Review of
DIGITAL COMMUNICATION

Review of DIGITAL COMMUNICATION

'State of the Art' in Digital Signalling, Digital Switching and Data Networks

---

J. Das
Department of Electrical Engineering,
Indian Institute of Technology
Kanpur, India

JOHN WILEY & SONS
NEW YORK CHICHESTER BRISBANE TORONTO SINGAPORE

First published in 1988 by
WILEY EASTERN LIMITED
4835/24 Ansari Road, Daryaganj
New Delhi 110 002, India

Distributors:

Australia and New Zealand:
Jacaranda-Wiley Ltd., Jacaranda Press,
JOHN WILEY & SONS, INC.
GPO Box 859, Brisbane, Queensland 4001, Australia

Canada:
JOHN WILEY & SONS CANADA LIMITED
22 Worcester Road, Rexdale, Ontario, Canada

Europe and Africa:
JOHN WILEY & SONS LIMITED
Baffins Lane, Chichester, West Sussex, England

South East Asia:
JOHN WILEY & SONS, INC.
05-05 Block B, Union Industrial Building
37 Jalan Pemimpin, Singapore 2057

Africa and South Asia:
WILEY EASTERN LIMITED
4835/24 Ansari Road, Daryaganj
New Delhi 110 002, India

North and South America and rest of the world:
JOHN WILEY & SONS, INC.
605, Third Avenue, New York, NY 10158, USA

Library of Congress Cataloging in Publication Data

ISBN 0-470-20221-1 John Wiley & Sons, Inc.
ISBN 0-85226-151-9 Wiley Eastern Limited

Printed in India at Rajkamal Electric Press, Delhi.

To

My Students

Past, Present and Future

Preface

"We have now reached a stage when virtually anything we want to do in the field of communications is possible: the constraints are no longer technical, but economic, legal or political. Thus if you want to transmit the Encyclopaedia Britannica around the word in one second, you can do so."

—Arthur C. Clarke

Looking around us and observing the technological developments in electronics and telecommunications all over the world, one is inclined to agree with the futurologists that we have almost reached the Information age and the Information revolution—a revolution much more significant than the Industrial revolution of the last century. The changing social needs and the development of new technologies are bringing the evolution of new communication capabilities in the form of videotex, electronic mail, high-speed FAX, teleconferencing and interactive CATV. These are also leading to automated offices and wired homes, where most of the future information will be generated, processed and utilized for running the industries, business and the government, and for education and recreation. At the same time, the spectacular developments in solid state electronics, computers, digital switching and transmission, have brought in the Digital revolution —leading to universal digital connectivity. It is also evident that communication is getting more and more integrated with computers and data networks, and eventually, we shall have a fairly integrated general purpose computer-communication (C & C) network. These C & C networks with their flexibility and multiservice facilities will then lead to the Integrated Services Digital Network (ISDN), which is the desired goal of the current developments in telecommunications.

The futuristic ISDN will not only provide the global digital connectivity through digital switches and digital transmission, but also the required signal processing, data compression, data protection and storage. Technological advances in devices may eventually result in the complete 'system-on-a-chip', requiring completely new approaches to the system design. As a consequence of the above wide ranging developments, the academicians will have to develop new unified theories for end-to-end information systems, including models of humans as information processors. We shall thus need in future more information and communication theorists than we need now; and these theorists will have to be trained in a wide variety of disciplines and technologies.

Most of the technological developments in digital communication has been very rapid and has taken place only during the last two decades. As a result, the senior professionals and academics have not been able to keep pace with these developments and therefore, have an urgent need to update their knowledge in these areas. Moreover, it is very necessary that our electrical engineering students, specialising in communication, must have a strong base in digital communication systems as well. Unfortunately, a comprehensive book giving the 'state-of-the-art' information on Digital Communication is not available so far. To fulfil the need of such a book, the present volume has been written, presenting a review of the recent developments in important areas of digital communication.

The book consists of ten chapters, and discusses such topics as: principles of digital modulation, source encoding, data transmission through cables and optical fibres, digital radio including satellite communication, data networks and digital switching, information theory and coding, survival of communication including spread-spectrum techniques, and future trends including ISDN. Conceptually, a system point of view has been taken in discussing the various topics, and the total range of digital signal processing necessary in a digital network, has been brought out in subsequent chapters, thus presenting a continuity of thought from end-to-end. The main emphasis has been on the 'state-of-the-art' and to discuss important results in the context of present-day developments in digital communication networks. Most of the information contained in the book are now available only in published journals. The book also contains an exhaustive bibliography.

The book is an outgrowth of the author's teaching and research activities in the area of digital communication over the last three decades. Further, the author has been called upon, over the years, to give many series of lectures in various international and national workshops. The book is actually compiled out of these lecture notes, presently edited and updated. It is hoped that the book will fulfil the need of a reference volume for the faculty and the professionals; and the exhaustive bibliography would help the reader to continue further reading in the area of his interest. The book may also be used in teaching courses in digital communication systems by selectively choosing the topics of interest. The contents of the book is mainly complementary (but in a few cases overlapping) to the topics in 'Principles of Digital Communication' by Das, Mullick and Chatterjee. By choosing selectively from the two books, two/three semester courses at the senior level may easily be organised. Although the book is primarily addressed to the professionals, it is no doubt that the senior students and academics will also find it useful and informative.

In writing the book, the author has drawn largely from the published materials in recent journals and some of the recent books in communication. As such, the author is indebted to a large extent to his predecessors and the researchers; and the acknowledgements have been made in proper places in the text. The author has spent a lifetime at the Indian Institute of Technology, Kharagpur and recently at the Indian Institute of Technology, Kanpur, and as such, he is largely indebted to his innumerable students, some of them colleagues later on, for the intellectual enrichment he derived through their interaction in class rooms, laboratories and research projects over all these years.

The author is grateful to the authorities of IIT, Kanpur, for providing financial support to the project through Quality Improvement Program. The author also wishes to thank Mr. J.C. Verma for excellent artwork, Mr. Joseph John for correcting the manuscript with care and patience, and Mr. C.M. Abraham for excellent typing of the manuscript and for general secretarial help.

Special thanks are due to my wife, Bela, for her encouragement and enthusiastic support during the long hours, spent in preparing the manuscript.

J. DAS

IIT, Kanpur
India

Contents

CHAPTER 1

Introduction

In ancient civilizations, the development of signs, symbols, codes and languages led to better communication between member units of a society, and the art of printing, with the availability of books, widened the area of communication between man and man at great distances. Although Joseph Henry invented the telegraph in 1831 and Samuel Morse followed with his code in 1832, the major breakthrough in long-distance telecommunication was due to: (i) the invention of the telephone by Graham Bell in 1875; (ii) the successful production and detection of electromagnetic waves by Heinrich Rudolf Hertz in 1888; and (iii) G.M. Marconi's success in the transatlantic radio experiment in 1901. It is indeed an interesting exercise to correlate the growth of the modern civilization with the landmarks of discoveries/inventions in electronics and telecommunication, as given in Table 1.1. The range of human communication has expanded through many orders because of the successful developments in telegraph, telephone, radio, TV and, now, satellite and fibre-optic systems. The present status of telecommunication is easily visualized by going through the progress made in last two decades.

The status of telecommunication varies widely in different countries [1]. From the statistics of 1974, it is observed that the world total number of telephones was 336×10^6, and, with an average growth of 20% per year, the expected total in 1984 was 10^9 (approximately). However, the telephone density in the developed First World is about 40; in the socialistic Second World, the density is 1.76; and in the developing Third World, the density is 1.15 only, for every 100 people. It has been also analysed that countries with high GNP have a high telephone density and, at the same time, the growth rate is also high in developed countries. The traffic growth and the related network development depend on various factors, such as: economic activity, service improvements, telecommunication maturity and service reliability. Thus, it becomes more difficult for developing countries to expand their telecommunication facilities because of their dependence on capital, technical knowhow, and a much higher demand for more telephones. The growth of the international telecommunication network depends on the traffic behaviour, and that on economic activity and social needs. The traffic growth during the last decade for different countries varies from 10% to

Table 1.1 Some landmarks in electronics and communication

Mediterranean civilization—(Egyptian)—pictograph, ideograph and hieroglyphs
Phonetic writing—tachygraphy by Greeks—400 B.C
True shorthand—Tyro, 60 B.C. (for recording the speech of Cisero)
Cipher—Roger Bacon's scripts (A.D. 1214-1294)
Bush telegraph—talking drum of Congo tribes (2-state code)
Ogam script—by ancient Celts, A.D. 400
Universal language (artificial)—by Descartes 1629
John Wilkins—1668, invention of new language with special grammar—supported by the Royal Society
Lingua Franca—1921 (League of Nations)
Electrical signalling—Roger Bacon 1267, Watson 1746
Morse code—1832; Joseph Henry—telegraph—1831
Submarine telegraph cable—1850
Maxwell's theory of EM field—1864
Quadruplex telegraphy—Heaviside and Edison—1875
Graham Bell—magneto telephone—1875; Microphone—1878
Gramophone/phonograph—Edison—1877
Voice modulation of a light beam by microphone—Bell—1878
First manual exchange—1878
Piezoelectric crystals (Pierce *et al.*)—1880
Edison effect—1883
Hertz's experimental production and detection of EM waves at $\lambda = 5$ m and 50 cm; reflectors, 1888
Strowger automatic switch—1889
Generation and detection of EM wave ($\lambda = 6$ mm)—1895 by J.C. Bose
Marconi-transmitted coded message at 1.75 miles—1895
Marconi's transatlantic radio experiment—1901
Heaviside and Kennedy—discovery of ionosphere—1902
I.T.U.—1865 in Paris, International Radiotelegraph Union—1906
First telegraphic transmission of picture from Munich to Numbering—Korn—1904
Electron (Thomson)—1897
Vacuum diode (Fleming)—1904
Cathode ray tube—1905
Triode valve—De Forest—1906
Auto exchange—1907
Radio transmitter—Marconi—1916
Wave filters—Campbell (birth of FDM)—1914-18
Ultrasonics (submarine detection)—1914-18 (World War I
Superhet radio receiver (Armstrong)—1918
Crossbar switch—1919
Printing telegraphy—1920
Experimental radar—1922
FM theory—John Carson—1922
Electronic TV picture tube (Zworykin)—1923
Telegraph theory—Nyquist—1924
Carrier and VFT on lines—1925
Information theory (Hartley)—1928
Television (Jenkins)—1925, Fransworth—1929
Commercial transatlantic radio telephone—1926
Negative feedback amplifier (Black)—1927

Photoelectric devices—1929
MW transmission experiment—1931 (Clavier)
Vocoder (Dudley)—1936
Coaxial multichannel circuits—1933
Radio telemetering (radiosonde)—1937
Bell Lab. Crossbar Exchange No. 1—1938
Magnetron, Klystron, Radar—1939 (World War II)
Control and telemetry—WW II
First Computer at Harvard University—1944
Pulsed Radar—14 MW; 10 cm—1949
Information theory—Gabor—1946; Shannon—1948; Wiener—1948
Transistor—Bardeen—1947; Shockley—1949
Audio submarine cable—1950
Transatlantic coaxial cable—1955
ESS—1955-60
Maser (Gordon *et al.*)—1955-56
Laser (Schawlow and Townes)—1958-60
Tunnel diode—Esaki—1957
Integrated circuits—1960
Man-on-Moon—Apollo Program—1961-66
ESS No. 1—1964
Microprocessor—1970
ESS No. 5—1980
Fibre-optic communication—1966-76

Satellite—1957-80	*Digital communication*
Sputnik—1957	PCM (Reeves)—1938
Score—1958	Oliver, Pierce, Shannon—1948
Courier—1960	T_1-carrier—1962
Telstar—1962	DM—(CC Cutler)—1952
Relay—1963	F. De Jeager—1952
Syncom—1963	Greefkes—1968
Early Bird—1964	
Mariner IV—1965	*Coding*
Mariner V—1966	R.W. Hamming—1950
Intelsat—(I-IV)—1968-74	P. Elias—1955
ATS-F—1974	BCH—1959-60
Intelsat V—1979-80	Woozencraft—1957 (sequential decoding)
INSAT IA—1982	Viterbi—1967 (M-L decoding)

45%; the highest growth rate being that of Brazil, and the world average is 16%. Major investments for international telecommunication have been made in submarine cables and satellite system networks, and the total was approximately $ 4000 million till 1978. Since the media diversity for international network is provided by cable and satellite paths in a complementary way, equal importance is now given for the development of both the systems. It is estimated that approximately $ 1000 million for satellites and $ 1000 million for cables have been invested after 1978 [2]. Thus, telecommunication, both national and international, supports a very large industry, indeed!

1.1 PROGRESS IN NETWORKS AND SYSTEMS

Over the decades, the telecommunication network in many countries has grown from a small regional system to a worldwide one. At present, a long-distance telephone call is processed through a central office (CO) and a toll exchange; and for international calls, an additional gateway exchange is involved. The transmission media available are subscribers' cables, coaxial cables and MW LOS links, and, recently, satellite and fibre-optic links are being used extensively. A large telecommunication network using many exchanges and various media is shown in Fig. 1.1. Till 1960, the communication networks including exchanges were basically analog and the services handled were mostly voice and telex circuits. With the success of Pulse code modulation (PCM) and Time division multiplexing (TDM) through Bell System T_1-carrier in 1962, all transmission circuits, first short-haul trunks and then long-haul circuits, are fast becoming all-digital. Concurrently, digital exchanges using PCM-TDM techniques are also being introduced in communication networks. In the early days, transoceanic circuits were either through narrow-band submarine cables or HF radio; but with growing demands for more circuits, wideband transmission systems using MW links, coaxial cables, and now, satellite and fibre-optic systems are being used. The demand for more radio spectrum for expanding communication is further growing, and this will soon lead to the utilization of millimetric waves and optical communication. At the same time, the national and regional networks are maturing into all-digital global networks through the use of MW

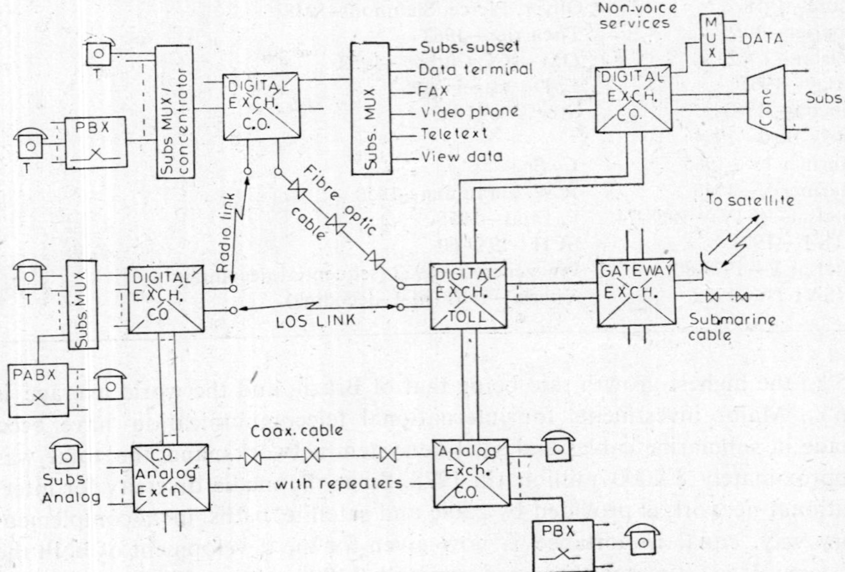

Fig. 1.1 Scheme of a large telecommunication network including analog/digital exchanges serving both voice and non-voice subscribers. Transmission media are: MW LOS links, satellite links, coaxial/fibre-optic cables and submarine cables (C.O.: central office, SUBS.: subscribers).

links, satellites, submarine cables and optical fibres, incorporating digital techniques.

1.1.1 Switching and Data Networks [3, 4, 5]

Historically, the switching networks have developed through manual to automatic and then digital exchanges. The automatic exchanges first developed were Strowger and Panel systems, and in the late 1930's, the Crossbar exchanges were introduced. This system was improved further till early 1960, when the concept of PCM-digital exchanges matured. An operational electronic switching system (ESS) was first installed in France (known as RITA) in 1968. Then, of course, the progress in ESS was very rapid, and, today, all major telecommunication administrations are committed to the futuristic all-digital switching networks.

In the context of the development of all-digital switching and transmission systems, the major contributions have been by the Bell Labs, ITT, CNET of France, and NEC/HITACHI of Japan. ITT has proposed the Network 2000 as the concept of their telecommunication network for A.D. 2000. The ITT system-12 is the first generation of digital exchanges developed with the concept of Network 2000, using distributed processing and fibre optics for transmission. In Japan, similarly, digital switching systems, such as, HDX-10 of Hitachi, NEAX-61 of NEC, are being installed using latest LSI technologies to achieve a compact, economical and fully electronic switching system. In UK, the system-X, an integrated digital switching and transmission system, has been developed for both data and telephone switching.

Bell Labs have developed the No. 4 ESS for toll switching and a newer version No. 5 ESS for local offices. Both the systems use stored-program-controlled (SPC) digital time-division switching and provide operational savings and ease in system growth. The digital switch consists of four major subsystems: (a) the central processor; (b) a message switch; (c) a time-multiplex switch; and (d) interface modules. The ESS uses fibre-optic links as trunks, microprocessor-controlled interface modules, and a time-space-time switching network. The SPC is distributed among the central processor and interface module units. The No. 5 ESS has a subscriber capacity of 12,000 lines and the No. 4 ESS has a capacity of 10^5 trunks. Both use the PCM format for incoming lines and the information rate of 32.768 Mb/s through fibre-optic links. Modular hardware, custom LSI chips and high-level programming languages are the key factors in these ESS.

Concurrently, the computer architecture has gone through an evolution, and today's distributed processing system has developed through point-oriented computers in the 1950's, the family series in the 1960's, large computers in the early 1970's and then, area-oriented horizontally distributed types in the late 1970's and supercomputers in the 1980's. The distributed system is intended to carry out all possible processing at the site, where information is generated and used. Instead of a single ultra-large computer, the new system applies horizontal integration of several smaller computers to increase the system's capabilities, resulting in a better flexibility of operation and maintenance, and in greater economy. Hardwarewise, the single-chip processing power is approximately doubling each year. Bell Lab's BELLMAC-32 A, a recently completed single-chip 32-bit microcomputer, has a computing power almost equal to some of the popular minicomputers. The trends in microcomputer development suggest that the micros, minis and the maxicomputers, all three may be expected to process about 2×10^7 instructions per second in the early 1990's. Except for a few special purpose supercomputers, the computer-on-a-chip will, by then, serve most of the processing requirements in our telecommunication networks. The optical disc, magnetic bubbles and semiconductor memories will meet the needs of large data bases at much lower costs.

The digital data system, as proposed by Bell Systems and others, uses data multiplexers and concentrators, and the resulting high rate data (64 Kb/s or 1.544 Mb/s) is processed through the digital switches for onward transmission. The digital hierarchy is developed by using higher order multiplexing of the primary MUX data streams of 1.544/2.048 Mb/s, to generate the ultimate rate of 120/274/400/800 Mb/s. These are transmitted through coaxial cables/waveguides/radio relay/optical fibres, using multilevel DC-free codes. In satellite communication, 50/60 Mb/s data is transmitted per transponder, giving 700/1000 speech channels. Newer TDMA systems are being developed for 100/350 Mb/s data rate in the 20/30 GHz band. Over and above the national common-carrier systems, independent high-capacity data networks are being developed in many countries, e.g., in USA and Japan, using satellite and LOS links. Regarding the network protocol, the packet switching has been so far the most popular technique.

1.1.2 Transmission Systems [6, 7, 8]

As in the case of switching and data networks, the transmission media also have gone through many stages of development: from open-wire lines, through carrier quads, and coaxial cables, to the most wideband fibre-optic cables on the one hand, and from HF radio, through VHF, UHF and MW, to infra-red and laser beams, as shown in Table 1.2. Till the 1960's, the multichannel carrier systems used analog FDM techniques both in cable and radio media. But with the successful application of PCM-MUX in short-haul trunks, more and more MUX systems were designed with digital TDM techniques for both short-haul and long-haul transmission. The bandwidths and number of speech channels in both analog and digital MUX systems are also indicated in Table 1.2.

Wideband transcontinental communication is mainly achieved through LOS links, satellites, submarine cables and optical fibres. Traditionally, the earlier systems were analog, but digital techniques have been introduced in recent Digital Radio, TDMA satellites and fibre-optic communication systems. The channel capacities now achieved are: 200 Mb/s for digital LOS;

60 Mb/s for satellites; and 400/800 Mb/s for optical fibres. It is expected that with single-mode fibre cables, the data rate will go up to a few Gb/s.

Table 1.2 Transmission media: stages of development

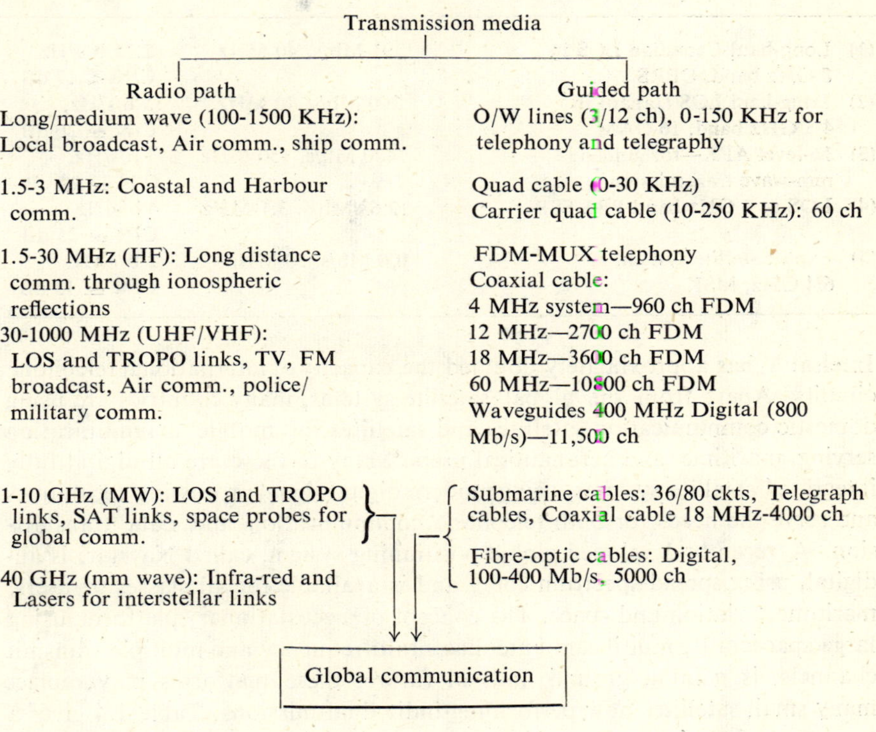

In Digital Radio, the most popular QPSK technique of modulation is now being supplemented with newer techniques, e.g., MSK, QPRS, higher-order APK and MA-MSK, giving better spectral efficiencies. Use of DFE (decision-feedback equalizer), adaptive multipath cancellers and phase-adaptive diversity techniques, has made the systems quite robust and resistant against fading, interference and noise. Presently, systems are in operation having capacities of: (a) 91 M bits, QPRS with spectral efficiency of 2.25 b/s/Hz and channel CNR < 17 dB; (b) 200 Mb/s-16 QAM, spectral eff. of 5 b/s/Hz and CNR $\simeq 20$ dB; (c) 100 Mb/s-MSK, spectral eff. of 2 b/s/Hz; and CNR $\simeq 13$ dB, as given in Table 1.3.

In satellite communication, spectacular progress has been made from the early Sputnik-Score-Courier-Synchom experiments to the commercial Intelsat IV and V satellites, providing 25,000 and 50,000 international telephone channels at a cost of $ 5000 per year per circuit only. Intelsat IV system provided two-thirds of the world's transoceanic telecommunications and consisted of 12 satellites and 300 earth stations in 125 different countries.

Table 1.3 Summary of LOS systems

System description	*System capacity and RF-BW*	*Spectral efficiency and CNR for $P_e \simeq 10^{-4}$*
(1) Long-haul Canadian LOS in 8 GHz band, QPRS	91 Mb/s, 40 MHz	2.25 b/s/Hz, C/N < 17 dB
(2) Long-haul LOS (Japan) in 4/5 GHz band, 16-QAM	200 Mb/s, 40 MHz	5 b/s/Hz, C/N $\simeq$ 20 dB
(3) 16-level APK—for guided mm-wave transmission	400 Mb/s, 150 MHz	3 b/s/Hz, C/N $\simeq$ 20 dB
(4) LOS at 2 GHz band PRS-SSB	12.63 Mb/s, 3.1 MHz	4 b/s/Hz C/N $\simeq$ 25 dB
(5) Expt. Satellite system, at 6/4 GHz, MSK	100 Mb/s, 50 MHz	2 b/s/Hz C/N $\simeq$ 13 dB

Intelsat V has approximately doubled the capacity of international telephone circuits. Apart from the global satellite systems, many countries are using domestic communication satellites and satellites for mobile communication serving maritime and aeronautical users. Many of these are all-digital fully integrated satellite systems, using modern digital signal processing techniques and serve the needs for both telephone communication and data transmission. A recently developed global positioning system, called Navstar, is all-digital, using spread-spectrum codes and is available for all classes of users, maritime, aviation and space. The concept of a geostationary platform, using large spacecrafts, multibeam antennas, multifrequency and multiple-transmit channels, is gaining ground and, in future, these platforms may replace many small satellites now performing individual missions. Table 1.4 gives a summary of the important satellite systems.

In the last two decades, the progress in optical fibre technology has been remarkable, and a very wideband digital communication system is now available for short-haul and long-haul trunk circuits. The optical source, generally laser diodes, at λ of 0.85, 1.3 and 1.55 μm, is used for transmission of data bits at 100–400 Mb/s rate. The path loss is about 0.3 to 1 dB/km, giving a repeater spacing of 20–30 km. Long-haul circuits are built up with regenerative repeaters and a total length of 2500 km or more is possible. All major telephone laboratories in the world are working on single-mode long-wavelength fibre-optic systems, and it is expected that the repeater spacing will increase further, say, to more than 100 km, by using coherent detection, at $\lambda = 1.6$ μm, and optical amplification techniques. At present, USA, Canada and Japan have long-haul circuits operating at wavelengths of 0.8–0.85 and 1.3 μm, and at bit rates of 44.7, 91 and 400 Mb/s. For transoceanic communication, single-mode multifibre cables at $\lambda = 1.3$ μm and associated regenerative repeaters are being designed with a channel capacity of 36,000 at a bit rate of 274 Mb/s. Developments are also in progress in the areas of WDM (wavelength division multiplexing), integrated optical IC's, optical amplification and coherent detection. From cost considerations, it is estimated that digital fibre cables with a transmission capacity of 100–1000 Mb/s will be cheaper than digital coaxial cables, at least by an order of magnitude. It is further predicted that, at the turn of the century, all fixed services will be by guided wave systems, specially fibre optics, and satellites will be mainly used for mobile and deep-space communications.

Table 1.4 Summary of satellite systems

(1) Commercial systems: INTELSAT IV, 6/4 GHz, 25,000 channels,
12 Satellites and 300 earth stations
INTELSAT V, 6/4 GHz, and 14/12 GHz;
50,000 channels, growth rate 20% per year

(2) Domestic satellites: USA— 4 Satellite systems using 1000 earth stations
Japan—JBS system
Canada—CTS system
India—INSAT-I/II

(3) All-digital domestic systems:

(a) Satellite Business System (SBS) of USA
RF—14/12 GHz, TDMA with DA, 5.5 and 7.6 m Roof-top antennas,
data rate 2.4 Mb/s to 6.3 Mb/s.
Channel capacity—14,000 voice or 8000 data circuits (= 500 Mb/s)

(b) FECS and RICS of Japan:
RF: 30/20 and 6/4 GHz, TDMA, 60 and 100 Mb/s
No. accesses: 4, QPSK
FEC-R7/8, data rate : 2.4 to 6.3 Mb/s
Ch. cap : digital TV : 32 Mb/s, or, 2 CTV + 192 voice ckts

(4) Mobile communication: Mari Sat, Fleet Sat, Lea Sat, Inmar Sat
(5) Global positioning system: Navstar
(6) Geostationary platform (futuristic) : 5000 kg Geoplatform, consisting of:

(a) 30/20 GHz system
(b) Large-aperture (12 m) multibeam 6/4 GHz system
(c) Satellite-switched 14/11 GHz system
(d) 12 GHz broadcast sat. system

The international telecommunication network, at present, is based on two principal media, submarine cables and satellites, which bridge the long distances over the seas. Submarine cables were the forerunner in the development of the global network and the satellites came into operation some ten years later. Contrary to the early predictions that satellites would rapidly supersede cable systems, it is now proved that they complement each other, and both the technologies have been instrumental in generating a revolution in international communication. An outline of the progress in both the technologies is given in Table 1.5, where it is seen that the traffic carried by satellites is a large share of the total volume. Figures, 1.2 and 1.3 show the relative growth and cumulative investments for both cable and satellite systems [2]. It is now easily seen that the progress in both the systems is quite balanced, and it is expected that the trend will continue till some new breakthrough occurs. The telecommunication administrations are now committed to the balanced use of both the systems, and, as a result, the new demand will be met by further expansion and development in terms of wider bandwidths (the relative cost per circuit mile has been reduced by 4 : 1 by increasing the cable bandwidth from 5 to 50 MHz) using fibre-optic cables and multibeam satellite-switched networks.

Table 1.5 Progress in international communication [2]

*Submarine cables**	*Year*	*Satellites*
TAT-1 (48)	1956	—
CANTAT-1 (80)	1961	—
COMPAC (80)	1962	—
TAT-3 (128)	1963	—
TRANSPAC-1 (128)	1964	—
—	1965	Early Bird (240)
SAT-1 (360)	1966	INTELSAT-II (240)
—	1968	INTELSAT-III (1200)
TAT-5 (845)	1970	—
—	1971	INTELSAT-IV (6000)
CANTAT-2 (1840)	1974	—
TAT-6 (4000)	1976	INTELSAT-IV A(6000 +2TV)
PANCAN-3 (5520)	1977	—
New cables (20, 940)	1980 } 1981	INTELSAT-V (12,000 + 2TV)

*Figures in parentheses indicate the number of circuits.

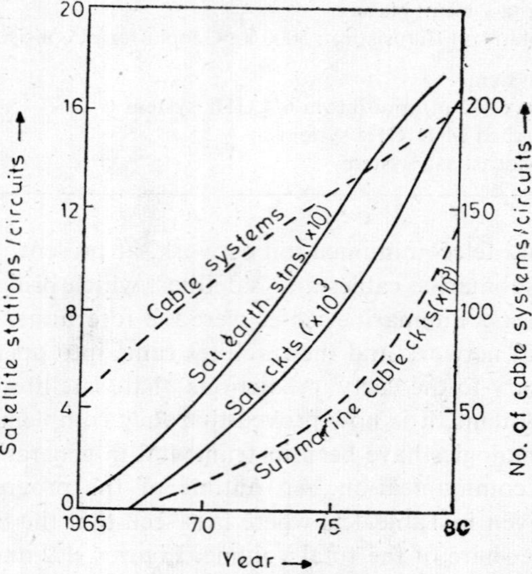

Fig. 1.2 Relative growth of satellite and submarine cable systems

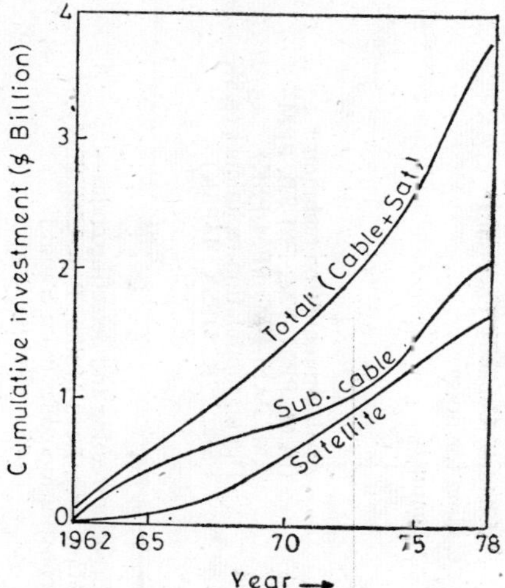

Fig. 1.3 Cumulative investment in submarine cables and satellites

1.1.3 Signal Processing

The major subsystems in a long-distance communication link are shown in Fig. 1.4, where the traditional information sources are: telegraph, speech, facsimile (FAX), TV and telemetry signals. For the purpose of matching these signals to the noisy channel, it is necessary to encode and transform the signals in a suitable form, and to this end, various techniques of signal representation and signal design are used, as shown in Table 1.6. The analog signals, e.g., speech and TV, are digitized through sampling and A-to-D conversion by PCM or delta modulation (DM), and if required, the signals may be compressed in bandwidth through various band-compression techniques, e.g., LPC, APC, and vocoder. For protection against noise in the channels, the analog signals are modulated through wideband schemes, e.g., FM, PPM, etc., and the digital signals are coded through error-correcting codes (FEC) or orthogonal codes, both of which use wider bandwidths, thus complying with the results of Shannon's channel capacity theorem. Some of the currently used robust communication systems utilize many coding techniques simultaneously, e.g., FEC/CDM/PSK, for their survival in a hostile environment.

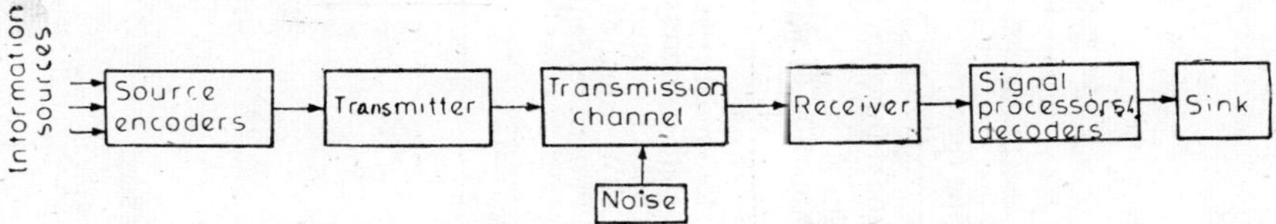

Fig. 1.4 A long-distance communication link

In digital exchanges, basically digital data are switched, and, as such, there is a need as well as possibilities of digital signal processing to optimize

Table 1.6 Signal representation and signal design

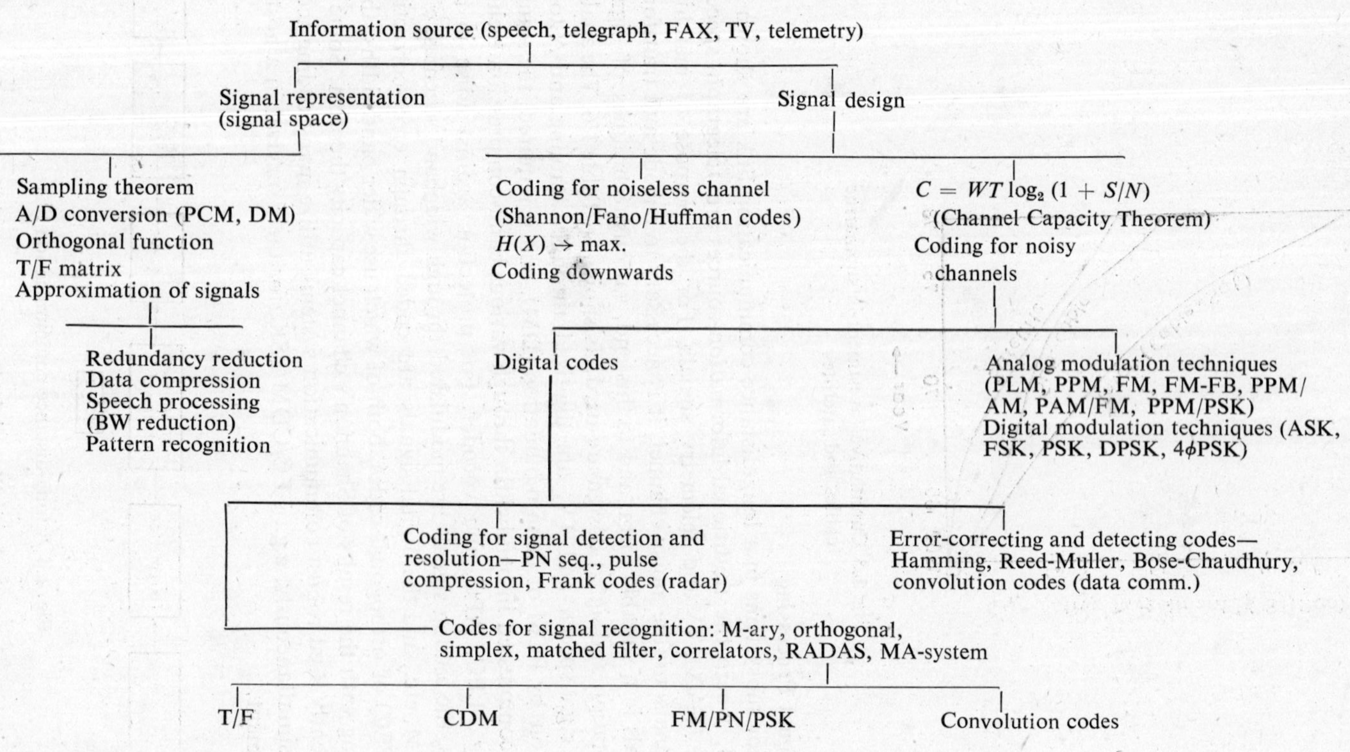

the data formats and their information content. In Fig. 1.5, some of the current data-processing techniques used for different voice and non-voice services are shown. These processing modules are mostly microcomputer controlled and, sometimes, operate in a time-sharing mode. These will be further discussed in Chapter 3.

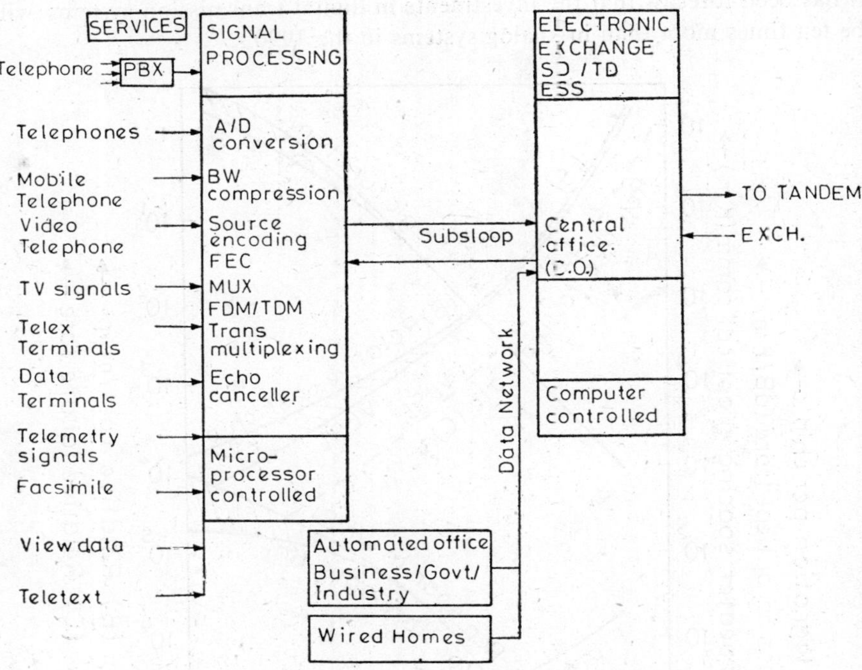

Fig. 1.5 Signal processing in a digital exchange

1.2 INTEGRATED DIGITAL NETWORKS

We have discussed above the current developments in switching networks and transmission, specially the use of digital techniques in switches and transmission systems. Concurrently, the progress in LSI and VLSI has been spectacular, and the trends in cost reduction, improvement in reliability and the volume of integration per chip, as shown in Fig. 1.6, indicate a very favourable bias towards digital networks. However, the customer services are, at present, partly digital and partly analog as seen in Fig. 1.5, and new services, e.g., videotex, electronic mail, teleconferencing and interactive CATV, are also analog/digital. But all signals may be converted to the digital form by efficient use of A/D converters, and modern techniques of signal processing may be utilized for source encoding, bandwidth compression, error correction, data formating, etc. The new services may also be made compatible with other digital streams through complex signal processing, if necessary. As the information moves along, the data/messages are switched through local exchanges (ESS), short-haul/long-haul trunks, toll exchanges and, ultimately, to a broadband medium for intercontinental transmission.

Since all signal processing and switching functions are now computer-controlled, it is economical to have an all-digital network. Although the long-haul trunks, till date, were cheaper for analog transmission, recent developments in digital carrier through digital radio, TDMA satellites and fibre-optic cables (see Fig. 1.6), have made digital transmission economical. As a result, it has been forecast that the investments in digital transmission systems will be ten times more than in analog systems in the 1990's.

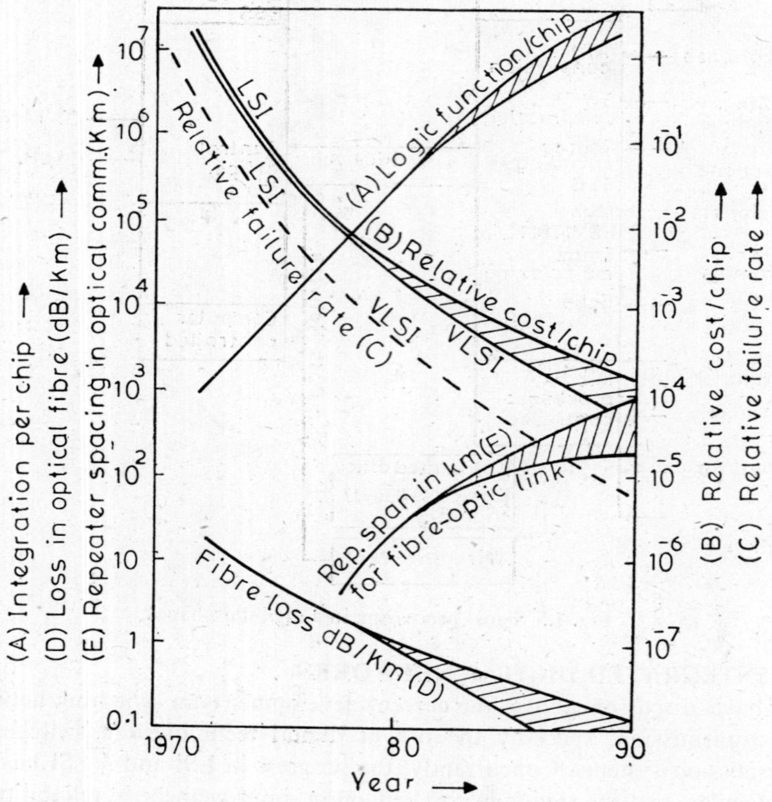

Fig. 1.6 Progress/forecast in specific areas during 1970-90

It has been discussed earlier that, over the decades, the computer architecture has gone through an evolution, while switching and transmission are gradually becoming all-digital. At present, all ESS's are computer-controlled, major signal processing modules are μP-controlled, large memories are in the form of VLSI chips, and the customer services are also processed through digital techniques; all these lead to a fairly integrated general purpose 'computer and communication' (C and C) network, which will ultimately lead to the ISDN (Integrated Services Digital Network), as designated by CCITT [9].

The concept of the integrated switching and transmission network emerged some time in late 1960's. Before the emerging of all-digital techniques, it was difficult to conceive a single switching centre for digital/analog signals of different bandwidths, e.g., voice, data, FAX, videophone and TV, and, at the same time, it was realized that the interfacing of different signals at the regional communication centre would be very complex in a hybrid environment. The integration of switching and transmission through digital techniques, however, automatically results in an integrated network. But the problem remains that the digital stream for videophone and TV would be at much higher bit rates (2 Mbits and above) than that required for speech. It was, therefore, suggested that two groups of users, one for narrow-band signals and the other for wide-band signals, should be processed in two separate switching centres, and the overall processing for long-distance connections should be done in a superswitching centre [10]. In this process, ESS will gradually replace the older exchanges and we shall have a homogeneous all-digital network. An alternative proposal was an integrated system [11], where a set of unique networks are constructed for different types of signals, and the overall control of these networks is given to the ESS, as shown in Fig. 1.7. Thus, a common control and a common signalling system are provided for the overall network, where the networks themselves remain somewhat individual. The futuristic integrated network, as proposed above, however, has now taken shape in the form of all-digital PCM-TDM switches and PCM transmission, where data rates as high as 8 Mb/s are easily processed. It is expected that higher data rates, including digital TV data, will soon be accommodated in the integrated digital network (refer to Chapter 10).

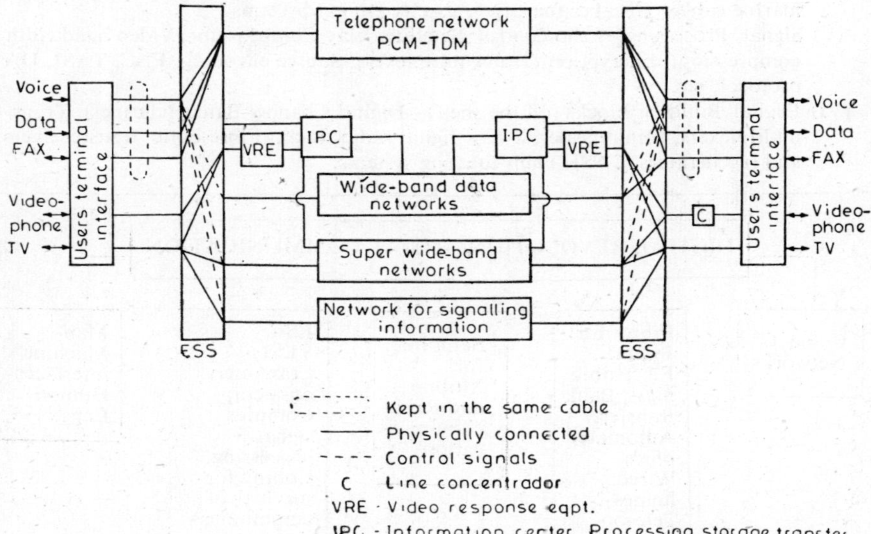

Fig. 1.7 A futuristic integrated network

The reasons for this rapid progress towards all-digital networks have been discussed by many experts, and Aaron [12] has analysed the trends in digitization of various telecommunication services in different countries. It is shown that the growth has been rather linear during the 1968-78 period in almost all countries, and this has been mostly in short-haul digital trunks. USA has the maximum ckt-km of digital trunks installed (10^8 km in 1976), followed by Japan (10^7 km in 1976), where the present growth is faster. The continental countries are following the same pattern of growth. The factors which are contributing to the spectacular progress of digital network, leading to the 'Digital Revolution' of this decade, are indicated in Fig. 1.8. The net result of this almost universal digital connectivity would be the rapid development of business communication and home electronics, and the 'C and C' offices will be paperless; but at the time, this will ensure accurate decision making through the efficient use of the office equipment. The home-office, the reception room, the decision room, and the executive's office will all be interconnected through video terminals and data networks; and customers and executives will mostly discuss business through computer terminals. Thus, C and C, the systematic merging of computers and communications, is a key to the success and universality of future communications.

(I) Services—Digital audio/video Telephone, Digital Telemetry, Telex, Data, Facsimile, View data, Teletext, Digital TV (32 Mb/s), Automated home and office.
(II) Digital ESS—Distributed processing.
(III) Devices—LSI cheaper, nonlinear optical and millimetre wave devices favour digital.
(IV) Media—Digital radio, TDMA—Satellite, Digital Fibre optics—including Submarine cables; Cheaper for large-capacity Digital Systems.
(V) Signal Processing—Echo control, Satellite delay Compensator, Video bandwidth compression, Encryption, Transmultiplexers, Source encoding, FEC, TASI, DA protocol, etc.
(VI) Digital Building Blocks (off the shelf)—Digital Channel Banks (cheaper), Transmultiplexers, Minirepeater, 2-way digital radio rack, Fibre-optic transreceivers and regenerators, TASI, Fault-locating systems.

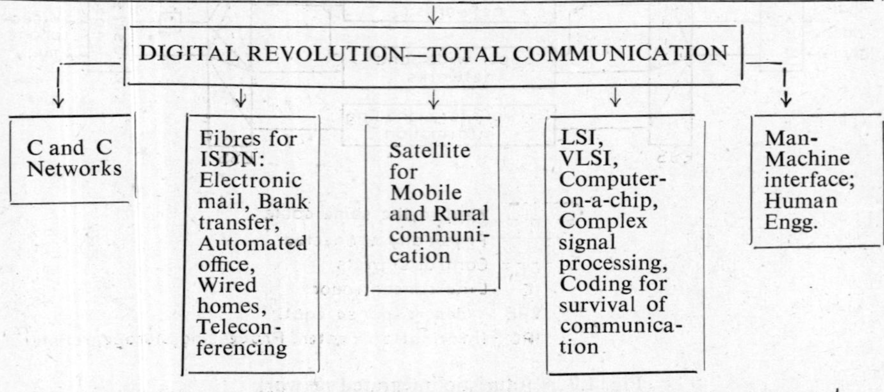

INCREASED PRODUCTIVITY; BETTER LIFE-STYLE; AND MORE RELAXED SOCIETY

Fig. 1.8 Digital Revolution; input-output

Hopefully, the Digital Revolution will also lead to increased productivity, better life-style and more relaxed society [13].

As a cosequence of the above wide-ranging developments in 'C and C' networking, the academicians will have to develop a new unified theory for end-to-end information systems involving Shannon's theory, theory of computational complexity and theory of multiuser communication networks, including link reliability and cryptographic protection. Since man-machine interfaces will be an important element of future networks, it will be necessary to study the model of humans as information processors. In the opinion of Prof. E.C. Posner of JPL, Caltech [14], "the upshot of all this leads to the conclusion that: we will need more information and communication theorists in A.D. 2000 than we need now. Not only that, but these theorists will be more broadly educated and trained in an even wider variety of disciplines and technologies than we are trained in now".

1.3 OUTLINE OF THE BOOK

The organization of the book follows, to a large extent, the signal flow in a digital communication network, as shown in Fig. 1.4, and the important topics, selected for discussion, belong to the class of signal representations and design, as given in Table 1.6. However, the overall approach, has been from the system point of view, and, as such, typical systems have been described after the suitable introduction of their principles. Nevertheless, basic principles and theory are included wherever necessary.

In Chapter 2, principles of data transmission and reception are discussed. Binary and multilevel modulation techniques including their error-rate performances have been explained, and necessary auxiliary circuits for bit and carrier recovery using PLL have been also included.

In Chapter 3, digital encoding systems, using PCM, DM and their improved versions, for speech and television signals, have been explained in detail. Speech-band compression techniques, e.g., vocoder, APC and LPC, are discussed. Recent developments in the encoding of wide-band FDM and television signals have been included. Special digital circuits, such as TASI and echo canceller, have been also discussed.

Base-band data transmission techniques using voice grade cables, coaxial cables and optical fibres are discussed in Chapter 4. Signal shaping, using raised cosine filters, partial response technique and line codes, is explained along with the techniques of equalization. Principles and current practice in base-band data multiplexing and the currently used hierarchy are also included. Some of the current voice-grade and high-rate data transmission systems are also illustrated.

In Chapter 5, one of the major chapters in the book, all important techniques related to digital ratio, including MW LOS and satellite links, are introduced and illustrated. For LOS links, fading and multipath problems and their solution through adaptive receivers have been discussed. In satellite communication, basic principles as well as some system design techniques have been covered, and a review of current satellite systems has been given. Multiaccessing in radio systems has been explained and this leads to the important and currently popular spread-spectrum (SS) technique which has been discussed in some detail. This chapter along with Chapter 4 discusses important aspects of transmission media and techniques for both low- and high-rate data.

From the point of view of digital networking, computer communication networks and electronic switching systems (ESS) are two most important topics and they are discussed in Chapters 6 and 7, respectively. Techniques of circuit/packet switching have been explained and a review of currently available cable data and packet radio networks have been given in Chapter 6. Principles of ESS, interface circuits and switch design are the basic topics in Chapter 7. A review of large ESS along with a discussion on their mutual synchronization problems is also included.

Chapter 8 introduces Information theory and coding, as applied to communication systems, and discusses error control techniques and their performances. Shannon's fundamental theorems are explained and the channel capacity theorem is used to evaluate the efficiencies of several communication systems.

Survival of communication, a system-oriented approach to the problems of coexistence of several communication networks in a common environment, is the topic of Chapter 9. Techniques for suppression of RFI and mutual interference, and for data security have been discussed. The application of SS techniques for anti-jamming and anti-multipath purposes has been also included.

In the concluding chapter, i.e., Chapter 10, the futuristic ISDN has been elaborated upon with reference to the emerging non-voice services and their compatibility with the developing all-digital networks. Such innovations, as the automated offices, wired homes, personal information processing and communication systems (PICS), etc., have been introduced and their impact on the emerging ISDN discussed.

Thus, the content of the book summarizes the present status of digital communication, along with applications, and indicates the direction in which further developments and innovations will take place.

REFERENCES

1. Das, J., 'Telecommunication and Social Objectives', *JIETE*, vol. 26, pp. 555-560, Nov. 1980.
2. Dawidziuk, B.M. and Preston, H.F., 'International Communications, Network Developments and Economics', *Electrical Communication*, vol. 54, No. 2, pp. 127-138, 1980.
3. Cookson, A.F., 'Network 2000 Concept—Oversew', *Electrical Communication*, vol. 53, No. 3, pp. 158-162, 1979.
4. Special Issue on No. 4 ESS, *BSTJ*, vol. 56, No, 7, 1977.
5. Issue on No. 5 ESS, *Bell Labes Record*, Dec. 1981.
6. Special Issue on Digital Radio, *IEEE Trans. on Communication*, vol. Com-27, No. 12, Dec. 1979.

7. Satellite Communication Issue, *IEEE Comm. Magazine*, vol. 18, No. 5, 1980.
8. Koyoma, M. and Itoh, T., 'Present and Future of Large Capacity Optical Fibre Transmission', *IEEE Comm. Magazine*, vol. 19, No. 3, pp. 4-11, 1981.
9. Kobayashi, Koji, 'Telecommunication in the Future', *IEEE Comm. Magazine*, vol. 18, No. 4, pp. 23-27, 1980.
10. Finet, A.E., 'Telecommunications Intergrated Network', *Trans. IEEE*, Com-21 No. 8, p. 916, 1973.
11. Ogata, K., 'General Trend in Electrical Communication in Japan', *Trans. IEEE*, Com-20, No. 8, p. 689, 1972.
12. Aaron, R., 'Digital Communications—The Silent Revolution', *IEEE Comm. Magazine*, vol. 17, No. 1, pp. 16-26, 1979.
13. Das, J., 'The Information Age and Digital Revolution', *JIETE*, vol. 28, pp. 633-642, 1982.
14. Posner, E.C., 'Information and Communication in the Third Millennium', *IEEE Comm. Magazine*, vol. 17, No. 1, pp. 8-15, 1979.
15. Das, J., Mullick, S.K. and Chatterjee, P.K., *Principles of Digital Communication*, Wiley Eastern, New Delhi, 1986.

CHAPTER 2

Principles of Data Transmission

A modern digital communication system consists of many subblocks, as shown in Fig. 2.1. The overall system may be grouped into the following subsystems:

(a) Source encoder including A/D conversions and data reduction.
(b) Signal-design processor including error-correcting/detecting codecs and modulation hardware.
(c) Matched filter receiver.
(d) Decoders.

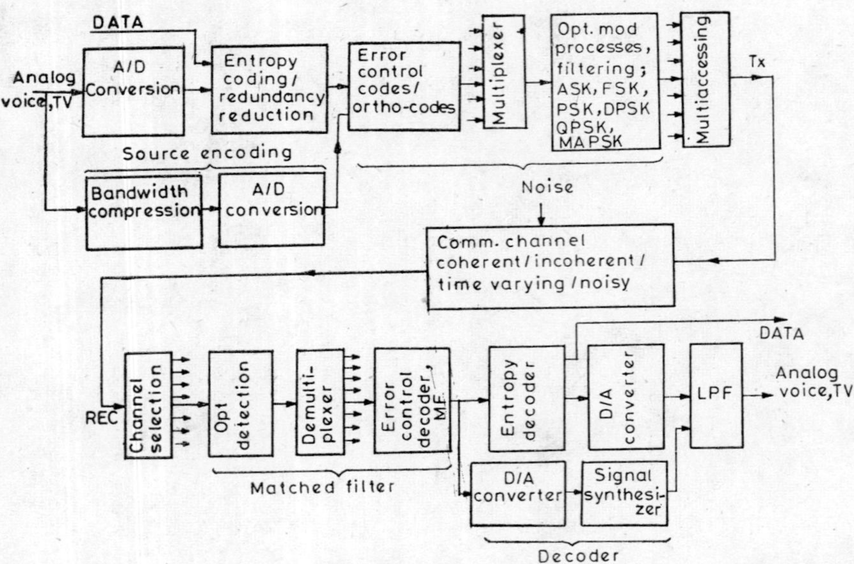

Fig. 2.1 Digital communication system model

Source encoding and error protection will be the topics in later chapters, but modulation and demodulation techniques including optimum detection will be discussed in this chapter. Since the digital data, either from the A/D codec or from a computer peripheral, has a low-pass spectrum and the transmission medium is generally band-pass, it is necessary to match the data with the medium through a modulation process. These modulation techniques are mainly: Amplitude Shift Keying (ASK/on-off), Frequency Shift Keying (FSK), and Phase Shift Keying (PSK). For binary inputs, the speed of transmission is generally 1 bit/Hz/s. But for higher speeds of transmission, the binary data may be converted to multilevel bauds and each baud then modulates the carrier leading to multilevel M-ASK, multifrequency M-FSK, multiphase M-PSK and multilevel QAM (APK). There are other spectrally efficient modulation schemes, e.g., CPFSK, OFFSET-QSK, MSK and PRS, which will be discussed in Chapter 5. In this chapter, we shall mainly discuss the above modulation schemes, their implementation, the detection process and the overall error rates at the receiver output.

2.1 MODULATION TECHNIQUES

In an ideally low-pass channel, unipolar (1/0) or bipolar (± 1) data may be directly transmitted and received. However, a random data sequence occupies a large bandwidth and it is necessary to use a low-pass filter after the data generator to restrict the required bandwidth. Moreover, the low frequency response in a long line is restricted due to the use of coupling transformers and other device in the repeater amplifiers. Thus, in a voice-grade data circuit which is a band-pass medium, one of the modulation schemes has to be used. Even in coaxial cable systems, which are typically wideband low-pass media, the low frequency response is poor and such techniques as AMI, pseudo-ternary and PRS, are used to overcome this problem.

The bandwidth requirement of a data transmission system may be calculated from the following considerations. The power spectral density (PSD) of a random binary signal is given by:

$$G(f) = \frac{\sigma^2}{T} \mid S(f) \mid^2 + \frac{a_0^2}{T^2} \sum_{k=-\infty}^{\infty} \left| S\left(\frac{k}{T}\right) \right|^2 \delta(f - kf_r), \tag{2.1}$$

where $S(f)$ is the Fourier transform of the pulse waveform $s(t)$, σ^2 = variance of the data symbols, a_0 = mean amplitude of the symbols and $f_r = 1/T$ = repetition rate of the pulses. For binary RZ pulses shown in Fig. 2.2(a),

$$\mid S(f) \mid^2 = (A\tau)^2 \left(\frac{\sin \pi f\tau}{\pi f\tau}\right)^2, \qquad \tau = \text{pulse width} \leqslant T,$$

and for a bipolar symmetrical $[p(1) = p(0)]$ RZ random pulse train with $a_0 = 0$, $G(f)$ is

$$G(f)_{[\mathrm{BSR}]} = \frac{A^2}{T}\left(\frac{\sin \pi f\tau}{\pi f}\right)^2. \tag{2.2}$$

It is thus seen that $G(f)_{[\mathrm{BSR}]}$ has the same spectral density as $\mid S(f) \mid^2$ within a constant, and is shown in Fig. 2.2(b). For maximum energy transfer through the channel, the channel bandwidth [BW(ch)] has to include at

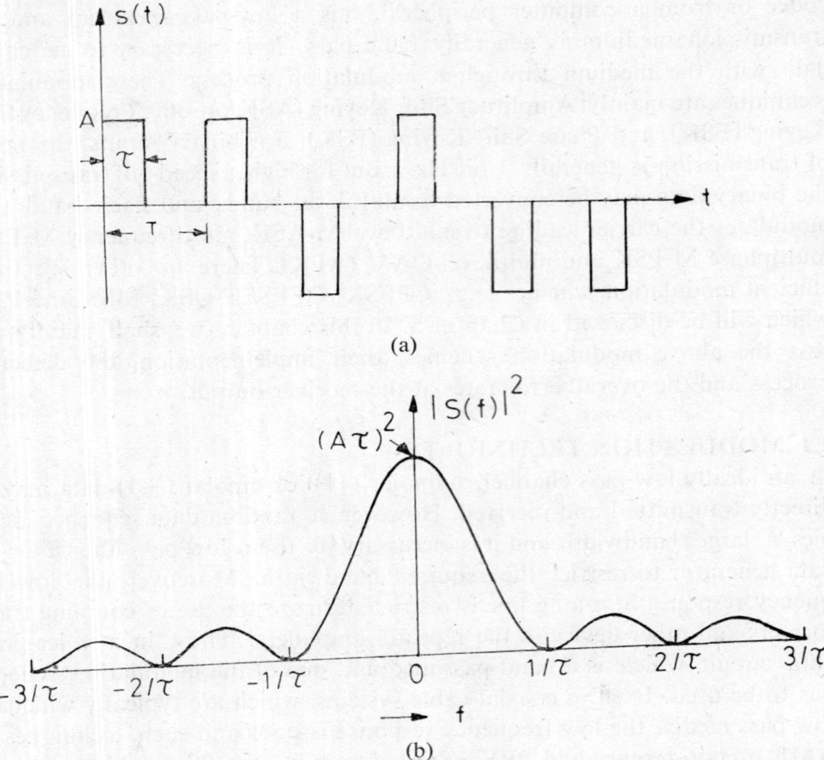

Fig. 2.2 (a) *RZ* pulse train and (b) The energy-density spectrum of a single pulse

least the main lobe of $G(f)$, and the BW(ch.) $\geqslant 1/\tau$. Since $\tau \leqslant T$, the minimum BW(ch.) is $1/T$ Hz. On the other hand, the sampling theorem tells us that random narrow pulses, $\tau \ll T$, can be transmitted through an ideal lowpass channel if the channel cutoff frequency $f_c = 1/2T$ Hz. The ideal lowpass filter produces sin x/x type of impulse response with zeros at $t_k = k/(2f_c) = kT$, and the successive pulses can be distinguished without any error. Since ideal filters are not physically realizable, the zeros of the pulse response will no longer be at $t_k = kT$, and this gives rise to the well-known intersymbol interference (ISI) problem in practical systems. As a partial solution to this, the signalling pulses are shaped with raised cosine or PRS filters such that the oscillatory tail of the response is reduced, thereby reducing the ISI. Consider a raised-cosine filter given by the amplitude response [1]:

$$A(\omega) = \frac{1}{2}\left[1 + \cos\frac{\pi\omega}{2\omega_c}\right] = \cos^2\frac{\pi\omega}{4\omega_c}, \qquad \omega \leqslant 2\omega_c;$$

and its impulse response is:

$$h(t) = \left(\frac{\tau\omega_c}{\pi}\right)\frac{\sin 2\omega_c t}{2\omega_c t[1 - (2\omega_c t/\pi)^2]}, \tag{2.3}$$

where $\tau \leqslant T/2$. The above $A(\omega)$, shown in Fig. 2.3(a), is said to have a roll-off factor $\alpha = 1$, since $A(\omega)$ extends up to $2\omega_c$. In practice, $0.25 \leqslant \alpha \leqslant 1$ is used. Figure 2.3(b) shows the impulse response $h(t)$ corresponding to $\alpha = 0.5$ and 1. For finite pulse width τ comparable to T, $A(\omega)$ is modified by the factor $(\omega\tau/2)/\sin(\omega\tau/2)$ to obtain the same responses as in Fig. 2.3.

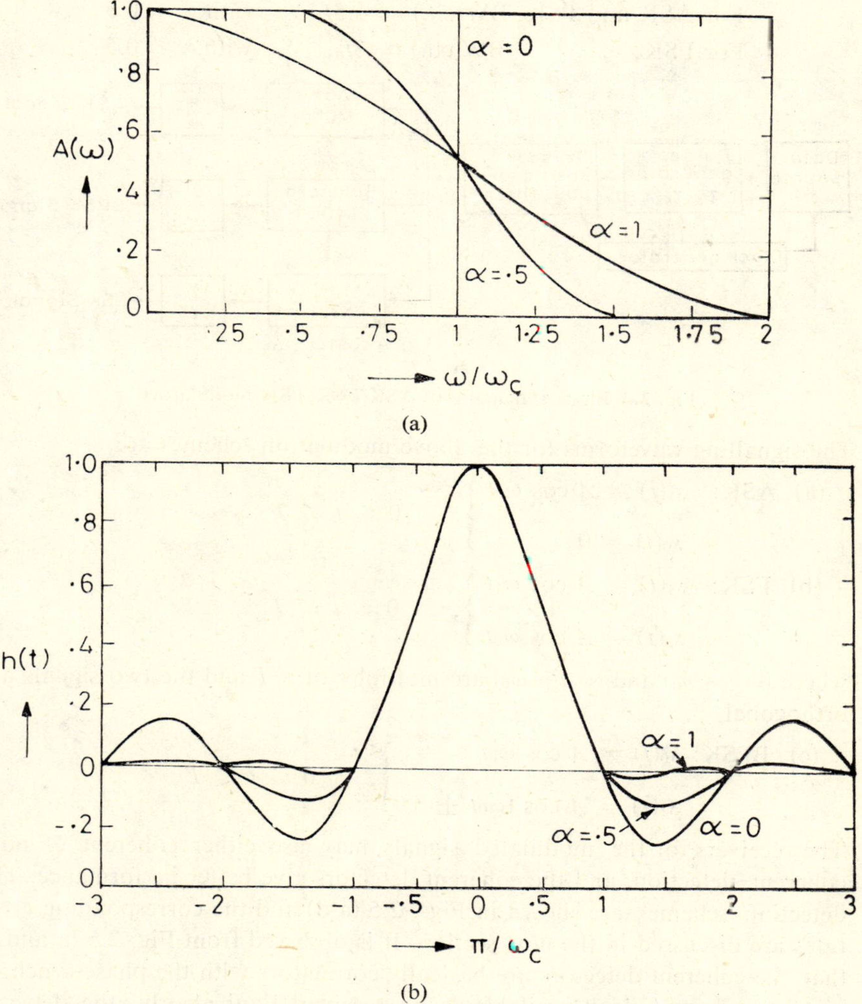

Fig. 2.3 Raised cosine spectral shaping: (a) filter characteristics and (b) impulse responses

In practice, a random pulse train with $\tau = T/2$ is transmitted through the modified raised-cosine filter and the overall ISI due to the filter plus medium is corrected by proper equalization at the receiver. An alternative technique of spectrum shaping is the Partial Response Signalling (PRS), which will be discussed in Chapter 4.

For band-pass channels, the base-band data are shifted to the required pass band through ASK/FSK/PSK modulation, as shown schematically in Fig. 2.4. The low-pass filter is optional, since the overall spectrum shaping may be done by a properly designed band-pass filter as well. The overall bandwidth required now is approximately:

For ASK and PSK, BW(ch.) $\simeq 1.5 f_r$, with $\alpha = 0.5$

For FSK, BW(ch.) $\simeq 3 f_n$, with $\alpha = 0.5$

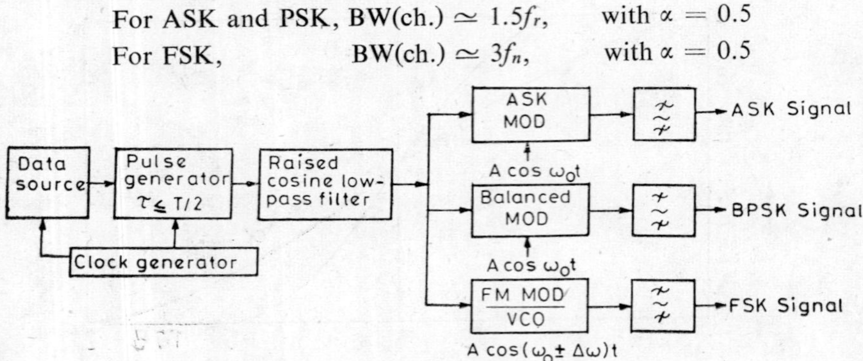

Fig. 2.4 Block schematic of ASK/FSK/PSK modulators

The signalling waveforms for the above modulation schemes are:

(a) ASK: $$\left.\begin{aligned} s_1(t) &= A \cos \omega_0 t \\ s_0(t) &= 0 \end{aligned}\right\}, \quad 0 \leqslant t \leqslant T$$

(b) FSK: $$\left.\begin{aligned} s_1(t) &= A \cos \omega_1 t \\ s_0(t) &= A \cos \omega_0 t \end{aligned}\right\}, \quad 0 \leqslant t \leqslant T$$

where $(\omega_1 - \omega_0)$ and $(\omega_1 + \omega_0)$ are multiples of π/T and the two signals are orthogonal

(c) BPSK: $$\left.\begin{aligned} s_0(t) &= A \cos \omega_0 t \\ s_1(t) &= A \cos (\omega_0 t \pm \pi) \end{aligned}\right\}, \quad 0 \leqslant t \leqslant T$$

The receivers for the modulated signals may use either coherent or non-coherent detection, and the coherent detectors give better performance. The detection schemes are shown in Figs. 2.5 (a-d) and the corresponding error rates are discussed in the next section. It is observed from Fig. 2.5 (a and c) that the coherent detectors are basically correlators with the phase-synchronized carrier and clock available at the receiver. Equivalently, the detector may be realized through a matched-filter (MF) correlator, as discussed below. It may be noted that in Fig. 2.5(a), the BPSK signal will give bipolar (± 1) output, whereas for ASK, the output will be unipolar (1/0) only. The incoherent detection for BPSK signals is realized by using the phase of $(k - 1)$th pulse as the reference for the kth pulse in the coherent detector, as shown in Fig. 2.6b. To obtain the correct data bits at the receiver, the transmitter data are precoded differentially, generating DPSK signals,

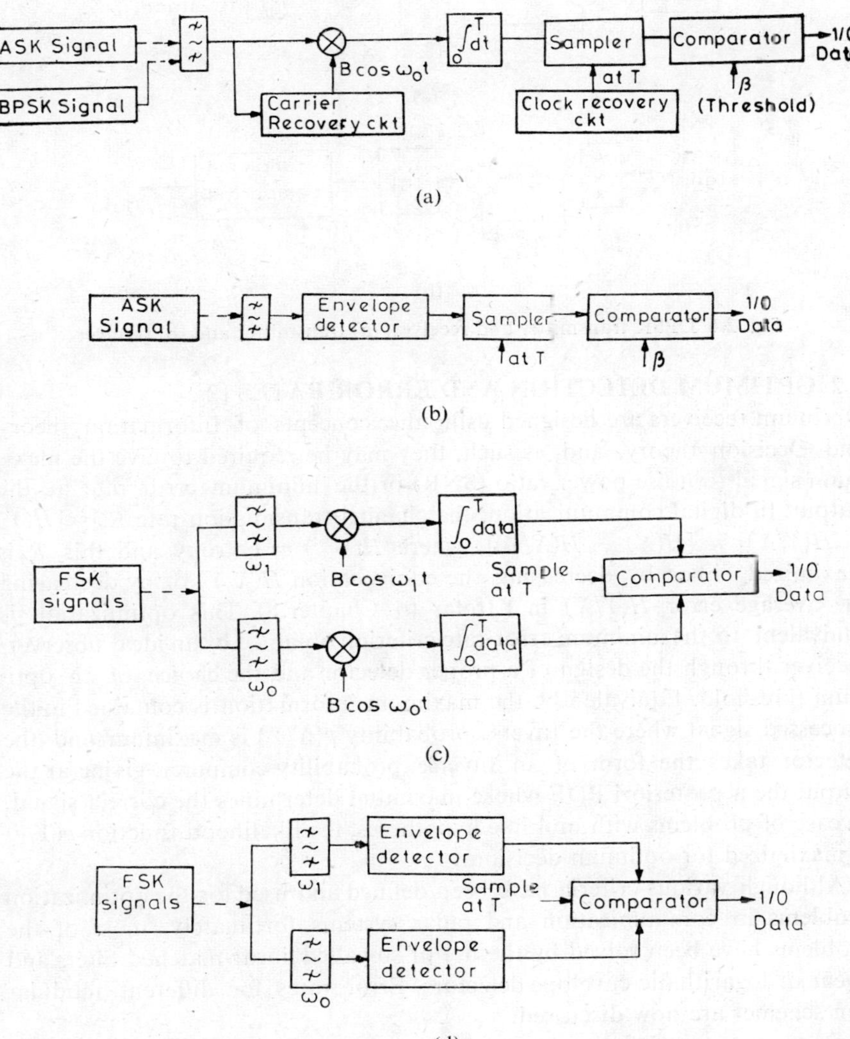

Fig. 2.5 Coherent/incoherent detectors for digital signals: (a) coherent detector for ASK; (b) incoherent detectors for ASK; (c) coherent detector for FSK; and (d) incoherent detector for FSK

also shown in the Fig. 2.6(a). It may be checked that the message bits: {0010100110 . . .} produces {00011000100 . . . } at the precoder output and the transmitted phase are: $\{000\pi\pi000\pi00 \ldots\}$. Now the detector output gives the same data bits as the original message. This avoids the difficulties of carrier synchronization at the receiver, but the error performance deteriorates by 1 dB only as compared to that of CPSK.

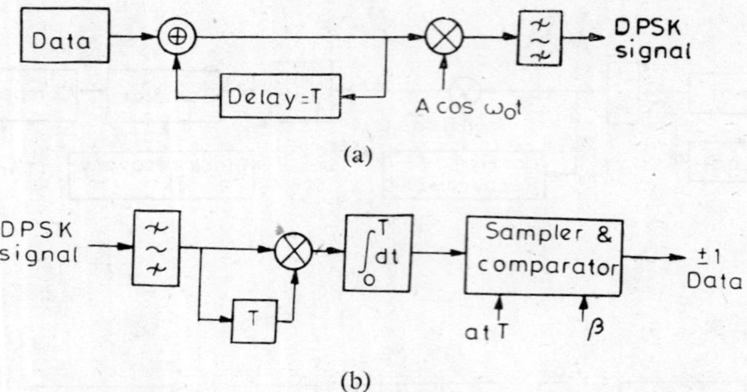

Fig. 2.6 DPSK transmitter and receiver: (a) transmitter and (b) receiver

2.2 OPTIMUM DETECTION AND ERROR RATES [2]

Optimum receivers are designed using the concepts of Information theory and Decision theory, and, as such, they may be required to give the maximum signal-to-noise power ratio (SNR) or the minimum error rate at the output. In digital communication, the channel transmission rate $R = [H(Y) - H(Y/X)] = [H(X) - H(X/Y)]$, where $H(\quad) =$ entropy and this R is maximized either by decreasing the equivocation $H(X/Y)$ or by decreasing the average error $H(Y/X)$ in Y (refer to Chapter 8). This optimization is equivalent to the minimum error-rate criterion obtained by an ideal observer/receiver through the design of a proper detector and the choice of an optimum threshold. Equivalently, the maximum information is contained in the processed signal where the Inverse probability $p(X/Y)$ is maximum and the detector takes the form of an Inverse probability computer, giving at the output the a posteriori PDF whose maximum determines the correct signal. In case of problems with multiple hypotheses, the likelihood function $p(Y/\theta)$ is maximized for optimum decision.

Although various criteria have been defined and used for the optimization problems in communication and radar systems, fortunately, most of the problems have been solved by the use of so-called linear matched filters and linear or logarithmic envelope detectors. Error rates for different modulation schemes are now discussed.

2.2.1 Weighted Probabilities an Hypothesis Testing [3]

Consider a simple model of 1/0 binary signals, designated by A_1/A_0 volts and disturbed by Gaussian noise in the channel as shown in Fig. 2.7. For a single sample or for a single digit transmission, the conditional probabilities at the receiver would be

$$p(y/0) = \frac{1}{\sqrt{2\pi}\sigma} \exp\,[-(y - A_0)^2/2\sigma^2]$$
$$p(y/1) = \frac{1}{\sqrt{2\pi}\sigma} \exp\,[-(y - A_1)^2/2\sigma^2] \tag{2.4}$$

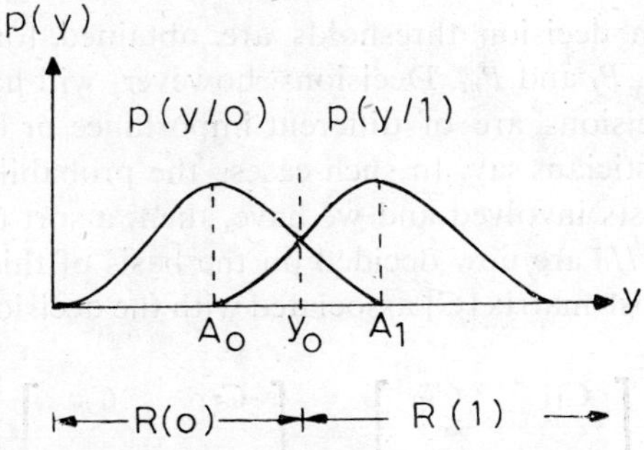

Fig. 2.7 PDF of A_1/A_0 in binary transmission

If the source probabilities are $p(0) = P_0$, $p(1) = P_1$, and $P_0 + P_1 = 1$, and if the decision at the receiver is taken such that the hypothesis H_1 is true (i.e., y is called '1') if the received signal exceeds the threshold y_0 and the hypothesis H_0 (i.e., y is called '0') is true otherwise, as shown by the regions $R(1)$ and $R(0)$, respectively, in the figure, then the error probabilities P_f (= false alarm probability) and P_m(= probability of miss) are given by:

$$P_f = P_0 \int_{R(1)} p(y_0/0)\, dy = p_0 \int_{y_0}^{\infty} p(y/0)\, dy \tag{2.5}$$

$$P_m = P_1 \int_{R(0)} p(y/1)\, dy = P_1 \int_{-\infty}^{y_0} p(y/1)\, dy$$

The total error probability $P_e = (P_m + P_f)$ is now minimized by choosing y_0 such that the likelihood ratio

$$\lambda_1 = \frac{p(y/1)}{p(y/0)} > \frac{P_0}{P_1} \text{ for hypothesis } H_1$$

and

$$\lambda_0 = \frac{p(y/0)}{p(y/1)} > \frac{P_1}{P_0} \text{ for hypothesis } H_0. \tag{2.6}$$

Such a receiver is called an 'ideal observer'. The method is also equivalent to the 'maximum a posteriori detection' procedure which maximizes the inverse prabablities $p(1/y)$ and $p(0/y)$ simultaneously. The optimum decision threshold is now given by:

$$y_0 = \frac{A_1 + A_0}{2} + \frac{\sigma^2}{A_1 - A_0} \log_e (P_0/P_1) \tag{2.7}$$

and for the Binary symmetric channel with $P_0 = P_1$, $P_f = P_m$, we have

$$y_0 = \frac{A_1 + A_0}{2} \tag{2.8}$$

The Bayes' Criterion

The above optimum decision thresholds are obtained for the unweighted probabilities P_0, P_1, P_f and P_m. Decisions, however, will have to be biased if the different decisions are of different importance or have different risk values, as the statisticians say. In such cases, the probabilities are weighted according to the costs involved and we have, then, a sort of a cost matrix. The hypotheses H_1/H_0 are now decided on the basis of this cost matrix.

In general, the cost matrix $[C]$ associated with the decision matrix $[H]$ may be written as

$$[C] = \begin{bmatrix} C_{11} & C_{10} \\ C_{01} & C_{00} \end{bmatrix} = \begin{bmatrix} C_{11} & C_m \\ C_f & C_{00} \end{bmatrix}$$

for

$$[H] = \begin{array}{c|cc} X \backslash Y & 1 & 0 \\ \hline 1 & H_1 & H_0 \\ 0 & H_1 & H_0 \end{array}$$

where C_{ij} is the cost of choosing H_j when actually H_i was true. In practical situations, the cost C_{11} and C_{00} for correct decisions H_1 and H_0 may be taken as zero (except when some credit is given for the correct decisions) and the $[C]$ reduces to

$$[C] = \begin{bmatrix} 0 & C_m \\ C_f & 0 \end{bmatrix}$$

for the error probabilities

$$[P] = \begin{bmatrix} 0 & P_m \\ P_f & 0 \end{bmatrix}$$

Using the cost matrix, the average risk per decision is now given as:

$$\begin{aligned} \bar{C}(y_0) &= P_0 \left[C_{00} \int_{R(0)} p(y/0)\,dy + C_{01} \int_{R(1)} p(y/0)\,dy \right] \\ &\quad + P_1 \left[C_{01} \int_{R(0)} p(y/1)\,dy + C_{11} \int_{R(1)} p(y/1)\,dy \right] \\ &= C_f P_0 \int_{y_0}^{\infty} p(y/0)\,dy + C_m P_1 \int_{-\infty}^{y_0} p(y/1)\,dy, \quad \text{if } C_{11} = C_{00} = 0 \end{aligned} \tag{2.9}$$

The average risk $\bar{C}(y_0)$ is now minimized by choosing the optimum value of y_0, which may be obtained by setting $\frac{d}{dy}[\bar{C}(y_0)] = 0$. The results, thus, obtained are:

$$\begin{aligned} &\text{choose } H_1 \text{ for} \quad \lambda_1(y) > \lambda_t \\ &\text{choose } H_0 \text{ for} \quad \lambda_1(y) < \lambda_t \end{aligned} \tag{2.10}$$

where

$$\lambda_t = \frac{P_0(C_f - C_{00})}{P_1(C_m - C_{11})} = \frac{P_0 C_f}{P_1 C_m}, \quad \text{if } C_{11} = C_{00} = 0$$

$$= \frac{\exp[-(y_0 - A_1)^2/2\sigma^2]}{\exp[-(y_0 - A_0)^2/2\sigma^2]}$$

The corresponding decision threshold is

$$y_0 = \frac{A_1 + A_0}{2} + \frac{\sigma^2}{A_1 - A_0} \log_e \left(\frac{P_0 C_f}{P_1 C_m}\right) \tag{2.11}$$

Thus, the threshold is biased towards the decision costing more, i.e. $C_f > C_m$, $P_0 = P_1$, then $y_0 > (A_1 + A_0)/2$ = the mean value of the signal.

This decision strategy, which minimizes the average risk $\bar{C}$, is known as the Bayes' criterion, and the minimum average risk $\bar{C}_{min}$ obtained by using eq. (2.11) for y_0 in eqn. (2.9) is known as Bayes' risk. When the relative costs $(C_{01} - C_{00}) = (C_{10} - C_{11})$, the Bayes' criterion is equivalent to the ideal observer of eqn. (2.6). The Bayes' criterion, defined by eqn. (2.10), is also true when the decision is based on measurements of more than one variable. It is necessary, however, to have a priori knowledge of the relative costs and the source probabilities P_0 and P_1 (resulting in simple hypotheses H_1 and H_0). The concept of the Bayes' strategy demands the possibility of a large number of observations or trials.

2.2.2 Binary Communication Using Ideal Observer

For an ideal observer, the requirement that the average error rate $H(Y/X)$ is to be minimized leads to the condition:

$$H(Y/X) = -\sum_x p(x_i) \cdot \sum_y p(y_i/x_i) \cdot \log_2 p(y_i/x_i) \simeq 0$$

If $p(y_i/x_i)$ approaches unity, then $H(Y/X)$ tends to zero. From Bayes' theorem on inverse probability, we have

$$\sum_x p(x_i/y_i) = Kp(x_i) \sum_y p(y_i/x_i),$$

where $p(x_i)$ and $p(y_i)$ are known. The ideal observer has to maximize both the 'inverse probability' and the 'likelihood function' with the help of optimum detection and filtering.

If the cost matrix is modified as

$$[C] = \begin{bmatrix} 0 & 1 \\ 1 & 0 \end{bmatrix}$$

then the Bayes' criterion is equivalent to the ideal observer. The 'maximum a posteriori detection' procedure, where the hypotheses H_1/H_0 are chosen on the basis of: $p(1/y) > p(0/y)$ or otherwise, also gives the same results and optimum thresholds for the ideal case of sampled functions or digits used as the transmitted symbols. For the unipolar (1/0) binary transmission and reception, the optimum threshold $y_0 = A/2$, where $A = 1$, $p(0) = p(1) = 0.5$, and the error probabilities are (for Gaussian noise)

$$P_e = P_f + P_m = 2P_f = \int_{A/2}^{\infty} \frac{1}{\sqrt{2\pi}\sigma} \exp(-y^2/2\sigma^2)\, dy$$

$$= \int_{A/2\sigma}^{\infty} \frac{1}{\sqrt{2\pi}} \exp(-z^2/2)\, dz,\ z = y/\sigma$$

$$= \operatorname{erfc}(A/2\sigma) \tag{2.12*}$$

where

$$\operatorname{erf}(x) = \int_{-\infty}^{x} \frac{1}{\sqrt{2\pi}} \exp(-z^2/2)\, dz,$$

and $\quad \operatorname{erfc}(x) = [1 - \operatorname{erf}(x)].$

For the bipolar transmission with $\pm A/2$ volts signals, $y_0 = 0$, and

$$P_e = \operatorname{erfc}(A/2\sigma) \tag{2.13}$$

which is the same as eqn. (2.12). But the signal power S for unipolar transmission is $A^2/2$, whereas S for bipolar is $A^2/4$, thus showing an advantage of 3 dB in the SNR required for bipolar signalling for the same P_e.

To improve the visibility of radar echoes or to decrease P_e in binary communication, the signal may be repeated n times or the pulse width made much larger than the inverse of the receiver bandwidth B. In such cases, the signal plus noise is sampled n times, the sampling interval being just greater than $1/B$, and the decision H_1/H_0 is based on the joint PDF of $Y(= y_1, y_2, \ldots, y_n)$ exceeding the threshold or not. For coherent detection and integration at the receiver, the sampled signals are independent and the decisions are baseed on

$$\lambda_1(Y) = \frac{\exp\left[\sum_{j=1}^{n} - (y_j - A_1)^2/2\sigma^2\right]}{\exp\left[\sum_{j=1}^{n} - (y_j - A_0)^2/2\sigma^2\right]} > \frac{P_0}{P_1} \cdots$$

$$\text{for hypothesis } H_1 \tag{2.14}$$

and H_0 otherwise, where $y_j = j$th sample value and A_1/A_0 stands for 1/0 digits. The optimum decision threshold is now given by

$$y_0 = \left[\frac{n(A_1 + A_0)}{2} + \frac{\sigma^2}{A_1 - A_0} \log_e (P_0/P_1)\right] \tag{2.15}$$

---

*An alternative definition of erf (x) is:

$$\operatorname{erf}_*(x) \triangleq \frac{2}{\sqrt{\pi}} \int_0^x \exp(-z^2)\, dz \text{ and } \operatorname{erfc}_*(x) = 2 \operatorname{erfc}(\sqrt{2}x).$$

Using this value of $\operatorname{erf}_*(x)$, $P_e = \frac{1}{2} \operatorname{erf}_*[A/(2\sqrt{2}\sigma)]$. However, eqn. (2.12) will be used in the present text.

which is the same as eqn. (2.7) for $n = 1$. H_1 is now decided upon if $\sum_n y_j \geqslant y_0$ and H_0 for $\sum_n y_j < y_0$. The optimum receiver, for the above operations, consists of a coherent detector, an integrator and threshold device.

2.2.3 Exactly Known Signals of Any Shape

Consider that the received signal $Y = S + N$, with transmitted signal present and $Y = N$, with zero transmission, where

$$S(t) = \{S_1, S_2, \ldots, S_{TW}\} \qquad \text{for } 0 < t < T \tag{2.16*}$$
$$= 0 \qquad \text{otherwise}$$

Then, $Y(t) = \{y_1, y_2, \ldots, y_{2TW}\}$ and the joint PDF's $p(Y/S)$ and $p(Y/N)$ are represented in the same $2TW$-dimensional vector space. Assuming further that the sampled values are independent, the likelihood ratio $\lambda_s(y)$ is given by [similar to eqn. (2.14)]

$$\lambda_s(Y) = \frac{p(Y/S)}{p(Y/N)} = \frac{\exp\left[-\sum_{i=1}^{2TW} (y_i - S_i)^2/2\sigma^2\right]}{\exp\left[-\sum_{1}^{2TW} y_i^2/2\sigma^2\right]} \tag{2.17}$$

and the decisions are to be taken as $\lambda_s(Y) > \lambda_t$ for hypothesis H_S, and H_N otherwise. λ_t here is normally unknown as the a priori probabilities are not known.

The above condition is further simplified as [3]:

$$u \triangleq \frac{2}{n_0}\int_0^T y(t)\cdot S(t)\,dt > \beta \triangleq \left(\frac{E}{n_0} + \ln \lambda_t\right)$$
$$\text{for hypothesis } H_S \tag{2.18}$$

*[Refer to Appendix A]. $S(t)$ is represented here by its sample values $\{S_1, S_2, \ldots, S_{2TW}\}$, where T and W are the time duration and bandwidth of the signal. The signal energy E is then given by:

$$E = \int_{-\infty}^{\infty} S^2(t)dt = \frac{1}{2W}\sum_{1}^{2TW} S_i^2$$

and the signal power

$$P = \frac{1}{2WT}\sum_{1}^{2TW} S_i^2$$

Further, the band-limited white Gaussian noise of duration T is also completely specified by the $2TW$ samples which are statistically independent. The joint PDF of the samples $\{n_1, n_2, \ldots, n_{2TW}\}$t is given by:

$$p(n_1, n_2, \ldots, n_{2TW}) = \left(\frac{1}{2\pi\sigma^2}\right)^{TW} \cdot \exp\left[-(n_1^2 + n_1^2 + \ldots + n_{2TW}^2)/2\sigma^2\right].$$

where $E = \int_{-\infty}^{\infty} S^2(t)\,dt$, $\sigma^2 = Wn_0$, $n_0/2 =$ two-sided spectral density of white noise. The optimum receiver is then, an I.F. cross-correlator followed by a sampler-comparator, as shown in Fig. 2.8, given that the signal is completely known in amplitude, shape and time of arrival. The values of λ_t and

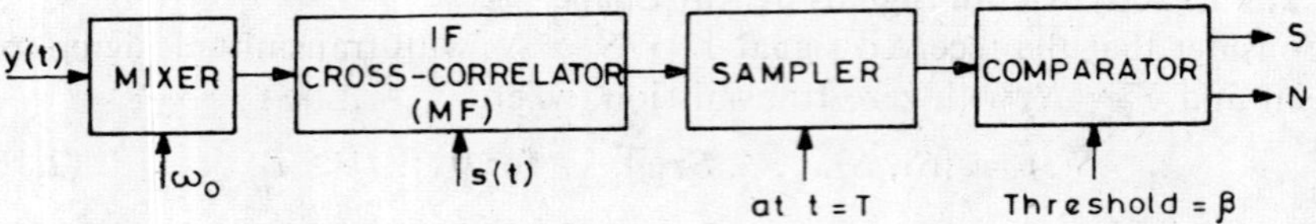

Fig. 2.8 Optimum receiver for an exactly known signal

β depend on the criteria to be used, and for the Neyman-Pearson test*, they are calculated from the allowable value of P_f. The conditional PDF's $p(u/N)$ and $p(u/S)$ are obtained from $p(Y/N)$ and $p(Y/S)$, respectively, and

$$p(u/N) = \sqrt{\frac{n_0}{4\pi E}} \cdot \exp\left(-n_0 u^2/4E\right)$$

and

$$p(u/S) = \sqrt{\frac{n_0}{4\pi E}} \cdot \exp\left[-n_0(u - 2E/n_0)^2/4E\right] \tag{2.19}$$

These PDF's are Gaussian and, hence, the error probabilities are obtained as

$$P_f = \int_{\beta}^{\infty} p(u/N)\,du = \text{erfe}[\beta\sqrt{n_0/2E}]$$

$$P_m = \int_{-\infty}^{\beta} p(u/S)\,du = \text{erfc}[(\beta - 2E/n_0)\sqrt{n_0/2E}] \tag{2.20}$$

Then, the detection probability is simply, $P_d = (1 - P_m)$.

If, however, the received signals are of varying amplitudes, but of constant shape, i.e.,

$$S_j(t) = A_j S_0(t)$$

where $S_0(t)$ is of constant energy and shape, then, we have to evaluate the average likelihood ratio $\lambda_s^{av}(Y)$. The condition for this compositehy pothesis testing is now modified as

$$u \triangleq \frac{2}{n_0}\int_0^T y(t)\cdot S_0(t)\,dt > \beta \triangleq \left(\frac{E}{n_0} + \ln \lambda_t^{av}\right), \tag{2.21}$$

*When the a priori probabilities and the cost matrix are unknown, the Bayes' criterion cannot be applied. In such situations, as in radar search problems, the observer preselects a value of P_f which he can afford, and seeks a decision strategy that maximizes the detection probability $P_d = (1 - P_m)$, keeping P_f constant. This strategy is known as the Neyman-Pearson criterion.

where $\lambda_t^{av} = (3/A_0^2)\lambda_t$, assuming the least favourable distribution of A_j, i.e., $p(A_j) = 1/A_0$, $0 < A_j < A_0$. The optimum receiver is now the same as of Fig. 2.8., except that the threshold of the comparator is only modified.

2.2.4 Detection of Two Exactly Known Signals of Arbitrary Shape

The basic problem here is to find the optimum way to distinguish between either of two waveforms, $S_1(t)$ and $S_2(t)$, defined over the interval 0-T sec. As an extension of the results of the previous section, the likelihood ratios for the bandlimited signals $S_1(t)$ and $S_2(t)$, corrupted by white Gaussian white noise are

$$\lambda_{S_1} = \frac{p(Y/S_1)}{p(Y/S_2)} = \frac{\exp\left[-\sum_{i=1}^{2TW} (y_i - S_{i1})^2/2\sigma^2\right]}{\exp\left[-\sum_{i=1}^{2TW} (y_i - S_{i2})^2/2\sigma^2\right]}$$

$$> \lambda_t \text{ for hypothesis } H_{S_1} \tag{2.22}$$

$$\lambda_{S_2} = \frac{p(Y/S_2)}{p(Y/S_1)} > 1/\lambda_t \text{ for hypothesis } H_{S_2},$$

where $S_1(t)$ and $S_2(t)$ are represented by $2WT$ independent samples. Simplifying further, the condition for minimum P_e is

$$2\sigma^2 \ln (\lambda_{s1}) > 2\sigma^2 \ln (P_2/P_1) \ldots \quad \text{for } H_{S_1}$$

which reduces to

$$\int_0^T y(t)(S_1 - S_2)\, dt > \left(\frac{n_0}{2} \ln \frac{P_2}{P_1} + \frac{E_1}{2} - \frac{E_2}{2}\right) \tag{2.23}$$

where

$$\Sigma S_{i1}^2 = 2WE_1, \qquad \Sigma S_{i2}^2 = 2WE_2, \qquad P_1 = p(S_1),\ P_2 = p(S_2),$$

and

$$\Sigma 2y_i(S_{i1} - S_{i2}) = 4W \int_0^T y(t)(S_1 - S_2)\, dt$$

The optimum receiver is then given by two matched filters, matched to $S_1(t)$ and $S_2(t)$, respectively, such that $h_1(t) = S_1(T - t)$ and $h_2(t) = S_2(T - t)$. The outputs of two filters are compared at $t = T$, as shown in Fig. 2.9, and

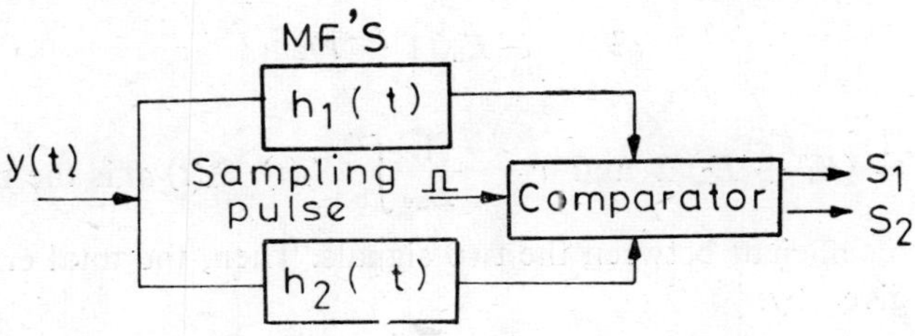

Fig. 2.9 Optimum receiver for two exactly known signals

the decision is made in favour of the larger output. Alternatively, a single matched filter, with $h(t) = [S_1(T - t) - S_2(T - t)]$, may be used along with a threshold in the comparator. For the special case of $E_1 = E_2, P_1 = P_2 = 0.5$, the threshold would be zero.

To determine the error probabilities, it is seen from eqn. (2.23), that an error of deciding H_{S_2} for H_{S_1} occurs when

$$u \triangleq \int_0^T n(t)(S_1 - S_2)\,dt < \left[\frac{n_0}{2}\ln(P_2/P_1) - \frac{1}{2}\int_0^T (S_1 - S_2)^2\,dt\right] \triangleq u_1 \tag{2.24}$$

where $y(t)$ for S_1 transmitted is $[S_1(t) + n(t)]$. Since $n(t)$ is Gaussian with zero mean, then u is also Gaussian with zero mean, and the variance σ_u^2 is

$$\sigma_u^2 = \frac{n_0}{2}\int_0^T (S_1 - S_2)^2\,dt$$

The PDF $p(u/S_1)$ is now given by

$$p(u/S_1) = \frac{1}{\sqrt{2\pi}\sigma_u}\exp(-u^2/2\sigma_u^2)$$

The errors occur when $u < u_1$ and the total probability of deciding H_{S_2} for H_{S_1} is

$$P_m = \int_{-\infty}^{u_1} p(u/S_1)\,du = \operatorname{erfc}(Z_1) \tag{2.25}$$

where

$$Z_1 = -\frac{u_1}{\sigma_u} = \frac{a}{2} + \frac{1}{a}\ln(P_1/P_2) \text{ and } a^2 \triangleq \frac{2}{n_0}\int_0^T (S_1 - S_2)^2\,dt$$

Similarly, the total probability of deciding H_{S_1} for H_{S_2} is

$$P_f = \operatorname{erfc}(Z_2), \qquad Z_2 = \left[\frac{a}{2} + \frac{1}{a}\ln(P_2/P_1)\right] \tag{2.26}$$

The overall probability of error in the binary communication is then given by

$$P_e = P_1P_f + P_2P_m \tag{2.27}$$

Because of the symmetrical form of P_f and P_m, it is seen that P_e depends only on P_1/P_2 and a^2. For binary signalling, $P_1 = P_2 = 0.5$, and $P_e = P_f = P_m$. Now, the parameter a^2 may be written as:

$$a^2 = \frac{4}{n_0}E_{av}(1 - \rho),$$

where $E_{av} = (E_1 + E_2)/2$ and $\rho = \frac{1}{E_{av}}\int_0^T S_1(t)\cdot S_2(t)\,dt$ is the normalized correlation coefficient between the two signals. Then, the total error probability P_e is given by:

$$P_e = \operatorname{erfc}\sqrt{(1 - \rho)E_{av}/n_0} \tag{2.28}$$

P_e then depends on the average signal energy E_{av}, n_0 and the cross-correlation coefficient ρ. Apparently P_e does not depend on the signalling waveform since an MF/correlation receiver is used. For minimum P_e, a^2 has to be maximum and this is possible for $\rho = -1$ only [i.e., $S_1(t) = -S_2(t)$]. Then the optimum signalling waveforms are also necessary for minimum P_e. Finally, the optimum matched filter receiver, discussed above, corresponds to minimum P_e, maximum SNR and also to the least mean-squared error at the output. All the three criteria lead to the same solution of the matched filter (or cross-correlator) as the optimum receiver.

2.3 PRACTICAL BINARY SYSTEMS [4, 5]

Various binary communication systems, viz., unipolar 1/0 and bipolar ± 1, DSB-AM, FSK and PSK, as discussed in Sec. 2.1, are being used in practice. In most systems, again, there are possibilities of non-coherent and coherent detection. The results of eqns. (2.25) to (2.28) are applicable to these systems using coherent techniques, as the matched filtering is fundamentally equivalent to coherent reception. The results for non-coherent systems may be extrapolated from Rice and Rayleigh distributions of the envelope detector outputs in non-coherent receivers.

In case of baseband 1/0 and DSB-AM transmission and reception through coherent detectors or matched filters, eqn. (2.27) is modified as

$$P_e = \text{erfc}\sqrt{E/2n_0} \tag{2.29}$$

where $E_1 = E$, $S_2 = 0$, $E_2 = 0$, $a^2 = 2E/n_0$, $P_1 = P_2 = 0.5$, and the optimum threshold $= E/2$. The case for non-coherent reception with on-off signalling has been discussed in [5], where the opimum threshold is shown to be $\sqrt{2 + E/2n_0}$ and the error probability for high CNR is approximately (using optimum threshold) given by

$$P_e \cong \frac{\sqrt{n_0}}{2\sqrt{\pi E}} \exp\left(-\frac{E}{4n_0}\right) + \frac{1}{2} \exp\left(-E/4n_0\right)$$

$$\cong \frac{1}{2} \exp\left(-\frac{E}{4n_0}\right) \ldots \qquad \text{for } \frac{E}{n_0} \gg 1 \tag{2.30}$$

Equation (2.29), on the other hand, approximates, for large CNR, to

$$P_e \cong \sqrt{n_0/\pi E} \exp\left(-E/4n_0\right) \tag{2.31}$$

Thus, at high CNR, the two detection schemes perform almost equally well, the superiority of the coherent reception being only at lower CNR's.

For the binary FSK systems, shown in Fig. 2.5, again there are the possibilities of coherent and non-coherent reception. For the coherent case using orthogonal signalling, the error probability is modified as

$$P_e = \text{erfc}\left(\sqrt{E/n_0}\right)$$

$$\cong \sqrt{\frac{n_0}{2\pi E}} \exp(-E/2n_0) \ldots \qquad \text{for } E/n_0 \gg 1, \qquad (2.32)^*$$

where $E_1 = E_2 = E$, $S_1(t)$ and $S_2(t)$ are orthogonal and $a^2 = 4E/n_0$ since $\int_0^T S_1 S_2 \, dt = 0$.

For the case of non-coherent reception using filters and envelope detectors

$$P_e = \frac{1}{2} \exp\left(-\frac{E}{2n_0}\right) \qquad (2.33)$$

which again gives error rates similar to that of the coherent case for high CNR.

In case of base-band bipolar signals and coherent binary phase-shift keying (PSK) systems, the parameters of eqn. (2.27) are: $E_1 = E_2 = E$, $S_1(t) = - S_2(t)$, $P_1 = P_2 = 0.5$ and $a^2 = 8E/n_0$. Hence, the overall error probability is given by

$$P_e = \operatorname{erfc}(\sqrt{2E/n_0}) \qquad (2.34)$$

$$\cong \frac{1}{2}\sqrt{\frac{n_0}{\pi E}} \cdot \exp(-E/n_0) \ldots \qquad \text{for high CNR}$$

When there is a phase error $\Delta\theta$ between the transmitter and receiver carriers, P_e deteriorates to erfc $[\sqrt{(2E)/n_0} \cos \Delta\theta]$ and 1 dB loss in CNR to achieve the same P_e occurs when $\Delta\theta \simeq 25°$. For the DPSK signals of Fig. 2.6, the overall error rate is now found to be

$$P_e = \frac{1}{2} \exp(-E/n_0) \qquad (2.35)$$

DPSK is seen to have 3 dB lesser CNR requirements than the non-coherent FSK and gives performances similar to that of the coherent PSK at high CNR. It should be noted that for systems using the ideal Nyquist bandwidth, the CNR $\gamma = E/n_0$; but for larger BW with $\alpha > 0$, the effective noise power in the receiver increases by $(1 + \alpha)$. Then, for the same P_e, the signal power has to be increased proportionately, thus making the effective CNR $\gamma = (1 + \alpha)E/n_0$ [i.e., the ratio C/n_0 is increased by $(1 + \alpha)$, keeping P_e and n_0 constant].

All the different error rates have been plotted in Fig. 2.10, where it is seen that the coherent PSK gives the minimum P_e since it has the maximum value of a^2. This is achieved with ideal synchronization and matched-filtering at the receiver. At high CNR, most of the binary systems perform within a few dBs of each other, if constant amplitude signals and Gaussian white noise are assumed. A reference error rate $P_e = 10^{-5}$ is obtained for $E/n_0 = 9.6$ dB in PSK, for $E/n_0 = 12.6$ dB in FSK and for $E/n_0 = 15.6$ in ASK, using coherent detection. For incoherent detection, the CNR requirement is only 1 dB more in respective systems.

---

*For coherent binary FSK signalling, if the frequency separation $\Delta\omega$ between the two carriers is made equal to $1.4\pi/T$, then the minimum P_e is obtained (since $\rho \simeq -1$).

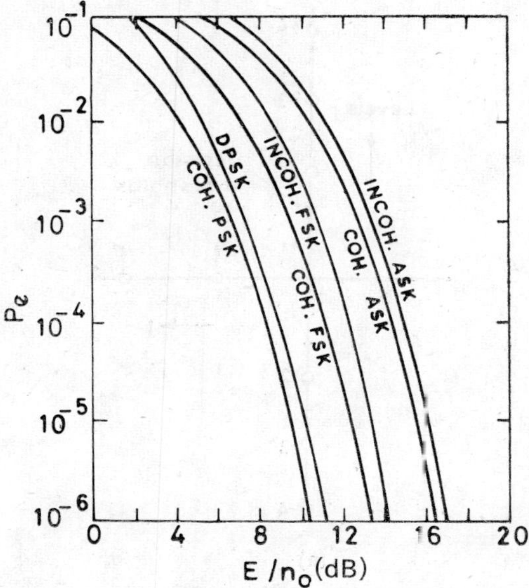

Fig. 2.10 Error rates in binary communication systems

2.4 M-ARY SIGNALLING—M-ASK, M-FSK, M-PSK and QAM [4]

In binary signalling, as discussed above, the speed of transmission is approximately 1 bit/s/Hz. For conserving bandwidth and for a higher rate of transmission, a few of the binary digits may be grouped together and converted to a multilevel symbol (baud) through D to A conversion. This multilevel baud may now be used to generate multilevel M-ASK, multifrequency M-FSK, multiphase M-PSK, and multilevel QAM (APK) signals, thereby reducing the channel BW by a factor k, if k bits are grouped to produce $M = 2^k$ level baseband signal. This BW reduction is achieved at a penalty of the required CNR for a desired P_e, as will be now discussed.

The signal constellations for M-ASK and M-PSK are shown in Fig. 2.11. The corresponding signalling waveforms are:

(a) M-ASK : $S_j(t) = A_j(a_0 \cos \omega_0 t), \quad 0 \leqslant t \leqslant kT_b = T, \qquad (2.36)$

where $1/T_b =$ data rate, $a_0 \cos \omega_0 t$ is the basic signalling waveform and A_j is the level corresponding to the k-bits of source alphabet with $j = \{1, 2, \ldots, M\}$, $M = 2^k$.

(b) M-PSK : $S_j(t) = A_0 \cos \left[\omega_0 t + \frac{2\pi}{M}(j-1)\right], \quad 0 \leqslant t \leqslant kT_b = T, \qquad (2.37)$

where $j = \{1, 2, \ldots, M\}$ and E_j, the signal energy, is the same for all j.

Referring to eqn. (2.1) and the above equations, it is observed that the PSD $G(f)$ for the signalling waveforms are the same as in eqn. (2.1) with

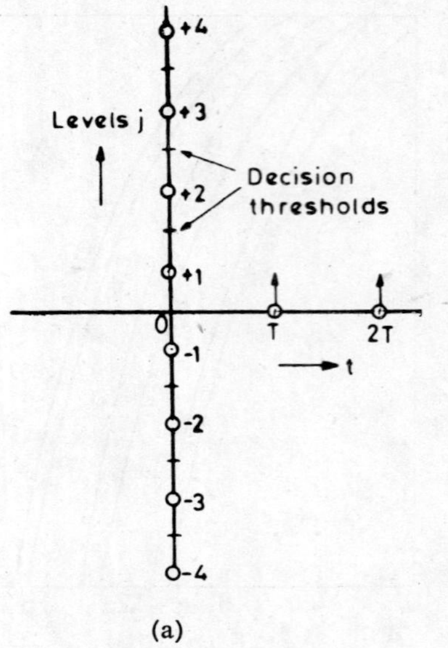

(a)

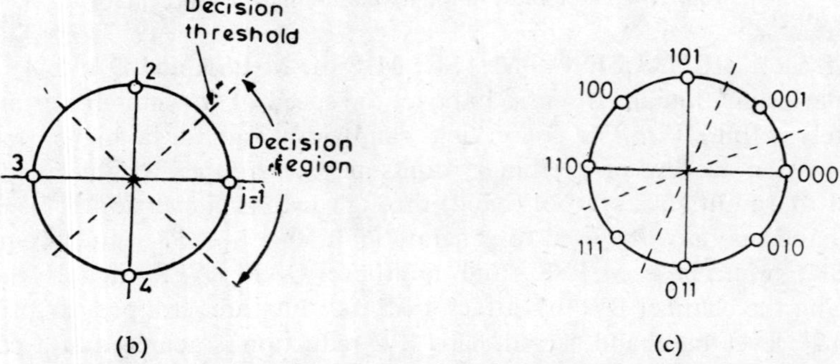

(b) (c)

Fig. 2.11 Signal constellations for M-ASK and M-PSK; (a) M-ASK, $M = 8$; (b) QPSK, $M = 4$; and (c) MPSK, $M = 8$

$T = kT_b$. Figure 2.12 shows the PSD's for both M-ASK and M-PSK signals and the reduction of channel BW with larger M is evident. The block schematic of the transmitters and receivers for M-ASK are shown in Fig. 2.13, where it is seen that the modulation and detection of signals are straightforward. Of all M-PSK systems, QPSK, $M = 4$, is the most popular technique as it provides a signalling rate of 2 bits/s/Hz, and requires the same E_b/n_0 for a given P_b as for the BPSK signals. The block schematics of QPSK transmitter and receiver are shown in Figs. 2.14a and b, where two quadrature carriers are each multiplied by half the data bits, thus making $T = 2T_b$ and the channel BW half that required for a BPSK modulator. The receiver is complementary to the modulator and uses coherent detection of the quadrature signals. The conceptual modulator and demodulator for M-PSK, $M = 8$,

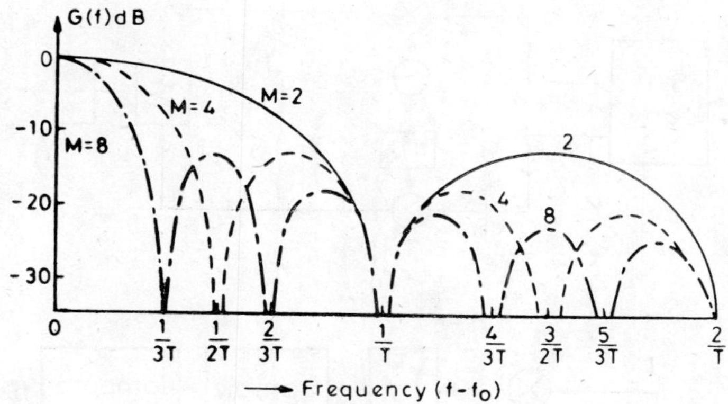

Fig. 2.12 $G(f)$ for M-ASK and M PSK signals

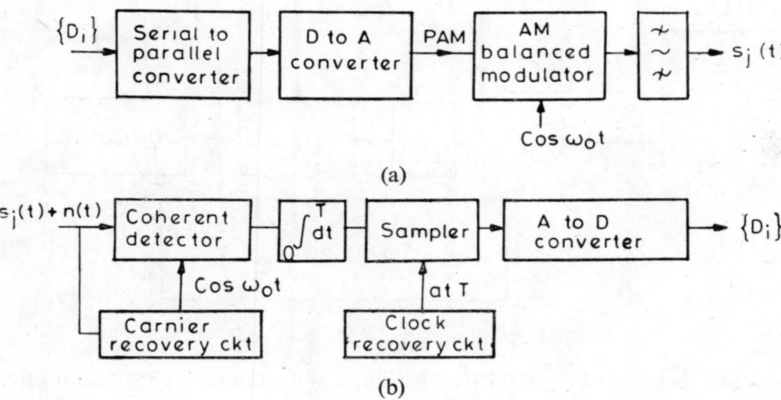

Fig. 2.13 Block schematic of M-ASK transmitter and receiver (coherent)

16, . . . , are shown in Figs. 2.15(a) and (b), where the k-bit data are converted to a multilevel PAM signal and this PAM phase modulates the carrier. The demodulator is complementary, but has the problem of sign ambiguity and non-linear quantization is required. Both the modulator and the demodulator may be simplified by using the quadrature carriers as in QPSK and modulating these carriers with suitable amplitude functions as required by the signal constellation shown in Fig. 2.11(c) for $M = 8$. Table 2.1 gives the coefficients of the I-Q vectors required for M-PSK, $M = 8$; and a coefficient generator after the S/P converter generates these amplitude functions. The M-PSK modulator is then the same as the QPSK modulator with this modification. The corresponding demodulator has I/Q channels whose outputs are quantized and decoded according as the coefficient values of Table 2.1, as shown in Fig. 2.15(c).

The receiver may be further modified using a second quadrature reference at $+\pi/4$ and $-\pi/4$, thus giving another set of vectors y_A and y_B. The combinations of y_I, y_Q, y_A and y_B may be used to decode the data bits [6].

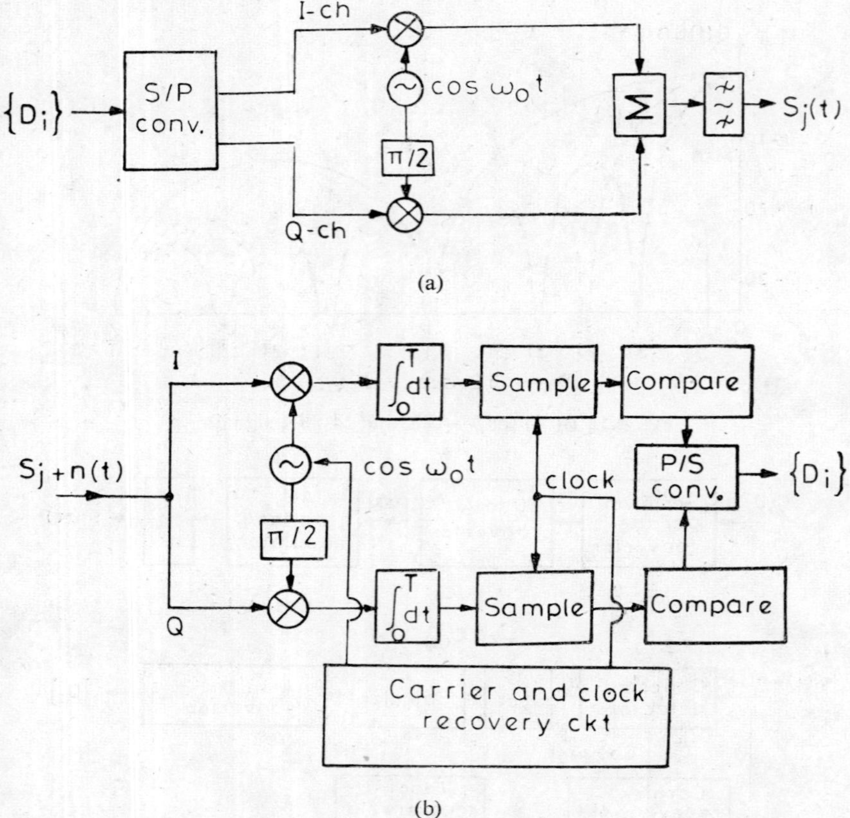

Fig. 2.14 Block schematic of QPSK transmitter and receiver: (a) TX, and (b) Rec.

Table 2.1

Data words (Gray code)	Quadrature coefficients		Quantized output	Composite signal
	y_I	y_Q		
000	1.0	0	10	$\cos \omega_0 t$
001	0.707	0.707	11	$\cos (\omega_0 t + \pi/4)$
101	0	1.0	01	$\cos (\omega_0 t + \pi/2)$
100	—0.707	0.707	—11	$\cos (\omega_0 t + 3\pi/4)$
110	—1	0	—10	$\cos (\omega_0 t + \pi)$
111	—0.707	—0.707	—1—1	$\cos (\omega_0 t + 5\pi/4)$
011	0	—1	0—1	$\cos (\omega_0 t + 3\pi/2)$
010	0.707	—0.707	1—1	$\cos (\omega_0 t + 7\pi/4)$

As in the case of binary DPSK, the incoherent phase comparison detection is also possible for M-PSK. At the transmitter, the weightage of the symbol is conveyed by the phase transitions between successive pulses rather than as the absolute phases of the pulses. In the receiver, the differential phase

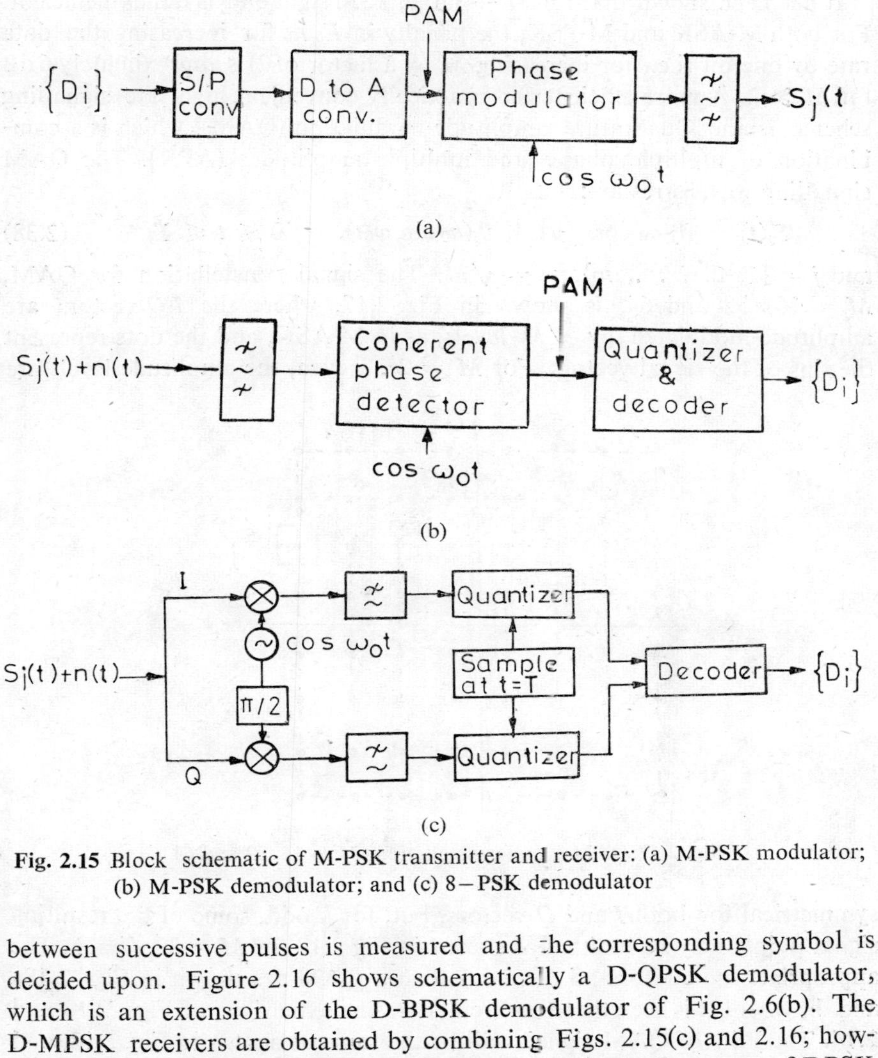

(a)

(b)

(c)

Fig. 2.15 Block schematic of M-PSK transmitter and receiver: (a) M-PSK modulator; (b) M-PSK demodulator; and (c) 8 – PSK demodulator

between successive pulses is measured and the corresponding symbol is decided upon. Figure 2.16 shows schematically a D-QPSK demodulator, which is an extension of the D-BPSK demodulator of Fig. 2.6(b). The D-MPSK receivers are obtained by combining Figs. 2.15(c) and 2.16; however, there is a 3 dB degradation in the P_e vs E_b/n_0 performance of DPSK receivers for $M > 2$.

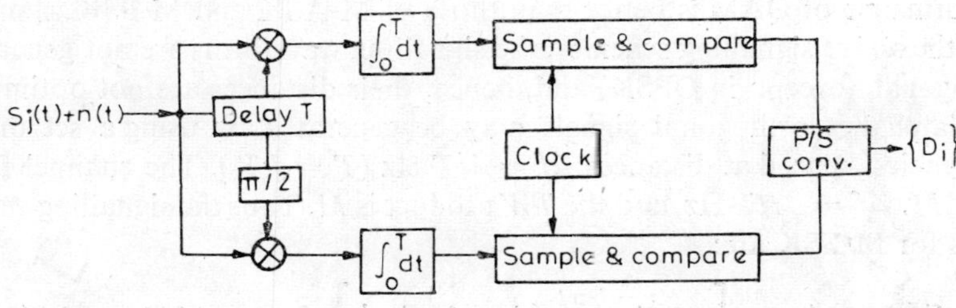

Fig. 2.16 D-QPSK receiver

It has been shown that for $M \rightarrow$ large, PSK signalling is rather inefficient. For both M-ASK and M-PSK, the penalty in E_b/n_0 for increasing the data rate by one bit (i.e., for increasing M by a factor of 2) is approximately 6 dB for $M \gg 2$. A more efficient and practically convenient high rate signalling scheme is the Quadrature amplitude modulation (QAM), which is a combination of multiple phases and multiple amplitudes (APK). The QAM signalling waveforms are:

$$S_j(t) = A_j(a_0 \cos \omega_0 t) + B_j(a_0 \sin \omega_0 t), \qquad 0 \leqslant t \leqslant T, \tag{2.38}$$

and $j = \{1, 2, \ldots, m\}$, $m = \sqrt{M}$. The signal constellation for QAM, $M = 16$, 32 and 64, is shown in Fig. 2.17, where the I-Q vectors are amplitude-modulated by $\sqrt{M}$ levels, as in M-ASK, and the dots represent the tips of the signal vectors. For $M = 2^k$, k even, the amplitude levels are

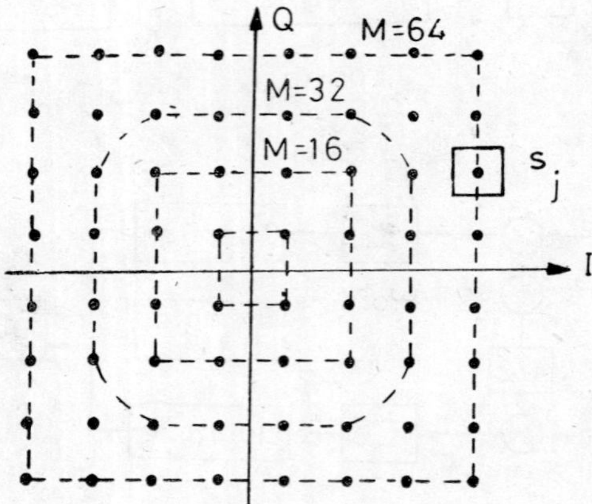

Fig. 2.17 QAM signal constellation, $M = 16$, 32 and 64

symmetrical for both I and Q vectors, but, for k odd, some of the resulting signal points may be omitted as shown in Fig 2.17 for $M = 32$, $k = 5$. The appropriate modulator and demodulator for QAM are shown in Figs. 2.18a and b, where it is seen that the circuits are the extensions of Figs. 2.13 and 2.14. The PSD of QAM signals will be the same as shown in Fig. 2.12, and the effective channel BW is reduced by a factor of k, when k bits of data are grouped together to form the multilevel baseband signal. Moreover, the error performance of QAM is better than those of M-ASK and M-PSK signals.

In the above signalling schemes, the signalling waveforms are not generally orthogonal (except in QPSK) and, hence, their distances are not optimum. Multialphabet orthogonal signals may be generated by using a set of M frequencies, spaced at distances $\Delta f = 1/T$ Hz ($T = kT_b$). The channel BW, $W = M.\ \Delta f = M/T$ Hz, and the TW product is M. Thus the signalling waveforms for M-FSK are:

$$S_j(t) = A_0 \cos \left[\omega_0 t + \frac{2\pi t}{T}(j-1) + \phi \right], 0 \leqslant t \leqslant kT_b = T \tag{2.39}$$

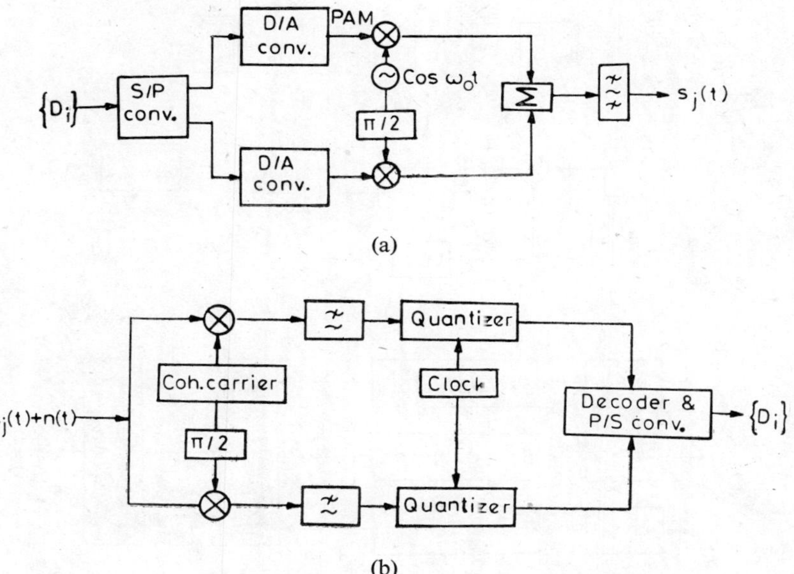

Fig. 2.18 Block schematic of M-QAM transmitter and receiver: (a) modulator and (b) demodulator

and $j = \{1, 2, \ldots, M\}$.

Theoretically, it is possible to transmit and detect waveforms coherently, if the phase references are available at the receiver. In practice, however, such phase references are difficult to maintain because of the multiplicity of frequencies and a non-coherent receiver is generally used. A simplified modulator and demodulator for M-FSK are shown in Figs. 2.19a and b, where a PLL loop is used as the demodulator. A more efficient detector, related to the optimum receiver for M-ary orthogonal signals (discussed in Chapter 8), consists of M matched filters, followed by M envelope detectors and the decisions are based on the largest envelope output which is recognized by a 'Greatest of' decision circuit, as shown in Fig. 2.19(c).

2.4.1 Error Rates and Comparisons [3]

For M-ASK, the error occurs whenever the received vector is outside the decision region shown in Fig. 2.11(a). Assuming that all levels are equiprobable, the symbol error probability is calculated as:

$$\text{M-ASK (coherent): } P_e(M) = \frac{2(M-1)}{M}\,\text{erfc}\sqrt{E_0/2n_0}, \quad E_0 = \frac{a_0^2 T}{2}$$

$$= \frac{2(M-1)}{M}\,\text{erfc}\sqrt{\frac{3}{M^2}\cdot\frac{2E_{av}}{n_0}} \qquad (2.40)^*$$

*For large x, $\text{erfc}(x) \simeq \frac{1}{\sqrt{2\pi}x}\cdot\exp(-x^2/2)$.

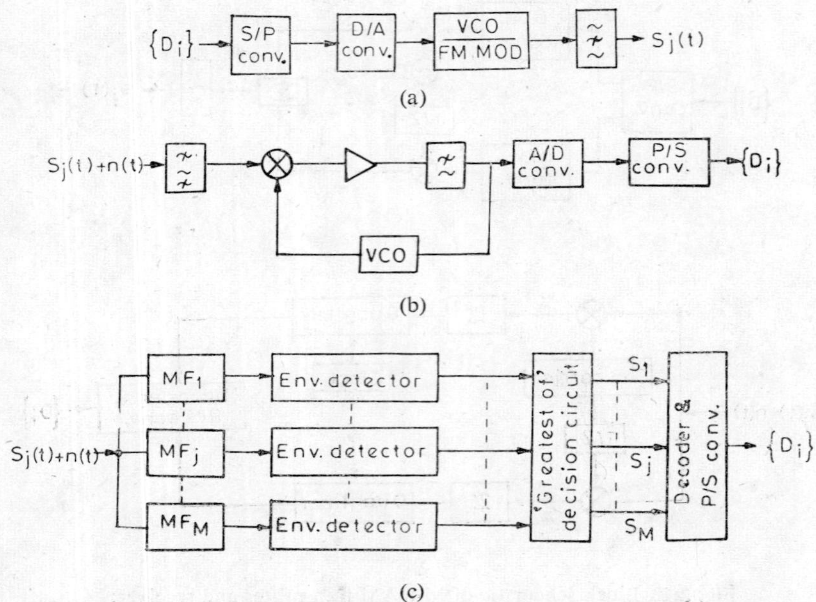

Fig. 2.19 Block schematic of MFSK transmitter and receiver: (a) modulator; (b) demodulator using PLL; and (c) optimum incoherent receiver

where E_{av} per symbol $= (M^2/12)\, E_0$. Since each symbol carries $\log_2 M$ bits of information, the bit error rate $P_b \simeq P_e(M)/\log_2 M$.

For M-PSK signals, using the coherent phase detector of Fig. 2.15(b), the symbol error rate is estimated as (for $M \geqslant 4$):

$$\text{M-PSK} : P_e(M) \simeq 2\text{erfc}[\sqrt{2E_s/n_0} \cdot \sin(\pi/M)]$$

$$\simeq \frac{1}{\sqrt{\pi E_s/n_0}\, \sin \pi/M} \cdot \exp[-(E_s/n_0) \cdot \sin^2(\pi/M)] \quad (2.41)$$

where E_s, is the symbol energy $= (A_0^2 T/2)$. For $M=4$, eqn. (2.41) reduces to

$$P_e(4) = 2\text{erfc}\sqrt{E_s/n_0}.$$

Since $E_s = 2E_b$, BER in QPSK is:

$$P_b(\text{QPSK}) = \text{erfc}\sqrt{2E_b/n_0} \quad (2.42)$$

which is the same as eqn. (2.34) for BPSK signals.

For the differentially encoded M-PSK signals, the error rate is obtained as:

$$\text{D-MPSK} : P_e(M) \simeq 2\ \text{erfc}\ [2\sqrt{E_s/n_0} \cdot \sin(\pi/2M)] \quad (2.43)$$

It will be seen that DPSK results are worse by 2.3 dB for $M = 4$, and by approximately 3 dB for $M \geqslant 8$.

Multilevel QAM provides the best results, and for k even, $m = \sqrt{M}$, the error rates are:

$$P_e(m) = \frac{2(m-1)}{m}\ \text{erfc} \sqrt{\frac{E_0}{2n_0}}, \qquad E_0 = \frac{a_0^2 T}{2} \quad (2.44)$$

For k odd, $M = 8, 32, 128$, etc., a bound on P_e is:

$$\text{M-QAM}: P_e(M) \leqslant 4\,\text{erfc}\sqrt{E_0/2n_0}, \qquad k \geqslant 3$$
$$\simeq \frac{4\sqrt{Mn_0}}{\sqrt{6\pi E_{av}}} \exp\left[-\frac{3E_{av}}{2Mn_0}\right], \tag{2.45}$$

where E_{av} for both vectors is $(m^2E_0)/6 = kE_b$, and $P_e(M) \simeq kP_b$.

Finally, for orthogonal M-FSK, using an incoherent optimum receiver, the error rate is:

$$\text{M-FSK}: P_e(M) = \sum_{n=1}^{M-1} \frac{(-1)^{n+1}}{n+1} \binom{M-1}{n} \times \exp\left[-\frac{nE_s}{n_0(n+1)}\right] \tag{2.46}$$

and BER $P_b \simeq \frac{1}{2} P_e(M)$ for $M \gg 1$. It may be shown that with $M \to \infty$,

$$P_e(M) \leqslant 2^k \cdot \exp\left[-kE_b/2n_0\right]$$
$$= \exp\left[-kE_b/(2n_0) - k \ln 2)\right] \tag{2.47}$$

and in the limit $k \to \infty$, $M \to \infty$, $P_e(M) \to 0$, if $E_b/n_0 > 2 \ln 2$.

This lower limit of E_b/n_0 for error-free communication is 3 dB higher than the idealized Shannon's limit of $E_b/n_0 \geqslant \ln 2$, as discussed in Chapter 8.

The various error rates as given by the above equations are plotted in Figs. 2.20, 2.21, 2.22, and 2.23, for $M = 4, 8, 16$ and 32. For M-ASK, it is

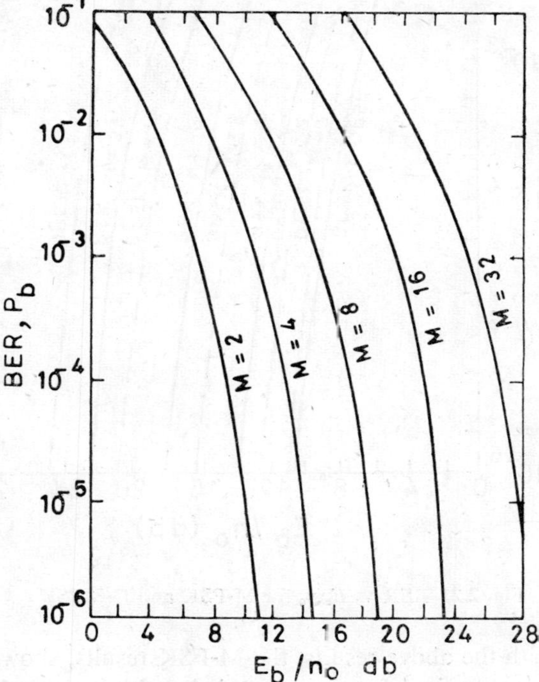

Fig. 2.20 BER vs E_b/n_0 for M-ASK

seen from Fig. 2.20 that the penalty in E_b/n_0 for the same P_b increases as M increases; for $M = 4$, the loss $\simeq$ 4 dB, but for $M \geqslant 8$, the loss $\simeq$ 6 dB for increasing M by a factor of 2. For M-PSK, it is observed from Fig. 2.21, that P_b for QPSK and BPSK are almost the same, since the 3 dB advantage in signalling rate compensates for the excess CNR required in QPSK. For higher values of M, the penalties in CNR are approximately: 4 dB for $M = 8$, 6 dB for $M \geqslant 16$. Further, DPSK results are seen to be poorer than M-PSK results approximately by 3 dB for $M \geqslant 8$. The QAM results of Fig. 2.22, are, however, the best among the bandwidth efficient communication systems. It is seen that E_b/n_0 requirement of M-QAM for $P_b = 10^{-5}$ is less, as compared to that of M-PSK, by: 4 dB for $M = 16$; 7 dB for $M=32$; and 10 dB for $M=64$. Moreover, the hardware for QAM is rather straightforward, and the technique has been used in many recent digital radio systems.

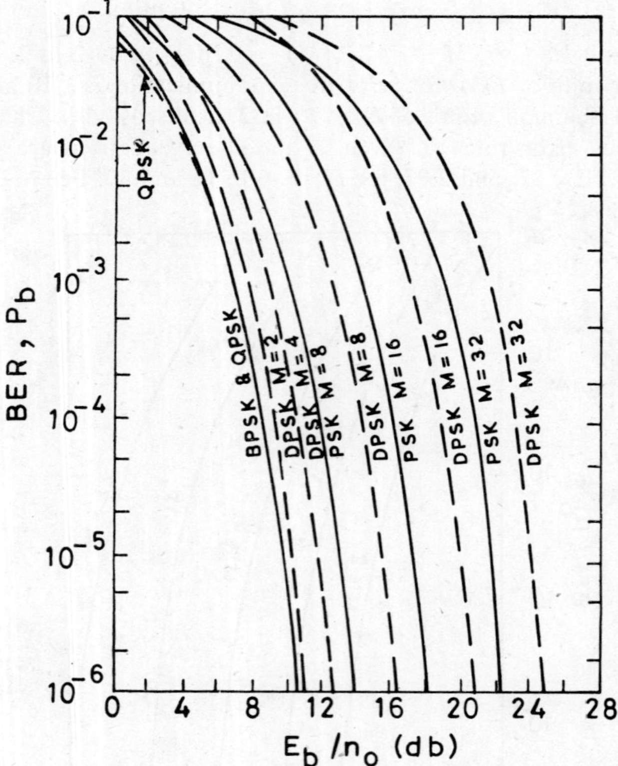

Fig. 2.21 BER vs E_b/n_0 for M-PSK and D-MPSK

In contrast with the above results, the M-FSK results, shown in Fig. 2.23, indicate that E_b/n_0 required for a given P_b is less for larger values of M. It should be noted that in M-FSK, the bandwidth is exchanged for CNR and

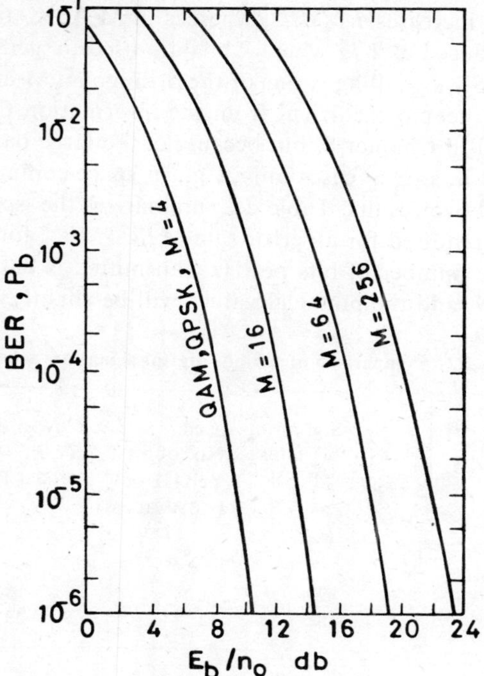

Fig. 2.22 BER vs E_b/n_0 for M-ary QAM

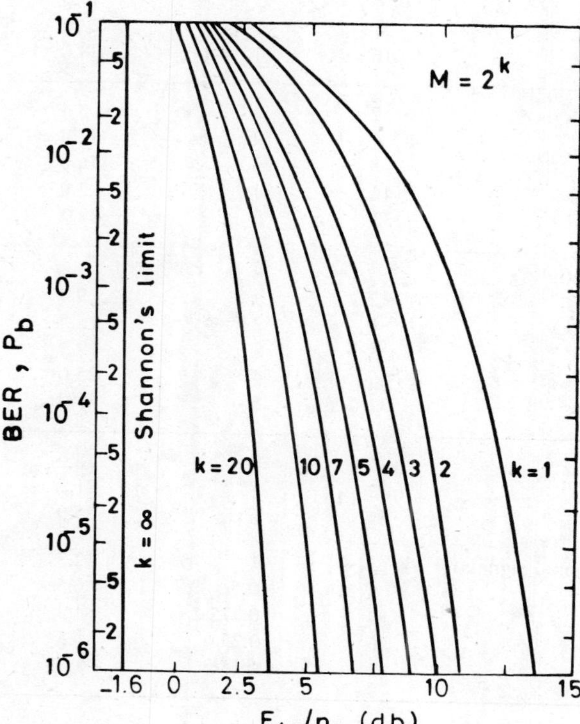

Fig. 2.23 BER vs E_b/n_0 for M-FSK signals

the channel BW increases as M/k, whereas in M-ASK, and M-QAM, the channel BW is reduced as $1/k$, where k bits are used to generate the M-ary signals. The M-FSK signalling is one of the orthogonal M-ary systems which are proved to be the most efficient from the information theoretic point of view, as discussed in Chapter 8, but because of the large bandwidth requirement, this is used in special cases only, e.g., in space communication.

Based on the above results, Table 2.2 summarizes the spectral efficiency, CNR and E_b/n_0 required for an error rate of $P_b = 10^{-4}$ for different signalling schemes. The number of bits per Hz transmitted is based on the idealized Nyquist BW and the practical values will be slightly lower. Similarly,

Table 2.2 Comparison of various digital signalling schemes

Signalling scheme	Signalling speed: No. of states per symbol	Signalling speed: No. of bits per Hz of RF BW (ideal)	C/N in dB for $P_b = 10^{-4}$ (ideal BW)	E_b/n_0 in dB for $P_b = 10^{-4}$
ASK-(DSB-AM) and envelope detection	2	1	15	15
	4	2	23	20
ASK-(DSB-SC) coherent detection	2	1	9.0	9.0
	4	2	16.0	13.0
	8	3	22.0	17.0
	16	4	28.0	22.0
PSK (coherent)	2	1	8.4	8.4
	4	2	11.4	8.4
	8	3	16.8	11.8
	16	4	22.0	16.3
	32	5	28.0	21.0
D-PSK (differentially coherent)	2	1	9.4	9.4
	4	2	13.5	10.5
	8	3	19.3	14.3
	16	4	25.0	19.0
	32	5	31.0	24.0
QAM (APK)	16	4	18.0	12.0
	32	5	21.5	14.5
	64	6	25.0	17.0
M-FSK (orthogonal signals)	2	1	12.3	12.3
	4	0.5	6.5	9.5
	8	0.375	4.3	8.3
	16	0.25	1.5	7.5
	32	0.156	−1.75	6.25

the relation between E_b/n_0 and C/N is based on the Nyquist BW, i.e., $\frac{C}{N} = \left(\frac{kE_b}{n_0}\right)$ for the bandwidth-reduced systems and $\frac{C}{N} = \frac{E_b}{n_0}\cdot\frac{k}{M}$ for M-FSK. It is easily seen from Table 2.2 that for multialphabet signalling, QAM is the most efficient; but if a large bandwidth is available, M-FSK will require minimum E_b/n_0 for efficient signalling.

2.5 CARRIER AND CLOCK SYNCHRONIZATION [7, 8]

In coherent detection, as discussed above, it has been assumed that a synchronized carrier and a synchronized sampling pulse (for bit recovery) are available at the receiver. Since in both ASK and PSK, suppressed carrier modulation (through the use of multiplier/balanced modulator) is used to minimize the transmitted power, a ready carrier reference is not available in the received signal $y(t) = s(t) + n(t)$. For ASK and BPSK, the carrier, however, may be regenerated through a simple squaring circuit and a local VCO may then be locked to this frequency through a phase-locked loop (PLL). In M-ASK and M-PSK, more complex circuits are necessary as now discussed.

In a PLL, as shown in Fig. 2.24(a), the noisy carrier at the input is multiplied by a locally generated reference carrier (form a VCO) to give the error voltage $\epsilon(t)$ as:

$$\epsilon(t) = \frac{AB}{2}\,[\sin(\theta - \tilde{\theta}) + \sin(2\omega_0 t + \theta + \tilde{\theta})] \tag{2.48}$$

and the loop filter output $e(t)$ is simply the low frequency component of $\epsilon(t)$. This is approximated as:

$$e(t) = \frac{AB}{2}\sin(\theta - \tilde{\theta}) \simeq \frac{AB}{2}(\theta - \tilde{\theta})$$

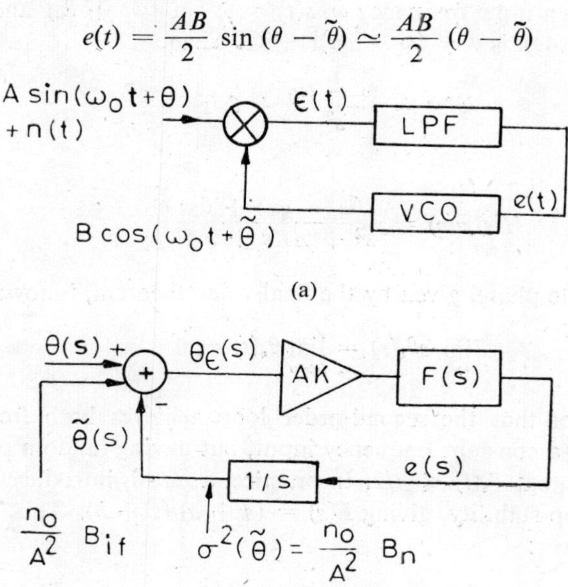

Fig. 2.24 (a) a PLL circuit and (b) Linearized equivalent of a PLL

for $$(\theta - \tilde{\theta}) = \theta_\epsilon \ll \pi/2 \tag{2.49}$$

Thus, for small errors, the PLL is a linear feedback circuit, where $e(t)$ controls the VCO output in such a way that $\epsilon(t) \to 0$ and $\tilde{\theta} \to \theta$, the true value of the carrier phase. Since the VCO output phase $\psi(t)$ is given by:

$$\psi(t) = \omega_0 t + K_1 \int e(t)\,dt = \omega_0 t + \tilde{\theta}, \tag{2.50}$$

the VCO basically behaves like an integrator and the overall loop now behaves as a second-order loop [assuming that the loop filter has a transfer function $F(s) = 1 + a/s$]. Neglecting the constant phase $\omega_0 t$ in the product terms and using eqns. (2.48) to (2.50), the linearized PLL is shown in Fig. 2.24(b) and its governing equations in the transform domain are:

$$\begin{aligned} \theta_\epsilon(s) &= \theta(s) - \tilde{\theta}(s) = [1 - H(s)]\theta(s) \\ H(s) &\triangleq \frac{\tilde{\theta}(s)}{\theta(s)} = \frac{AKF(s)}{s + AKS(s)} \end{aligned} \tag{2.51}$$

where $H(s)$ is the closed loop transfer function and $AK = ABK_1/2$.

For a second-order loop, with $F(s) = (1 + a/s)$, we have

$$H(s) = \frac{AK(s + a)}{s^2 + AKs + aAK}.$$

For an unknown input frequency ω, $s(t) = A \sin(\omega t + \theta_0)$ and the input θ, as in eqn. (2.49), is $\theta = (\omega - \omega_0)t + \theta_0$. Then,

$$\theta(s) = \frac{\omega - \omega_0}{s^2} + \theta_0/s$$

and

$$\theta_\epsilon(s) = [1 - H(s)]\theta(s) = \frac{(\omega - \omega_0) + \theta_0 s}{s^2 + AKs + aAK} \tag{2.52}$$

The steady-state phase, given by the final-value theorem, is now:

$$\lim_{s \to 0} s\theta_\epsilon(s) = \lim_{t \to \infty} \theta_\epsilon(t) = 0 \tag{2.53}$$

It is thus seen that the second-order loop achieves both frequency and phase lock for a constant frequency input, but having random phase, under the condition that $\theta_\epsilon(t) \ll \pi/2$. In practice, a zero is introduced in $F(s)$ for the ease of loop stability, giving $F(s) = (s + a)/(s + b)$. This results in a steady-state error:

$$\lim_{t \to \infty} \theta(t) = \lim_{s \to 0} s\theta_\epsilon(s) = \frac{b(\omega - \omega_0)}{aAK} \tag{2.54}$$

But by making the loop gain large, the error is minimized. In case of continuously varying input frequencies, say, for continuous Doppler shifts or FM waveforms, the equivalent input phase $\theta = \alpha t^2 + (\omega - \omega_0)\,t + \theta_0$. In such cases, the loop with the imperfect integrator gives unbounded phase error at $t \to \infty$, but with an ideal integrator, $\theta_\epsilon(t) = \alpha/aAK$ for $t \to \infty$. Thus, the loop gain has to be large; alternatively, a third order PLL is used to make $\theta_\epsilon(t) \to 0$.

The transient and steady-state behaviour of the PLL is generally obtained by analysing the non-linear differential equations of higher orders through phase-plane techniques. Some of the important results of this analysis are given in Table 2.3. The most important property of the loop is its pull-in range, i.e., the maximum $\Delta\omega$ that may be permitted so that the loop locks on to the input ω_0 and θ. This lock-on operation is possible with reasonable $\Delta\omega$, because the difference frequency $\Delta\omega$ produces a d.c. voltage at the phase detector output which pushes the average VCO frequency towards ω_0

Table 3.3 Characteristics of a second-order PLL

Parameter	Values
Noise BW, B_n	With perfect integrator: $B_n = (AK + a)/4$ With imperfect integrator: $B_n = \dfrac{AK(AK+a)}{4(AK+b)}$
Pull-in range $= \Delta\omega$ (rad/sec)	$\Delta\omega$ very large with T_{acq} large
*Frequency lock range $= \Delta\omega_m$	For perfect integrator, $\Delta\omega_m = 2.6\omega_n$, $\xi = \dfrac{1}{\sqrt{2}}$ With imperfect integrator, $\Delta\omega_m < 2\omega_n\sqrt{\xi\omega_n b + 1}$
Hold-on range	$\simeq AK$ rad/sec
With linearly varying input frequency, $\omega = \omega_0 + Dt$, the search rate $D = \dfrac{2\Delta\omega}{T_0}$, where $2\Delta\omega$ = freq. range to be searched; T_0 = time period	$D \leqslant 2B_n^2$, $\xi = \dfrac{1}{2}$, with perfect integrator; with imperfect integrator, $D \simeq 2B_n^2\left(1 - \dfrac{1}{\sqrt{\text{loop SNR}}}\right)$
Lock-on time T_{acq} (acquisition time)	$3.5(\Delta f)^2/B_n^3$

*Natural frequency of the loop $\omega_n = \sqrt{aAK}$; damping constant $\xi = AK/2\omega_n = \dfrac{\omega_n}{2a}$ (=0.707 in general).

in small increments, say, $\delta\omega$, and ultimately VCO locks to ω_0 after a large acquisition time T_{acq}. The change of VCO frequency in steps of $\delta\omega$ results in 'cycle skipping' (or cycle slipping) and once the VCO is in lock, the loop stops skipping cycles and begins to pull into phase lock with $\theta_\epsilon \to 0$. The value of T_{acq} is then calculated based on the value of $\delta\omega$ at each step and the time taken for reaching the phase lock after the frequency lock. As an example, for a PLL with $B_n = 5$ Hz, $\Delta f = 10$ Hz and initial $\theta_\epsilon = -\pi/2$, it is calculated that:

Time required for frequency lock $\simeq$ 3 sec, and the time for phase lock $\simeq$ 0.5 sec, thus giving, $T_{acq} \simeq 3.5$ sec.

So far, the PLL has been discussed neglecting the input noise in $y(t)$. It is easily shown that the equivalent noise power entering the linearized loop is $n_0 B_{if}/A^2$, as shown in Fig. 2.24(b), where $n_0/2$ is the double-sided noise spectral density of $n(t)$ and B_{if} is the IF BW. The variance of $\tilde{\theta}$ is then calculated as:

$$\sigma^2(\tilde{\theta}) = \frac{n_0}{2A^2} \int_{-\infty}^{\infty} | H(j\omega) |^2 \frac{d\omega}{2\pi} = \frac{n_0}{A^2} B_n, \tag{2.55}$$

where B_n is the noise bandwidth of the loop. In general, B_n should be large enough so that the loop is able to track efficiently the variation of the input phase, i.e., $B_n \geqslant$ phase noise BW, but $B_n \ll$ the system BW, which is wide enough to allow the sidebands of the modulated signal. Thus, the loop behaves as a lowpass filter for $n(t)$, thereby improving the loop SNR considerably as compared to the CNR at the RF/IF level of the receiver (SNR gain $\simeq B_{if}/B_n$). It has been further shown that the frequency of skipping cycles becomes quite large if the loop error $\sigma^2(\tilde{\theta}) > 0.25$. The frequency of skipping cycles $\simeq (4B_n/\pi) \exp[-2/\sigma^2(\tilde{\theta})]$, which gives the value of $10^{-4}B_n$ approximately for $\sigma^2 = 0.25$. When a single cycle is skipped, a transient of 2π occurs in $\tilde{\theta}$ and the loop behaviour deteriorates. This results in the so-called 'threshold' of the input CNR below which the loop SNR decreases rapidly and the linear approximation of the loop does not hold.

2.5.1 Carrier Synchronization

The second-order PLL, as discussed above, is generally used for tracking carrier frequencies, since its steady-state phase error tends to zero at $t \to \infty$, if the initial frequency offset is within $\pm\omega_n$. Consider first the binary ASK and PSK systems where the received $s(t)$ is a suppressed carrier signal. But the carrier may be regenerated through a squaring loop as shown in Fig. 2.25, where the BPF at $2\omega_0$ drives the PLL and the VCO output is divided and phase-shifted to give the synchronized carrier $B \sin(\omega_0 t + \tilde{\theta})$. To maintain a constant input to the PLL, since the loop gain $\propto A$, a bandpass limiter is used to remove the input level variation. This results in a SNR degradation of 1 dB only for small CNR's and gives 3 dB advantage for large CNR's.

Further, to simplify the multiplier circuit of the PLL, the VCO output is hardlimited and a switched multiplier is used without any degradation of the loop performances.

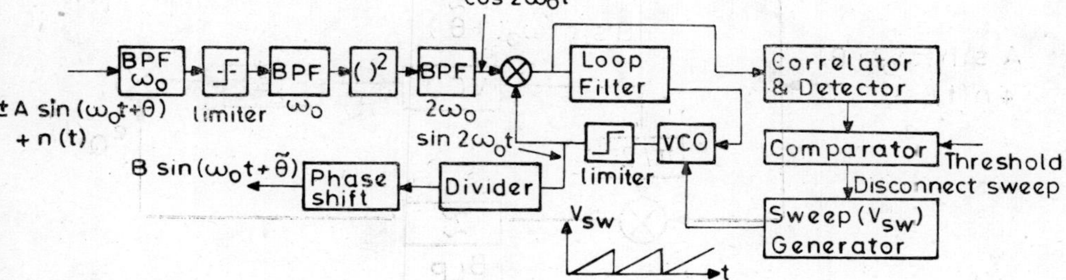

Fig. 2.25 Carrier acquisition and tracking circuit

The initial acquisition of the unknown input frequency may be obtained by using a swept VCO, as shown in Fig. 2.25, where the maximum sweeping rate is proportional to aAK and also to the square of the noise BW, B_n of the loop. In case where the initial frequency offset Δf is not too large, PLL may be operated in both acquisition and tracking mode, by using a large B_n for acquisition and a small B_n for tracking. Since $T_{acq} \simeq 3.5(\Delta f)^2/B_n^3$, the acquisition time is reduced by making B_n large. Once the acquisition is obtained, the loop is switched to the tracking mode by making B_n small, which reduces the tracking error $\sigma^2(\tilde{\theta})$. It may be shown that the variance of the phase error is increased due to the squaring circuit and, as compared to eqn. (2.55), the new variance is:

$$\sigma^2(\tilde{\theta}) = \frac{n_0 B_n}{A^2}\left(1 + \frac{n_0 B_{if}}{2A^2}\right) \tag{2.56}$$

where B_{if} = BW of the BPF. Since $A^2/(2n_0B_{if}) = \gamma$ = CNR, the phase variance is increased by a factor of $[1 + 1/(4\gamma)]$; and for $\gamma \simeq 10$ dB, there is hardly any deterioration of the loop SNR.

An alternative sheme is the well-know Costas loop [6], shown in Fig. 2.26, where the error voltage $e(t)$ is the product of the $I-Q$ signals given by:

$$e_I(t) = \frac{AB}{2}[\cos(\theta - \tilde{\theta}) \cos(2\omega_0 t + \theta + \tilde{\theta})]$$

$$e_Q(t) = \frac{AB}{2}[\sin(\theta - \tilde{\theta}) + \sin(2\omega_0 t + \theta + \tilde{\theta})]$$

Neglecting harmonic terms, $e(t)$ is given by:

$$\begin{aligned} e(t) &= \frac{(AB)^2}{4} \sin(\theta - \tilde{\theta})\cdot\cos(\theta - \tilde{\theta}) \\ &= C \sin 2(\theta - \tilde{\theta}), \qquad C = \text{constant} \\ &\simeq 2C(\theta - \tilde{\theta})\ldots \qquad \text{for } (\theta - \tilde{\theta}) \ll \pi/2 \end{aligned} \tag{2.57}$$

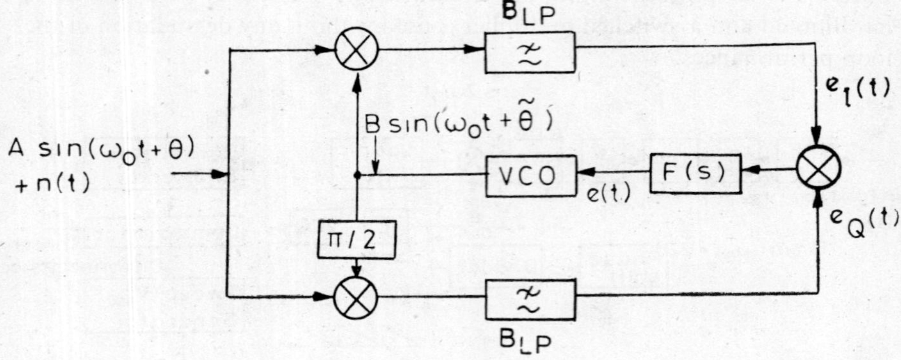

Fig. 2.26 Costas loop

This error voltage [where the A^2 term removes the data modulation from $e(t)$], controls the VCO as in a PLL to make $e(t) \to 0$. The Costas loop, then, acquires the synchronized carrier as in the squaring loop of Fig. 2.25 and the variance of the phase error is given as:

$$\sigma^2(\tilde{\theta}) \simeq \frac{n_0 B_n}{A^2}\left(1 + \frac{n_0 B_{LP}}{A^2}\right) \tag{2.58}$$

where B_{LP} = the data filter BW.

A somewhat superior carrier acquisition circuit is the decision feedback PLL shown in Fig. 2.27, where the usually correct data are multiplied with the delayed error $\theta_\epsilon(t - T_b)$ giving $e(t)$ as (the delay T_b compensates for the delay in the data recovery arm):

$$e(t) = \text{Low frequency component of}\left\{\frac{A^2 B}{B}\,[\sin(\theta - \tilde{\theta}) + \sin(2\omega_0 t + \theta + \tilde{\theta})]\right\}$$

$$\simeq C \sin(\theta - \tilde{\theta}), \qquad C = \text{constant.} \tag{2.59}$$

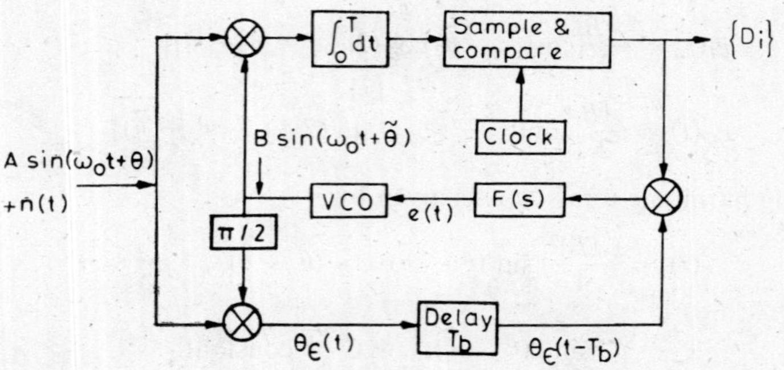

Fig. 2.27 A decision feedback PLL

The data removal from $e(t)$ through the term A^2 is done somewhat differently in this scheme and assuming that the error in $\{D_i\}$ is less than 10^{-2}, the input noise now disturbs only one arm instead of the two arms in the Costas loop. Hence, the phase error is less in this circuit, and its variance is:

$$\sigma^2(\tilde{\theta}) \simeq \frac{n_0 B_n}{A^2}\left(\frac{\pi}{4}\ \frac{n_0}{A^2 T_b}\right)$$

$$\simeq \frac{\pi}{4}\ \frac{B_n T_b}{\gamma^2}, \tag{2.60}$$

where the input CNR $\gamma = (A^2 T_b)/n_0$. The equation is also valid at low input CNR.

The above circuits for carrier recovery in binary systems may now be extended for M-ary systems. For M-PSK signals, the received signal is passed through an Mth power-law device which removes the data and generates the Mth harmonic Mf_0. This frequency is locked through a PLL and f_0 obtained from a divider circuit, as shown in Fig. 2.28. For $S_j(t)$ given by eqn. (2.37), the output of the Mth power-law device is:

$$Z_j(t) = [S_j(t)]^M$$

$$= \frac{A_0^M}{2^M - 1} \cos\,[M\omega_0 t + (j-1)2\pi] + \text{(other harmonics excluding } M\omega_0) \tag{2.61}$$

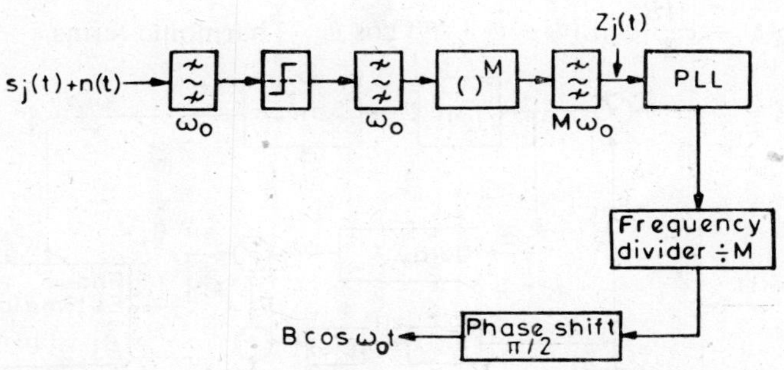

Fig. 2.28 Carrier acquisition for M-PSK using Mth power loop

The $M\omega_0$ — BPF now selects the Mth harmonic and locks the PLL. The PLL output is now B sin $(M\omega_0 t)$, which is divided and phase-shifted to obtain B cos $\omega_0 t$. The Mth power-law devices introduce more noise in $Z_j(t)$, but by making the $M\omega_0$-BPF narrow band, the PLL SNR is improved.

An alternative method of carrier acquisition in M-PSK is the generalization of the binary Costas loop to an M-phase loop, as shown in Fig. 2.29. The input signal is multiplied by B cos $[\omega_0 t + (\pi/M)(j-1)]$ in M multipliers to generate data signals in each path. All data outputs are then multiplied to obtain the error signal $e(t)$ and cancel the data modulation. The $e(t)$ controls the VCO to generate ω_0 with correct phase. A simpler method is to extend the decision feedback PLL of Fig. 2.27 for the M-PSK signal, and

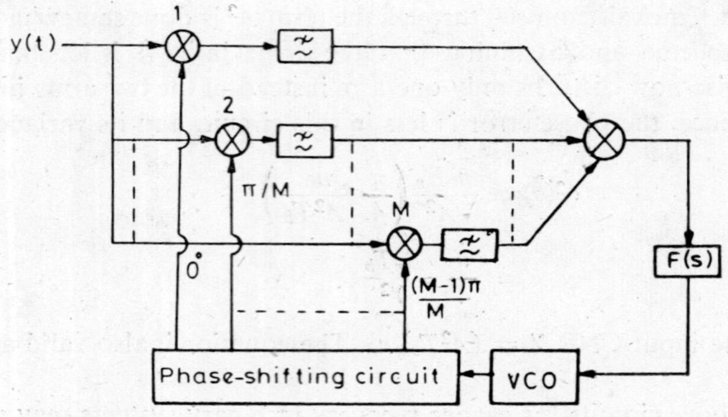

Fig. 2.29 *M*-phase Costas loop

the technique is also suitable for multilevel QAM signals. The generalized decision feedback PLL is shown in Fig. 2.30, where the modulation phase $\theta_j = (2\pi/M)(j - 1)$ of $S_j(t)$ is estimated from the *I-Q* components a and b as $\hat{\theta}_j = \tan^{-1}(b/a)$. Then, multiplying the delayed data signals by $\cos \hat{\theta}_j$ and $-\sin \hat{\theta}_j$ as shown, the error components $e_I(t)$ and $e_Q(t)$ are:

$$e_I(t) = \frac{AB}{2} \cos (\theta - \tilde{\theta} + \theta_j)(-\sin \hat{\theta}_j) + \text{harmonic terms}$$

$$e_Q(t) = \frac{AB}{2} \sin (\theta - \tilde{\theta} + \theta_j) \cos \hat{\theta}_j + \text{harmonic terms.}$$

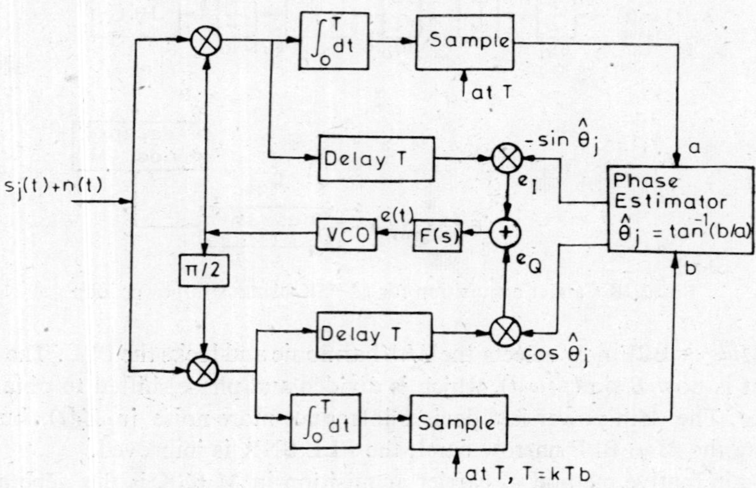

Fig. 2.30 Decision feedback PLL for M-PSK and QAM

Summing and neglecting harmonic terms (assuming $\theta_j = \hat{\theta}_j$),

$$e(t) = \frac{AB}{2} \sin (\theta - \tilde{\theta}) \tag{2.62}$$

This error signal $e(t)$ now controls the VCO and $\theta_\epsilon = (\theta - \tilde{\theta})$ tends to zero. This circuit gives a better performance than the circuit using Mth-law devices.

2.5.2 Clock Synchronization [6, 7]

The problem of clock (bit or symbol) synchronization for digital modulation systems is somewhat similar to that of the carrier synchronization. Although sophisticated techniques, such as an MAP estimator, may be used for special systems, simple squaring or similar loops operate quite efficiently with reasonably acceptable CNR's at the receiver input. A few of these suboptimal circuits are shown in Fig. 2.31. The squaring loop of Fig. 2.31(a) is similar to that of Fig. 2.25 and has an equivalent performance. In Fig. 2.31(b), $d(t)$ multiplied by $d(t - T_b/2)$ produce a line spectrum at $f_c = 1/T_b$ and this may

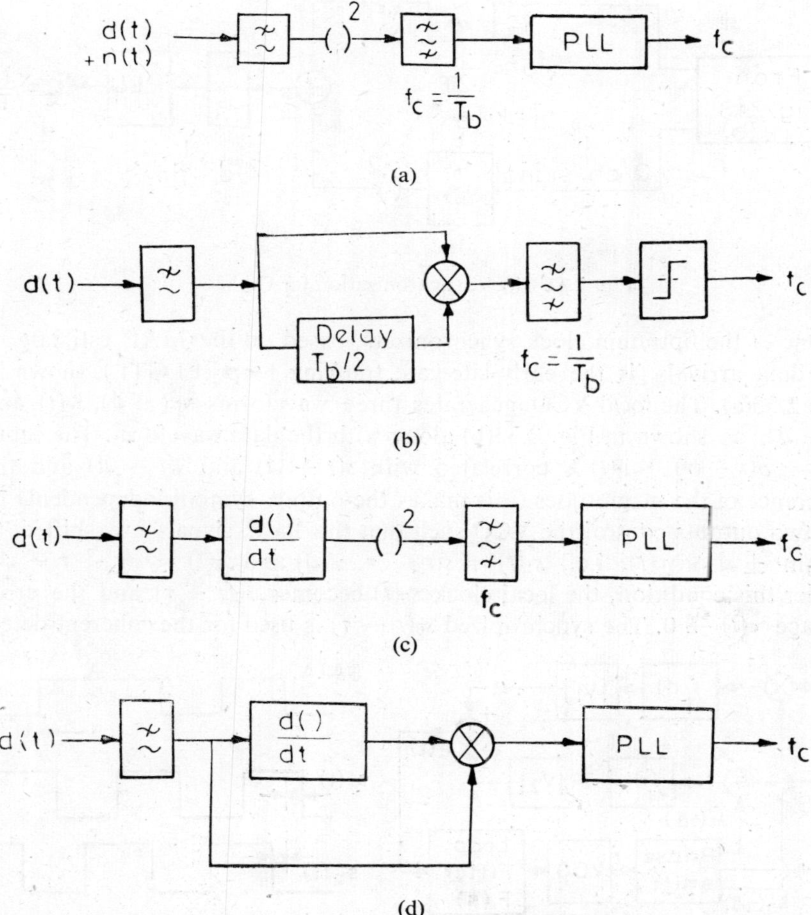

Fig. 2.31 Few suboptimal clock recovery circuits: (a) squaring loop; (b) delay and multiply bit synchronizer; (c) differentiate and square circuit and (d) differentiate and multiply circuit

be filtered to give the required clock frequency. Similarly, the circuits of Figs. 2.31(c) and (d) produce line spectra at f_c and they are tracked by PLL. The variance of the timing error in each case is worse by only a few per cent as compared to the optimum MAP synchronizers, but the circuits have the advantage of much simpler implementation. Specifically, the error variance for the squaring loop is:

$$\sigma_\epsilon^2/T_b^2 = \alpha \frac{n_0 B_n}{A^2}, \qquad 0.5 < \alpha < 1. \tag{2.63}$$

For QAM signals, a similar synchronizer is shown in Fig. 2.32, where the outputs of the quadrature arms are filtered, squared, added and then tracked through a PLL. This behaves as a modified squaring loop of Fig. 2.31(a).

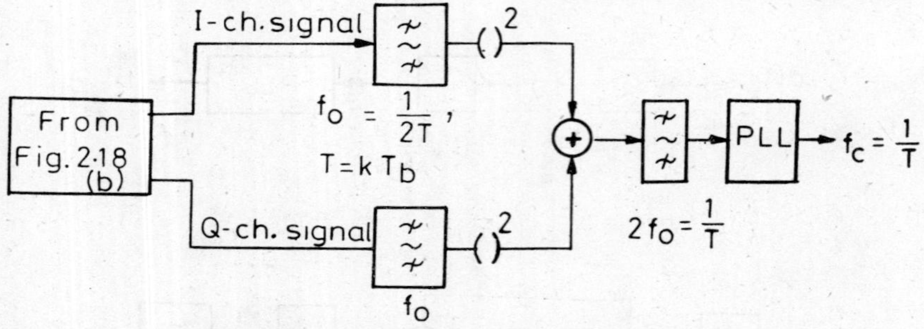

Fig. 2.32 Clock synchronization for QAM

One of the optimum clock synchronizers, based on the MAP estimate of the data arrivals, is the early-late-gate tracking loop (ELGTL), shown in Fig. 2.33(a). The local VCO generates three waveforms $s_1(+\Delta)$, $s_0(t)$ and $s_2(-\Delta)$, as shown in Fig. 2.33(b) along with the data waveform. The input $y(t) = d(t - \tau) + n(t)$ is correlated with $s(t + \Delta)$ and $s(t - \Delta)$ and the difference of the magnitudes (this makes the output symbol-independent) of the two outputs control the VCO such that the local signals are shifted to within $\pm\Delta$ of $y(t)$, i.e., $s_1(t) = s(t - \tau + \Delta)$ and $s_2(t) = s(t - \tau - \Delta)$. Under this condition, the local clock $s_0(t)$ becomes $s_0(t - \tau)$ and the error voltage $\epsilon(t) \to 0$. The synchronized $s_0(t - \tau)$ is used for the coherent detec-

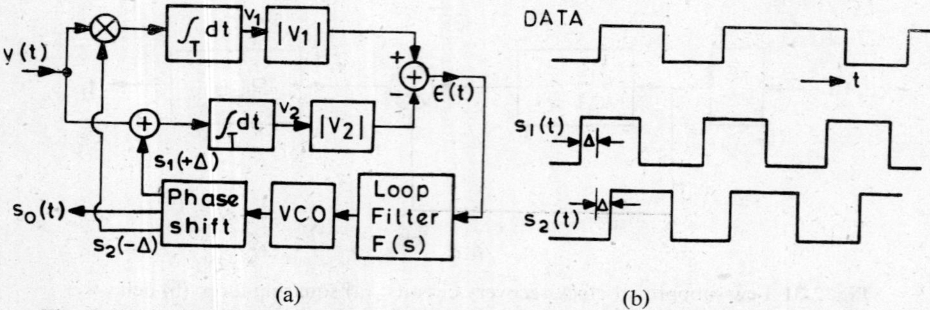

Fig. 2.33 (a) Early-late-gate bit synchronizer, (b) Waveform of data $s_1(t)$ and $s_2(t)$ as used in the cross-correlators

tion of the data. It has been shown that $\Delta = T_b/4$ is optimum and the corresponding mean-square timing error is:

$$\frac{\sigma_\epsilon^2}{T_b^2} = \frac{1}{8}\frac{B_n T_b}{E_b/n_0} = \frac{1}{8}\frac{n_0 B_n}{A^2} \tag{2.64}$$

This result is better than that of eqn. (2.63) and the circuit gives near-optimum performance.

2.6 INTRODUCTION TO ANALOG MATCHED FILTERS [9]

In the discussions on optimum detection in Sec. 2.2, it has been shown that the matched filter or a cross-correlator is the heart of the optimum detector, and, as such, this has to be carefully designed for optimum system performances. In Figs. 2.8 and 2.9, it is required to obtain the integral

$$\int_0^T y(t)\cdot s(t)\,dt \tag{2.65}$$

which may be instrumented either by a cross-correlator or by a matched filter having the impulse response $h(t) = s(\tau_0 - t)$. A linear filter having this impulse response is optimum from the points of view of different criteria used in the optimum detection and estimation, viz., SNR criterion, likelihood ratio criterion and inverse probability criterion, and hence this linear filter is called a 'matched filter', whose two important properties are:

$$\begin{aligned} h(t) &= k\cdot s(\tau_0 - t) \\ H(j\omega) &= k\cdot S^*(j\omega)\exp(-j\omega\tau_0) \end{aligned} \tag{2.66}$$

where k and τ_0 are constants. It is easily shown that for a signal $s(t)$ disturbed by white Gaussian noise $n(t)$, the peak signal-to-noise power ratio at the matched filter output is

$$(S/N)_{\text{pk}} = 2E/n_0, \tag{2.67}$$

where E = signal energy and $n_0/2$ = two-sided noise spectral density. The above (S/N) value is maximum for the class of linear filters that may be used for signal processing.

The philosophy of the matched filter may be discussed on the basis of the signal generation and reception of Fig. 2.34, where the impulse response of the signal generating filter is $s(t)$ and that of the matched filter is $s(-t)$ (neglecting the constants). The received signal and the filter output are now given by

$$\begin{aligned} y(t) &= s(t) + n(t) \\ y_0(t) &= y_s(t) + y_n(t) \end{aligned} \tag{2.68}$$

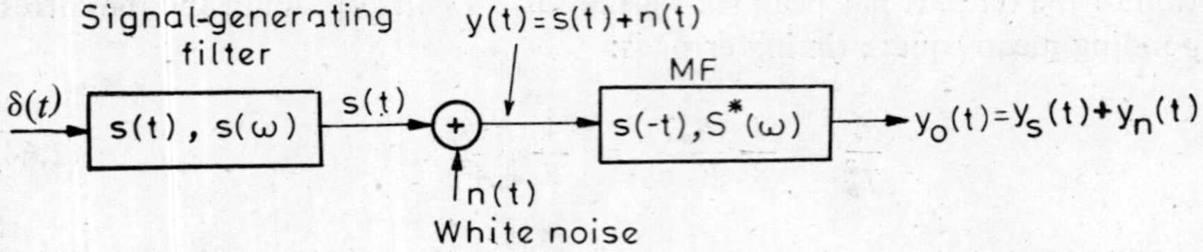

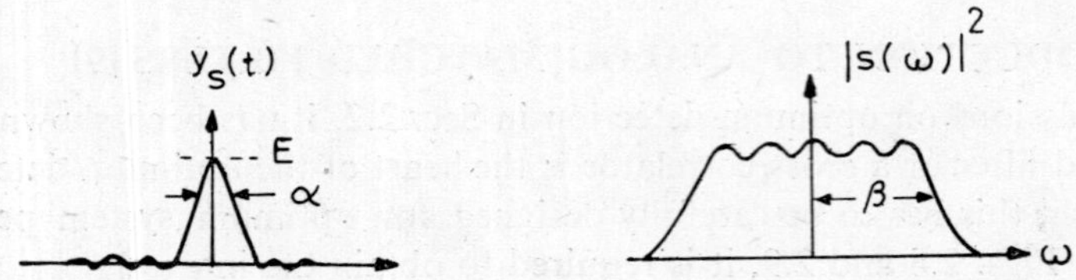

Fig. 2.34 Properties of the MF

where $y_s(t)$ is the signal component and $y_n(t)$ is the noise component of the output. Further,

$$y_s(t) = \int_{-\infty}^{\infty} s(-\tau)\cdot s(t-\tau)\, d\tau = y_s(-t)$$

and

$$y_s(0) = \int_{-\infty}^{\infty} s^2(-\tau)\, d\tau = E \tag{2.69}$$

Also,

$$S/N = \frac{y_s^2(0)}{\langle y_n^2(0)\rangle} = E^2 \bigg/ \left(\frac{n_0 E}{2}\right) = \frac{2E}{n_0}, \tag{2.70}$$

and noise power at MF output $= n_0E/2$.

The spectrum of $y_s(t) = |S(j\omega)|^2$, whose nature is shown in Fig. 2.34. The effective widths of $y_s(t)$ and $|S(\omega)|^2$, denoted by α and β, are such that

$$\alpha\beta \geqslant \pi.$$

Thus, it is seen that $\alpha \geqslant \pi/\beta$, but one would expect that the linear filter output should be an impulse of large height so that the signal detection and its parameter estimation (say, the delay τ_0) are possible with minimum error. To obtain an impulse at the output, the receiving filter may be an inverse filter with the transfer function $1/S(j\omega)$, but the gain of such a filter would be very large as $\omega \to \infty$, since $s(t)$ would always be a band-limited signal. Such an inverse characteristic would then increase $y_n(t)$ infinitely and, hence, the signal $y_s(t)$ would be lost in the noise. The compromise solution is then used, where the signal output is made as close to an impulse as possible, but the noise is suppressed outside the signal band. The transfer function $H(j\omega)$ of the matched filter should now have $|H(j\omega)| = |S(j\omega)|$, but the phase of $H(\omega)$ should be the inverse of that in $S(\omega)$, as required by the inverse filter. Thus,

$$H(j\omega) = S^*(j\omega) \tag{2.71}$$

The output peak signal-to-noise power ratio ($=2E/n_0$) is now independent of peak power, time duration, bandwidth and waveshape, but depends only on the signal energy E. For the purpose of radar detection and binary communication systems, all waveforms having the same energy are equally effective, given that the noise is white and Gaussian and the matched filter reception is used. For the estimation of signal parameters and for optimum signal resolution, however, the nature of the waveform is also important, and we have for such purposes, various pulse-compression techniques and coded waveforms to achieve the optimum results.

The output SNR ($= 2E/n_0$) has been calculated on the basis of the true white Gaussian noise of infinite total power. But in cases where the disturbing noise is band-limited and of fixed total power, the output SNR is shown to be

$$(S/N)_{pk} = 2BT(S/N)_i = \frac{2EB}{N_i} \tag{2.72}$$

where B = matched filter bandwidth, T = signal duration and $(S/N)_i = P_i/N_i = E/(TBn_0)$ = input signal-to-noise power ratio. Clearly then, the output SNR may be improved by spreading the signal energy more and more, keeping the total noise power fixed, and, of all signals with the same energy, the one with the largest bandwidth is the most useful. Thus the time-bandwidth product = (BT) of the signal or its matched filter is a very important parameter of the filter and the complexity of the filter is also given by the product BT. Since this signal may be represented by $2BT$ independent samples, the corresponding matched filter requires no more than $2BT$ elements or parameters to be specified for its complete synthesis.

2.6.1 Practical Matched Filters

The simplest matched filter required in a radar receiver is the IF filter used for the reception of isolated rectangular RF pulses. Since a rectangular RF pulse has $(\sin x)/x$ type of spectral distribution, and it is difficult to design a practical filter with such $|H(j\omega)|$, the practical IF filters generally have either Gaussian, approximately rectangular or the tuned circuit type of $|H(j\omega)|$. The loss in the peak output SNR of these filters is within 1 dB only as compared to the ideal matched filter, and a few (4 to 6) cascaded single-tuned IF stages gives the best performance.

In the consideration of the synthesis of more elaborate matched filters, as required to optimize the accuracy, resolution and ambiguity of the detected signals (as used in radar signals), the constraints of physical realizability and practical realizability have to be kept in view. As such, the usual approach in the matched filter design has been to synthesize such networks whose impulse response is the desired signal optimized from the point of view of accuracy, resolution and ambiguity, and to use the same in a complementary way to receive the signal. Because of the complementary functions of signal generation and signal reception being done by the same network, the problems of realizability are automatically considered. Delay lines and all-pass networks have been extensively used in sophisticated matched filter design. A simple example of signal generation and reception of an arbitrary waveform is shown in Figs. 2.35 (a and b), where the outputs of the impulse-excited tapped delay line (TDL) are weighted by factors $\{a_0, a_1, \ldots, a_n\}$, and the summed output is

$$s_1(t) = a_0\delta(t - \tau_0) + a_1\delta(t - \tau_1) + \ldots + a_n\delta(t - \tau_n) \qquad (2.73)$$

where τ's are the tap delays and the filter output will be $s(t)$ as shown. If now $s(t)$ is introduced at the right-hand side of the delay line, then the summed output will be a sharp pulse, giving the required cross-correlated output of eqn. (2.69).

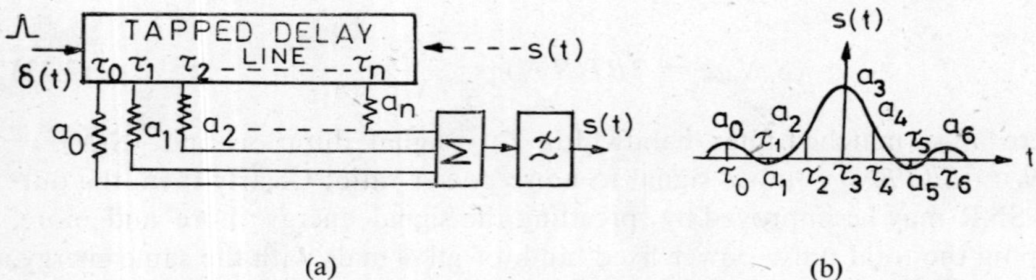

Fig. 2.35 Signal generation by TDL: (a) delay line MF and (b) generated signal $s(t)$

In a more general case of signal generation, the weighting functions may be of the form $H_i(j\omega)$ and the transfer function of the terminating filter $F(j\omega)$, thus giving the output

$$S(j\omega) = F(j\omega) \cdot \sum_{i=0}^{n} H_i(j\omega) \exp(-j\omega\tau_i) \qquad (2.74)$$

The transfer function of the corresponding matched filter is now

$$\begin{aligned} H(j\omega) &= F^*(j\omega) \sum^{n} H_i^*(j\omega) \exp[-j\omega(\tau_n - \tau_i)] \\ &= S^*(j\omega) \exp(-j\omega\tau_n) \end{aligned} \qquad (2.75)$$

The various coded waveforms, as required in orthogonal and spread-spectrum signalling, are generated and received through such filters incorporating tapped delay lines, and by assigning amplitude, phase, frequency or complex functions to the weighting networks.

The use of cascaded all-pass networks also produces a signal-generating and matched-filter pair. The overall transfer function of the cascaded all-pass networks, shown in Fig. 2.36(a), is of the form $\prod^{n} K \exp(-j\theta_i)$, having constant magnitude and an approximately linear phase $\theta_i(f)$ over the useful bandwidth W. If the networks are grouped into two filters A and B such that,

$$\prod^{A} K_A \exp(-j\theta_i) \cdot \prod^{B} K_B \exp(-j\theta_i) = K_A K_B \exp(-j\omega\tau_0) \quad (2.76)$$

then the filters A and B are matched to one another as shown in Fig. 2.36(b). The signal generated by the filter A may be processed through the matched filter B to produce a sharp pulse at the output with only a constant delay τ_0.

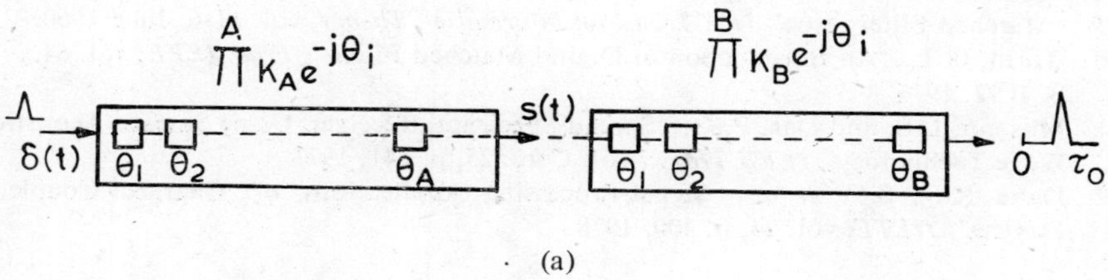

(a)

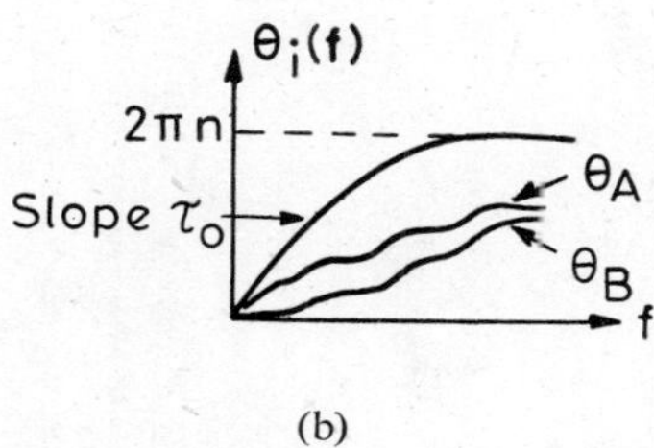

(b)

Fig. 2.36 All-pass network as an MF: (a) cascade of all-pass networks and (b) phase function of the networks

Such all-pass networks are easily designed by using simple bridged-T sections in cascade. A similar type of matched filter giving constant amplitude but linearly increasing (or decreasing) group delay over the useful band is used to process the 'chirp' signal, which has its amplitude constant but the frequency linearly varying with time, giving the overall output as a sharp IF pulse. Such FM pulse-compression techniques have been widely used in radar systems to obtain better time resolution.

Recent developments in MF-device techniques have led to the availability of (a) digital MF's using LSI technology [10] and (b) SAW and CCD devices [11, 12]. These can be configured into low-pass and band-pass TDL's and are now extensively used for various signal-processing requirements. They are discussed in some detail in Appendix B.

REFERENCES

1. Sunde, E.D., 'Theoretical Fundamentals of Pulse Transmission', *BSTJ*, vol. 33, pp 721-788, May 1954 and pp. 987-1016, July 1954.
2. Das, J., 'Optimum Detection and Communication in Presence of Noise', Pts. I, II, III and IV, *ITE Students Journal*, vol. 14, pp. 27, 106, 156, 212, 1973.

3. Helstorm, C.W., *Statistical Theory of Signal Detection*, Pergamon, Press, NY, 1960.
4. Das, J., Mullick, S.K. and Chatterjee, P.K., *Principles of Digital Communication*, Wiley Eastern, New Delhi, 1986.
5. Schwartz, M., Benett, W.R. and Stein, S., '*Communication Systems and Techniques*, McGraw-Hill, NY, 1966.
6. Bellamy, J., *Digital Telephony*, John Wiley & Sons, NY, 1982.
7. Lindsay, W.C., *Synchronization Systems in Communications*, Prentice-Hall, NJ, 1972.
8. Holmes, J.K., *Coherent Spread-Spectrum Systems*, John Wiley & Sons, NY, 1982.
9. 'Matched Filter Issue', *IRE Trans. on Information Theory*, vol. IT-6, June 1960.
10. Turin, G.L., 'An Introduction to Digital Matched Filters', *Proc. IEEE*, vol. 64, p. 1092, 1976.
11. Milstein, L.B. and Das, P.K., 'Spread-Spectrum Receiver Using Surface Acoustic Wave Technology', *IEEE Trans.*, vol. Com. 25, p. 841, 1977.
12. Datta Roy, S.C. *et al.*, 'Signal-Processing Applications of Charged Coupled Devices', *JIETE*, vol. 24, p. 400, 1978.

CHAPTER 3

Digital Encoding Systems

Till recently, all telecommunication networks, involving switching and transmission, were analog, and, all signal processing, e.g., modulation, multiplexing, filtering, bandwidth compression etc., was done through analog techniques. Necessarily, the requisite accuracy and reproducibility suffered. With the progress in analog-to-digital (A/D) conversion techniques and the availability of microprocessor (μP) chips for performing complex mathematical operations, it is now possible to execute complex signal processing using digital techniques; and reproducible modules are easily designed and fabricated. As such, we have now μP-controlled equalizers for correction of line distortion, error-correcting codecs to provide immunity to noise, linear-predictive codecs (LPC) for speech compression and automatic systems for speech recognition. In some cases, e.g., for A/D conversion, LPC, and equalization, custom-made LSI's are now commercially available; they are in the form of either single-chip signal processor or multichip processors.

As a prelude to the digital communication systems described in later chapters, we discuss in this chapter some of the digital techniques for A/D conversion, LPC and speech recognition. Also, the digital signal processing for transmultiplexer, TASI and echo canceller has been included.

3.1 A/D CONVERSION OF SPEECH SIGNALS

Starting with the development of the pulse code modulation (PCM) [1], in post-war years, continuous efforts have been made to further improve the now-classical PCM codec through adaptive and differential techniques, and to devise new techniques, e.g., delta modulation (DM) and its modified forms. While PCM and DM are based on waveform coding and are applicable to any arbitrary, band-limited signal including telemetry and television signals, special codecs have been developed for speech signals using the parametric properties of the signal. Such parametric speech codecs are also quite a few in number and Table 3.1 gives a summary of the codecs useful for speech encoding [2, 3].

Table 3.1

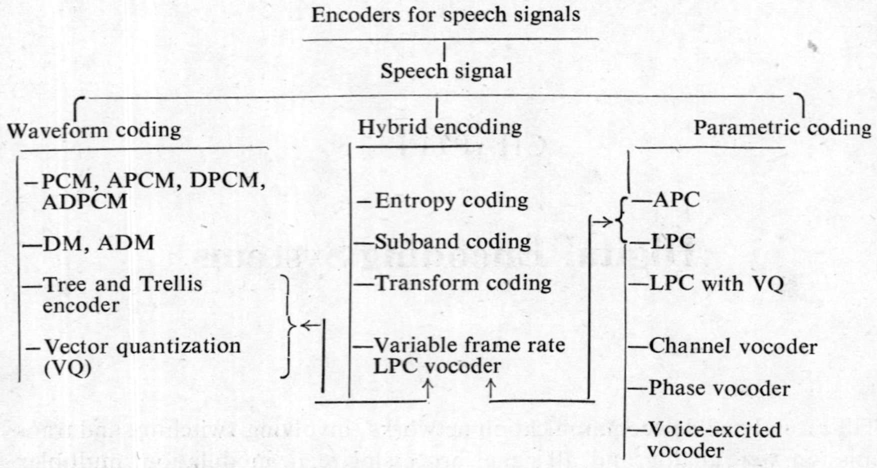

3.1.1 Pulse Code Modulation (PCM) [4, 5]

The sampling theorem [1] provides efficient representation of a continuous waveform through discrete samples, taken at intervals of $\tau \leqslant 1/2W$, where W is the low-pass bandwidth of the signal. The waveform is reconstructed by multiplying the samples with an interpolating function, say, $\sin x/x$ and the reconstructed signal is given by

$$\tilde{x}(t) = \sum_{-N}^{N} x\left(\frac{n}{2W}\right)\frac{\sin \pi(2W\,t - n)}{\pi(2W\,t - n)}, \qquad N = TW \tag{3.1}$$

where the number of samples $= (2TW + 1)$ in T secs, should be large and the $\sin x/x$ filter is represented by an ideal low-pass filter of bandwidth W. In the frequency domain, the spectrum of the sampled signal has sidebands around the harmonics of $f_s = 1/\tau$ and to avoid aliasing error $f_s \geqslant 2W$. For band-pass signals having a bandwidth $W = (f_2 - f_1)$, the sampling rate f_s varies between $2W$ and $4W$, depending upon ratio of f_2/W.

The representation of an analog waveform in the digital form requires quantization both in time as well as in amplitude. Thus, the sample values have to be quantized and coded, usually in the binary form, to produce the equivalent digital bit stream. The PCM is the most popular method of digitizing analog signals. The PCM signals may be generated by different techniques: (a) counter type ADC, (b) feedback type coder, (c) parallel coder, and (d) cascade coder. Of these, the most commonly used type is the feedback coder which uses a D/A converter (DAC) in the feedback circuit, as shown in Fig. 3.1(a). The DAC consists of an $R - 2R$ ladder network which produces binary-weighted currents in response to the switch positions controlled by the steering pulses, generated at the comparator output as shown in Fig. 3.1(b). The signal sample is now compared repeatedly with the summed output of the DAC, and the comparator output gives the PCM bits which are the binary equivalent of the signal sample. In the PCM receiver, a complementary DAC with pulse-steering circuits is used to generate the quantized approximation $\hat{x}(t)$. The ADC and DAC are now available as LSI chips.

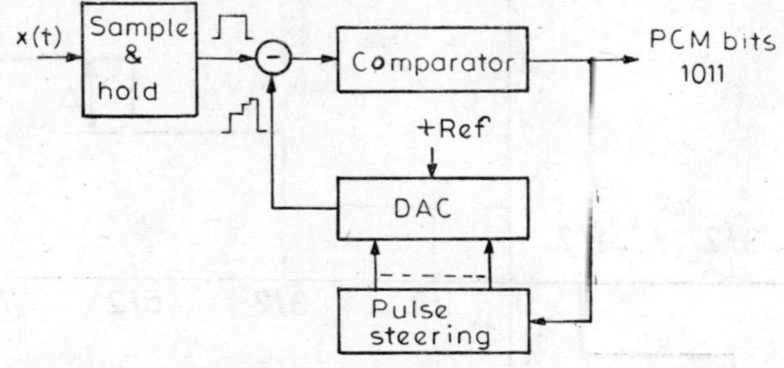

Fig. 3.1(a) Feedback type A-to-D coder (ADC)

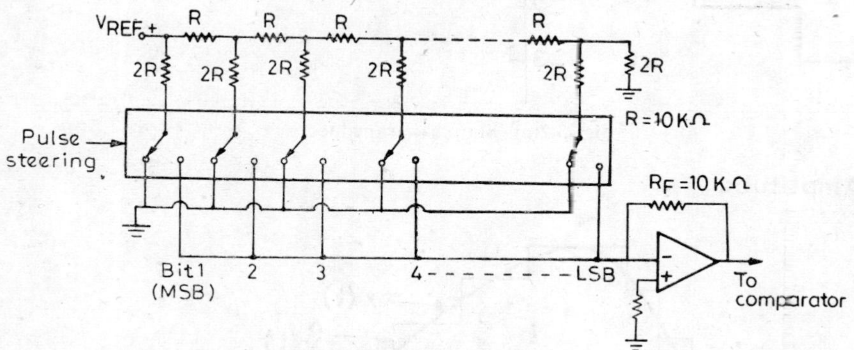

Fig. 3.1(b) Schematic of DAC

The digital representation of an analog waveform cannot be exact and there will be residual error called quantization error which is, however, small. The input-output characteristics of a linear (fixed-step) quantizer is equivalent to that of a 'staircase transducer', as shown in Fig. 3.2(a) and the corresponding error waveform $e(t)$, while approximating a smooth analog signal, is shown in Fig. 3.2(b). It is seen that the error waveform is generally of a triangular shape and the error amplitudes are within $\pm\Delta/2$, where Δ is the step size of the quantizer. Further, all error amplitudes within $\pm\Delta/2$ may be assumed to be equiprobable, giving $p(e) = 1/\Delta$, and the mean error power is now given by

$$\epsilon^2 = \Delta^2/12 \tag{3.2}$$

This gives the signal-to-quantizing noise power ratio as [5]:

$$\begin{aligned} S/N_q &= (6B - 7.2)\ \text{dB}, && \text{for } 4\sigma \text{ loading with random signals} \\ S/N_q &= (6B + 3),\ \text{dB}, && \text{for sinusoidal signals} \end{aligned} \tag{3.3}$$

where B is the number of bits in PCM words, giving the peak-to-peak signal level $L = \Delta \cdot 2^B$.

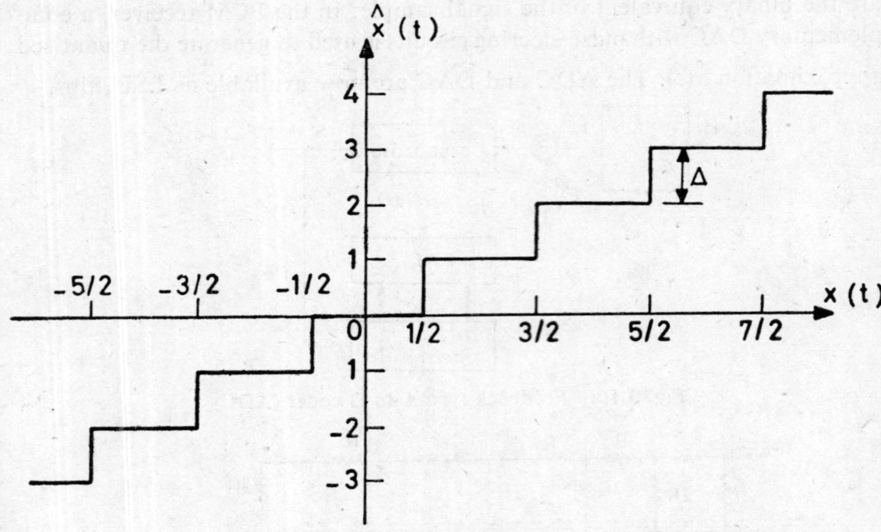

Fig. 3.2(a) Staircase transducer

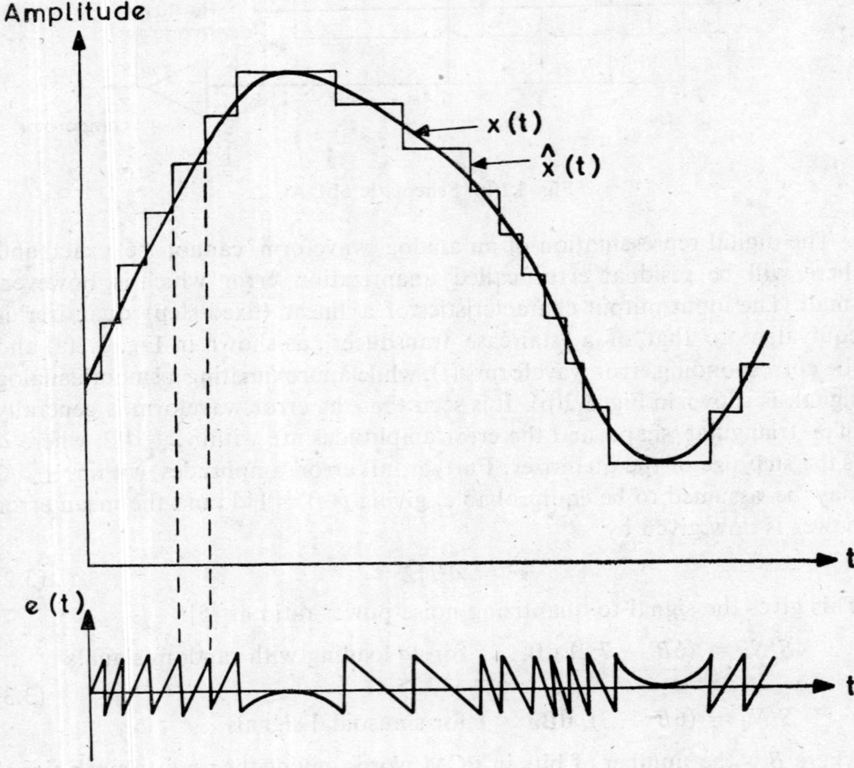

Fig. 3.2(b) Approximation $\hat{x}(t)$ of a smooth waveform and the corresponding $e(t)$

The error spectrum, however, depends upon the number of bits used for coding, and the higher the bit rate, the wider is the error spectrum. But because of the reconstruction of the waveform through quantized sample values, the total error power ($= \Delta^2/12$) is folded back in the signal bandwidth and hence the quantizing noise is only dependent on the quantization step Δ. There is some decrease in the error power within the signal bandwidth if the sampling rate f_s is made much larger than the Nyquist minimum [5]. But PCM always uses the Nyquist sampling rate, whereas in delta modulation, sampling is done at a much higher rate.

The values of S/N_q given by eqn. (3.3) are for 0 dB reference input When the signal level decreases, the S/N_q value also deteriorates proportionately, thus giving a very poor dynamic range of the codec. To overcome this disadvantage, a non-linear quantization is preferably used in PCM codecs, where smaller signal levels are quantized with smaller step sizes and the larger values with larger step sizes. Such a non-linear quantization is obtained through the use of a compressor before the quantizer and a matching expander at the decoder output as shown in Fig. 3.3(a). In the log-PCM codec, the compression law, shown in Fig. 3.3(b), is given by

$$y = \frac{\log_e (1 + \mu \mid x \mid)}{\log_e (1 + \mu)}, \qquad \mu > 0 \quad \text{and} \quad \mid x \mid \leqslant 1 \tag{3.4}$$

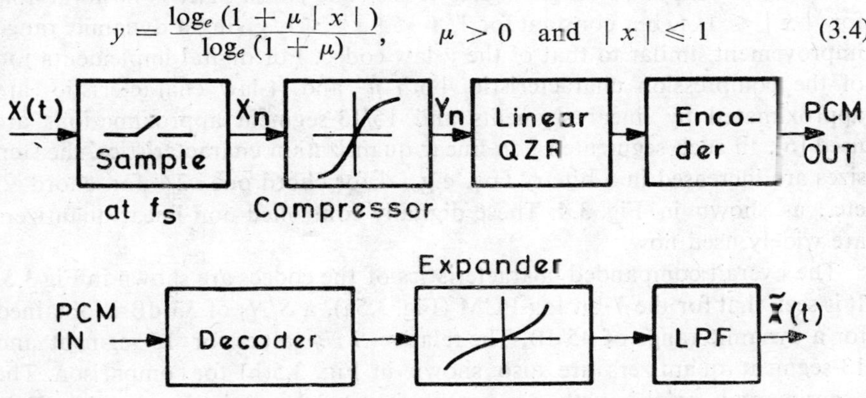

Fig. 3.3(a) Block diagram of a companded PCM codec

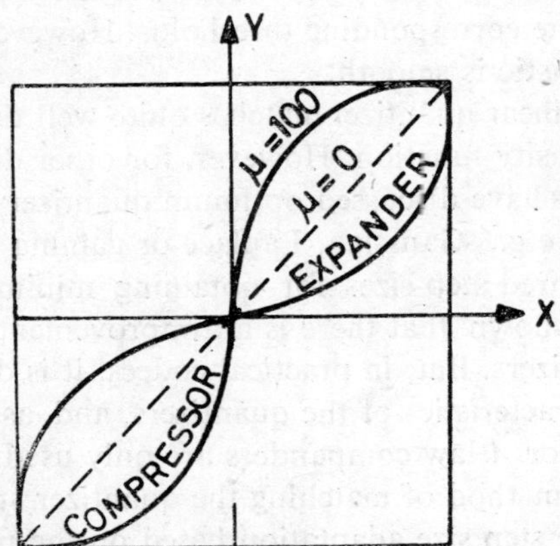

Fig. 3.3(b) Compressor-expander characteristics

It may be shown that for an exponential distribution of $x(t)$, the SNR of the codec is $S/N_q = 12/\Delta^2$. The SNR is now independent of the signal level and normally 4σ loading, i.e., $X_{max} = 4\sigma_x$, σ_x^2 = variance, is a satisfactory condition to avoid overload. The companding advantage (i.e., the increase in the dynamic range) of the μ-law (usually, $100 < \mu < 200$) is approximately 26 dB for a 7-bit codec. The compressor is realized by using temperature-compensated diode networks and the expander uses a complementary circuit consisting of diodes and resistances.

In some PCM multiplexed system, e.g., in 30-channel PCM, a different compression law, known as A-law (with $75 < A < 150$) is used and its equation is given as:

$$y = \begin{cases} A\,|\,x\,|/(1 + \log_e A), & 0 < |\,x\,| \leqslant \dfrac{1}{A} \\[2ex] \dfrac{1 + \log_e (A\,|\,x\,|)}{1 + \log_e A}, & \dfrac{1}{A} < |\,x\,| \leqslant 1 \end{cases} \tag{3.5}$$

A-law gives a linear input-output characteristic for low signal levels and a compressed output for high signal levels. The codec SNR is now variable for $|\,x\,| \leqslant 1/A$, but constant for $1/A < |\,x\,| \leqslant 1$, giving a dynamic range improvement similar to that of the μ-law codec. For digital implementation of the compression characteristic, both μ- and A-law characteristics are approximated by linear segments and 15/13-segment approximations are used [6]. In such segmented non-linear quantization characteristics, the step sizes are increased in a binary law, e.g., Δ for chord one, 2Δ for chord 2, etc., as shown in Fig. 3.4. These digitally controlled non-linear quantizers are widely used now.

The overall companded characteristics of the codecs are shown in Fig 3.5. It is seen that for the 7-bit log-PCM (Fig. 3.5a), a S/N_q of 33 dB is obtained for a dynamic range of 45 dB. The relative S/N_q values for 15-segment and 13-segment quantizers are also shown in Fig. 3.5(b) for comparison. The experimental results with sine wave inputs show ripples in the SNR characteristics because of the 'jerky' use of higher chords in quantizers as the input crosses the corresponding thresholds. However, for random inputs, the SNR characteristic is smooth.

The above non-linear quantizer matches quite well the analog signals with the exponential density function. However, for other distribution of signals, Max [7] and others have discussed optimum quantizers based on the signal density functions, e.g., Gaussian, Laplace or gamma functions. They have calculated the required step sizes for obtaining minimum N_q for different signals and have shown that there is an improvement of a few dB in S/N_q for optimum quantizers. But, in practical codecs, it is difficult to implement the non-linear characteristics of the quantizers, and, as such, in commercial systems, the μ-law or A-law companders are only used.

A more flexible method of matching the quantizer step size to the signal power is to use the step size adaptation based on the magnitude of previous

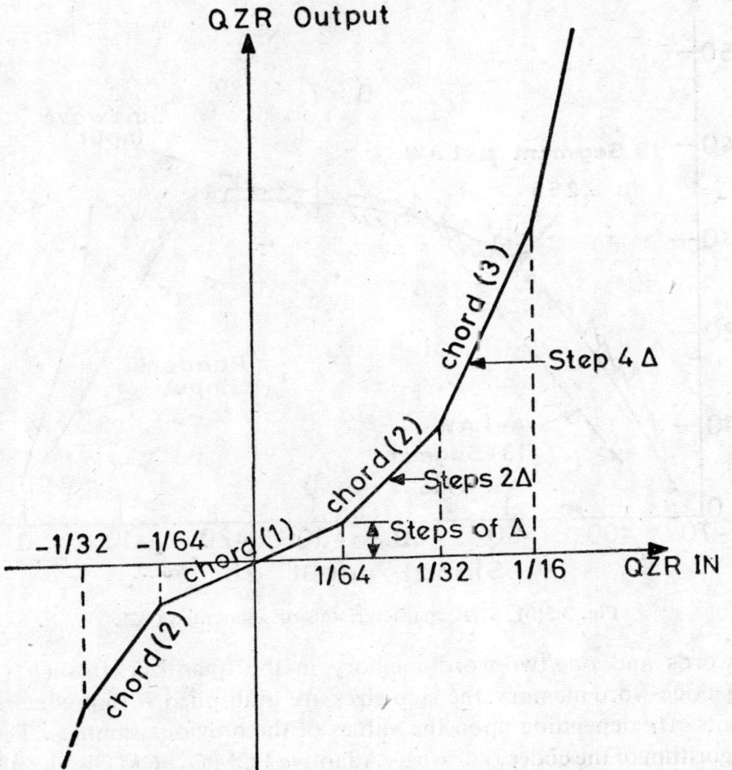

Fig. 3.4 Segmented QZR characteristics

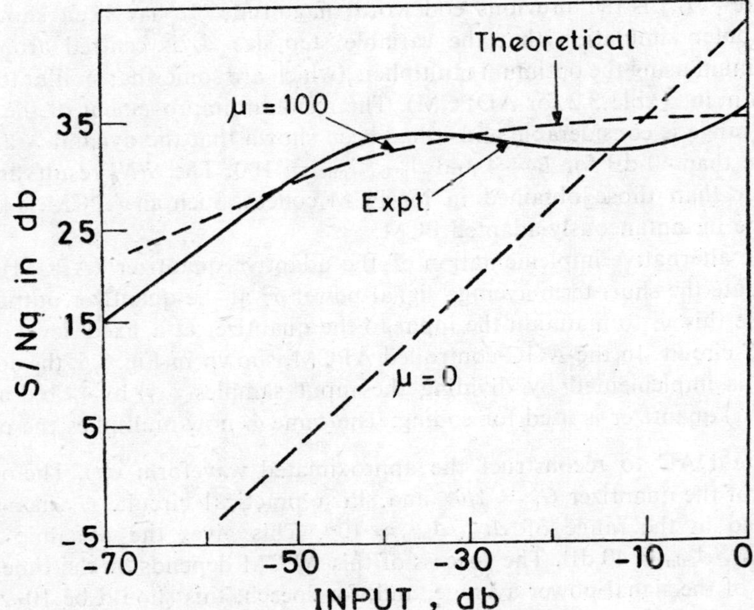

Fig. 3.5(a) SNR values for 7-bit log-PCM codec

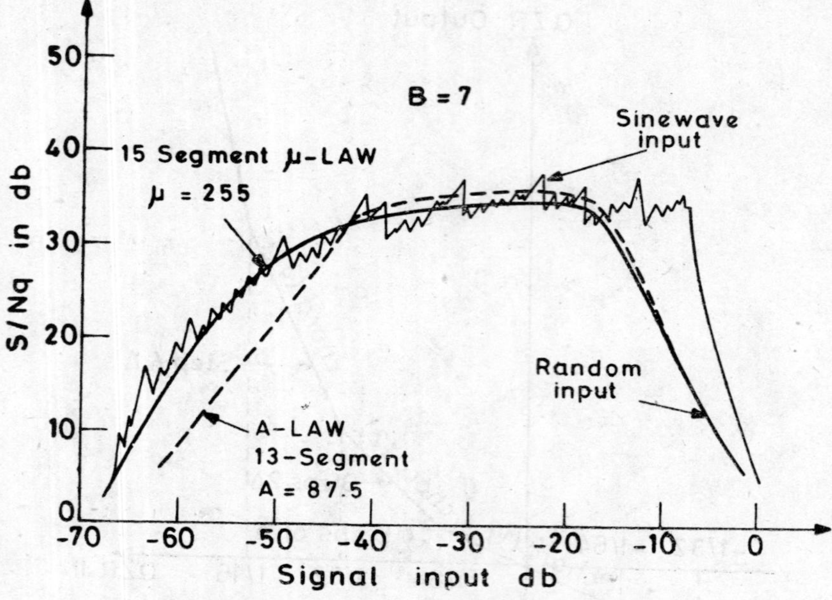

Fig. 3.5(b) SNR characteristics of segmented QZR

code words and one/two-word memory in the quantizer. In such a coder having a one-word memory, the step-sizes are multiplied with predetermined constants M_i depending upon the values of the previous samples. The step size algorithm of the codec, known as Adaptive PCM (APCM), is given by [4]:

$$\Delta_{r+1} = \Delta_r \cdot M_i \mid H_r \mid , \tag{3.6}$$

where $\mid H_r \mid$ is the previous codeword magnitude. It has been shown by computer simulation that the variable step size Δ_r is centred around an optimum using the optimum multipliers (which are somewhat similar to those shown in Table 3.2 for ADPCM). The resultant improvement of the dynamic range is considerable and it has been shown that the dynamic range is more than 40 dB for $B = 4$ and $\Delta_{max}/\Delta_{min} = 100$. The S/N_q results are also better than those obtained in log-PCM codecs. Such an APCM is known as the instantaneously-adapted PCM.

An alternative implementation of the adaptive quantizer (APCM) is to estimate the short-term average signal power $\hat{\sigma}_r^2$ at the quantizer output and to use this $\hat{\sigma}_r$ to maintain the input to the quantizer at a fixed level by an AGC circuit. In the AGC-controlled APCM, shown in Fig. 3.6, the adaptation is implemented by dividing the input samples $\{x_r\}$ by $\hat{\sigma}_r$ and a fixed (linear) quantizer is used for coding. The same $\hat{\sigma}_r$ now multiplies the output of the DAC to reconstruct the approximated waveform $\tilde{x}(t)$. The overall gain of the quantizer $G_r = 1/\hat{\sigma}_r$, and, in a practical circuit, G_r has to be limited in the range of $\Delta_{max}/\Delta_{min} \simeq 100$. This gives the dynamic range $= \Delta_{max}/\Delta_{min} = 40$ dB. The success of this APCM depends on the time constant of the signal-power average, and, for speech, this should be 10–20 ms.

This APCM is now called the syllabically-adapted PCM. A comparison of the performances of log-PCM and APCM shows that the APCM gives an SNR improvement of 4-5 dB, even with $B = 3$, over log-PCM for both Gaussian and Laplacian PDF of $x(t)$. At the same time, APCM gives low idle-channel noise in the system.

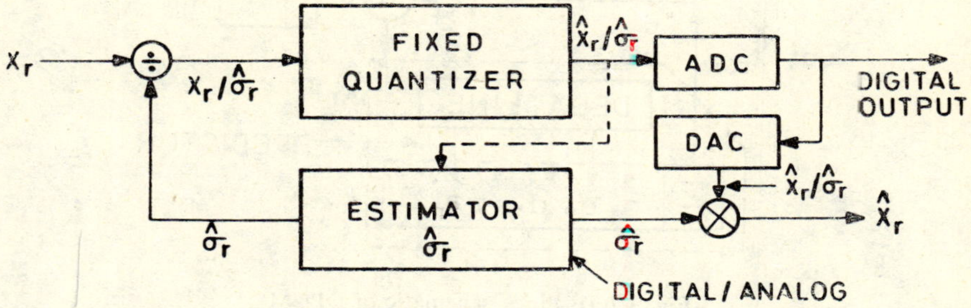

Fig. 3.6 AGC-controlled adaptive QZR

3.1.2 Differential PCM (DPCM) [4]

Since many of the naturally occurring information signals, e.g., speech and TV, exhibit significant correlation between successive samples at the Nyquist rate, the variance of the first difference $\langle d_r^2(1) \rangle$ is smaller than the variance σ_x^2 of the signal itself. Such a differential input to a quantizer will behave more as a random signal and an optimized quantizer may be used.

Assuming the signal correlation between x_r and x_{r-1} to be ρ_1, we have

$$\begin{aligned} \langle d_r^2(1) \rangle &= \langle (x_r - x_{r-1})^2 \rangle \\ &= 2\sigma_x^2(1 - \rho_1) \\ &< \sigma_x^2 \text{ for } \rho_1 > 0.5 \text{ (as in speech signals).} \end{aligned}$$

It will then be advantageous to use $d_r(1)$ as the input to the quantizer and reconstruct the signal through an integrator. However, a better approximation to $\{x_r\}$ is provided through a feedback predictor, and the difference $\langle d_r^2 \rangle$ is further minimized, as shown in Fig. 3.7(a). For a single-tap (a_1) predictor, the difference signal

$$\langle d_r^2(a_1) \rangle = \langle x_r^2 \rangle (1 - \rho_1^2), \qquad \text{with } a_1 = \rho_1$$

and the SNR of the DPCM quantizer is given by

$$S/N_q = \frac{\langle x_r^2 \rangle}{\langle (\hat{x}_r - x_r)^2 \rangle} = \left(\frac{\sigma_x^2}{\sigma_d^2}\right)\left(\frac{\sigma_d^2}{\sigma_e^2}\right), \tag{3.7}$$

where, $\sigma_e^2 = \langle (\hat{d}_r - d_r)^2 \rangle$; $\sigma_d^2 = \langle d_r^2 \rangle$. Since $(S/N)_Q$ for the quantizer $= \sigma_d^2/\sigma_e^2$, the gain due to DPCM is:

$$G_p = \frac{\sigma_x^2}{\sigma_d^2} = \frac{\langle x_r^2 \rangle}{\langle (x_r - \tilde{x}_r)^2 \rangle} \tag{3.8}$$

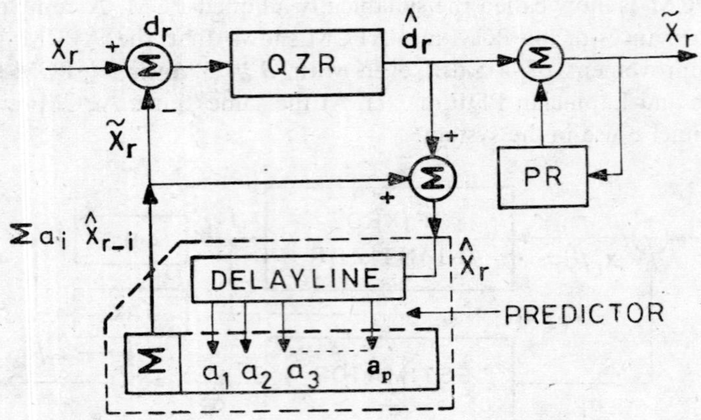

Fig. 3.7(a) Block schematic of DPCM

To maximize G_p, σ_d is to be minimized and this may be done by using a multitap predictor, where the delayed-weighted sum of previous quantized samples forms the linearly predicted values $\{\tilde{x}_r\}$ for the incoming samples $\{x_r\}$. Thus,

$$\tilde{x}_r = \sum_{i=1}^{P} a_i \hat{x}_{r-i},$$

where the predictor transfer function

$$P(z) = \sum_{i=1}^{P} a_i Z^{-i},$$

and its output is $\tilde{x}_r$ with input $\hat{x}_r$, as shown in Fig. 3.7(a). The optimum coefficients, a_i's, are obtained from the matrix relation:

$$[A_{\text{opt}}] = \Gamma^{-1}[\rho], \tag{3.9}$$

where

$$[A] = \begin{bmatrix} a_1 \\ a_2 \\ \cdot \\ \cdot \\ \cdot \\ a_p \end{bmatrix}; \; [\rho] = \begin{bmatrix} \rho_1 \\ \rho_2 \\ \cdot \\ \cdot \\ \cdot \\ \rho_p \end{bmatrix}; \; \Gamma = \begin{bmatrix} \rho_0 & \rho_1 & \cdots & \rho_{p-1} \\ \rho_1 & \rho_0 & \cdots & \rho_{p-2} \\ \cdot & & & \\ \cdot & & & \\ \cdot & & & \\ \rho_{p-1} & \rho_{p-2} & \cdots & \rho_0 \end{bmatrix}; \; \rho_0 = 1$$

and $[\rho]$ is the normalized autocorrelation matrix. The resulting optimum gain over PCM is:

$$G_{\text{opt}} = [1 - \sum_{i=1}^{p} a_i \, \rho_i]^{-1} \tag{3.10}$$

It is known that two-sample correlation $\rho_1 > 0.5$ for speech, and this gives an SNR improvement of about 4 dB. With the weighting a_1, a_2, etc., the SNR gain over PCM saturates at 12 dB. With a multitap predictor, the variation of the SRN gain with p is show in Fig. 3.7(b), where the spread of the curve is due to different speakers. It is seen that the gain saturates with 2 or 3 coefficients only, and with a large p, $G_{opt} \leqslant 12$ dB.

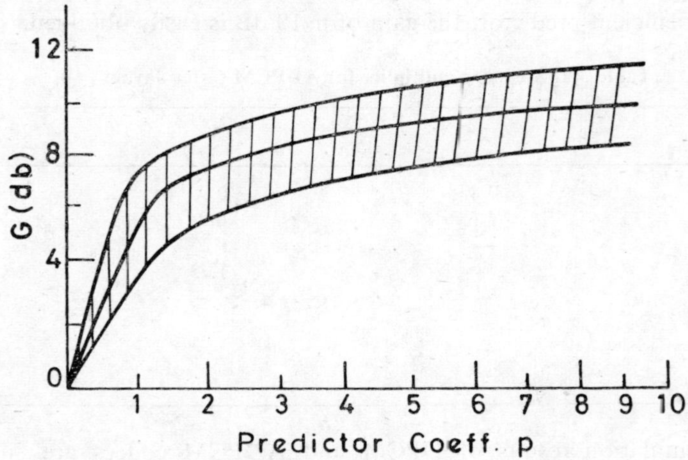

Fig. 3.7(b) SNR gain in DPCM over PCM (after Jayant [4])

More sophistication in the DPCM codec can be introduced by adding the step size control as in APCM. Such an ADPCM codec is schematically shown in Fig. 3.8. Here the step size control is obtained through $\hat{\sigma}_q$, where $\hat{\sigma}_q = \sqrt{\langle q_r^2 \rangle}$, as in Fig. 3.8. For instantaneous adaptation using one-word

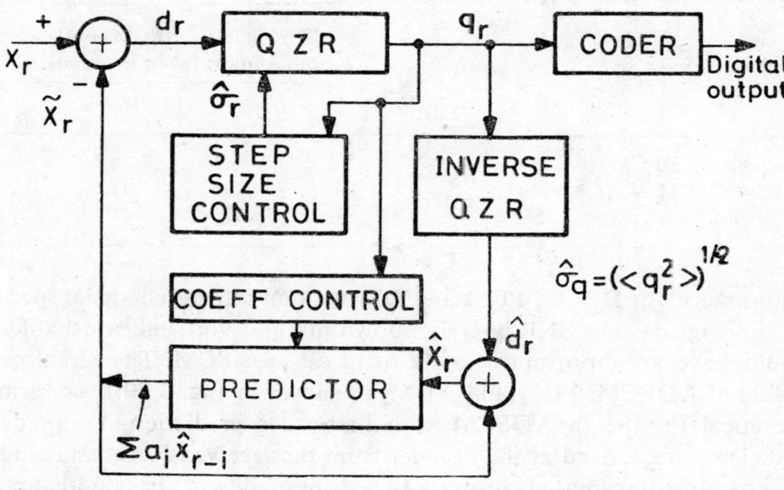

Fig. 3.8 Block schematic of ADPCM

memory, the step size multipliers for 2/3/4-bit coding are given in Table 3.2, and these values are somewhat similar to those used for APCM. Since on-line calculation of A_{opt} is not generally practicable, the coefficient values are up-dated only after every 4–32 ms, and, during this time, the correlation matrices of speech samples are recalculated. In most ADPCM codecs, the step size control is adaptive, but the optimum tap weights are maintained constant for a particular input signal whose statistics have been predetermined. With such a fixed coefficient predictor, the gain of 8–10 dB is easily obtained.

Table 3.2 Step size multiplier for ADPCM (after Jayant [4])

$B \rightarrow$	2	3	4
M_1	0.8	0.9	0.9
M_2	1.6	0.9	0.9
M_3	—	1.25	0 9
M_4	—	1.75	0.9
M_5	—	—	1.2
M_6	—	—	1.6
M_7	—	—	2.0
M_8	—	—	2.4

The simulation results of DPCM and ADPCM codecs are shown in Table 3.3, which gives the SNR gains over PCM. It is seen that adaptive DPCM with a frame time T_{AD} of 4 ms only (= 32 samples of speech) gives an improvement of 13 dB and further increase of the frame time does not help. Since for differential encoding (for DPCM and DM), the correlation between the signal samples are assumed, O'Neal [9] determined the bounds of SNR

Table 3.3 DPCM performance (after P. Noll [8])

Type of codec	SNR gain over PCM in dB with no. of taps (p) in the predictor			
	1	3	5	10
Non-adaptive DPCM	5 4	8.4	8.6	9.0
Adaptive DPCM, T_{AD} = 4 ms	5.6	10.0	11.5	13.0
Adaptive, DPCM, T_{AD} = 32 ms	5.6	9.6	11.1	12.6

performance for DPCM, PCM and DM with an integrated signal spectrum, e.g., TV signals. The SNR bounds, shown in Fig. 3.9(a), indicate that DPCM should have an improvement of 12 to 13 dB over PCM. The actual perfor-mances of ADPCM, PCM and ADM, as shown in Fig. 3.9(b), confirm the theoretical results, In ADPCM with first-order prediction, the prediction coefficients are derived at the decoder from the received bit stream and are not separately transmitted. But in a two-stage predictor, for adaptive-predic-tive coding (APC), the predictor coefficients are also transmitted along with

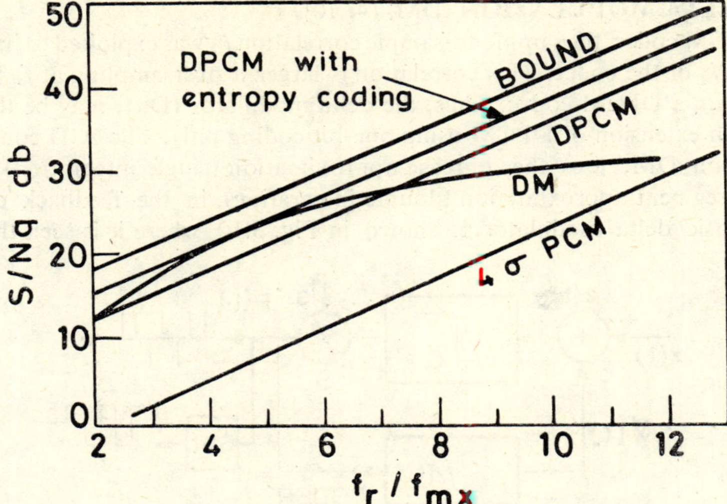

Fig. 3.9(a) SNR bounds (after O'Neal)

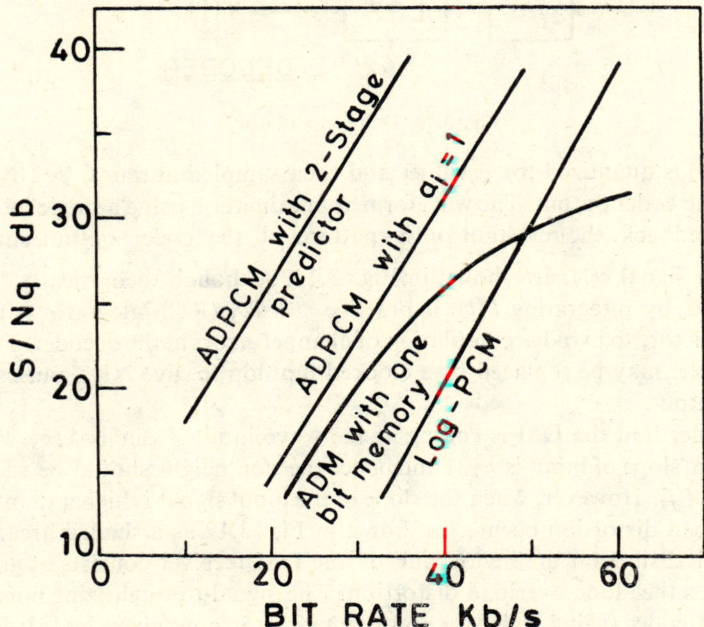

Fig. 3.9(b) SNR for different codecs (with speech input)

quantizer outputs. It has been shown that the advantages of adaptive prediction and adaptive quantization in ADPCM are additive, and as the prediction improves with $p > 2$, the quantizer input tends to become Gaussian and small variations in a_i's do not change the codec characteristics.

3.2 DELTA MODULATION (DM) [4, 10]

In DPCM codec, the sample-to-sample correlation ρ_1 was exploited to improve the S/N_q of the codec. This correlation is larger if oversampling at $f_s \gg 2W$ is used in a DPCM codec. Thus, the delta modulator (DM) may be thought of as an extension of DPCM using one-bit coding only. The A/D converter, based on DM, uses the staircase approximation (single integration) or the linear segment approximation (double integration) in the feedback circuit. The basic delta modulator is shown in Fig. 3.10, where it is seen that the

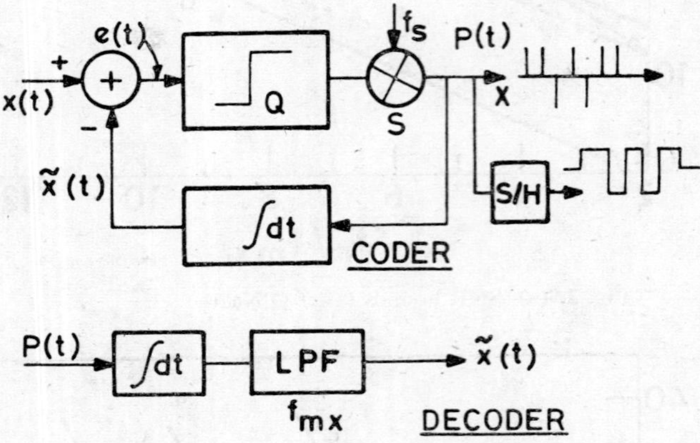

Fig. 3.10 Delta modulator

error $e(t)$ is quantized to $\pm$ values and then sampled at rate $f_s \gg 2W$, giving $P(t)$ as the coder output. The waveform approximation using a single integrator in the feedback, the resultant pulse pattern at the coder output and also the error signal $e(t)$, are shown in Fig. 3.11. Although theoretically, $\tilde{x}(t)$ is generated by integrating $P(t)$, in practice, a leaky (RC) integrator is used as the predictor, to avoid accumulation of channel errors at the decoder. Further, the sampler may be replaced by a clocked flip-flop to give NRZ pulses at the coder output.

Consider that the DM is coding a sine wave input $A \sin(\omega_m t + \theta)$. The maximum slope of input is $\omega_m A$ and hence the step height should be such that $\Delta \geqslant \omega_m A/f_s$. However, when the slope of the input signal is higher than above, an overload distortion occurs, as shown in Fig. 3.12 as a shaded area. Thus, the total distortion at the output of the DM receiver consists of granular noise plus the slope-overload distortion. The signal-to-quantizing noise ratio of a DM codec (using staircase approximation) is now given by [5]:

$$S/N_q \simeq 0.05K^2 f_s^3/(f_{\max} f_m^2), \qquad \ldots K \leqslant 1 \tag{3.11}$$

It is now seen that S/N_q improves with f_s at a rate of 9 dB/oct only as compared to the exponential improvement in PCM. and the SNR deteriorates with increasing signal frequencies. The above SNR is for sinusoidal imputs and for random white Gaussian signal, the S/N_q is approximathly 4 dB worse

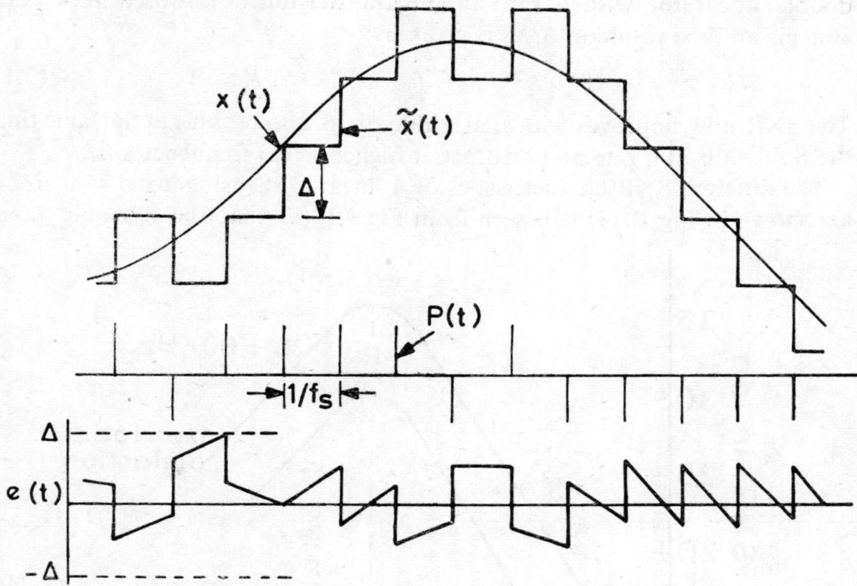

Fig. 3.11 Waveform approximation and $e(t)$ in DM codec

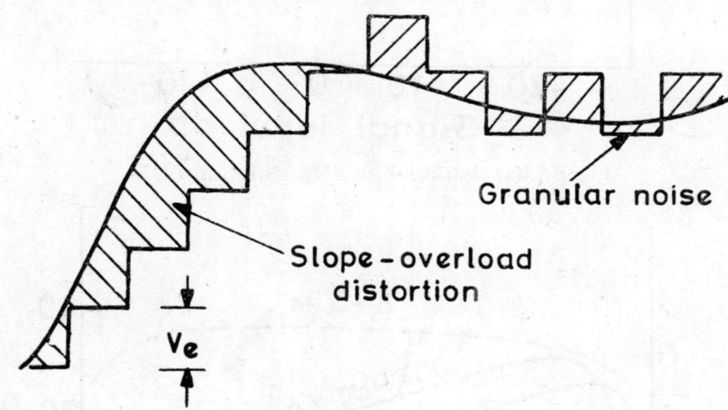

Fig. 3.12 Overload distortion in DM

with $f_m = f_{\max}$. However, for mean speech frequencies ($\simeq f_{\max}/3$), S/N_q is better by 10 dB, showing that the DM codec matches very well for signals with an integrated spectrum, e.g., speech and TV signals. The optimum step height for random signals has been given by Abate [11] as:

$$\Delta_{(\text{opt})} = (\langle (x_r - x_{r-1})^2 \rangle)^{1/2} \cdot \ln (f_s/f_{\max}) \qquad (3.12)$$

If one uses a double integrator in the feedback circuit of a DM codec, then the slope-overload distortion decreases; but at the same time, there is a problem of overshoot and instability in the codec. As a result, a modified double integrator with a zero in the transfer function is used in practical systems and the resultant S/N_q is given by:

$$S/N_q \simeq 6 \times 10^{-4} \cdot K^2 \cdot f_s^5/(f_m^4 f_{\max}), \quad \ldots K \leqslant 1 \tag{3.13}$$

The SNR now improves with f_s at a rate of 15 dB/oct, but, at the same time, the SNR falls at a rate of 12 dB/oct at higher signal frequencies f_m.

The different SNR characteristics of a linear DM (using fixed step size Δ) are shown in Fig. 3.13. It is seen from Fig 3.13(a) that the dynamic range

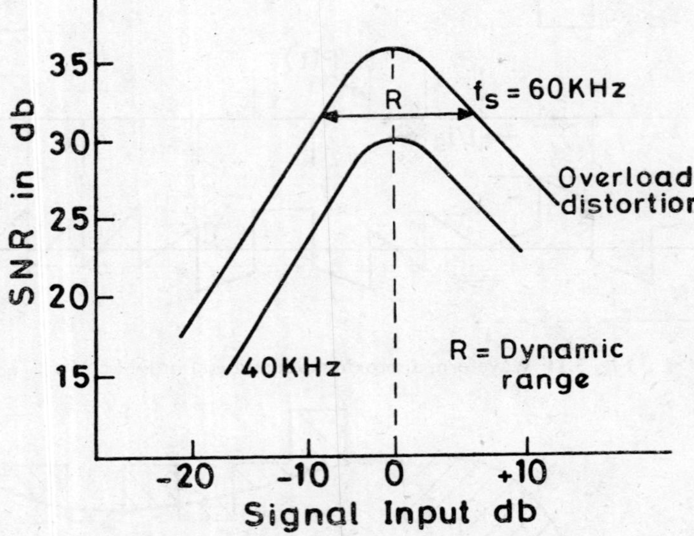

Fig. 3.13(a) Variation of SNR with S in LDM

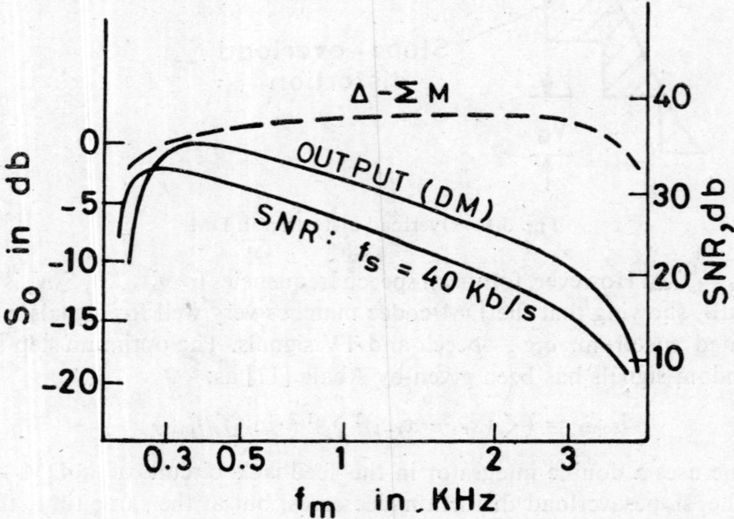

Fig. 3.13(b) Variation of output and SNR with f_m in LDM

is poor, as the SNR decreases for both lower and higher input signals. Although, a peak SNR of 35 dB is obtained for $f_s = 60$ kHz as given by eqn. (3.11), the effective dynamic range R is hardly 20 dB. This problem is similar to that obtained in PCM. The frequency response characteristic of an LDM, as shown in Fig. 3.13(b), also deteriorates at higher input frequencies. However, the frequency response can be made uniform over the required bandwidth by using a Delta–ΣM codec where the predictor in the feedback consists of a low pass filter only. As a result, the peak SNR of D–ΣM is lower than that of an LDM [10].

3.2.1 Adaptive DM (ADM) [5]

The improvement of the dynamic range had been the concern of many workers for the last decade and two types of adaptive DM have been developed, viz., instantaneous and syllabic ADM. In the instantaneously adapted DM (somewhat similar to ADPCM), proposed by Jayant [4], a one-bit memory is used to indicate the continuous increase/decrease of the slope of the input signal and the step sizes are adapted according to the algorithm:

$$\Delta r = \begin{cases} P \cdot \Delta_{r-1}, & \text{if Sgn } e_r = \text{Sgn } e_{r-1} \\ Q \cdot \Delta_{r-1}, & \text{if Sgn } e_r \neq \text{Sgn } e_{r-1} \end{cases}$$

and

$$P \cdot Q \simeq 1, \qquad 1 < P_{\text{opt}} < 2 \tag{3.14}$$

The codec is schematically shown in Fig. 3.14(a), where the probable values of P and Q are: $P = 1.5$, $Q = 0.66$. In Fig. 3.14(b), the reduction of slope-overload noise is indicated due to instantaneous adaptation of the step sizes. It has been shown that the codec has a very large dynamic range and an SNR improvement of 10 dB is obtained at a comparatively low f_s. Responses of the codec to large bandwidth triangular inputs have been shown to be excellent.

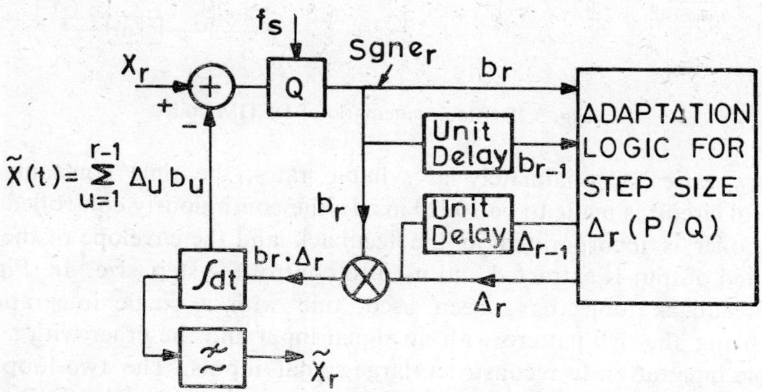

Fig. 3.14(a). Block diagram of ADM with one-bit memory

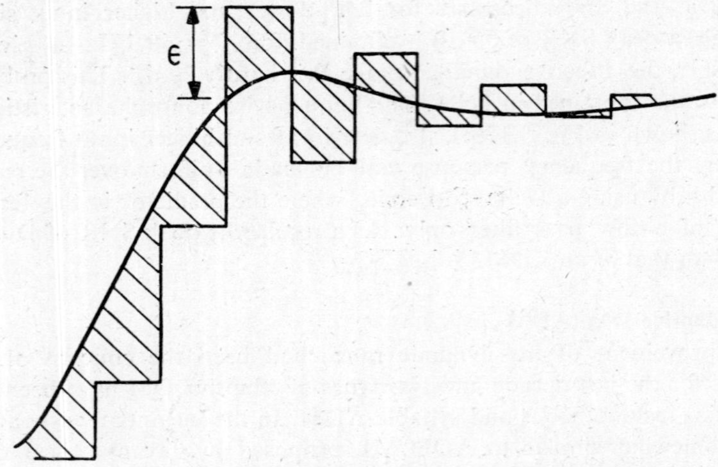

Fig. 3.14(b) Effect of instantaneous companding in ADM

In the syllabic ADM, either digital or analog control may be used. In the digitally controlled DM (DCDM), proposed by Greefkes [12], a 4-stage shift register is used to indicate the sharp rise in the slope of the signal and the step size is increased continuously if continuous ones are obtained at the codec output, as shown in Fig. 3.15. The step size starts decreasing when the codec output no longer contains series of one's. Since the speech signal

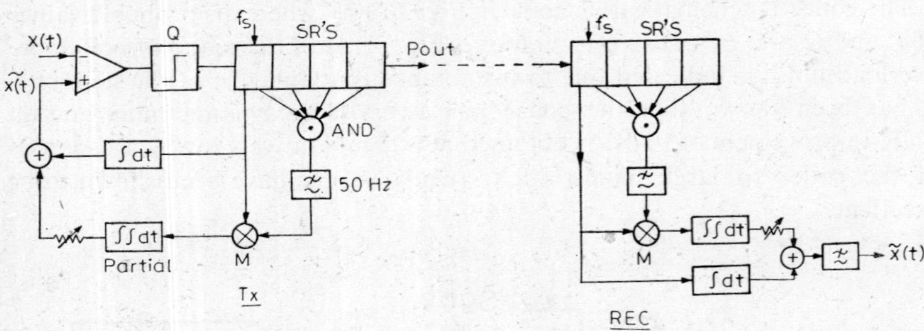

Fig. 3.15 Block schematic of DCDM codec

energy varies approximately at syllabic rates, the time constant of the control circuit is made to be 10–20 ms. In the continuously controlled ADM, a decoder is incorporated in the feedback and the envelope of the reconstructed output is extracted and used to control the step size. In Fig. 3.15, two feedback loops have been used, one with a single integration for stabilizing the 1/0 pattern with no signal input and the other with a partial double integration to reconstruct large-signal inputs. The two-loop codec gives a better performance than the signle-loop codec [13]. The SNR characteristics and the dynamic range of the DCDM are shown in Fig. 3.16, where it is seen that with $f_s = 40$ kHz, the $S/N_q = 30$ dB, and a dynamic range of

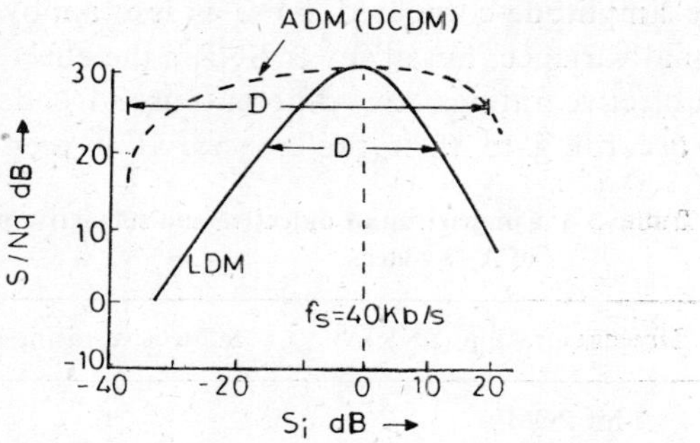

Fig. 3.16 SNR and dynamic range of DCDM codec

40 dB is obtained. The coder matches with the speech spectrum and the speech transmission quality test has shown that the 40 Kb/s DCDM has a similar quality as given by 7-bit log-PCM. The coder is now available as an LSI chip. A comparison of the variation of SNR with bit rates has been shown in Fig. 3.9, where it is seen that the ADM performance is better than that of log-PCM for bit rates less than 50 Kb/s. However, it has been shown that ADPCM gives a still better performance at low bit rates.

3.2.2 Effect of Channel Errors

One of the important parameters to be considered in digital transmission is the effect of channel errors at the receiver output [12, 14]. It is known that for PCM, $S_0/N_e = 1/(4P_e)$, where N_e = equivalent noise power at the decoder output and P_e = the channel error rate; and P_e worse than 10^{-4} is not acceptable for good quality speech. In ADM, $S_0/N_e \simeq 1/(2.5P_e)$, but it has been experimentally observed that for an ADM with $f_s = 20$ Kb/s, a 95% intelligibility score is obtained even with 5% error in the received bit stream. The intelligibility scores of ADM codecs with $f_s = 14.4$, 10 and 7.2 Kbits remain constant at 92%, 83% and 80%, respectively, with $P_e \leqslant 0.05$ [14]. The reason for this is that the adaptive control voltage in ADM is a very low frequency signal (of bandwidth approximately 50 Hz) and this signal does not respond to error rates less than 5%. As a result, the overall reconstructed signal level remains stable. On the other hand, in PCM, because of the different weightages on different bits of the word, occasional errors in the MSB give a disturbing noise at the codec output. Thus, in a noisy channel, say, in mobile digital radio communication, ADM performs much better than PCM.

3.2.3 Subjective Rating

Recent literature recognizes that the SNR characteristic alone is not the complete measure of the performance of the codecs, specially at low bit rates [4]. Because of the signal-dependent error waveform at the output of the quantizer, the perceptual quality of the coder varies with the amount of correlation between the signal amplitude and the error waveform. At higher

SNR's, the amplitude-correlated noise is less annoying than the additive noise of equal variance, but at lower SNR's the effect is the reverse. On the basis of subjective ratings, the different useful codecs are catalogued in Table 3.4 according to their preferences. It is seen that the ADM and

Table 3.4 Comparison of objective and subjective performances of A/D codecs

Objective rating (SNR)	Subjective rating (preference)
7-bit PCM	
ADM-40 Kbit	ADM-40 Kbit
	7-bit PCM
ADM-32 Kbit	ADM-32 Kbit
6-bit PCM	
4-bit ADPCM	4-bit ADPCM
	6-bit PCM
5-bit PCM	
3-bit ADPCM	3-bit ADPCM
	5-bit PCM
4-bit PCM	4-bit PCM

ADPCM codecs are subjectively assessed to be better than the corresponding PCM coders having similar SNR's. It has been further shown from the spectrograms of speech and quantization error in PCM and ADPCM that, in ADPCM, the error spectrum is falling with frequency and is matched to the long-term spectrum of speech itself. There is a certain speech-related buzziness in ADPCM noise and the quantizing noise is less in ADPCM than in log-PCM during the silent intervals of speech. In ADM at low bit rates, the slope-overload distortion constitutes the major component of the total error power and this is mostly signal-correlated. Thus, the perceptual quality of the ADM is largely controlled by the granular noise, and the slope-overload distortion is less annoying. It has been then suggested that the perceptual quality of ADM codecs may be improved by 'slow' adaptive strategies (by operating at a certain overload) and by making error-power spectra decrease at higher frequencies.

3.3 DATA REDUCTION

It is well known that speech and television signals are highly redundant, but much of the redundancy is eliminated when they are encoded into digital streams by DPCM or ADM codecs. Further analysis of the quantizer input signals in DPCM codecs shows that the PDF of the inputs is somewhat Laplacian (two-sided exponential) and, as such, some correlation between input samples still remains. O'Neal [15] then suggested additional coding of the DPCM output by using Entropy coding techniques (Huffman coding or Shannon-Fano coding; refer to Chapter 8). This resulted in an improvement in the S/N_q ratio by about 5 dB without increasing the bit rate (say, 32 Kb/s for speech) of the codec. This is equivalent to a saving of approximately one bit per sample in a DPCM codec, and larger saving in bits/samples may be possible using entropy coding with sufficient buffer memory in the codec.

Once a sufficiently accurate digital representation of a signal, say, speech, has been obtained, it is possible to use discrete forms of Orthogonal transforms, viz., Fourier, Hadamard and Karhunen-Loeve (K-L), to reduce the bit rate and thus reduce the redundancy of the digital signal [16]. A study on this has shown that the K-L transform provides a reduction in the bit rate by 13.5 Kbit/s, Fourier by 10 Kbit/s and Hadamard by 7.5 Kbit/s, as compared with the 56 Kb/s rate required in the companded PCM. Alternatively, the SNR improvement obtained by the transform techniques are: K-L—9 dB; Fourier—6 dB; and Hadamard—3 dB, as compared to the SNR's of the log-PCM at the respective bit rates. These improvements in SNR are somewhat independent of the bit rate of the codec, at least in the range of 20-56 Kb/s.

Many of the present-day techniques in data compression and bandwidth reduction use digital signal processing to reduce the redundancy in the digital data obtained through standard A/D conversion. For such digital processing, many techniques, e.g., adaptive sampling, prediction, interpolation and statistical encoding, have been used [17]. For telemetry signals, data converters using either predictors or interpolators, have been designed to give compression ratios of 10 to 50 with an error tolerance of 0.5% to 2%. The adaptive predictive coders, e.g., ADPCM, ADM, discussed earlier have these data reduction techniques somewhat integrated within the coder hardware, and, as such, little gain may be obtained by further entropy encoding of the bit streams from these coders.

3.4 PARAMETRIC SPEECH CODERS [2, 3]

The A/D codecs discussed in earlier sections are based on waveform coding and can code various signals, e.g., speech, TV and telemetry signals, equally well. However, some redundancy exists at the digital output, and this can be removed if the parametric descriptions of the signal and its source are available. Using this knowledge, a different class of codecs has been designed, specially for speech signals. It has been shown that the speech generation process can be approximately modelled as a linear system driven by a quasi-periodic sequence of pulses or by noise as shown in Fig. 3.17(a). An all-pole filter is quite sufficient to represent the effect of the vocal tract and subsequent transmission medium of the signal. The equivalent transfer function can be represented by a predictor as shown in Fig. 3.17(b). Since the voiced speech has a long-term correlation and the vocal tract transfer function gives a short-term correlation, the speech model can be represented by a two-stage predictor with a single pulse as the input as shown in Fig. 3.17(c). As a result, an inverse filter should be able to remove the short-term and long-term redundancy of the speech signal. Such an inverse filter and its equivalent parallel form are shown in Figs. 3.18(a and b). The predictor

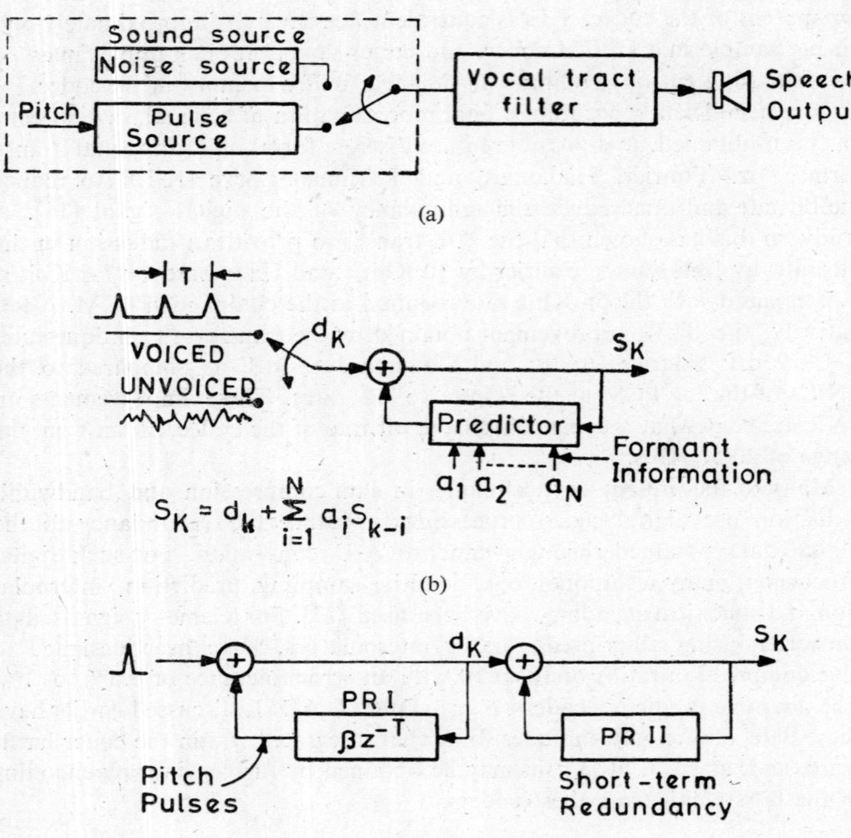

Fig. 3.17 Speech models: (a) source-system models; (b) linear predictor model; and (c) voiced speech mode

coefficients β and T are obtained by minimizing the function:

$$\sum_{k=1}^{n} V^{k^2} = \sum_{k=1}^{n} (S_k - \beta \hat{S}_{k-T})^2 \tag{3.15}$$

where $n \simeq$ 75–100 and $T \simeq$ 3–20 ms. T is selected to maximize the normalized correlation:

$$\rho_T = \frac{\sum^{n} S_k \cdot S_{k-T}}{(\sum^{n} S_k^2 \cdot \sum S_{k-T}^2)^{1/2}} \quad \text{and} \quad \beta_{\text{opt}} = \frac{\sum^{n} S_k S_{k-T\text{opt}}}{\sum^{n} S_{k-T\text{opt}}^2} \tag{3.16}$$

And for the second predictor, $[a_i] = [A_{\text{opt}}] = \Gamma^{-1} \cdot [\rho]$ as in eqn. (3.9).

Based on the above two-stage predictor model, Atal and Schroeder [18] have developed an adaptive-predictive codec (APC) whose performances at low bit rates are excellent. In the system block diagram shown in Fig. 3.19, the step adaptation control and the predictor adaptation control are obtained

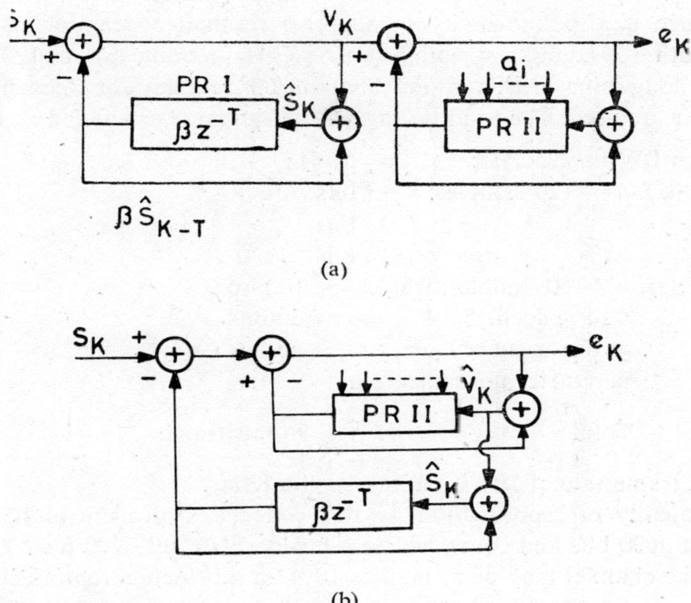

Fig. 3.18 Speech parameter extraction: (a) inverse filter and (b) nested inverse filter

from the quantizer output and the values of a_i's, β and T are obtained from the matrix eqns. (3.9) and (3.16). These parameters are transmitted to the receiver along with the quantized residual output q_k and the signal reconstructed in a complementary decoder. The approximated signal $\{\hat{S}_k\}$ is now given as:

$$\hat{S}_k = e_k[1 - \beta Z^{-T}]^{-1}[1 - \Sigma a_i Z^{-i}]^{-1}, \qquad e_k = \sigma_r \cdot q_k \cdot Q^{-1}$$

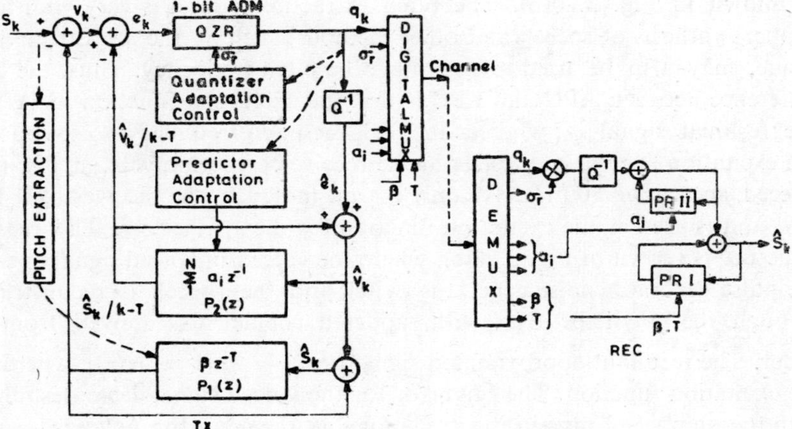

Fig. 3.19 Block diagram of the APC codec

The codec is also known as an adaptive residual coder. For a speech of bandwidth 3 kHz and a sampling rate of 6 kHz, a frame time of 120 samples for the long-term predictor and that of 60 samples for the short-term predictor are used. The total bit rate of the system is calculated as follows:

Speech BW = 2950 Hz: $f_s = 6$ kHz
For $P_1(z)$, $n = 120$ samples: β—3 bits
T—7 bits
σ_r—4 bits
For $P_2(z)$, $n = 60$ samples; frame = 10 ms
4 of a_i coeff. 5×4 = 20 bits
q_k (error signal) = 60 bits
parameter normalization = 2 bits

Total = 96 bits/frame

For a frame rate of 100/s, bit rate = 9.6 Kbits.

The quality of reproduction by this codec is equivalent to 4.5-bit log-PCM, at 9600 bits and equivalent to 4-bit log-PCM, at 7200 bits. The error rate in the channel may be as high as 10^{-3} for satisfactory reproduction and, with an error rate $= 10^{-2}$, the system performance is tolerable. This codec is one of the best low-bit rate codec for speech. However, calculation of the predictor coefficients and their adaptation cannot be done on-line and it requires a mini-computer to calculate these coefficients.

The above APC codec has been further modified by Atal and Hanauer [19] and the modified codec is known as Linear Predictive Codec (LPC). In this codec, shown in Fig. 3.20, the speech is again modelled as the sum of weighted-delayed samples and the predictor coefficients are obtained in the same manner as in the case of APC by using the correlation matrix of the input signal. The excitation function of the speech is now obtained by a peak-picker circuit after removal of the short-term redundancy of the signal, as shown in Fig. 3.20(b). The pitch extraction circuit is very vital to the proper synthesis of speech and other methods, such as the Cepstrum technique, may also be used to determine the pitch frequency. Thus, the basic difference between APC and LPC is that the excitation function in APC is the residual signal q_k, whereas in LPC, the pitch frequency/noise source is the excitation for the synthesizer at the receiver. Alternatively, a low-pass-filtered speech of 500 Hz BW only is transmitted to the receiver by 1 Kbit DM and is used as the excitation function for the synthesizer. The receiver structure is shown in Fig. 3.20(c), where the voiced/unvoiced signal operates the pulse generator/noise generator switch and the speech is reconstructed through the predictor $H(z)$ with updated coefficients received from the coder. The resultant approximated signal is: $\hat{S}_n = e_n/[1 - H(z)]$, where e_n is the excitation function. The rms error for the voiced signal decreases rapidly with the number of taps in the predictor, but the error for unvoiced sound varies slowly with the number of predictor taps. In general, 10 to 12 predictor taps are used. The total number of bits required for acceptable-quality

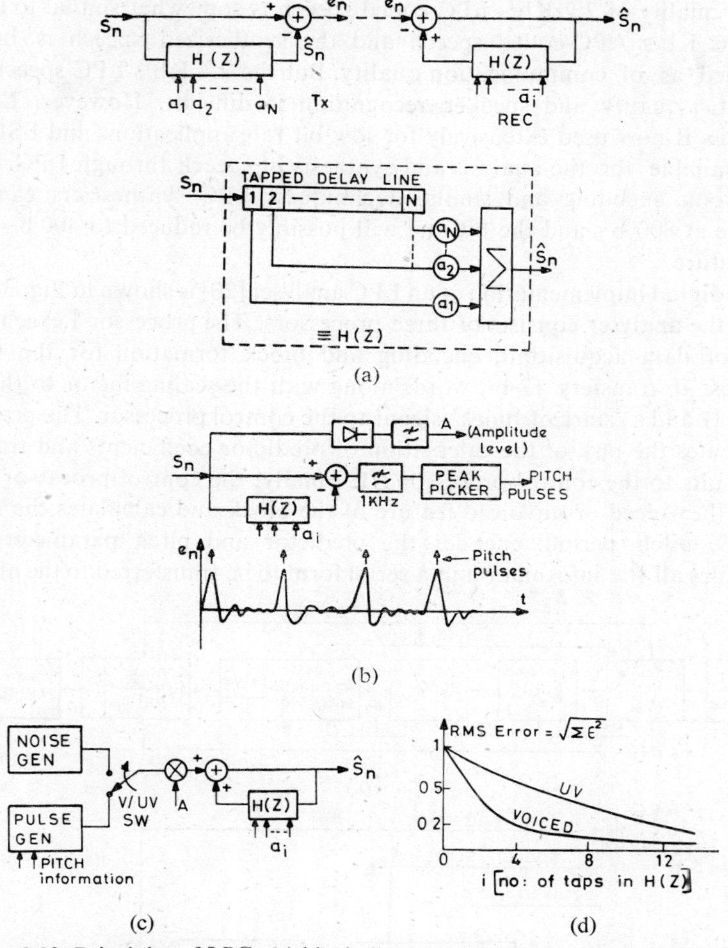

Fig. 3.20 Principles of LPC: (a) block diagram of LPC; (b) pitch extraction in LPC; (c) receiver for LPC; and (d) RMS error vs. predictor length

speech is given as follows:

Parameter	*Bits required*
Pitch	6
V/UV	1
A (rms values)	5
Frequency and BW of poles of $[1 - H(z)]$; $(N = 10\text{-}12)$	60
Total	72 bits/frame

Bit rate for 10 ms frame 7200
Bit rate for 33 ms frame 2400

The quality of 7.2 Kb/s LPC-coded speech is somewhat similar to that of the 9.6 Kb/s APC-coded speech and the synthesized speech is broadly classified as of communication quality. But the 2.4 Kb/s LPC speech is of synthetic quality and speaker recognition is difficult. However, LPC at 2.4 Kb/s is now used extensively for low-bit rate applications and LSI chips are available for the analysis and synthesis of speech through LPC. Using code-book encoding and similar techniques, LPC synthesizers can now operate at 800 b/s and the bit rate will possibly be reduced to 400 b/s in the near future.

The digital implementation of an LPC analyser [20] is shown in Fig. 3.21(a), where the analyser consists of three processors. The processor I executes the tasks of data acquisition, encoding and block formation for the speech samples; it transfers 12-bit words along with the scaling factor to the processor II and a 'start of block' signal to the control processor. The processor II executes the task of the calculation of predictor coefficients and transfers the results to the control processor III. Finally, the control processor determines the voiced or unvoiced feature of the block and calculates the energy and the pitch period; encodes the predictor and pitch parameters, and assembles all the information in a serial form to be transferred to the modem.

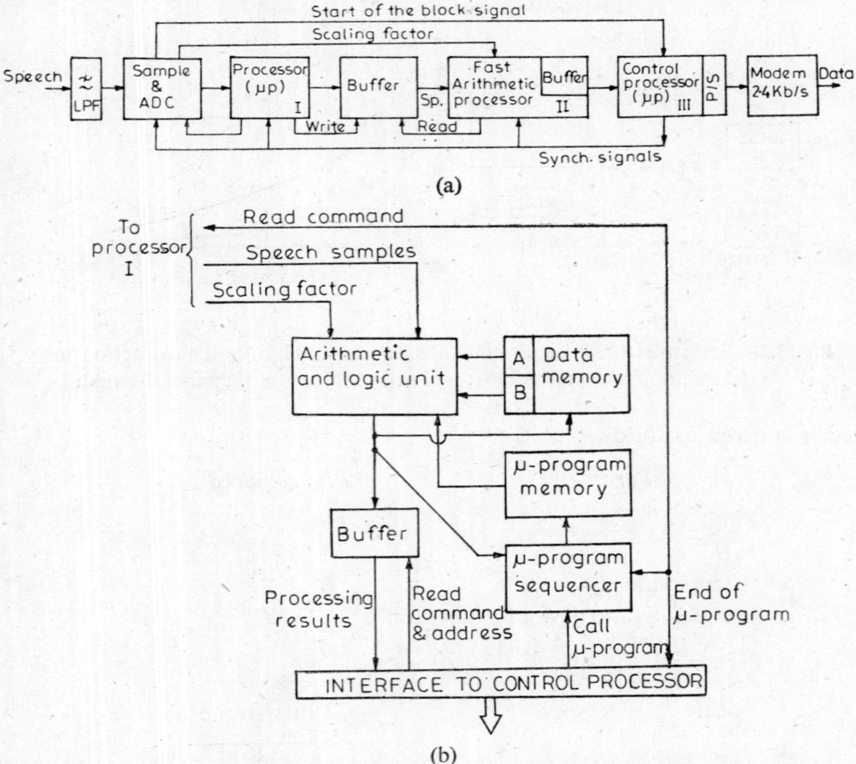

Fig. 3.21 (a) Block schematic of an LPC analyser; Sp-speech sample to arithmetic unit and (b) block diagram of the fast arithmetic processor II

The modem operates at 2.4 Kb/s and uses CCITT V. 24 interface. The pre-processor I and the control processor III are based on general-purpose μP's, so that the main characteristics of the LPC (e.g., block length, number of predictor coefficients, format of the serial data, etc.) may be altered merely by modifying the software. The fast arithmetic processor II, as shown in Fig. 3.21(b), performs a series of repetitive operations at high speed, and, as such, it is custom-made with bipolar technology. The processor is organized around an arithmetic and logic unit which carries out the commands in microinstruction form sent by a microprogram memory, the addressing of which is controlled by a sequencer. A command from the central processor initiates the execution of each microprogram; the corresponding processing is then done by the arithmetic unit, at its own speed, until it has been completed. The A and B memories are used as the data tables by the arithmetic and logic unit. The overall control of the LPC and the synchronization of various units are with the control processor.

The LPC decoder/synthesizer performs the complementary functions, as shown in Fig. 3.22. The broad functions are: (a) interfacing and demultiplexing the received serial data; (b) calculation of the signal samples from the excitation and prediction parameters; and (c) D/A conversion to analog signals. The control and the arithmetic processors are similar to the one used in the analyser. But the number of operations performed by the arithmetic unit is much smaller, hence, the complexity of the unit is less.

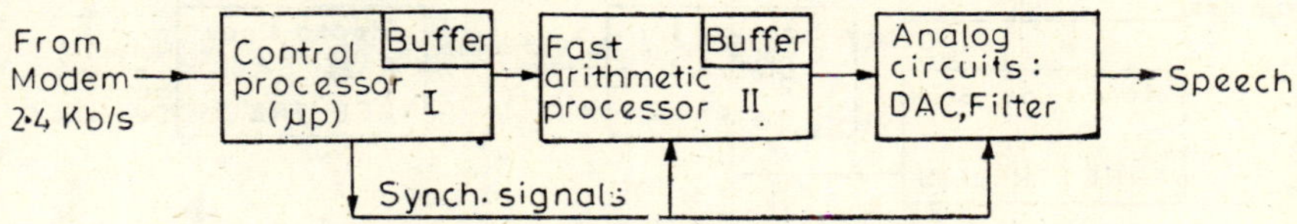

Fig. 3.22 Block schematic of an LPC synthesizer

The parameters and the number of bits per frame required for the digital LPC are as follows:

Frame length	22.5 ms
Sampling rate, f_s	8 kHz
Frame rate	44.4 frames/s
Number of input samples, n	180
Number of coefficients N	
—for voiced signal	10
—for unvoiced signal	4
Computation method	Covariance
Bits for a_i	41
Pitch bits	7
Gain (A) bits	5
V/UV	1
Total bits/frame	54
Transmission rate	2.4 Kb/s

Detailed tests of the analyser/synthesizer have shown that the intelligibility and speaker recognition in the synthesized speech are very satisfactory. Thus, the LPC codecs will find wide applications in the telecommunication systems and networks.

3.5. RECENT DEVELOPMENTS IN SPEECH CODING

A general description of various established digitization techniques has been given in earlier sections. Recently, some new techniques, e.g., Tree, Trellis and Vector quantization, spectrum shaping of the quantization noise, variable frame rate transmission of narrow-band LPC parameters, etc., have been developed [21]. Some of these approaches have been indicated in Table 3.1.

3.5.1 Multipath Search Coding [22]

The conventional coding schemes, e.g., PCM, DPCM, ADM, are based on instantaneous decision of the quantizers, and, as such, are called single-path coders. In multipath search coding (MSC) schemes, on the other hand, future values of input samples x_{n+1}, x_{n+2}, . . . , are considered as well before a (delayed) decision is made about the optimum channel code C_n to be transmitted. In Fig. 3.23, showing the structure of MSC schemes, the input samples $\{x_n\}$ are fed into the input buffer of length N. The encoder compares

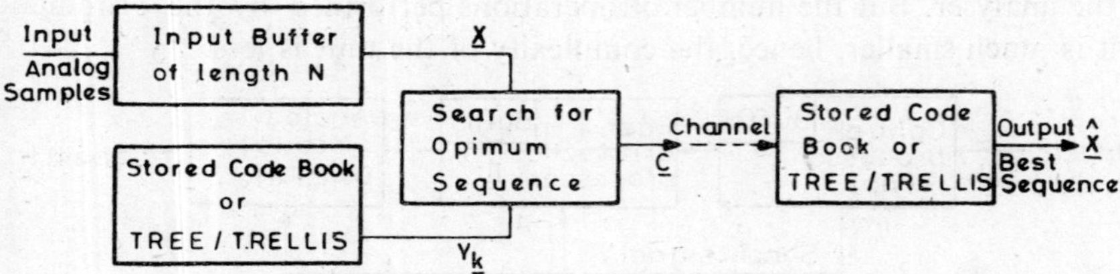

Fig. 3.23 Structure of multipath search coding (MSC) schemes

the buffered vector $\underline{X}$ with a collection of possible output sequences $\underline{Y}_k$, $k = 1, 2, \ldots, 2^N$. The optimum output sequence is the nearest neighbour sequence based on the squared error criterion:

$$\epsilon_k^2 = (\underline{X} - \underline{Y}_k)^T(\underline{X} - \underline{Y}_k) \rightarrow \text{minimum} \tag{3.17}$$

The decoder is informed about the chosen output sequence by a binary channel sequence C_n, and on its basis, the decoder outputs the corresponding output sequence approximating $\{x_n\}$.

MSC coding strategies are based on:

(1) Code-book coding, also known as list coding/vector quantization (VQ).
(2) Tree coding.
(3) Trellis coding.

In the code-book coding schemes, the set of possible output sequences $\underline{Y}_k$, $k = 1, 2, \ldots, 2^N$, is arranged in a code-book, whose elements are multi-valued, giving 'typical' output sequences. When the optimum output sequence (with minimum ϵ_k^2) has been found, the corresponding index of that sequence is transmitted as the channel sequence $\underline{C}$ in a binary format using N bits, as shown in Fig. 3.24. In tree and Trellis coding schemes, the output sequences of length L are arranged in the form of a tree or trellis of depth L, as in Fig. 3.24. Its branches are weighted with reconstruction (multilevel) values, and the different sequences have a number of common elements.* Each

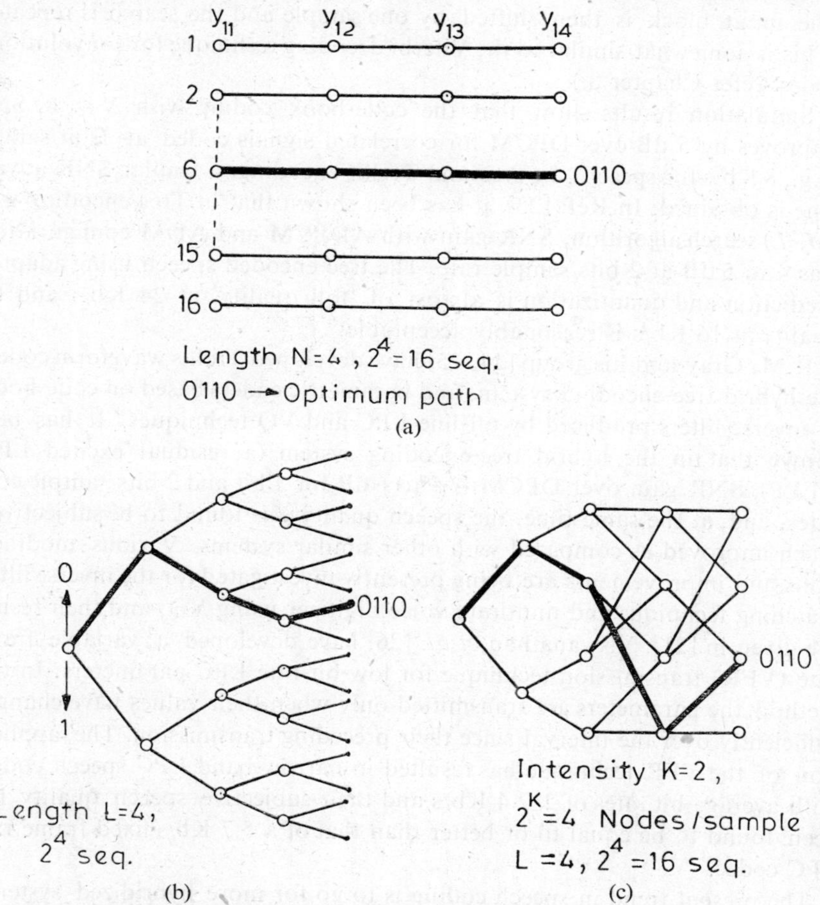

Fig. 3.24 MSC coding: (a) code-book coding; (b) tree encoding; and (c) trellis coding

*The scheme for allotting different weightages to different elements of a code in the code-book or to different branches in a tree or trellis, so that 'typical' (i.e., highly probable) output sequences result, may be either deterministic or probabilistic. In the deterministic schemes, the elements may be generated successively on a sample-by-sample basis by a given algorithm known to the encoder and decoder (as in ADPCM or ADM). Thus, the channel code not only defines a path map but also the weightage of the elements from an alphabet size of $2B$, where each source sample contains B bits of information. In the probabilistic scheme, the weightages are less restricted and may be obtained from known probabilistic models or from a training sequence of data. It is also possible to pick the elements of the code-book as generated by a PN sequence and to find a good code-book by a trial-and-error method.

sequence has a path through the Tree or Trellis, and the channel sequence then provides information regarding the path map and the equivalent input to the decoder. MSC schemes may work in a block mode, i.e., a complete L-sample source sequence is released after the optimum $\underline{Y}_k$ has been found; or in an incremental mode, where only a single symbol is released at a time. The input block is then shifted by one sample and the search is repeated. (This is somewhat similar to the Viterbi decoding technique for convolutional codes, refer Chapter 8.)

Simulation results show that the code-book coding with $N = 8$, SNR improves by 5 dB over DPCM for correlated signals coded at 1-bit/sample (e.g., 8 Kb/s for speech). In Tree and Trellis encoding, a similar SNR advantage is obtained. In Ref. [23], it has been shown that in Tree encoding with (M, L) search algorithm, SNR gain with ADPCM and ADM configurations was 4 to 5 dB at 2 bits/sample rate. The tree-encoded speech using adaptive prediction and quantization is almost of 'toll quality' at 24 Kb/s and the quality at 16 Kb/s is reasonably acceptable.

R.M. Gray and his group [24, 25] have developed Trellis waveform coders, the hybrid tree-encoding system, and parametric coders based on code-books of inverse filters produced by off-line LPC and VQ techniques. It has been shown that in the hybrid tree-encoding system (a residual excited LPC-RELP), SNR gain over DPCM is 5 to 6 dB for 1 bit and 2 bits/sample code rates, and, at the same time, the speech quality was found to be subjectively much improved as compared with other similar systems. Various modifications and improvements are being presently investigated for the inverse filter-matching technique and multirate voice digitizer using VQ, and their results are given in [21]. Viswanathan *et al.* [26] have developed a variable frame rate (VFR) transmission technique for low-bit-rate LPC parameters. In this method, the parameters are transmitted only when their values have changed sufficiently over the interval since their preceding transmission. The application of the VFR technique has resulted in narrow-band LPC speech codecs with average bit rates of 2–2.4 Kb/s and their subjective speech quality has been found to be equal to or better than that of a 5.7 Kb/s fixed frame rate LPC codec.

The present trend in speech coding is to go for more hybridized systems, where various established techniques, e.g., APC/LPC, ADPCM/ADM, Tree/VQ and variable rate transmission, are combined to optimize the performance, thus leading to low-bit-rate digitized speech in the range 1.0 to 10 Kb/s.

3.5.2 LPC-VQ Codecs [27]

It has been shown that the digital LPC at 2.4 Kb/s requires 41 bits/frame to encode the spectral data (equivalent to predictor coefficients) and this results in a family of over 2.10^{12} spectral shapes, which are far in excess of the number of shapes present in ordinary speech. As such the LPC bit streams have some redundancy, and the data can be further compressed if some data reduction techniques, say, code-book (VQ) encoding, is applied. Such a hybrid codec, where the LPC bit streams are further compressed by code-book encoding, is shown in Fig. 3.25. In the codec, the spectral data are encoded by the VQ technique, while the pitch and gain data are encoded differentially. For vector quantization, the distortion measure chosen is the

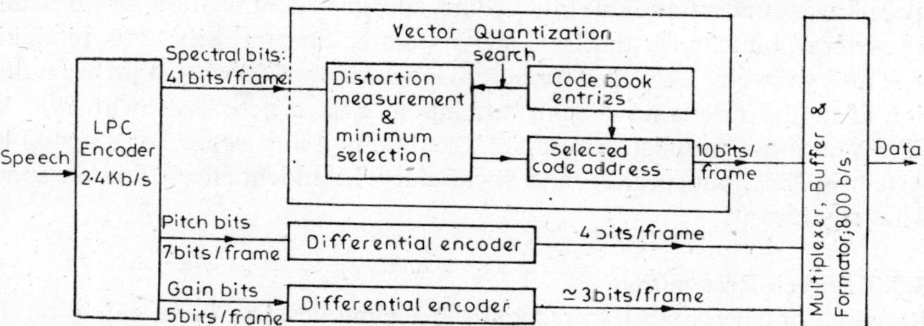

Fig. 3.25 A hybrid codec at 800 b/s using LPC and VQ

likelihood ratio distortion measure (LRDM), which masures the similarity in power spectral density between the input vector and the code-book (template) vector. Economy in the code-book entries is made in two ways: spectra representing non-speech signals are not included in the library, and groups of spectra which are audibly indistinguishable are compressed into a single representative vector. As a result, 1024-member code-book, equivalent to a 10-bit code-word per frame, instead of 41-bit word in 2.4 Kb/s-LPC, has been found to be adequate to give the required fidelity. The LRDM for different sizes of the code-book is shown in Fig. 3.26, where it is seen that the qualitative difference between a 4096-member code-book and a 1024-member code-book is small and no significant subjective difference (in

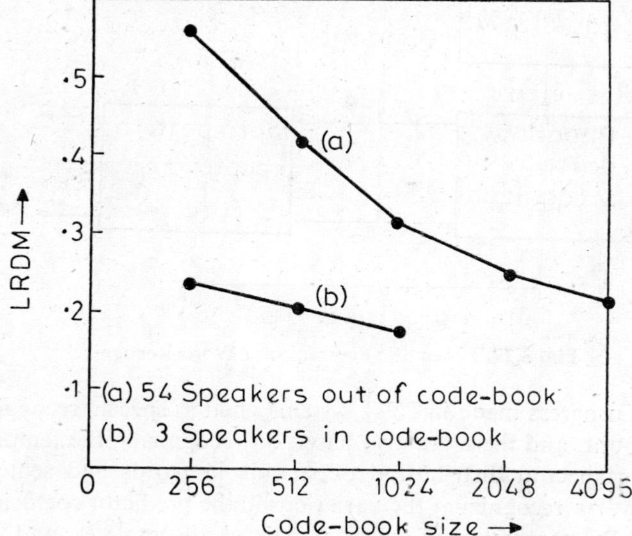

Fig. 3.26 Distortion measure with the size of the codebook

intelligibility score ≅ 80%) can be recognized. However, there is some significant difference in LRDM for speakers in the code-book (i.e., for those who helped to produce the code-book [templates) and the speakers out of the code-book, as shown in Fig. 3.26. Even with speakers out of the code-book, speech was understandable and some speaker identification was possible. The overall bit rate for the LPC-VQ system is: Spectral bits —10; pitch bits —5; and gain bits —2.6 per frame, giving a total of 800 b/s. Further reduction of the bit rate is now being attempted and experiments with 400 b/s LPC-VQ show promising results. The intelligibility score for the 400 b/s system is 72%, but speaker- and vocabulary-dependent effects become somewhat significant.

3.5.3 Speech Recognition

Research on speech pattern recognition techniques has been going on for more than three decades. In one of the earliest phonetic pattern recognizers, designed by Dudley and Balashek [28], ten spoken digits from a single speaker were recognized through spectrum matching. Recognition of the words was done at two stages: the first step was to match the power spectra of the ten phonemes with their stored spectral distributions and then to recognize the word by matching the duration of such phonetic patterns. A block diagram of the recognizer is shown in Fig. 3.27, which gave reasonable results for a single speaker but the system was highly speaker-dependent.

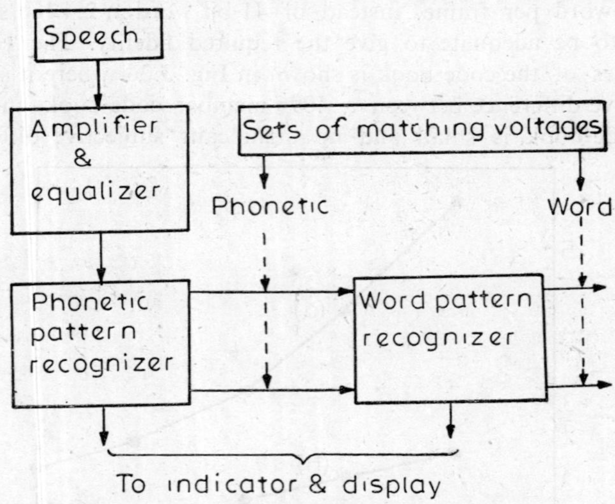

Fig. 3.27 Schematic diagram of a Word Recognizer

With the progress made in LPC systems, better speech recognizers have now been built, and these are also based on recognizing the temporal variations in the spectral distribution of sounds in words and sentences, and equivalently, on recognizing the variations in the predictor coefficients of the LPC. The LPC parameters of words are stored as templates and compared with those of the spoken words. The spoken words are recognized through the minimum distortion measure, and are used for deriving command signals, say, in dialling telephone numbers. Such a Voice Dialer [29] has been designed and is in operational use in a limited way. The block schematic of the voice dialer is shown in Fig. 3.28, where two modes of operation are possible: (a) spoken (command) input of subscriber names to dial subscriber

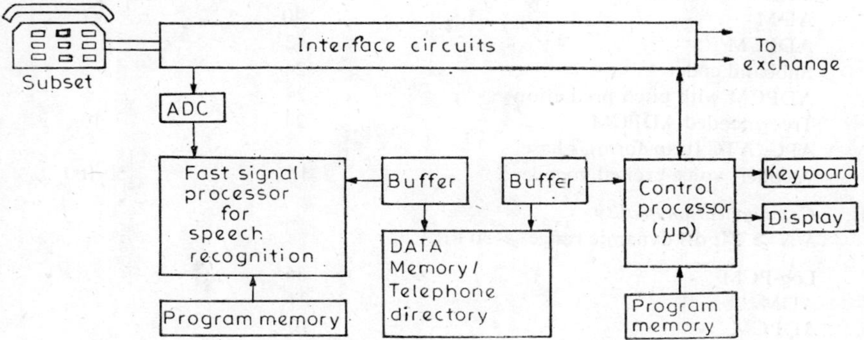

Fig. 3.28 Block schematic of a Voice Dialer

numbers stored in an electronic telephone directory; and (b) spoken input of telephone numbers to replace normal push-button or rotary dialling. The basic hardware of the system consists of a speech recognition unit, a control unit and data memories for an electronic telephone directory and the reference templates. Speech recognition is mainly performed by a fast signal processor chip and a standard 16-bit μP executes less time-critical parts of the algorithm, which need complex decision logic. The control processor is primarily used for overall system control including the supervision of telephone line status, handling of the keyboard input and generation of the display output. The system, however, is broadly speaker-dependent. The vocabulary of a speaker-independent speech recognition system has to have a large amount of data and the template matching needs considerable processing power. As such, speaker-independent recognizers may be used at common locations, say, at exchanges, only; but the speaker-dependent recognizers are able to service a limited number of users and generally have a vocabulary size of 30–50 words/names. Such user-friendly voice dialers are going to be popular in future.

3.5.4 Quality and Complexity

Flanagan *et al.* have discussed the quality and complexity of different codecs in [3] and their assessment of the codecs is shown in Table 3.5. It is seen that the quality of waveform coders deteriorates significantly at bit rates lower than 16 Kbits, while the vocoder (including LPC) performs optimally at bit rates of 2.4 to 4.8 Kb/s. In the range of 1.2 Kb/s, the codecs are still useful for information storage, as used in voice response systems, although their quality needs to be improved further.

Table 3.5 Status of A/D codecs: speech codecs (0-3.4 kHz) (after Flanagan *et al.* [3])

	Types of codecs	Kb/s	Relative complexity
A.	*Toll-quality transmission*		
	$S/N \geqslant 32$ dB: dynamic range $\geqslant$40 dB		
	Log-PCM (7-bit)	56	3
	ADM	40	1
	ADPCM	32	1
	Subband coder	24	5
	ADPCM with pitch prediction	24	5
	Tree-encoded ADPCM	24	10
	APC, ATC (transform), phase vocoder, voice-excited vocoder	16	50-100
B.	*Communication quality*		
	$S/N \geqslant 24$: dB dynamic range $\geqslant$ 30 dB		
	Log-PCM	36	
	ADM	24	
	ADPCM	16	
	Subband	9.6	
	APC, ATC, PV, VEV	7.2	
	VFR-LPC	2.4	
C.	*Synthetic quality*		
	Channel vocoder and LPC	2.4	
	LPC with orthogonal coefficients	1.2	
	Formant vocoder	0.5	

3.6 PICTURE ENCODING

For a high quality AD conversion of TV signals, it is necessary to have the bit rate of the coder from 100 to 150 Mbit/s. Although, the standard log-PCM coder may be used at higher speeds, it becomes difficult to have the conversion time of the order of 100 ns with the standard circuitry. A new PCM terminal [30], developed at Comsat Labs, for colour TV, uses digital circuits only and no operational amplifiers. The AD loop contains a digital feedback circuit which corrects the quantization errors in the front stage(s), thus reducing the required settling time and required accuracy of the first-stage approximation. The converter operates in either parallel only or series-parallel modes. The 8-bit coder, as shown in Fig. 3.29, gives an SNR of 45 dB with a full-load sinusoidal input.

It is well known that the bandwidth requirement in TV signals is much in excess of the total information content in the signal. It has been estimated that TV signals contain only 1.1-1.5 bits of information per pixel. As such, it should be possible to encode TV signals efficiently with 1.5 to 2 bits per sample. To this goal, many fruitful investigations [31] have been carried out over the last two decades. The redundancy reduction techniques in TV are: statistical coding, differential coding, dual-mode systems and two-dimensional transform coding .A large class of literature giving the ultimate entropy

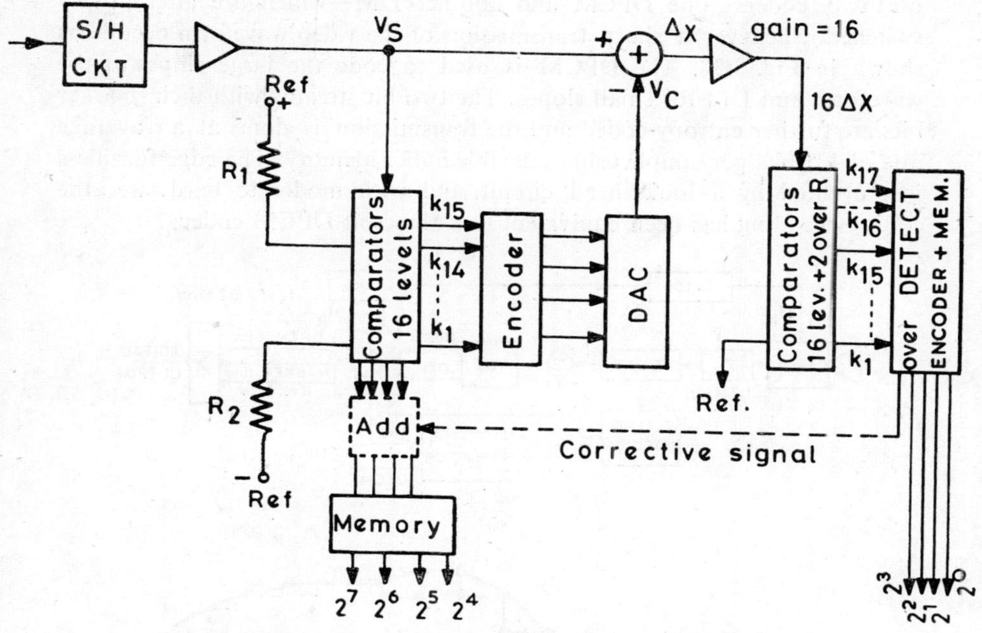

Fig. 3.29 Two stage parallel ADC with digital feedback loop

of TV signals is available [17]. Limb [32] has shown that if the source and the receiver are properly matched for TV encoding, then high quality pictures may be obtained with 0.8-2.0 bits per picture element on an average. He has demonstrated this by utilizing an adaptive non-linear coder and differential feedback. Many modified DPCM coders have been developed recently with 3 or 4 bits per sample. For 1 MHz picture-phone signals, a DPCM cadec has been developed at 6.312 Mbit/s to be transmitted in the Bell system T2 digital facility [33]. This codec effectively uses 4 bits per sample, but transmits approximately 3 bits on an average by utilizing the horizontal blanking period and a small buffer store. It has also been shown that there is a certain amount of redundancy in the output of the 3/4-bit DPCM codec and further economy in bits/s may be obtained by a suitable entropy coding. Alternatively, a coder with 24 levels rather than 8 may be used and the 24 levels are coded with a variable word-length code of the Shannon-Fano type. With a buffer memory of about 10,000 bits, this gives a substantial improvement in the picture quality of the videophone using the standard 6.312 Mbits transmission. Another important development has been the use of dither signals for picture coding. The pseudorandom quantization now produced an uncorrelated error spectrum and the picture quality is improved for 3/4 bits per sample coding.

An important result in picture coding has been the development of an adaptive dual-mode codec for TV signals, where high-quality pictures have been produced at the low bit rate of 1.5 bit per sample [34]. This codec uses two coders—one DPCM and another DM—which are alternatively switched whenever a sharp transmission of the video waveform occurs, as shown in Fig. 3.30. The DPCM is used to code the large slopes of the waveform and DM for small slopes. The two bit streams with their marker bits are further entropy-coded and the transmission is done at an average rate of 1.5 bits per sample using a flexible buffer memory. The edge fuzziness is prevented by a look-ahead circuit, and with moderate hardware, the coding-decoding has been equivalent to a three-bit DPCM codec.

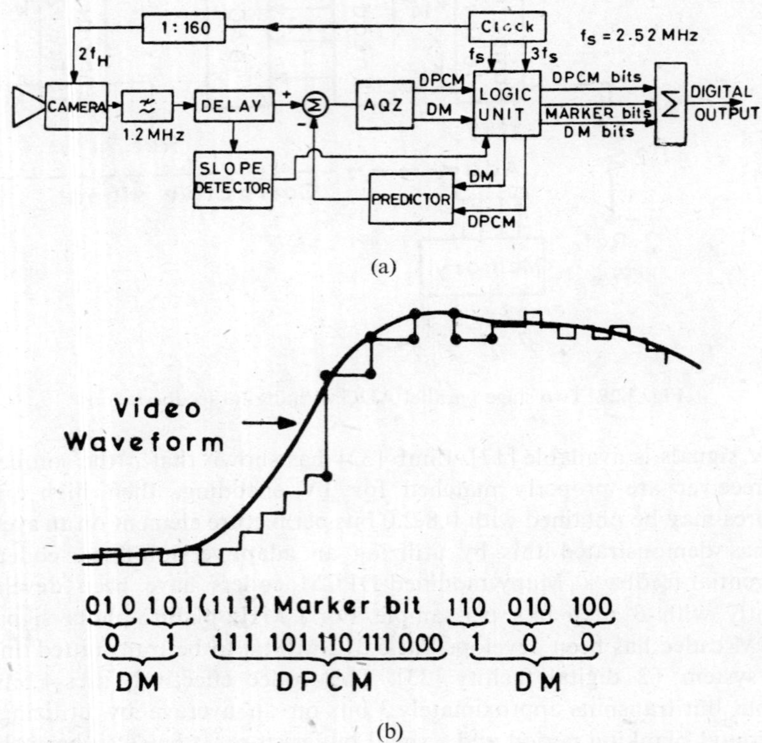

Fig. 3.30 (a) Dual-mode coder for picture phone signals; (b) Example of a transition from DM to DPCM and back to DM (after Frei *et al.* [34])

Experimental results with various images have indicated that sampled images can be modelled rather accurately by a third-order Markov process, and as such, a third-order predictor is quite efficient for DPCM. However, two-dimensional (2-D) predictors are generally better than 1-D predictors [35]. The points adjacent to $S_0(\simeq \Sigma a_i S_i)$ in various spatial directions are shown in Fig. 3.31, where it is seen that the storage of earlier lines is necessary to take advantage of 2-D correlation in pictures. Both DPCM and ADM with 2-D predictors have been designed [36] and the experimental results show that the rendition of vertical edges is now significantly improved and gives better visual perception. Three-dimensional prediction may be obtained by including some points in the earlier frames as well as from the present frame, and, for scenes with low detail and small motion, this frame

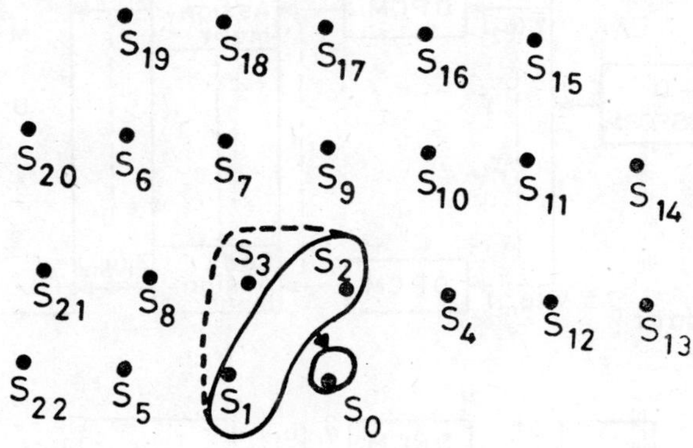

Fig. 3.31 Showing pixels adjacent to S_0. Usually S_1, S_2 and S_3 are used to predict S_0. Alternatively S_1 and S_2 are used to predict S_0

difference prediction appears to be the best. But in scenes with higher detail and motion, field difference prediction does better than frame difference prediction [37]. (Both adaptive quantization and adaptive prediction are used in 2-D or 3-D DPCM and ADM codecs.) However, more successful adaptive predictors for frame-to-frame coding are the ones that take into account the motion of objects, and 20-70% saving of the DPCM bit rate is possible.

Two-dimensional transform techniques, using K-L, Fourier, Hadamard and other transforms, have been widely used to reduce the bit rate of the digitized picture [38]. These coding techniques achieve a better quality of pictures at low bit rates, say, less than 2 bits/pixel, and the associated noise and quality degradation introduced in the processing, as well in the channel, are less objectionable to the human observer. The DPCM and ADM, on the other hand, produce good quality pictures at rates greater than 2 bits/pixel. Since the memory requirement and the processing time are very large in 2-D transform techniques, a hybrid technique using a combination of 1-D transform and DPCM coding has been found to be efficient [35]. The coder takes a 1-D transform of each line of the picture and then operates on each column of the transformed data through DPCM (or ADM), as shown in Fig. 3.32.

Symbolically, one-dimensional unitary transformation of the data and its inverse are given by:

$$[Y] = [A]\cdot[X]$$
$$[X] = [A]^{-1}\cdot[Y], \tag{3.18}$$

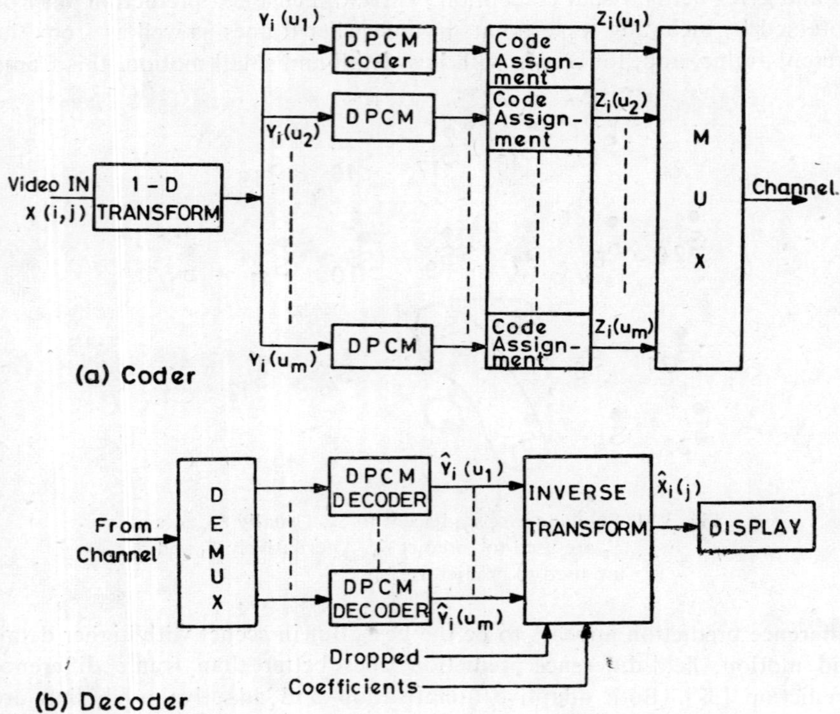

Fig. 3.32 Block diagram of a hybrid codec using 1-D transform

where the sampled data $\{x(i, j)\}$ are divided into arrays of $M \times N$ pixels; $[A]$ is the transformation matrix and $A = A^{-1}$. Equivalently,

$$\left.\begin{aligned} y_i(u) &= \sum_{j=1}^{M} x_i(j) \cdot A(j, u), \qquad & i &= 1, 2, \ldots, N \\ & & j &= 1, 2, \ldots, M \\ x_i(j) &= \sum_{u=1}^{M} y_i(u) \cdot A^{-1}(u, j) & u &= 1, 2, \ldots, M \end{aligned}\right\} \tag{3.19}$$

where $A(j, u)$ is the forward transformation kernel. The column vectors $\{y_i(u)\}$ are DPCM coded to $\{Z_i(u)\}$ for transmission in the channel, and recovered in the receiver through appropriate decoders. Then the inverse transformation of the estimated $\{\hat{y}_i(u)\}$ gives the estimated pixels $\{\hat{x}_i(j)\}$.

The equipment complexity and the number of computational operations are considerably less now than in the case of 2-D transforms. The theoretical and experimental results indicate that acceptable quality pictures are obtained using 1 bit/pixel or less and the degradation due to channel errors with $P_e \leqslant 10^{-2}$ is negligible. Effective SNR's for the Markov data have been calculated and they are approximately:

Bit rate/pixel	1	2	3	4
SNR dB	30	37	42	47

The other efficient technique for bandwidth reduction of TV signals is the interframe coding (also known as conditional frame-to-frame coder), where the high frame-to-frame correlation of the pixels is used to predict the future values of pixels. One method of coding video data is simply to transmit the gray levels of the elements that have changed in successive frames with proper addressing information. The receiver then generates the successive frames by replenishing the previous frames with the transmitted information. Naturally, less information needs to be transmitted, if only the elements that change beyond a particular threshold level are updated. The use of an appropriate value for the threshold eliminates the granular noise in the background without noticeably degrading the encoded video data. Experiments with picture-phone signals using the 'conditional frame replenishment' technique have indicated that good pictures are obtained with 1 bit/pixel. The system generates information at an uneven rate, and, as such, a large memory and variable length codes for transmission of data are used [39].

A particularly efficient picture signal digitizer, based on interframe coding, has been developed by NEC, Japan [40]. In the codec, shown in Fig. 3.33, the digitized picture signal is encoded by an interframe coder, in which, the differential between the present and the previous frame is only encoded at an average rate of 2-3 bits per sample. To further reduce the redundancy of the differential signal, this signal is coded through a variable length coder (entropy coding). The variable rate digital output is then equalized through

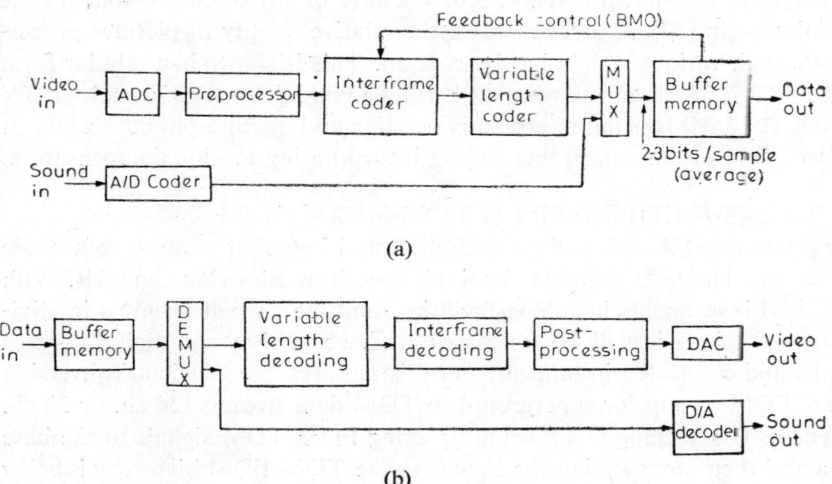

Fig. 3.33 Block schematic of an interframe codec

a buffer memory of 1 Mbit, and the buffer overflow is controlled through the feedback of the buffer-memory-occupancy value (BMO) to the adaptive quantizer in the interframe coder. Thus, for signals with a high information rate (due to excessive motion of the picture), the quantization is made coarser, giving a lesser number of bits per-sample, so that the buffer overflow does not occur, and this degrades the SNR marginally. In an NTSC colour signal, the colour subcarrier phase-reverses in successive frames and this generates a large frame difference, even when the picture is perfectly still. To avoid this, a frame-to-frame phase inversion of the chrominance component of the composite signal is carried out in the preprocessor after the ADC. The overall bit rate of the codec for an excellent (broadcast) quality picture is 20-30 Mb/s, and the bit rate for picture-phone signals may be 1.5–3 Mb/s only. It is possible to have a variable bit rate at the coder output, say 17–27 Mb/s, depending on the relative motion of the picture. Three or more such variable rate digital signals may be multiplexed into a constant rate digital channel (e.g., 60 Mb/s channel for 3 signals), by using the adaptive bit sharing technique (similar to TASI). Experiments with three-channel TV transmission using a 60 Mb/s satellite channel (giving an average bit rate of 20 Mb/s per signal) have shown excellent rcsults. Further improvements in the quality of the picture transmitted through noisy channels have been achieved through using forward-error-correcting (FEC) codes, e.g., the 239/255 double error correcting BCH code. Experiments have shown that the mean error-free time is longer than 1 hr at a channel BER $= 10^{-5}$, and is about 5 sec for BER $= 10^{-4}$, showing the FEC gain of the order of 10^5. Further progress in coding of teleconferencing signals is being made and it will soon be possible to transmit these signals digitally through the T_1-carrier at 1.5 Mb/s only.

In summary, the approaches that have been used in picture coding may be classified as shown in Table 3.6. We have briefly discussed some of the techniques. In [37], an assessment of the relative quality of pictures processed through various codecs has been given. This is shown in a tabular form in Table 3.7. The trend shows that it will be possible to digitize colour TV signals at 15 Mb/s using interpolative coding and picture-phone signals at 1.5 to 3 Mb/s using interframe coding incorporating motion compensation.

3.7 TRANSMULTIPLEXERS (TRANS-MUX) [41, 42]

The A/D and D/A codecs have been discussed in earlier sections, but these codecs are basically suitable for single speech or television channels. With the initial investments in analog facilities, and the current progress in digital transmission, it will be necessary to have analog and digital systems coexist and connected in tandem; and this requires the efficient conversion of an FDM group (or supergroup) to TDM data streams (24 ch. or 30 ch. systems). The technique of demultiplexing FDM/TDM signals to the base band and then reconverting the signals to the TDM/FDM hierarchy is being widely used; however, a more economical solution with better system performance may be achieved by using a digital interface for the conversion of

Table 3.6

Methods of picture coding

- PCM
  - Log-PCM
  - APCM
- Predictive
  - Non-adaptive (DPCM)
  - Adaptive
  - Conditional replenishment
  - Delayed (tree) encoding
- Transform
  - K-L, Hadamard
  - Cosine, Haar, Slant
  - Adaptive transforms
  - Coefficient selection
  - Quantization
- Statistical coding
  - Huffman
  - Shannon-Fano
- Interpolative and extrapolative
  - Subsampling, spatial and temporal
  - Adaptive
- Other methods
  - Contour
  - Run length
  - Bit plane

Hybrid coding (Predictive and Transform)

Table 3.7 (After Netraveli and Limb [37])

Picture codecs: A. Broadcast TV signals, BW ≃ 4.5 MHz
B. Picture-phone signals, BW ≃ 1 MHz

Type of codec	Bit-rate Mb/s for the quality					
	Excellent		Good		Fair	
	A	B	A	B	A	B
Log-PCM	95	20–40	50	—	—	—
DPCM	45	10	32	6	16	4
ADM	60	10	30	6	16	4
Hybrid Transform/ DPCM	40	8	25	6	15	3
Intraframe coding (adaptive)	50	30	35	20	25	10
F-F conditional replenishment	30	6	25	3	20	1.5
Interpolative coding (futuristic)	20	—	15	—	10	—
F-F coding with motion compensation	—	3	—	1.5	—	1

FDM to TDM and vice versa. Such a digital interface has been designated as a 'transmultiplexer'. Transmultiplexers have been designed for both the FDM group band and the FDM supergroup band; their specifications being: (a) 24 ch. T_1-carrier at 1.544 Mb/s is converted to two FDM groups in the band of 60–108 kHz; (b) two 30–ch. (CEPT-PCM-30) digital carriers at 2.048 Mb/s are converted to one FDM supergroup in the band of 312-552 kHz; (c) FDM to TDM conversions follow the same frequency/bit rate plan; and (d) allowed signal impairments are:

Maximum idle channel noise with reference to peak signal:	−80 dB
Maximum rms non-linear distortion with reference to peak signal	−40 dB
Minimum cross-talk attenuation between any two channels:	60 dB

Frequency response: 500 to 3000 Hz: within ± 0.5 dB

$$\left.\begin{array}{r}\text{200 to 500 Hz}\\ \text{and 3000 to 3400 Hz}\end{array}\right\}: \text{ within } +0.5, -3 \text{ dB}$$

The block schematic of a transmultiplexer for T_1-carrier is shown in Fig. 3.34, where SSB modulators/demodulators are based on the analog principles of Hartley's or Weaver's method. It is seen that in the TDM to FDM direction, the demultiplexed channel signals are converted to linearly

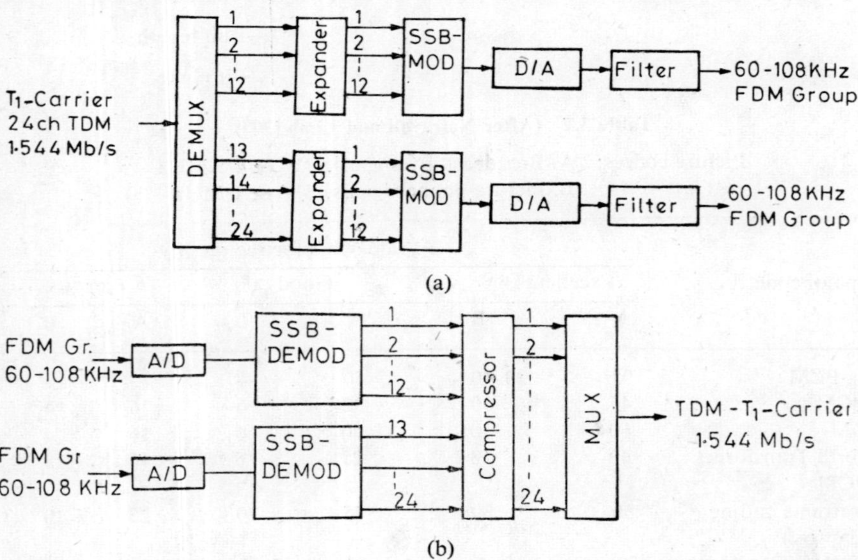

Fig. 3.34 Block diagram of a 24 ch. transmultiplexer: (a) TDM-FDM conversion and (b) FDM-TDM conversion

encoded samples through the μ-law expander. The FDM signal is then digitally generated by means of a bank of SSB modulators, and the analog

FDM group signal is obtained by D/A conversion and analog filtering. The FDM to TDM conversion is obtained by complementary processing, and as such it is only necessary to consider in detail the operation of the digital SSB modulators. The basic problem in the digital SSB MOD is the sampling rate at which the digital filtering and other operations may be made. As, for example, the minimum sampling rate for the FDM group is 112 kHz (using bandpass sampling and keeping a guard band of 8 kHz between signal images), and this is close to the 96 kHz rate required if the signal is converted to baseband (0–48 kHz) first. But for a digital bandpass filter of order 16, as required in an SSB MOD, the number of machine cycle/s required is approximately $4 \cdot 10^6$; and this will be multiplied for 12 channels. Thus, the straightforward bandpass method of SSB generation requires a very large number of computations and the associated hardware. The major effort, then, has been to process the digital signals at lower sampling rates and thus reduce the computation rate and the hardware.

3.7.1 Digital Signal Interpolation

A signal $x(t)$, with a bandwidth 0–W Hz, is usually represented through the Nyquist theorem by samples $x(n)$ at $t = n/2W = nT$, where T is the sample period. If now $x(t)$ is sampled at a faster rate, say, at M/T, then a new sequence $y(m)$ is obtained. In the frequency domain, $x(n)$ has a spectrum that is periodic with period $1/T = 2W$ Hz, but $y(m)$ has a spectrum of width $\pm W$ around zero and M/T Hz only, as shown in Fig. 3.35. This process of creating a new sequence at a faster rate is known as digital interpolation and the inverse process of sample rate reduction is known as decimation. In principle, $y(m)$ may be generated from $x(n)$ by simply converting $x(n)$ to the baseband by a DAC and then resampling $x(t)$ at a rate M/T samples/s. The process can be carried out digitally if a low pass recursive digital filter with cut-off at $1/2T$ Hz [as shown in Fig. 3.35(a)], but operating at the higher sample rate M/T Hz, is used to filter the low rate sequence $x(n)$ (intervening input samples are considered to be zero). The resultant output sequence will be the required $y(m)$ at the rate M/T Hz, and its spec-

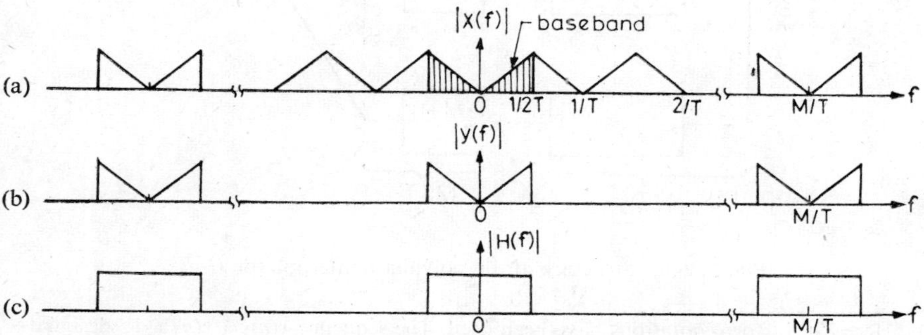

Fig. 3.35 (a) Periodic spectrum of $x(n)$; (b) spectrum of interpolated $y(m)$; and (c) transfer function $H(f)$ of low-pass digital filter

trum will be the same as that of Fig. 3.35(b). Thus, the filter simply removes unwanted signal images and leaves a spectrum that is obtained by sampling $x(t)$ at the higher rate M/T Hz. In the dual process of decimation, the high rate sequence $y(m)$ is filtered through the same low pass digital filter having the characteristics of Fig. 3.35(c) and the unwanted signal images between $1/2T$ Hz and $M/2T$ Hz are removed. Then, retaining every Mth output sample only and discarding the rest, gives $x(n)$ and the corresponding periodic spectrum of Fig. 3.35(a). Mathematically, the last operation performs a step-and-repeat process (with step interval $1/T$ Hz) on the spectrum of Fig. 3.35(b) to obtain that of Fig. 3.35(a).

Conceptually, the above digital interpolation technique may be generalized to perform bandpass interpolation (filtering) (even for single sidebands instead of double sidebands as above), and this leads to the bandpass method of digital SSB generation. However, it has been already mentioned that this simple technique would require approximately $5 \cdot 10^7$ machine cycles/s for computational purposes. Thus, it is necessary to execute the digital interpolation at a lower sampling rate, say, at 8 kHz, instead of at 112 kHz (for FDM group signals) calculated earlier. The digital interpolation at lower sampling rate is possible by using polyphase interpolators, whose structure is shown in Fig. 3.36(a).* The network consists of all-pass phase-shifting networks $H_i(z^M)$ and the digital delays z^{-i}. An approximate high rate sequence $y(m)$ may be obtained by shifting the sequence $x(n)$ by successive delays of (iT/M), $i = 1, 2, \ldots, M\text{-}1$, and then interleaving the

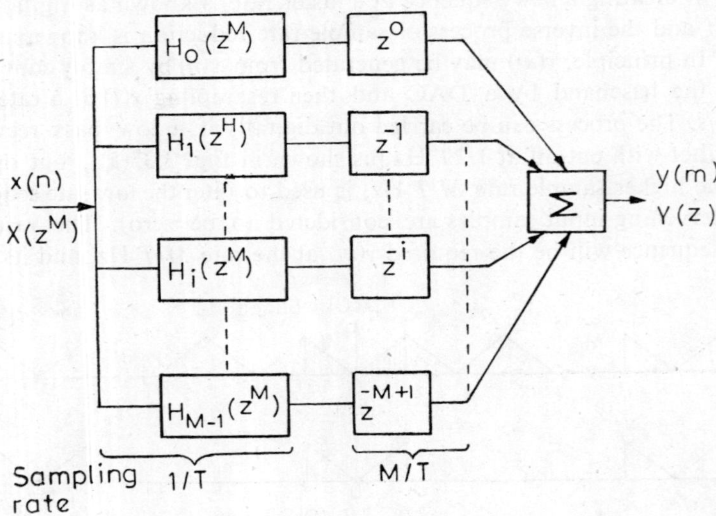

Fig. 3.36(a) Structure of the polyphase interpolator $H(z)$

*Here Z-transform notations have been used. The sequence $x(n) = x(nT)$ is designated as $X(z^M)$, and the sequence $y(m) = y(m\,T/M)$ as $Y(z)$. If $1/T = 8$ kHz and $M = 14$, then the digital delays z^{-1} are at the sampling rate of 112 kHz but $H_i(z^M)$ is operated at the sampling rate of 8 kHz only.

delayed sequences in a summer to produce $y(m)$. However this interleaved sequence will not have the spectral property of Fig. 3.35(b), and to have the correct low-pass characteristic, the additional phase-correcting networks in the form $H_i(z^M)$ are added (but operated at the lower sample rate of $1/T$ Hz). It may be shown that the proper phase characteristics ϕ_i of the branch filters are in the form of descending steps which can be decomposed into a linear component ϕ_{li} and a periodic component ϕ_{pi}, as shown in Fig. 3.36(b).

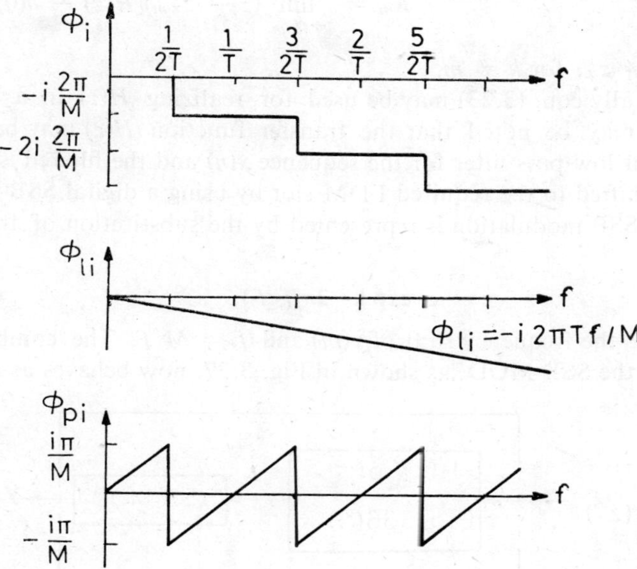

Fig. 3.36(b) Decomposition of phase characteristics of the branch filters

The linear component ϕ_{li} is realized by z^{-i}, and the periodic component by $H_i(z^M)$. It is now seen from Fig. 3.36(a) that the transfer function of the polyphase interpolator is given by:

$$H(z) = \frac{Y(z)}{X(z^M)} = \sum_{i=0}^{M-1} z^{-i} H_i(z^M) \tag{3.20}$$

The transfer function $H(z)$ may be realized in either of the following forms:

(a) Polynomial form: $$H(z) = \frac{\sum_{k=0}^{K} a_k z^k}{\sum_{m=0}^{M} b_m z^m}, \qquad K \leqslant M \tag{3.21}$$

(b) Pole-zero form: $$H(z) = a_k \frac{\prod_{k=1}^{K} (z - z_{0k})}{\prod_{m=1}^{M} (z - z_{\infty m})}, \qquad K \leqslant M \tag{3.22}$$

(c) Residue form (parallel form):

$$H(z) = h(0) + \sum_{m=1}^{M} \frac{R_m}{(z - z_{\infty m})} \tag{3.23}$$

where

$$h(0) = \lim_{z \to \infty} H(z)$$

and

$$R_m = \lim_{z \to z_{\infty m}} (z - z_{\infty m})[H(z) - h(0)],$$

but $z_{\infty m} \neq z_{\infty k}$ for $k \neq m$.

Specifically eqn. (3.23) may be used for realizing $H(z)$ in a convenient form. It may be noted that the transfer function $H(z)$ may be used as a convenient low-pass filter for the sequence $x(n)$ and the filtered signal $y(m)$ may be shifted to the required FDM slot by using a digital SSB MOD. The complex SSB modulation is represented by the substitution of the variable z as:

$$z \to z \cdot \exp(-2\pi j f_c/f_s), \tag{3.24}$$

where f_c is the frequency shift of $y(m)$ and $f_s = M/T$. The combination of $H(z)$ and the SSB MOD, as shown in Fig. 3.37, now behaves as a bandpass

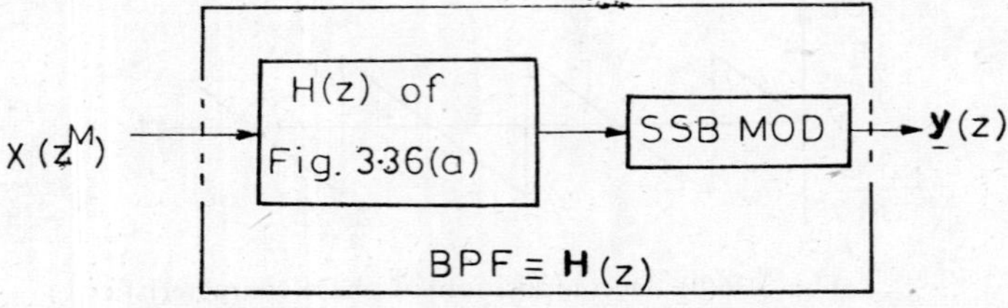

Fig. 3.37 Equivalent bandpass filter using a combination of a polyphase network and an SSB MOD

filter, and a bank of bandpass filters may be obtained by shifting their pass band by the frequency $(1/T)$ with respect to each other. The overall transfer function $\underline{H}'(z)$ of the network is given by:

$$\underline{H}'(z) = H(z \exp(-2\pi j \Omega_c)), \qquad \Omega_c = f_c/f_s = f_c T/M$$

$$= \sum_{i=0}^{M-1} z^{-i} \cdot \exp(2\pi j \Omega_c i) \cdot \underline{H}_i(z^M \cdot \exp(-2\pi j M \Omega_c)) \tag{3.25}$$

The real SSB signal is now obtained by multiplying $X(z^M)$ by the transfer function

$$\underline{H}(z) = 2Re[\underline{H}'(z)] = 2Re[\underline{H}(z \exp(-2\pi j \Omega_c))] \tag{3.26}$$

Thus, the frequency-shifted SSB signal is given by

$$\underline{Y}(z) = \underline{H}(z) \cdot X(z^M) \tag{3.27}$$

which is realized by digital interpolation and digital SSB modulation.

3.7.2 Transmultiplexing Algorithms [41]

Transmultiplexing algorithms are classified into four groups: (a) bandpass filter bank; (b) low-pass filter bank; (c) Weaver structure; and (d) multistage modulation. We discuss below the techniques of implementation of some of these algorithms.

The bandpass filter bank algorithm is implemented by using the realization given by eqns. (3.20), (3.26) and (3.27). As indicated in Fig. 3.34, the N-channel TDM sequence is demultiplexed, and each channel sequence $x_r(n)$ (where $0 \leqslant r \leqslant N-1$) is SSB-modulated to the required band in the FDM format. Since $x_r(n)$ has a periodic spectrum, as shown in Fig. 3.35(a), the FDM signal may be generated by suitable filtering in a bandpass filter bank, whose pass bands are shifted by the frequency $(1/T)$ with respect to each other. The filter bank outputs are summed to give the FDM signal.* An economic realization of the filter bank is now given by eqn. (3.26), and its block schematic is shown in Fig. 3.38. The digital FDM output $\Sigma \underline{Y}_r(z)$ is

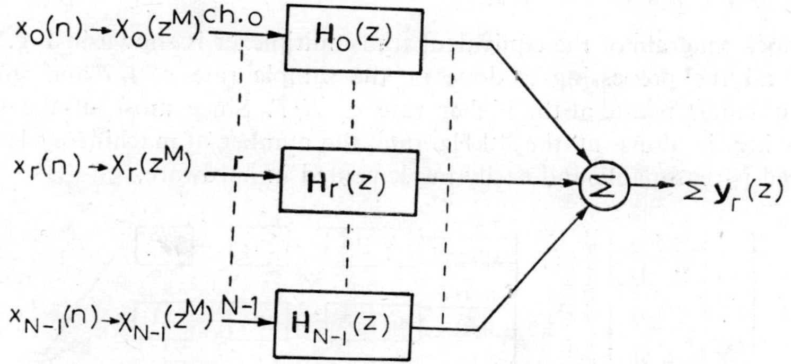

Fig. 3.38 Block schematic of the filter bank realization

$$\Sigma \underline{Y}_r(Z) = \sum_{r=0}^{N-1} \underline{H}_r(z) \cdot X_r(z^M) \tag{3.28}$$

and

$$\underline{H}_r(z) = 2Re\left[\sum_{i=0}^{M-1} z^{-i} \exp(2\pi jir/M) \cdot \underline{H}_{i,r}(z^M)\right] \tag{3.29}$$

where it is assumed that the frequency shift $f_{cr} = rfs/M = 2rW$, $\Omega_{cr} = r/M$, and $\exp(-2\pi jM\Omega_{cr}) = \exp(-2\pi jr) = 1$. Since the same transfer function $H(z)$ of Fig. 3.36(a) is generated in each filter branch r, then $H_{i,r}(z^M) = H_i(z^M)$. Using this and eqn. (3.29) in eqn. (3.28), the real-valued FDM

*For channels with $N/2 \leqslant r \leqslant N-1$, the resulting spectra are in the inverted positions. This is corrected by premodulating these channel signals by a carrier $f_s/2$ and then processing the shifted (in frequency) signals through the bandpass interpolators. This modulation process is particularly simple and is equivalent to the multiplication of the samples $x(nT)$ by $\{(-1)^n\}$. The FDM signal then exhibits the correct frequency for all channels.

output is given as:

$$\sum_{r=0}^{N-1} Y_r(z) = \sum_{r=0}^{N-1} 2Re\,[\sum_{i=0}^{M-1} z^{-i} \exp(2\pi jir/M) \cdot \underline{H_i}(z^M)] \cdot X_r(z^M)$$

Interchanging the order of summation and rearranging, we get

$$\sum_{r=1}^{N-1} Y_r(z) = \sum_{i=0}^{M-1} z^{-i} \cdot 2Re[\underline{H_i}(z^M) \sum_{r=0}^{N-1} X_r(z^M) \cdot \exp(2\pi jir/M)] \tag{3.30}$$

If now $M = N$, then the term

$$\sum_{r=0}^{N-1} X_r(z^N) . \cdot \exp\left[j\left(\frac{2\pi}{N}\right) ir \right]$$

represents the inverse discrete Fourier transform (IDFT) for N points at the $(1/T)$ sampling rate. The eqn. (3.30) is now rewritten as:

$$Y(z) = \sum_{r=0}^{N-1} Y_r(z) = \sum_{i=0}^{N-1} z^{-i} \cdot 2Re[\underline{H_i}(z^N) \cdot \text{IDFT}\{X_r(z^N)\}] \tag{3.31}$$

The block diagram of the equivalent transmultiplexer is shown in Fig. 3.39, where all the processing is done at the sample rate of $1/T$ and only the delay circuits operate at the higher rate of N/T. Since most of the signal processing is done at the 8 kHz rate, the number of machine cycles now required is reasonable and easily implemented in hardware.

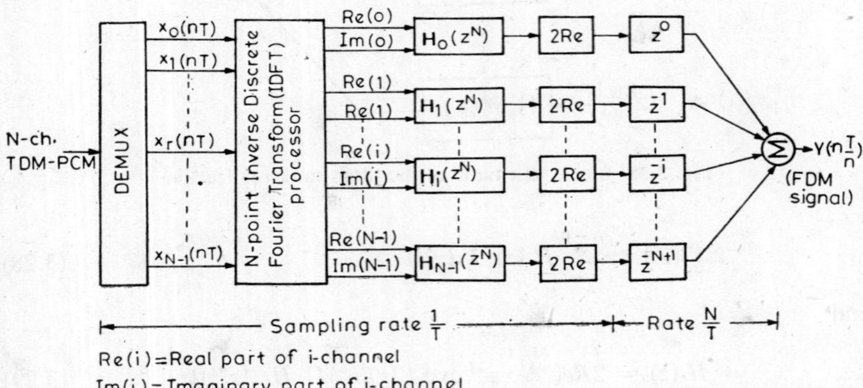

Fig. 3.39 Block diagram of a transmultiplexer using eqn. (3.31)

An interesting simplification of the Trans-Mux is possible if N is chosen as $N = 2^k$. Then the Trans-Mux may be constructed by summing up the outputs of k modules, where each module is a 2-channel Trans-Mux, as shown in Fig. 3.40. With $N = 2$, no imaginary parts occur in the DFT and $H_0(z^2)/H_1(z^2)$ represents filters with real coefficients. Further, by combining the processors of all modules, we obtain a 'Hadamard processor', where only trivial multiplications with (± 1) have to be carried out [43]. Such a Trans-Mux will indeed be a simple one.

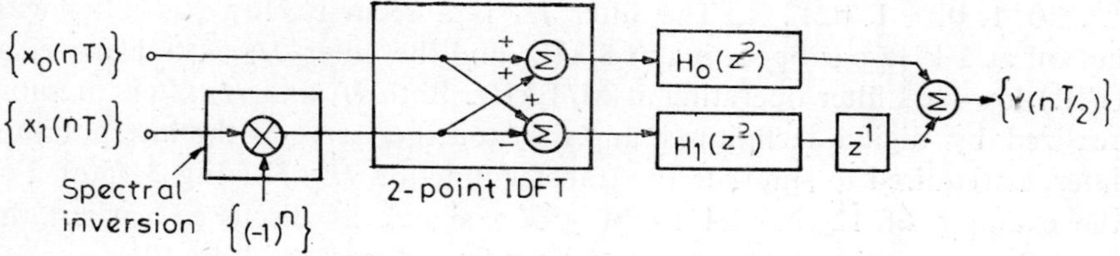

Fig. 3.40 Transmultiplexer for $N = 2$ using eqn. (3.31)

In Fig. 3.39, the polyphase network has to process complex signals, and as such, the network is also complex. To avoid this, a modified Trans-Mux has been developed using SSB conversion of the digital signals (after Demux) and then processing the complex outputs through IDFT and $2N$ low-pass filters (simulated by the polyphase network). This low-pass technique has been studied by Bonnerot *et al.* [44] and others [41].

3.7.3 The Weaver Structure [42]

The classical Weaver's method of SSB generation may be modified to realize a digital Trans-Mux. The amount of circuitry is kept low by ingenious partitioning of the low-pass filters into recursive and transversal sections, operating at different sampling rates. The block diagram of the basic converter is shown in Fig. 3.41, where the speech signal sampled at 8 kHz ($= 1/T$) is modulated by an auxiliary carrier of 2 kHz and the filtered signal (through H_1 and H_2) now occupies the frequency band of ± 2 kHz. This band is then translated to the required band to generate the FDM signal. The low-rate modulation is very simple, as the sampled carrier is given by the sequence

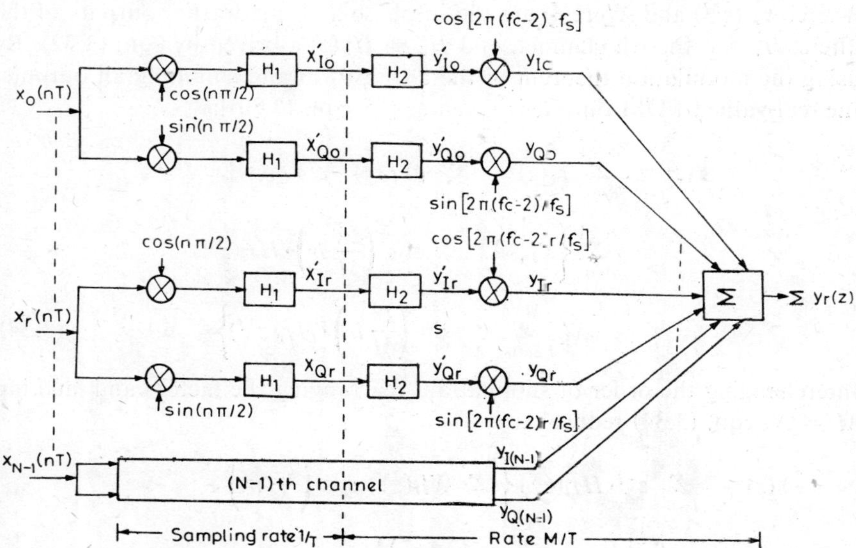

Fig. 3.41 Block schematic of Trans-Mux using the Weaver method

$\{\ldots 0, 1, 0, -1, 0, \ldots\}$. The filter H_1 is a recursive low-pass filter with cut-off at 2 kHz but operating at 8 kHz, and the filter H_2 is a transversal (FIR) low-pass filter operating at M/T kHz. Both H_1 and H_2 filters are now realized by digital techniques, and they together serve as the interpolation filter, as required to simulate the transfer function $H(f)$ of Fig. 3.35(c). For the example of 12-channel FDM group signal, as discussed earlier, the overall computation rate per channel is now reduced to 1.4×10^6/s, i.e., a reduction by a factor of three has been achieved. Further reduction of the computation rate is obtained by using the IDFT technique, as now discussed.

The transfer function of the low-pass filter cascade $H(z) = H_1(z) \cdot H_2(z)$ may be factored into

$$H(z) = [H_R(z)]^{-1} \cdot H_T(z)$$

where $[H_R(z)]^{-1}$ represents the recursive part and $H_T(z)$ the transversal part of the transfer function. Using eqn. (3.20), we get

$$H_T(z) = \sum_{i=0}^{M-1} z^{-i} H_{Ti}(z^M) \tag{3.32}$$

and H_2 is now configured as the polyphase interpolator of Fig. 3.36(a). The right-hand section of the Trans-Mux can be simplified further by noting that for any channel r, the interpolator outputs are:

$$\begin{aligned} Y'_{Ir}(z) &= X'_{Ir}(z^M) \cdot \sum_{i=0}^{M-1} z^{-i} \cdot H_{Ti}(z)^M \\ Y'_{Qr}(z) &= X'_{Qr}(z^M) \cdot \sum_{i=0}^{M-1} z^{-i} \cdot H_{Ti}(z^M) \end{aligned} \tag{3.33}$$

where, $X'_{Ir}(z^M)$ and $X'_{Qr}(z^M)$ are the in-phase and quadrature outputs of the filters H_1 for the rth channel, and $H_2 = H_T(z)$ as given by eqn. (3.32). By using the modulation theorem of the z-transform and summing all outputs, the real-valued FDM output is given as [*cf.* eqn. (3.30)]:

$$\begin{aligned} Y(z) &= \sum_{r=0}^{N-1} Y_r(z) = \sum_{r=0}^{N-1} [Y_{Ir}(z) + Y_{Qr}(z)] \\ &= \sum_{i=0}^{N-1} \left\{ X'_{Ir} \sum_{i=0}^{M-1} z^{-i} \cos\left(\frac{2\pi}{M} ir\right) \cdot H_{Ti}(z^M) \right. \\ &\quad \left. + X'_{Qr} \sum_{i=0}^{M-1} z^{-i} \sin\left(\frac{2\pi}{M} ir\right) \cdot H_{Ti}(z^M) \right\} \end{aligned} \tag{3.34}$$

Interchanging the order of summation, rearranging the factors and making $M = N$, eqn. (3.34) reduces to

$$\begin{aligned} Y(z) = \sum_{i=0}^{N-1} z^{-i} \cdot H_{Ti}(z^N) &\left\{ \sum_{r=0}^{N-1} X'_{Ir}(z^N) \cos\left(\frac{2\pi}{N} ir\right) \right. \\ &\left. + \sum_{r=0}^{N-1} X'_{Qr}(z^N) \cdot \sin\left(\frac{2\pi}{N} ir\right) \right\} \end{aligned} \tag{3.35}$$

Using a cosine and a sine processor, the required Trans-Mux is realized [45], and is shown in Fig. 3.42. Since all circuits except the final delay circuits operate at 8 kHz rate, the computation rate of the the Trans-Mux is now reduced to 5×10^5/s per channel, giving an overall reduction by a factor of nine approximately.

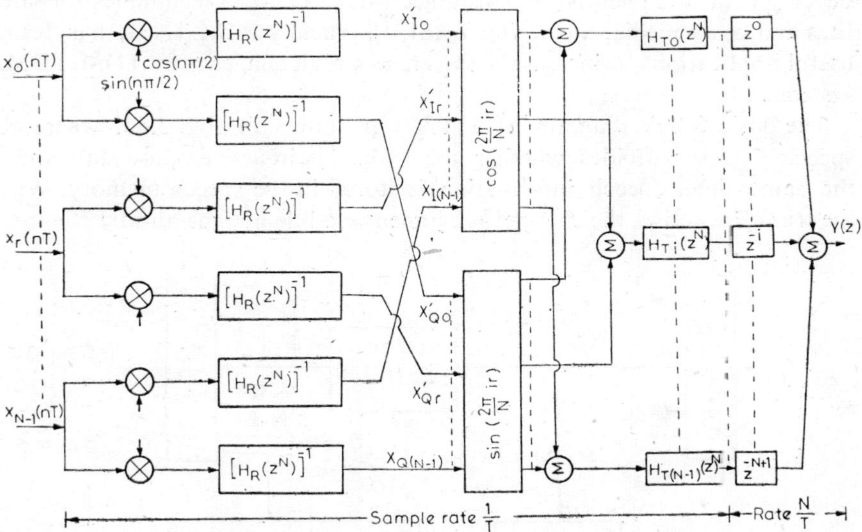

Fig. 3.42 Block diagram of Trans-Mux using eqn. (3.35)

The fourth method of realizing a Trans-Mux is the multistage modulation method, where both the sampling rate and the spectral positions of the signals to be processed are changed step by step. Various intermediate FDM signals are generated and the number of channels multiplexed increases from stage to stage. In the final stage, the FDM group/supergroup signal is formed. A particular advantage of this method is that it only requires a few types of filters and they may be easily realized. An efficient realization of a supergroup (60-channel) Trans-Mux using this method has been given in [46].

The application of Trans-Mux is not yet universal and no standard has been evolved so far for the technique and the performances of the system. One of the serious problems in a Trans-Mux under looped condition (in $2W/4W$ circuits) is the stability; and except for the multistage modulation method, all other methods are subject to parasitic oscillations under adverse conditions. However, attempts are being made to solve this problem and to evolve an universally accepted design technique for future Trans-Mux systems.

3.8 DIGITAL TASI [47, 48]

In two-way speech transmission, advantage may be taken of the fact that the speech activity α each way is approximately 0.4 and the idle time of the channel may be used to transmit speech signals from other trunks. The technique of multiplexing signals from N trunks through M transmission channels, $N > M$, is known as Time Assignment Speech Interpolation (TASI)*, and is somewhat similar to the random multiplexing technique and the Demand Assignment (DA) technique. TASI has been extensively used in long-distance submarine cables using analog techniques [47], but with the development of Time-division Multiaccessing (TDMA) techniques for satellites and other media, the digital TASI, specifically PCM-TASI, has found useful applications to increase the average circuit efficiency in TDM/TDMA systems.

The basic block diagram of a TASI is shown in Fig. 3.43, where the speech detector decides whether the trunk-i is in active/pause state and at the same time, speech information is stored in the speech memory. When the trunk-i is active, the channel assignment module assigns an idle channel-j

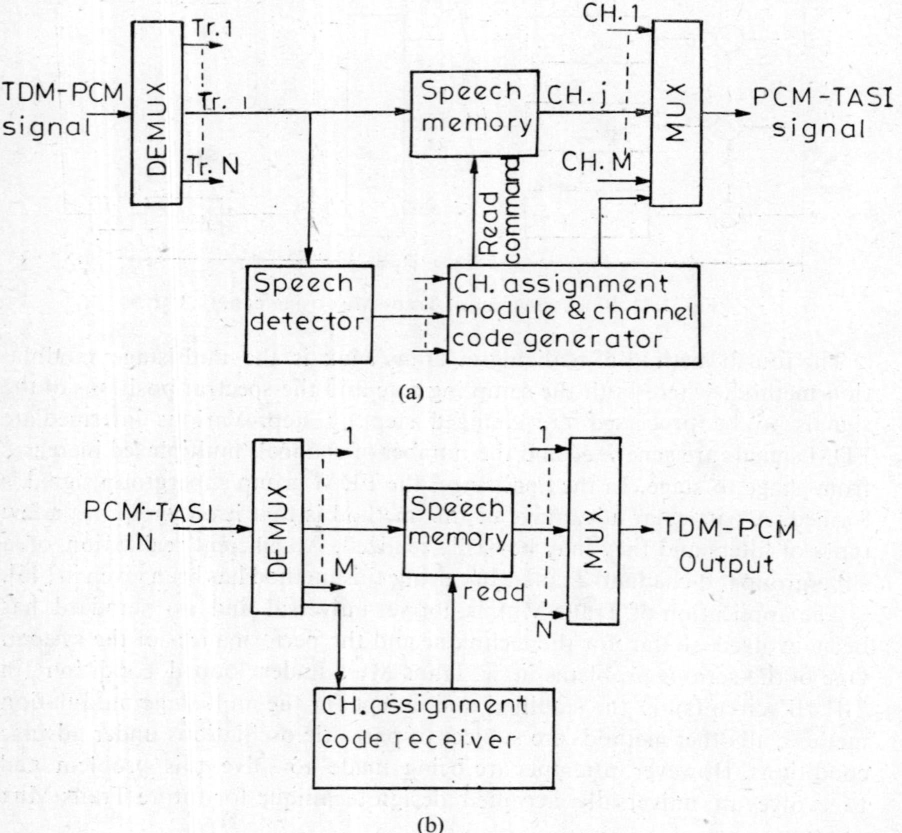

(a)

(b)

Fig. 3.43 Block diagram of PCM-TASI

*Here a 'trunk' means the connection from the switching exchange to TASI and a 'channel' means the connection from the TASI to the distant receiver. Both trunk and channel signals are in TDM formats.

to the trunk-i and the information from the speech memory is transferred to the channel after suitable speed conversion/time compression. The assignment module also generates the trunk identification code and transmits the same through the MUX circuit. At the receiver, the channel assignment signal is decoded, and the channel signals are redistributed to N trunks under the control of the assigned codes. The speech memory is a temporary store and also carries out the speed conversion/time expansion as required.

The speech detector circuit, as shown in Fig. 3.44, operates on the principles of energy detection or integral detection of the speech samples, Since the instantaneous amplitude detection is not reliable due to circuit noise, the speech samples are subjected to two thresholds, one for amplitude and another for minimum duration, as obtained through the comparator and the counter. The comparator output is counted in a reversible counter and the integrated output above a certain threshold is used to signal the channel assignment module that a channel should be allotted to the incoming trunk. The circuit shown in Fig. 3.44 is for per channel detection; however, the detector may be organized on a TDM-basis as well. In an experimental equipment, the performance of the circuit was measured with the amplitude threshold = the 4th step of a 7-bit PCM word (companded through logarithmic μ-law, μ = 100), equivalent to a level of -45 dBm. The detection

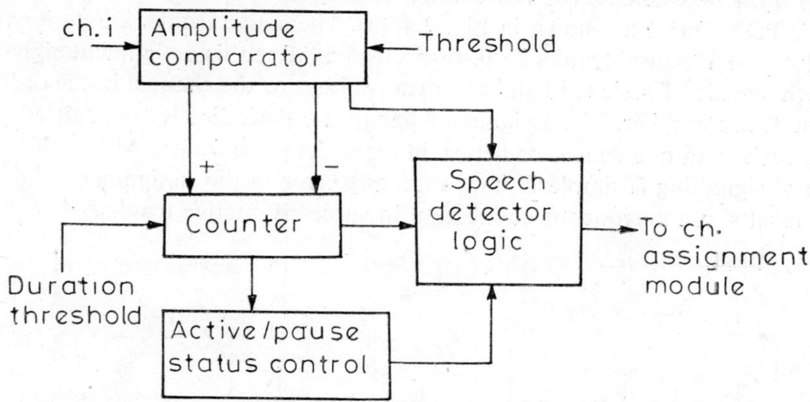

Fig. 3.44 Single-channel speech detector

time in ms vs.duration threshold for various d(=threshold level/input signal level) is shown in Fig. 3.45, where it is seen that with $d = -20$ dB, and duration threshold = 4, the detection time is less than a msec. The listening test showed that for an amplitude threshold = 4th step and duration = 4, no clipping could be recognized at the receiver output. However, when a trunk is disconnected from a channel during a pause, the listener feels a void and to overcome this, a small amount of noise is inserted into the 'pause' trunks. Further, a simple echo suppressor may also be incorporated in the circuit to improve the overall performance.

The trunk assignment procedure may be of two types:

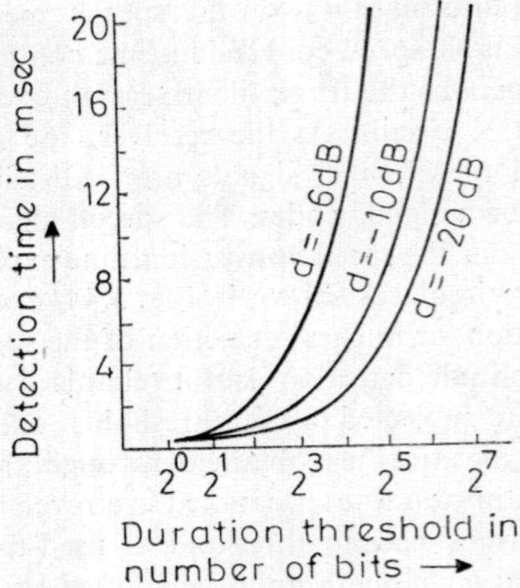

Fig. 3.45 Detection characteristics of the speech detector; d = threshold level/input signal level (after Amano and Ota [48])

(*a*) *Type I*—where N trunks are designated by an N-bit code, as apreamble in the TDM frame as shown in Fig. 3.46(a). The active trunks are designated by 1 and 'pause' trunks by 0, respectively. After this assignment signal, speech signals of assigned trunks are transmitted to the channel in the order of the trunk number. The assignment signals are periodically transmitted to the receiver in one frame, or better, in more than one frame. Although this type of signalling is simple in hardware, any error in the assignment signal bits results in an erroneous connection in successive active trunks.

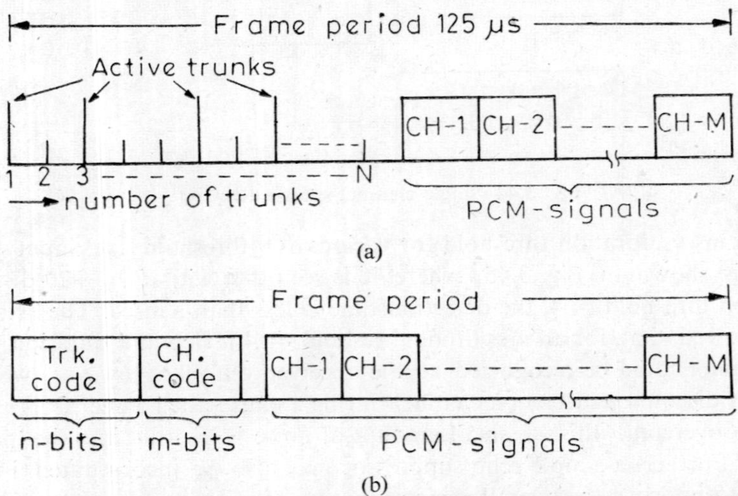

Fig. 3.46 Types of channel assignment signal: (a) channel allocation scheme type I and (b) channel allocation scheme Type II

(*b*) *Type II*—where the preamble consists of $(n + m)$ bits, as shown in Fig. 3.46(b). N trunks are designated by n bits and M channels by m bits. The assignment signal is generally transmitted whenever a new assignment is made, and the confirmation signal is transmitted during the idle time of the preamble slot. If an error occurs in the trunk/channel code, then the error is confined to two trunks and the erroneous connection continues for one talk-spurt only.

To avoid errors in the assignment signal, an error control procedure is used, e.g., the signal is repeated twice or thrice and the coincidence/majority decision is made in the receiver. Analysis and experiments for the following cases have been made and the results are shown in Table 3.8. The cases studied are:

Type I: (a) No error protection.
(b) Assignment signals are repeated twice and bit-by-bit coincidence is checked at the receiver. The assignment is not changed for the trunk corresponding to the bit not coincided.
(c) The assignment signals are repeated three times and a majority decision is made at the receiver.

Type II: (a) The assignment signals are repeated three times and a majority decision is made at the receiver.
(b) The assignment signal is repeated five times and a majority decision is made.

Table 3.8 Value of F and t_c for Type I and Type II assignment procedures
$N = 64(n = 6)$, $M = 31(m = 5)$, $k_a = 8$, $\alpha = 0.4$, $\tau = 1.0$ s, $k_s = 8$ for Type I and 5 for Type II

Assignment procedure type	Theoretical results		Experimental results F for
	F	t_c (ms)	$p_b = 10^{-4}$; and 10^{-5}
I (a)	236	2.25	—
I (b)	$445p_b + 0.76$	4.25	0.8
I (c)	1.97×10^3p_b	6.25	10^{-1}; 10^{-2}
II (a)	1.84×10^4p_b	1.50	1; 10^{-1}
II (b)	$1.02\times10^5\,p_b^2$	2.50	10^{-3}; 10^{-5}

Assuming that: p_{as} = the error rate in the assignment signal, p_b = BER and s = number of frames required to transmit a set of complete assignment signals, the relative impairment factor F may be defined as:

$$F = \frac{\text{speech impairment due to errors in the assignment signal}}{\text{speech impairment due to errors in non-TASI systems}}$$

$$= p_{as}\cdot s/p_b \tag{3.36}$$

In Table 3.8, the value of F and the worst case time t_c required to connect a trunk-to-a trunk are shown for the five cases with the following parameters: $N =$ number of trunks; $M =$ number of channels; $\alpha =$ average activity; $\tau =$ average talk-spurt length; $T =$ frame period; $k_a =$ number of bits of assignment signal per frame; $k_s =$ number of bits for synchronization of the assignment signal.

It is seen from Table 3.8 that the impairment due to errors is rather negligible for BER $\leqslant 10^{-4}$, for Types I(c), II(a) and II(b) procedures. But the values of t_c are less for Type II systems. On the overall basis, the type II(a) signalling procedure is the suitable one for TASI.

The channel capacity gain η for TASI can be calculated by assuming the Poisson distribution of speech activity α and an exponential distribution for talk-spurt length τ. The values of η now depend on the values of s, as larger values of s economized on the channel capacity. Further, if α increases, then there is a possibility of blocking of trunk signals and, as such, the blocking probability (also called 'freeze out') P_B should be kept low, say, $P_B \leqslant 10^{-2}$. With the parameter values as: $P_B = 5 \cdot 10^{-3}$, $\alpha = 0.4$ and $\tau = 1.0$ s, the TASI gain η vs. N is shown in Fig. 3.47. It is seen that for a large number of trunks, say, $40 \leqslant N \leqslant 80$, and $s = 8$, the gain η is between 1.8 and 2. Thus, TASI provides a useful digital technique to double the channel capacity of long-distance transmission links.

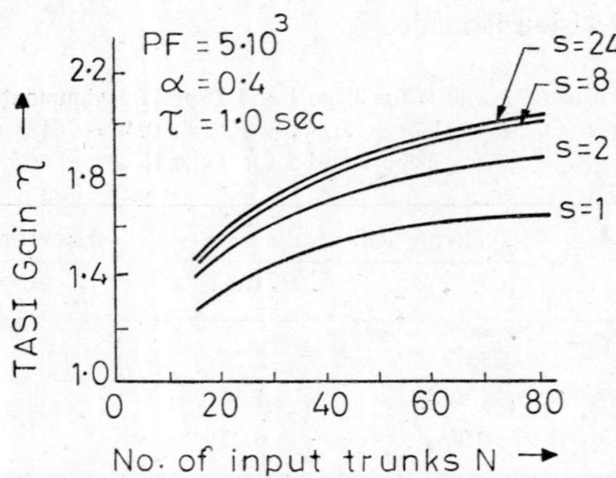

Fig. 3.47 TASI gain vs. number of trunks N (after Amano and Ota [48]

3.9 ECHO CONTROL TECHNIQUES [49]

In a long-distance telephone circuit, as shown in Fig. 3.48, echoes are generated due to the mismatch of impedances at the hybrid coupler (required for $2W/4W$ conversion) and such other mismatches in the terminals. The echo signal, as shown in the figure, travels back to the talker after certain delay and the delayed echo (with delay > 10 ms) is very disturbing unless suppressed to a relative level of -40 dB or less. The classical solution for echo control has been to use a voice-actuated echo suppressor at each hybrid,

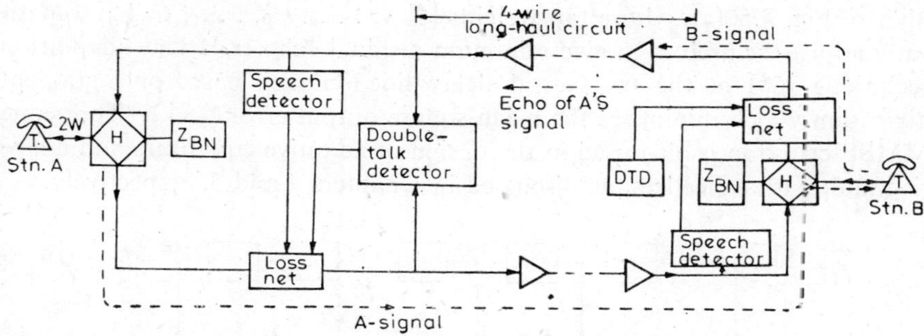

Fig. 3.48 A long-distance telephone circuit showing generation of echoes and their elimination using echo suppressors. T—telephone; H—hybrid transformer, Z_{BN}—balance network; and DTD—double-talk detector

as shown in Fig. 3.48. Assuming that Stn. A is talking, the speech detector output (in Stn. B) actuates a loss network in the transmit path of the hybrid B; and this blocks the return echo signal and also the signal from the talker B. Thus the system requires that only one talker should be active at a time. To avoid blocking of both way signals, when the other talker is active, the double-talk detector (DTD) is incorporated and this disables the suppressor circuit temporarily. Echo suppressors have been successfully used in long-distance circuits having moderate round trip delay, say, less than 100 ms. It has, however, a tendency to 'chop' during back-and-forth conversation, and, sometimes, clip out a segment at the beginning of an utterance. As such, the echo suppressors have not found favour for use in circuits having longer delay, say, 200 ms or more, as in satellite-routed circuits.

The solution for echo suppression in long-delay circuits has been found in an echo canceller, which simulates the estimated echo and subtracts this from the desired signal, as shown in Fig. 3.49. Since the magnitude, phase

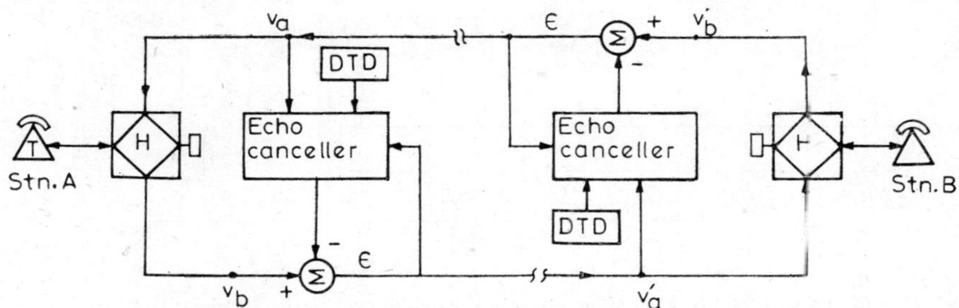

Fig. 3.49 Telephone circuit with echo cancellers at both ends

and delay of the echo signal v_b, leaking through the hybrid and other circuits, depend on the termination of the hybrid, a fixed filter cannot simulate the expected echo. As such, an adaptive filter is used to generate the estimated echo, and the transfer function $H(f) = F(v_b/v_a)$, where $F(\;)$ = the Fourier transform, is best simulated by a non-recursive transversal filter of Fig. 3.50(a), whose tap weights $\{C_{-N}, \ldots, C_0, \ldots, C_N\}$ automatically adjust themselves to give minimum residual echo $\{\epsilon_n\}$. The adaptation technique [50] of the transversal delay line (TDL) is based on a gradient algorithm which minimizes the mean-square output error $\langle \epsilon_n^2 \rangle$. The same MMSE criterion is also used in the design of adaptive equalizers and adaptive multipath cancellers, as discussed in Chapters 4 and 5, respectively.

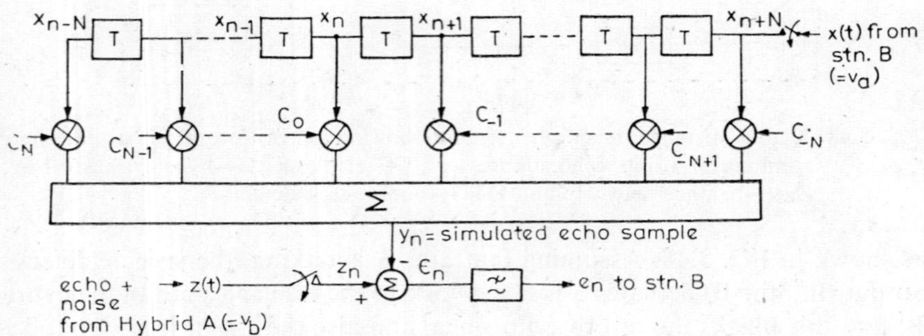

Fig. 3.50(a) Simulation of echo samples by a transversal filter; $\{x_n\}$ are the samples taken at a rate $> 2f_{\max}$

As shown in Fig. 3.50(a), the echo signal $x(t)(=v_a)$ is sampled as $\{x_n\}$ at a rate $1/T \geqslant 2\, f_{\max}$, $f_{\max}$ = maximum signal frequency, and fed to the TDL, whose tap weights $\{C_N, \ldots, C_0, \ldots, C_{-N}\}$ are obtained by correlating $\{\epsilon_n\}$ with the stored echo samples $\{x_{-N}, \ldots, x_{n-1}, x_n, x_{n+1}, \ldots, x_{n+N}\}$, as shown in Fig. 3.50(b) (for the tap C_{-N} only). The weighted outputs of the TDL are summed to form the replica echo y_n and this is subtracted from

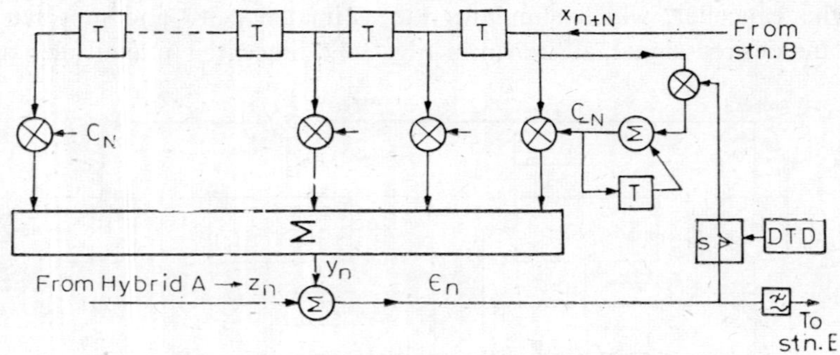

Fig. 3.50(b) Echo canceller showing one tap-adjustment circuit (C_{-N}). (Double-talk detector (DTD) disables the circuit when both talkers are active)

$\{z_n\}$ [the sample values of $z(t) = v_b + n(t)$] to form the error $\{\epsilon_n\}$. The error samples $\{\epsilon_n\}$ are filtered and allowed to reach the hybrid B along with the signal from talker A. The MMSE strategy to minimize $\langle \epsilon_n^2 \rangle$ requires that the tap weights $C_j(n+1)$ at the instant $(n+1)T$ be related to the weights

$C_j(n)$ at the earlier instant nT by

$$\underline{C}(n+1) = \underline{C}(n) - \mu\nabla \langle \epsilon_n^2 \rangle \tag{3.37}$$

where $\underline{C}(n)$ is the tap-weight vector $\{C_N, \ldots, C_0, \ldots C_{-N}\}$ at time nT, μ is the adaptation constant (step size), and $\nabla \langle \epsilon_n^2 \rangle$ is the gradient of the mean ϵ_n^2 at time nT with respect to $\underline{C}(n)$. As a first approximation, using μ small, one may compute $\nabla\epsilon_n^2$ instead of $\nabla \langle \epsilon_n^2 \rangle$, and the component of this gradient vector are given as:

$$\nabla\epsilon_n^2 = \left\{\frac{\partial\epsilon_n^2}{\partial C_N}, \ldots, \frac{\partial\epsilon_n^2}{\partial C_0}, \ldots, \frac{\partial\epsilon_n^2}{\partial C_{-N}}\right\} \tag{3.38}$$

The gradient vector is evaluated by using

$$\frac{\partial\epsilon_n^2}{\partial C_j(n)} = 2\epsilon_n\cdot\frac{\partial\epsilon_n}{\partial C_j(n)} = -2\epsilon_n\cdot x_{n-j} \tag{3.39}$$

since, from Fig. 3.50(a),

$$\epsilon_n = z_n - y_n = z_n - \sum_{j=-N}^{N} C_j(n)\cdot x_{n-j},$$

and

$$\frac{\partial\epsilon_n}{\partial C_j(n)} = -x_{n-j},$$

where, $\underline{X}_n = \{x_{n-N}, \ldots, x_n, \ldots, x_{n+N}\}$ = stored echo samples. Thus in vector form, $\nabla\epsilon_n^2 = -2\epsilon_n\cdot\underline{X}_n$ and the eqn. (3.37) reduces to

$$\underline{C}(n+1) = \underline{C}(n) + 2\mu\epsilon_n\underline{X}_n \tag{3.41}$$

The error surface $\langle \epsilon_n^2 \rangle$ is parabolic and has a single minimum and eqn. (3.41) always converges to this minimum. The above gradient algorithm is quite robust in the sense that any direction-preserving variation on the correction term $\mu\epsilon_n\underline{X}_n$ works almost as well. As a result, one may use only sgn (ϵ_n) in place of ϵ_n in eqn. (3.41) and this reduces the hardware complexity of Fig. 3.50(b); but, at the same time, the convergence is guaranteed (though the rate of convergence is somewhat reduced). Such simplified hardware has been used in many commercial designs.

The performance of echo cancellers depends on the length of the TDL (with reference to the time spread of the echo), on the value of μ (larger values of μ give larger steady-state errors) and on the particular algorithm used for approximating $\{z_n\}$. The gradient algorithm is widely used, but it requires an FIR-TDL. Other adaptive algorithms, e.g., least-square (Kalman), adaptive lattice, may also be used. These have a faster rate of convergence as compared to that of the gradient algorithm. In long-distance telephone circuits, the echo return loss due to the hybrid is approximately 15 ± 3 dB, and the echo canceller provides an additional loss of 30 dB; thus, the echo is suppressed by more than 40 dB as required. Here also, double-talk detectors (DTD) are provided, as shown in Fig. 3.50(b), to freeze the tap weights C_j, when the other talker is active. Otherwise, the other talker's signal will act as a large additive noise in ϵ_n and disturb the tap-weight settings considerably. Further improvements in the performance of the echo cancellers are obtained by blocking the residual echo and noise using centre-clipping devices (which only pass signals above a small threshold value) in conjunction with the cancellers. Such echo cancellers are now widely used in satellite-based circuits [51].

3.9.1 Echo Cancellation for Duplex Data Transmission [52]

As compared to the $2W/4W$ telephone circuits, the full-duplex data transmission has two special problems:

(a) The near-end echoes are as disturbing as the far-end echoes (in telephone circuits, the near end echoes are in the form of a desirable side tone and the far-end echoes only are cancelled). Moreover, the far-end echo may have a very large delay, e.g., for a satellite link, the delay is 500 ms approximately. The echo canceller has to cancel both the near-end and the far-end echoes.

(b) In full-duplex operation, double-talking is implied, and, as such, the tap weights of the canceller cannot be frozen as in Fig. 3.50(b).

To provide an echo-free interface between the 4-W modem and the 2-W telephone line, the hybrid couplers and echo cancellers are provided at station locations for duplex data transmission systems, as shown in Fig. 3.51. Data signals are scrambled to eliminate the possible correlation between the transmitted and received signals. The gradient algorithm is now realized digitally using a μP and a RAM. The correlation coefficients are calculated four times during a bit interval (one calculation per bit interval will also suffice, but the error ϵ_n will be larger) and are stored in the RAM. The μP updates the tap weights after each calculation. The summed output $\{y_n\rangle$ is now the replica echo and is subtracted from the received signal to give ϵ_n which is then used to calculate the tap weights in the next cycle. In tap-weight calculations, only the sgn (ϵ_n) is used and a dither signal in the form of random noise with rectangular distribution is added to ϵ_n to improve the estimation of the echoes.

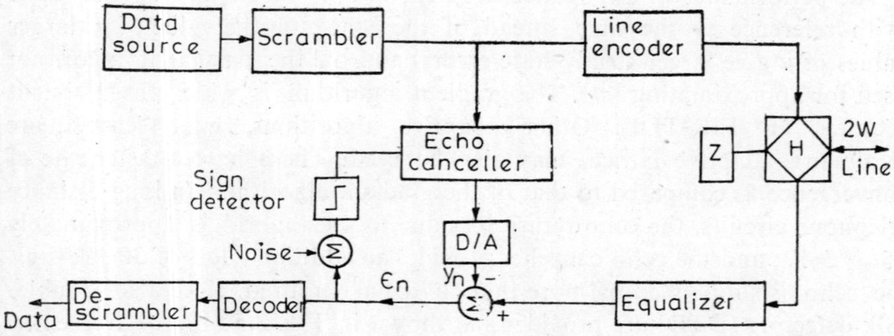

Fig. 3.51 Block diagram of a duplex data transmission system.

The near-end echo has a short delay but a high signal level. Since the typical hybrid attenuation is of the order of 10 dB and the line attenuation can be 40 to 50 dB, then the echo signal level is higher than the received signal by 30 to 40 dB. As such, the echo canceller provides an echo suppression of 50 to 60 dB to give an SNR of at least 20 dB. The far-end echo is, however, a low-level signal, but with a large delay. This can be minimized only by having a long impulse response of the simulating filter, say, by using an IIR filter. For duplex data transmission systems, it is convenient to use a combination of a FIR (transversal) filter and an IIR filter, shown in Fig. 3.52(a), such that the FIR filter cancels the strong undelayed near-end

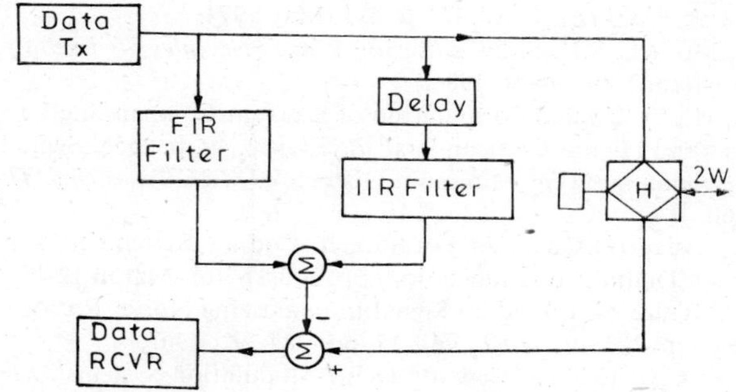

Fig. 3.52(a) Echo canceller usig FIR and IIR filter.

echo and the IIR filter with a longer impulse response cancels the weaker far-end echo [53]. An alternative solution is to provide a bulk delay between sections of the active transversal filter, as shown in Fig. 3.52(b). This drastically reduces the number of taps required in the filter, but it will be necessary to sound the echo channel, before data transmission, to estimate the delay of the far-end echo. Although it is preferable to operate the TDL in the canceller at the Nyquist rate, it is possible to operate the TDL at the data rate as well; but in the latter case, the two transmitters at stations A

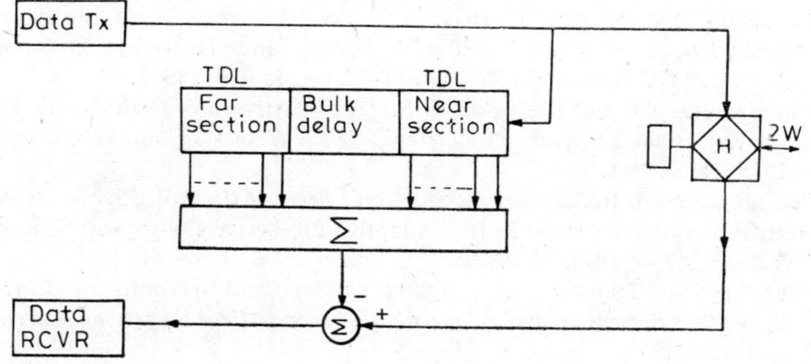

Fig. 3.52(b) Echo canceller using a bulk delay in the TDL

and B must operate in full synchronism. Regarding the problem of double-talk, a special algorithm (averaged-gradient algorithm) is used to obtain the convergence, although at a rather slow rate.

REFERENCES

1. Oliver, B.M., Pierce, J.R. and Shannon, C.E., 'The Philosophy of PCM', *Proc. IRE*, vol. 36, pp. 1324-1331, 1948.
2. Flanagan, J.L., *Speech Analysis, Synthesis and Perception,*, Academic Press, N.Y. 1965.
3. Flanagan, J.L. *et al.*, 'Speech Coding', *IEEE Trans.*, vol. Com-27, No. 4, p. 710, 1979.
4. Jayant, N.S., 'Digital Coding of Speech Waveforms: PCM, DPCM and ADM Quantizers', *Proc. IEEE*, vol. 62, p. 611, May 1974.
5. Das, J., Mullick, S.K. and Chatterjee, P.K., *Principles of Digital Communication*, Wiley Eastern, New Delhi, 1986.
6. Kaneko, H., 'A Unified Formulation of Segment Companding Laws and Synthesis of Coders and Digital Companders', *BSTJ*, vol. 49, p. 1555, Sept. 1970.
7. Max, J., 'Quantizing for Minimum Distortion', *IRE Trans. Inf. Th.*, vol. IT-6, pp. 7-12, 1960.
8. Noll, P., 'Adaptive Quantizing of Speech Coding Systems', *Proc. IEEE*, Zurich Seminar on Digital Communication. pp. B-3(1)-3(6), March 12-15, 1974.
9. O'Neal, J.B. Jr., 'A Bound on Signal-to-quantizing Noise Ratios for Digital Encoding Systems', *Proc. IEEE*, vol. 55, pp. 287-292, 1967.
10. Das, J., 'A Critical Review of Delta Modulation Systems', *Electro-Technology*, vol. 16, p. 41, March 1972.
11. Abate, J.E., 'Linear and Adaptive Delta Modulation', *Proc. IEEE*, vol. 55, pp. 298-308, 1967.
12. Greefkes, J.A., 'A Digitally Companded Delta Modulation Modem for Speech Transmission', *Proc. IEEE*, Int. Conf. Communications, pt 7, pp. 33-48, June 1970.
13. Chakravarty, C.V. and Faruqui, M.N., 'Two-loop Adaptive Delta Modulation', *Trans. IEEE*, Comm. vol. Com-22, pp, 1710-1713, Oct. 1974.
14. Melnick, M., 'Intelligibility Performance of a Variable Slope Delta Modulation', *Proc. IEEE*, vol. 59, p. 1382, 1971.
15. O'Neal, J.B. Jr., 'Entropy Coding in Speech and Television Differential PCM Systems', *IEEE Trans. Inf, Th.*, vol. IT-17, pp. 758-761, 1971.
16. Campanella, S.J. and Robinson, G.S., 'A comparison of Orthogonal Transformations for Digital Speech Processing', *IEEE Trans.*, Com. Tech. vol. Com-19, p. 1045, Dec. 1971.
17. Special issue on 'Redundancy Reduction', *Proc. IEEE*, vol. 55, No. 3, 1967.
18. Atal, B.S. and Schroeder, M.R., 'Adaptive Predictive Coding of Speech Signals', *BSTJ*, vol. 49, p. 1973, 1970.
19. Atal, B.S., and Hanauer, S.L., 'Speech Analysis and Synthesis by Linear Prediction of the Speech Wave', *Journal of the Acoustical Society of America*, vol. 50, pp. 637-655, 1971.
20. Grassot, F. and Voillaume, J.C., 'Low Bit Rate Speech Transmission', *Elec. Comm.*, vol. 55, No. 4, pp. 316-322, 1980.
21. Special issue, 'Bit Rate Reduction and Speech Interpolation', *IEEE Trans.*, vol. Com-30, No. 4, April 1982.
22. Fehn, H.G. and Noll, P., 'Multipath Search Coding of Stationary Signals with Applications to Speech', *IEEE Trans.*, vol. Com-30, No. 4, April 1982.
23. Jayant, N.S. and Christensen, S.A., 'Tree-encoding of Speech Using the (M, L) Algorithm and Adaptive Quantization,' *IEEE Trans.*, vol. Com-26, p. 1376, 1978.
24. Stewart, L.C., Gray R.M., and Linde, Y., 'The Design of Trellis Waveform

Coders', *IEEE Trans.*, Com-30, p. 702, April 1982.
25. Matsuyama, Y. and Gray, R.M., 'Voice Coding and Tree-encoding Speech Compression Systems Based upon Inverse Filter Matching', *IEEE Trans.*, Com-30, p. 711, April 1982.
26. Viswanathan, V.R. *et al.*, 'Variable Frame Rate Transmission: A Review of Methodology and Application to Narrow-band LPC Speech Coding', *IEEE Trans.* Com-30, p. 674, April 1982.
27. Carmody, J. and Rothweiler, J., 'Speech Coding at 800 and 400 Bits per Second', *Elec. Comm.*, vol. 59, No. 3, pp. 260-265, 1985.
28. Dudley, H. and Balashek, S., 'Automatic Recognition of Phonetic Patterns of Speech', *Jour. Acoust. Soc. America*, vol. 30, pp. 721-732, 1958.
29. M. Immendörfer, 'Voice Dialer', Elec. Comm., vol. 59, No. 3. pp. 281-285, 1985.
30. Honra, O.A., 'A 150 MBPS A/D and D/A Conversion System', *Comsat. Tech. Review*, vol. 2, No. 1, p. 39, Spring 1972.
31. Schreiber, W.F., 'Picture Coding', *Proc. IEEE*, vol. 55, p. 320, March 1967.
32. Limb, J.O., 'Source-Receiver Encoding for TV Signals', *Proc. IEEE* vol. 55, p. 364, 1967.
33. Abbott, R.P., 'A Differential Pulse Code Modulation Codec for Videotelephony Using Four Bits per Sample', *IEEE Trans.*, Com. Tech. vol. Com-19, p. 907, Dec. 1971.
34. Frei, A.H., Schindler, H.R. and Vettiger, P., 'An Adaptive Dual-Mode Coder/Decoder for Television Signals', *Trans. IEEE*, Com. Tech. vol. Com-19, p. 933, Dec. 1971.
35. Habibi, A. and Robinson, G.S., 'A Survey of Digital Picture Coding', *Computer*, p. 22, May 1974.
36. Schilling, D.L. *et al.*, 'Video Encoding Using Adaptive Delta Modulation', *IEEE Trans.*. Com-26, p. 1682, 1978.
37. Netraveli, A.N. and Limb, J.O., 'Picture Coding: A Review', *Proc. IEEE*, vol. 68, p. 366, 1980.
38. Faruqqui, M.N., 'Bandwidth Compression of Image Signals', *JIETE*, vol. 26, p. 29, 1980.
39. Haskell, B.G. *et al.*, 'Interframe Coding of Video Telephone Pictures', *Proc. IEEE*, vol. 60, p. 792, 1972.
40. Kaneko, H. and Ishiguro, T., 'Digital Television Transmission Using Bandwidth Compression Techniques', *IEEE Communication Soc. Mag.*, vol. 18, No. 4, pp. 14-22, 1980.
41. Scheuermann, H. and Göckler, H., 'A Comprehensive Survey of Digital Transmultiplexing Methods', *Proc. IEEE*, vol. 69, pp. 1419-1450, 1981.
42. Freeny, S.L., 'TDM/FDM Translation as an Application of Digital Signal Processing', *IEEE Com. Soc. Mag.*, vol. 18, pp. 5-15, Jan. 1980.
43. Claasen, T.A.C.M. and Mecklenbräuher, W,F.G., 'A Generalized Scheme for an All-Digital Time-division Multiplex to Frequency Division Multiplex Translator', *IEEE Trans. Circuit Syst.*, vol. CAS-25, pp. 252-259, 1978.
44. Bonnerot, G. *et al.*, 'Digital Processing Techniques in the 60-channel Transmultiplexer', *IEEE Trans. Com. Tech.*, vol. Com-25, pp. 698-706, 1978.
45. Darlington, S., 'On Digital Single Sideband Modulators', *IEEE Trans.*, vol. CT-17, pp. 409-414, 1970.
46. Tsuda, T. *et al.*, 'Realization of a TDM-FDM Transmultiplexer with Multistage Structure Using ROM Multipliers'. *Fujitsu Scientific Tech. Jour.*, pp. 17-37, March 1979.
47. Bullington, K. and Fraser, J.M., 'Engineering Aspects of TASI'. *BSTJ*, vol. 38, pp. 353-364, 1959.
48. Amano, K. and Ota, C., 'Digital TASI System in PCM Transmission', *Int. Comm Conf.*, Boulder, June 1969.

49. Weinstein, S.B., 'Echo Cancellation in the Telephone Network', *IEEE Comm Soc. Mag.*, vol. 15, pp. 9-15, June 1977.
50. Widrow, B. *et al.*, 'Adaptive Noise Cancelling; Principles and Applications', *Proc. IEEE*, vol. 63, pp. 1692-1716, 1975.
51. Suyderhoud, H.G. *et al.*, 'Results and Analysis of a Worldwide Echo Canceller Field Trial', *COMSAT Tech. Rev.*, vol. 5, pp. 253-974, Fall 1975.
52. Alvestad, T. and Eriksen, T.J.C., 'Echo Canceller for Two-wire Data Modems', *Elec. Comm.*, vol. 59, No. 3, pp. 333-337, 1985.
53. Gilsanz, M. *et al.*, 'Adaptive Echo Cancelling for Baseband Data Transmission', *Elec. Comm.*, vol. 59, No.. pp. 338-344, 1985.

CHAPTER 4

Digital Cable Systems

Introduction [1, 2]

The rate of data transmission in a given low-pass (baseband) channel is governed by the Nyquist-Shannon theorem, giving an ideal transmission rate of 2 symbols (bauds)/s/Hz. However, all practical channels, e.g., speech and cable circuits, HF and RF links etc., are non-ideal, having attenuation and phase distortion, and thus the rate of transmission is always less than the Nyquist rate. Moreover, many such channels show strong low frequency attenuation (due to coupling transformers) and delay distortion, thus making the circuit virtually a bandpass channel. This problem is solved by using d.c.-free codes, e.g., AMI, pseudoternary and PRS, and also by shifting the data spectrum through ASK, FSK and PSK modulation. For the purpose of increasing the spectral efficiency of data modems, multilevel signalling using PAM, FM, multiphase PSK and multilevel QAM are used. Characteristics of these modulation schemes have been discussed in Sec. 2.4, and Table 2.2. gives a summary of their error performances. It has been shown that M-QAM with coherent detection, is the most efficient. However, there is considerable degradation in their performance due to various channel imperfections, e.g., amplitude and delay distortions, time dispersions etc., and this leads to interference (ISI), as discussed later.

In this chapter, we shall discuss voice-grade data circuits and also high-rate digital transmission systems using wide band cable media, e.g., coaxial cables and optical fibres. The associated problems of channel equalization, synchronization and TDM multiplexing will also be discussed.

4.1 CHANNEL CHARACTERIZATION [1, 3]

A data communication system generally consists of a transmitter, the channel and a receiver. The problem of error-free detection of the signal is attempted to be solved by suitably choosing matched filters and detection techniques. The channel may be linear or non-linear (generally considered to be somewhat linear) and time-invariant or time-varying. Many of the common types of channels, such as voice-band and wide band telephone systems, high-frequency radio systems, troposcatter systems and space probes, exhibit somewhat similar basic characteristics, differing mainly in the degree to which the various imperfections are present. Thus, the channel may be modelled for the simplest case by a linear transfer function followed by an additive Gaussian noise source. This model takes into account the deterministic impairment, dispersion, as well as the most prevalent Guassian noise.

A transmission channel may be characterized either in the frequency domain or in the time domain. For a transmitted signal $s(t)$, the received signal, in Fig. 4.1, is given by:

$$r(t) = f\{s(t)\} + n(t) = r_0(t) + r_n(t)$$

and in the frequency domain:

$$R(\omega) = H(\omega) \cdot S(\omega) + N(\omega) \tag{4.1}$$

where $r_0(t)$ is due to $s(t)$ only. The response $r(t)$ is given by the convolution

$$r(t) = \int_{-\infty}^{\infty} h(\tau)\, s(t-\tau)\, d\tau + r_n(t) \tag{4.2}$$

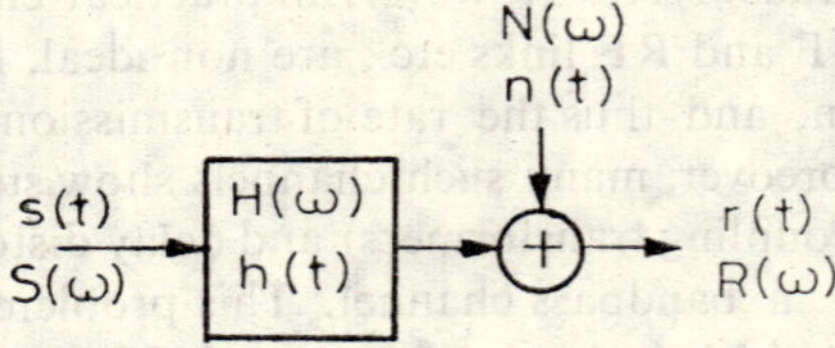

Fig. 4.1 An equivalent channel

For an algebraic time-invariant channel, $H(\omega)$ is a gain constant g which may be a random variable. The PDF of g is usually either Rayleigh or Rician, and such channels are referred to as Flat-Flat fading channels. If $g(t)$ is time-varying, then the signal will be dispersed in frequency by an amount equal to the bandwidth of $g(t)$, and the channels are now called Frequency Dispersive or Time-Selective channels. One such example is the Rayleigh frequency dispersive fading channel, as observed for high-frequency radio circuits, where $g(t)$ is wide sense stationary. For fading channels, $h(t)$ is a simple function of a random process usually Gaussian and the signal component of the received signal is of the form:

$$r_0(t) = \sum_j b_j\, s(t - \tau_j)$$

$$\simeq \int_{-\infty}^{\infty} h(\tau)\, s(t-\tau)\, d\tau \tag{4.3}$$

If the frequency domain characteristics $H(\omega)$ is not ideal, then this frequency selective channel gives rise to a dispersion of the impulse response $h(t)$ and the channel is now called a time-dispersive channel. Noting the dispersion of the impulse response in the channel, a channel may now be modelled as a tapped-delay line (TDL), as shown in Fig. 4.2. where the tap

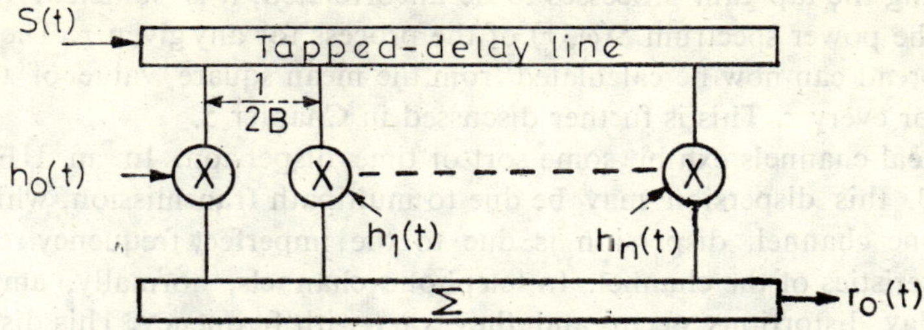

Fig. 4.2 Tapped-delay line model of a channel

gains are given by the sample values of the impulse response taken at the Nyquist rate. The representation now becomes:

$$s(t-\tau) = \sum_{-\infty}^{\infty} s(t - \frac{n}{2B}) \frac{\sin 2\pi B (\tau - \frac{n}{2B})}{2\pi B (\tau - \frac{n}{2B})}$$

where B is the bandwidth of $s(t)$, and

$$\begin{aligned} r_0(t) &= \int_{-\infty}^{\infty} \Sigma s(t - \frac{n}{2B}) h(\tau) \frac{\sin x}{x} d\tau \\ &= \Sigma s(t - n/2B)\, h_n(t) \end{aligned} \tag{4.4}$$

where $\quad x = 2\pi\, B(\tau - n/2B) \quad$ and $\quad h_n(t) = \int_{-\infty}^{t} h(\tau) \frac{\sin x}{x} d\tau$

The tap gain of the delay line, in Fig. 4.2, is now $h_n(t)$ and the length of the line extends over the significant dispersion of the impulse response. The sampled values of $r(t)$ may also be written as:

$$r_k = \Sigma h_n\, s_{k-n} + \eta_k \tag{4.5}$$

where η_k is the sampled noise.

4.1.1 Time-varying Channel [3]

If the impulse response of the channel is time-varying, the signal component of the output is given by

$$r_0(t) = \int_{-\infty}^{\infty} h(t, \tau) s(t - \tau)\, d\tau \tag{4.6}$$

where $h(t, \tau)$ is the response at t of the linear time-varying channel for an impulse applied τ sec earlier. A troposcatter channel falls into this category. The channel may now be modelled with the above tapped-delay line of Fig. 4.2, where tap gains $h_k(t)$ are sample functions of random processes, usually zero-mean Gaussian. The channel is now completely characterized by the tap-gain correlation function given by

$$\begin{aligned} R_h(t_1, t_2, \tau, \eta) &= E[h_1(t_1, \tau) h_2(t_2, \eta)] \\ &= R_h(\Delta, \tau, \eta), \quad \text{with} \quad \Delta = (t_1 - t_2) \end{aligned} \tag{4.7}$$

Assuming the tap gain processes to be uncorrelated, it is sufficient to only know the power spectrum $S(\omega, \tau)$ of the process for any given τ. The multipath spread can now be calculated from the mean square value of the tap gains for every τ. This is further discussed in Chapter 5.

All real channels exhibit some sort of time dispersion. In an HF radio channel, this dispersion may be due to multipath transmission, while in a telephone channel, dispersion is due to the imperfect frequency response characteristics of the channel. In telephone channels, normally, amplitude and delay distortions occur and they vary with frequency. This distortion results in intersymbol interference (ISI), which is somewhat deterministic in short-haul circuits. In theory, it is possible to remove this intersymbol interference by using sophisticated equalization techniques. However, there is always some non-linearity, such as frequency offset, phase jitter and gain saturation, which sets a limit to the ultimate symbol rate that can be supported by a real channel.

4.2 MULTILEVEL SIGNALLING [1]

The optimum receivers and signalling formats for data communication through an additive Gaussian noise channel have been discussed in Chapter 2. However, the effect of ISI on the error performances has to be estimated and the necessary equalization of the channel implemented to evolve an efficient data communication system. The baseband PAM channel will be considered for the purpose and the results may be extended to the FSK and PSK systems by considering the equivalent baseband signals and low-pass channels.

Consider a PAM baseband data transmission system using an ideal low-pass channel perturbed by Gaussian noise as shown in Fig. 4.3. Assume that the allowed transmitter levels are $\pm d, \pm 3d, \ldots \pm (L-1)d$, where $2d$ is the distance between adjacent levels. The input symbols $\{a_n\}$ appear as a series of amplitude-modulated, overlapping pulses $\{a_n\, x(t)\}$ at the output of the receiving filter $G_R(\omega)$. The pulse waveform $x(t)$ is given by the inverse Fourier transform as:

$$x(t) = \frac{1}{2\pi} \int_{-\infty}^{\infty} G_T(\omega) \cdot H(\omega) \cdot G_R(\omega)\, e^{j\omega t}\, d\omega \tag{4.8}$$

and with the addition of noise, we get

$$y(t) = \sum_n a_n\, x(t - nT) + n(t)$$

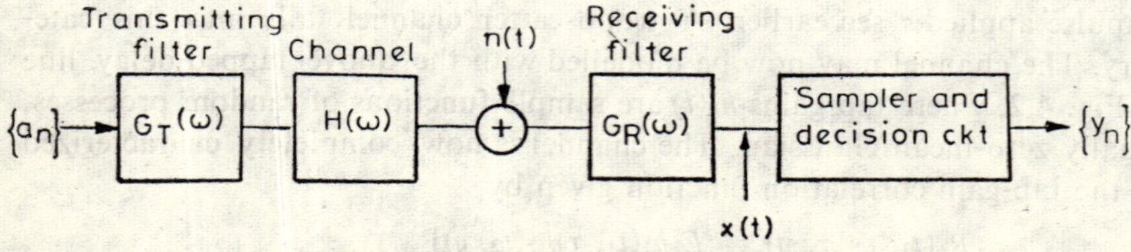

Fig. 4.3 A PAM system using threshold detection

The sampled values of the output $y(t)$ may be written as:

$$y_k = \sum_n a_n x_{k-n} + \eta_k \tag{4.9}$$

where $\quad y_k = y(kT + t_0)$; $t_0 =$ system delay.

The normalized desired output may now be written as:

$$y_k/x_o = a_k + (\frac{1}{x_0} \sum_n a_n x_{k-n} + \eta_k/x_0), \qquad n \neq k \tag{4.10}$$

For no error to occur, the following condition should be fulfilled:

$$| \sum_n a_n x_{k-n} + \eta_k | < x_0 d, \qquad n \neq k \tag{4.11}$$

It is thus seen that the ISI and the noise, given respectively, by the 2nd 3rd terms of eqn. (4.10), affect the error performance in a cumulative way. The problem of the design is to minimize the ISI and the probability of error.

For minimizing the effect of channel noise, the transmit and receive filters are optimized so that the two filters have complementary characteristics given by [1]:

$$G_R(\omega) = \frac{\sqrt{| X(\omega) |}}{\sqrt[4]{N(\omega)}} \quad \text{and} \quad G_T(\omega) = \frac{X(\omega)\cdot\sqrt[4]{N(\omega)}}{\sqrt{| X(\omega) |}} \tag{4.12}$$

where $X(\omega) =$ the spectrum of $x(t)$, as defined in eqn. (4.8).

The filters may be identical, each accomplishing 'half' of the signal shaping if the noise spectral density $N(\omega)$ is flat and the signal spectrum $X(\omega)$ is real. With the above optimization, the probability of symbol error in a PAM channel is given by [*cf.* eqn. (2.40)]:

$$P_e = 2(1 - \frac{1}{L})\ erfc[d/\sigma_N] \tag{4.13}$$

where

$$erfc(x) = \frac{1}{\sqrt{2\pi}} \int_x^\infty \exp\left(-\frac{t^2}{2}\right) dt \text{ and } \sigma_N^2 = \text{noise variance.}$$

If the average transmitted power is P_s and the noise power in the channel is $P_N = n_o/T$, then the symbol error probability is given by:

$$P_e = 2(1 - \frac{1}{L})\ erfc\left[\left(\frac{3}{L^2 - 1} - \frac{P_s}{P_N}\right)^{1/2}\right] \tag{4.14}$$

and the bit error $P_b \simeq P_e/(\log_2 L)$.

This error probability has been shown in Fig. 2.20 for various values of $L\ (= M)$. This result may also be expressed in terms of the rate of transmission per Hz for a fixed error rate (say, $P_e = 10^{-5}$) and is shown in Fig. 4.4, where

$$R/W = \frac{2 \log_2 L}{1 + \alpha}, \qquad W = \text{channel BW}$$

and

$$P_s/P'_N = \frac{L^2-1}{3} \cdot \frac{(4.42)^2}{1+\alpha} \tag{4.15}$$

with α as the roll-off parameter of the channel filter and P'_N is the tolal noise in BW = W.

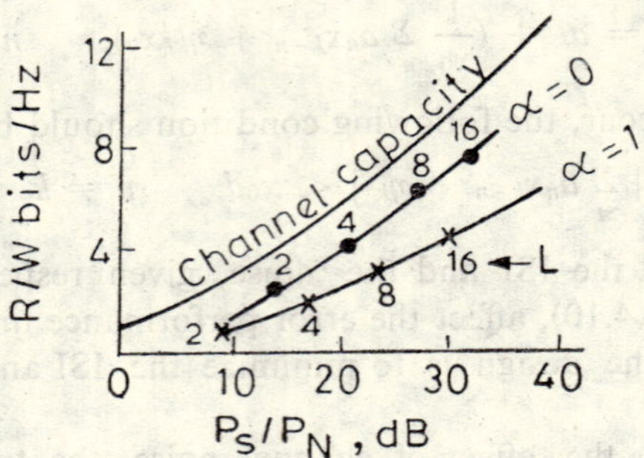

Fig. 4.4 Required signal-to-noise ratios for $P_e = 10^{-5}$ (after Lucky *et al.* [1])

4.2.1 The Effect of Intersymbol Interference

The effect of ISI may be considered in terms of the closure of the eye pattern or in terms of the rms error. Recalling that an error occurs in detecting the correct pulse whenever the sum of ISI and noise exceeds the distance to the nearest decision threshold as given by eqn. (4.11), the overall error probability may be shown to be:

$$P_e = \frac{1}{L^{2k}} \sum_{i=1}^{L^{2k}} \left\{ erfc \left[\frac{x_0 d - D(i)}{\sigma_N} \right] + erfc \left[\frac{x_0 d + D(i)}{\sigma_N} \right] \right\} \tag{4.16}$$

where the dispersion of x_n does not exceed k symbols on each side of the main pulse, and $D(i)$ is the ISI for each sequence. This equation, however, is not of much practical use for numerical computation. A more useful relation is obtained by using the mean-square eye closure measure. If

$$\epsilon^2 = \frac{1}{x_0^2} \sum_n{}' x_n^2, \quad (n \neq 0)$$

is the mean-square distortion in the system impulse response, then the error probability is given by

$$P_e = 2erfc \left\{ \left[\frac{L^2-1}{3} \epsilon^2 + \frac{e_N^2}{x_0^2 d^2} \right]^{1/2} \right\} \tag{4.17}$$

wheye ISI has been assumed to be approximately Gaussian.

There is considerable difficulty in actually visualizing the effect of ISI on error probability. However, a comparison of P_e vs. SNR is shown in Fig. 4.5, for binary data, having peak distortion $D = 1.2$ for 10 dB attenuation distortion and $\alpha = 0.6$ [1]. The curve (a) of the figure was obtained from

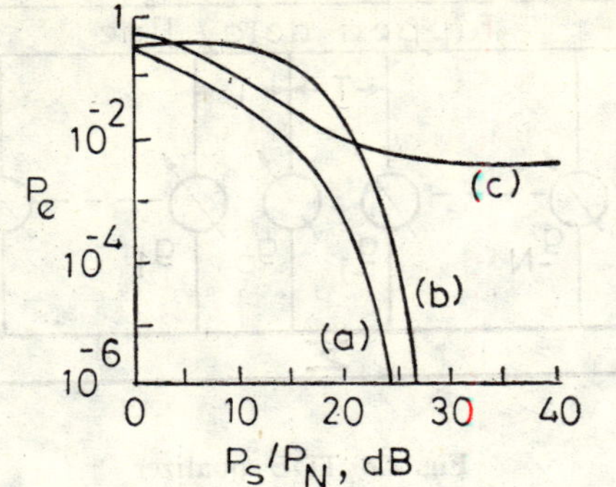

Fig. 4.5 Probability of error P_e for binary data having attenuation distortion = 10 dB, α = 0.6 (after Lucky *et al.* [1])

eqn. (4.16) using $k = 5$ and a sequence length of 11. The curve (b) was obtained from the upper bound of the relation for P_e using the peak distortion D as a parameter, and the curve (c) was obtained from eqn. (4.17) assuming Gaussian ISI. It is seen that curve (a) is close to the true value of P_e. Comparing Figs. 4.4 and 4.5, it is seen that for $P_e = 10^{-5}$, the P_s/P_N required in a binary system, having an ISI of $D = 1.2$ and $\alpha = 0.6$, is 10 dB more than in a binary system having zero ISI.

4.3 EQUALIZATION TECHNIQUES [4, 5]

The deterioration of the error performances due to ISI has been shown in the earlier section, specifically for PAM transmission. It is, therefore, necessary to minimize ISI by using a delay equalizer. Adjustable LC delay equalizers have been in use in long-distance transmission systems for a long time. However, these lumped-parameter equalizers are difficult to adjust on line and are not suitable for automatic operation. The tapped-delay line characterization of a general transmission medium, as shown in Fig. 4.2 has led to the development of a transversal filter as an adjustable equalizer. This versatile filter was invented by Wiener and Lee in 1935 and first described in literature by Kallman in 1940. In theory, an inverse transversal filter should be able to make the ISI zero. But this would require an infinite length of the TDL and a practical compromise has to be made. It is also possible in practice to combine matched filtering, necessary for noise reduction, and equalization in the same TDL, if the taps are spaced at the Nyquist intervals rather than at symbol intervals. Such versatile transversal filters have been constructed by various workers.

Considering a TDL with the tap gains as shown in Fig. 4.6, the impulse respose of the TDL is:

$$h(t) = \sum_{-N}^{N} g_n \delta(t - nT)$$

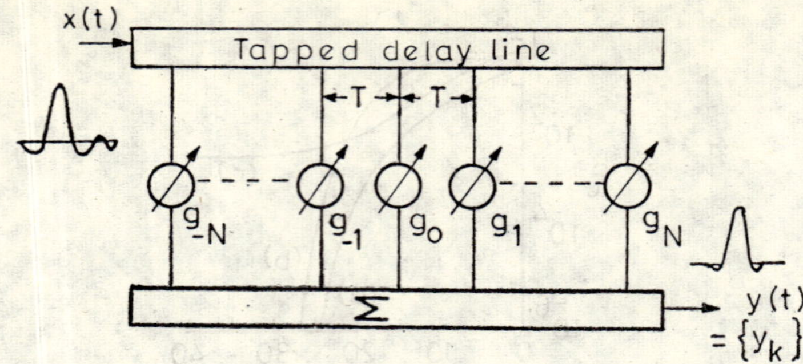

Fig. 4.6 TDL equalizer

while its frequency characteristic is:

$$H(\omega) = \sum_{-N}^{N} g_n e^{-jn\omega T}$$

For an input $x(t)$, the output $y(t)$ is given by:

$$y(t) = x(t)*h(t) = \sum_{-N}^{N} g_n x(t-nT)$$

and (4.18)

$$y_k = \sum_{-N}^{N} g_n x_{k-n}$$

The condition for the ISI to be zero is that the tap coefficients should be

$$\{g_n\} = \frac{1}{\{x_n\}} = \frac{T}{2x}\int_{-\pi/T}^{\pi/T} \frac{g_0 T}{X_1(\omega)} \cdot e^{jn\omega T}\, d\omega \tag{4.19}$$

where $X_1(\omega) = T \sum_n x_n e^{jn\omega T}$.

This requires an infinite length of the TDL, except in trivial cases. Considerable literature is available on the different practical techniques for the design of the TDL equalizer. The preset equalizers are normally used in short-distance circuits, where the channel parameters are more or less time-invariant. For long-distance circuits and for radio channels, it is necessary that the equalizer should be automatic with initial training facilities or adaptive, so that the time-varying characteristics of the channel can be accommodated in the equalizer operation. Various equalizer-adjustment algorithms also have been developed. They are based on either the peak-distortion criterion or the mean-square distortion criterion. The peak distortion D is defined as:

$$D = \frac{1}{y_0}\left\{ \sum_k |y_k| - y_0 \right\} = \frac{1}{y_0} \sum_k{}' |y_k|, \qquad (k \neq 0)$$

Using eqn. (4.18) and normalizing $y_0 = 1$, we get

$$D_0 = \sum_{\substack{k=-\infty \\ k\neq 0}}^{\infty}{}' \left| \sum_{-N}^{N} g_n x_{k-n} \right| \tag{4.20}$$

It has been shown that the peak distortion is a convex function of the $2N$ variables g_n, $|n| \leqslant N \neq 0$. Further, if $D_0 < 1$, then the minimum distortion D must occur for those $2N$ tap-gain settings which cause $y_k = 0$, $k \neq 0$. Thus, to find the tap gains g_n, it is sufficient to solve $2N$ linear equations, and any minimum of D found by systematic search must be the absolute (global) minimum of distortion.

The mean-square distortion of $y(t)$ is defined as:

$$\epsilon^2 = \frac{1}{y_0^2} \sum_n{}' \; y_n^2 \tag{4.21}$$

It is now necessary to solve simultaneously $(2N + 1)$ linear equations and, subsequently, scale the tap-gain values obtained to achieve the proper pulse height. This set of equation has the interesting property that the optimum setting of the equalizer will force the cross-correlation between the error signal and the input pulse to be zero at each sample point within the range of the equalizer. This interesting property is utilized in designing the hardware of the adaptive equalizer.

It is thus seen that the receiver in a data communication system has to have a matched filter to minimize the channel noise and a delay equalizer to minimize the ISI. In practical systems, the noise performance of the conventionally equalized system is somewhat poorer than for the ideal linear receiver. This is more so because of the finite length of the TDL. However, the overall effect is usually negligible when the length of the equalizer is sufficient to ensure the suppression of the ISI and the SNR of the channel is reasonably good. On the other hand, the effect of the matched filter on the ISI may be large, since the MF will square the amplitude characteristic and eliminate the phase distortion of the channel. For certain types of ISI, e.g., for parabolic delay, the MF will minimize the ISI but for the slope—attenuation distortion, the MF will make matters worse. The transversal filter equalizer has to be more elaborate to do both the jobs of noise reduction and the minimization of ISI.

4.3.1 Practical Equalizers [5, 6, 7]

Practical equalizers may generally be classified as manual, preset/automatic, adaptive and decision beedback equalizers. The classical equalizers have been designed mostly on the basis of LC filter techniques but all modern equalizers are designed on the basis of the TDL as discussed earlier. The actual delay lines may be LC delay lines having the required bandwidth and delay, but, at present, active delay lines using 'sample and hold' techniques are more popular and easier to construct. The new series of delay lines using bucket-brigade techniques and charge-coupled devices are also available now. It is possible now to completely integrate an adaptive equalizer with mostly digital ICs. An adaptive equalizer is shown in Fig. 4.7, where the transversal filter has been used as the TDL [7]. The system is designed for multilevel PAM transmission and the A/D converter after the summing

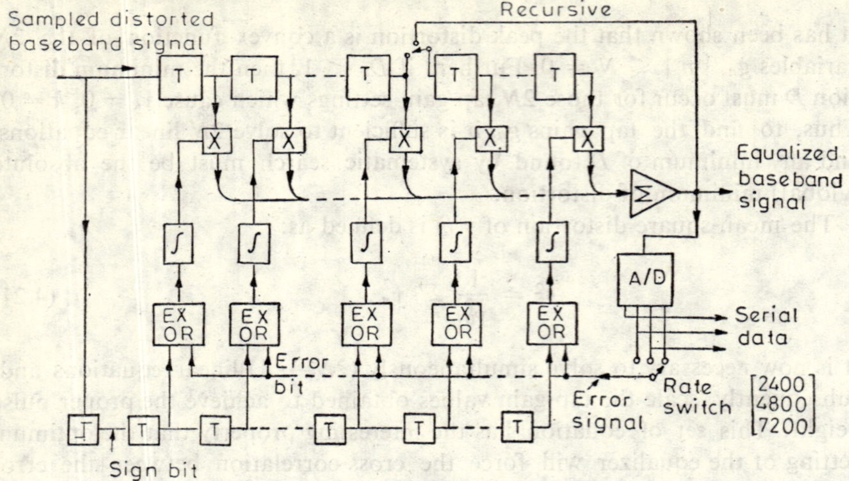

Fig. 4.7 An adaptive equalizer

amplifier converts the data into serial form and an extra bit is added to the converter to give the error signal. This signal is correlated with the delayed sign bit of the signal through EX-OR circuits and the control voltages for the tap-gain settings are generated by integrating EX-OR outputs. The transversal filter can be altered to a recursive filter by means of a strap change, and the second half of the filter operates in a feedback mode. This increases the efficiency of the equalizer; however, the recursive mode is unstable for certain types of line distortion. The eye pattern of a 4-level PAM signal before and after equalization is shown in Fig. 4.8.

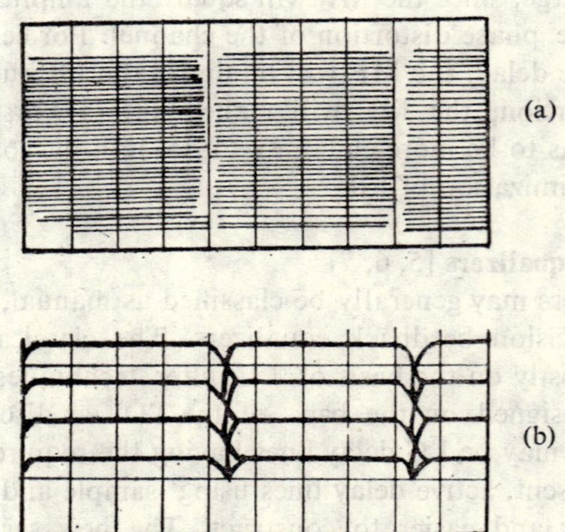

Fig. 4.8 Eye pattern at the output of the equalizer: (a) before equalization and (b) after equalization

It has been shown that in a noisy dispersive time-varying channel, an adaptive decision feedback equalizer [6] performs better than the linear equalizer. As shown in Fig. 4.9, the decision feedback circuit uses the past decisions about the digits to minimize the ISI of the future digits by coherently subtracting the interference from previously detected digits. It also automatically adapts the equalizer parameters to the changes in channel characteristics. This equalizer has been shown to be insensitive to the

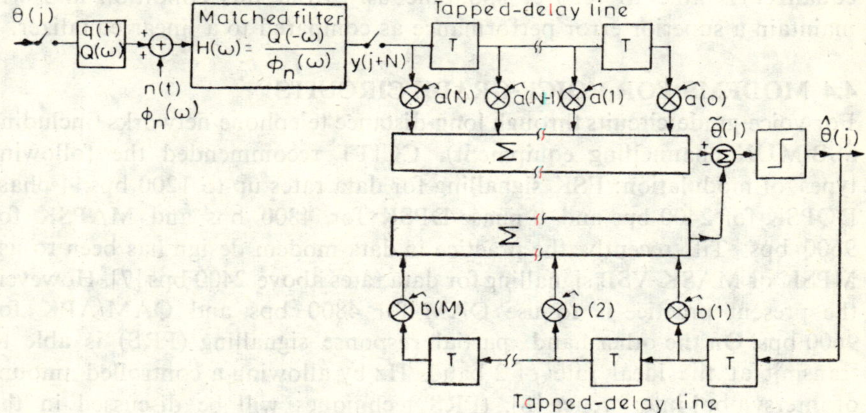

Fig. 4.9 A decision feedback equalizer

quantization of the input signal and the quantization and adjustment of its own parameters, and so can be constructed at a reasonable cost. The performance of this equalizer and some comparisons with a linear equalizer and a Bayesian demodulator are shown in Fig. 4.10. It is seen that the performance of this equalizer is close to that of the optimum Bayesian receiver

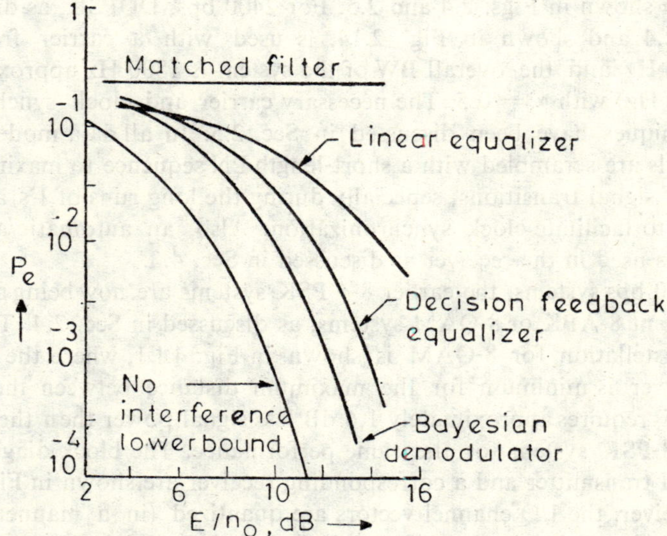

Fig. 4.10 Comparison of P_e vs E/n_0 for various equalizers and an ideal receiver

and is considerably superior to that of the linear equalizer. However, the Bayesian receiver is impractical, whereas the decision feedback equalizer is easily instrumented with digital hardware. The only drawback of the equalizer is that there is a possibility of error propagation, and, sometimes, the errors at the output may occur in bursts. Simulation studies have shown that the equalizer is able to recover spontaneously from this condition and will maintain a superior error performance as compared to a linear equalizer.

4.4 MODEMS FOR VOICE-GRADE CIRCUITS

For voice-grade circuits through long-distance telephone networks (including SSB-MUX channelling equipment), CCITT recommended the following types of modulation: FSK signalling for data rates up to 1200 bps, 4-phase DQPSK for 2400 bps and 8-phase DPSK for 4800 bps and MAPSK for 9600 bps. Till recently, the practice in data modem design has been to use MPSK or MASK-VSB signalling for data rates above 2400 bps [7]. However, the present practice is to use QAM for 4800 bps and QAM/APK for 9600 bps. On the other hand, partial response signalling (PRS) is able to transmit at the ideal rate of 2 bauds/Hz by allowing a controlled amount of intersymbol interference [8]. (PRS techniques will be discussed in the next section.)

The nominal bandwidth available in voice-grade circuits is 300–3400 Hz, but it is practicable to use only the midband of the channel, say, 600–2400 Hz for data transmission, because of the non-linear delay distortion at the band edges. As such, 1200 bps-FSK systems use the signalling frequencies of 1200 and 2200 Hz and each of the receiving filters has a BW of 1800 Hz (with $\alpha = 0.5$) around each carrier. The modulator and demodulator for FSK have been shown in Figs. 2.4 and 2.5. For 2400 bps, DQPSK, as discussed in Sec. 2.4 and shown in Fig. 2.14, is used with a carrier frequency $f_c = 1600$ Hz and the overall BW of the system is 1800 Hz approximately (700–2500 Hz) with $\alpha = 0.5$. The necessary carrier and clock synchronization techniques have been discussed in Sec. 2.5. In all data modems, the data signals are scrambled with a short-length PN sequence to maximize the number of signal transitions, sepecially during the long runs of 1's and 0's, and thus to facilitate clock synchronization. Also, an automatic/adaptive equalizer is used in the receiver as discussed in Sec. 4.3.

For 4800 bps systems, the earlier 8-ϕ PSK systems are now being replaced by equivalent 8-APK or 8-QAM systems, as discussed in Sec. 2.4. The best signal constellation for 8-QAM is shown in Fig. 4.1 1, where the average signal power is minimum for the maximum distance between the signal points. This requires approximately 1.6 dB less signal power then the corresponding 8-PSK system for the same performance. The block diagrams of an 8-QAM transmitter and a corresponding receiver are shown in Fig. 4.12. In the receiver, the I/Q channel vectors are quantized (in a manner somewhat similar to Table 2.1) and then jointly decoded to give the corresponding data. The system uses $f_c = 1800$ Hz and a nominal BW of 2400 Hz

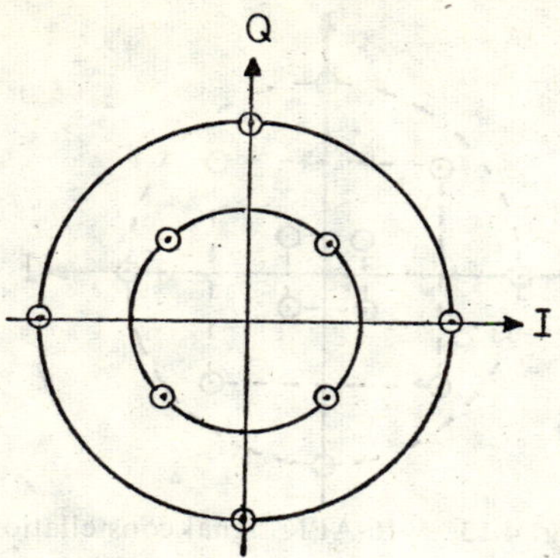

Fig. 4.11 Signal constellation for 8-QAM

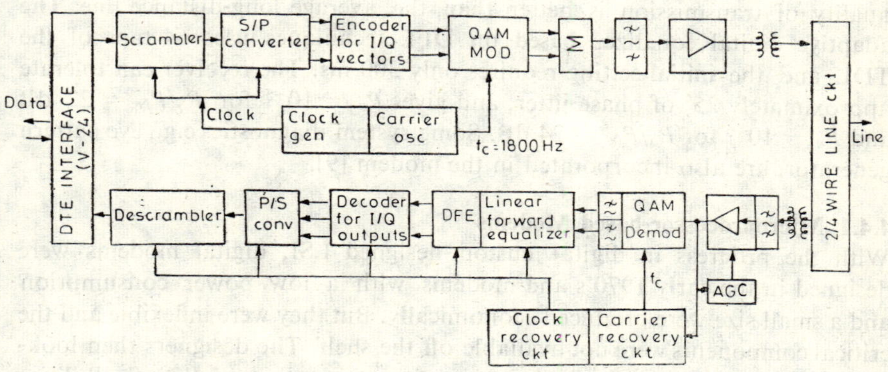

Fig. 4.12 Block diagram of a QAM modem for 4800 bps

(600–3000 Hz) with $\alpha = 0.5$. The receiver of the modem uses a two-section equalizer operated digitally and its initial set-up time is 300 ms approximately. QAM has the added advantage of its insensitivity to channel phase degradation e.g., phase jitter, as compared to an SSB or VSB system; and can easily tolerate 30° *p*-to-*p* phase jitter for $P_e \leqslant 10^{-6}$. The P_S/P_N requirement for $P_e \simeq 10^{-6}$ is 20 dB only, and can tolerate an envelope delay distortion of 3 ms over the useful frequency band (due to the use of the efficient DFE in the receiver path) [9].

In 9600 bps modems for voice-grade channels, it is necessary to transmit 4 bits per symbol, thus making the baud rate as 2400 baud/Hz. Again, QAM and 16-APK are the optimum modulation techniques. The signal constellation for 16-QAM has been shown in Fig. 2.17 and Fig 4.13 shows the signal points for 16-APK. The modem block diagram is similar to that of 4800 bps-modem shown in Fig. 4.12, except that the encoder and decoder now deal with 4 bits per symbol. With $f_c = 1800$ Hz, (or 1700 Hz in some systems) and the signal spectrum restricted to the voice band, the filter roll

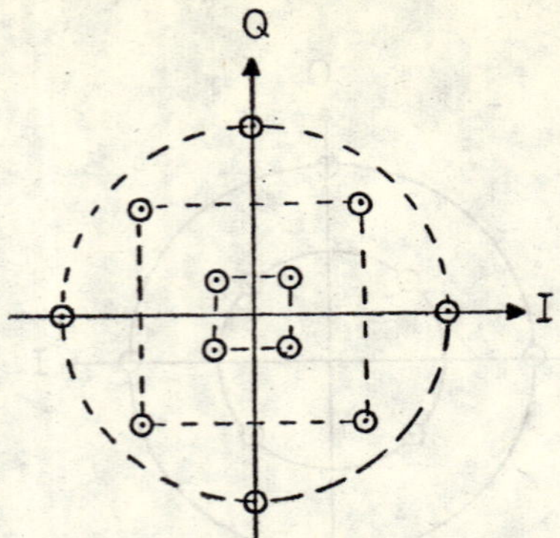

Fig. 4.13 A 16-APK signal constellation

off $\alpha \leqslant 0.2$ only, and the signal is restricted to 400 to 3200 Hz. The system operates over Type 3002, C 2-conditioned telephone circuits, where the quality of transmission is better than the average long-distance line. The adaptive digital equalizer based on DFE techniques uses 64 taps of the TDL and the initial setting requires only 300 ms. The receiver can tolerate approximately 25° of phase jitter, and gives $P_e = 10^{-5}$ for $P_S/P_N = 22$ dB and $P_e = 10^{-6}$ for $P_S/P_N = 24$ dB. Some system diagnostics e.g., eye pattern generator, are also incorporated in the modem [9].

4.4.1 Microprocessor-based Modems

With the progress in digital custom-designed LSI, digital modems were designed in the early 1970's and modems with a low power consumption and a small size were produced economically. But they were inflexible and the critical components were not available off the shelf. The designers then looked into the possibility of using low-cost microprocessors with their flexibilities for a more flexible design of data modems with built-in comprehensive diagnostic facilities. This resulted in a few successful μP-based data modems now available commercially.

All digital data modems were designed by replacing the analog filters and modulators by digital filters and digital multipliers. A new type of digital filter, consisting of a transversal part and a simple recursive network, simplified the hardware considerably. The modem was programmable by changing the contents of the memories and the same hardware could adapt itself to several modulation schemes. The LSI implementation of the hardware made the modems compact and economical [10]. A convenient conceptual model of a versatile data transmitter is shown in Fig. 4.14(a) and its functional diagram for digital implentation is shown in Fig. 4.14(b) [11]. It may be easily shown that the output signal of the transmitter is given by the sample values:

$$S(n\tau_s) = I(n\tau_s)\cdot\cos(\omega_c n\,\tau_s) + Q(n\tau_s)\cdot\sin(\omega_c n\tau_s), \tag{4.22}$$

where

$$I(n\tau_s) = D_I * h(n\tau_s), \qquad \tau_s = \frac{1}{f}$$

$$Q(n\tau_s) = D_Q * h(n\tau_s)$$

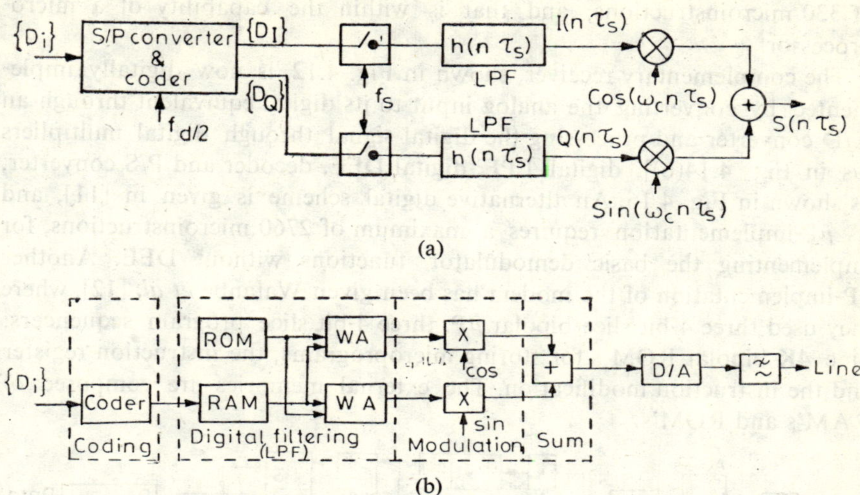

Fig. 4.14 (a) Conceptual model of a data lransmitter and (b) functional diagram of the versatile digital data transmitter

and cos $(\omega_c n\tau_s)$, sin $(\omega_c n\tau_s)$ are the samples of the I/Q carriers. After D/A conversion and simple low pass filtering, the line signal $S(t)$ is generated and transmitted. The technique may be used for QPSK/APK/QAM modulation for data rates of 2400, 4800 and 9600 bps, and the twin digital filters (DF) operate with samples at the rate of 14.4 kHz. The TDL-DF is so designed that there is no need for the output BPF. The digital version of the transmitter, as in Fig. 4.14(b) may, now be implemented with a microprocessor and a simplified block diagram (using Intel 3000) is shown in Fig. 4.15. The complete hardware consists of a CPA consisting of 6 CPE's, an MCU, and an MM of 157 bytes of 32 bits each. The external memory consists of an ROM for the storage of system parameters (356 bytes of 8 bits for filter coefficients, carrier samples, and encoding rules) and a small RAM each for storage of signal points (20 × 4 bits).

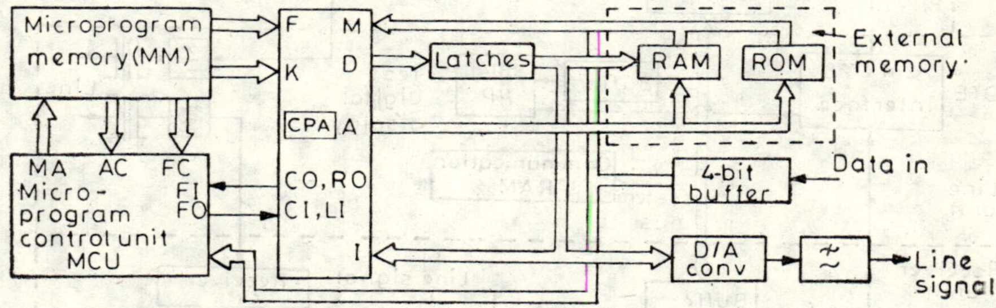

Fig. 4.15 μp-implementation of the data transmitter (CPA: Central processing array)

There are a few latches and a D/A converter which generates the required analog line signal. The calculation of an output sample requires a maximum of 330 microinstructions, and that is within the capability of a microprocessor.

The complementary receiver, shown in Fig. 4.12, is now digitally implemented by converting the analog input to its digital equivalent through an A/D converter and processing the digital signal through digital multipliers [as in Fig. 4.14(b)], digital LPF, digital DFE, decoder and P/S converter, as shown in Fig. 4.16. An alternative digital scheme is given in [11], and its μP-implementation requires a maximum of 2760 microinstructions, for implementing the basic demodulator functions without DFE. Another μP-implementation of the modem has been given Watanbe *et al.* [12], where they used three 4-bit slice bipolar μP, three 4-bit slice program sequencers, nine 4K bipolar ROM's for storing microprograms, the instruction register and the instruction modification. The external memories are composed of RAM's and ROM's.

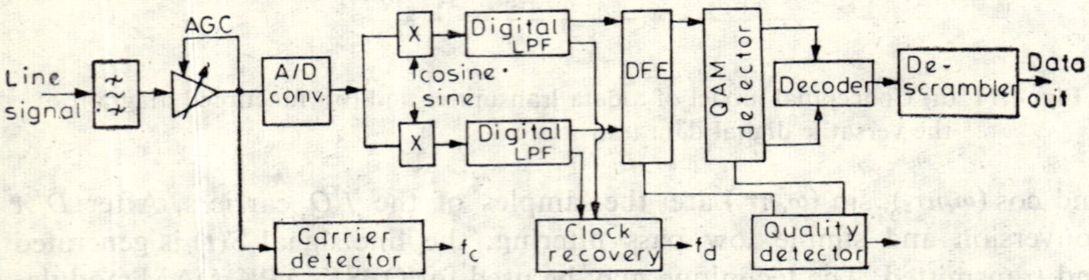

Fig. 4.16 A digital receiver for QAM signals of variable rates

In a recent development of a μP-based variable-rate QAM data modem [13], the main signal-processing functions of the receiver are carried out by a custom-designed array processor and other functions of the modem by three 8085/8048 μP's, as shown in Fig. 4.17. The array processor is built

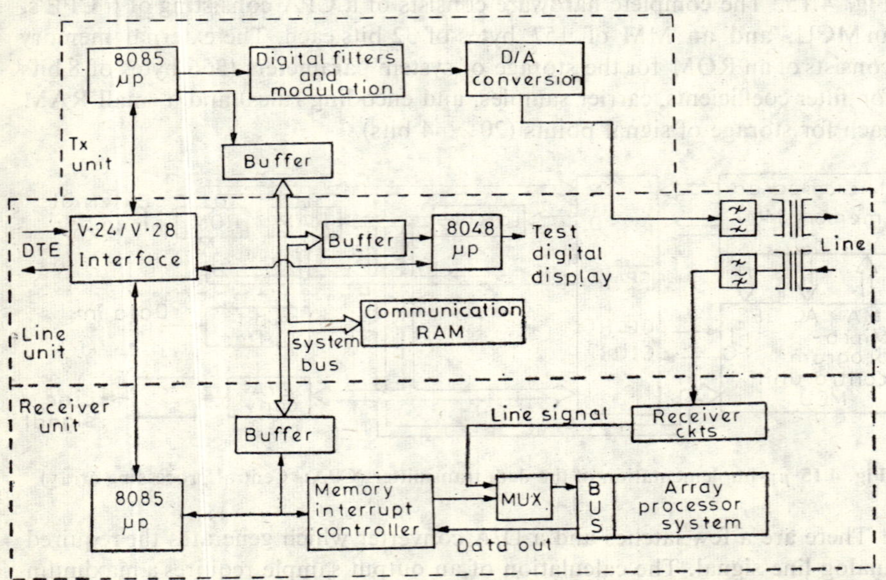

Fig. 4.17 Schematic structure and data flow in the μP-based modem

around four LSI circuits, capable of performing 1.5 million complex 16-bit multiplications per second. A powerful automatic adaptive equalizer using DFE allows data transmission at a high speed (up to 9600 bps) over unconditioned lines, including leased lines and switched lines. The modem operates virtually error-free at the input P_S/P_N of 22 dB at 9600 bps, 18.5 dB at 7200 bps, and 14.0 dB at 4800 bps. Comprehensive diagnostic facilities are built in, giving a 4-digit display for different types of tests, e.g., line levels, error counts, block count, etc. An optional eye pattern generator can be incorporated within the modem, thus providing further performance monitoring. The trend is then surely visible that single-chip modems will be developed soon and be available off the shelf.

4.5 PARTIAL RESPONSE SIGNALLING

In band-limited channels, controlled intersymbol interference may be introduced to reduce the ISI effect and the resultant techniques are known as partial response signalling (PRS) (also known as correlative level coding). Lender, Kretzmer and others [14, 15] have done extensive work on PRS and experimental data sets using PRS-SSB and PRS-FM having a transmission rate of 2 bits/Hz have been developed. Certain 3-level PRS codes, e.g., $(1+D)$, $(1-D^2)$, where D is the delay operator, have been successfully used in voice-grade channels to give maximum data rates and minimum distortion. Futher bandwidth compression of data signals may be obtained through higher-level codes, such as, $(1+2D+D^2)$ or $(1+2D+D^2-D^4-2D^5-D^6)$. The spectrum of PRS codes has a null at $1/2T$, instead of at $1/T$ for binary waveforms (T = bit period), and thus, the bandwidth required for PRS codes is the ideal Nyquist bandwidth. This leads to the promise of transmitting data at the rate of 2 bauds/Hz. Examples of useful low pass PRS codes are shown in Table 4.1, and their pulse shapes along with $F(\omega)$ are shown in Fig. 4.18.

Table 4.1 Low-pass PRS codes, their spectra and SNR characteristics

Sr. No.	$F(D)$	Spectrum: $F(\omega)$	No. of levels received	SNP. degradation over ideal binary in dB
1.	Duobinary$(1+D)$	$2T\cos\frac{\omega T}{2}$	3	2.1
2.	$(1+2D+D^2)$	$4T\cos^2\frac{\omega T}{2}$	5	6.0
3.	$(2+D-D^2)$	$T[2+\cos\omega T - \cos 2\omega T] + jT[\sin\omega T - \sin 2\omega T]$	5	7.2 (1.2 without precoding)
4.	Modified duobinary $(1-D^2)$	$2T\sin\omega T$	3	2.1

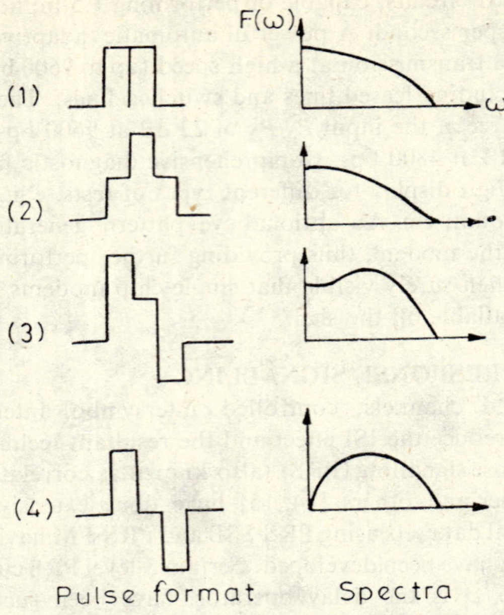

Fig. 4.18 Pulse shapes and corresponding spectra of PRS codes of Table 4.1

The coding process may be illustrated by considering the transmission of the data input $\{D_i\} = \{0110101110010\}$ by using the duobinary code $(1 + D)$. Since each pulse is represented as a twin pulse, the coded output is obtained through the linear sum $(d_n + d_{n-1})$, as shown in Fig. 4.19(a), and the resultant coded output is a 3-level signal given by:

$$\{C_i\} = \{01211112210110 \ldots\}$$

In the receiver, $\{C_i\}$ may be decoded through an inverse coder with a transfer function $H(D) = 1/(1 + D)$ as shown in Fig. 4.19(b). However, this feedback decoder propagates noise pulses, and, as such, a simple multilevel decision circuit is preferred as a decoder. But to have proper correspondence between 1/O data and the received levels, the data bits are precoded in the transmitter using a mod-2 coder, shown in Fig. 4.19(c), where the coder output

$$b_i = d_i \oplus b_{i-1}$$

Now the $\{b_i\}$ is coded with the $(1 + D)$ code, and the resultant $\{C_i\}$ is a 3-level code, whose (0/2) levels (or $\pm$ 1 levels if $\{C_i\}$ is made d.c.-free) correspond to 0's in the data bits $\{D_i\}$ and 1-level (or 0-level in the balanced case) corresponds to 1's of $\{D_i\}$. The complete codec is shown in Fig. 4.20, where the coded $\{C_i\}$ is filtered and transmitted to the line. In the receiver, a fullwave rectifier and a limiter recover the data $\{D_i\}$. The codecs for other codes of Table 4.1 are similar in principle, except that in 5-level codes

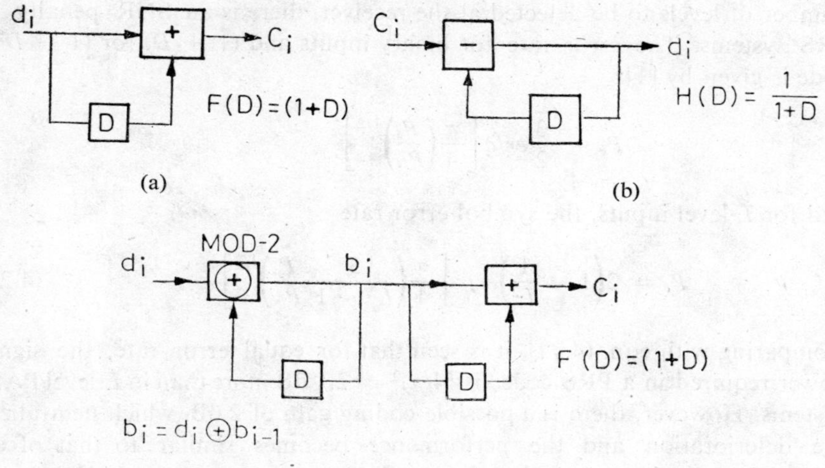

Fig. 4.19 (a), (b) and (c): Coders for the duobinary code

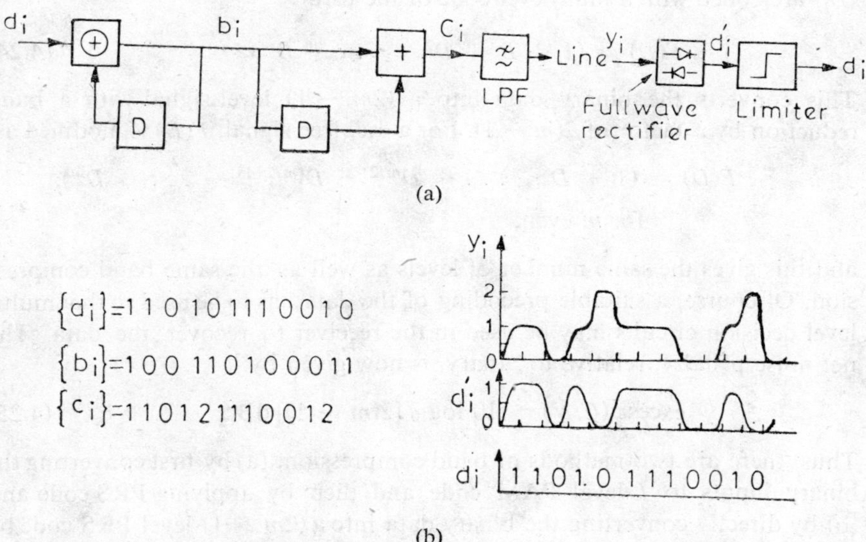

Fig. 4.20 (a) Baseband doubinary codec and (b) coded data bits and decoder waveforms

(Nos. 2 and 3), two successive fullwave rectifiers (or 5-level decision circuits) will be needed to recover $\{D_i\}$. A comprehensive discussion on precoders and spectra of multilevel PRS codes has been given in [16].

Duobinary $(1 + D)$ and other PRS codes can be easily extended to multi-level inputs. For an L-level input and duobinary code, the number of received levels will be $(2L - 1)$, e.g., for binary input $\{d_i\}$, the received $\{y_i\}$ are 3-level; for 4-level d_i's, y_i's have 7 levels and so on. Because of the larger number of levels to be detected at the receiver, there is an SNR penalty in PRS systems. The error rate for binary inputs and $(1 + D)$ [or $(1 - D^2)$] code is given by [1]:

$$P_e = \frac{3}{2} erfc \left[\frac{\pi}{4} \left(\frac{P_s}{P_N} \right)^{1/2} \right]$$

and for L-level inputs, the symbol error rate

$$P_e = 2\left(1 - \frac{1}{L^2}\right) erfc \left[\frac{\pi}{4} \left(\frac{3}{L^2 - 1} \cdot \frac{P_s}{P_N} \right)^{1/2} \right] \quad (4.23)$$

Comparing with eqn. (4.14), it is seen that for equal error rate, the signal power required in a PRS codec is $(4/\pi)^2 = 2.1$ dB more than in L-level PAM systems. However, there is a possible coding gain of 2 dB, which neutralizes the deterioration and the performance becomes similar to that of an L-level PAM system, with the advantage of spectrum economy [14].

Lender has further suggested that considerable bandwidth compression of data signals is possible by using the polybinary concept, where the data bits $\{D_i\}$ are coded with a multilevel code of the form:

$$F(D) = (1 + D + D^2 + \ldots + D^m) \quad (4.24)$$

This converts the binary data into a $(2m - 1)$ level signal with a band reduction by a factor of $2(m - 1)$. For a d.c.-free signal, $F(D)$ is modified as:

$$F(D) = (1 + D + \ldots + D^{m/2} - D^{(m/2+1)} - \ldots - D^m),$$
$$\text{for } m \text{ even,}$$

and this gives the same number of levels as well as the same band compression. Of course, a suitable precoding of the data has to be used so that multilevel decision circuits may be used in the receiver to recover the data. The net noise penalty, relative to binary, is now given by:

$$\text{excess } (C/N) = 10 \log_{10} [2(m - 1)] \text{ dB} \quad (4.25)$$

Thus, there are two methods of band compression: (a) by first converting the binary inputs to L-level PAM code and then by applying PRS code and (b) by directly converting the binary data into a $(2m - 1)$ level PRS code by using eqn. (4.24).

4.5.1 Application in Voice-grade Circuits

Although a combination of PRS and SSB provides an excellent method of data transmission at the Nyquist rate, such data sets are rarely used in the available voice-grade circuits, because of the delay distortion of these channels at both lower and higher ends of the band. It is, however, possible to transmit data with PRS coding directly through voice-grade circuits. Because of the bandpass nature of the channels, it is necessary that the PRS system polynomial be of the form $F(D) = (1 - D)^n \cdot (1 + D)^m$. Of these codes, the modified duobinary code $(1 - D^2)$ is the simplest and has a spectrum given by:

$$|F(\omega)| = |G(f)| \cdot 2T \sin \omega T, \qquad \omega \leqslant \pi/T,$$

where T = bit duration = $1/f_r$, f_r = signalling rate and $G(f)$ the spectrum of the random input. The other higher-order codes suitable for the voice-grade channels are given by:

(a) $F(D) = 1 - 2D^2 + D^4,$

$$= (1 - D^2)^2$$

$$|F(\omega)| = |G(f)| \cdot 4T \sin^2 \omega T$$

(b) $F(D) = 1 - 2D + 2D^3 - D^4,$

$$= (1 - D)^3(1 + D),$$

$$|F(\omega)| = |G(f)| \cdot 8T \sin \omega T \sin^2 \frac{\omega T}{2}$$

(c) $F(D) = 1 - 3D^2 + 3D^4 - D^6,$

$$= (1 - D^2)^3$$

$$|F(\omega)| = |G(f)| \cdot 8T \sin^3 \omega T \tag{4.26}$$

It is observed that the balanced d.c.-free codes are suitable for bandpass channels and, with the higher-order codes, the spectrum is further concentrated at the midband only, resulting in less distortion. However, the number of received levels now increases, and the decoding and equalizing processes are somewhat complex [16].

It has been shown theoretically and experimentally by various authors [14] that the P_e vs. SNR characteristics deteriorates for higher-order codes. Kabal and Pasupathy [17] have, however, shown that it is not absolutely essential to have precoded PRS signalling with bit-by-bit detection at the receiver and, in special cases, using a partial response decoder suggested by them and no precoding, the SNR degradation is less. It is particularly noted from their results that the SNR degradation can be minimized if the highest value sample of the PRS code can be taken as the reference and other lower-value samples are equalized in the receiver by using the technique of the decision feedback equalizer [18]. Such a decoder is shown in Fig. 4.21, where the pre-echoes are equalized first in the feed-forward section and then the trailing response is equalized through DFE. As an example, for the PRS coder using the 7-level code $(1 - 2D + 2D^3 - D^4)$, the noise margin deterioration with the PRS decoder of Fig. 4.21 is only 4 dB. For an equivalent 7-level PRS code using precoding and multilevel decision circuit as the decoder, the noise margin deteriorates by 7 dB [19].

Some interesting experimental results with a data transmission set using directly the PRS codes: $(1 - D^2)$, $(1 - 2D + 2D^3 - D^4)$ and $(1 - 3D^2 + 3D^4 - D^6)$, were obtained through voice-grade circuits (including SSB-MUX channelling equipment) [20]. The receiver used the decoder shown in Fig. 4.21 (without precoding of the transmitted signals) and an adaptive equalizer along with a compromise pre-equalizer. The decoder and pre-equalizer converted all higher-order codes to $(1 - D^2)$ code before being sampled for

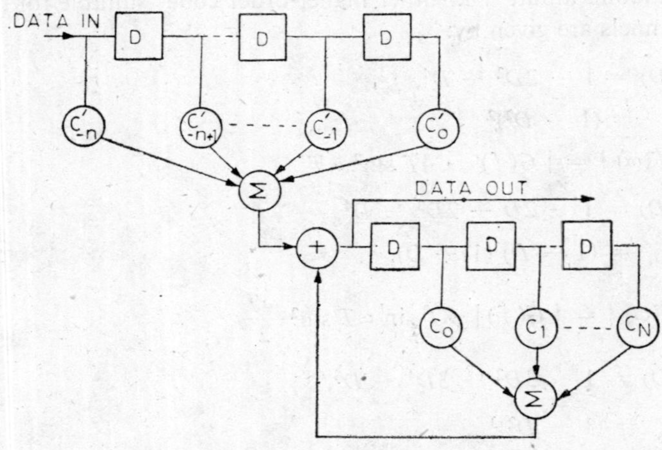

Fig. 4.21 Feed-forward-cum-feedback decoder for higher-order codes

adaptive equalization. This simplifies the decoder hardware and makes the receiver rather robust. The transmission rate realized with excellent performance was 6.4 Kbps and for 7.2 Kbps, the performances degraded marginally. Table 4.2 shows some of the experimental results where it is seen that an average of 20% tolerance in speed and phase is easily realizable in practical systems. The error rate performances are seen to be better than those shown by Lender [14]. It is thus seen that 3-, 7- and 9- level PRS coders are suit-

Table 4.2 Speed and phase tolerance, and P_e vs. P_S/P_N at 6.4 Kbps

Code $F(D)$	Equalization strategy	Mean-speed deviation (% 1/T) allowed for % eye closure of:		Mean-phase deviation (% T) allowed for % eye closure of:		Required P_S/P_N for the given P_e	
		50%	100%	50%	100%	P_e=10^{-3}	P_e=10^{-5}
		%	%	%	%	dB	dB
$(1 - D^2)$	Pre-equalized	15	25	14	25	14.4 15.0*	17.8 18.2*
$(1 - 2D + 2D^3 - D^4)$	Post-equalized	22	28	17	22	17.3	20.4
$(1 - 3D^2 + 3D^4 - D^6)$	Post-equalized	26	30	16	20	19.4	22.4

*at the rate of 7.2 Kbps

able for data transmission at the ideal Nyquist rate, even though the channel is of the bandpass type. Although the 3-level code is simple to implement, the higher-order codes, where the spectrum is further concentrated at the midband only, are more suitable for channels with higher high-pass cutoff, and give lesser distortion in the received code. The technique and codes suggested here will, then, be useful to provide optimum data rates through voice grade circuits without using any frequency translation of data signals as discussed in Sec. 4.4, thus simplifying the modem along with its synchronizing and equalizing circuits.

4.6 MULTIPLEXING

The traditional methods of sharing wideband channels among multiple subscribers have been circuit switching and message switching (with store and forward operation). Access to the common channel may be provided on the basis of (a) permanent assignment, (b) demand assignment and (c) random access. These are achieved mainly by using frequency division (FDM), time division (TDM) and code division multiplexing (CDM) techniques. Historically, the analog multiplexing technique using FDM has been perfected first. We have now a hierarchy of FDM systems going up to 10,800 circuits, using a bandwidth of 60 MHz approximately in the L_5 system of the Bell telephone [21]. The intermediate systems are: 3 MHz L_1 for 600 channels; 4 MHz P-4 MTr for 960 ch.; 12 MHz C-12 MTr for 2700 ch. and 18 MHz L_4 for 3600 ch. as shown in Fig. 4.22. Over the last two decades, there is a shift from the coaxial cable system to the radio relay system and, as of today, 62 per cent of the carrier circuit miles is through microwave relay in the 4, 6 and 11 GHz bands. It is now possible to utilize many of these wideband systems for transmitting Mbits of digital data using a suitable interface.

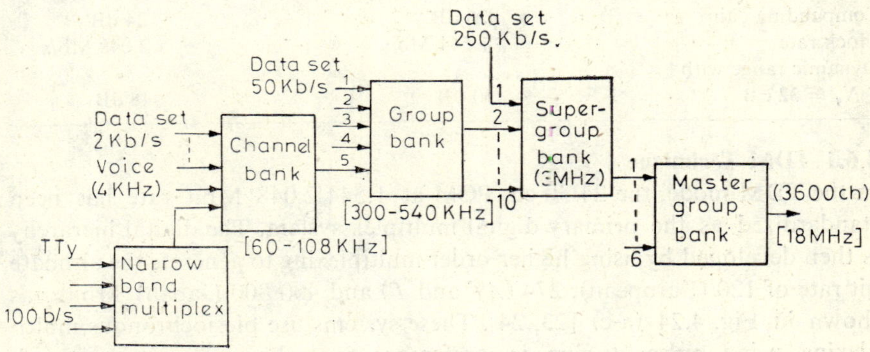

Fig. 4.22 Hierarchy of long-haul FDM transmission system for speech and data

For multiplexing digital voice, 24-ch./30-ch. log-PCM codecs have been developed. Such a 24-ch. codec is shown in Fig. 4.23, where two compressor-encoders are used alternatively for odd and even channels, thus avoiding cross-talk between channel samples [22]. The relevant parameters of the two

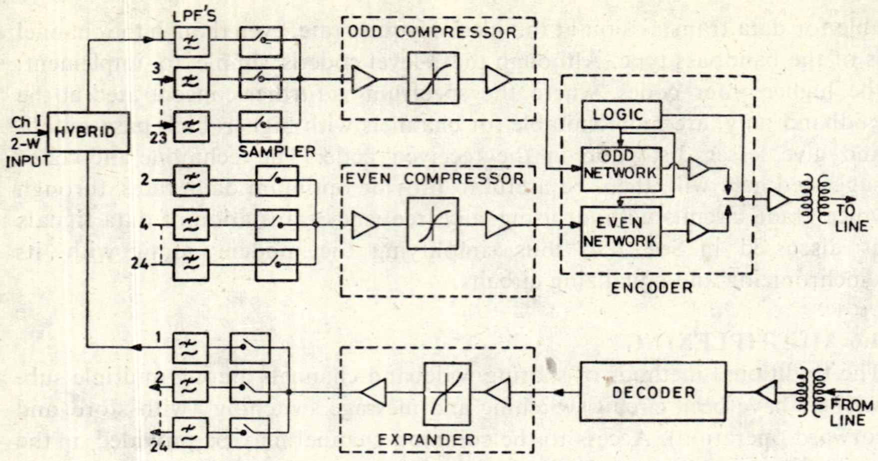

Fig. 4.23 24-channel log-PCM system

systems are shown in Table 4.3. The higher-order PCM-MUX systems are based on these basic MUX systems.

Table 4.3

System parameter	24-ch. PCM	30-ch. PCM
Sampling rate	8 kHz	8 kHz
Frame time	125 μs	125 μs
No. of bits per sample	7 + 1	8
Signalling	8th bit	31st channel
Synchronization	193rd bit in the frame	32nd channel
Companding	μ – law, $\mu = 100$	A – law, $A = 87.6$
Companding gain	26 dB	24 dB
Clock rate	1.544 Mb/s	2.048 Mb/s
Dynamic range with $S/N_q = 32$ dB	50 dB	48 dB

4.6.1 TDM Technique

In the TDM mode, the 24/30 ch. PCM at 1.544/2.048 Mbit rate has been standardized as the primary digital multiplex system. The digital hierarchy is then developed by using higher-order multiplexing to generate the ultimate bit rate of 120 (European), 274 (*AT* and *T*) and 400/800 (Japan) Mbits, as shown in Fig. 4.24 (a-c) [23, 24]. These systems use plesiochronous multiplexing using pulse stuffing in preference to techniques using network synchronization. A supergroup or master group FDM signal may be inserted into the digital hierarchy at an appropriate point using a suitable wideband codec, e.g., a transmultiplexer [25]. These codecs usually convert FDM signals to PCM having (10 + 3) or more digits and are of special design. Since binary codes are not suitable for transmission in cables or radio systems, d.c.-free codes, e.g., ternary or multilevel, are used.

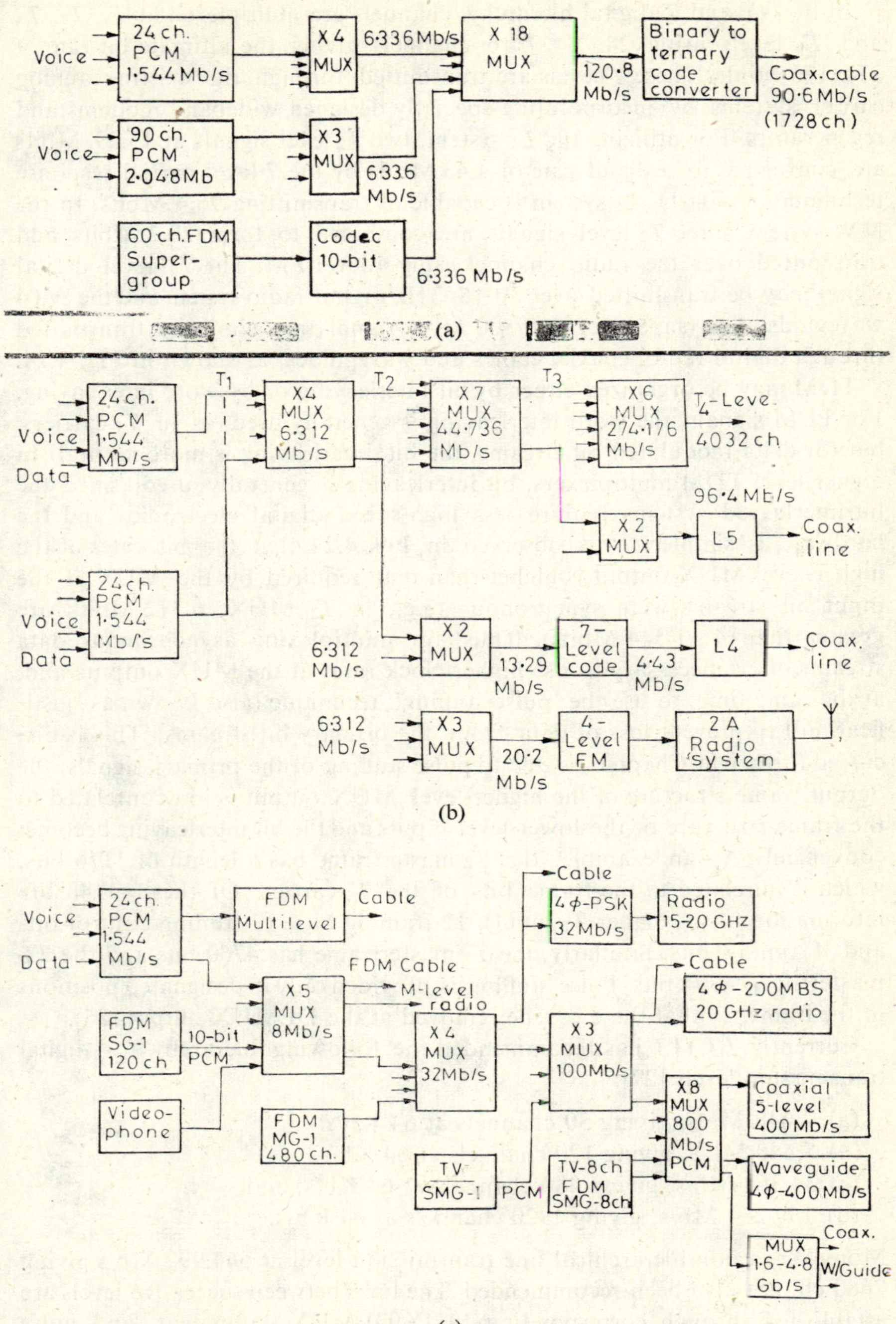

Fig. 4.24 Hierarchy of digitally multiplexed systems: (a) 120 Mb/s coaxial cable system; (b) *AT* and *T* digital hierarchy; and (c) digital hierarchy in Japan

In the AT and T digital hierarchy, channels are multiplexed at T_1, T_2, T_3 and T_4 levels using $24\times4\times7\times6$ channels giving the ultimate bit rate of 274·176 Mbits. These T-levels are transmitted through the existing analog carrier systems by incorporating specially designed wideband modems and regenerators. For utilizing the L_4 system, two T_2 level signals at 13.29 Mbits are converted to a baud rate of 4.43 MBds by the 7-level partial response technique. Similarly. L_5 system is capable of transmitting 96.4 Mbits. In the MW system, three T_2 level signals are combined to form 20.2 Mbits and transmitted over the radio channel using 4-level FM. The T_4 level digital signal may be transmitted over an 18 GHz digital radio system and the WT4 waveguide. The (Japanese) 400/800 Mbit signals are similarly transmitted through digital radio, coaxial cables and waveguides, as shown in Fig. 4.24.

TDM may be organized either by bit interleaving or by word interleaving. For PCM signals, the word interleaving is generally used, as in T_1-carriers; but for delta-modulated bit streams, the bit interleaving is more natural. In higher level TDM multiplexers, bit interleaving is generally used, since the bit-interleaved systems require less high-speed digital electronics and the hardware is simpler. It is observed in Fig. 4.24 that the bit rates at the higher-level MUX output is higher than that required by the MUX if the input bit streams were synchronous (e.g., in T_2 MUX, 6.312 Mbits are greater than 4×1.544 Mbits). Thus, for multiplexing asynchronous data streams, it is necessary to use higher clock rates at the MUX outputs and, at the same time, to use the 'pulse-stuffing' technique (also known as 'justification') to prevent loss of data from the primary bit streams. This is discussed further in Chapter 7. Due to pulse stuffing of the primary signals, the output frame structure of the higher-level MUX output is now unrelated to the frame structure of the lower-level inputs and the bit interleaving becomes convenient. As an example, the T_2-masterframe has a length of 1176 bits, which is unrelated to the frame bits of the T_1-carrier. Of these, 1148 are information bits (287 per T_1-input), 12 framing bits, 12 stuffing control bits and 4 timing bits. Similarly, the T_3-masterframe has 4760 bits and the T_4-masterframe 4704 bits. Pulse stuffing is provided only in designated positions in the frame, so that these can be removed at the DEMUX output [26].

Currently, CCITT has recommended the following hierarchy for digital transmission levels [27]:

(a) 2.048 Mb/s, giving 30 channels at 64 Kb/s;
(b) 8.448 Mb/s, giving 120 channels at 64 Kb/s;
(c) 34.368 Mb/s, giving 480 channels at 64 Kb/s; and
(d) 139.264 Mb/s, giving 1920 channels at 64 Kb/s.

Moreover, a non-hierarchical line transmission level at 564.992 Mb/s giving 7680 ch. has also been recommended. The links between successive levels are established through corresponding MUX/DEMUX equipment, and pulse stuffing is used as discussed above.

4.6.2 Synchronization

For optimum detection of pulsed (data) signals, it is necessary to have an accurate estimation of the carrier frequency and phase, data timing, etc. In general, the synchronization requirements are for:

(a) carrier phase and frequency;
(b) clock timing;
(c) word/code timing; and
(d) frame timing.

Carrier synchronization techniques are based on the classical theory of the phase-lock loop and the various modifications of the loop are: (a) squaring loop; (b) Costa's loop; (c) decision directed feedback loop; (d) baseband modulation reconstruction loop; and (e) data-aided loop. These have been discussed in Sec. 2.5.1.

Bit/symbol synchronization generally uses the principle of maximizing the a posteriori probability to obtain the estimate of the epoch or the time uncertainty from the start of the observation to the first bit transition point. These MAP synchronizers have structures known as: (a) Data Transition-Tracking loop (DTTL); (b) Differential Coincidence Type Bit Synchronizer (DCS); (c) Early-Late-Gate Bit Synchronizer (ELGS); and (d) Filter and Square Bit Synchronizer (FSTL) [28]. In terms of the tracking error, the more complicated circuits, as DTTL and ELGS, are superior to FSTL by only a few per cent, but FSTL has the advantage of much simpler implementation. Some of these synchronizers have been discussed in Sec. 2.5.2.

4.6.3 Line Codes

For efficient synchronization, it is necessary to have as many 1/0 transitions as possible. Moreover, repetitive data patterns generate line spectra which are undesirable from the interference point of view. It is, therefore, necessary to randomize the input data patterns, and this is accomplished by using scramblers at the input of the data modems and multiplexers. In a particular system, the scrambler used is a PN-sequence generator whose generator polynomial is:

$$f(X) = 1 + X^6 + X^7$$

This PNS multiplies the data at the transmitter and divides the received signal in a self-synchronized mode at the receiver to generate the original data.

Data scramblers, however, do not prevent long string of zeros in the transmitted sequence. They simply ensure that the relatively short repetitive patterns are randomized and avoid the line spectra in the cable/radio medium. To avoid long strings of zeros and to ensure the removal of d.c. from the transmitted signal, it is necessary to use further coding of the bit streams before transmission in the channel. The techniques, known as Line coding, normally insert a '1' whenever three or more zeros occur in succession. Further, binary to ternary encoding is used to obtain a balanced signal and also to compress the signal bandwidth. Some examples of line codes used in practice along with other specifications are given in Table 4.4 [26].

Table 4.4 Examples of line codes

System designation	Transmission rate (Mb/s)	Line codes	Media	Repeater spacing
AT and $T-T_1$	1.544	Bipolar (AMI)	Twisted pair	6000 ft
AT and $T-T_2$	6.312	B6ZS	Low capacitance twisted pair	14,800 ft.
AT and $T-T_4$	274.176	Polar binary (NRZ)	Coaxial cable	5700 ft.
30-channel CCITT	2.048	HDB3	Twisted pair	2 km
GTE-9148A	3.152	PRS $(1-D^2)$	Twisted pair	6000 ft.
Canada-LD-4	274.176	B3ZS	Coaxial cable	1.9 km
CCITT	8.448	HDB3	Low capacitance cable	—
,,	34.368	HDB3/4B−3T (MS43)	Coaxial cable	2 km
,,	139.264	CMI/4B−3T/ 6B−4T	,,	2/4.5 km
,,	564.992	AMI	,,	1.55 km
8TR 609 Phillips	140	CMI/4B-3T	Coaxial cable	2/4.5 km
39-STC	140	CMI/7B8B	Optical fibre	9 km

In the T_1-carrier, the d.c.-wander problem was solved by using the 'alternate mark inversion' (AMI) (also known as Bipolar coding), where an input binary data $\{D_i\}$ is converted to a pseudo-ternary code using the rule:

A logic '0' is encoded with zero voltage, while a logic '1' is alternately encoded with positive and negative voltages. Thus, an input $\{D_i\}$ = {101100101} is coverted to a ternary signal $\{T_i\} = \{+ 0 - + 00 - 0 +\}$. The coding also reshapes the signal spectrum such that there is a peak at $f = 0.5/T$ instead of at $f = 0$, which helps in recovering the clock easily. The 'bipolar violation' in the line also provides for monitoring the channel conditions.

The AMI encoding does not solve the problem of the undesirable long strings of zeros, and it is then necessary to use zero-substitution codes, known as binary N-zero substitution (BNZS) [29], to increase the number of transitions in the data streams. The BNZS, generally used are B3ZS, B6ZS, where three or six consecutive zeros are replaced by a ternary code. The BNZS codes are used along with AMI, and as such, bipolar violations occur occasionally. As an example of B3ZS, the string '000' is replaced by one of the codes $(00\,-)$, $(+\,0\,+)$, $(00\,+)$, $(-\,0\,-)$, depending upon the polarity of the preceding pulse and the number of bipolar pulses since the last substitution. Consequently, the $\{D_i\}$ = {10100011100001}, is converted by the B3ZS code as, $\{T_i\} = \{+ 0 - 00 - + - + 00 + 0 -\}$. Similarly in B6ZS, the six zeros are replaced by either {0 − + 0 + −} or {0 + − 0 − +), depending upon the polarity of the immediately preceding pulse. Another popular BNZS is the high density bipolar coding, known as HDB3, where strings of four zeros are replaced by one of the codes (000 −), (+ 00 +), (000 +), (− 00 −), depending upon the polarity of the preceding pulse and the number of bipolar pulses in between substitutions. This also produces bipolar violations in the resultant signal bits.

In the above AMI and BNZS coding, the higher transmission capacity of ternary codes is not utilized, but only the efficient timing information is used. To take advantage of the higher information rate in ternary codes, the 4B-3T and 6B-4T codes are used for transmission in coaxial cables as indicated in Table 4.4. In 4B-3T code, 4 binary bits are coded into 3 ternary bauds, thus saving 25% of the required bandwidth. Similarly, in the 6B-4T code, 6 binary bits are coded into 4 ternary bauds, giving a bandwidth saving of 33%. In all these teranry codes, care is taken to maintain the d.c. balance.

The other binary codes used in practice are Digital biphase (also known as diphase or Manchester) code and Coded mark inversion (CMI). In the diphase code, the binary bits are multiplied by a square wave of the period equal to the bit period T. Thus a given $\{D_i\} = \{110010110\}$ is coded as $\{C_i\}$ =101001011001101001}, where the C_i bits have a bit time = $T/2$. The code $\{C_i\}$ has strong timing transitions and there is no d.c.-wander. This code has been used in the Ethernet circuits. In CMI, as recommended by CCITT, the zeros are encoded as a half-cycle square wave of one particular phase, whereas the successive '1's are coded as NRZ pulses with opposite phase. As an example, $\{D_i\}$ = {110010110) is coded as: (C_i) = (110001011101001101) with the new bit duration = $T/2$. CMI is used as an interface code for 140 Mb/s systems.

4.6.4 Hierarchical Multiplexers

Based on the CCITT recommendations and the above discussions on line codes, standard MUX/DEMUX units are now manufactured by various telecommunication industries. The basic units are for 8/34/140 Mb/s and 565 Mb/s units are also available. The block schematic of a MUX/DEMUX unit is shown in Fig. 4.25, where the major emphasis is on the synchronizing circuit using 'pulse-stuffing' technique [27]. The multiplexing is based on bit interleaving and the 1536-bit frame contains 'housekeeping' bit groups including the frame alignment code and the pulse-stuffing control bits. The line code for the 2/8/34 Mbit streams is HDB 3 and for the 140 Mbit stream, the code is CMI as shown in Table 4.4. The line interface circuits perform the decoding and coding accordingly, and the crystal clock provides the necessary timings with tolerances of $\pm 10^{-5}$ approximately as required. At the receiver side, the frame alignment code is used to synchronize the composite binary signal with the receiver timing circuits. After demultiplexing at the synchronized bit rate, the individual bit streams are

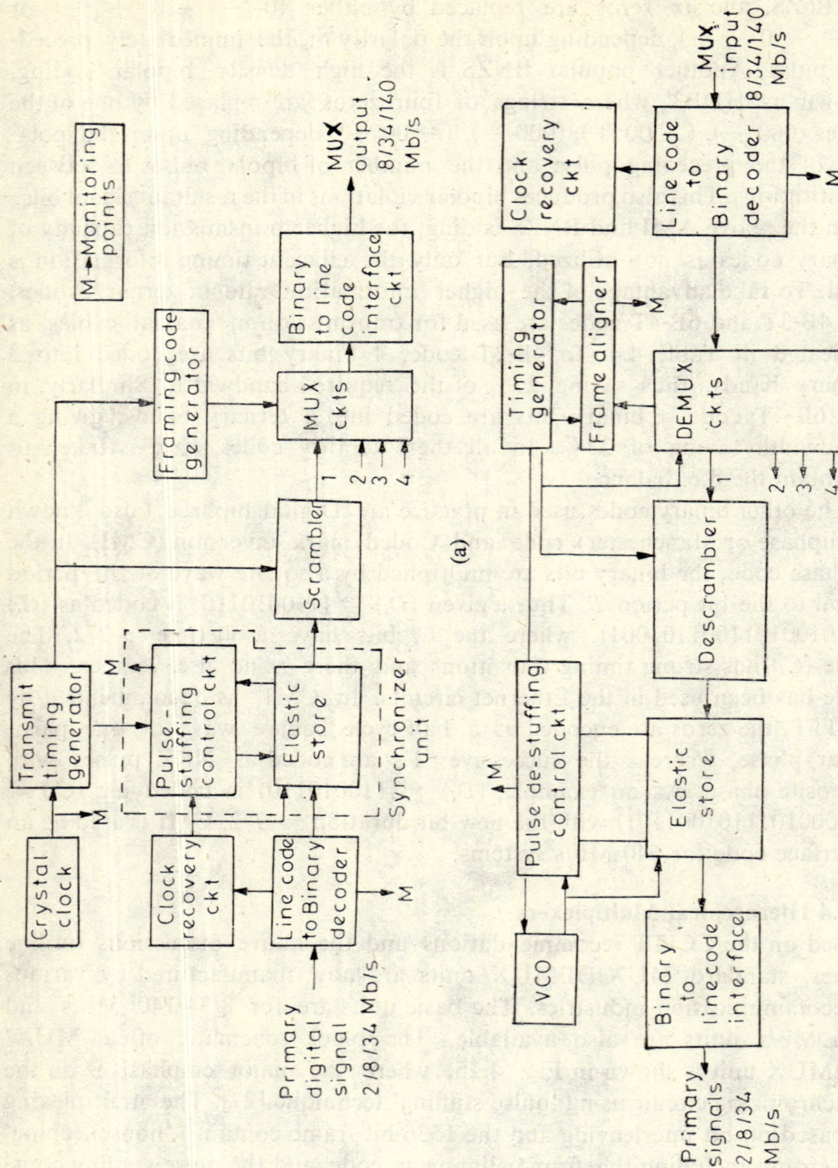

Fig. 4.25 MUX-DEMUX units for digital signals: (a) MUX unit and (b) DEMUX unit

de-stuffed in the elastic store and read out at a constant rate determined by the VCO. Finally, the recovered primary signals are converted to the HDB3 code.

The 140/565 Mbit MUX/DEMUX unit is similar to that shown in Fig. 4.25, except that the line code is CMI at the trans-input and AMI at the trans-output. The frame length is 2688 bits and contains the frame alignment and othe housekeeping bits. The frame alignment code is (111110100000), and is generated at the start of each frame. Because of the high frequencies involved, the major part of multiplex functions are implemented with LSI circuits and the circuits at higher frequencies (> 500 MHz) are implemented through 'macrocell' array IC's.

4.7 DIGITAL COAXIAL CABLE SYSTEMS [31, 32]

Historically, PCM transmission at 1.544 Mb/s was introduced for inter-exchange trunk traffic over the older audio cables with repeaters placed at the points where the loading coils were installed. With the growth of higher-order digital systems, it was necessary to introduce coaxial cables for digital transmission and, quite often the cables, used for high-capacity analog FDM systems, were reused for high capacity digital systems as well. Table 4.4 indicates that coaxial cable systems are now manufactured for transmission rates of 34–565 Mb/s, with repeater spacings of 2/4.5 km.

Two standard types of coaxial cables are normally used and their specifications are:

(a) 1.2/4.4 mm type with the loss α in dB $= (0.066 + 5.15\sqrt{f} + 0.0047f)$ dB/km and a temperature coefficient of 0.202% per °C.

(b) 2.6/9.5 mm type with α in dB $= (0.013 + 2.305\sqrt{f} + 0.003f)$ dB/km and a temperature coefficient of 0.19% per °C (the frequency f is in MHz).

It is easily estimated that the smaller coaxial cables gives a loss $\alpha \simeq 84$ dB for a 2-km length of the cable at 52.2 MHz which is the Nyquist bandwidth of the 140 Mb/s system. Thus. the repeater spacings are 2 km for smaller cables and 4.5 km for larger cables. For the 565 Mb/s system, the repeater spacing using the larger cable is 1.55 km (as used for 60 MHz analog FDM systems). A micro-coaxial cable with dimensions 0.8/2.9 mm, is used for the 34 Mb/s system and gives a repeater spacing of 2 km. If the 1.2/4.4 mm cable is used for the 34 Mb/s system, then the repeater spacing is 4.2 km.

The cable loss vs. frequency and the phase vs. frequency characteristics are well defined for coaxial cables (except for small changes due to temperature), and a common equalizer easily corrects these deviations. The equalizers are sometimes designed using tapped-delay lines and the overall frequency response is made to have a 100% raised-cosine spectrum. This ensures minimum ISI at the detector input. The cross-talk performance of the cables is extremely good; the far-end cross-talk ratio and near-end cross-talk attenuation are greater than 140 dB for 1–1000 MHz. This enables the same composite cables to be simultaneously used for digital as well as analog MUX systems. Underground coaxial cables are also effectively screened against external electrical interference. The presently installed analog cables have minor degradations due to impedance irregularities, arising out of tail cable mismatch, defects in joints, gas seal, connectors, etc., and these lead to some extra ISI. However, it has been estimated that with proper joints, connectors, etc., these cables can support digital transmission up to 565 Mb/s. In general, the degradations are less in larger cables and at lower transmission rates.

A simplified block diagram of a 140 Mb/s coaxial cable system is shown in Fig. 4.26. The 140 Mb/s CMI-coded multiplexed signal is reconverted to binary, scrambled through a 7/9-stage shift register sequence and then recoded with a 4B-3T ternary code. The transmitted baud rate is now 104.448 Mbaud/s. In a 6B-4T ternary coded system, the transmission rate is 92.8 Mbauds. The transmitter also includes power-feed filters and a monitoring system to indicate any fault in the chain of the coders. In the receiver, the received signal is first amplified and equalized through automatic equalizers. Then complementary decoding and recoding generate the 140 Mb/s CMI signal which is now fed to the DEMUX equipment. The receiver has elaborate monitoring circuits to locate faults on an in-service basis. In the 140 Mb/s system, the repeater spacing is either 2 km for smaller cables or 4.5 km for larger cables and the maximum loss allowed between repeaters is 84 dB at 52.224 MHz. The AGC in the receive amplifier compensates for this loss and the accompanying gain and phase distortions are corrected. The transmit output is ± 6 V raised-cosine pulses. The overall error rate in the system is kept better than 10^{-10} with a noise margin of 5 dB.

In the 34 Mb/s system, used in micro and smaller coaxial cables, the input from the lower-level MUX is HDB3-coded and the line code used is 4B-3T. Thus, the transmission rate is 25.776 Mbauds. The nominal distances between repeaters are 2.25 km and 9.3 km for different cables. However, only the micro and smaller coaxial cables are used for 34 Mb/s systems. The general scheme of the line terminal is similar to that of Fig. 4.25, except that the interface code is HDB3 and not CMI. The transmit level is $\pm$ 3 V and AGC of the receiver compensates for the line loss between 56 and 84 dB. Supervisory facilities for monitoring and fault location are provided.

The 565 Mb/s coaxial system has been introduced in continental countries recently. The system is used along with the larger 2.6/9.5 mm cable with a nominal repeater spacing of 1.55 km, giving a loss of 57–69 dB at 282.5 MHz. The line terminal configuration is similar to that shown in Fig. 4.25, except that the interface and line codes used are AMI, which possesses very low spectral energy at low frequencies and simplifies error detection.

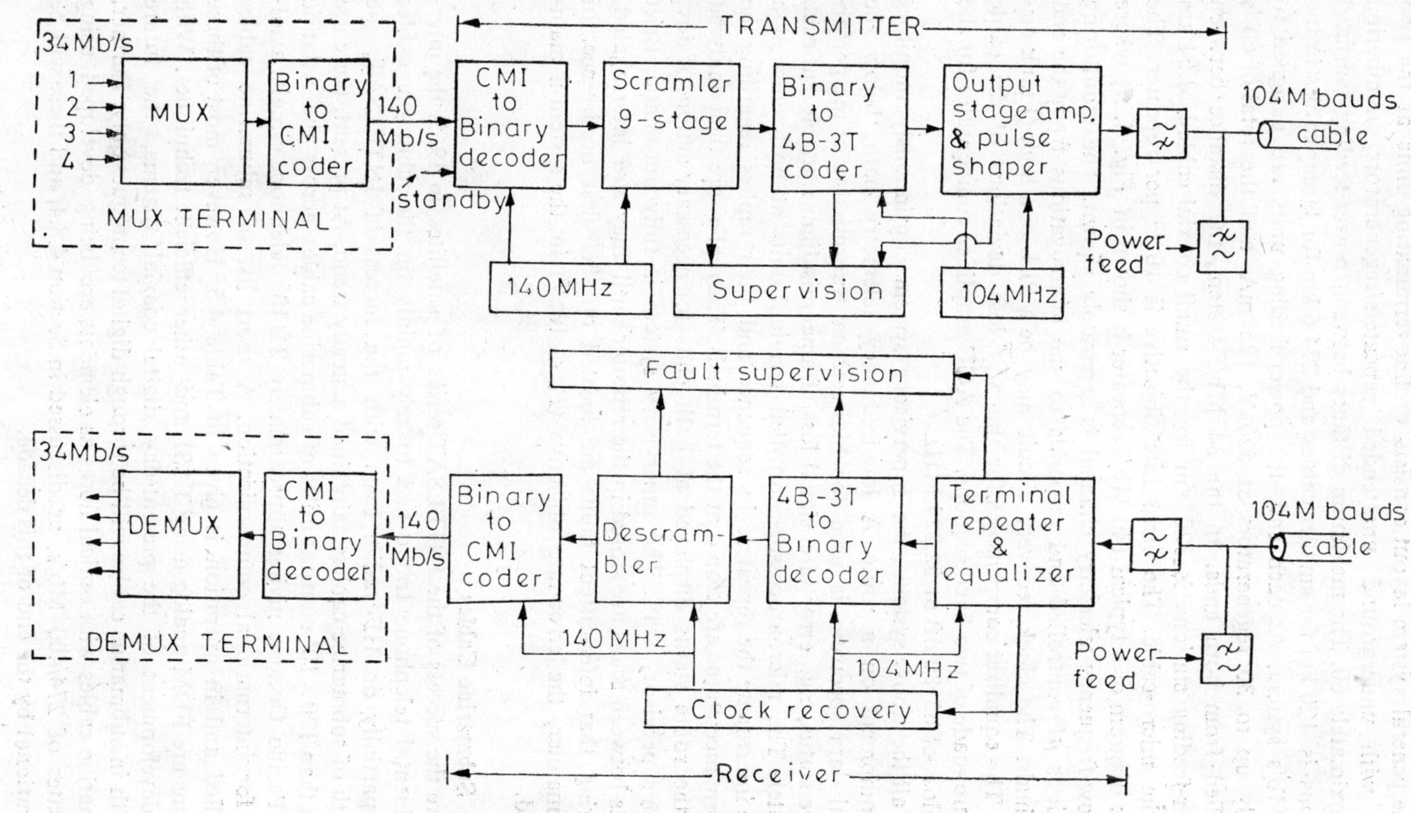

Fig. 4.26 Block schematic of a 140 Mb/s coaxial system

The general system layout consists of line-terminating units at the two ends with underground unattended repeaters/regenerators, numbering approximately 60. The maximum distance between power-feeding (surface) stations is 120 km for smaller cables and 274.6 km for larger cables in the 140 Mb/s system. At each end, the power-feeding units are designed to supply up to 30 regenerators at 520 V, 125 mA, and the total of 60 is supplied from both ends. In the 34 Mb/s system, the distance between power-feeding stations is 120 km for the small coaxial cable and 54 km for the micro-cable. The error rate objective is 10^{-10} per repeater. The block diagram of a typical 140 Mb/s repeater is shown in Fig. 4.27, where the low-frequency telemetry channel is separately shown. The monitoring circuit is μP-controlled and responds to the interrogations from the end terminals. The clock recovery circuit may be a PLL or a high-Q tank circuit. The equalizer compensates for the $\sqrt{f}$ loss deviations of the cable and are made of two fixed sections. The AGC amplifier compensates for the path loss of 65–80 dB at 52.224 MHz.

In all the above systems, the supervisory system continuously monitors all equipment on a route. A low-frequency telemetry path, below the digital data spectrum, is used for bidirectional transmission of the performance status, e.g., error rate, signal loss, frame misalignment, etc., at each repeater. The microprocessor-controlled terminal units at both ends of a route interrogate the repeaters in sequence and their replies regarding the performance status are stored at the terminals. The replies are then analysed and the route status displayed at both ends for necessary action, if any, by the supervisory staff. The supervisory system is fully automatic. Brief events between scans are stored in the repeater until they are interrogated, ensuring that intermittent faults are located on their first occurrence. In this manner, the error-rate objective ($P_e < 10^{-10}$) in the system is maintained.

4.7.1 Submarine Cables

Before the success of the INTELSAT series of satellites in 1965, the intercontinental telephone traffic had to rely mainly on the submarine cables and partially on HF radio. Even with the success of INTELSAT, the growth of submarine cables maintained a steady trend. At present, there are more than 170×10^3 nautical miles of submarine cables across the Atlantic and Pacific Oceans, giving approximately 2×10^5 telephone circuits available for international communication. A short list of submarine cables installed and in operation is given in Table 4.5. However, most of these systems are FDM analog using TASI and other efficient techniques. With the development of the present-day digital coaxial systems, the future growth in submarine cables will be through digital techniques only. Further, submarine cables with optical fibres as elements are being developed using bit rates of 274/400 Mb/s, as discussed in Section 4.8.4, and these will be operational by the end of this decade.

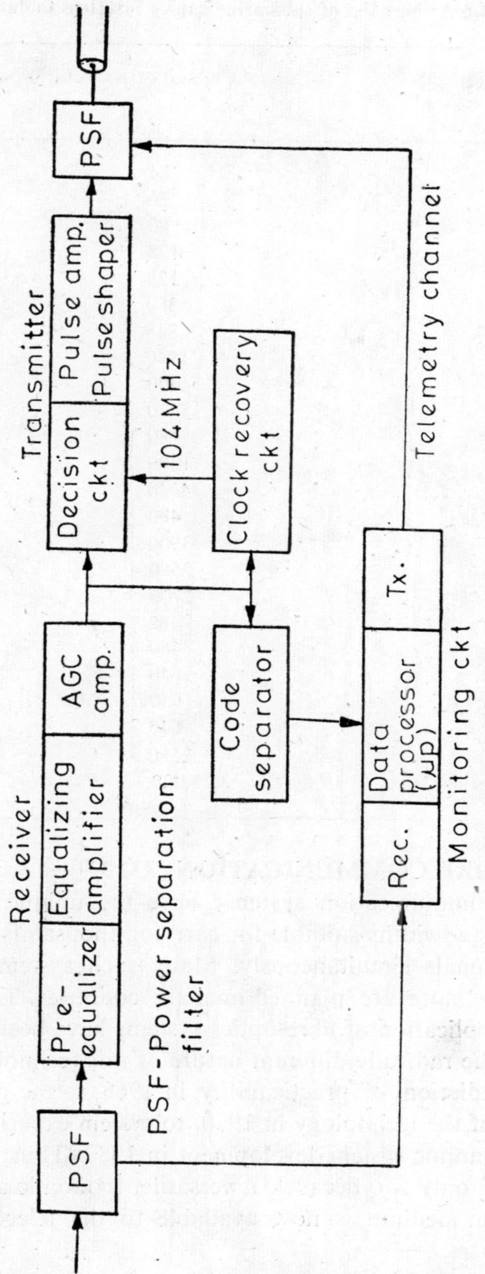

Fig. 4.27 Block diagram of a typical repeater at 140 Mbauds

Table 4.5 A short list of submarine cables installed to date

Cable designation	Capacity (in circuits)	Year of Completion
TAT-1	84	1956
CANTAT-1	80	1961
COMPAC	80	1962
TAT-3	128	1963
TRANSPAC-1	128	1964
SAT-1	360	1966
TAT-5	845	1970
CANTAT-2	1840	1974
TAT-6	4000	1976
PENCAN-3	5520	1977
Ivory Coast–Nigeria	480	1980-81
East–West Malaysia	1200	1980-81
UK–Spain No. 3	4140	1980-81
Singapore–Indonesia	480	1980-81
UK–Denmark No. 3	3900	1980-81
France–Algeria No. 3	2580	1980-81
Japan–Korea	2700	1980-81
Syria–Greece	480	1980-81
IOCOM	480	1980-81
Virgin Islands–Brazil	640	1980-81
Virgin Islands–Venezuela	640	1980-81
Guam–Taiwan	640	1980-81
France–Greece	2580	1980-81
TAT-8 (digital fibre optics)	36,000	1988

4.8 OPTICAL FIBRE COMMUNICATION [33, 34]

The optical fibre communication systems have the unique advantage of having very large bandwidths suitable for carrying thousands of voices and tens of television signals simultaneously. Many such systems are already installed and many more are planned in many countries. The pace of the development and application of fibre-optics systems have been rapid, particularly in view of the radically different nature of this technology; from the first theoretical prediction of practicability in 1966, to the glimmer of the possible feasibility of the technology in 1970, to system experiments in 1976 and to practical economic system development in 1980. Thus, it is seen that in a short span of only two decades, a versatile, economic and very wide-band communication medium is now available to the telecommunication planners and designers.

The fibre-optics communication systems have quite a few technical advantages over the convention systems using metallic (screened pair and coaxial) cables. They are:

(a) Small dimensions and weight of the cable.
(b) Long repeater spacing.

(c) Very wide bandwidth.
(d) Freedom from electromagnetic interference and cross-talk.
(e) More economical for many applications.

As such, these systems are useful for many applications, e.g., submarine communications, intercity long-haul communications, intracity trunking networks, subscriber loops and military communications. It is predicted that at the turn of the century, the statellite systems for fixed services will be replaced by guided-wave systems, specially optical fibres. Satellites, then, will be mainly used for mobile and deep-space communications.

A basic fibre-optics system consists of the transmitter, the fibre-medium and the receiver, as shown in Fig. 4.28. In the transmitter, the coded binary signal modulates, through a driver circuit, the intensity of the output of the optical source (LED or Laser). The source output is coupled into the fibre and the light-wave propagates along it. At the receiver, the optical pulses are converted by a photodetector (PIN or APD) into current pulses which are amplified, equalized and detected as binary signals. This binary signal is then decoded as the received data. In the source drivers, direct current modulation, through a differential amplifier or an emitter follower, is used up to bit rates of 140 Mb/s. At 565 Mb/s or more, the laser source can be driven by a voltage amplifier with a low output impedance ($\simeq$50 ohms). Lasers, available at present, have low threshold currents and a reduced sensitivity to temperature, and, as such, a single control loop on the average emitted power is used for both output stabilization and alarm purposes. In the receiver, APD photodiodes are generally used to achieve the best SNR, and a feedback circuit is used to control the thermal drift and the avalanche gain. Only at the lower bit rates of, say, 1–10 Mb/s, PIN detectors are used to simplify the receiver. The transimpedance amplifier is the preferred

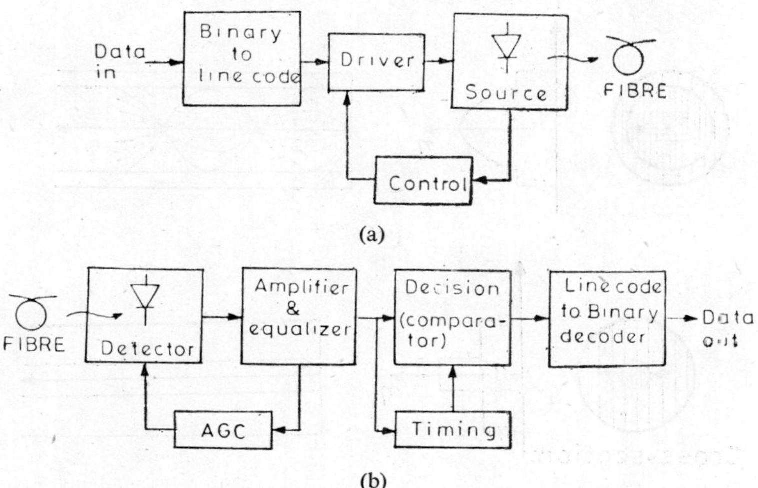

Fig. 4.28 An optical fibre digital communication system (a) Transmitter and (b) Receiver.

choice for signal bandwidths up to 150 MHz; above this value, the PIN-FET voltage amplifier is the simplest and cheapest solution. In contrast with the coaxial systems, the linecoders used are binary only and they are of the *m*B-*n*B family, where *m* bits are converted to *n* bits in the coder ($n > m$), thus introducing some redundancy in the signal for error monitoring and spectrum shaping. 3B-4B, and 5B-6B, codes are widely used; also, scrambled binary signals with parity digits are sometimes used. Some experiments have been done with multilevel codes, e.g., HDB3 and PRS [33].

4.8.1 Optical Fibres [35]

The main components of the fibre-optics system are the optical fibres, light-emitting sources and photodetectors. An optical fibre for communication has a cylindrical core of very high purity glass, surrounded by a cladding of a glass having a lower refractive index η_1. In other less demanding applications, plastic materials may be used for the cladding. There are three main types of optical fibres, viz., (a) multimode step index (SI), (b) multimode-graded index (GI) and (c) monomode (or single-mode) (SM) fibres as shown in Fig. 4.29. In the multimode step index fibre of Fig. 4.29(a), the core is of

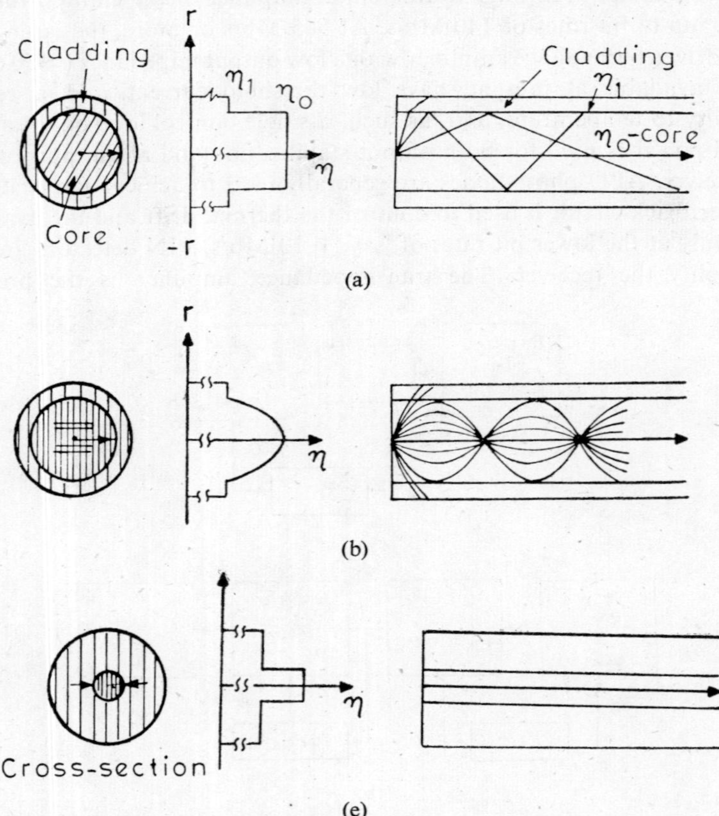

Fig. 4.29 Three types of optical fibres: (a) multimode step index; (b) multimode-graded index and (c) single mode

optically homogeneous material of refractive index η_0, and the change from η_0 to η_1 is abrupt. In the multimode-graded index fibre of Fig. 4.29(b), the η value changes smoothly from the cladding to the centre of the core, where η is maximum. The single-mode fibre has a profile behaviour similar to that of the step index one, but has a much smaller core, as shown in Fig. 4.29(c). The behaviour of η with the radial coordinate r may be approximated as:

$$\eta^2(r) = \eta_0^2 - \Delta^2\left(\frac{r}{a}\right)^g \tag{4.27}$$

where $\eta_0 = \eta$ at the centre of the fibre, a = core radius, and $\Delta = \sqrt{\eta_0^2 - \eta_1^2}$. With $g = 2$, the η-profile is parabolic and with $g \to \infty$, the profile is a step as in Figs. 4.29(a) and (c). The typical dimensions of the fibre are: (a) for multimode fibres, core diameter is 50 to 75 μm and the cladding diameter is 125 μm; and (b) for the single-mode fibres, the cladding diameter is similar to the above, but the core diameter is 4 to 10 μm.

A simlified geometrical approach as shown in Fig. 4.30 for a SI-fibre, may be used to determine the conditions for proper guidance of the optical waves through the fibre. It may be shown by using Snell's law that if ϕ_i is larger than the critical angle $\phi_c = \sin^{-1}(\eta_1/\eta_0)$, then the rays in a SI-fibre will be totally reflected into the core and then guided by the fibre through a large number of subsequent reflections. If, on the other hand, $\phi_i < \phi_c$, then a large number of rays will be refracted into the cladding and ultimately lost after a small number of reflections. It is further seen that the maximum angle of incidence θ_m at the input of the fibre for which the light will be guided by the fibre itself is given by:

$$\theta_m = \sin^{-1}(\Delta/\eta_e) \tag{4.28}$$

where $\eta_e = \eta$ of the fluid in which the fibre is immersed ($\eta_e = 1$ for air).

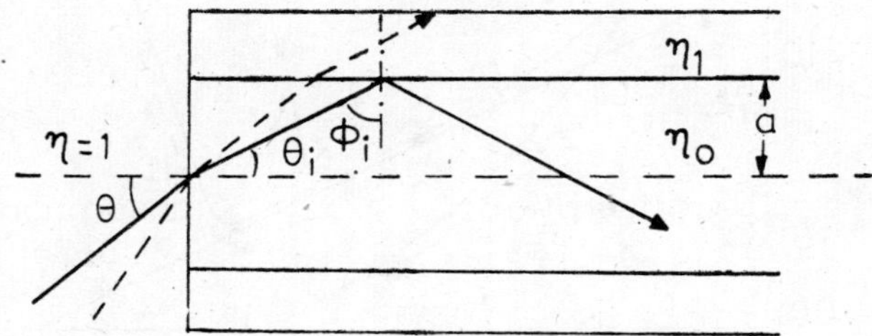

Fig. 4.30 Propagation of light waves in an SI fibre

The quantity θ_m is normally called the numerical aperture (NA) of the fibre. Further, for a length of the fibre = L, the actual path length $L_p(\theta_i)$ for different rays is given by:

$$L_p(\theta_i) = L \sec \theta_i \tag{4.29}$$

Similar results are also calculated for GI-fibres, where the rays are continuously refracted due to the parabolic variation of η. These rays travel along quasi-sinusoidal paths, as shown in Fig. 4.29(b), and all rays go across the fibre axis, while a few are skew rays (which do not cross the fibre axis).

The basic material of which both the core and the cladding of the fibres are made is silica (SiO_2), with the addition of suitable dopants in order to obtain the required η-profile behaviour. The intrinsic causes of attenuation (α) of signals in the fibres are: (a) material absorption; and (b) material scattering. The material absoption has two main components: (a) the ultraviolet (UV) absorption due to stimulation of electron transitions which decreases exponentially with the wavelength λ and is negligible above 1–1.2 μm; and (b) the infrared (IR) absorption, due to interactions of photons with molecular vibrations, and is negligible for $\lambda < 1.6$ μm. With germania (GeO_2) as the dopant, the above attenuation characteristic is not changed basically. The overall α-characteristic of the silica fibres is shown in Fig. 4.31, where it is seen that the region whith $0.8 < \lambda < 1.6$ μm forms an useful window suitable for telecommunication purposes. In addition to the above, the presence of impurities (e.g., Sc, Ti, V, Cr, Mn, Fe, Co, Ni, Cu, etc.) gives broad absorption peaks. The attenuation at $\lambda = 0.84$ μm due to these impurities ranges from 20–2500 dB/km/ppm (part per million) and it is necessary to limit the impurities to less than 1 ppb (part per billion). But the absorption due to hydroxil ion (OH^-) is present in most of the commercial fibres and this gives α-peaks at 1.37, 0.95 and 0.72 μm. However, with controlled manufacturing techniques, the only disturbing α-peak is at 1.37 μm with a value of 10–12 dB/km, as shown in Fig. 4.31.

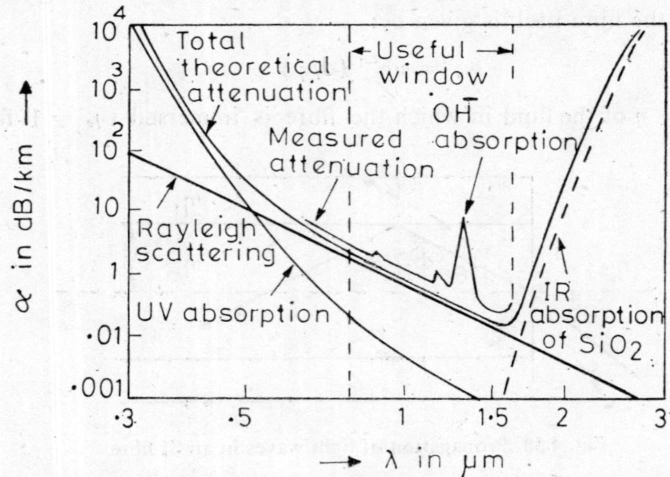

Fig. 4.31 Theoretical and practical attenuation in SiO_2 fibres

The second fundamental cause of attenuation in fibres is the material scattering, which is due to local fluctuations of density and η occurring at distances which are small compared to λ of light. This scattering is mainly

Rayleigh scattering and follows the $1/\lambda^4$ law, as shown in Fig. 4.31. Thus, it is seen that there are three distinct windows suitable for long-distance communication and they are: (a) $0.8 < \lambda < 0.9$ μm; (b) $\lambda \simeq 1.3$ μm; and (c) $\lambda \simeq 1.55$ μm. These are being successfully exploited for future communication systems. In addition to the above intrinsic losses, some loss in cables may occur due to bending, which is partially compensated by having a large value of Δ/λ, or a large cladding core diameter ratio.

The transmission rate (in Mb/s). of a fibre is limited by the pulse spreading mainly due to the dispersion of three types, viz., modal, material and waveguide dispersions. The modal dispersion is typical of multimode fibres, where the path differences and hence the propagation time between different rays having different values of θ_i, as given by eqn. (4.29), produce a spread in the output pulse. Fortunately, in the GI-fibres, a self-equalization of the transit times occurs due to the fact that the rays with longer paths travel in regions with lower values of η and have higher speeds. The maximum transit time differences per km is given by:

$$\begin{aligned} \tau_s &= \frac{\Delta^2}{2\eta_0 C_g}, \qquad \text{for SI-fibres} \\ \tau_g &= \frac{\Delta^4}{8\eta_0^3 C_g}, \qquad \text{for GI-fibres} \end{aligned} \tag{4.30}$$

where C_g = velocity of light in the fibre and Δ is as defined in eqn. (4.27). For a typical value of $\Delta = 0.2$, $\tau_s \simeq 45$ ns/km and $\tau_g \simeq 0.2$ ns/km, i.e., $\tau_s/\tau_g > 200$, showing a large improvement in GI-fibres.

The other factor for bandwidth limitation is the material dispersion caused by the variation of η and C_g with different colours (λ's) present in the transmitted signal. Since all existing light sources (LEDs and laser) emit radiation with a spectrum, material dispersion occurs in all types of fibres, but for silica fibres, this dispersion is zero at a $\lambda \simeq 1.3$ μm. Thus, systems operating at $\lambda = 1.3$ μm can have longer repeater spacings than for $\lambda = 0.85$ μm. The third factor in pulse spreading is the wavelength dispersion which is mainly due to the λ-dependence of the V-number $[= (2\pi a\Delta)/\lambda]$ of the fibre. The λ-dispersion is generally appreciable only in SM fibres, where fortunately, the material dispersion and λ-dispersion cancel each other at $\lambda \simeq 1.3$ μm, as shown in Fig. 4.32. At such wavelengths, SM fibres can have a bandwidth of 100 GHz.km. It is also possible to design SM fibres with zero dispersion at $\lambda \simeq 1.5$ μm, where other losses are minimum. Since the overall dispersion is λ-dependent, it is natural that the sources with smaller spectral width, e.g., lasers, will have lesser pulse-spread and hence larger capacity in terms of GHz.km. This is shown in Fig. 4.33, where it is seen that multimode dispersion effects control the fibre capacity in SI and GI fibres, when used in conjunction with lasers having spectral widths $\simeq$ 1 nm. When LED's are used, the material and waveguide dispersions (jointly known as chromatic dispersion) control the bandwidth in

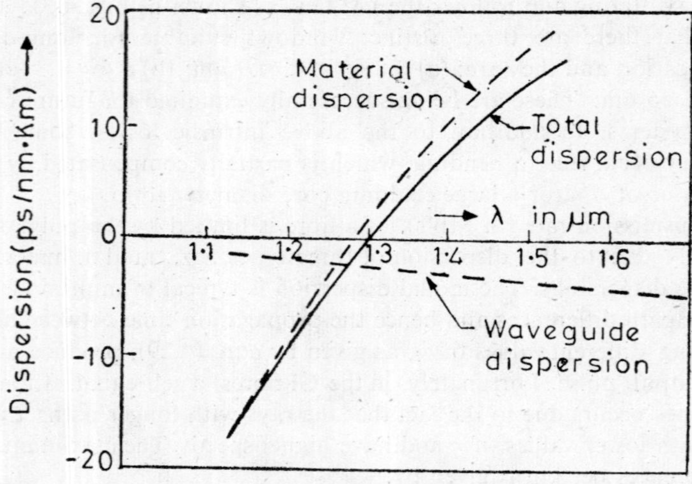

Fig. 4.32 Dispersion in SM fibres

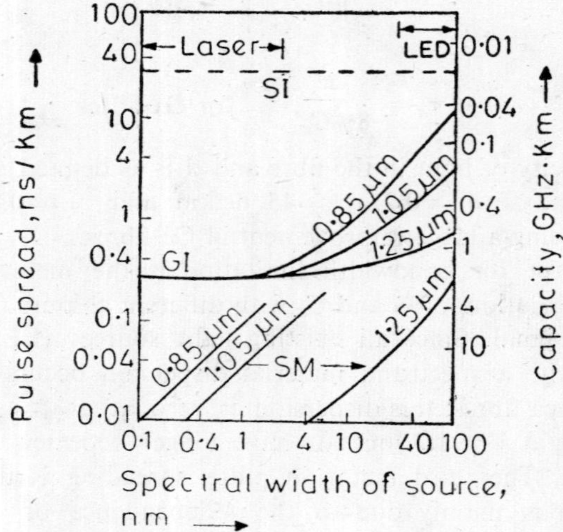

Fig. 4.33 Pulse spread and transmission capacity vs. source spectral width

GI and SM fibres. In SM fibres, the chromatic dispersion effects are apparent even with the use of lasers. A summary of the properties of different fibres is given in Table 4.6 [36].

Commercial glass fibres are made using one of the four techniques, viz., outside vapour-phase oxidation (OVPO) of the Corning Glass Works, modified chemical vapour deposition (MCVD) of the Bell Labs, vapour-phase axial deposition (VAD) of the Nippon Telegraph and Telephone Corporation and the multi-component glass-fibre technique of the British Post Office. The first three techniques make fibres whose composition is mostly SiO_2. Small

Table 4.6 Properties of optical fibres

Type of fibre and core diameter (μm)	Materials	Loss in dB/km for λ (μm) 0.85	1.05	1.5	1.5	Transmission capacity in GHz. km
SM, 1–10	Core and cladding: Silica-based glass	2	1	0.38	0.2	50–100
Multimode SI, 50–200	Core and cladding:					
	(a) Silica-based glass	2	1	0.5	0.2	0.005 to 0.02 (for (a)–(c))
	(b) Core: silica glass Cladding: plastic	2.5	1.5	High	High	
	(c) Core: multicomp. glass Cladding: same	3.4	6	High	High	
Multimode GI, 50–60	Silica glass	2	1	0.5	0.2	1
	Multicomp. glass	3.5	10	High	High	0.4

amounts of Ge, P and sometimes B are added as dopants to control the η and to lower the working temperature below that of SiO_2. All the three techniques produce fibres with low loss and wide bandwidths. The MCVD and OVPO methods are essentially batch processes where a 'preform' is first prepared and then drawn into a finite length, up to 10 km, of fibre.

Cables

For application in long-distance communication, the optical fibres have to be inserted in cables for physical protection. Since the fibres have fragile physical characteristics and their transmission properties get altered by microbending, special care has to be taken while designing and manufacturing fibre-optic cables. Normally, the optical fibres are protected with a primary coating during the drawing process and while cabling, a secondary coating is provided in the form of either a loose or a tight plastic jacket of diameter 0.4–1 mm. A few fibres are then bunched together and a plastic outer sheath is provided. Underground cables are further protected by an armouring (steel wires or tapes; kevlar) below the plastic sheath. The general structures of the cables are mainly of three types: (a) 'classical' stranded structure; (b) grooved cylindrical structure; and (c) ribbon structure. The details of the stranded structure are shown in Figs. 4.34 (a-c) and the other two structures are shown in Fig. 4.35 (a and b). The stranded structure is similar to that of the conventional copper cables and is widely used in many countries. The helically grooved cylindrical structure is used in France and Japan, and the ribbon structure is used in the USA. This structure is based on a plastic ribbon containing 12 fibres and several ribbons (up to 12) are superposed and twisted to form the cable core. This is then protected in the usual manner. A typical fibre cable with 48 fibres and the stranded structure has the following characteristics:

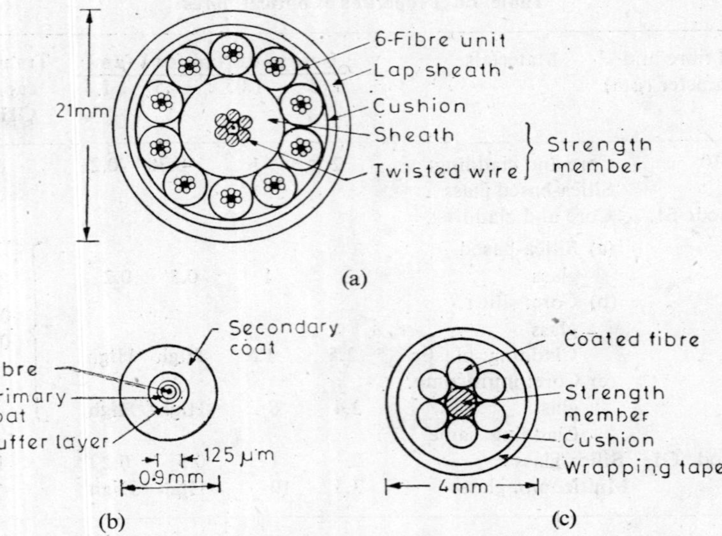

Fig. 4.34 Stranded structure of a fibre cable: (a) composite fibre cable; (b) single-coated fibre and (c) 6-fibre unit

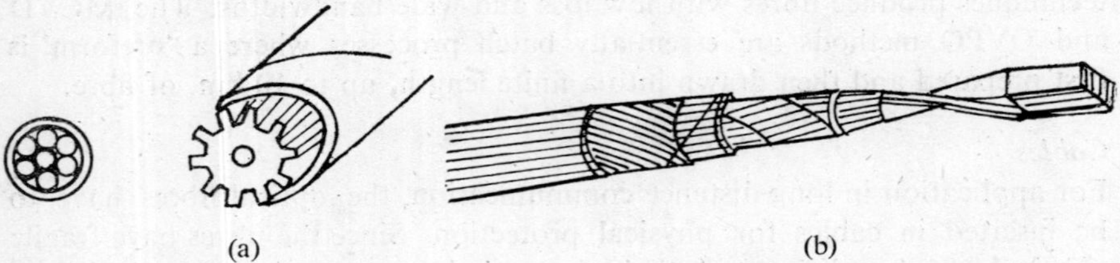

Fig. 4.35 Two structures of fibre cables: (a) grooved cylindrical structure and (b) ribbon structure

Weight	740 kg/km
No. of interstitial pairs	8
Cable diameter	30 mm
Unit cable length	1 km, max.
Loss at $\lambda = 0.85\ \mu m$	Av. 2.78 dB/km
Bandwidth (0.85 μm)	Av. 800 MHz. km
Repeater spacing	Av. 5 km × 4
Transmitted signal	32/100 Mb/s, Analog TV

Connectors, Splices and Couplers

In a complete optical fibre system, fixed connections (i.e., splices) and removable connectors are required to join fibres and connect fibres to devices and other components. These connectors introduce a considerable amount of loss and, therefore, their design and method of use are quite critical. Fusion and adhesive bonding are the two basic techniques to splice individual fibres. The average splicing loss is 0.1–0.2 dB and an additional loss of 0.3–0.4 dB may occur with fibre mismatch. There are two methods for designing connectors for fibres, each giving a loss of 1 dB. In the first method, the prepared fibre ends are aligned in a compound lens composed of a moulded plastic biconical centre element with an optical fluid in each of the two concave cavities. The other design employs a double-elbow four-rod guide to bias the prepared fibre ends into one of the interstices between the rods. Making connectors for SM fibres is really critical.

In a distributed fibre optics network, multiport devices known as couplers, are required to divide or combine the transmission paths. The simplest coupler is the tap, which is a 3-port device and splits the incoming signal into two outgoing channels. Taps can be used as combiners and splitters. There are three designs, viz., (a) the biconical tapered coupler, (b) the beam splitter tap, and (c) the selfoc lens splitter. The technology of these components is rapidly evolving and new design concepts are expected in future.

4.8.2 Sources and Detectors

The optical sources, at present used for fibre-optics communication, are light-emitting diodes (LED) and laser diodes (LD), based on *p-n* junctions of suitable semiconductor materials. The emission radiation from a semiconductor is due to the recombination of electrons and holes, occurring when the material is excited out of its thermodynamic equilibrium through the injection of a current of electrons. The recombination can take place with direct transition, giving rise to the emission of a photon with a frequency corresponding to the energy gap between the conduction and valence bands. In the indirect transition, a phonon (lattice vibration), in addition to the photon, is emitted or absorbed, but this is a slower process with a much lower efficiency and occurs in Si and Ge materials. The sources are, therefore, implemented using direct transition materials, e.g., GaAs, GaAlAs and InGaAsP emitting at 0.9, 0.8–0.9 and 1.2–1.6 μm respectively.

LED's are based on spontaneous emission, i.e., when the device is forward biased, a diffusion of electrons occurs through the *p-n* junction and in the *p*-region there is a recombination of electrons and holes causing emission of photons, at a wavelength depending on the materials used. The emission spectra of LED's are approximately 30–100 nm wide, and the direction of emission of the photons is completely random, leading to an uniform distribution of the light emission in all directions (in the surface-emitter type). This leads to high coupling losses (up to 95%) while injecting the optical power into the fibre which has a limited NA. In the surface-emitting LED's, it is possible to inject into the device a current density up to some tens of KA/cm^2, and these LED's can couple into the 50/125 μm fibre a power of 50–100 μW. In the edge-emitting LED's, which have a structure similar to the one normally used in laser diodes, the emission is in the plane of the junction and, due to a better optical confinement, is more directive, leading to a higher coupling efficiency. The optical power emitted by an LED varies almost linearly with the injection current I_D, as shown in Fig. 4.36, and is suitable for analog modulation as well with little distortion. LED's with capability of generating several milliwatts of optical power are available, but only a fraction of it can be coupled to the fibre. LED's have achieved good reliability results well ahead of lasers, and lifetimes up to about 10^8 hours for GaAlAs LED's in the 0.8–0.9 μm wavelength have been reported.

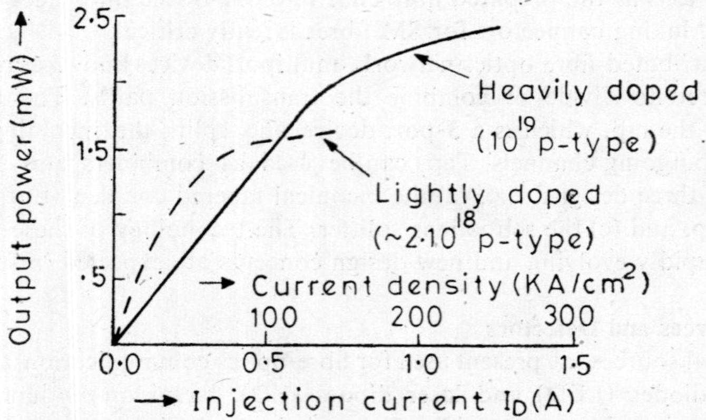

Fig. 4.36 Optical power output vs. injection current in LED's (GaInAsP/InP)

LED's are simpler, more economical and structurally more reliable devices. But the signalling rate is restricted to 100 Mbit-km due to the wider spectral width and lower coupling efficiencies. Thus, LED-based fibre-optics systems are quite suitable and cost effective in short-haul circuits; and with the development of low-loss fibres in 1.3 μm band, LED's will be a viable source for long-haul high data rate links also. Some estimates of maximum link lengths and data rates are shown in Fig. 4.37, where it is seen that it will be possible to use LED's for 140 Mb/s systems using GI fibres up to a length of 30 km at $\lambda = 1.3\ \mu$m.

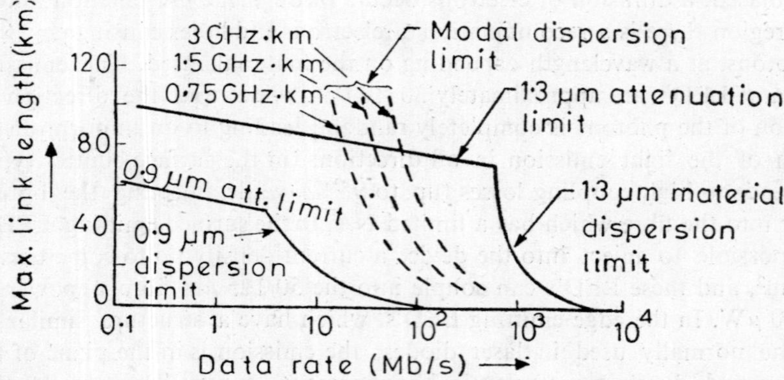

Fig. 4.37 Estimates of maximum link lengths and data rates for LED sources at 0.9 and 1.3 μm

Laser diodes are based on stimulated emission, which takes place when a photon with suitable energy causes the recombination of an electron and a hole, with the emission of another photon having the same direction and phase, and the process is repeated in such a way that an amplification of light is obtained. Using a resonant cavity, a portion of the light travels back and forth and is amplified at each trip. When the light thus produced overcomes the light absorbed, diffused or emitted out of the device, the laser effect is obtained. The stimulated emission is obtained through the injection of very strong localized current densities, which, in turn, modifies the thermodynamical equailibrium of the laser material. The optical power output of a laser diode is large only above a certain current threshold I_{th}, as the stimulated emission takes place with $I_D > I_{th}$. At the same time, for $I_D > I_{th}$, the emitted spectrum is of the order of 2 nm, and is much narrower than that in LED's.

The laser diodes are fabricated on several layers of semiconductor materials, suitably doped (double heterostructure), ensuring a confinement of electrons and holes in a very thin region (about 0.2 μm). Moreover, this region is surrounded by two materials having a lower refractive index, giving rise to a guiding structure, ensuring the confinement of the light. In the widely used 'stripe geometry' structure, the use of a suitable stripe contact limits the active region to a width of 10–20 μm, thus ensuring a better coupling with the fibre and reducing the threshold current $I_{th} \simeq 100$ mA. More sophisticated structures, e.g., Buried Heterostructure (BH) or Transverse Junction Stripe (TJS), permit a substantial reduction of I_{th}, with the best $I_{th} \simeq 4$ mA. Using special structures, it is also possible to obtain monomode lasers, having a very narrow spectrum of width $\geqslant 10^{-4}$ nm (less than 100 MHz).

For fibre communication systems, direct bandgap double heterojunction devices, with λ in the range of 0.8–1.6 μm operating at high current densities of 10^3 A cm^{-2} or more, are generally used. The material system consisting of $Al_xGa_{1-x}As$ on the GaAs substrate has been used for optical sources working in the 0.8–0.9 μm band. For longer wavelengths, e.g., in the 0.92–1.65 μm band, $Ga_x In_{1-x}P_y As_{1-y}$ grown on the InP substrate is currently used. The power available from a semiconductor laser is generally 1–20 mW and coupling efficiencies of 30%–50% can be easily obtained with 50/125 μm fibres with NA = 0.2. Special structures have been developed to obtain very high powers and 40–80 mW of continuous power have been obtained. This can couple at least 20 mW of power to a fibre and, at the same time, be modulated at very high speeds, say, a few Gbits/sec. In contrast with LED's, lasers are temperature-sensitive, and the outputs usually decrease with the temperature, as shown in Fig. 4.38. The threshold current $I_{th} \propto \exp(T/T_0)$, where T is the absolute temperature of the device, and T_0 is a constant ($\simeq$ 120–190°K for GaAIAs lasers and 50–80°K for InGaAsP lasers). Because of the sharp threshold in the output characteristics, lasers are not suitable for analog modulation. In general, lasers are more expensive than LED's and have lower stability and reliability. The problem of reliability has been practically solved for GaAlAs devices and extrapolated lifetimes of 10^5–10^6 hours have been obtained. But for the longer wavelength devices, the problem is still being investigated. The channel capacity of a laser-excited system is much more than that of an LED-excited system, and at $\lambda = 1.3$ μm with single-mode fibres, it is possible to transmit data at 400 Mb/s rate over 30 km or so. With further improvements in fibres and lasers, data rates up to 2 Gb/s over 50 km can be expected in the near future. It is also expected that at $\lambda = 1.55$ μm, data rates up to 1 Gb/s over distances of more than 100 km will be feasible soon and will be used effectively in submarine cables.

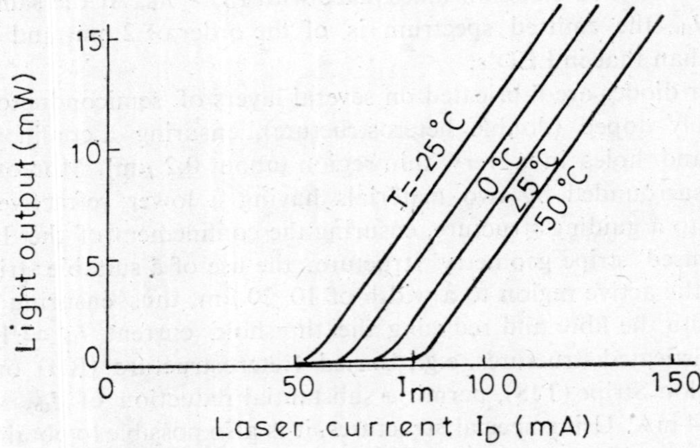

Fig. 4.38 Typical laser outputs vs. I_D for a semiconductor laser at various temperatures

Photodetectors

Two types of optical detectors are normally used in telecommunication systems, viz., (a) unity gain PIN photodiodes and (b) avalanche photodiodes (APD). The semiconductor materials used for these are Si in the 0.8–0.9 μm band and Ge or In GaAsP (or InGaAs) in the 1.2–1.6 μm band. The PIN diode is based on three layers of semiconductor material: a *p*-doped layer, an intrinsic layer and a *n*-doped layer, as shown in Fig. 4.39(a). With a reverse bias (approximately 20 V for Si), a strong electric field is generated

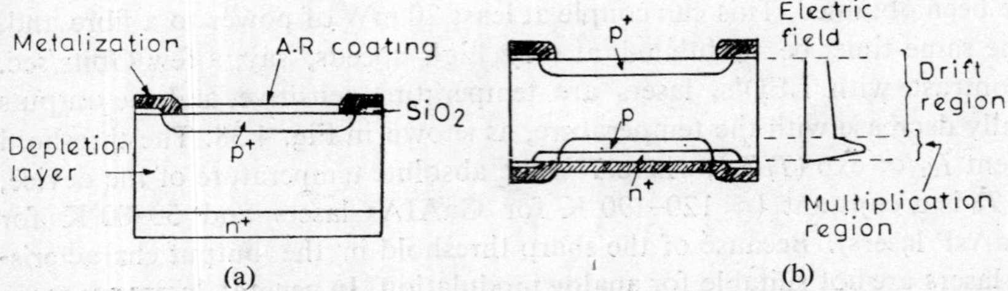

Fig. 4.39 Structures of photodiodes: (a) PIN structure and (b) APD structure

in the i-layer, which can separate the electron-hole couples generated by the photons impinging on the device. In such a way, an electrical current, proportional to the optical power absorbed by the device, is obtained. By the addition of a further layer, with a low p-doping, as shown in Fig. 4.39(b), it is possible to obtain in a region of the device, a very strong electric field, which can accelerate the electrons at very high speeds; such electrons are then able to ionize the lattice atoms, extracting other electrons. In such a way, an 'avalanche' effect is produced and this gives rise to an internal amplification in the device. APD's require much higher bias voltages (100–400 V) and they are temperature-sensitive, so that the thermal drift and the avalanche gain have to be controlled through proper feedback circuits.

The commercially available Si-PIN diodes have sensitivities in the range of −50 dBm (with BER $\simeq 10^{-9}$ at 10 Mb/s) and quantum efficiency of 85% (in the 0.8–0.9 μm band), but the APD's have a higher sensitivity in the range of − 60-to −65 dBm. An important paramater, which limits the sensitivity of the photo diodes, is the dark current (a few nA), i.e., the noise current present in the diodes in the absence of any optical signal. Further, the APD's have an excess noise because of the avalanche multiplication, resulting in an optimum gain, above which the excess noise predominates over the signal gain. Si-APD's have low excess noise and their optimum gain is in the range of 50–150. Thus, Si-APD's are used for higher bit rates, although Si-PIN diodes are more economical and stable for lower bit rates. At longer wavelengths (1.2–1.6 μm), Si-APD's have less favourable performance, and the most commonly used detectors are Ge-APD's and InGaAsP (or InGaAs)-PIN diodes along with a FET amplifier. (InGaAsP-APD's are not available as yet). The sensitivities of the diodes very with the signalling rate and they are approximately:

(a) Ge-APD . . . −45 to −30 dBm for bit rates of 50–500 Mb/s.
(b) PIN-FET . . . −45 to −35 dBm for bit rates of 100–500 Mb/s.

4.8.3 Fibre-Optics Transmission Systems

A simple block diagram of an optical fibre communication system has been shown in Fig. 4.28. Such systems at 8/34/140/400 Mb/s have been designed and installed in many countries. One such typical system at 140 Mb/s is shown in Fig. 4.40 [37, 38]. Its overall specifications are:

System capacity: 140 Mb/s (1920 telephone channels or 2 digital TV channels).

Electro-optical components:
Source: Semiconductor Laser $\lambda = 0.85$ μm, with power coupled = −3 to 0 dBm
Medium: a pair of GI multimode fibres with 3.5 dB/km loss and 1 ns/km dispersion at $\lambda = 0.85$ μm
Detector: APD with a gain of 100 and sensitivity of −48 dBm
Maximum allowable path loss: 32 dB
Overall receiver gain: 50 dB

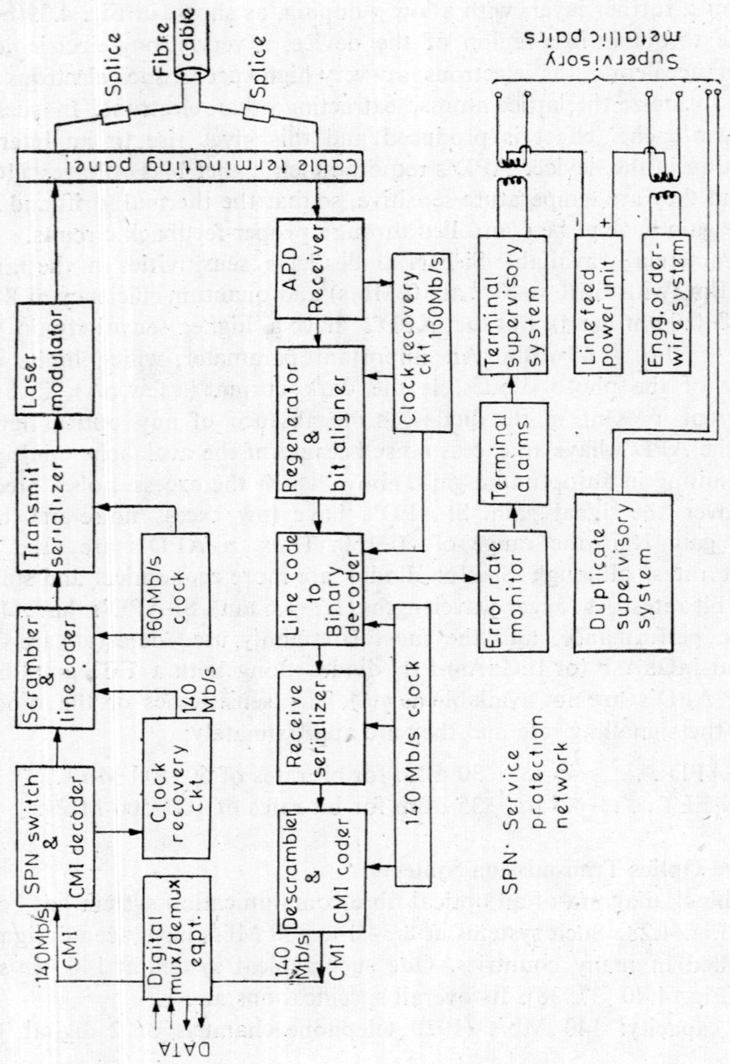

Fig. 4.40 Block diagram of a 140 Mb/s optical transmission system

Repeater spacing: 9 km
Power-feeding span: 40 km
Maximum path length: 2300 km
Overall error rate: better than 10^{-9}
Line code: input CMI; output 7B 8B
Transmission rate: 160 Mbaud
System supervision provided for 256 terminals/repeaters.
Alarms provided for signal failure, excessive error rate, laser drive failure, excessive laser power and laser out of limits.

In the transmit path of Fig. 4.40, the processing of signals is somewhat similar to that in coaxial cable systems, except that the decoded binary data are first scrambled with an 11-stage PNS genetator and then converted to a redundant binary line code, 7B-8B, which includes mark parity control. This facilitates the error detection in the receiver by detecting digital sum violations or mark parity violations. The transmission rate now is increased to 160 Mbaud, and the laser is modulated to give a sequence of half-width light pulses for transmission via the fibre. The terminal receiver consists of the APD receiver, regenerator, line code decoder, descrambler and the CMI coder. The APD receiver, whose details are shown in Fig. 4.41, has the APD closely coupled to a low noise transimpedance preamplifier followed by an AGC amplifier. Variations in the received signal level are compensated by both the APD bias control and the AGC gain, giving a total control of 25 dB. The APD bias is derived from a simple DC/DC converter as the current required is only about 1 μA. The signal pulses, distorted due to the fibre dispersion, is reshaped to give raised-cosine pulses through a manually controlled variable equalizer. Reshaped pulses are now amplified and regenerated into the digital form for further processing in the line code decoder and descrambler. Finally, the pulses are CMI-coded at 140 Mb/s and sent to the MUX/DEMUX unit. A PLL circuit converts the 160 Mbaud line signal to the data signal at 140 Mb/s. This circuit also acts as a jitter reducer with a bandwidth of 50 Hz and a maximum jitter gain of 0.5 dB.

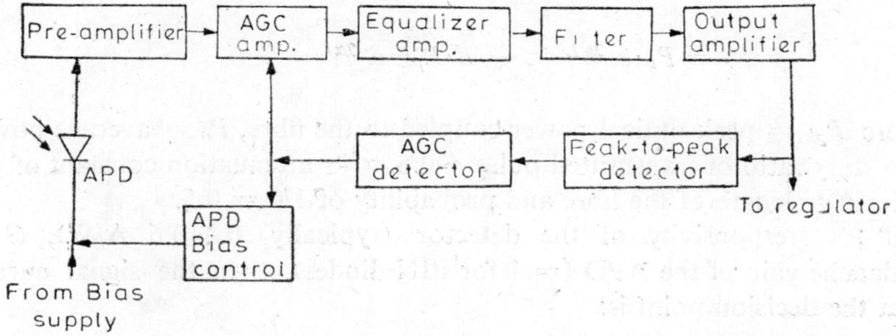

Fig. 4.41 Block diagram of an APD receiver

The metallic pairs available in the fibre cable are used for transmitting and receiving supervisory signals and also for power feed to the unattended repeaters. Both-way repeaters, as shown in Fig. 4.42, are installed at nominal intervals of 9 km and the total path length may be of the order of 2000 km.

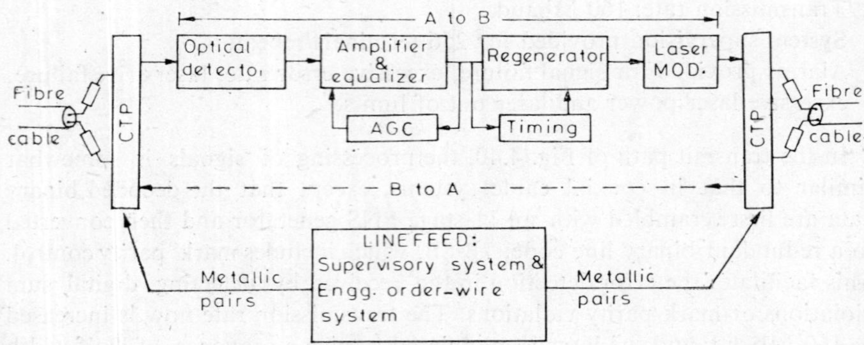

Fig. 4.42 Block diagram of an optical repeater

Except the coding and scrambling equipment, which are not required in the repeaters, the basic units in a repeater are the same as those of the terminal equipment. An elaborate monitoring system, based on a μ-processor, memories and interfaces, is provided to supervise the transmission quality of the signal along the route and the necessary alarms/indicators ensure the overall BER in the total system. Engineering order-wire circuits are also provided for the ease of maintenance and continuous supervision.

Link Calculations

The calculation of maximum repeater spacing possible, with a given source, fibre and the detector, depends on the optical power P_0 injected into the fibre, the fibre loss, the detector sensitivity and the overall noise power at the decision point. These are related by the following considerations:

P_o for LED's is around 20 to 100 μW (equivalent to -17 to -10 dBm) and for lasers around 0.3 to 2 mW (-5 to $+$ 3 dBm].

The mean received power P_r at the receiver is:

$$P_r = P_o \cdot e^{-2\alpha L} = \frac{1}{2} \cdot d \cdot P_{pk} \cdot e^{-2\alpha L} \tag{4.31}$$

where P_{pk} = peak optical power coupled to the fibre, P_o = average power, d = duty ratio of transmitted pulse $\sim$0.5, α = attenuation constant of the fibre, L = length of the fibre and probability of 1/0 = 0.5.

If ρ = responsivity of the detector (typically 0.4–0.6 A/W), G = avalanche gain of the APD (= 1 for PIN-diodes), then the signal current I_p at the decision point is:

$$I_p = 2\rho G P_r \tag{4.32}$$

The noise power generated in the receiver is due to the usual thermal noise I_{th}^2 and the quantum noise I_{qn}^2 due to the photodetector (assuming that the dark current of a few nA is negligible). These are given as:

$$I_{th}^2 = \frac{4KT\,B_n F}{R} \tag{4.33}$$

$$I_{qn}^2 = 2q\rho G^2\, F_G B_n P_r \tag{4.34}$$

where K = the Boltzmann constant, T = absolute temperature, B_n = noise bandwidth of the detector amplifier, F = NF of the amplifier, R = input resistance of the amplifier, q = electron charge and F_G = NF of the amplifier taking into account the excess noise due to the random nature of the avalanche process (F_G = 1 for PIN diodes). Thus, the signal-to-noise ratio at the decision point is given by:

$$S/N = \frac{I_p^2}{I_{th}^2 + I_{qn}^2} \tag{4.35}$$

For a given BER at the receiver output, the value of S/N required is obtained from the decision theory results (Ref. Chapter 2), and is approximately 20 dB for a BER $= 10^{-10}$. From the above equation, the required value of P_r may be calculated for a given S/N. Such sample calculations are shown in Fig. 4.43 for a PIN detector and also for a APD detector with optimum gain. It is observed from the curve (a) that for lower values of S/N as required for digital signalling, the thermal noise controls the required P_r; but for larger values of S/N, as required for analog modulation P_r is controlled by the quantum noise. For the APD detector, as given by (b), P_r required is less because of the optimum gain.

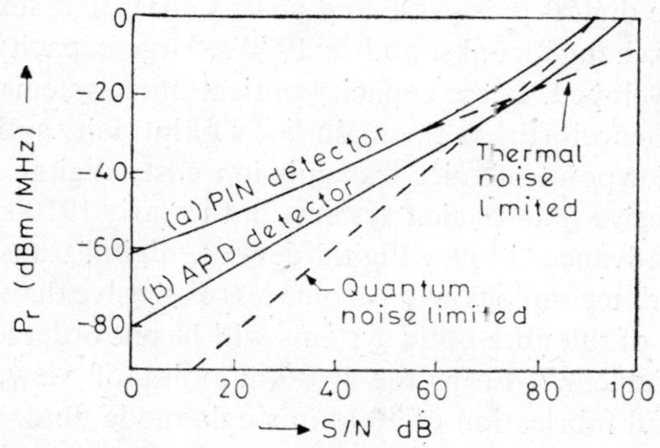

Fig. 4.43 Optical mean power P_r vs. S/N for two detectors

With the required P_r obtained as above, the repeater spacing L can be simply calculated from

$$L = \frac{P_o\,(\text{dB}) - P_r\,(\text{dB})}{\alpha\,(\text{dB/km})} \tag{4.36}$$

This assumes that the fibre bandwidth is much larger than the signal bandwidth, and B_n is independent of L. However, for higher transmission rates, B_n is also a function of L; and there is a power penalty to compensate for the signal dispersion in the fibre. Further, losses due to connectors, splices and ageing of the equipment have to be considered and adequate margins are to be provided while designing the total system. With the current development of fibres, sources and detectors, the possible repeater spacings at various λ's and data rates are shown in Fig. 4.44, where it is seen that the expectations with the 1.55 μm system are indeed very encouraging.

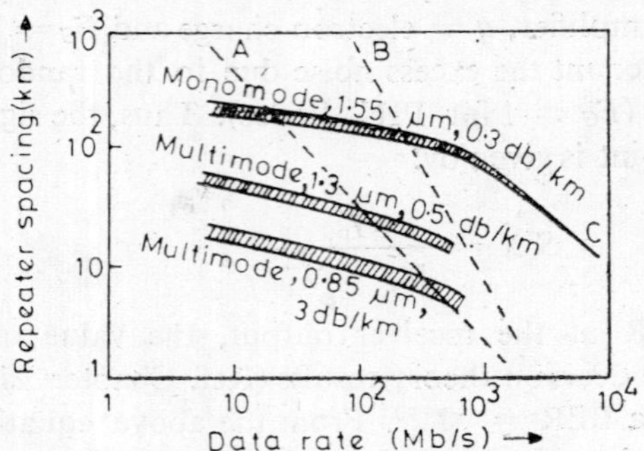

Fig. 4.44 Projected repeater spacings with various fibres at λ = 0.85, 1.3 and 1.55 μm. (A and B give the limits due to modal dispersion, assuming 0.3 ns/km and 0.3 ns/√km respectively for pulse broadening. C gives the limit due to laser emission instability)

4.8.4 Current Trends

In the 1960's the digital pair cable and analog coaxial systems were developed for interexchange trunks, and in 1970's, large-capacity digital coaxial systems were developed. Large capacity optical-fibre systems will be used in the 1980's and, henceforth, as the main body of intracity and intercity trunking. From the viewpoint of voice transmission cost, digital coaxial systems were more expensive than analog systems in the early 1970's, but because of the remarkable advances in new digital devices, digital transmisslon systems using optical-fibre technology will become less expensive than analog systems. The overall cost of the fibre-optic systems will be one order less by the 1990's as is shown in Table 4.7. From the loss/km point of view, recent reports indicate successful fabrication of 100 km single-mode fibre, whose losses are as low as 0.2 dB/km at 1.55 μm. The present development in fibre technology indicates a loss of 0.3 to 1 dB/km for the wavelength range of 1 to 1.8 μm. However, the possible wavelengths presently used or available in the near future are 0.85, 1.3 and 1.55 μm; this limitation being mainly due to the non-availability of laser diodes operating at other wavelengths. An estimate of the transmission bit rate and repeater spacing for long-haul circuits strongly depends upon the fibre type (step-index multimode, graded-index multimode, and single-mode) and the wavelength (0.85, 1.3 and 1.55 μm) used. For 100–400 Mb/s, a single-mode fibre with longer wavelengths (1.3μm presently and 1.55 μm in future) gives a repeater spacing of 20–30 km. For long-haul systems, regenerative repeaters, using a laser diode and a photodetector, are being built in hybrid IC forms and, in future, will become more integrated in a few IC chips.

Table 4.7 Trends in transmission cost reduction in cables

Year	Type of cables	Bit rates	Repeater spacing	Relative transmission cost
1960-70	Pair cable (digital)	64 Kb/s	1→ 2 km	70→20
	,, (analog)	–	1→ 2 km	15→5
1970-80	Coaxial cable (digital)	274 Mb/s	1→ 2 km	20→5
1980-90	Fibre cable (digital)	100–400 Mb/s	10→ 30 km	10→2
1990-2000	,, ,,	1000 Mb/s	50→100 km	5→1

Considerable research efforts are being made to extend the laser technology to the ultralong wavelengths in the 2–10 μm range and this will reduce the fibre losses to 0.01 dB/km. In this wavelength range, a completely new set of materials will have to be developed for sources, detectors and fibres. Glasses based on heavy metal fluorides are promising materials for the purpose. A whole new technology of material processing and fabrication techniques has also to be developed. Further, the 2–10 μm lasers will probably require cooling; however, portable coolers of liquid-helium temperature are already available. The expected performance parameters for 2–10μm optical communication systems are given in Table 4.8 [39].

Table 4.8 Expected performances of 2-10 μm optical systems

Laser source: expected life	$5 \cdot 10^6$ hrs
Fibre loss	0.01-0.05 dB/km
Repeater spacing	200-80-800 km
Bandwidth-distance	10^3 GHz-km

All major telephone laboratories, e.g., Bell Labs, GTE-Lenkurt in USA, Bell Northern in Canada, Nippon in Japan and European telephone administration labs, are working on single-mode, long-wavelength fibre-optic systems. However, at present, the working systems are operating at 44.7, 90 and 140 Mb/s using mostly first-generation 0.8–0.85 μm components. The Bell system is committed to a policy of converting all trunk lines and local loops into fibre optics, whenever this is economical. Bell Canada has already installed 3600 km of fibre cables (44,000 km of fibres) for the Canadian telephone network. They have also the largest fibre-optic system, where a broad-band network links exchanges, which are from 100 to 200 km apart and require 3200 km of 12-fibre cable. It is expected that within five years, large-capacity optical fibre systems will be put into wide commercial use. In Japan, a field trial of the long-haul system has been in progress: The system parameters are: Bit rate $\simeq$ 400 Mb/s, repeater spacing $\simeq$ 20 km, total path length $\simeq$ 2500 km, single-mode fibre cable having 0.7 dB/km loss, wavelength $\simeq$ 1.3 μm and overall error rate $< 10^{-8}$ for 2500 km [40].

Both in USA and Japan, development efforts are going on for a multi-fibre cable for undersea telephone transmission that will increase the repeater spacing to 35 km (from the present 9.4 km for coaxial cables) and have two-way voice channel capacity of 36,000 (from the current 4200 channels in coaxial systems). The Trans-Atlantic service is expected to start in 1988 with transmission of streams at 274 Mb/s on a single-mode cable operating at 1.3 μm. The cable will be 6500 km long and reach ocean depths of 6.5 km. Nippon T and T in Japan is developing a fibre-optic system due to become operational in 1985. This will transmit 5000 to 18,000 voice channels at 400 Mb/s, using 1.3 μm laser sources and a repeater spacing $>$ 25 km. With the development of the 1.6 μm system, the repeater spacing may be 100 km or more.

Although optical-fibre technology has already proven its versatility as the next generation communication system, we may expect further improvement and drastic changes in the forthcoming era. Possible realizable developmets are: WDM (wavelength division multiplexing) [41], bilateral transmission in a fibre, integration of repeater electronics, high-speed repeater technology, optical IC's, optical amplification, coherent carrier transmission and fibre-optic transducers. Various WDM techniques are being investigated to transmit video, voice and data over fibre-optic telephone systems for the home. A particular implementation of WDM is the subscriber loop (under trial at a cost of $ 6.3 million) now installed and operating in the rural town of Elie in Canada, where some customers (150 homes and businesses) receive their video, FM radio, data and telephone services from a 0.83 μm laser and transmit telephony, data and video switch-control information upstream using an LED at 0.92 μm. These two sources transmit bidirectionally on the same fibre, using a WDM coupler. The researchers see a growing demand for WDM techniques in both local loop and trunk systems. The repeaters for fibre-optic systems may use WDM techniques, where ten optical channels may be allocated in bilateral directions, similar to the technique used in guided millimeter-wave systems. It is believed that optical IC's will play an important role for the WDM technique and more than 50,000 telephone channels or 160 digitized colour TV channels could be handled by a single fibre. Use of the coherent detection of an optical carrier will increase the repeater spacing further, say, to more than 100 km, and numerous cost and reliability benefits will be obtained, specially for transoceanic transmission systems. The progress so far and the progress expected in optical fibres and optoelectronic components are broadly shown in Fig. 4.45 [39], and it is seen that, by the turn of the century, the optical-fibre systems will be the cheapest and the most reliable medium for both short-haul and long-haul communication circuits.

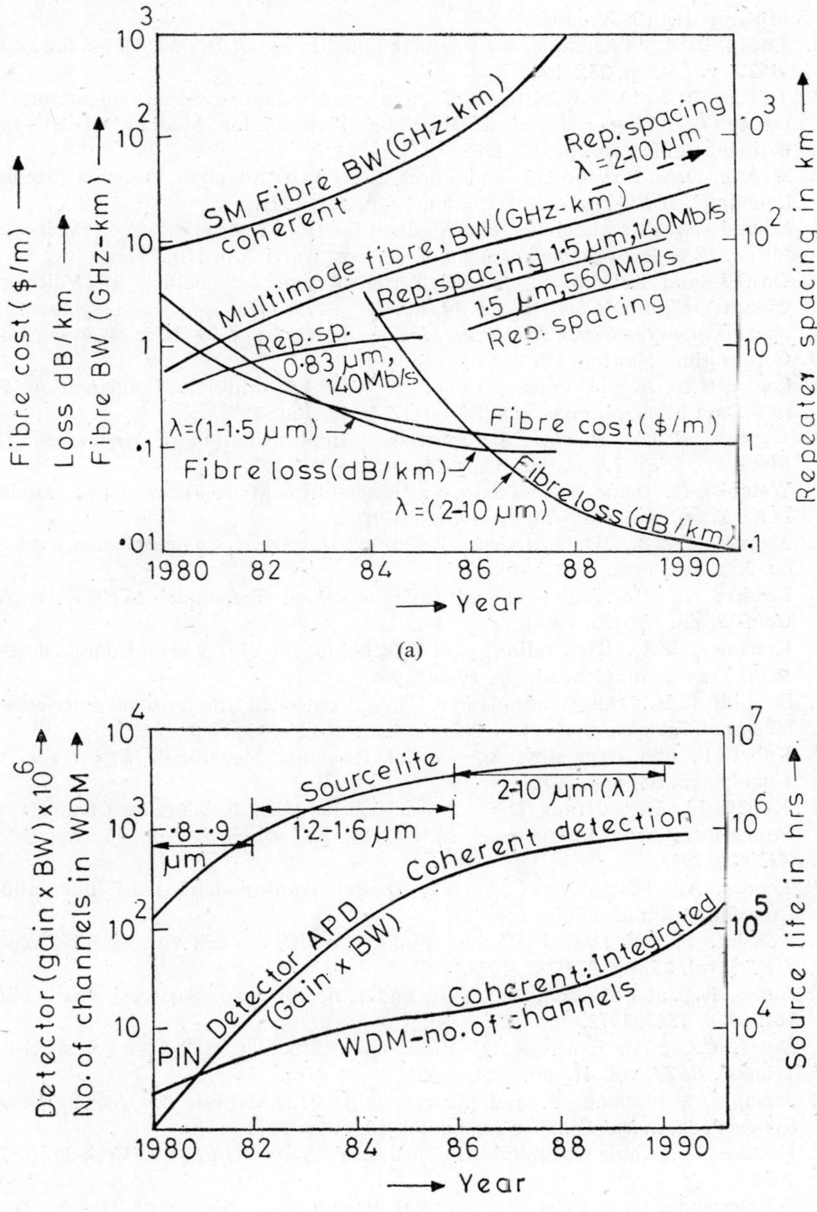

Fig. 4.45 Progress in (a) optical-fibre performances and (b) optoelectronic components

REFERENCES

1. Lucky, R.W., Salz, J. and Weldon, E.J. Jr., *Principles of Data Communication*, McGraw-Hill, N.Y., 1968.
2. Bennett R.W. and Davey, J.R., *Data Transmission*, McGraw-Hill, N.Y., 1965.
3. Schwartz, M., Bennett, W.R. and Stein, S., *Communication Systems and Techniques* McGraw-Hill, N.Y., 1966.
4. Lucky, R.W., 'Tcchniques for Adaptive Equalization of Digital Communications, *BSTJ*, vol. 45, p. 255, 1966.
5. DiToro, M.J., 'A New Method of High-speed Adaptive Serial Communication through Any Time-variable and Dispersive Transmission Medium', IEEE International Conference, p. 763, 1965.
6. George, D.A., Bowen, R.R. and Storey, J.R., 'An Adaptive Decision Feedback Equalizer', *IEEE Trans.*, vol. Comm. 19, p. 281, 1971.
7. Edvardsson, K. and Nymen, H., 'Modems for Data Transmission on Voice-grade Lines', *Electrical Communication*, vol. 48, nos. 1 and 2, p. 110, 1973.
8. Das, J. and Rakshit, S., '3-level Partial Response Signalling for Voice-grade Circuits', *Electronic Letters*, vol. 14, no. 18, p. 579, 1978.
9. *Specifications for Codex 9600 Data Modem and Codex 4800 Data Modem*, Codex Corporation, Newton, USA.
10. Logan, H.L., Jr. and Forney, G.D., 'A MOS/LSI Multiple Configuration 9600 BPS Data Modem', *Proc. ICC'76*, no. 12, p. 48, June 1976.
11. Gerwan, P.J.V. *et al.*, 'Microprocessor Implementation of High-speed Data Modems', *IEEE Trans.*, vol. Com-25, pp. 238-250, 1977.
12. Watanbe, K., Inaue, K. and Sato, Y., 'A 4800 Bit/s Microprocessor Data Modem', *IEEE Trans.*, vol. Com-26, pp. 493-498, 1978.
13. Nyman, H. *et al.*, 'Data Modem Evolution', *Electrical Communication*, vol. 57, no. 3, p. 187, 1982.
14. Lender, A., 'Correlative Digital Communication Technique', *IEEE Trans.*, vol. Com-12, pp. 128-135, 1964.
15. Kretzmer, E.R., 'Generalization of a Technique for Binary Data Communication, *IEEE Trans.*, vol. Com-14, pp. 67-68, 1966.
16. Schmidt, K.H., 'Data Transmission Using Controlled Intersymbol Interference', *Electrical Communication*, vol. 48, nos. 1-2, pp. 121-133, 1973.
17. Kabal, P. and Pasupathy, S., 'Partial Response Signalling', *IEEE Trans.*, vol. Com-28. pp. 921-934, 1975.
18. Kobayashi, H. and Tang, D.T., 'A Decision Feedback Receiver for Channels with Strong Intersymbol Interference', *IBM J. Res. and Develop.*, vol. 17, no. 5, pp. 413-419, 1973.
19. Lender, A., 'Seven-level Correlative Digital Transmission', IEEE International Convention, Philadelphia, June 1976.
20. Rakshit, S. and Das, J., 'Some Studies on PRS through Voice-grade Circuits', *JIETE*, vol. 25, pp. 278-285, 1979.
21. James, R.T. and Mueneh, P.E., 'A.T. and T. Facilities and Services', *Proc. IEEE*, vol. 60, p. 1342, 1972.
22. Davis, C.G., An Experimental Pulse Code Modulation System for Short-haul Trunks', *BSTJ*, vol. 41, pp. 1-24, 1962.
23. Jessop, A., Norman, P. and Waters, D.B., '120-Megabit per Second Coaxial Systems', *Electrical Communication*, vol. 48, p. 79, 1973.
24. Kumagi, D., 'Cable Communication in Japan', *Trans. IEEE*, vol. Com-20, p. 707, 1972.
25. Scheuermann, H. and Gockler, H., 'A Comprehensive Survey of Digital Transmultiplexing Methods', *Proc. IEEE*, vol. 69, p. 1419, 1981.
26. Bellamy, J., *Digital Telephony*, John Wiley and Sons, N.Y., 1982.

27. Barbetta, A. and Natens, M., 'Digital Multiplexers for Rates from 2 to 565 $Mbits^{-1}$', *Electrical Communication*, vol. 57, No. 3, p. 251, 1982.
28. Lindsay, W.C., *Synchronization Systems in Communication and Control*', Prentice-Hall, N.J., 1972.
29. Buchner, J.B., 'Ternary Line Codes', *Phillips Telecom. Rev.*, vol. 34, pp. 72-86, 1976.
30. Mayo, J.S., 'Experimental 224 Mb/s PCM Terminals', *BSTJ*, vol. 44, pp. 1813-1842, 1965.
31. Kolb, W. *et al.*, 'Coaxial Line Equipment for 34 to 565 Mbits', *Electrical Communication*, vol. 57, p. 243, 1982.
32. Bax, W.G. and Wagenmakers, J., '140 Mb/s Coaxial Transmission System 8TR 609', *Phillips Telecom. Rev.*, vol. 37, p. 144, 1979.
33. Sandbank, C.P., *Optical Fibre Communication Systems*, John Wiley and Sons, N.Y., 1980.
34. Barnoski, M.K., '*Fundamentals of Optical Fibre Communications*', Academic Press, N.Y., 1982.
35. Tosco, F., 'Lecture Notes on Optical Communication Systems', at the Workshop on Phyisics of Communication, ICTP, Trieste, Nov. 1983.
36. Maskara, S.L., 'Optical Fibre Communication Systems for Voice and Video Signals—State of the Art', *IETE Tech. Rev.*, vol. I, No. 4, pp. 35-50, 1984.
37. Special Issue on 'Optical Communication', *Philips Telecomm, Rev.*, vol. 40, No. 2, July 1982.
38. Thomas, D.S., '140 Mbit s^{-1} Multimode Optical System', *Electrical Communication*, vol. 57, pp. 201-207, 1982.
39. Kao, C.K., 'Rapid Evolution of Fibre Optics', *Elec. Comm.*, vol. 58, pp. 112-114, 1983.
40. Koyoma, M. and Itoh, T., 'Present and Future of Large Capacity Optical Fibre Transmission', *IEEE Comm. Mag.*, vol. 19, No. 3, p. 4, 1981.
41. Special Issue on 'Optoelectronics', *Hitachi Rev.*, vol. 33, No. 4, Aug. 1984.

CHAPTER 5

Digital Radio Systems

The demand on the radio frequencies for long-distance communication led to the development of UHF and MW links as early as the 1930's. Over the following decades, long-haul multichannel communication systems have been built up using repeatered MW links, and line-of-sight (LOS) propagation. The frequency bands normally used for MW LOS links are: 1.7–2.3, 3.4–4.2, 3.925–8.5; and 10.5–13.25 GHz with channel bandwidths of 3.5/20/30/40 MHz. The multiplexing and modulation technique utilized is FDM-FM and 600–1200 voice circuits are accommodated in most of the systems. The LOS repeater spans are usually 15 to 80 km and a total link length of 5000 km or more may be bult up with suitable repeaters. Using high-power transmitters (50 kW or more), longer distances over the horizon (say, 100 to 1000 km) may be covered using the path diffraction or the troposcatter mode of propagation. Both the LOS and Tropo links suffer from fading and multipath transmission, resulting in the loss of signals during the trough of the fade-cycle (10–100 Hz). To achieve a 99.99% reliability, a 40 dB fade margin is generally provided for while designing the overall link. Moreover, frequency/space diversity reception is used to improve the reliability and performance of long-haul systems. For short-haul and mobile communication, VHF/UHF frequencies are used and they also suffer from fading and multipath problems, requiring fade margin and diversity techniques.

LOS and Tropo systems so far were based on analog modulation schemes, mostly FDM-FM; however, during the last decade or more, digital modulation techniques have proved to be more efficient and economical. The growing introduction of digital switching in the communication network has also led to the development of digital radio techniques to form integrated digital networks. Satellite communication is also becoming all digital. The trend in USA and Canada, as also in Japan and European countries, clearly indicates the predominance of the digital approach in new long-haul transmission facilities. In the forecast for 1990, it is shown that terrestrial microwaves, satellite communication and fibre-optic systems will be predominantly all-digital networks, the ratio between digital to analog being approximately 10 : 1, if not more.

In the background of the above, this chapter discusses digital LOS systems including the techniques of improving their performances through diversity, DFE and adaptive receivers, and digital satellite communication systems including the recent progresses. Multiaccessing in radio systems, with special emphasis on spread-spectrum techniques, is also discussed.

5.1 ANALOG MW SYSTEMS [1]

A typical LOS-MW system consists of terminal transmitters and receivers, and a number of repeaters, spaced at a distance of 30 to 60 km, as shown in Fig. 5.1. The transmit-receive antennas may be combined into one using diplexers, and if necessary, four frequencies may be used instead of two as shown, to avoid cross-talk between the 'GO' and 'RETURN' channels

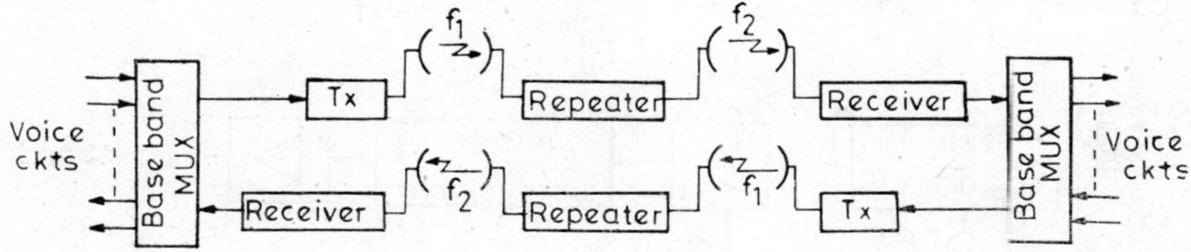

Fig. 5.1 A typical LOS system

More detailed block diagrams of the transmitter, the repeater and the receiver are shown in Fig. 5.2. The repeaters* are usually non-demodulating, and have a power output of a few watts, but their input/output frequencies are different. In the receiver, an effective AGC circuit is used to maintain the signal level during the fade cycle. The most commonly used antennas are the parabolic reflectors with a waveguide feed or a feed horn at the focus. Their diameters are 4 to 5 m for multihop systems (having an approximate gain of 30 dB) and 1 to 2 m for mobile systems. For multifrequency use, horn reflector antennas are more efficient. In some cases, the antennas are energized with two planes of polarization simultaneously for transmission and reception on nearby frequencies. For Tropo systems, the RF power output of 1–50 kW and antennas having a diameter of 10–30 m (with gain $\simeq$ 50–70 dB) are used.

Because of the random fading of the signal, it is necessary to use space/frequency diversity or both. A typical quadruple diversity system using both space and frequency diversity along with dual polarization is shown in Fig. 5.3. Two parabolic antennas, suitably spaced and with multiple feeds for simultaneous transmission at one frequency and reception at two other frequencies, are used; further, one transmits in a horizontal polarization and the other in a vertical polarization. Thus, four distinct signals are received by the four receivers at the receiving station, thereby giving a fourfold diversity to the system. The signals from the four receivers are suitably

*In short-haul systems, the demodulating type of repeaters is used to facilitate channel dropping at intermediate stations.

combined, as shown, and a large diversity gain is obtained. There are mainly three methods of diversity combining, viz., (a) Selection combining; (b) Maximal ratio combining; and (c) Equal gain combining. Of these, maximal ratio (m.r.) combining gives the maximum gain, and the average signal-to-noise ratio CNR improves by 7 dB with quadruple diversity. However, for an outage rate of 0.01%, as required in a practical system, the CNR gain is greater than 30 dB for quadruple and 20 dB for dual diversity.

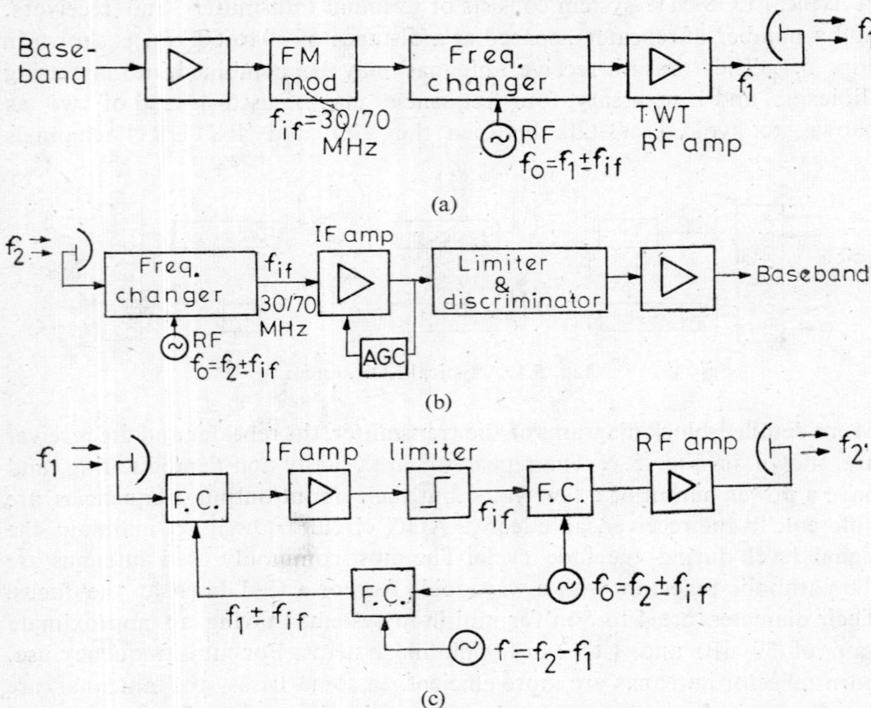

Fig. 5.2 Typical block diagrams of a MW-LOS communication system: (a) terminal transmitter; (b) terminal receiver and (c) typical MW repeater

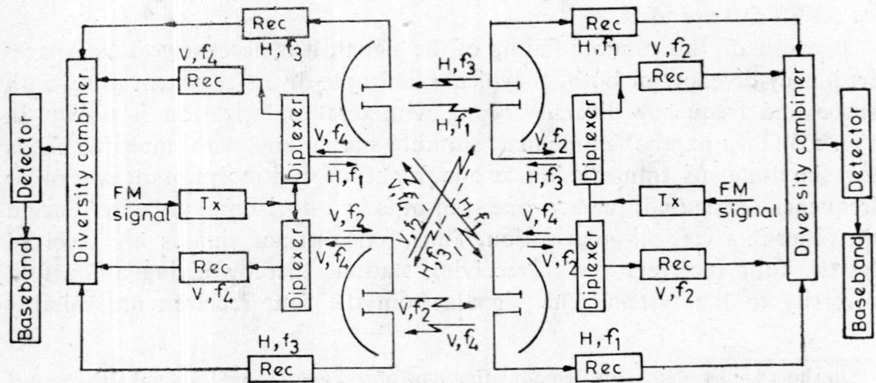

Fig. 5.3 Quadruple diversity configuration with spaced antennas and crossed polarization

5.1.1 Link Calculations

Consider a hypothetical 2500 km LOS link with 50 repeaters, each at a mean distance of 50 km, and the link provides 6C0 audio channels (60 kHz–2.54 MHz) with a Ch. BW = 30 MHz.

According to CCIR recommendations, the total weighted noise power allowed at the output of the circuit is 7500 pW, of which 3750 pW may be due to thermal noise and the other 3750 pW due to intermodulation and other noise. The reference sinusoidal test tone signal at the output is 0 dBm (0 dB at 1 mW) and the psophometric weighting is 2.5 dB. We shall now calculate the transmitter power required for the above link with and without fading.

Considering first the case without fading, the weighted noise power N_0 allowed per hop = 3750/50 = 75 pW = −71.25 dBm. The corresponding weighted SNR at the voice output is:

$$S_0/N_0 \text{ (weighted)} = 0 - 71.25 = 71.25 \text{ dB}$$
$$S_0/N_0 \text{ (unweighted)} = 71.25 - 2.5 = 68.75 \text{ dB} \tag{5.1}$$

The thermal noise N_i at the receiver input is given by

$$N_i = kT_0B_{if}\,(\text{NF}) \tag{5.2}$$

where $k = 1.38 \times 10^{-23}$ watt/sec/°K, $T_0 = 293$°K, $B_{if} = 30$ MHz and NF = 10 dB (assumed). Then, $N_i = -89.2$ dBm.

Assuming that FDM-FM modulation is used, the SNR improvement in the receiver output is calculated as.

$$\text{Noise Improvement Factor (NIF)} = 10 \log \left[\frac{B_{if}}{2\delta f}\left(\frac{F_d}{f_0}\right)^2\right] \tag{5.3}$$

where δf = audio channel BW, F_d = peak frequency deviation per channel, and f_0 = the mean frequency of the top channel. For the above system, NIF ≃ 18 dB. Then, combining the eqns. (5.1), (5.2) and (5.3), the required C_i (= received power P_r) is:

$$C_i = 68.75 - 89.2 - 18 = -38.45 \text{ dBm}.$$

Using the free-space path loss equation and the antenna gains of the transmitter and the receiver of a single hop, the received power P_r (= C_i) is given by:

$$P_r \text{ (in dBm)} = P_t \text{ (in dBm)} + (G_t + G_r - A_f) \text{ dB} - 20 \log\left(\frac{4\pi d}{\lambda}\right) \text{ dB}, \tag{5.4a}$$

where P_t = transmitted power, dBm; G_t* = tx. antenna gain; G_r = receiver

---

*Antenna gain for parabolic reflectors is:

$$G = 0.54\left(\frac{\pi D}{\lambda}\right)^2, \qquad D = \text{diameter}$$
$$= 20 \log f + 20 \log D - 42.3 \text{ dB} \tag{5.4c}$$

and the antenna beam width (between 3 dB points) is:

$$\theta \simeq 70\lambda/D = \frac{21{,}000}{fD}, \quad \text{where } D \text{ is in meters and } f \text{ in MHz.}$$

antenna gain; d = distance between the antennas; λ = wavelength of transmission; A_f = RF circuit losses and

$$\text{free-space path loss} = 20 \log \left(\frac{4\pi d}{\lambda}\right) \tag{5.4b}$$

For a single hop of 50 km using transmission at 2 GHz (λ = 15 cm), path loss = 132 dB. Assuming 3 m antennas, $G_t = G_r = 33$ dB. Usually, RF circuit losses $\simeq (5 + 5) = 10$ dB; then from eqn. (5.4a), we get

$$P_r = (P_t + 66 - 132 - 10) \text{ dBm}$$

With the required $P_r = C_i = -38.45$ dBm, the required P_t is:

$$P_t = -38.45 + 76 = 37.55 \text{ dBm} = 5.7 \text{ W}$$

For a fading channel with a required fade margin of 40 dB, the required P_t will be 57 KW, indeed a very large power.

Considering the problem from a different point of view, we note that the FM receiver should be operated above threshold, i.e., $C_i/N_i \geqslant 10$ dB approximately even at the trough of the fading cycle. Then,

$$C_i \geqslant (-89.2 + 10) \text{ dBm}.$$

With a mean received carrier power $P_r = -38.45$ dBm, as calculated above, the fade margin of $(-38.45 + 79.2) = 40.75$ dB is automatically provided. However, the worst SNR at the receiver output (at the midband) will be:

$$S_0/N_0 \text{ (mean)} = 10 + 18 + 3 = 31 \text{ dB}$$

instead of 68.75 dB. as given in eqn. (5.1).

Multihop Fading

For multihop fading, CCIR's recommendations are: With reference to 0 dBm test tone signal, the total noise power allowed at the output of the circuit $\leqslant$ 75,000 pW (weighted) for 99.99% of time. Since in an LOS link, the output carrier level C_0 is maintained constant through AGC during the trough of the cycle, the noise power N_0 is also increased by the same gain. If the fade depths are of the order of 40 dB, then for a single-hop fade, the allowable N_0 per hop is $(75{,}000/10^4) = 7.5$ pW. In multihop fading, it is normally assumed that 30% of the hops may fade simultaneously. This requires a further reduction of N_0 by 15 dB approximately; thus, in a general fading situation, the median $N_0 \leqslant 0.25$ pW $= -96$ dBm (weighted) $= -93.5$ dBm (unweighted). Using eqn. (5.1), the mean $S_0/N_0 = 93.5$ dB, and the worst-case $S_0/N_0 = (93.5 - 55) = 38.5$ dB. The required C_i is now obtained, as in the non-fading case, as:

$$C_i = \frac{S_0}{N_0} + N_i - \text{NIF} = -13.7 \text{ dBm}.$$

With reference to the value of $C_i = -38.45$ dBm in the non-fading case, the excess power required is $(38.45 - 13.7) \simeq 25$ dB, which is generally known as the Fade margin of the link. The transmitted power P_t now required is 62.55 dBm (=1.8 kW) (*cf.* the value of 57 kW obtained earlier).

Because of the large power requirement to maintain a reasonable SNR at the output of the circuit during the troughs of the fade cycles, it is necessary to use a quadruple diversity reception, which provides a 30 dB gain for 99.99% of time (a dual-diversity system will have a gain of 20 dB only). Adding the baseband equalization gain of 5 dB in FM, the total gain obtained is 35 dB. Thus, the above power requirement is now reduced to: $P_t = 62.55 - 35 = 27.55$ dBm $\simeq 0.57$ W only, and this provides for 55 dB fade margin, giving $S_0/N_0 \geqslant 38.5$ dB for 99.99% of time. The power requirement decreases with increase of the frequency of transmission as the antenna gains increase with frequency. Table 5.1 shows the important parameters at different frequencies for the above 2500 Km LOS link.

Table 5.1 Parameters of a 2500 Km, 600-ch LOS Link with 50 Hops

Total allowable noise power at the output $\leqslant$ 75,000 pW; RF BW = 30 MHz; 3 m antenna; RF losses = 10 dB; Ch. C_i/N_i = 20.5 dB; C_i = −13.7 dBm; $S_0/N_0 \geqslant 38.5$ dB with circuit reliability of 99.99%; fade margin = 55 dB.

Parameter	Frequency in GHz			
	0.9	2.0	4.0	8.0
Path loss (dB)	125	132	138	144
Required Tx. power without diversity (dBm)	68.55	62.55	56.75	50.65
,, ,, (KW)	7.2	1.8	0.45	0.1125
Required Tx. power with diversity and equalization (dBm)	33.55	27.55	21.75	15.65
,, ,, ,, (Watts)	2.28	0.57	0.14	0.035

In a troposcatter link, only a single hop with a fade depth of 40 dB may be considered. However, the path loss is large and, hence, the power requirement is also large. As an example, consider a Tropo link with a hop of 500 Km, transmission frequency = 2 GHz, RF BW = 30 MHz and parabolic antennas with diameter = 30 m. The antenna gain is now 54 dB. The path loss, as in [2] = 235 dB. For a single hop with a fade depth of 40 dB, the maximum allowable noise power at the receiver output is 7.5 pW; then $N_0 \leqslant -78.5$ dBm (unweighted). Using

$$\text{NIF} = 18 \text{ dB}, C_i = (78.5 - 89.2 - 18) = -28.7 \text{ dBm}.$$

Therefore, the transmitter power

$$P_t = (-28.7 + 235 + 10 - 108 - 35) = 73.3 \text{ dBm} = 21.4 \text{ KW},$$

which is the usual range of power used in Tropolinks.

In digital LOS/Tropolinks, the CNR at the receiver input is related to BER at the receiver output, and depending upon the type of modulation used, the minimum $C_i/N_i \geqslant 10$ dB for a BER $= 10^{-5}$. The fade margin required is the same, and the median $C_i/N_i \geqslant 50$ dB. The use of diversity reduces the requirement considerably; however, the diversity technique alone is not sufficient to maintain the required BER, and the decision feedback equalization or other adaptive techniques have to be used to realize efficient digital LOS/Tropolinks, as discussed in later sections.

5.2 DIGITAL MODULATION

An important aspect of digital radio transmission is the type of modulation used, in relation to its efficient use of the frequency spectrum, its sensitivity to noise and interference, its simplicity of practical equipment design and its resulting overall economy. The classical modulation techniques, e.g., ASK, FSK, PSK and DPSK as discussed in Chapter 2, have served well so far, but it has been now necessary to develop more efficient techniques for bandwidth conservation and to improve their relative sensitivity to noise, interference and fading. Extensive efforts have been made in recent years to develop new modulation schemes, known as continuous phase FSK (CPFSK), minimum phase-shift FSK (MSK), off-set QPSK (O-QPSK), Q-PRS, higher-order amplitude-phase keying (APK/QAM) and MA-MSK. QAM and PRS signalling have been discussed in Chapters 2 and 4 respectively. For channel transmission, the signalling waveforms are normally shaped through either raised-cosine filters (with $0.33 < \alpha < 1$), as discussed in Sec. 2.1, or PRS waveshaping shown in Fig. 4.18. The signal sets of PSK, MSK etc., are now modified as shown in Fig. 5.4, where it is seen that the number of signal vectors increases from 4 to 9 in case of Q-PRS and from 16 to 49 in case of 16-MA-MSK. This, of course, results in some degradation in error performance of PRS-filtered signals, as discussed in Chapter 4.

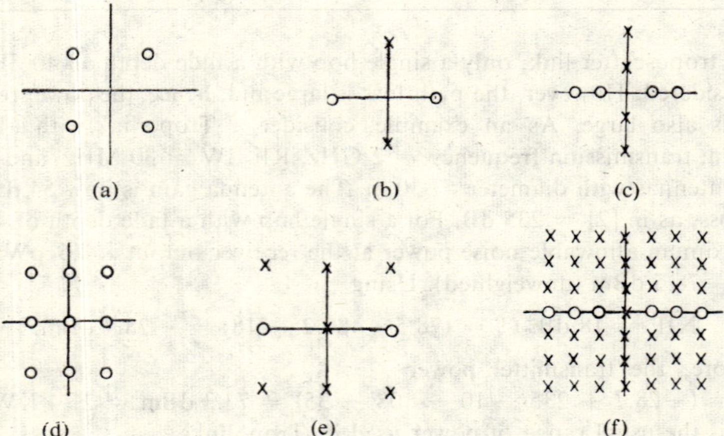

Fig. 5.4 Signal set description with raised-cosine and PRS filtering: (a) Q-PSK, R-cosine; (b) MSK, O-QPSK, R-cosine filtering; (c) MA-MSK, O-QPSK 4 level, R-cosine; and (d) QPSK-PRS; (e) MSK, O-QPSK, PRS-filtering; and (f) MA-MSK, 4-level PRS

The modulation waveforms of QPSK, O-QPSK and MSK are shown in Fig. 5.5, where it is seen that the Q-channel-modulating signal of MSK is

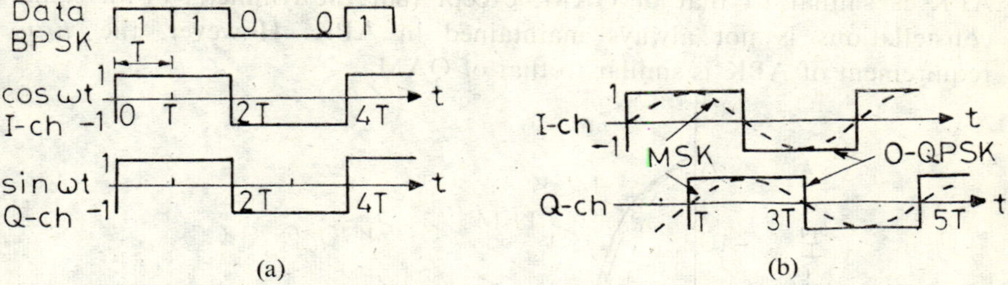

Fig. 5.5 Modulation waveforms of QPSK, O-QPSK and MSK: (a) QPSK and (b) MSK and O-QPSK

shifted by T sec and, at the same time, shaped to give a sharp spectral roll-off. The basic symbol shape in MSK is given by [3]

$$p(t)\,(\text{MSK}) = \sin\frac{\pi t}{2T}, \qquad 0 \leqslant t \leqslant 2T \tag{5.5}$$

and the resultant power spectral density $G(f)$ of an MSK signal is:

$$G(f)\,(\text{MSK}) = \frac{8T(1 + \cos 4\pi fT)}{\pi^2(1 - 16T^2f^2)} \tag{5.6}$$

where $f =$ frequency offset from the carrier. The rapid spectral roll-off of MSK is due to the fact that it avoids the abrupt phase changes at bit transition instants ($|\,\phi\,| \leqslant \pi/2$ due to the delay of T in the Q-component) unlike that in QPSK. In the Tamed-FM (TEM) [4], the phase transition is made even smoother, as shown in Fig. 5.6. The resultant spectra $G(f)$ of QPSK,

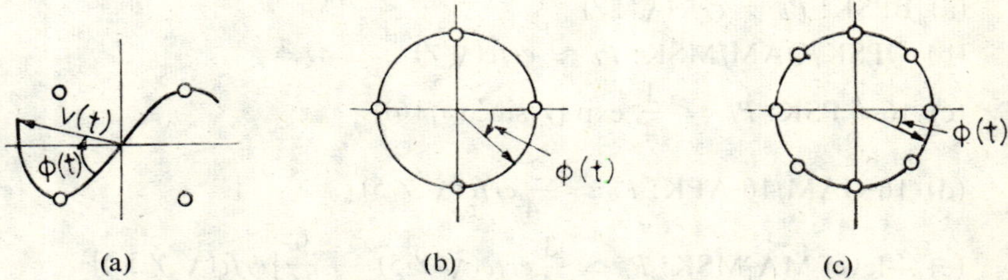

Fig. 5.6 Phase transition in QPSK, MSK and TFM; (a) QPSK, $|\,\phi\,| \leqslant \pi$; (b) MSK $|\,\phi\,| \leqslant \pi/2$; and (c) TFM, $|\,\phi\,| \leqslant \pi/4$

MSK and TFM are shown in Fig. 5.7, where it is seen that the spectra roll-off is sharper in MSK than in QPSK. Although the MSK spectrum has a wider main lobe (the first null is at $f = 0.75/T$) than QPSK, 99% of the total energy is contained in a bandwidth $B \simeq 1.2/T$ in MSK, but $B \simeq 8/T$ in QPSK. Thus, MSK is spectrally more efficient and may be used in digital FDM systems with very little adjacent channel interference. The error performance of MSK is within 1 dB of that in QPSK; and its demodulation and synchronization circuits are simple. The multiamplitude MA-MSK is a higher-order signalling format similar to APK/QAM. The signal format of APK is similar to that of QAM, except that the symmetry of the signal constellations is not always maintained in APK. However, the E_b/n_0 requirement of APK is similar to that of QAM.

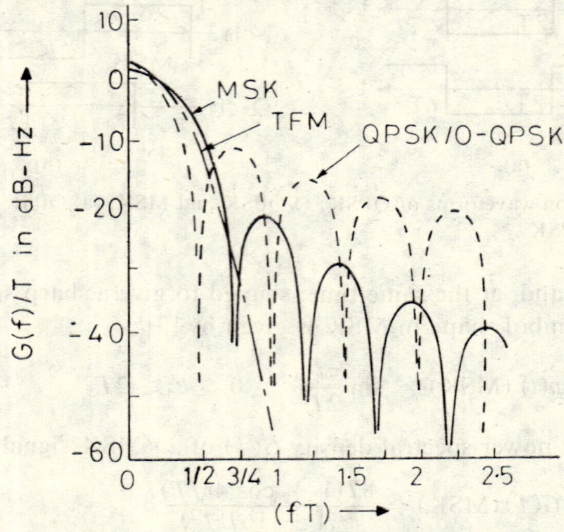

Fig. 5.7 Spectrum of MSK, TFM and QPSK

For the purpose of obtaining higher spectral efficiencies, multilevel PAM/QAM, multiphase PSK, higher-order APK and MA-MSK signalling have been used in LOS systems. Error-rate performances of these are shown to be [5, 6, 7].

(a) BPSK: $P_b = erfc(\sqrt{2\gamma})$

(b) QPSK/QAM/MSK: $P_b \simeq erfc(\sqrt{\gamma})$

(c) 16-ϕ PSK: $P_b \simeq \frac{1}{4} \exp [\gamma \sin^2 (\pi/16)]$

(d) 16-QAM/16-APK: $P_b \simeq \frac{3}{4} erfc(\sqrt{\gamma/5})$

(e) 4-level MA-MSK: $P_b \simeq \frac{3}{4} erfc(\sqrt{\gamma/5}) - \frac{9}{16}[erfc(\sqrt{\gamma/5})]^2$ (5.7)

where P_b = BER, $\gamma = C/N$, C = average carrier power and $N = \sigma^2$. P_b vs γ curves for various modulation schemes are shown in Fig. 5.8. It is seen that APK/QAM are more efficient in terms of E_b/n_0 as compared to multiphase PSK. MSK, QPSK and O-QPSK give similar performances, except that C/N required for $\alpha < 1$ is more because of ISI problems, as shown in Fig. 5.9. In Q-PRS, C/N requirement is 2.1 dB more, but with carrier-phase offset, the deterioration is more, as shown in Fig. 5.10. The overall bit-rate efficiency for different modulation schemes using the Nyquist bandwidth ($\alpha = 0$) and raised-cosine filtering ($\alpha = 0.33$ and 1), are shown in Figs. 5.11(a) and (b). It is evident that the simple modulation schemes

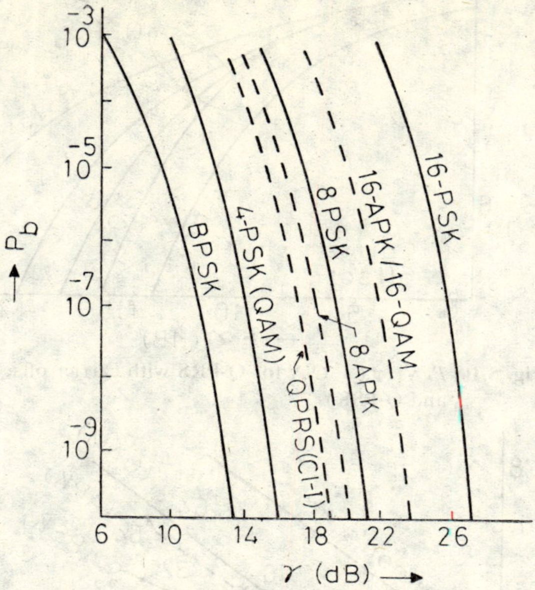

Fig. 5.8 P_b vs. γ for different digital modulation systems. RMS γ is specified in the double sided Nyquist BW.

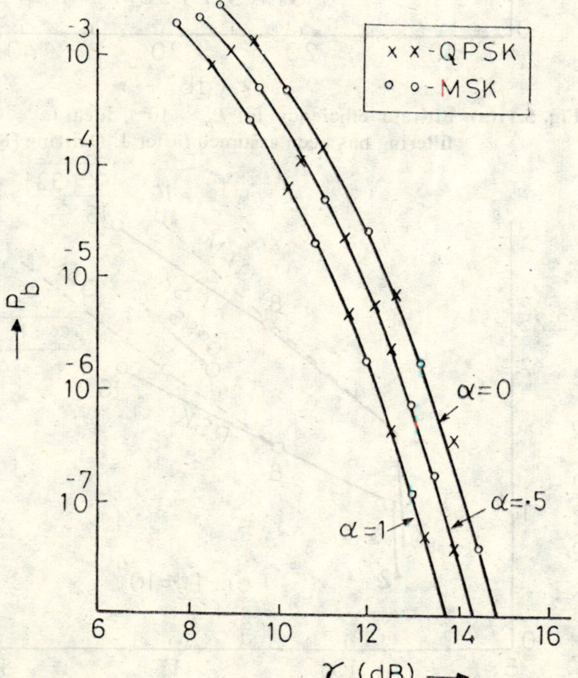

Fig. 5.9 P_b vs. γ: same for MSK and O-QPSK

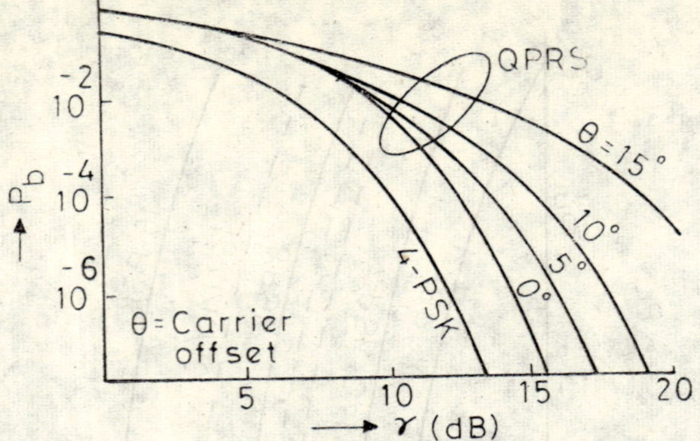

Fig. 5.10 P_b vs. γ (= C/N for Q-PRS with carrier offset and Q-PSK)

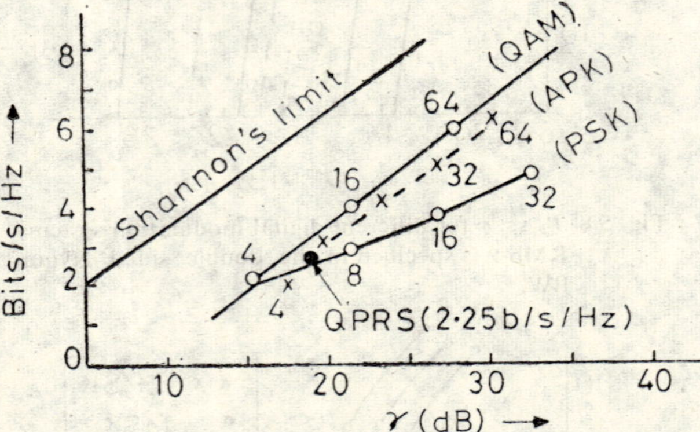

Fig. 5.11(a) Bit-rate efficiency for $P_b = 10^{-8}$, ideal ($\alpha = 0$) filtering has been assumed (after J. Oeitting [8])

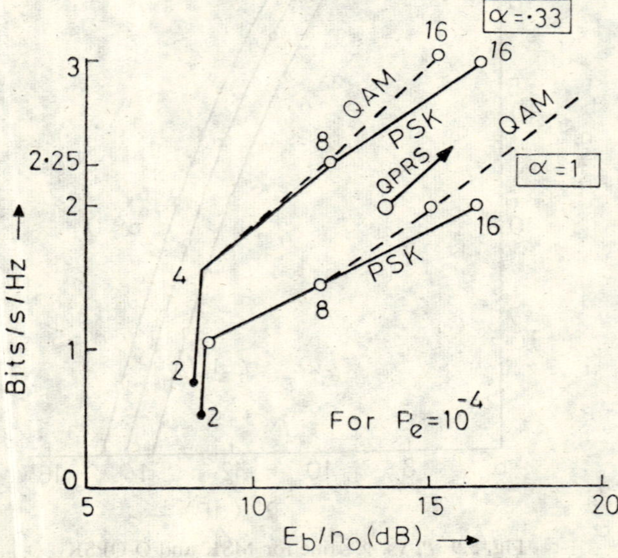

Fig. 5.11(b) Bit-rate efficiency for $P_b = 10^{-4}$, $\alpha = 0.33$ and 1.0 (after J. Oeitting [8])

cannot achieve the ideal Shannon's bit-rate efficiency, and there is a gap of 8 to 10 dB, which can be partially bridged by efficient coding techniques. The useful characteristics of APK/FSK/PSK systems are given in Table 5.2 [8], where the bit rate efficiency (in practical BW), E_b/n_0, and the excess power requirement in the presence of CW interference (I) have been shown. These data will be useful for designing high efficiency systems.

Table 5.2 Relative Performances of Digital Modulation Schemes (after Oeitting [8]

Type	Modulation	Speed bits/s/Hz (RF BW)	E_b/n_0 (dB) for $P_b = 10^{-4}$	CW interference: E_b/n_0 (dB) reqd. for $P_b = 10^{-4}$	
				$C/I = 10$ dB	$C/I = 15$ dB
AM	OOK-coherent	0.8	12.5		
	OOK-env. det.	0.8	13.0	~ 20	14.5
	QAM	1.7	9.5		
	QPRS	2.25	11.7 (coding gain 2 dB)	12.5	11.7
FM	PSK-noncoherent, $D = 1$	0.8	11.8	14.7	13.3
	CP-FSK-coherent, $D = 0.7$	1.0	8.9*		
	MSK, $D = 0.5$	1.9	9.4	~ 11.4	
	MSK- diff. encoding $D = 0.5$	1.9	10.4		
PM	BPSK-coh.	0.8	9.4	10.5	9.5
	D-CPSK	0.8	9.9	11.0	9.9
	DPSK	0.8	10.6	12.0	10.6
	QPSK	1.9	9.9	12.2	9.9
	D-QPSK	1.8	11.0	> 20	14.0
	8ϕ-PSK-COH	2.6	12.8	~ 20	15.8
	Off-set QPSK	1.8	10.4		
	16-ϕ-PSK-COH	2.9	17.2		> 24
AM/PM	16-PAK	3.1	13.4		18.0
	4-level MA-MSK, O-QASK (raised cosine)	3.6	14.7		
	4-level MA-MSK O-QASK (PRS)	4.2	16.8		

(*Assumes 3-bit integration time)
D = FM Mod. Index.

5.3 ILLUSTRATIVE SYSTEM DESCRIPTIONS [9]

Using the recently developed techniques of modulation, many efficient LOS systems have been designed and experimented with. Some brief descriptions of a few such systems are given as follows:

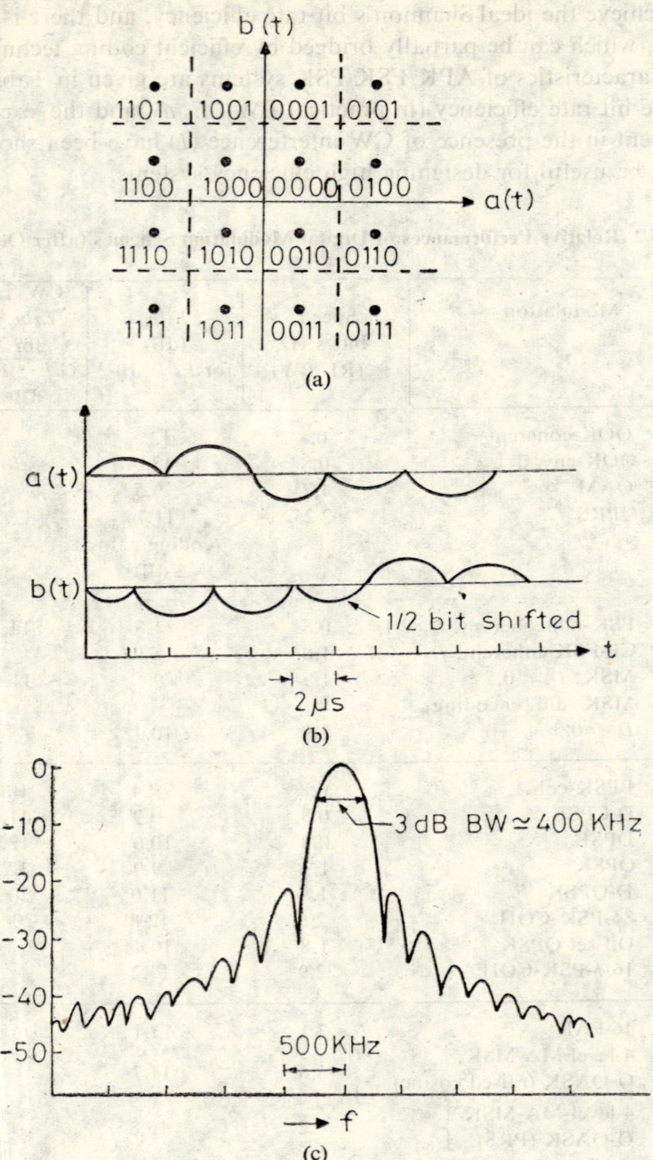

Fig. 5.12 (a) 4-level QASK represented as 16 discrete carrier amplitudes and phases (Gray-coded); (b) waveforms $a(t)$ and $b(t)$ in quadrature baseband channels for MA-MSK; and (c) spectrum of the 4-level MA-MSK signal (1 Mb/s)

(a) 91 Mb/s Digital Radio with Q-PRS [10]

This is a long-haul, 8 GHz, Canadian system having the characteristics:

RF BW = 40 MHz, 2.25 bits/s/Hz, $C/N < 17$ dB for $P_e = 10^{-4}$

Although both Q-PRS and 8-PSK satisfy the requirement of 2.25 b/s/Hz spectral efficiency, Q-PRS-complexity is the same as that of Q-PSK only, and much less than that of 8-ϕ PSK. Q-PRS has the 2 dB processing gain and the eye-opening is similar to that of 8-PSK for amplitude and delay distortions. DFE is used for intersymbol interference correction.

Practical performance: $C/N \rightarrow 1$ dB less than ideal E_b/N_0 ($\simeq$11.7 dB).

C/N degradation $\leqslant 1$ dB for $C/I \simeq 6$ dB.

(*b*) *4-level MA-MSK at 1 Mb/s* [7]

Basically, the system is 16-QASK, with 4 levels of I and Q carrier amplitudes. The base band pulse shaping and offset quadrature components, similar to that used in MSK, was chosen to produce desirable spectral properties. The signal set for the system is shown in Fig. 5.12(a) and I-Q channel waveforms are shown in Fig. 5.12(b). Spectrum for 1 Mb/s data signals is shown in Fig. 5.12(c). The BER, as given by eqn. (5.7) and also the experimental results are shown in Fig. 5.13. The system achieves substantial band width reduction and improved performance over M-ϕ PSK. The modulator/demodulator circuits used are similar to those used in 16-QASK, except that the modulating waveforms are shaped as shown in Fig. 5.12(b). Necessary phase tracking, AGC and symbol synchronization are incorporated in the system. The system operates within 0.5 dB of the theoretical performance.

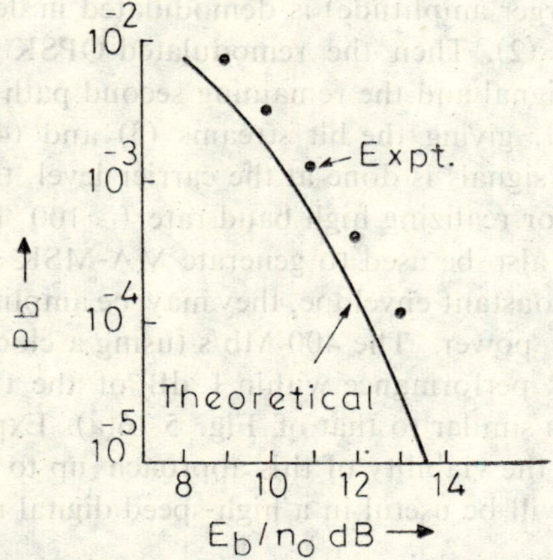

Fig. 5.13 P_b vs $\cdot E_b/n_0$ for 1 Mb/s MA-MSK system

(*c*) *200 Mb/s, 16-QAM System* [11]

The specifications of the system are:

Operating frequency 4.4–5.0 GHz; Channel capacity: 200 Mb/s;

Modulation: 16-QAM; Differential encoding; Coherent detection;

RF channel spacing: 40 MHz; Symbol rate: 50 Mbauds/s;

Spectral efficiency: 5 b/s/Hz, $\alpha = 0.5$; Repeater spacing: 50 km;

Regeneration at every repeater.

The block diagrams of the transmitter and receiver are shown in Fig. 5.14, where the signal set generated is similar to that of the 4-level MA–MSK of Fig. 5.12(a). For achieving 5 b/s/Hz efficiency, differential encoding and special binary transversal filters (BTF) have been used. The resultant signal spectrum is shown in Fig. 5.15, and the first null is seen to be at $f < 1/T_s$, T_s = symbol time. The side lobes outside $f = 40$ MHz are reduced by further filtering. In BER performance, there is a 1.5 dB degradation from the theoretical, as shown in Fig. 5.16(a) and the equivalent $E_b/n_0 \simeq 17$ dB for $P_e = 10^{-5}$. The effect of CW interference is shown in Fig. 5.16(b), where there is a 2 dB deterioration for $C/I = 30$ dB. Also, a 5 dB back-off is necessary in the TWT amplifiers to maintain the linearity of the signal set and this may lead to a further 1–2 dB degradation in C/N for a given P_b. However, the system is the most efficient in terms of b/s/Hz

(d) 400 Mb/s, 16-APK System [12]
The 16-APK has the same signal set as that of Fig. 5.12(a), but has a novel modem configuration, as shown in Fig. 5.17. In the modulator, the APK signal is generated as the sum of the outputs of two QPSK modulators, each driven by the two bits of the 4-bit APK word. The output level of Mod-I is 6 dB higher than that of Mod. II, and the resulting signal constellations are shown in Fig. 5.18, which is the same as in Fig. 5.12(a). The demodulation is performed again in parallel as shown in Fig. 5.17(b). The first path QPSK signal (of larger amplitude) is demodulated in detector I, giving the bit streams (1) and (2). Then the remodulated QPSK signal is subtracted from the 16-APK signal and the remaining second path signal is demodulated in detector II, giving the bit streams (3) and (4). Since the logical translation of 4-bit signals is done at the carrier level, the circuit is simple and is very useful for realizing high baud rate ($\geqslant 100$ Mb/s) transmission. The technique may also be used to generate MA-MSK signals and because the signals have a constant envelope, they may be amplified non-linearly and summed at a high power. The 400 Mb/s (using a clock rate of 100 MHz) modem gave a BER performance within 1 dB of the theoretical, and the P_b vs. C/N curve is similar to that of Fig. 5.16(a). Experiments at higher clock rates showed the viability of this approach up to 1.6 Gbits/s rate of transmission, and will be useful in a high-speed digital mm-wave communication system.

(e) 4-bit/s/Hz-PRS-SSB Digital Radio at 2 GHz [13]
By using 7-level PRS signalling, 192 PCM channels corresponding to a bit-rate of 12.63 Mb/s are accommodated in an RF BW of 3.1 MHz only. Initially, 2 data bits are used to form 4-level PAM signals and then each PAM pulse is multiplied by the modified duobinary code ($1\text{-}D^2$), resulting into 7-level signals and, at the same time, the channel capacity improves to 4 b/s/Hz. PRS signals at 3.156 MHz are now converted to SSB signals at 2 GHz. The BER characteristics show: $P_b = 10^{-5}$ for $C/N = 25$ dB,

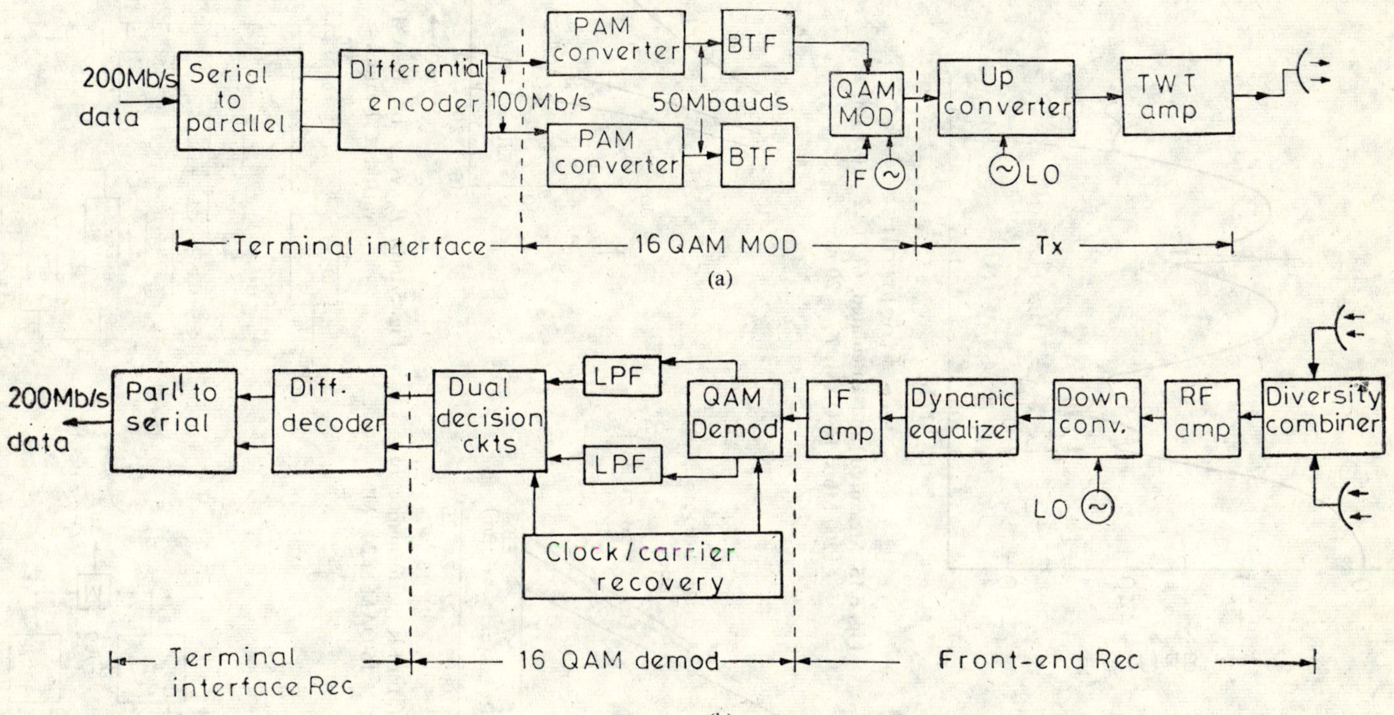

Fig. 5.14 16-QAM transmitter and receiver; (a) transmitter and (b) receiver and demodulator

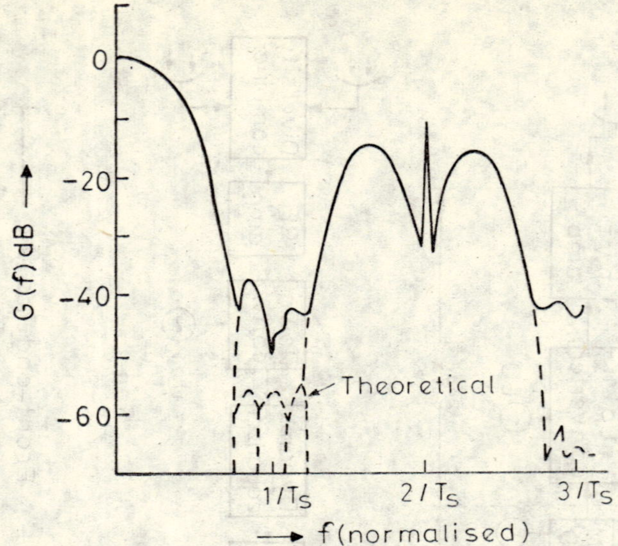

Fig. 5.15 Measured and calculated $G(f)$, $\alpha = 0.5$ for 200 Mb/s, 16-QAM, $T_s = 20$ ns

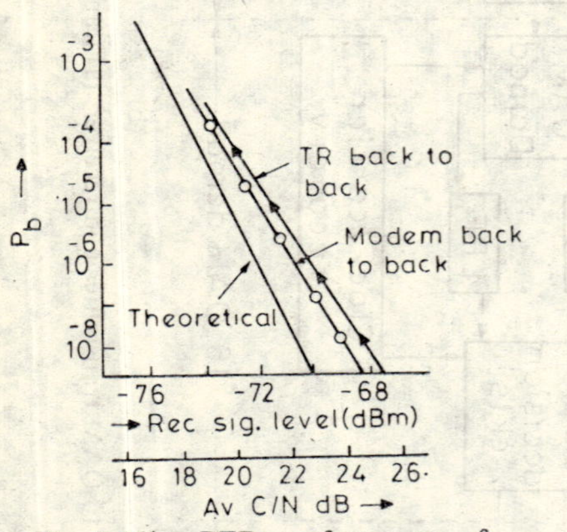

Fig. 5.16(a) BER performance of 16-QAM at 200 Mb/s

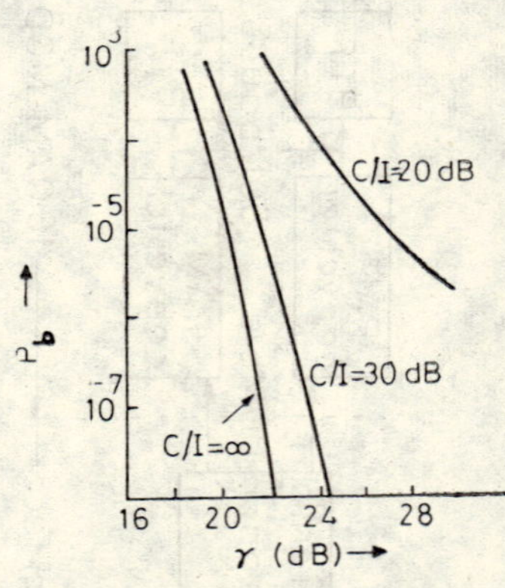

Fig. 5.16(b) P_b vs. γ in 16-AMQ for various C/I values

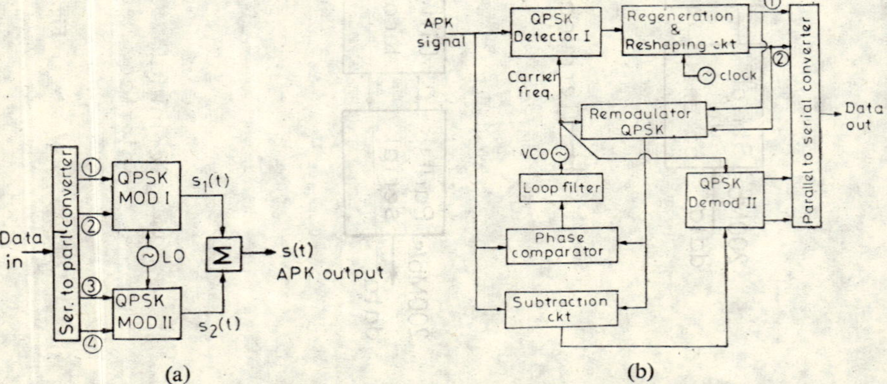

Fig. 5.17 16-APK modulator and demodulator: (a) APK parallel modulator and (b) APK parallel demodulator

$P_b = 10^{-6}$ for $C/N = 27$ dB, which is somewhat worse than the values of the 16-QAM system [Fig. 5.16(a)].

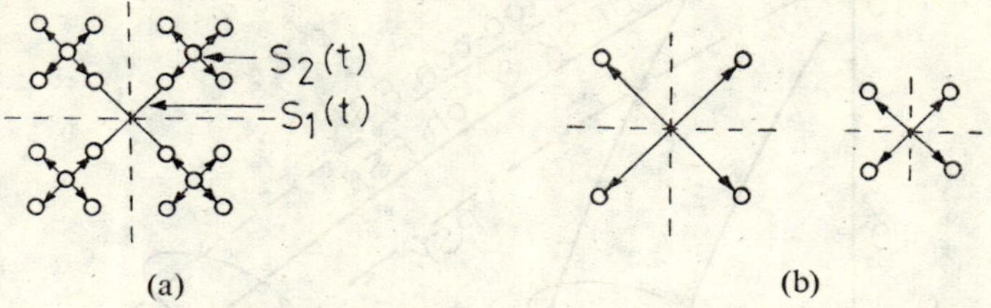

Fig. 5.18 Signal constellation of Fig. 5.17(a)

5.4 FADING AND MULTIPATH

LOS and Troposcatter channels exhibit fading and multipath phenomena and the digital ratio systems are adversely affected. Assuming Rayleigh fading, the average error rates for different signalling schemes deteriorate to [2]:

(a) Coh. ASK, $\overline{P}_b$ with optimum threshold $\rightarrow \frac{1}{\gamma_0}$, for large γ_0.

(b) Non-coh. FSK, $\overline{P}_b = \frac{1}{2 + \gamma_0} \rightarrow \frac{1}{\gamma_0}, \quad \gamma_0 \gg 1.$

(c) Coh. FSK, $\overline{P}_b = \frac{1}{2}\left[1 - \frac{1}{\sqrt{1 + 2/\gamma_0}}\right] \rightarrow \frac{1}{2\gamma_0}, \quad \gamma_0 \gg 1.$

(d) DPSK, $\overline{P}_b = \frac{1}{2 + 2\gamma_0} \rightarrow \frac{1}{2\gamma_0}, \quad \gamma_0 \gg 1.$

(e) Coh. PSK, $\overline{P}_b = \frac{1}{2}\left[1 - \frac{1}{\sqrt{1 + 1/\gamma_0}}\right] \rightarrow \frac{1}{4\gamma_0}, \quad \gamma_0 \gg 1, \quad (5.8)$

where $\gamma_0 = \langle\gamma\rangle$ = average CNR over a fading cycle. The results are shown graphically in Fig. 5.19.

Based on these results, the E_b/n_0 requirements of different modulation schemes for $\overline{P} = 10^{-2}$ with Rayleigh fading have been calculated and are given in Table 5.3 [8]. By using Error control coding, $\overline{P}_b$ can be improved to the desired value. The effect of delay distortion, mainly due to the filters in the transmit-receive chain, is also shown in the Table, for both quadratic and linear delay with the maximum differential delay $d = T_s$, the symbol duration. It is seen that in QAM, MSK and QPSK, the E_b/n_0 requirements are similar for both Rayleigh fading and delay distortion, but in higher-order signalling alphabets (e.g., in 16-PSK and 16-APK), the penalty is more.

Using (space/frequency) diversity combining, the error rate improves, depending upon the order of diversity M. Of the three techniques, viz., selectioncom bining, maximal ratio combining and equal gain combining.

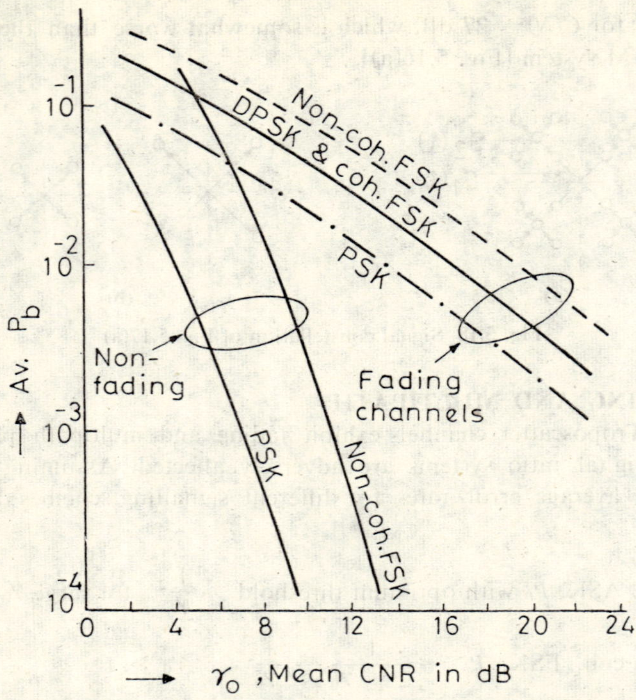

Fig. 5.19 Average P_b for FSK and PSK systems in Rayleigh fading

Table 5.3 Performance of Digital Modulation Schemes in Presence of Rayleigh Fading and Delay Distortion (after Oeitting [8])

Type	Modulation	Average E_b/n_0 (dB) for $P_b = 10^{-2}$ and Rayleigh fading	E_b/n_0 (dB) for $P_b = 10^{-4}$ for delay distortion $d/T_s = 1$	
			Quadratic	Linear
	OOK-coh	17 } Assumes	12.8	12.4
	OOK-env. det.	19 } opt. threshold	13.3	16.9
AM	QAM	14	9.8	15.8
	QPRS	16	—	> 25
	FSK-Non-coh.	20	13.5	16.0
	CP-FSK coh, $D = 0.7$	13 } Assumes	8.8	8.6
	CP-FSK-Non-Coh.,	} 3-bit integration		
FM	$D = 0.7$	18 } time	10.2	12.7
	MSK, $D = 0.5$	14	9.8	15.8
	MLK-diff. encoding,			
	$D = 0.5$	17	10.8	16.8
	BPSK-coh	14	9.8	9.6
	D-CPSK	17	10.3	10.1
	QPSK	13.5	9.8	15.8
PM	D-QPSK	20	16.3	—
	Off-set QPSK	13.5	9.8	15.8
	8-ϕ PSK	16.5	<25	~ 25
	16-ϕ PSK	21.0	—	—
AM/PM	16-APK	18	—	—

the maximal ratio (m.r.) combining provides the maximum improvement.* Assuming $\gamma_{0i} = \gamma_{0j} = \gamma_0$, i.e., the same mean CNR in all branches, the average error rates in maximal ratio combining are given by [2]:

For non-coherent FSK and DPSK:

$$\overline{P}_b(\text{m.r}) = \frac{1}{2} \prod_{j=1}^{M} \left[\frac{1}{(1+\alpha\gamma_0)} \right], \quad \text{where} \begin{cases} \alpha = 1/2 \ \ldots \ \text{NC-FSK} \\ \alpha = 1 \ \ldots \ \text{DPSK} \end{cases}$$

$$\simeq \frac{1}{2} \prod_{j=1}^{M} \left[2P_{ej} \right], \quad \text{for } \gamma_0 \geqslant 1 \tag{5.10}$$

For coherent FSK and PSK,

$$\overline{P}_b(\text{m.r}) \simeq \frac{1}{2\sqrt{\pi}} \times \frac{1}{(\alpha\gamma_0)^M} \times \frac{\left(M - \frac{1}{2}\right)!}{M!} \tag{5.11}$$

where γ_0 is small; and $\alpha = 1/2$ for C-FSK aud $\alpha = 1$ for PSK. These results are shown in Fig. 5.20, where it is seen that the implicit

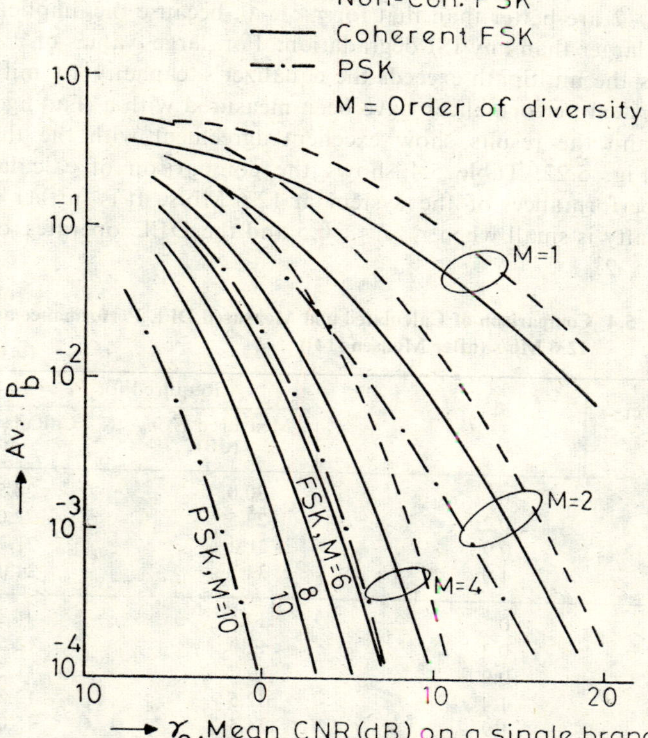

Fig. 5.20 Error rates with maximal ratio combining for FSK/PSK systems in Rayleigh fading (after Schwartz *et al.* [2])

*In the m.r. combiner, the diversity branch weights a_j are given as:

$$|a_j| \propto |S_j/\sigma^2| \quad \text{and} \quad \angle a_j = -\angle s_j(t),$$

where $s_j(t)$ is the signal with power S_j in the jth branch. This combiner then gives an instantaneous output CNR as:

$$\gamma = \sum_{j=1}^{M} \gamma_j \tag{5.9}$$

diversity gain potential is large for large M and the system gives CNR gains more than in a non-fading LOS case.

5.4.1 Experimental Results [14, 15]

Monsen [14] and others have made extensive studies on fading multipath channels, using diversity-combining and decision feedback equalizers (DFE). The system configuration for a data rate of 12.6 Mb/s, using a QPSK modem, DFE and quadruple diversity, is shown in Fig. 5.21(a). Because of the multipath, the pulsed signal is spread to the extent of $(T + \tau_m)$, where τ_m is the multipath spread and shown in Fig. 5.21(b); and this results in a considerable ISI at the receiver input. The precursors of the ISI are now eliminated by 3-tap forward filters and the postcursors are corrected by the common backward filter (DFE). The theoretical performance of the system is shown in Fig. 5.22. The results are somewhat similar to those of Fig. 5.20, except that BER varies with τ_m. It is also seen that BER for small values of τ_m/T are better than that for $\tau_m = 0$, because the implicit diversity gain is larger than any ISI degradation. For large values of τ_m/T, BER increases as the multipath exceeds the equalizer's capacity to mitigate the ISI. The system performances have been measured with a Fading channel simulator and the results show excellent agreement with the theoretical results of Fig. 5.22. Table 5.4 shows the comparison of calculated and measured performances of the system at 12.6 Mb/s. It is further seen that the ISI penalty is small when $\tau_m/T \leqslant 0.5$ and the DFE operates efficiently with $\tau_m/T \leqslant 2$.

Table 5.4 Comparison of Calculated and Measured DFE Performance at 12.6 Mb/s (after Monsen [14])

Order of diversity, M	τ_m/T	Required for $\bar{P}_b = 10^{-6}$	
		Measured $\bar{E}_b/n_0$ (dB)	Calculated $\bar{E}_b/n_0$ (dB)
2	0	30.0	28.5
2	0.2	22.5	22.0
2	0.9	21.0	21.7
2	1.9	>33	24.0
4	0	14.5	14.2
4	0.2	12.0	12.0
4	0.9	10.7	11.5
4	1.4	12.5	12.5
4	1.9	14.5	15.0

The system with DFE was also tested in a field test facility [15] over a 168-mile Tropolink, using 4.5 GHz and 900 MHz channels. The test was made at data rates 3–12.6 Mb/s, using $M = 2$ or 4, and the values of τ_m, the multipath spread, ranged from 0.163 μs to 0.41 μs. The test results are shown in Fig. 5.23. The effectiveness of DFE and diversity for digital Tropolinks is well established by the above results.

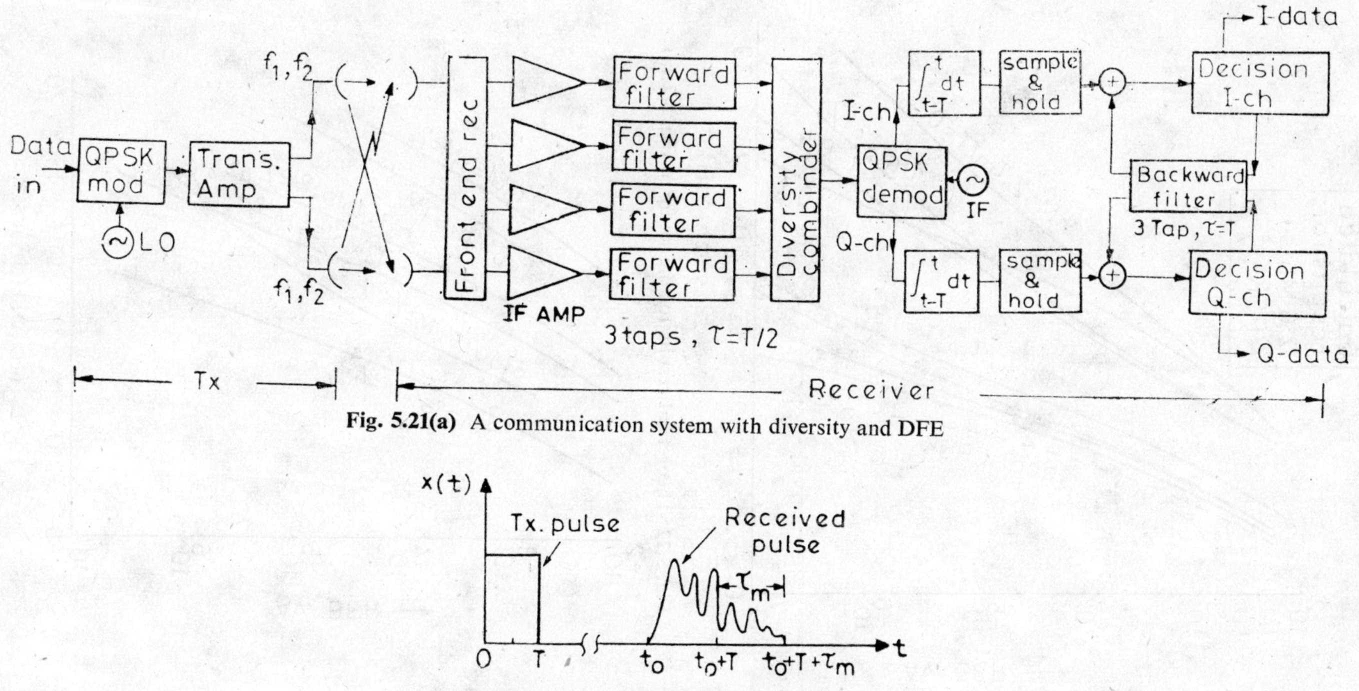

Fig. 5.21(a) A communication system with diversity and DFE

Fig. 5.21(b) Response of a multipath channel

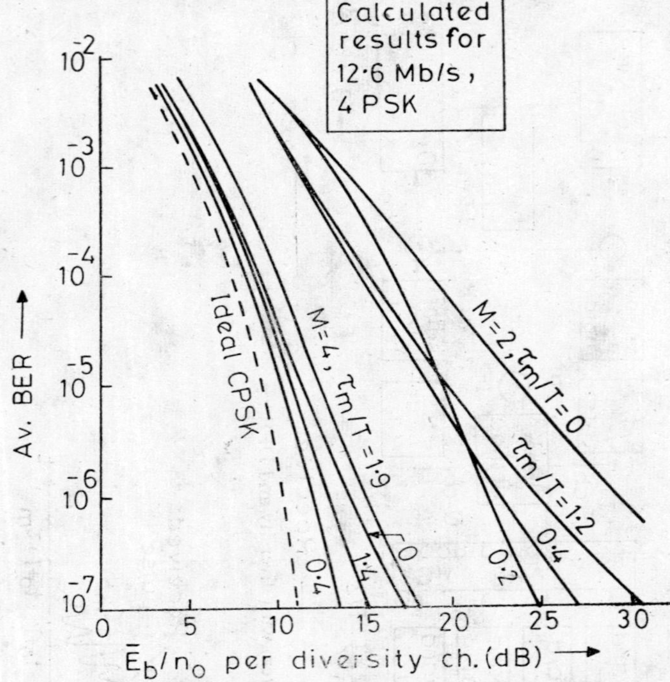

Fig. 5.22 Theoretical results for the set-up of Fig. 5.21(a)

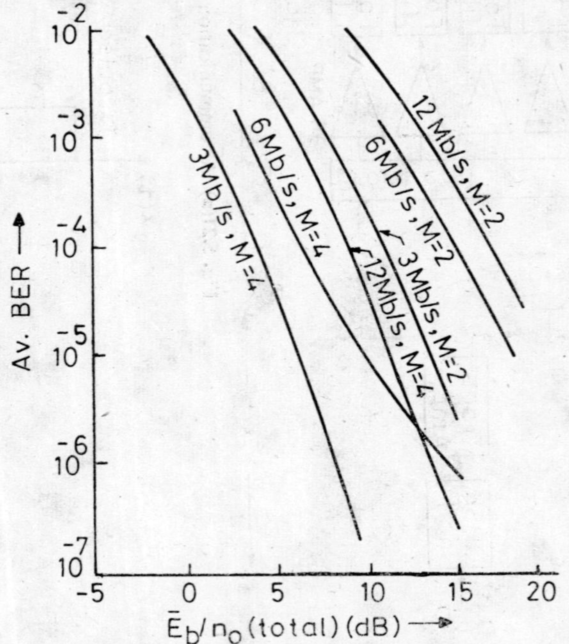

Fig. 5.23 Experimental test results for the set-up of Fig. 5.21(a) (after Ehrman and Monsen [15])

5.4.2 Recent Results

Apart from Monsen's experiments, Barnett [16] in his recent investigations has shown that the traditional average power fade depth was a poor indicator of BER performance, as shown in Fig. 5.24(a). Modest in-band amplitude dispersion, say, 0.2 dB/MHz, was sufficient to cause BER to exceed 10^{-3} and thus amplitude dispersion was found to be a good indicator of BER performance. The excellent correlation between BER and the in band frequency dispersion is shown in Fig. 5.24(b).

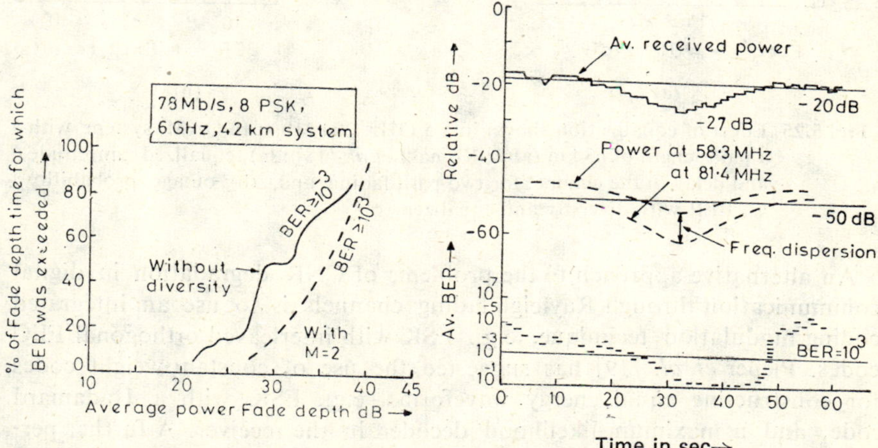

Fig. 5.24(a) BER as a function of Fade depth, with and without diversity

Fig 5.24(b) Dispersion and BER vs. time (sec) of the system of Fig. 5.24 (a) without diversity (after Barnett [16])

Anderson *et al.* [17] have also concluded, from their experiments with the 91 Mb/s, QPRS, RD-3 system, as follows:

Multipath induced outage is much higher than would be predicted from the measured flat-fade margin of the system. The primary cause of the outage is the in-band distortion caused by the frequency selectivity of the multipath fading process. Phase-adaptive space-diversity combining is very effective in reducing the severity of the in-band distortion for a given fade depth. A simple adaptive linear equalizer in conjunction with diversity provides an improvement factor $\simeq 20$, thus reducing the multipath outage to an acceptable limit (per hop outage probability $< 10^{-6}$).

In the diversity experiments with the 200 Mb/s, 16-QAM system [18], the above results were confirmed and it was found that a 4-dB amplitude dispersion occurrence probability, within a ± 20 MHz bandwidth, corresponds to the outage probability. A dynamic equalizer was used to correct the delay/amplitude dispersion and the results are shown in Fig. 5.25(a). The outage probabilities of the system, corresponding to no-diversity, with diversity and with dynamic equalizer, are 2.5%, 0.45% and 1.7%, respectively, as shown in Fig. 5.25(b). By using the dynamic equalizer in conjunction with diversity, sufficient improvement ($\simeq 10$) is expected.

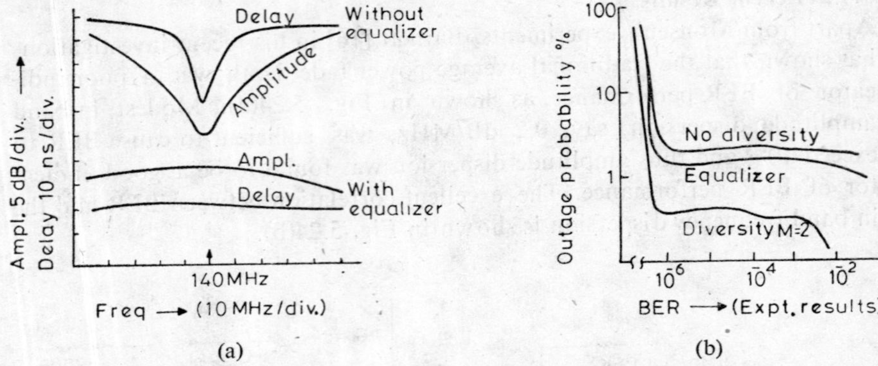

Fig. 5.25 Effect of equalization shown in a 5 GHz, 200 Mb/s, 16-QAM system with a path length of 63 km (after Komaki *et al.* [18]): (a) equalized amplitude and delay in the channel for two-path fading and, (b) outage probability vs. BER with diversity and equalizer

An alternative approach to the problems of CNR degradation in digital communication through Rayleigh fading channels is to use an integrated coding/modulation technique, e.g., FSK with interleaved orthogonal/FEC codes. Pieper *et al.* [19] has suggested the use of constant-weight codes for constructing equal energy waveforms, e.g., FSK with a Hadamard code, and a maximum-likelihood decoder in the receiver. A further performance gain is achieved by the concatenation of the RS code and the Hadamard code. Interleaving the coded data over part of the fading cycle before transmission over the channel avoids the error bursts due to interference normally present. The block diagram of such a coded system and its theoretical performances using different codes are shown in Fig. 5.26. For comparison, $\bar{P}_b$ for two-diversity FSK systems is also shown in the figure. It is observed that the coding gain (with concatenation) can be as high as 10 dB, assuming, of course, that proper interleaving/deinterleaving has been done.

A digital Tropo system for tactical communication [20], using dual diversity and interleaved FEC, has been designed by Marconi Ltd. The system uses QPSK with R 1/2 modified two-error correcting Hamming code, and the receiver C/N in the range of 8–30 dB. With 128 channels of 16 Kb/s ADM, the transmission bit rate is approximately 4 Mb/s, and the receiver BER $< 10^{-3}$. The codec requires an interleaving delay of 25 ms, equivalent to 100 Kb/s, and the storage has been accomplished through RAM's and PROM's. The system has given excellent field performance.

From above results, it may be concluded that:

(a) The propagation impairments on LOS systems, i.e., the effect of fading and multipath, are serious when digital radio is implemented in predominantly analog frequency bands. More spectrally-efficient modulation schemes exhibit an increased susceptibility to multipath and rain and,

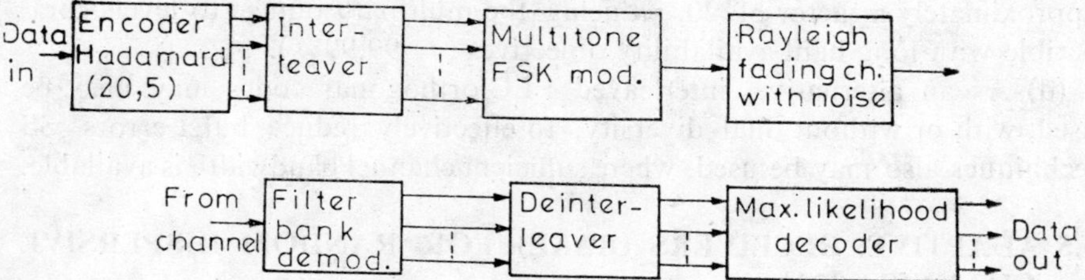

Fig. 5.26 (a) Block diagram of a coded communication system

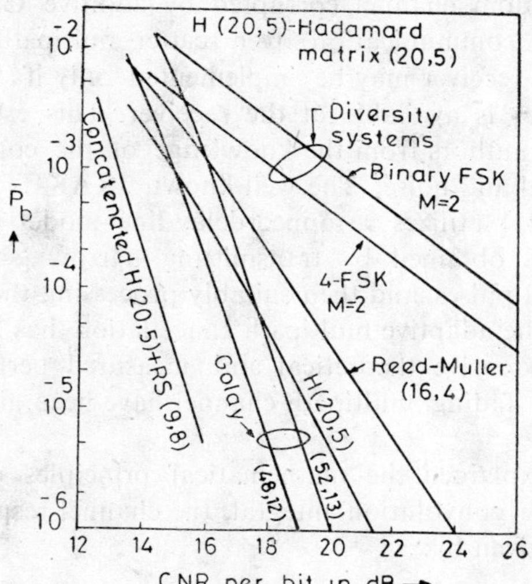

Fig. 5.26 (b) BER performances of several codes and conventional diversity (after Pieper *et al.* [19])

as such, they require special considerations. The traditional method of providing fade margins in FM analog systems does not provide full protection in digital systems and the commonly used average-power fade-depth is a poor indicator of the error rate performance. But the in-band amplitude and delay dispersions have a good correlation with the error-rate performance. In all experimental investigations, it was found that the primary cause of outage is the in-band distortion caused by the frequency selectivity of the multipath fading process.

(b) The phase-adaptive space-diversity combining is very effective in reducing the effect of fading. In addition, it affords some increase in the effective fade margin of the signal, i.e., it reduces the severity of in-band distortion for a given fade depth. However, diversity only does not guarantee the system performance within allowable limits.

(c) It is necessary to use an adaptive linear amplitude-cum-delay equalizer along with the diversity to obtain satisfactory performance. It has been shown that the combined improvement due to diversity and equalizer is approximately a factor of 20, reducing the multipath outage to levels compatible with long-haul availability objectives ($\simeq$ 99.99%).

(d) As an alternative, interleaved FEC/orthogonal codes may also be used, with or without dual diversity, to effectively reduce burst errors. SS techniques also may be used, where sufficient channel bandwidth is available.

5.5 ADAPTIVE RECEIVERS (RAKE) FOR RANDOM DISPERSIVE CHANNELS [21]

It is well known that the MF/cross-correlator receiver is the optimum. But in a 'Gaussian' random channel corrupted by additive Gaussian noise, a situation typical in communication over scatter-multipath channels, the cross-correlator/MF receiver may be implemented only if the estimate of the channel statistics is available at the receiver. This estimate has been obtained by various authors from the knowledge of the correlation matrix of the received signal and noise. The well-known 'RAKE' receiver, based on time-diversity [21] utilizes a tapped-delay-line model of the channel; and the estimate is obtained by transmitting narrow 'sounding' pulses along with the signal pulses, and then suitably processing the delay line output. Alternatively, the adaptive multipath cancellation has been suggested by Morgan [22]. Extensive theoretical and measured performances of a DFE modem for a fading multipath channel have been given by Monsen and others [14].

Kailath [23] has explained the mathematical principles of the channel estimator. Using the convolution integral, the channel response of a time-varying medium is given by:

$$Z(t) = \int_0^t a(\tau, t) \cdot x(t - \tau)\, d\tau \tag{5.12}$$

where $a(\tau, t)$ = time-varying impulse response of the medium, $a(\tau, t) = 0$ for $\tau < 0$ and $x(t)$ = transmitted signal, $x(t) = 0$, for $t < 0$.

In the discrete analog form, the channel output may be written as:

$$Z(m) = \sum_{k=0}^{m} a(k, m) \cdot x(m - k),$$

and in the matrix form, $[Z] = [A] \cdot [X]$, where $X_t = \lfloor \underline{x(0)\ x(1)\ x(2) \ldots} \rfloor$ is a row vector representing the discrete input signal $\{x^{(k)}\}$ and X_t = transpose of X. With the additive Gaussian noise in the channel, the received signal is:

$$[Y] = [A] \cdot [X^{(k)}] + [N] = [Z^{(k)}] + [N] \tag{5.13}$$

where all variables are Gaussian.

The channel response, given by $[A]$, may be closely represented by a discrete filter, say, a TDL whose tap functions $a_i(t)$ are sample functions from a Gaussian process. Assuming the correlation matrix ϕ_{AA} to be known, $\phi_{ZZ}^{(k)}$, $\phi_{YY}^{(k)}$ may be estimated. The conditional PDF $p(Y/X^{(k)})$ is now given by:

$$p(Y/X^{(k)}) = \left(\frac{1}{\sqrt{2\pi}}\right)^n \cdot \frac{1}{|\phi_{YY}^{(k)}|^{1/2}} \cdot \exp\left[-1/2\{Y_t \cdot |\phi_{YY}^{(k)}|^{-1} Y\}\right] \tag{5.14}$$

The optimum receiver then computes the set of a posteriori probabilities $\{p(X^{(k)}/Y)\}$ by using the Bayes' rule. One simplifies the calculation by computing the quadratic form:

$$\Lambda'^{(k)} = Y_t[\phi_{YY}^{(k)}]Y.$$

which reduces to:

$$\begin{aligned}\Lambda^{(k)} &= Y_t\phi_{nn}^{-1}\cdot H^{(k)}Y \\ &= Y_tH^{(k)}Y = (Y)_t(H^{(k)}Y)\end{aligned} \tag{5.15}$$

where ϕ_{nn} = Identity matrix for white Gaussian noise with $n_0 = 1$, and

$$H = \phi_{zz}\phi_{yy}^{-1} = I - (\phi_{nn}\phi_{yy}^{-1}),$$

The optimum receiver, shown in Fig. 5.27, consists of an estimating filter $H^{(k)}$ followed by a cross-correlator, whose output is simply related to $p(X^{(k)}/Y)$. For non-white Gaussian noise in the channel, the receiver may be modified by including a 'whitening' filter along with the estimating filter, and the basic structure remains the same.

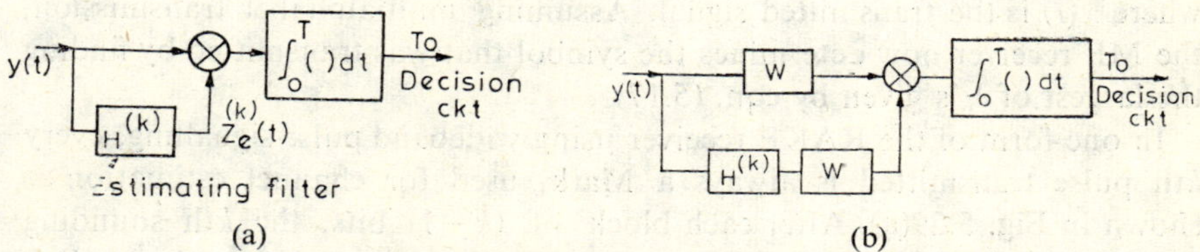

Fig. 5.27 An optimum receiver (a) For white noise in the channel, (b) For non-white noise

The solution given by eqn. (5.15) is computationally complex; moreover, the solution has to be updated coninuously because of the time-varying nature of the channel. Alternatively, the parameters of the estimating filter $H^{(k)}$ may be obtained by transmitting repetitive narrow sounding pulses (where the pulse width $\tau_s \simeq 1/W \ll \tau_m$, the multipath spread, $W =$ ch·BW) through the channel and processing the response through a TDL with taps at distances of $1/W$. If the tap outputs of the TDL are averaged over a number of sounding pulses, i.e., over a sufficiently large time (but small compared with the reciprocal of the channel fading rate), then the resultant response is the (almost) noiseless estimate $\{\hat{a}_m(t)\}$ of the channel impulse response. This estimate may now be used in the correlation of Fig. 5.27 to coherently detect the time-varying noisy signal.

The time-varying channel may now be modelled as a TDL, shown in Fig. 5.28 with coefficient multipliers, . . . $a_{-1}(t)$, $a_0(t)$, $a_1(t)$ . . . which are obtained through the estimating filter as above. The response of this TDL to the kth sounding pulse $x_{0k}(t)$, band-limited to W Hz, is

$$\hat{Z}_k(t) = \frac{1}{W} \sum_{m=-\infty}^{\infty} a_m(t)x_{0k}\left(t - \frac{m}{W}\right) \tag{5.16}$$

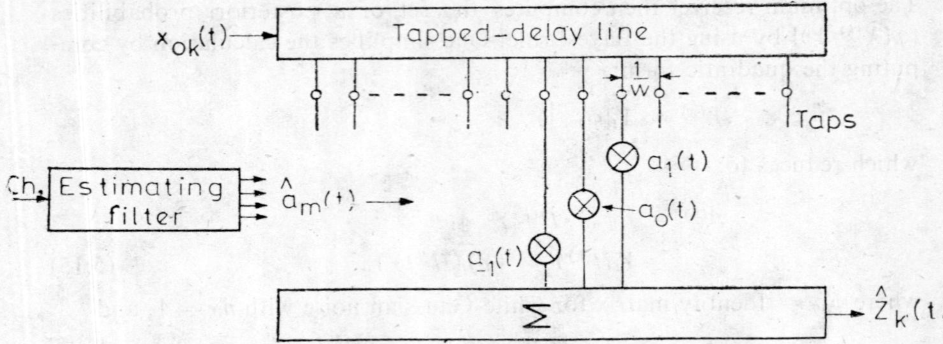

Fig. 5.28 TDL model of a multipath fading channel

The complementary TDL is the equivalent MF for optimum receiver operation and the response of the MF to a noisy received signal is given by:

$$\lambda_k = \frac{1}{W} \sum_{p=-\infty}^{\infty} \hat{Z}_k^*(p/W)[x(p/W) + n(p/W)] \tag{5.17}$$

where $x(t)$ is the transmitted signal. Assuming multialphabet transmission, the MF receiver now determines the symbol that was transmitted by finding the largest of λ_k's given by eqn. (5.17).

In one form of the RAKE receiver using wideband pulse signalling, every kth pulse transmitted is always a Mark, used for channel estimation as shown in Fig. 5.29(a). After each block of $(k-1)$ bits, the kth sounding bit is inserted. In the receiver, shown in Fig. 5.29(b), the received sounding pulse is separated through the TDM-switch and processed through the estimating filter having a large time constant. Thus, the estimates are derived

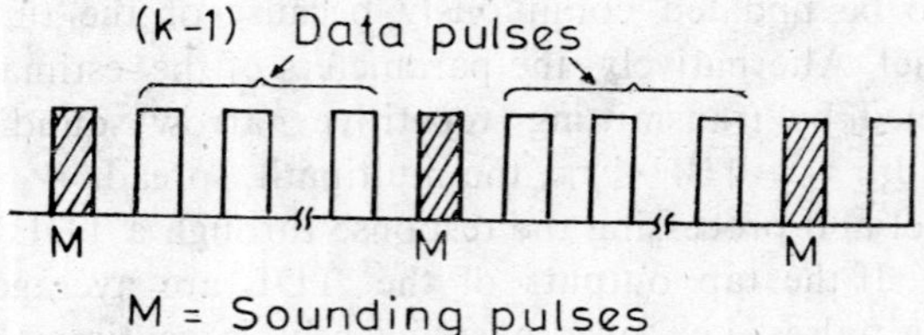

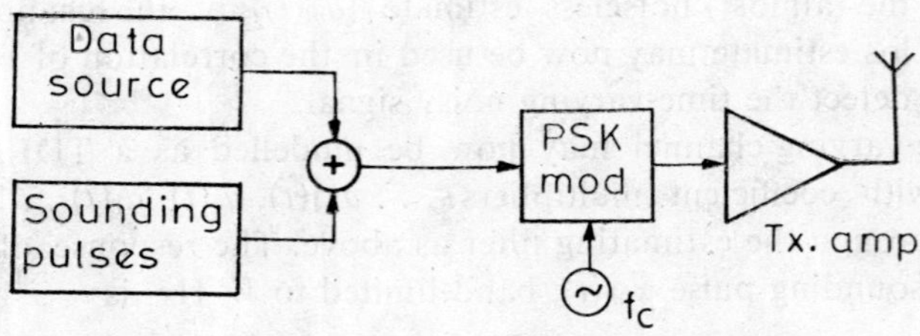

Fig. 5.29(a) A RAKE transmitter showing sounding pulses

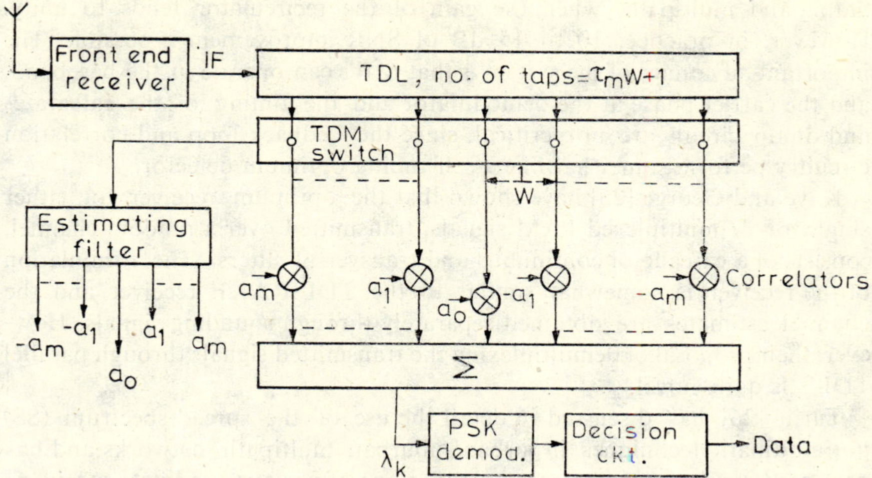

Fig. 5.29(b) A RAKE receiver showing TDL and estimator-correlator

with a higher SNR (say, by 10–15 dB) than the individual information bits. Assuming that the estimates are effectively noiseless, the correlator output at each tap represents a weighted verson of $[x(t) + n(t)]$ suitable for maximal ratio combining. These are summed and the output detected at the appropriate instant once per information pulse time T_b. This time-diversity receiver provides a larger diversity gain [the equivalent order of diversity $M \simeq (\tau_m W + 1)$] than in space-frequency diversity systems, and a higher bit rate of transmission is now possible.

An interesting and practically useful modification of the RAKE receiver has been proposed by Sussman [24]. In the proposed system shown in Fig. 5.30,

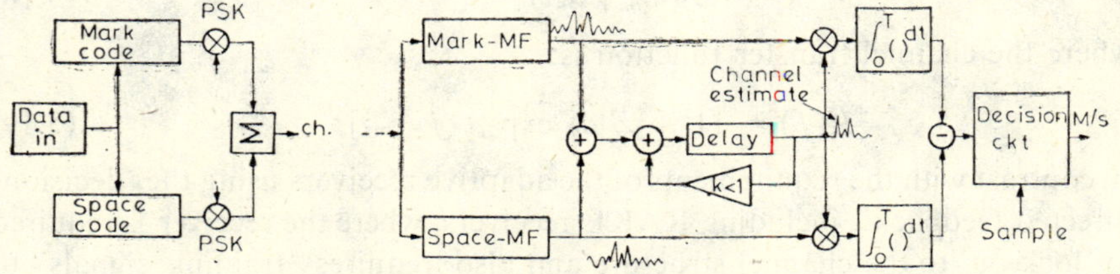

Fig. 5.30 A modified RAKE receiver

two orthogonal wideband codes are transmitted for mark/space and the MF receivers decode the noisy signal and resolve the multipaths. Since the channel response without noise is not available at the receiver, the estimate of the channel response is obtained through an auxiliary delay line, where the MF outputs are recirculated to obtain a more accurate (less noisy) estimate. The recirculated 'cleaned up' response will show the multipath pattern and give an excellent estimate of the channel characteristic. Ideally, the SNR at the input to the decision circuit, $(S/N_0) = E_b/n_0$ independent of fading and multipath, when the gain of the recirculator tends to unity. However, in practice, 10 to 15 dB of SNR improvement is possible. The important advantage of the system is that MF's can operate at the baseband, and the carrier phase at the demodulator and the timing of the integrate-and-dump circuit are not critical, since the feedback loop and correlation circuitry perform somewhat like a self-tuning optimum detector.

Kaye and George [25] have shown that the optimum receiver for either single or M-multiplexed PAM signals, transmitted over a random channel, consists of a cascade of continuous and transversal filters. The formulation of the receiver is somewhat similar to the TDL-RAKE receiver and the channel estimates are obtained separately through sounding signals. However, their proposal of demultiplexing the transmitted signals through parallel TDL's is quite novel.

Turin [26] has discussed in detail the use of the spread-spectrum (SS) antimultipath techniques in urban/suburban multipath networks and has presented results of analysis and simulations of various candidate receivers. His results show that Digital RAKE receivers, using MF's and DPSK, perform almost according to theory and give a time-diversity gain of 6 dB or more with reference to a non-fading LOS path. Even at a high data rate (above 1 Mbit/s), the intersymbol interference degrades the performance of DRAKE by 2–3 dB only.

A novel technique of eliminating interference due to multipath in digital transmission, have been suggested by Morgan [22]. His 'adaptive multipath canceller' (AMC) uses the configuration of an adaptive filter, shown in 5.31(a), with the constraint that the transfer function of the multipath channel should be minimum phase, i.e., the $\hat{a}_m(t)$ of eqn. (5.16), should be constrained as:

$$\Sigma \mid \hat{a}_m \mid < 1 \tag{5.18}$$

where the channel transfer function is:

$$H(f) = [1 + \Sigma\, \hat{a}_m \exp\{-j\omega\tau_m\}] \tag{5.19}$$

In contrast with the requirements of the adaptive receivers using the 'decision-directed feedback', including RAKE receivers, where the receiver is required to 'lock-on' to the channel structure and also requires 'training signals' to guarantee the initial convergence, the AMC requires only the knowledge of the transmitted power spectrum. This technique does not use decision-directed feedback, has guaranteed convergence and stability, and is simple to implement. As shown in Fig. 5.31(a), the adaptive weights of the transversal filter are derived by correlating the tap signals with the canceller output and this process adaptively cancels any delayed component of the received signal that is correlated with the direct signal, thus tending to cancel multipath components. However, the initial delay t_1 has to be properly optimized for proper cancellation. The theoretical and simulation performance of the AMC has been given by the author and it has been shown that the frequency-selective nature of the received spectra through multipath is very effectively corrected and made uniform through AMC as seen in Fig. 5.31(b), thus cancelling the multipath. By optimum adjustments of the tap delays, the effective ISI decreases by 10 to 15 dB and the RMS eye closure improves from 60% to 10% as seen in Fig. 5.31(c). The overall ISI improvement is shown in Fig. 5.31(d). In the presence of noise, the AMC improves the system performance to the limiting value of the data SNR. The performance advantage of 5 to 20 dB is rather insensitive of tap spacings, except the initial delay in the system. With narrow-band interference, the technique is very successful in eliminating the interference from the output. Particularly significant is the potential of this adaptive technique to acquire and track the multipath without any a priori knowledge of specific parameters.

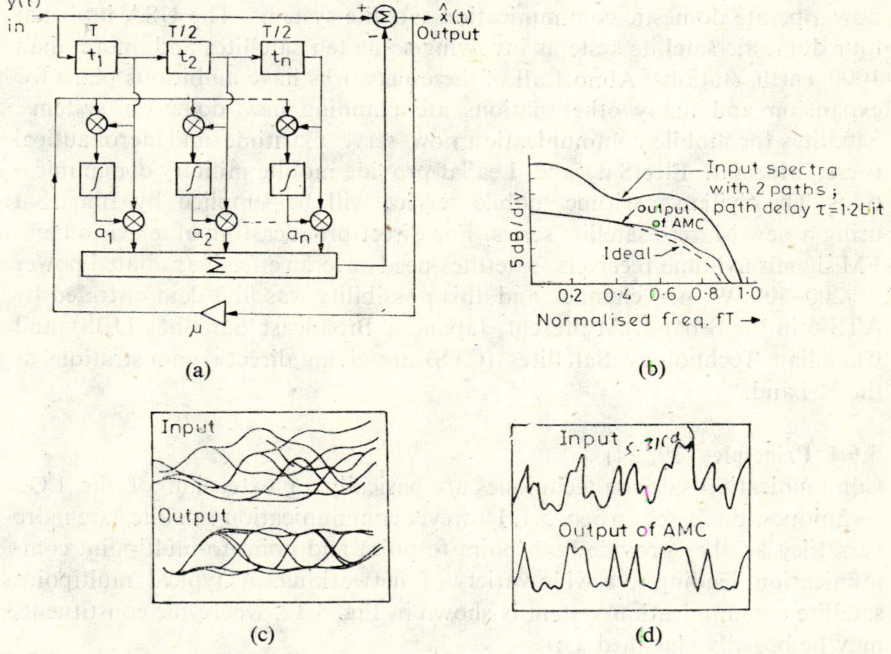

(a) (b)

(c) (d)

Fig. 5.31 Adaptive multipath cancellor (AMC) (after Morgan [22]); (a) AMC using correlation cancelling loops:
(b) Theoretical AMC power spectra for rectangular pulse, 50% attenuation. $WT = 1$, $S/N = \infty$ Frequency dispersion eliminated by AMC
(c) Eye pattern with and without AMC, $\tau = 1.2$ bit, 50% attenuation
(d) Overall ISI improvement

5.6 SATELLITE COMMUNICATION [27]

In the last 20 years, statellite technology has matured into a powerful communication technique providing a large capacity, but having a nodal network topology which haa an enormous potential for control, flexibility, processing, switching and growth. Starting with the launching of Sputnik in

1957, a series of experimental satellites, e.g., Score (1958), Echo (1960), Courier (1960), Telstar (1962), Relay (1961) and Syncom (1963), proved the techniques of maintaining continuous communication with earth stations. This resulted in the first international commercial communication satellite, Intelsat I in 1965. Today, the Intelsat IV provides approximately 25,000 circuits at a reduced cost of 5000 dollars per year per circuit. Intelsat V will now provide the reuse of the spectrum by polarization diversity and will also use higher frequency assignments for some routes. The Intelsat IV satellite system provides two-thirds of the world's transoceanic telecommunications and consists of 12 transponders linked to 300 earth stations in 125 different countries. Intelsat V has approximately doubled the capacity of international telephone circuits.

Many countries, e.g., the USA, Canada, Russia, Japan, Indonesia, India, now operate domestic communication satellite systems. The USA itself has four domestic satellite systems involving some ten satellites and more than 1000 earth stations. Almost all of these networks have ambitious plans for expansion and many other nations are planning new domestic systems. Satellites for mobile communication now serve maritime and aeronautical users. MariSat, FleetSat, and LeaSat provide mobile military communications. The civilian maritime mobile service will be supplied by InmarSat using a new Marecs satellite series. For direct broadcasting of television and FM signals to home receivers, satetlites need have an effective radiated power of 200–500 W per channel and this possibility was first demonstrated by ATS-6 in the S-band. At present, Japanese Broadcast Satellites (JBS) and Canadian Technology Satellites (CTS) are giving direct demonstrations at the K-band.

5.6.1 Principles [29, 34]

Communication satellite techniques are basically an extension of the LOS techniques, discussed in Sec. 5.1. However communication satellites are more versatile, as they provide both point-to-point and point-to-multipoint communication, leading to a wide variety of networking. A typical multipoint satellite communication system is shown in Fig. 5.32, where the constituents may be broadly classified as:

(a) The space segment consisting of the satellites and the corresponding control stations.
(b) The ground segment consisting of large/small earth stations, mobile vehicular stations and mobile data-collecting platforms.
(c) Terrestrial interface (tail-end) linking the ground stations to the national communication network.

To be able to communicate at large distances between the ground and the spacecraft (a distance of 36,000 km between the earth and the geo-stationary satellite with a free-space path loss of 200 dB approximately), the ground stations use large powers, large directive antennas and very sensitive

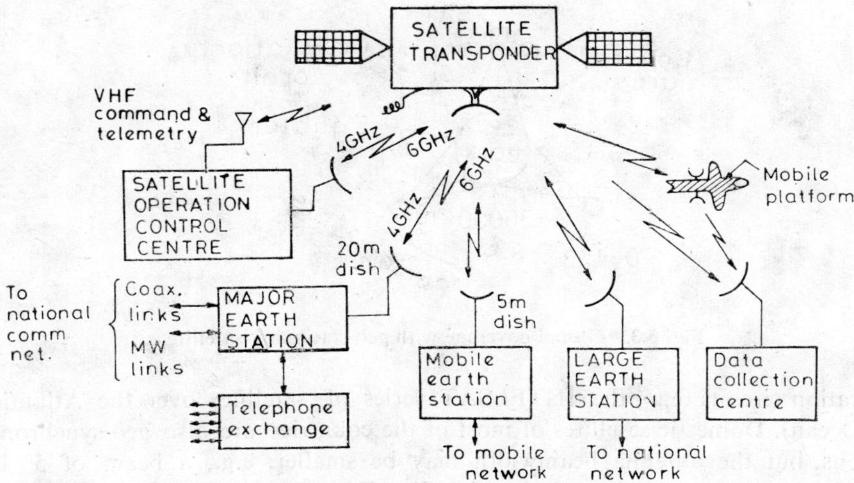

Fig. 5.32 A typical satellite communication system

receivers. Assuming the specifications of a typical transponder in a geo-stationary spacecraft as:

EIRP (equivalent isotropic radiated power): 30 dB W
Up/Down link frequency: 6/4 GHz
Link BW: 40 MHz
Antenna gain: 15 dB
Receiver noise temperature: 1200 °K
Sensitivity G/T: -15 dB/°K

it may be shown that the ground station capability for a single TV channel or 1000 speech channels has to be approximately:

EIRP: 90 dB W
Antenna gain: 52 dB
$G/T > 30$ dB/°K

We shall discuss the link calculations in detail in Sec. 5.7.

Satellites could be placed in (a) a circular orbit, (b) an elliptical orbit or (c) a geo-synchronous orbit, depending upon the specific user requirement. Although earlier experiments were made with satellites in elliptical orbits (e.g., in the Telstar experiment), the present-day communication satellites are predominantly geo-stationary, placed at a height of 35,786 km (=19,323 n.mi) over the earth's surface, as shown in Fig. 5.33. It is seen that the global coverage is possible with only three such satellites with antenna beamwidths of 17.3°. However, to provide orbit redundancy and greater coverage at higher altitudes, four or more satellites have to be used for global coverage. Moreover, the channel capacity per transponder in a satellite is restricted by bandwidth and power, and many satellites are parked at nearby positions to cater to the total demand of the international communi-

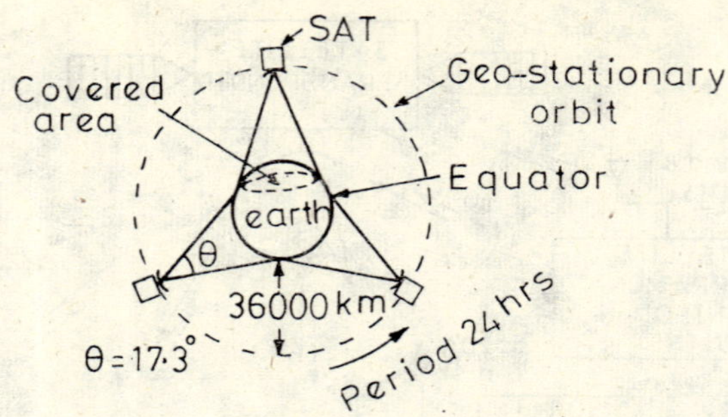

Fig. 5.33 Global coverage with geo-stationary satellite

cation circuits (e.g., the INTELSAT series of satellites over the Atlantic Ocean). Domestic satellites of most of the countries are also geo-synchronous, but the antenna beamwidth may be smaller, e.g., a beam of 5° is adequate to cover the whole of India. For countries at higher latitudes in the Northern Hemisphere, it is preferable to operate satellites in elliptical orbits. One such example is the Russian Molniya satellite system, which operates with an elliptical orbit having perigee = 1000 km, apogee = 39,375 km and a 12 hr period, as shown in Fig. 5.34. The system requires two or three satellites with the provision for station keeping and operating on the same orbit, so that a tracking antenna can maintain a continuous communication between two stations.

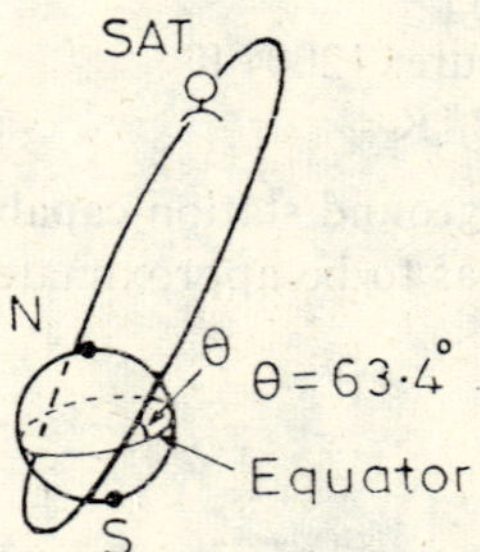

Fig. 5.34 Orbit of Molniya satellite; perigee = 1000 km, Apogee = 39,375 km, $\theta = 63.4°$; $T = 718.19$ minutes $\simeq$ 12 hr

To place the communication satellites in a geo-stationary orbit, it is necessary to have powerful launch vehicles, e.g., Thor-Delta, Atlas Centaur, Titan III of U.S.A., Ariane of the European Space Agency and Proton SL-12 of USSR. The Space Shuttle developed by NASA, USA, is now being used for placing satellites in specific orbits. Japan and China also have developed suitable launch vehicles for their satellites. For satellites in elliptical orbits, smaller rockets may be used. In general, all these vehicles employ multistage configurations using liquid and solid-fuelled stages, and the spacecraft payload is placed in the initial elliptical transfer orbit of 200 km by 36,000 km inclined to the equator (by 10.5°) by using the on-board guidance and control. Then a self-contained apogee propulsion system is used to carry out the orbit transfer to attain the final 36,000 km circular equatorial orbit. Figure 5.35 shows the typical flight sequence followed by a

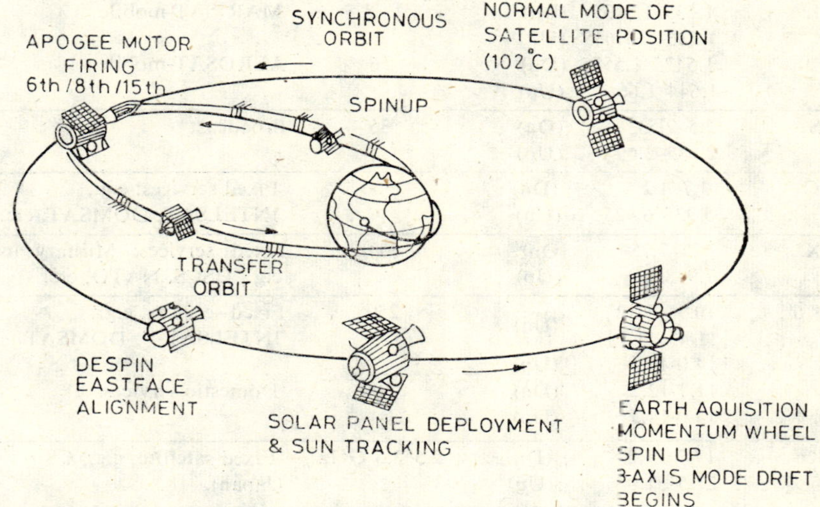

Fig. 5.35 A typical flight sequence of a synchronous satellite

spacecraft to attain the geo-stationary orbit. In case of space shuttle, the spacecraft payload along with a payload-assist module is placed in a 260 km circular parking orbit, and, at the appropriate equatorial crossing, the payload-assist module fires and transfers the spacecraft to the 260 km by 36,000 km transfer orbit. The apogee propulsion system in the spacecraft then performs the circularization and inclination correction operations. Once the satellite is placed in the geo-stationary orbit, the station-keeping techniques are used to maintain the orbit within ±0.1° in both north-south and east-west directions. Except for the space shuttle, all other vehicles are expendable, but the reusable space shuttle has now reduced the launch cost to a much lower value.

Based on the considerations of Galactic noise, rain attenuation, ionospheric scintillation and atmospheric attenuation as well as the state of the art in the hardware design, the frequency bands allotted for satellite communication are listed in Table 5.5. The international circuits, e.g., in the INTELSAT system, and most of the domestic circuits operate in the popular band of 6/4 GHz. However, some new systems in the 14/11 GHz band are also being operated, and possibly more systems will be engineered in this band.

The satellite communication system consists of the transmitter at the earth station, the satellite repeater (or transponder) and the receiver at the

Table 5.5

Band designation	Frequency range in GHz		Bandwidth in MHz	Type of services
Military UHF	0.24–0.3286 (Up and Down)		—	Mobile-Satellite LES, MARISAT, Fleet-SAT
L	1.535–1.5435	(Dn)	8.5	MARISAT-mobile
	1.636–5–1.645	(UP)		
	1.5425–1.5585	(Dn)	16	AEROSAT-mobile
	1.644–1.66	(Up)		
S	2.5–2.535	(Dn)	35	Broadcast
	2.635–2.69	(Up)		
C	3.7–4.2	(Dn)	500	Fixed services: e.g., INTELSAT, DOMSAT, etc.
	5.925–6.425	(Up)		
X	7.25–7.75	(Dn)	500	Fixed services: Military use, e.g., DSCS, NATO, etc.
	7.9–8.4	(Up)		
Ku	10.95–11.2, 11.45–11.7	(Dn)	250	Fixed services, e.g., INTELSAT-V, DOMSAT
	14.0-14.5	(Up)		
	11.7-12.2	(Dn)	500	Domestic services
	14.0-14.5	(Up)		
Ka	17.7–21.2	(Dn)	2.5–3.5 GHz	Fixed-satellite, e.g., CS (Japan).
	27.5–31.0	(Up)		

destination earth station, as indicated in Fig. 5.32. The earth station has an antenna subsystem, a power amplifier subsystem, a low-noise receiver subsystem and a ground communication subsystem, as shown in Fig. 5.36. The performance of an earth station is mainly specified by the transmitter EIRP and the receiver G/T, where G = receiver antenna gain and T = the noise temperature of the receiver front-end including the antenna. In the earlier satellite systems including the INTELSAT IV, the voice channels were multiplexed on the basis of SSB–FDM and the IF was FM-modulated by the multiplexed signal. In the later satellite systems, the digitized speech signals, in either PCM or DM format, are multiplexed on the basis of TDM and the TDM signal

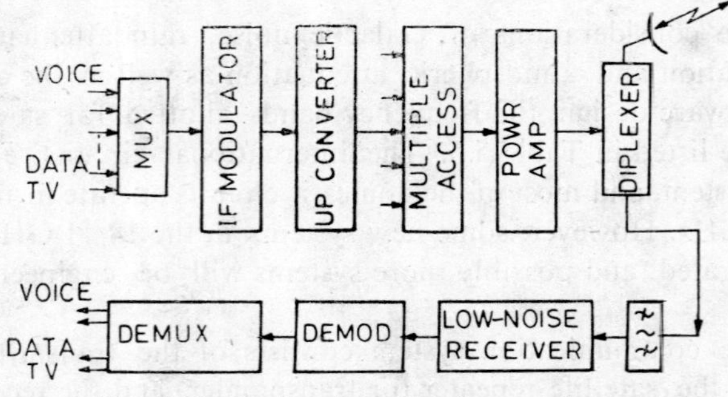

Fig. 5.36 Block schematic of an earth station

phase-modulates the IF using BPSK or QPSK. In this digital modulation scheme, it is possible to use error-correcting codes for improving the system performances. Assuming a 40 MHz slot (effective BW = 36 MHz) allotted to a station, it is possible to multiplex 900 speech channels using FDM-FM and 1000 channels usig PCM-QPSK modulation. However, when multiaccessing along with FDM-FM (i.e., multiple FM-modulated carriers are transmitted through the same power amplifier and antenna) is used, the EIRP has to be reduced to avoid excessive intermodulation noise: but with TDM-PSK modulation, the effect is negligible.

The satellite transponder is usually a frequency-translating repeating amplifier, as shown in Fig. 5.37. The uplink and downlink frequencies are

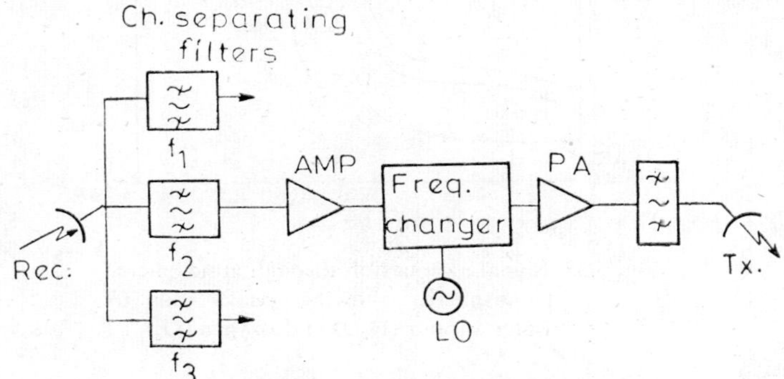

Fig. 5.37 Schematic of a satellite repeater

different in order to avoid instability in the amplifiers, and usually lower frequencies are used for the downlink. While multiaccessing with FDMA, sufficient back-off has to be given in the TWT-amplifiers to avoid intermodulation distortion. Since the antenna gain is small and the receiver noise temperature is large, the G/T of the transponder is poor ($\simeq -15$ dB/°K). Also, the EIRP of the repeater is small ($\simeq$ 20 to 30 dBW). Thus, the satellite transponder parameters are the limiting factors for the overall channel capacity of the total communication system.

5.6.2 Propagation Loss and Noise [29]

The radio signal from the earth station to the satellite (and vice versa) suffers the usual free-space path loss given by [see eqn. (5.4a)]*:

$$L_{FS} = 20 \log \left(\frac{4\pi R}{\lambda}\right) \text{ dB}$$

*The actual range of a satellite depends on the elevation angle θ of the ground station antenna, and the slant range R is given by:

$$R^2 = (h + r_e)^2 + r_e^2 - 2r_e(h + r_e) \cos \phi$$

where r_e = earth's radius (6378 km), h = satellite altitude (35,930 km for geo-stationary satellite), $h + r_e$ = 42,230 km, and ϕ = angle subtended by the satellite and the ground antenna at the centre of the earth.

For synchronous satellite, R varies from 36,000 km to 41,600 km with $90° \geqslant \theta \geqslant 0$. For $\theta = 5°$, 10° and 20°, R values for synchronous satellites are: 40,960; 40,480; and 39,360 km, respectively.

where R is the range of the satellite. Moreover, there are further losses due to absorption in the atmosphere (and troposphere), ionospheric scintillation and rainfall attenuation. At frequencies above 10 GHz, the attenuation due to water vapour and oxygen present in the atmosphere is considerable, as shown approximately in Fig. 5.38. It is seen that water vapour causes peaks

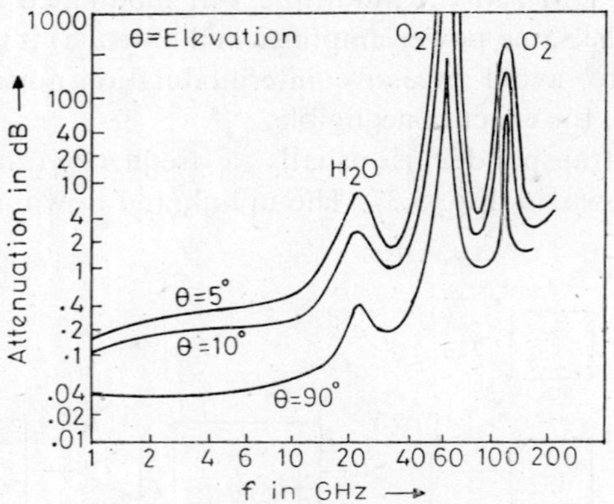

Fig. 5.38 Signal attenuation through atmosphere/troposphere, showing peaks due to water vapour (H_2O) and oxygen (O_2)

of attenuation at 22 and 183 GHz; and oxygen gives attenuation peaks at 60 and 119 GHz, approximately. These absorption losses are more at lower elevation angles and may be more than 100 dB at 60 MHz and above. Thus, systems operating in the Ku and Ka bands have to be designed with sufficient margins to compensate for the above losses.

The ionospheric effects, generally, are of importance only at frequencies below 1 GHz. The ionospheric scintillation, mainly caused by the F-layer irregularities, results in some fading of signals at UHF frequencies; but the effect is negligible at frequencies above, say, 4 GHz. Measured data give the approximate attenuation as: 22 dB, 2 dB and 0.5 dB at $f = 250$ MHz, 2.5 GHz and 7.5 GHz, respectively.

The attenuation due to rainfall depends on the rain rates in mm/hr, the effective path length through the rain and the angle of elevation of the earth station antenna. The rain attenuation can be severe at frequencies above 10 GHz and more so in tropical countries. For example, at $f = 20$ GHz, a signal may have a loss of 48 dB in heavy rain for an effective rain-path length (l_{eff}) of 4.8 km. The rain attenuation A_R is approximated as:

$$A_R = \alpha l_{eff}$$

$$\alpha = aR_p^b \tag{5.20}$$

where α is the specific attenuation in dB/km, R_p is the point rain rate in mm/hr, and a, b are constants depending on the frequency and the rain rate.

For $10 \leqslant f \leqslant 100$ GHz, a and b are in the range of:

$$0.12 \leqslant a \leqslant 1.08 \quad \text{and} \quad 0.74 \leqslant b \leqslant 1.19.$$

Figure 5.39 shows the approximate variation of α with frequency for $R_p \simeq 100$ mm/hr, 50 mm/hr and 10 mm/hr. The effective rain-path length is calculated from the freezing height and the elevation angle. Experiments have indicated that $l_{eff} \simeq 4.8$ km gives a good approximation to the rainfall attenuation at the zenith. Using the values of α from Fig. 5.39, and $l_{eff} \simeq 4.8$ km,

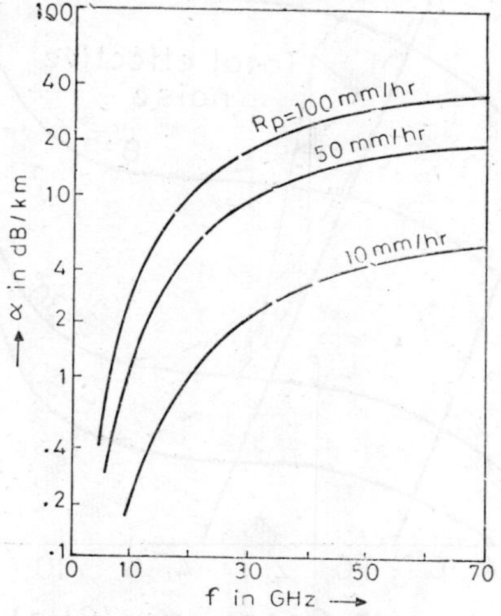

Fig. 5.39 Variation of rain attenuation coefficient α with frequency for different values of R_p

the approximate peak attenuation at different frequencies may be calculated. However, such extreme losses occur infrequently (say, for less than 0.1% of time) and are very localized. Sufficient system margins are then provided in designing satellite links at $f > 20$ GHz.

The effective (C_i/N_i) at the receiver depends on the effective noise temperature T_{eff} at the receiver input. T_{eff} mainly depends on the antenna temperature and the noise figure NF of the receiver front end. The antenna temperature T_a is a function of its radiation pattern, the physical temperature of the surroundings and the noise received from space. T_a is derived from the 'clear' sky temperature due to galactic noise, microwave background, and oxygen and vapour losses. The overall sky noise temperature for various elevation angles are shown in Figure 5.40, where it is seen that the effect of galactic noise in negligible above 2 GHz, but the noise due to H_2O and O_2 absorption is considerable for $f > 1$ GHz, specially at low elevation angles. It is further seen that the signal attenuation curves of Figure 5.38 are highly correlated with the noise temperature curves of Figure 5.40. Moreover, the rainfall attenuation also produces an increase in the antenna temperature,

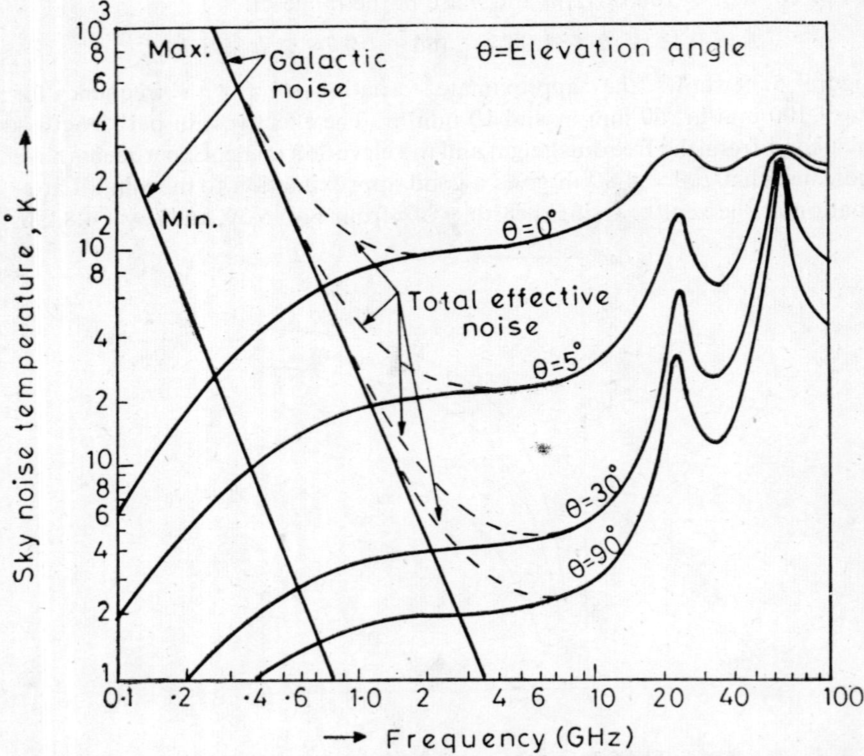

Fig. 5.40 Sky noise temperature due to galactic noise and atmospheric noise (due to oxygen and water vapour absorption)

which may be as high as 100 °K relative to the 'clear' sky temperature. The total noise power N_i at the receiver input is now given as:

$$N_i = kT_{\text{eff}}B_{\text{if}} \cdot (\text{NF}), \tag{5.2}$$

where k = Boltzman's constant $= 1.38 \cdot 10^{-23}$ joules/°K
$= -228.6$ dBW/°K/Hz

An overall communication link from one ground transmitter to the other ground receiver is characterised by the uplink and downlink carrier-to-noise (C/N) ratios, the intermodulation distortion in the repeater and the RF interference (I) from other systems, if any. Thus the composite C/N may be defined as:

$$(C/N)_T^{-1} = (C/N)_{\text{Up}}^{-1} + (C/N)_{\text{Dn}}^{-1} + (C/N)_{\text{IM}}^{-1} + (C/I)^{-1} \tag{5.21}$$

Generally $(C/N)_{\text{Dn}}$ is the limiting factor in obtaining the required SNR at the demodulated and demultiplexed signal level. Moreover, $(C/N)_{\text{Dn}}$ has to be above the threshold of the FM demodulator (say, 10 dB), and for QPSK, $(C/N)_{\text{Dn}} \geqslant 13$ dB. The allowable bandwidth of the RF signal and, hence, the channel capacity of the link are now calculated on the basis of the required SNR and the available $(C/N)_T$. To provide flexibility in the networking of the telecommunication links through satellites, various multiaccessing techniques, viz., FDMA, TDMA and CDMA, are utilized, and a versatile global network is configured.

5.7 SATELLITE LINK CALCULATIONS [35, 36]

Based on the eqns. (5.2)–(5.4) and the discussions above, the one-way satellite link equations may be written as:

$$\begin{aligned}
&\text{(a)}\ C_i\ (\text{dBW}) = P_T\ (\text{dBW}) + (G_T + G_R)_{\text{dB}} - (L_{FS} + L_M)_{\text{dB}} \\
&\text{(b)}\ N_i\ (\text{dBW}) = kT_eBF = n_0BF \\
&\text{(c)}\ (C_i/N_i)\ (\text{dB}) = (G_R/T_e)\ \text{dB} + (P_T + G_T)\ \text{dBW} \\
&\qquad\qquad - (L_{FS} + kB + F + L_M)\ \text{dB} \\
&\text{(d)}\ G_R/T_e\ (\text{dB}/^\circ\text{K}) = (C_i/N_i)\ \text{dB} - (\text{EIRP})_{\text{dBW}} \\
&\qquad\qquad + (L_{FS} + L_M + kB + F)_{\text{dB}} \\
&\qquad\qquad = (C_i/n_0)_{\text{dB-Hz}} - (\text{EIRP})_{\text{dBW}} + (L_{FS} + L_M + k)_{\text{dB}} \\
&\text{(e)}\ C_i/n_0\ (\text{dB}-\text{Hz}) = [(C_i/N_i) + \text{BF}]\ \text{dB}-\text{Hz}
\end{aligned} \tag{5.22}$$

where P_T = transmitter power, G_T = transmitting antenna gain, G_R = receiving antenna gain, L_{FS} = free-space loss in dB, L_M = miscellaneous losses due to coupling, etc., B = RF bandwidth, F = noise figure of the receiver, T_e = effective receiver temperature, C_i = received signal energy, N_i = received noise energy, n_0 = received noise density, EIRP = $P_T \cdot G_T$ watt, and G_R/T_e = receiver sensitivity or the figure of merit.

Equations (5.22) are valid for both uplinks and downlinks, using proper values of the parameters with reference to the ground station and the satellite. Usually (C_i/n_0) is used as a design parameter, since (C_i/n_0) is proportional to the maximum information rate transmittable with a given carrier power independent of the bandwidth (refer to Chapter 8).

To avoid saturation in the transponder and the consequent IM distortion, a maximum saturation flux density ϕ_s at the transponder input required to produce a saturated transponder output P_{TS} is generally specified. The required (maximum) ground station EIRP is then calculated as:

$$\phi_s\ W/m^2 = (\text{EIRP})_G/4\pi R^2 = \frac{\text{EIRP}}{(4\pi R/\lambda)^2} \cdot \left(\frac{4\pi}{\lambda^2}\right)$$

and in dB's, $\phi_s\ \text{dBW/m}^2 = \text{EIRP}_{(\text{dBW})} - L_{FS} + (4\pi/\lambda^2)_{\text{dB}}$

or

$$(\text{EIRP})_G\ \text{dBW} = \phi_s + L_{FS} - (4\pi/\lambda^2)\ \text{dB} \tag{5.23*}$$

The uplink (C_i/n_0) may now be written in terms of ϕ_s as:

$$(C_i/n_0)_U = [(G/T)_s + \phi_s - (4\pi/\lambda^2) - (k + L_M) - BO_i]\ \text{dB}-\text{Hz} \tag{5.24a}$$

---

*For f = 6 GHz for uplink, λ = 5 cm and $(4\pi/\lambda^2)$ = gain of 1 m² aperture at the transponder centre frequency = 36 dB.

and downlink (C_i/n_0) is:

$$(C_i/n_0)_D = [\text{EIRP}_{(\text{sat})} - L_{FS} + (G/T)_G - (k + L_M) - BO_0] \text{ dB–Hz} \tag{5.24b}$$

where $(BO_i,\ BO_0)$* = TWT input/output backoff in dB relative to single-carrier saturation.

The overall $(C_i/n_0)_T$ for the combined uplink and downlink is now obtained by rewriting eqn. (5.21) as:

$$(C_i/n_0)_T^{-1} = (C_i/n_0)_U^{-1} + (C_i/n_0)_D^{-1} + (C_i/n_0)_{IM}^{-1} \tag{5.25}$$

Consider the typical INTELSAT IV satellite operating with a large earth station having the parameters as:

Satellite	*Earth station*
EIRP = 22.5 dBW	EIRP = 90 dBW
$(G/T)_s = -11.6$ dB/°K	$(G/T)_G = 40.7$ dB/°K
$\phi_s = -73$ dBW/m²	Downlink path loss = 195.6 dB
	Uplink path loss = 199.1 dB

Using eqn. (5.24) and neglecting L_M and BO_i, we get

$$(C_i/n_0)_U = 228.6 - 73 - 18.6 - 36 = 101 \text{ dB–Hz}$$

and

$$(C_i/n_0)_D = 228.6 + 22.5 + 40.7 - 195.6 = 96.2 \text{ dB–Hz}$$

Combining through eqn. (5.25), $(C_i/n_0)_T = 95$ dB–Hz.

For a typical bandwidth of 40 MHz, $(C_i/N_i)_T$ at the ground receiver will be 19 dB, which would support FM/FSK/PSK systems.

To appreciate the intermodulation problems in the multicarrier operation of a TWT amplifier (used in a satellite repeater), consider a typical TWTA characteristic shown in Fig. 5.41. It is seen that the TWT shows saturation and non-linearity near the maximum input level, and multicarrier operation reduces the maximum output by about 2 dB. To minimize the corresponding intermodulation distortion, the input to the TWT is reduced in the form of backoff BO_i, and the $(C_i/n)_{IM}$ improved by about 15 dB for $BO_i = 10$ dB However, with larger BO_i, $(C/N)_{IM}$ improves but, at the same time, $(C_i/N)_D$ deteriorates because of the decrease in the satellite EIRP; and a compromise has to be found. The optimum operating point for the TWT is now obtained from the graphical relation of Fig. 5.42, where $(C_i/n_0)_T$ is calculated using eqn. (5.25) and its peak value indicates the optimum point [e.g., in the graph, $(C_i/n_0)_T$ max $\simeq$ 90 dB-Hz for $BO_i = 6$ dB]. Usually, there are two TWT's used in the amplifier chain, and both BO_i and BO_0 have to be used to control the *IM*-distortion. Figure 5.42 gives approximate characteristics for single carrier operation and, for multiple carrier operation, more

*In the multicarrier operation of the repeater, it is necessary to reduce the input to the first TWT, given by BO_i and also to the second TWT given by BO_0, to minimize the intermodulation distortion, as discussed later in the section.

backoff ($BO_i + BO_0$) will have to be provided as indicated by $(C_i/N)_{IM}$ curves of Fig. 5.41 (e.g., in INTELSAT IV, $BO_i = 11$ dB for multiple carrier operation).

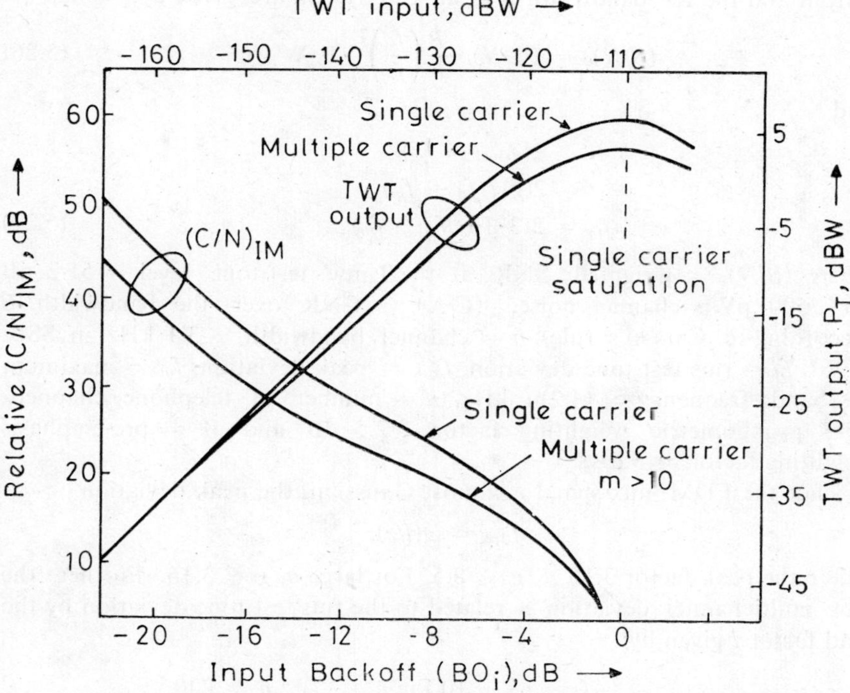

Fig. 5.41 TWTA input-output characteristics and $(C/N)_{IM}$ with backoff

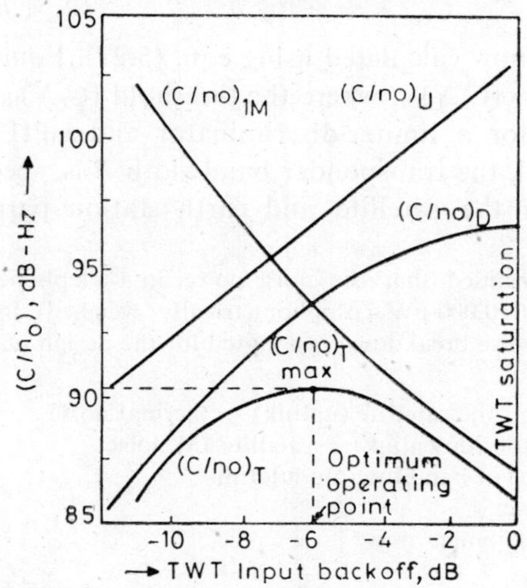

Fig. 5.42 Optimum TWT operation for single carrier

5.7.1 Channel Capacity [33, 36]

Consider first the single carrier operation with SSB/FDM multiplexed message signal and FM modulation of the carrier. The $(S/N)_0$ at the receiver output and the RF bandwidth B of an FM system are given by:

$$(S/N)_0 = (C/N)_i \cdot \frac{B}{b}\left(\frac{fd}{f_m}\right)^2 \cdot \mathrm{P} \cdot \mathrm{W} \tag{5.26}$$

and

$$\begin{aligned} B &= 2(f_{\mathrm{dpk}} + f_m) \\ &= 2(g \cdot l \cdot f_d + f_m) \\ &= 2(3 \cdot 16 \cdot l f_d + f_m) \end{aligned} \tag{5.27}$$

where $(S/N)_0^*$ = weighted SNR at the 1 mw test-tone level = 51.2 dB for 7500 pWp channel noise. $(C/N)_i$ = CNR over the bandwidth B (according to Carson's rule), b = channel bandwidth = 3.1 kHz in SSB/FDM, f_d = rms test-tone deviation, f_{dpk} = peak deviation, f_m = maximum baseband frequency $\simeq$ (4.2)n kHz, n = number of telephone channels, P = psophometric weighting factor = 2.5 dB and W = pre-emphasis weighting factor = 5 dB.

Since the FDM-mux signal is almost Gaussian, the peak deviation

$$f_{\mathrm{dpk}} = g(lf_d),$$

where the peak factor $3.16 \leqslant g \leqslant 8.5$. For large n, $g \simeq 3.16$. Further, the rms multichannel deviation is related to the rms test-tone deviation by the load factor l given by:

$$20 \log l = L = \begin{cases} -15 + 10 \log n \ldots & n \geqslant 240 \\ -1 + 4 \log n \ldots & 12 \leqslant n \leqslant 240 \end{cases} \tag{5.28}$$

For a given n, B is now calculated using eqn. (5.27). Equation (5.26) is valid only when $(C/N_i) > (C/N)_{\mathrm{th}}$, where the threshold $(C/N)_{\mathrm{th}}$ is approximately 13 dB and 10 dB for a limiter-discriminator and a PIL demodulator, respectively. In general, the transponder bandwidth B is specified and $(C/n_0)_T$ is calculated from the satellite and earth station parameters; then, the

---

*CCIR has recommended that the noise power in a telephone channel should not exceed a mean value of 10,000 pW (psophometrically weighed) in any hour. Accordingly, the following noise breakdown is adopted for the design of commercial satellite links:

(a)	Thermal noise at the satellite (uplink) + thermal noise at the earth station (downlink) + satellite IM noise	7500 pWp
(b)	Earth station out-of-band intermodulation	500 pWp
(c)	Interference noise	1000 pWp
(d)	Earth station equipment noise	1000 pWp
	Total	10,000 pWp

For an 1 mW test-tone audio output, this gives an output SNR = 51.2 dB.

channel capacity of the transponder using a single carrier is simply obtained from the above equations.

As an example, using the INTELSAT IV satellite and a standard earth station, as specified earlier, $B = 36$ MHz and $(C_i/n_0)_T = 95$ dB–Hz. Effective $(C/N)_i = (95 - 75.6) = 19.4$ dB and the required noise improvement factor (assuming $P + W = 6.5$ dB) $= 25.3$ dB. By an iterative process, the channel capacity is calculated to give $n \simeq 1000$ with $(S/N)_0 \simeq 52.5$ dB. It is observed that by decreasing $(C_i/n_0)_T$ to 85 dB-Hz, the threshold requirement of 10 dB using a PLL demodulator is satisfied, but $(S/N)_0 = 42$ dB with $n = 1000$. To improve $(S/N)_0$, one may reduce n and increase f_d to occupy the total available bandwidth. Thus, for $n = 600$, $(C_i/n_0)_T = 85$ dB-Hz; $(S/N)_0 \simeq 51$ dB. For $(C/N)_i < 10$ dB, threshold-extension demodulators, e.g., FMFB, may be used with a reduced number of channels, reduced RF BW, and with reduced values of $(S/N)_0$. Using such considerations, the number of channels available for various values of $(C_i/n_0)_T$ (using the INTELSAT IV transponder) is shown in Fig. 5.43. Since $(C_i/n_0)_T$ is linearly related to $(G/T)_G$, the curve of Fig. 5.43 may also be interpreted as the variation of n with $(G/T)_G$ for a required $(C_i/n_0)_T$. Smaller earth stations with $(G/T)_G < 41$ dB/°K give $(C_i/n_0)_T < 95$ dB-Hz; even then effective communication is established with a reduced number of channels and reduced RF BW.

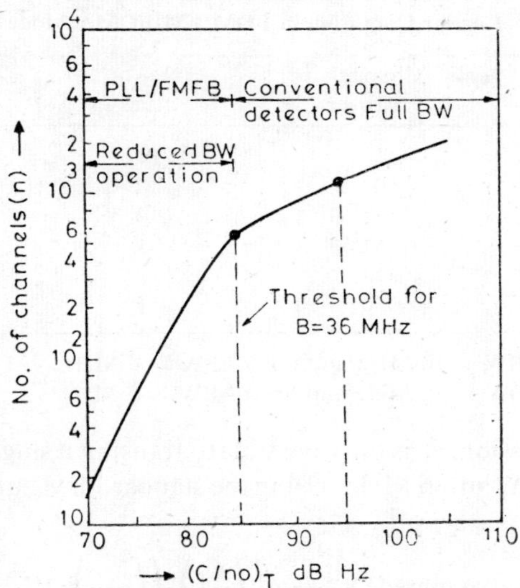

Fig. 5.43 Variation of channel capacity with $(C/n_0)_T$ for single carrier FDM operation

Modern satellites are sometimes provided with two types of downlink antennas, one for global coverage with a beamwidth of 17° and the other for lesser coverage with a beamwidth of 3.5° approximately, but with a higher gain (say, by 10–12 dB). They are known as global-beam and spot-beam antennas, respectively. The above calculations for the channel capacity have been made for the global-beam operation. With the spot-beam operation, the EIRP of the satellite increases by 10–12 dB and, for the same ground station, $(C_i/n_0)_T$ increases (almost) proportionately. As a result, the channel capacity (n) also increases. However, calculations show that for INTELSAT IV spot-beam operation, EIRP = 34.2 dBW, $(C_i/n_0)_T = 105$ dB-Hz; but $n = 1800$ only, giving an improvement of less than 3 dB in the spot-beam channel capacity.

The traffic requirements in a global network are such that a wide range of carrier capacities have to be accommodated by a commercial satellite. In INTELSAT IV, bandwidth units of multiples of 2.5 MHz were selected to allow a flexible interchange of carriers within a given frequency range and the groups of channels are accessed in the FDMA mode. The use of multiple FM carriers decreases the channel capacity in two ways: (a) the transponder EIRP is divided among the various input signals in a ratio proportional to their input powers and (b) the non-linearity in TWTA generates severe *IM*-distortion products, such that a sufficient backoff (e.g., in INTELSAT IV, $BO_i + BO_0 \simeq 11$ dB) is given to the TWTA to reduce *IM*-distortion. Thus, as the number of carriers increases, there is a sharp reduction in the channel capacity, as indicated in Table 5.6.

Table 5.6 Global-Beam Channel Capacity in INTELSAT IV for Multiple Carrier Operation in FDM/FM/FDMA Mode

RF BF per accessing carrier in MHz	Channels per carrier	Total no. of accesses	Total ch. capacity per transponder
2.5	24	14	336
5	60	7	420
10/5*	132/60	4	456
15/5**	252/60	3	564
18	320	2	640
36	960	1	960

*Three carriers of BW = 10 MHz and one of BW = 5 MHz.
**Two carriers of BW = 15 MHz and one of BW = 5 MHz.

For TV transmission, it is only possible to transmit a single carrier through the transponder BW of 36 MHz. Using the standard FM equation, the output SNR is given as:

$$(S/N)_0 \text{ (weighted)} = \frac{3}{2} g^2 \left(\frac{C_i}{n_0}\right)_T \left(\frac{f_{dkp}^2}{f_m^3}\right) \cdot W, \tag{5.29}$$

where $(S/N)_0$ = peak-to-peak luminance signal power/average noise power, g = crest factor for the luminance signal $\simeq 2$, f_{dpk} = peak FM deviation $\geqslant 10$ MHz, f_m = maximum video frequency $\simeq 5$ MHz and W = pre-emphasis noise weighting factor = 16.3 dB.

The SNR objective is: $(S/N)_0 \geqslant 52$ dB. Using the earlier value of $(C_i/n_0)_T = 95$ dB-Hz, $f_{dpk} = 13$ MHz (for $B = 36$ MHz) and $(S/N)_0 = 61$ dB. Thus, the system has a margin of 9 dB approximately.

5.7.2 Digital Signalling

The problems of *IM*-distortion due to multicarrier operation and the back-off in TWTA are very much minimized if digital signalling, instead of FM, is used for signal transmission. As such, PCM/TDM or DM/TDM signals are now being used in many of the satellites, and the most popular modulation technique used is QPSK. BPSK and D-PSK are also used in some cases. Referring to eqns. (2.34) and (2.42), the BER in both BPSK and QPSK is given by [eqn. (2.42)]:

$$P_b = erfc\sqrt{\frac{2E_b}{n_0}},$$

where E_b/n_0 is related to (C_i/n_0) by

$$E_b/n_0 = C_i/n_0 R_p \tag{5.30}$$

and R_p is the rate of information transmission in bit/s (for both BPSK and QPSK). For BER $= 10^{-5}$, E_b/n_0 required for QPSK is 10 dB; then using $(C_i/n_0)_T = 95$ dB–Hz for the composite link, as calculated earlier for the INTELSAT IV system, the rate of transmission R_p (power-limited) is calculated as:

$$\begin{aligned} R_p &= (C_i/n_0)_T \text{ dB–Hz} - E_b/n_0 \text{ dB} \\ &= (95 - 10) = 85 \text{ dB} = 300 \text{ Mb/s} \end{aligned} \tag{5.31}$$

The other constraint is the bandwidth B of the transponder, and for $B = 36$ MHz, $\alpha = 0.2$, the band-limited rate R_B is

$$R_B(\text{QPSK}) = 2B/1.2 = 60 \text{ Mb/s} \tag{5.32}$$

Thus, the above system is not power-limited, but band-limited. Using eqns. (5.24) and (5.31), the general equation for R_p may be written as:

$$R_p = [\text{EIRP}_{(\text{sat})} + (G/T)_G - (E_b/n_0 + L_{FS} + \kappa + L_M + BO_0)] \text{ dB} \tag{5.33}$$

The eqns. (5.32) and (5.33) are the two governing equations for calculating the actual rate of transmission R through the system, and R is given by the minimum of R_p and R_B. Thus,

$$R = \min\,[R_p, R_B] \tag{5.34}$$

Assume for a practical system, $(L_M + BO_0) = 5$ dB, then from eqn. (5.33), $R_p = 80$ dB $= 100$ Mb/s, but the bandwidth of 36 MHz will only support 60 Mb/s (which may be increased to 64 Mb/s with $\alpha = 1.125$) using QPSK. Since there is excess power available, it is possible to use higher-order modulation, say, 8-QAM, whereby R_B will increase.

If however, the system is power-limited due to lesser values of EIRP or G/T, or both, then it is possible to improve R by using error-correcting codes (refer to Chapter 8). As, for example, if, for the above example, $(G/T)_G = 30$ dB/°K, then $R_p = 10$ Mb/s only. But using Rate 1/2 convolutional codes, the required E_b/n_0 may be reduced to 5 dB, and R_p is now 30 Mb/s. The actual transmission rate through the channel is 60 Mb/s (due to

R 1/2 code) which matches R_B of eqn. (5.32). Thus, the advantage of low-rate FEC coding may be utilized for improving the channel efficiency for small earth stations (with $G/T < 40$ dB) and low satellite EIRP.

A special problem arises when emergency or remote area communication is established using a small earth station. say, a transportable terminal (TRACT). Assume that the TRACT parameters are: EIRP = 70 dBW (20 dB less than the standard earth station) and $G/T = 20$ dB/°K. Using a global beam transponder (similar to INTELSAT IV) and a standard earth station with EIRP = 90 dBW and $G/T = 41$ dB/°K, the channel capacity for the two-way link is calculated as:

(a) *From main to remote station*: SAT.EIRP = 22 dBW

$$(G/T)_G = 20 \text{ dB/°K; then, } (C_i/n_0)_T = 95 - 21 = 74 \text{ dB–Hz}$$

$$R_p \text{ (for } E_b/n_0 = 10 \text{ dB, } P_b = 10^{-5}) = (74 - 10) = 64 \text{ dB}$$
$$= 2.5 \text{ Mb/s}$$

Using BPSK, RF BW required = $1.2 \times 2.5 = 3$ MHz only. Since enough BW is now available R-1/3 FEC coding may be used to reduce E_b/n_0 to 4 dB for $P_b = 10^{-5}$. Then the increased rate of transmission R'_p due to 6 dB advantage in E_b/n_0 is:

$$R'_p = 2.5 \times 4 = 10 \text{ Mb/s}$$

and the RFBW using BPSK = $1.2 \times 10 = 12$ MHz only.

(b) *From remote to main station*: From eqn. (5.23), Flux density at the satellite input (through uplink) = $70 - L_{FS} + 36 = -93$ dBW/m².

Using the standard transponder with saturation flux density = -73 dBW/m² and EIRP = 22 dBW, the available EIRP is now $(22 - 20) = 2$ dBW only. Then, $(C_i/n_0)_T$ for the global beam = $95 - 20 = 75$ dB–Hz, and using eqn. (5.31), we set

$$R_p = 75 - 10 = 65 \text{ dB} = 3 \text{ Mb/s}$$

Using BPSK, RF BW required = $1.2 \times 3 = 3.6$ MHz only.

Again using R 1/3 coding, $R'_p = 12$ Mb/s, but the required BW is 43.2 MHz. This is more than the available BW of the transponder and the rate of transmission R'_p has to be limited to 10 Mb/s only. In general, it is not necessary that both-way links have almost the same R_p, as calculated above, and two transmission rates may vary widely depending upon the EIRP and G/T of the TRACT.

5.8 EARTH STATION AND TRANSPONDER [29, 34]

An earth station for satellite links may be a fixed, ground mobile, maritime, aeronautical or receive-only (for broadcast transmissions) terminal (see Fig. 5.36). A major earth station may be required to operate with different satellites and on a number of carrier frequencies simultaneously. As such, there are more than one transmit and receive chains in the transmitter and receiver of an earth station, as shown in Figs. 5.44 and 5.46. Usually a single large antenna is used for both transmit and receive signals, and automatic tracking facilities* are provided. The terrestrial link leading to national communication networks is connected after the baseband MUX-DEMUX equipment. An 'uninterrupted' primary power is also necessary for the continuous operation of the station. An earth station is usually specified by its EIRP, G/T and frequency of operation. A few typical characteristics of earth stations are given in Table 5.7.

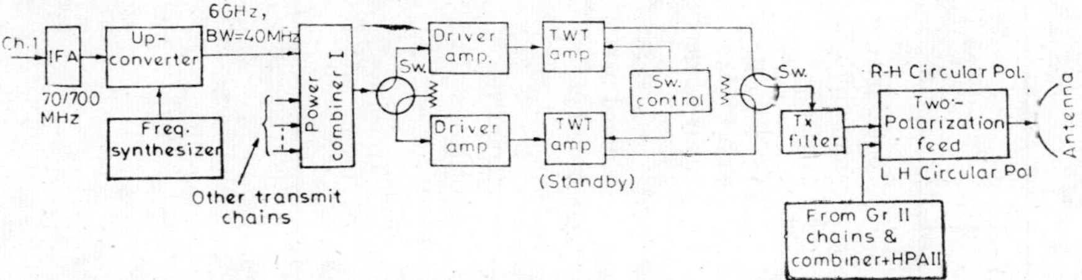

Fig. 5.44 Earth station transmitter using a common TWTA with redundancy

In single channel transmitters using a fixed 36-MHz band, a Klystron high-power amplifier (HPA) may be used. But in transmitters having multiple carrier chains, either a common wideband TWTA is used as shown in Fig. 5.44, or each channel uses a separate Klystron HPA, as shown in Fig. 5.45. A redundant TWTA is provided with a switch control in Fig. 5.44, but in Fig. 5.45, reliability is somewhat ensured by multiple HPA, whose

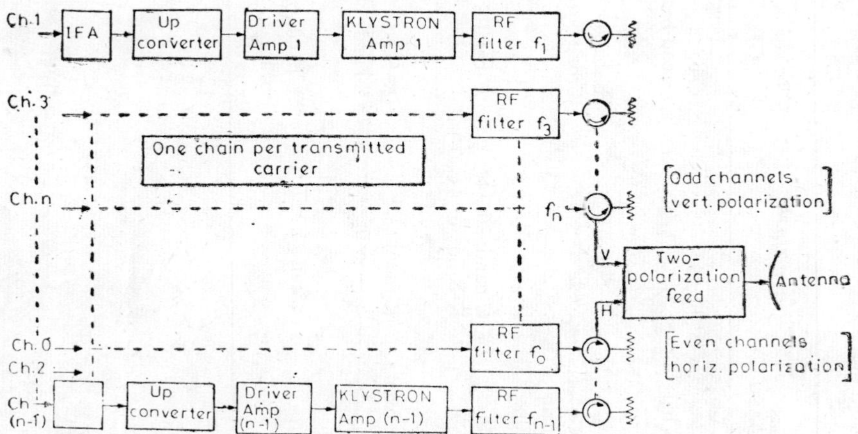

Fig. 5.45 Earth station transmitter for multicarrier operation using multiple Klystron HPA

outputs are combined through hybrids as shown. HPA output signals are heavily filtered to reduce undesired intermodulation and spurious components in the receiver band to avoid saturation in the receive amplifiers. The filtered signal is then passed through a waveguide to the polarizer and feed

*The principle is similar to that of Monopulse tracking radar. However, in simpler systems, only the antenna-pointing facilities are provided using motor drives for the elevation and azimuth control of the antenna mount.

Table 5.7 Typical Characteristics of Earth Stations

Types	Antennas	G/T, dB/°K	Frequencies, GHz	Transmitters	Receivers	Multiple access
INTELSAT Standard 'A'	29–35 m, auto-tracking	40–44	C-band broad beam	10 KW, up to 24 carriers	Cooled par-amp, up to 72 carriers	FDMA, TDMA
INTELSAT Standard 'B'	15–20 m, auto-track crossed linear polarization	35–50	K-band broad beam	2 KW, multiple carriers	Cooled par-amp LNA, many carriers	FDMA, TDMA, SCPC
Domestic systems	5–10 m, step-track	C-band: 30–34 K-band: 40–44	C-band K-band	1–8 KW, multiple carriers	Cooled par-amp LNA	FDMA, TDMA, SCPC
Direct broadcast receive-only terminals	0.5–1.5 m, fixed, pointable	K, 6–14 S, − 6–0	K-band S-band	—	Transistor LNA	FM, Video
Maritime mobile terminals	1–2 m, auto-track	−4.0	L-band, broad beam	100–200 W	Transistor LNA	SCPC, TDMA

diplexer, and the output with appropriate polarization illuminates the parabolic reflector. The non-linearity of the TWTA produces serious *IM*-distortion and a 7–10 dB backoff is used to avoid this. An alternative method of using feedback to reduce non-linearity is more efficient of power, but the system is more complex. In the multiple Klystron HPA also, there is a power loss ≃ 10 dB in the hybrid combiners, but the system is simpler and cheaper as compared to the wideband TWTA. A few typical HPA characteristics are shown in Table 5.8.

Table 5.8 Typical Characteristics of HPA

Parameters	C-Band	C-Band	K-Band
Type	Klystron	TWT	TWT
Frequency, GHz	5.925–6.425	5.925-6.425	14–14.5
Bandwidth, MHz	36	500	500
Power output, W	3.10^3	40/400	225/250
Gain, dB	76	45/70 ± 1	60
Harmonic output at saturated output, dB below carrier	−30	−60	−60
Residual AM, dBc	−40	−40	−40
3rd Order IM distortion	18 dB down at 3 dB BO_0; 24 dB down at 6 dB BO_0	20 dB down at 8 dB BO_0	—
Noise figure	—	35/37	37
Group delay:			
linear, ns/MHz	0.5	0.1	0.1
parabolic, ns/MHz²	0.05	0.05	0.05
ripple, ns/P-P	2	1.0	1.0
AM/PM conversion	4°/dB	5°/dB	5°/dB

The general block diagram of the receiver in an earth station is shown in Fig. 5.46, where the low noise amplifiers (LNA) (with a standby) mainly determine the sensitivity G/T of the receiver. With a common trans-receive antenna, the transmit to receive isolation has to be better than 90 dB.

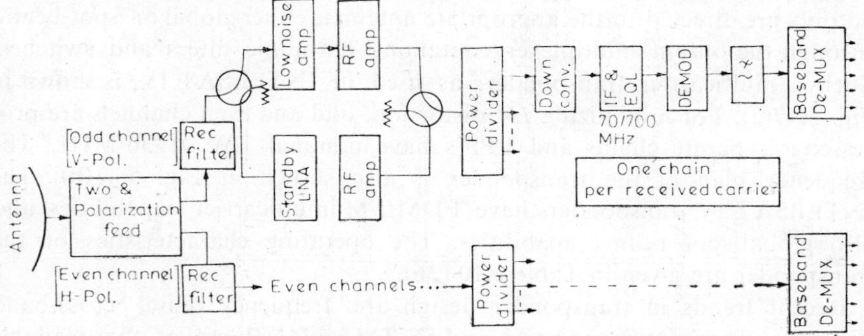

Fig. 5.46 Earth station receiver for multicarrier operation

Typically parametric amplifiers with cryogenic/thermoelectric cooling are used. Table 5.9 shows the typical characteristics of parametric LNA's. However, GaAs-FET amplifiers with noise temperature of 90–150°K (uncooled) and NF of 2.5 dB at 11 GHz and 4 dB at 18 GHz are also being introduced in modern earth stations. Down conversion is usually done in two stage (700/70 MHz) and the IF output is suitably equalized before FM demodulation. The baseband is either demultiplexed to feed the telecommunication network directly or the MUX signal is carried over a terrestrial link to the remote telecommunication centre.

Table 5.9 Typical Characteristics of LNA's

Parameters	C-band	C-band	K-band
Cooling	Thermoelectric	Cryogenic	Thermoelectric
Frequency, GHz	3.7–4.2	3.7-4.2	11.7–12.2
Bandwidth, MHz	500	500	500
Noise temp., °K	40	13–16	100–110
Gain, dB	55±0.5	60 ± 0.5	55 ± 0.5
VSWR input/output	1.2	1.25	1.2
IM distortion, dB	−64	−63	−64
Group delay (over 40 MHz):			
linear, ns/MHz	0.1	0.1	0.1
parabolic, ns/MHz²	0.03	0.01	0.03
ripple, ns/P-P	0.5	0.5	0.5
AM/PM conversion	0.5°/dB	0.5°/dB	0.5°/dB
Temperature range	0–50°C	0-50°C	0–50°C

5.8.1 Transponder

A simple frequency-translating satellite transponder has been shown in Fig. 5.37. As in the case of the earth station, multiple repeaters are necessary in a satellite for multiple carrier operation. In such a multiple carrier transponder, the uplink carriers are received by the antenna and the front-end receiver serves as an LNA and a broadband down converter. The broadband signal is now channelized into 12 repeater chains through band-pass filters (de-MUX). Each filter output is amplified by a redundant TWTA and the outputs are directed to the appropriate antenna, either global or spot-beam, through the output multiplexer containing harmonic filters and switches. Such a multicarrier transponder, as used in INTELSAT IV, is shown in Fig. 5.47(a). For minimizing *IM*-distortion, odd and even channels are processed in separate chains and BPF's have nominal BW of 36 MHz. The frequency plan of the transponder is also shown in Fig. 5.47(b). The INTELSAT IV transponders have FDM/FM multicarrier capabilities and also global/spot beam capabilities. The operating characteristics of the transponder are given in Table 5.10 [36].

Recent trends in transponder design are: frequency reuse, cross-band operation, beam interconnection and SS-TMA [34]. Reuse of the available 500 MHz band is made possible by providing dual-polarized transmit and

Table 5.11 Technical Characteristics of Satalliters

Sl. No.	System	Services	Frequency band	Mass (kg)	Primary power (W)	Coverage and no. of ant. beams	Antenna polarization	No. of transponders per satellite
1.	INTELAST IV-A	Telephone/TV	C	790	708	Global/Spot, 7/3	Circular	20
2.	INTELSAT V	Tel./TV	C, Ku	1020	1220	Global/Zonal/ Spot; 10/4	Circular and linear	15 + 6
3.	INTELSAT VI (for 1986)	TEL/TV/TTY/Data	C, Ku	2004	2260	Global/Zonal/ Spot; 14/4	Circular and linear	36 + 10
4.	STATSIONAR (USSR)	Teleehone	C	1250	700	Global/Spot, 2/4	Circular	6
5.	ANIK-D (Canada)	TEL/TV/TTY	C	635	1000	Canada/Northern USA; 5	Linear	24
6.	TDRS, Advanced Westar (USA)	Tel/Data	C, Ku	2132	1700	USA, 2/7	Linear and circulr	12 + 4
7.	France, Telecom-1	Tel/TV	C, Ku	1100	—	Semi-Global/ Spot/France 3/2	Circular and linear	4 + 6
8.	INSAT-1	Tel/TV/radiometry	S, C	1054	1250	India, 1/1	Linear	2 + 12
9.	COMSAT/MARISAT	Mobile Tel/TTY	L(ship) C(shore)	326	330	Global, 1	Circular	2

Sl. No.	Transponder bandwidth (MHz)	EIRP (dBW)	G/T dB/°K	Modulation	Multiaccess mode
1.	36	22/26/29	−18/−11.6	FDM/FM, SCPC, QPSK	FDMA, TDMA frequency reuse
2.	36/41/72/77 and 72/77/241	23.5–29 and 41.4–44.4	−18/−11.6/−8.6 and 0/3,3	FDM/FM, QPSK	FDMA, TDMA, frequency reuse
3.	36/41/72/77 and 72/77/150	23.5/28/31	−15/−8.5/−7/−1 and 1/4.3	—do—	—do—
4.	40	25–36	−15.8	FDM/FM,	FDMA
5.	36	36	−37.5/−27	FDM/FM, TV and SCPC	FDMA
6.	36/225	26/28/33 and 42–50.3	−7/−12.5 and −5/−4.4	FM, QPSK, (250 Mb/s)	FDMA, TDMA, Beam switching
7.	40/120 and 36	26/34.6 and 49	13.6 and 6.5	FDM/FM, BPSK	FDMA, SCPC, TDMA
8.	36	42 and 34	−4 and −4.5	FDM/FM, QPSK	FDMA
9.	4	20–29.5 and 18	−17 and −25	FM, BPSK	TDMA, FDMA

Fig. 5.36. At the destination, the receiver selects its own traffic signal through a suitable bandpass filter and then transfers the traffic to the telecommunication network after demodulation and demultiplexing. Multiple carriers from a single earth station or from multiple earth stations may now access the same transponder to fully load the repeater. As a result, of course, sufficient backoff has to be given to the repeater TWTA to reduce the IM-distortion to the acceptable level, thereby decreasing the overall channel capacity as discussed in Sec. 5.7 (see Fig. 5.41). However, the technique is commonly used for small and moderate traffic, and is more efficient than the broadcast mode of transmission.

The other methods of multiaccessing a transponder are:

(a) PCM (DM)/TDM/PSK/FDMA.
(b) PCM(DM)/PSK/FDMA, known as Single-Channel-per-Carrier System (SCPC).
(c) CFM/FDMA-SCPC.
(d) PCM (DM)/TDM/PSK/TDMA.
(e) Code Division Multiaccess (CDMA).

Instead of multiplexing traffic on the basis of SSB/FDM, as is done in FDM/FM/FDMA, the traffic, if digital, may be multiplexed on the basis of PCM (DM)/TDM (using digital TASI as well), where the analog signals are digitized through either PCM or DM coding. Each group of TDM signals may be transmitted by digital modulation of a carrier, either by PSK or QPSK, and multiple PSK-modulated carriers are now simultaneously processed in the common transponder, resulting in the PCM(DM)/TDM/PSK/FDMA mode of multiaccessing. However, this method is not a popular one, as the (somewhat similar) TDMA techniques are more efficient, as discussed below.

5.9.1 FDMA/SCPC [37]

In FDM/FM/FDMA mode of operation, the circuits between earth stations are usually preassigned, but the preassignment of circuits is efficient only for links with heavy traffic. As the number of circuits per link decreases, the link utilization for a given grade of service becomes increasingly inefficient. The solution to this problem is to share a pool of circuits among all earth stations having light traffic and in common view of the satellite. The circuits are then assigned on demand, forming temporary connection between any two stations and, on completion of the call, they are returned to the demand assignment pool. Such a flexible, multidestinational, fully variable, demand assignment service is provided by the Demand-assigned SCPC system. One form SCPC is the SPADE (Single-channel-per-carrier PCM Multiple-Access Demand-assigned Equipment) developed by COMSAT. In SPADE, the voice is digitized by 64 Kb/s-PCM (or by 32 Kb-DM as used in some continental SCPC systems) and the digital signal modulates the carrier through QPSK. The per-channel BW allowed is 38 kHz and the carriers are spaced at 45 kHz allowing a 7 kHz guard hand. The 36 MHz transponder BW is now divided into 800 channels including a common signalling channel (CSC) having a BW of 160 kHz, as shown in Fig. 5.49. In SPADE, the advantage is taken of low talker speech activity ($\simeq$38%), and the carriers are voice-activated, thus giving approximately 4 dB saving of the transponder power.

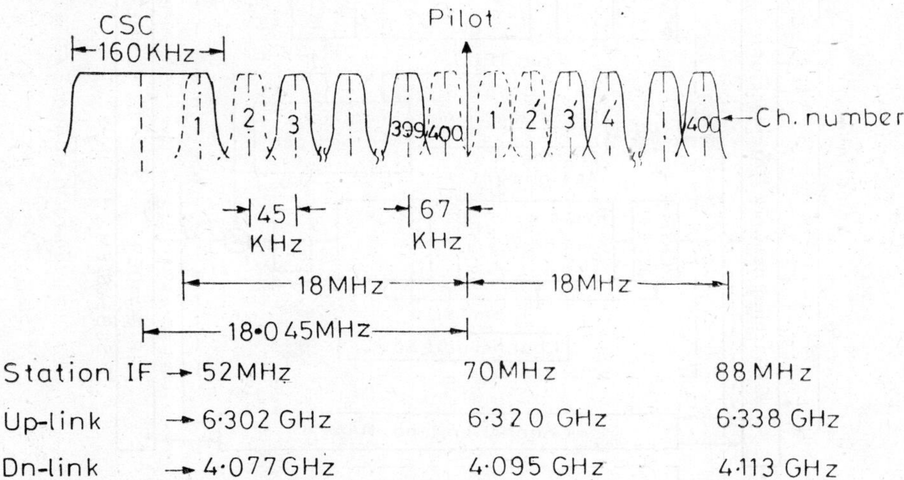

Fig. 5.49 SPADE frequency allocation scheme: dotted channels (1, 2, 1′, 2′ and 400) are normally not used

Using eqns. (5.31) and (5.33), it is easily calculated that $(C_i/n_o)_T$ per channel is now 66 dB-Hz and for E_b/n_o = 10 dB with $P_b = 10^{-5}$, R_p = 56 dB = 400 Kb/s. For the required R_p = 64 Kb/s = 48 dB, the available margin in the systems (for $L_M + BO_0$) is 8 dB and using voice-activation, the margin becomes 12 dB, which is reasonable in practice. With this margin, the channel capacity falls to 350, if voice-activation is not used.

The block diagram of a SPADE terminal is shown in Fig. 5.50, where the DASS unit controls the traffic through N channel units. On demand from the user interface unit, DASS assigns a free channel and a pair of unused (IF) frequencies to the user and transmits the information to the destination through the CSC. The channel unit digitizes the voice signal, and formats the bursts into packets of 256 bits including preamble and unique words. The voice detector switches the carrier to the QPSK modulator during voice bursts, and the modulated IF is summed in the IF subsystem. The IF is up-converted and amplified in the earth station terminal and fed to the antenna. In the channel receiver, the corresponding IF from the IF subsystem is demodulated and decoded in a complementary way and then transferred to the switching network. Each earth station monitors the CSC to maintain a log of the free frequency slots; and thus assigns at random the frequencies/channel to the calling user on demand. The CSC is a TDMA, 128 Kb/s, BPSK channel, and has a 50 msec frame with 1 msec access time. Thus, each of the terminals of a 50-terminal network can request a channel every 50 msec. The system has been found to be very effective in low-traffic routes.

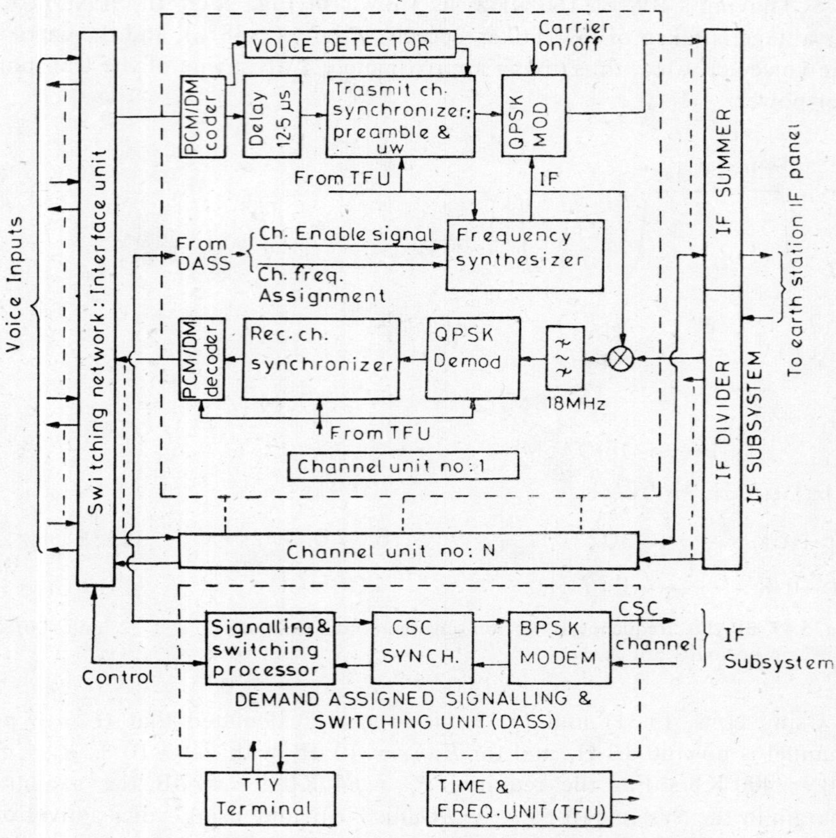

Fig. 5.50 Block diagram of a SPADE terminal

The alternative technique of SCPC is to use an individual FM channel per subscriber and assign the channels on demand from the common pool, as is done in SPADE. An FM-SCPC channel may take advantage of a voice-activated carrier as in SPADE and also of companding with pre-emphasis. Threshold extension demodulators, e.g., FMFB, and squelch detectors are also used to improve the overall SNR in the circuits. The general block diagram of an FM-SCPC is then the same as that of SPADE, except that the modem is now FM with companding and other improvements. Assuming that 800 channels are operated through the 36-MHz transponder, the available $(C/n_0)_T$ per channel is 66 dB-Hz, giving $(C/N)_i = 10$ dB. The value is critical, but a moderate FMFB/PLL demodulator will serve the purpose. The output SNR/channel is now obtained as:

$$(S/N)_0 \simeq 3D^3 + W + P + Q + (C/N)_i \tag{5.35}$$

$$\simeq 23 + 6.3 + 2.5 + 17 + 10 = 58.8 \text{ dB}$$

where D = FM deviation ratio, W = pre-emphasis noise weighting factor = 6.3 dB, P = phophometric weighting factor = 2.5 dB, Q = companding advantage = 17 dB. For an SNR objective of 51 dB, the available margin in the system is 8 dB (the margin is 12 dB with voice activation), as was calculated for SPADE. Without companding, however, the system performance would be very poor. It has been shown that with the INTELSAT IV system, the FM/SCPC may have a channel capacity of 1200 with the earth station G/T = 40.7 dB/°K and 800 for G/T=29.3 dB/°K [38].

5.9.2 TDMA [39, 40]

It is seen from above discussions that the channel capacity in FDM/FM/FDMA decreases considerably as the number of accesses to the transponder increases. In contrast, in TDMA, a single RF carrier is time-shared between a number of earth stations and the full output power of the TWTA may be utilized without any IM-distortion. Thus, the available EIRP, without backoff, remains constant independent of the number of accesses, and the channel capacity (≃ 1000) also remains approximately constant. Moreover, using digital TASI (refer to Chapter 3), the channel capacity may be approximately doubled.

In TDMA, the transmissions from earth stations are in the form of RF bursts within the PCM frame period of 125 μs. The format structure of a typical TDMA frame is shown in Fig. 5.51, where each burst consists of a

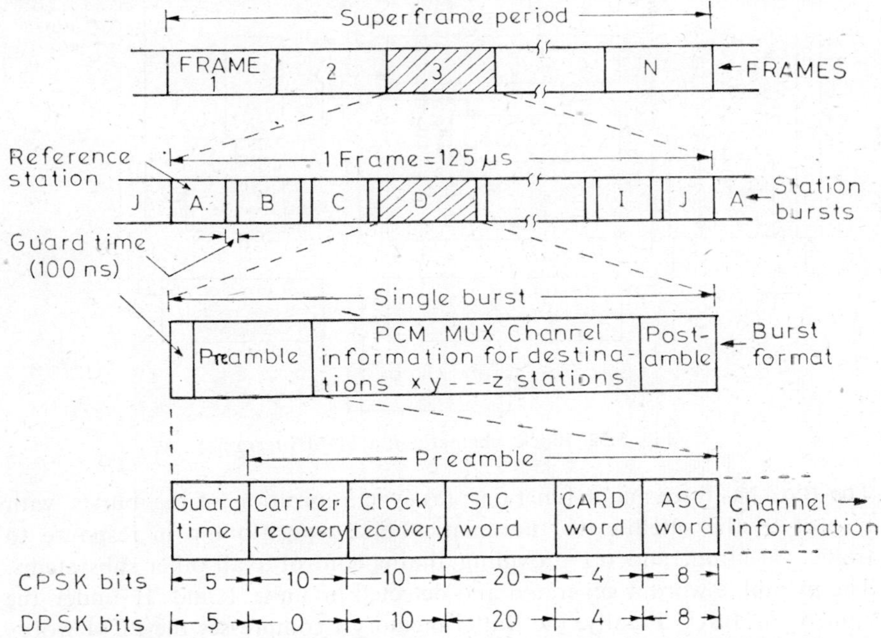

Fig. 5.51 Formal structure of a typical TDMA frame: CARD—channel allocation and routing data; ASC—access service circuit; and and SIC—station identification code

guard time ($\simeq$100 ns), preamble and the PCM-MUX information for destinations $x, y \ldots z$ stations. The preamble contains carrier and clock recovery bits, station-identification code (SIC), channel allocation and routing data (CARD) and an order-wire (voice) circuit (ASC). The bit allocations are as shown; but the actual CARD word is of 32 bits (21 bits of information +11 bits of parity using FEC, giving a BER $\simeq 2.10^{-24}$) and is transmitted in eight frames, 4 bits/frame. For DPSK modulation, carrier recovery bits are not necessary, and the preamble length reduces to 42 from 52 bits. The simplified block diagram of a TDMA terminal is shown in Fig. 5.52, where the Command Unit (CMU) and the Communication Control Unit (CCU) control the total operation of burst generation, reception, coding, decoding and synchronization. The CMU is a small stored-program computer which commands the operation of the system including burst allotment, initial acquisition, circuit assignment, signalling through CARD, and control in the event of faults. The CCU is the interface between CMU and other units and controls the CARD and ASC channels. Range measurement information is also processed through CCU to the Burst Synchronization Unit (BSU).

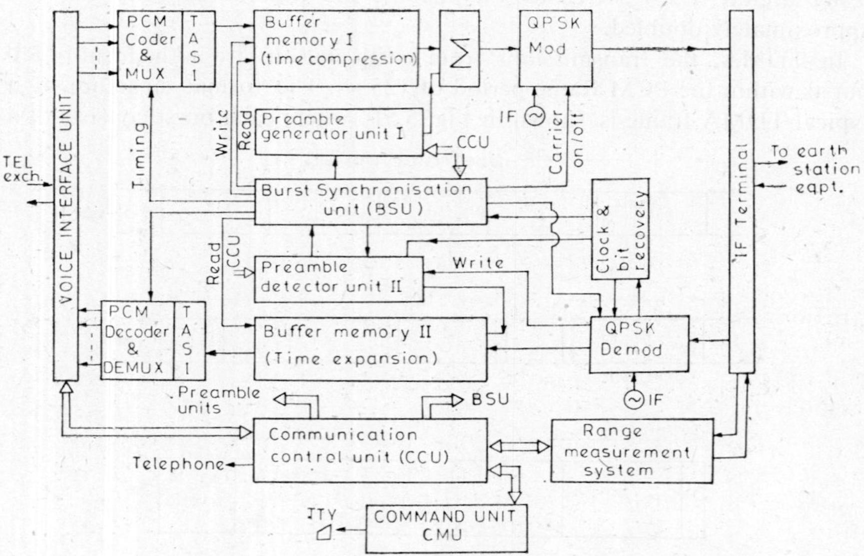

Fig. 5.52 Block schematic of a TDMA terminal

The BSU has three major functions: (a) synchronization of the bursts with respect to the satellite; (b) automatic burst repositioning in response to traffic variation; and (c) providing timing control to all other subsystems. The preamble word is generated and detected in units I and II under the control of BSU. Finally, the Buffer memory I compresses the PCM words to the required burst length (e.g., 125 μs PCM frame with 24 channels is compressed to 3 μs, for the transmission rate of 64 Mb/s) and the Buffer memory II expands the bursts in a complementary way.

When an earth station wants to transmit its bursts in the common digital channel to the satellite, the range data are measured and only the preamble word is emitted at the centre of the allotted slot with reference to the reference station's preamble (say, station A in Fig. 5.51). Then, the preamble word is moved to the predetermined position at the speed of 1 bit/frame. When the correct position is achieved, the synchronization circuit is locked and the information bits are added to the preamble for the normal operation. On demand from the interface unit, CCU allots the channel and sends the assignment and signalling information through CARD. Coded PCM words are now stored in Buffer I and added to the preamble. The composite burst is now PSK-modulated and transmitted through the HPA and the antenna. The satellite now receives the multiple bursts and retransmits them in a broadcast mode. At the destination, the station receives the desired preamble along with SIC (which also serves as a 'burst-start' signal), and the preamble is stripped out, decoded and distributed by the preamble unit II. PCM decoder and the interface unit, under the control of CCU, decodes and distributes the information according to the signalling information obtained through CARD. The QPSK modem should have an excellent burst operation capability, and is operated either in DCPSK or DPSK mode. The unique words (UW) used for SIC are specially selected so that SIC is detected with a tolerable error of 2 bits only. Experimental BER results of QPSK demodulators through the satellite loop are: BER $= 10^{-4}$, 10^{-5}, 10^{-6} for $(C/N)_i = 13, 14.5, 15.5$ dB, respectively. The probability of SIC misdetection is: $P_e = 10^{-4}, 10^{-5}, 10^{-6}$ for BER $= 10^{-3}, 10^{-4}, 10^{-5}$, respectively. The results are further improved by using a high rate code, e.g., R-7/8 convolutional codes for information bits.

For increasing the channel capacity in a TDMA system, both Demand assignment and TASI techniques have been proposed. In the TDMA/DA technique, which is somewhat similar to the DA technique of SPADE, the multidestinational slots in the information burst form a DA pool and they are assigned on-demand basis. This reduces the average length of the bursts and releases the idle slots for further use on demand, thereby increasing the overall channel capacity. Digital-TASI, on the other hand, is the well-known technique for increasing the capacity of a group of digital trunks (say, 40 and above). Using time preassignment of the channels and TASI, the traffic efficiency is almost the same as that of the DA mode. Experiments have shown that Digital-TASI almost doubles the channel capacity with reasonably large traffic per station.

To calculate the overall channel capacity of the system, it should be observed that the overhead bits per burst ($=57$ bits) are equivalent to 7 PCM words approximately. As such, for CPSK signalling, 7 voice channels per access are lost; and for DPSK signalling, only 6 voice channels ($=47$ bits of overhead) per access will be lost. The overall channel capacity for a 64 Mb/s system (equivalent to 1000 voice channels) is now given as:

$$\text{Channel capacity } n = 1000 - 7m, \qquad m = \text{number of accesses} \tag{5.36}$$

To improve the channel efficiency, it has been proposed that the frame time be increased from 125 μs to 750 μs or more, and the loss due to the overhead will decrease proportionately. The variation of n with the number of accesses, along with the channel capacity of the FDMA system, is shown in Fig. 5.53. The considerable improvement of n with a large frame time and also the superiority of the TDMA over FDMA are observed from the curves.

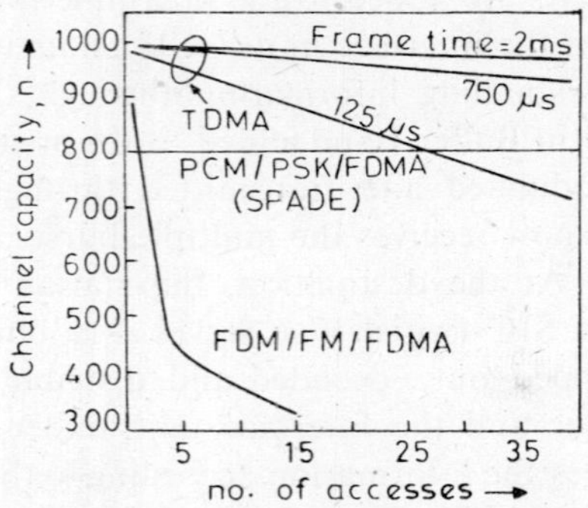

Fig. 5.53 Channel capacity variation with number of accesses for TDMA and FDMA

5.9.3 SS-TDMA [41, 42]

In a TDMA burst format; as shown in Fig. 5.51, the data for different destinations are multiplexed in the single burst and the destination receiver has to identify the particular slot in the frame which contains the required data. Consider the format of the burst shown in Fig. 5.54(a), where three slots of data (i, j, k) from the transmitting stations A, B and C are meant for the destinations i, j and k. If there is an electronic switch on-board the satellite and synchronized with the time slots, then it is possible to reconfigure the burst format such that the data for the destinations i, j and k are regrouped as shown in Fig. 5.54(b). This avoids the problem of searching for the destination data as required in the conventional TDMA. Assuming that the satellite has independent transmit and receive antennas, the switching matrix shown has to interconnect the antennas as required under the command from the ground. This technique is known as Satellite-Switched TDMA (SS-TDMA) and is being used in recent transponders, e.g., in

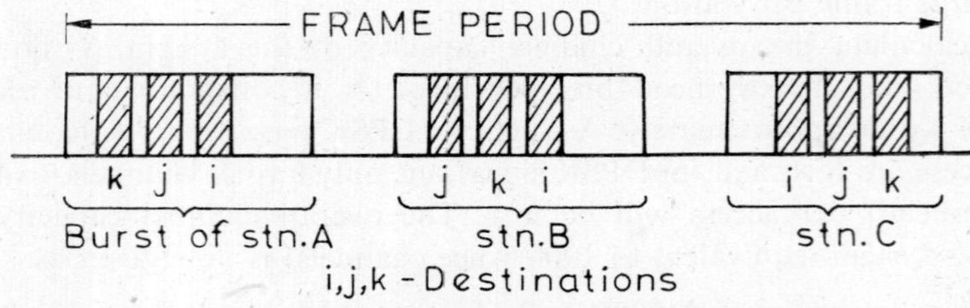

Fig. 5.54(a) TDMA burst input to satellite

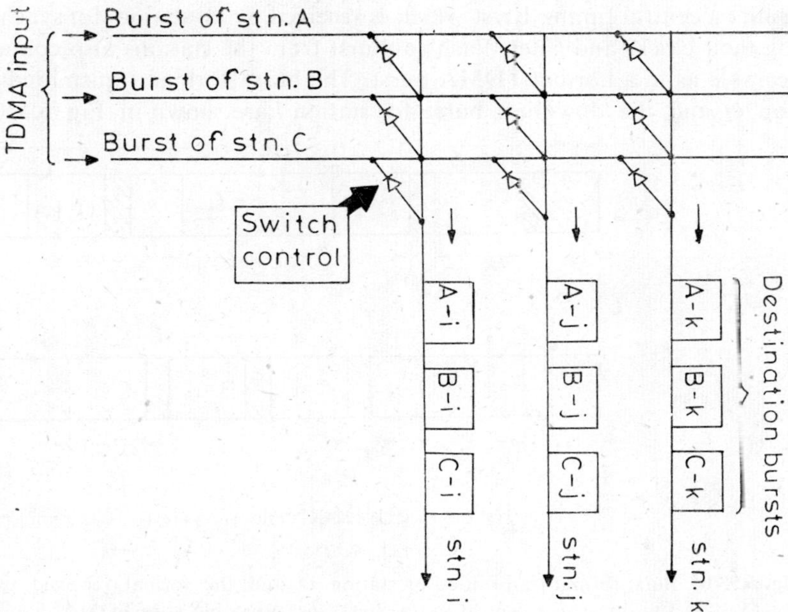

Fig. 5.54(b) Separation of subbursts by on-board matrix switch

Advanced Westar (USA). Figure 5.55(a) shows the overall scheme of the system, where the satellite has multiple spot beams and the on-board switching matrix switches the incoming data to the destination antenna as required. The reference station O (any other station also may be made as reference)

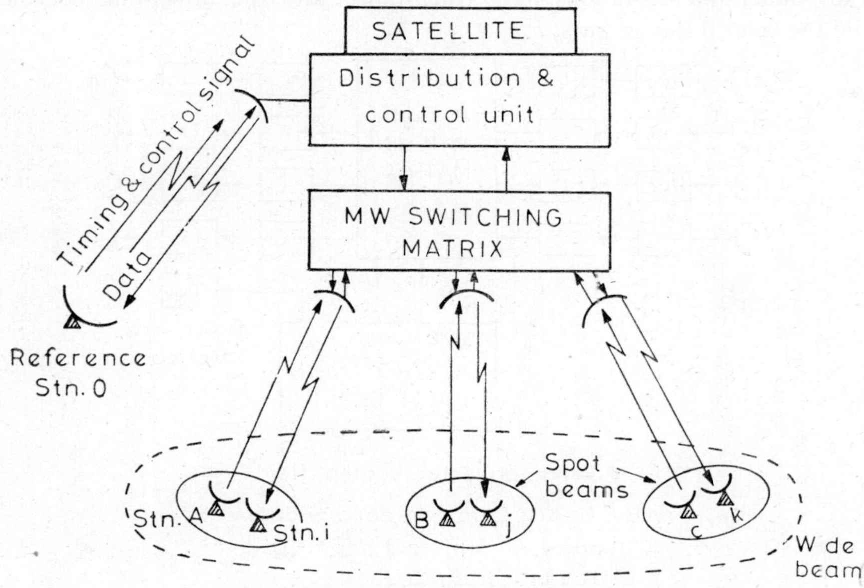

Fig. 5.55(a) Schematic of SS-TDMA

transmits a central timing burst which is received by all stations for synchronizing their clocks and gates. Each subburst from the stations also contains a preamble as in a normal TDMA burst. The burst format transmitted by station *O* and the downlink burst for station *i* are shown in Fig. 5.55(b).

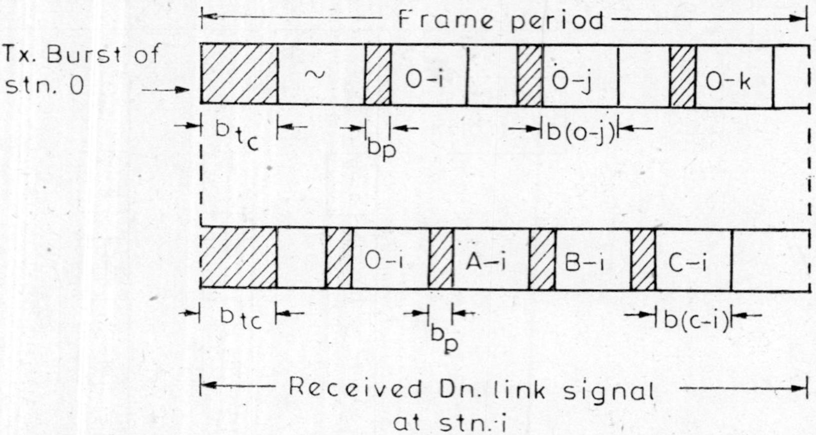

Fig. 5.55(b) Burst format transmitted by station *O* and the format received by station *i*; b_{tc} central timing burst, b_p: preamble as in TDMA; $b_{(o-j)}$ information bits from stn. *O* to stn. *j*

Uplink bursts from other stations have the same format as that of station *O*. Figure 5.56 shows the block diagram of a transponder using SS-TDMA. To avoid the uncertainty in switching instants due to the variable path delay of different stations, the reference station *O* is slaved to a master clock at the transponder and all stations now synchronize with the down-link beacon and the central timing burst b_{tc}.

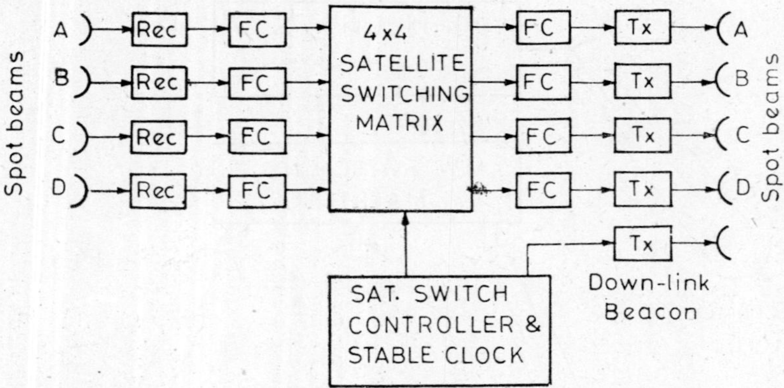

Fig. 5.56 Block diagram of a transponder using SS-TDMA

For high data rates, it is necessary to use fast microwave switches with redundant design for the switching matrix, and the spot beams should have sufficient isolation so that frequency reuse is possible. With these assumptions, the total data BW of the transponder with N spots is N times that of a transponder with a single wide beam antenna. The full power of the TWTA may now be used without any IM-distortion, leading to a further increase in the channel capacity. In recent designs, the transponder BW for SS-TDMA is made wider and it is possible to use the whole of the 500 MHz slot for one transponder with multiple spot beams. With a BW=400 MHz, the data rate $\simeq$ 600 Mb/s and for $N = 5$, the total channel capacity = 39,700, giving a 30% improvement over the conventional TDMA with the same number of spot beams [41]. It has been further suggested that SS-TDMA with an on-board regenerative repeater and baseband switching will have improved error-rate performance and power efficiency [42]. Using high-power PSK modulators, many TWTA may be replaced, thus saving weight and cost. It is reasonable to expect that the on-board signal processing including electronic switching will be incorporated in future satellites.

5.9.4 CDMA [43]

The basic philosophy in FDMA/TDMA has been that the signal/data for different destinations are orthogonal to each other either in the frequency domain or in the time domain. An alternative technique to achieve this is to use Code Division Multiple access (CDMA), where all uplink signals occupy the total available transponder BW; but each signal/data has its own orthogonal structure which is used to separate the desired signal/data from others. This is generally achieved by multiplying the signal/data by quasi-orthogonal PN sequences as unique words (since it is difficult to obtain a large number of purely orthogonal codes), and the composite signal is decoded through matched filters (MF)/correlation detectors. It is now possible to access the transponder randomly without time/frequency synchronization (but receivers have to acquire the address codes) and the CDMA systems are sometimes called Random Access Discrete Address (RADA) systems. The CDMA technique has been widely used in defence communication satellites, e.g., SKYNET (UK), DSCSII, DSCSIII (USA), NATOIII. Experimental civilian communication satellites using CDMA have also been developed by Japan and Australia, specially for remote area and emergency communication. The technique is discussed in detail in Sec. 5.11.2.

5.10 RECENT TRENDS IN SATELLITE COMMUNICATION [27]

In spite of the enormous growth in satellite communication technology, there are obvious limits to this growth. They are:

(a) The number of synchronous orbit assignments available is limited by interference between adjacent systems.
(b) The frequency bandwidth available for satellite systems is limited by nature and the regulatory process.

(c) The size of the satellite is limited by the current technology of launch vehicles, but this is being relaxed with the success of the Space Shuttle.

On the other hand, there are many promising techniques which will lead to long-term growth in spite of the above limitations. The important ones are:

(a) Frequency reuse (being used in INTELSAT V, COMSTAR etc.) and cross-linking (of specialized frequency band) between satellites in space.
(b) Multiple frequency band operation in a single large satellite.
(c) Network control technology for centralized automated control of earth stations.
(d) Solid-state amplifier technology.
(e) Multiple-beam ground antennas to have access to many satellites from one station.
(f) Multiple spot beams and SS-TDMA.
(g) Digital signal processing for voice, data and video.
(h) CDMA technique for small earth stations.

The problem of echo suppression in voice circuits (because of 0.5 sec delay in the path) has been solved by echo cancellers, now fabricated in a single VLSI chip at a lower cost. The effect of propagation delay in data transmission has been overcome by a satellite delay compensation unit, which permits much larger data block lengths (than in terrestrial circuits due to better error rates in satellite links) and increases the throughput efficiency by a factor of five. The video conferencing circuit has been made cheaper and more attractive by using the frame-differential band-compression technique. Even standard TV signals are now transmitted at 32 Mb/s rate only.

The application of Delta modulation, FEC, ARQ and M-ary QAM/MSK techniques improves the channel capacity manifolds and the band expansion due to A/D conversion of analog signals is more than compensated, resulting in better signal quality and lesser error rates with minimum satellite power. In TDMA using a frame time of 750 μs or more, the channel capacity of a transponder remains almost constant, even with a large number of accesses and FEC improves the channel capacity by approximately 6 dB for low G/T earth stations. VLSI and μ-processor techniques have greatly aided implementation of digital coding and data compression algorithms, and coupled with the forthcoming VHSIC technology, complex signal processors operating at very large bandwidths are being realized in a small number of chips. The processing of signals on-board the satellite leading to satellite-switched TDMA and regenerative repeaters are now possible using digital techniques and will lead to further improvement in the overall efficiency.

The all-digital, fully integrated domestic satellite systems, e.g., SBS of USA, FECS and RICS of Japan, use many of the digital processing techniques mentioned above. The Satellite Business System (SBS) has been designed to serve the full spectrum of communication needs of a large community of business and government organizations and others, and has the following characteristics:

Fully integrated digital-voice, data and image transmission capability; RF-12/14 GHz; TDMA with DA, voice activity compression (VAC); 5.5 and 7.6 m antennas for earth stations on customer premises; data 2.4 Kb/s to 6.3 Mb/s. Direct access to a switchboard wide-band communication network with reduced dependence on terrestrial access facilities. Equivalent useful capacity $\simeq$ 13,900 voice circuits or 8000 data circuits each of 56 Kb/s ($\simeq$ 500 Mb/s). The user of network services has available an assortment of advanced features, such as:

(a) flexible voice and data conference arrangements,
(b) multipoint distribution of digital data, including document distribution,
(c) teleconferencing, including multipoint video conferences,
(d) hot line and other priority connections,
(e) network access control, and
(f) electronic mail using a page-a-second communicating copier.

Japan has been developing domestic satellite communication systems since their first experiments with ATS-1 during 1967–68. They have now developed a medium capacity domestic satellite communication system using the 30/20 GHz and 6/4 GHz bands. This includes the Fixed earth station communication system (FECS) in the 30/20 GHz band, Remote island communication system (RICS) in the 6/4 GHz band and the Transportable earth station communication system (TECS) using both 30/20 and 6/4 GHz bands. Both for FECS and RICS, they are using TDMA, having 60 Mb/s and 100 Mb/s data rates. The 100 Mb/s system has the capability of transmitting and receiving two colour TV and 192 voice circuits, and has the following characteristics:

Bit rate 106.88 Mb/s, 2 CTV + 192 voice circuits, number of accesses 4; QPSK modulation with coherent detection; Network clock synchronization, FEC Rate-7/8 convolutional codes and threshold decoding; 1.544 Mb/s vioce and 32.064 Mb/s CTV signals. Data rates 6.4 Kb/s to 6.3 Mb/s.

Experiments are also being done with very small earth stations having one-to-three-channels capacity and using SSMA techniques in 6/4 GHz band. Another system for an integrated digital satellite communication for transmission of diverse bit rates of data, facsimile and video between small earth stations has been developed. TDMA-DA is being used for accessing the system.

Due to the congestion in the C-band, newer satellites are also utilizing Ku-band, e.g., in INTELSAT V and VI, Anik C, Advanced Westar, SBS, CS, etc. The use of the Ku band in domestic satellites will continue to grow. Since the ultimate capacity of K-band satellites is somewhat greater because of the easier implementation of frequency reuse, Japan has been carrying out a program of investigating the problems and possibilities of millimetre wave (Ka-band) satellite communication since 1967. Exhaustive study of rainfall attenuation, excess noise temperature due to rain, ionospheric scintillation and propagation delay at 20/30/32/35 GHz, has been made using CS, ECS and ETS-II experimental satellites. At Ka-band frequencies, it will be necessary to have site diversity incorporated using an auxiliary station to counteract the large rainfall attenuation at these frequencies. Network organization for such a system has also been studied. In general, the results have been favourable for implementing futuristic satellites in the Ka-band [44].

An excellent example of coordinated communication and tracking, using space communication techniques in the UHF, S and Ku-bands, is the communication and tracking system of the Space Shuttle Orbiter [45]. The composite system provides a transmission capability of digital voice, telemetry, television and data at a maximum rate of 50 Mb/s. An S-band link directly communicates with the ground station and another S/Ku-band link is routed through the Tracking and Data Relay Satellite (TDRS). In addition, 850 S-band channels are provided to communicate with other satellites or spacecraft using a variety of formats and modulation techniques. UHF is used for communication with extravehicular astronauts. During reentry and landing, additional L-band and C-band links are used as navigational aids.

The US defence department is developing a global positioning system, called NAVSTAR, which will serve most of the needs of USA's global navigation and position determination. This will provide for the first time a common system for all classes of users, maritime, aviation and space, and it may reverse the proliferation of special purpose navigation and positioning systems. The system is all-digital and the users determine their positions by tracking long digital codes transmitted from four NAVSTAR satellites. This is one of the unique usages for SS-codes and has a range measurement accuracy of 10 to 20 metres. The US defence considers NAVSTAR as the best way of keeping wide ranging forces. (including infantry) on target and potential civilian users are hopeful of getting this service as well.

The general trend towards larger spacecraft for satellite communication will lead to the support of large and complex antenna feed systems for multibeam, multifrequency use and the high power levels for multi transmit channels. Once large spacecrafts of 2500 kg or more are built for any one mission, their use for multiple mission will become obvious. Thus, the concept of the geostationary platform being realized, the platforms will replace many separate small satellites that are now performing individual missions. In addition to the economic benefits, platforms will be able to interconnect missions (e.g., international and maritime services) and thus conserve orbital arc and RF spectrum through the efficient use and reuse of several frequency bands. Initial studies show that a 5000 kg experimental geoplatform could accommodate several advanced communication payloads, e.g., a 30/20 GHz system, a large-aperture (12 m) multibeam 6/4 GHz system, a satellite-switched 14/11 GHz systems, and a 12 GHz broadcast satellite system. They would all be interconnected and could be used to accommodate a variety of user experiments and for commercial users, e.g., INTELSAT, as well.

In future, international services will continue to grow rapidly (at approximately 20% per year), and many new domestic systems will come into existence. New system topologies such as computer networking, broadcast and data collection, will be introduced, and new services such as video teleconferencing and electronic mail will become a part of satellite communication.

5.11 MULTIACCESSING IN RADIO SYSTEMS

Apart from the conventional FDM/TDM techniques, multiaccessing at the RF level is necessary for such common transponder systems, as used in mobile radio and satellite links. Traditionally mobile radio used FM terminals with frequency-sharing techniques and earlier satellites also used similar FDMA methods. To avoid the saturation problems in transponders leading to large intermodulation distortion, the more efficient TDMA techniques were then developed, as discussed in Sec. 5.9.2. For the purpose of random accessing, the Code Division Multiaccessing (CDMA) has been developed, specially for mobile networks. The CDMA techniques are now successfully used in many modern satellites, and hopefully this will be used more widely in mobile radio networks as well.

5.11.1 The Channel Assignment Problem

The established practice of assigning a wideband channel to multiple users by either frequency division multiplexing (FDM) or by time division multiplexing (TDM) is efficient only for heavy traffic densities. For low capacity traffic, these techniques are uneconomical and narrow-band channels are assigned to the subscribers on a pool basis having facilities for monitoring the channel. The two particular cases of such multiple assignments are (a) mobile radio communication and (b) low density static communication.

At present, there is a tremendous expansion of mobile radio communication for vehicular terminals in an urban environment and in future such communication will be carried out on a very large scale with inexpensive terminals. It is well known that if a common carrier is used by the entire set of unsynchronized vehicular terminals in a given city, the cross-interference will be very large. The problem is sought to be solved by dividing the city into regions, called cells, and each cell is assigned a separate carrier. Since it is not possible to assign an independent carrier to each of the cells, it is necessary to use multiple assignment of a carrier with proviso that no adjacent cells have the same carrier assigned. Since the communication will be through a master repeater in the city, this strategy will give minimum interference, the same carrier is repeated only with large distances between the cells (known as 'adjacency constraint'). Considerable theoretical work has been done on the methodology of the frequency usage and on the question of optimum cell shape and number. The desired 'adjacency constraint' is easily satisfied if more frequencies are available for assignment to different cells. But from the channel utilization point of view, one would like to use the minimum number of carriers and, at the same time, maintain the minimum interference required by the system. It has been proved earlier that only five frequencies (colours) will suffice to satisfy the desired 'adjacency constraint'. It has been recently shown by Minoli and Schneider [46] that only four frequencies also suffice to satisfy the desired 'adjacency constraint', the techniques being called the four-colour conjecture (4 CC). In a hypothetical case of 36 cells in a city, the frequency assignment plan for four carriers, F_1, F_2, F_3 and F_4, is shown in the Fig. 5.57, where it is seen that none of the frequencies has been assigned to adjacent cells. Such assignment can be organized through computer simulation as suggested by the authors. However, it will be necessary to evolve the design of the complete system systematically, considering the power requirement of the transmitters, attenuation factor, bandwidth and sensitivity of the receivers, etc., to satisfy the 'adjacency constraint'.

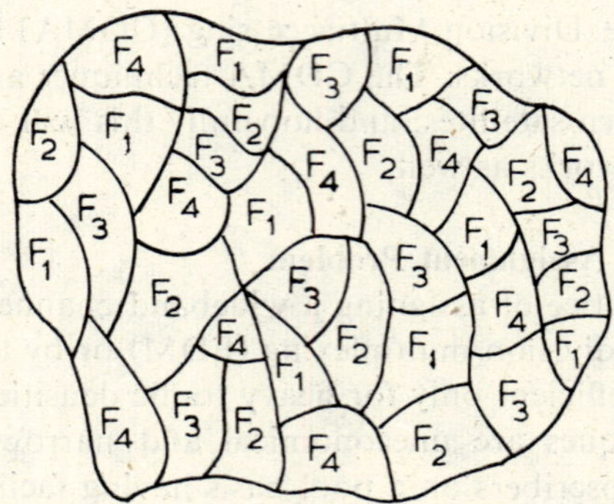

Fig. 5.57 Four-frequency assignment plan for a 36-cell city

For low-density traffic from static transmitters, the frequency assignment may be made on a shared-channel basis and it is interesting to calculate the capacity of such randomly used shared channels in a particular system configuration. The channel is divided into equal-width segments and the segments are assigned to more than one subscriber. With such multiple assignment, larger total bandwidth is available to each subscriber and thus larger average channel capacity results. It has been shown by Haber [47] that withsuch nonexclusivese gment assignment, the overall subscriber channel capacity is given by

$$C_{av} \simeq C_0 \frac{1}{p}[1 - p]^{(1/p-1)} = C_0 F \tag{5.37}$$

and

$$\begin{aligned} C_0 &= \text{channel capacity on an exclusive basis} \\ &= (B/m) \log_2 \; [1 + S/\{n_0(B/m)\}] \end{aligned} \tag{5.38}$$

where p is the probability of use by a given subscriber of his assigned k, k = number of segments assigned to each subscriber, B = channel bandwidth = nb, b = segment width, $n = B/b$ = number of segments, and

m = number of subscribers. The improvement factor in channel capacity is given by

$$F = \frac{1}{p}[1 - p)^{(1/p-1)} \tag{5.39}$$

As an example, with $p = 0.1$ and the reuse factor $r = 10$, $F = 4$. However, this improvement factor is dependent on the type of assignment for reuse. It has been further shown by the author that the assignment plan has to be as random as possible between different subscribers and this gives minimum variance of the overall channel capacity. Thus, it is shown that the average channel capacity is maximized when the probability of use p = inverse of the reuse factor r. This technique of frequency pooling with multiple assignment is quite an efficient method for accommodating low-density traffic.

5.11.2 CDMA Using the Spread Spectrum Technique (SSMA) [43, 48, 49, 50]

The network control in TDMA and FDMA is a complex problem and the systems are quite susceptible to jamming and interference. To avoid EMI/EMC problems present in these classical multiplexing systems, considerable attention has been given recently to the development of SSMA systems, where the total available spectrum is used commonly by all users and the users are identified by quasi-orthogonal addresses. The address coding is based on (a) CDMA [51] using direct sequence (DS) signalling and continuous transmission as in Fig. 5.58, (b) PAMA [52] using non-coherent frequency hopping

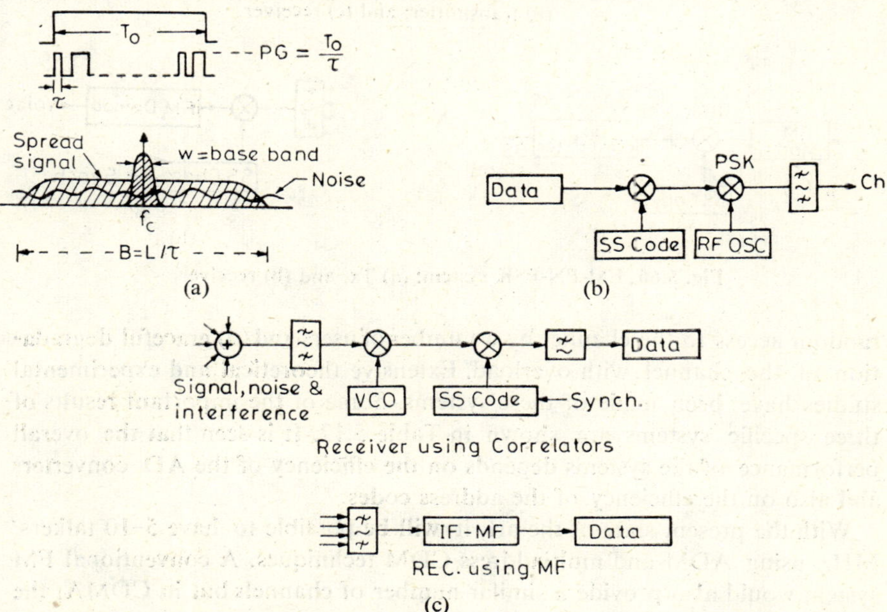

Fig. 5.58 Spread spectrum using DS modulation: (a) DS code and its spectrum; (b) transmitter; and (c) receiver

(FH) signal, also known as TF matrix coding as in Fig. 5.59 and (c) hybrid techniques [43], e.g., PPM-PN-FSK, FH-PN-PSK, FM-PN-PSK [53], as in Fig. 5.60. These wideband techniques have the advantages of (a) immunity to jamming, (b) security of information in the channel, (c) immediate

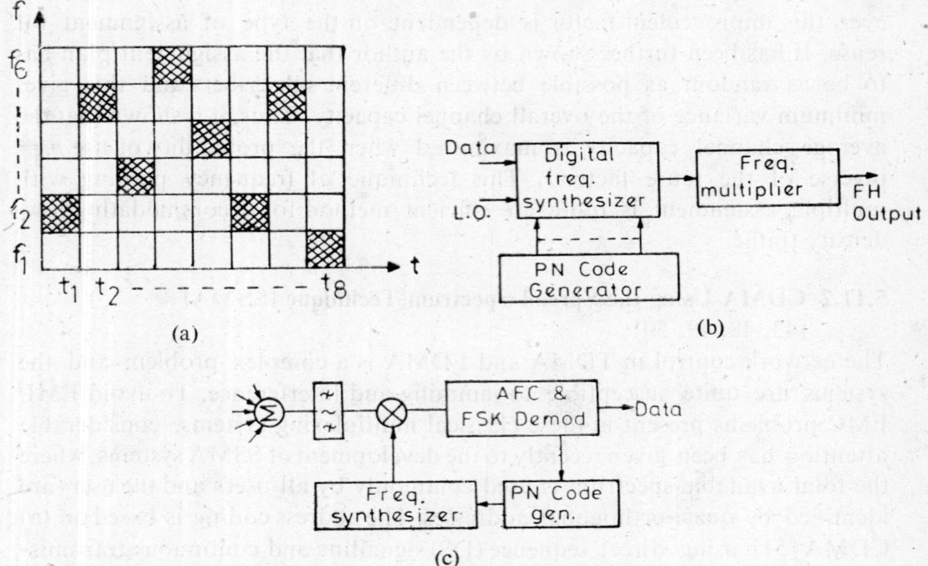

Fig. 5.59 PAMA: (a) FT-matrix coding; (b) transmitter; and (c) receiver

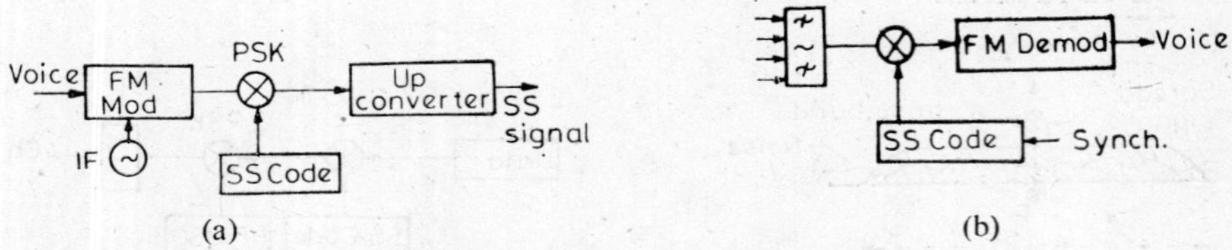

Fig. 5.60 FM-PN-PSK system: (a) Tx. and (b) receiver

random access to the channel by a number of users and (d) graceful degradation of the channel with overload. Extensive theoretical and experimental studies have been made of these systems. Some of the important results of three specific systems are shown in Table 5.12. It is seen that the overall performance of the systems depends on the efficiency of the AD converters and also on the efficiency of the address codes.

With the present state-of-the-art, it will be possible to have 5–10 talkers/MHz using ADM and multiaddress CDM techniques. A conventional FM system would also provide a similar number of channels but in CDMA, the accessing in random.

Table 5.12 SSMA performance: for speech=0–3.5 kHz, PSK, $K = \alpha K_{mx}$ = number of simultaneous talkers

Type of systems	System quality		K_0 = K/MHz		
	P_e	S_0/N_0	PAMA/ MA	CDM/PN/ PSK	PPM/PN/ FSK
(a) 7-digit PCM	10^{-5}	45 dB	3	3	18
(b) ADM 40 kB/S	10^{-5}	45 dB	3	4.5	25
(c) ADM 20 kB/S	10^{-3}	31 dB	5.5	16	30
(d) FM/PN/PSK [53]	. . .	. . .	k_0 = 5/MHz		
(e) PPM/Conv. Code/PSK [54]		. . .	k_0 = 15/MHz		

Because of the attractive performance of CDMA in mobile and other radio networks and also in satellite communication, many sophisticated systems have already been developed and are in use. The United States Army is using a RADA system, known as 'SMARTS'. The Skynet Satellite system of UK uses SSMA techniques for widebanding 2400 bits of data into a PN code at 10 Mbit rate, and using PSK, the 20 MHz band is utilized for a large number of accesses. The Australian Post Office has investigated the possibility of a small satellite system for distant cities and thinly populated areas and have come to the conclusion that the SSMA technique and DM-PSK system would give comparable performances [55]. Japan has developed a new satellite communication system using spread-spectrum random access (SSRA) [56]. In Germany, a CDMA system for an integrated network over glass fibres has been proposed and the advantage of the asynchronous integrated network based on CDMA has been shown [57].

As an example of remote area communication through satellite, it is observed that for TP channels of bandwidth = 120 Hz, it is difficult to have direct single channel transmission through satellite because of the Doppler shift and frequency stability problems. Alternatively, one may use 24-channel VFT (FDM-Mux) and digitize this to match 32 Kbit/s digital SCPC channel of the SPADE system. For the Symphonie Satellite, the available EIRP per SCPC channel = 8.2 dBW and it has been calculated that G/T required is 11 dB/°K in the first case (direct TP) and 15 dB/°K in the second case. If one uses SSMA technique for randomly accessing 24 TP subscribers, G/T required is only 11 dB/°K. The G/T requirement of small earth terminals along with the same transponder is shown in Table 5.13 for various simultaneous accesses.

5.11.3 Channel Capacity

Contrary to the general assumptions, Costas [58] has shown that the channel capacity of SSMA is comparable to that of FDMA and TDMA.

A comparison of the channel capacity of an SSMA system with that of a narrow-band system may be obtained by noting that a binary channel,

Table 5.13 SSMA for TP signals using 32 Kb/S transmission rate

M = No. of simultaneous accesses	EIRP in dBW per channel	G/T dB/°K
10	−1.8	3
21	−5	9
24	−5.6	11
30	−6.6	18
31	−6.9	24

having small P_e, has the capacity:

$$C_p = \frac{W}{\gamma} \cdot \frac{S}{N} \simeq \frac{1}{T_b} \leqslant W, \tag{5.40}$$

where W = channel BW, S = signal power, T_b = bit period, γ = SNR for small P_e and $N = n_o W$.

In the presence of jamming and co-channel interference $N' = n_o W + J$, and the error rate will be effected by

$$\gamma' = \frac{S}{N'} \cdot T_b W = \frac{ST_b}{(n_0 + J/W)} \tag{5.41}$$

which varies very slowly for large W. In order to effectively disrupt the circuit, the average jamming power required would be approximately $J_0 = n_0 W$. Thus, for a larger system bandwidth, the jamming power required is also large.

It is seen that the narrow-band channel capacity per circuit $C_{pN} \simeq B/K_{mx}$, but, using eqn. (5.40), the capacity per circuit for the broadband system $C_{pB} \simeq B/(\gamma \alpha K_{mx})$, where B = total bandwidth of the system, K_{mx} = maximum number of talkers, α = average fraction of time each talker transmits (αK_{mx} = average number of active talkers) and the effective N' in the broadband system $= \alpha K_{mx} S$. Then the ratio of the two channel capacities is:

$$C_{pB}/C_{pN} = 1/\alpha\gamma \tag{5.42}$$

For a practical case of $\alpha = 0.1$ and $\gamma = 10$ for $P_e = 10^{-5}$, it is seen that the two systems give the same channel capacity under the ideal assumption of strict spectrum allocation and constant transmission conditions in the narrow-band case. But in a congested band and with hostile jamming present, the broad-band system will normally far outperform the narrow-band system.

False Address Probability

The error probabilities of different SSMA systems have been calculated by various authors [51, 53, 59]. For DS-PSK signalling and correlator/MF receiver, as shown in Fig. 5.58, P_e for large BT is shown to be [51] [using eqns. (2.34) and (2.82)]:

$$P_e = erfc\sqrt{\frac{4BT_b}{K-1}} \simeq \frac{1}{2}\exp\left(-\frac{2BT_b}{K-1}\right) \quad \text{for large SNR} \quad (5.43)^*$$

where K = number of simultaneous talkers. The results of eqn. (5.43) are shown in Fig. 5.61(a) for various code lengths $L = BT_b$.

For single-address PAMA signalling, the false address probability, assuming Poisson's distribution of the occurrence of pulses, is given by [52]:

$$P_e = (1 - e^{-\lambda\tau})^q, \quad (5.45)$$

where

$$\lambda\tau = \frac{qK}{BT_b} = \frac{qKd}{MN};$$

and d = duty ratio $\leqslant 1$, q = number of slots used per address out of the MN slots in the TF-matrix, τ = duration of each time slot = M/B, M = number of frequencies and N = number of time slots.

The results of eqn. (5.45) are shown in Fig. 5.61(b) for various values of q; and $\lambda\tau = 0.7$ is the optimum for a given MN. The size MN of the matrix does not directly influence P_e, but only determines the number of independent addresses that may be obtained in the system.

The spectrum utilization factor η is poor in single-address PAMA systems and Chesler [52] has shown that a multi-address PAMA system, using one address for each k-bit word of the subscriber, is more efficient. The error probability is now given by:

$$P_e = \sum_{K_{mx}=0}^{\infty} 2^k \left[1 - (1 - q/Q)^{K_{mx}}\right]^q \cdot \frac{(K)^{K_{mx}}}{K_{mx}!} e^{-K}, \quad (5.46)$$

where Q is the matrix size, and the spectrum utilization factor now improves to:

$$\eta = \frac{\text{total information rate}}{B} = k \cdot K/Q \quad (5.47)$$

---

*Assuming that the received signals from the talker transmitters are of equal power and they add coherently, eqn. (5.41) is modified as:

$$\gamma' = \frac{ST_bB}{n_0B + (K-1)S} \simeq \frac{BT_b}{K-1}, \quad \text{neglecting } n_0 \ll \frac{K-1}{B} \quad (5.44)$$

It is to be noted that in writing γ', the processing gain of the MF of length BT_b has been introduced as per eqn. (2.82), and γ' is the predetection (E_b/n_0). However, due to the frequency and phase differences as well as due to code-phase differences of the received signals, the interfering signals do not add coherently. In [51] it is shown that the total interfering signal $J = \frac{(K-1)S}{2}$ and in [59], it is shown that $J = (2/3)(K-1)S$. It is also known that, by invoking the central limit theorem, the interfering signal J has an approximate Gaussian distribution for K large, and the peak-to-average power ratio of $J \simeq 10$ dB. Equation (5.28), which give the loading factor in a multichannel FM system, also indicate that the peak-to-average power ratio of random signals $\geqslant 10$ dB for large K. In view of the above, a conservative value of $\gamma' = (2BT_b)/(K-1)$ has been used in eqn. (5.43), but in practice γ' will be better than this value.

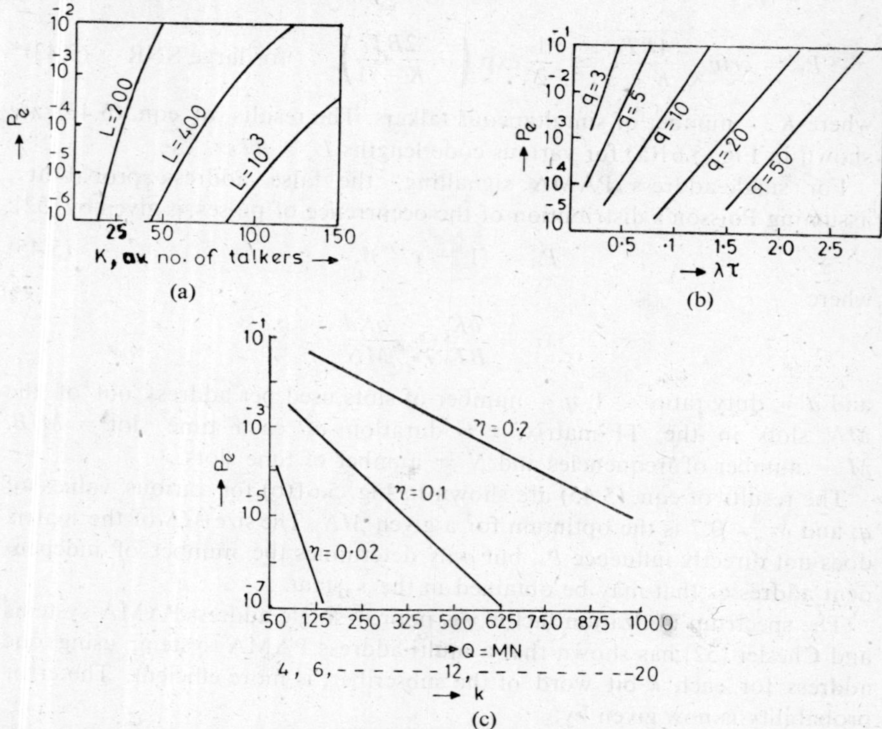

Fig. 5.61 Error probabilities in different SS-systems: (a) P_e in DS-PSK signalling; (b) P_e in single address PAMA system; and (c) P_e vs·Q in MA-PAMA system

The results of eqns. (5.46) and (5.47) are shown in Fig. 5.61(c), where it is seen that η improves considerably with k and Q.

In the FM-PN-PSK system of SSRA [53], the IF is FM-modulated by speech and further PSK-modulated by a PN-code at 16 Mb/s. This composite signal is upconverted to 6 GHz for transmission to the satellite. In the receiver, a complementary process recovers the voice signal, as shown in Fig. 5.60. The resultant $(S/N)_0$ at the receiver output is given by:

$$(S/N)_0 = (C/N)_i \left[\frac{3}{2}\,\frac{B_{if}}{f_{mx}} D^2\right] = (C/N)_i \times (\text{FM-gain}) \tag{5.48}$$

where D = deviation ratio of FM, B_{if} = FM BW, f_{mx} = speech BW and $(C/N)_i$ = SNR at FM demodulator input

$$\simeq (\alpha/K)\left[\frac{1}{1/\text{CNR (Up)} + 1/\text{CNR (Dn)} + 1/\text{PG}}\right],$$

$$\simeq (\alpha/K)\ \text{PG, neglecting the repeater and receiver noise} \tag{5.49}$$

where α = limiter effect coefficient $\simeq \pi/4 = -1$ dB:

$$\text{PG} = \text{Processing Gain} = \frac{\text{RF BW}}{\text{FM BW}};$$

CNR (Up) = Uplink CNR; CNR (Dn) = Downlink CNR.

For a standard transponder with 36 MHz BW, chip rate used was 16 Mb/s, FM BW = 30 kHz, $D \simeq 3$, $f_{mx} = 3.4$ kHz, FM threshold with FMFB $\geqslant 6$ dB, and $PG \simeq 30$ dB. The number of accesses possible $\simeq 200$ with $(S/N)_0 \geqslant 26$ dB.

The values in Table 5.12 have been calculated on the basis of the above equations. However, it is seen that η in each case is worse than in synchronized FDMA/TDMA.

5.11.4 Schemes for Improving the Spectral Efficiency in SSMA

In an ideally synchronized TDMA system, each user is assigned a time slot $= T_b/K$, where $T_b =$ information bit time and $K =$ number of active users. The spectrum utilization factor (SUF) η for PSK transmission of TDM signals is:

$$\eta = \frac{\text{Total information rate}}{\text{Channel bandwidth}} = \frac{K}{\text{BWE}} \simeq 1, \tag{5.50}$$

where BWE = bandwidth expansion in the channel. On the other hand, it has been shown that for PAMA systems [54], $\eta \leqslant \ln 2$ (= 0.695) and, ordinarily, for single-address PAMA system, $\eta \leqslant 0.1$. Chesler [52] has shown that multiaddress PAMA systems, using one address for each k-bit word and requiring $2^k = M$ addresses per user, provide better spectral efficiency than the single-address PAMA. The SUF now is given by:

$$\eta = \frac{kK}{Q},$$

where $Q =$ size of the TF matrix = BWE in the system.

His results show that the word error rate $P_w \simeq 10^{-5}$ is obtained with $k=20$ and $Q = 10^3$ for $\eta = 0.2$, and $K = 10$, whereas for single-address PAMA, $\eta = 0.05$ for similar error rates.

In CDMA systems using PSK transmission and matched filter/correlator receiver, the error-rate P_e is given by eqn. (5.43) and may be written as:

$$P_e \simeq \frac{1}{2} \exp\left(-\frac{2L}{K-1}\right)$$
$$\simeq \frac{1}{2} \exp\left(-\frac{2\text{BWE}}{K}\right); \quad \ldots K \text{ large}, \tag{5.51}$$

where $L =$ length of the address code $= BT_b =$ BWE. The predetection effective $(E_b/n_0)_{\text{eff}}$ is now equal to $2L/(K-1)$. For $P_e = 10^{-5}$, $(E_b/n_0)_{\text{eff}} \simeq 10$, and $\eta = K/(BT_b) \simeq 0.2$, which is much less than that obtained in a synchronized TDMA system.

Haber [59] has suggested the use of orthogonal multistate baseband signals (codes) to designate k-bit words (as used in M-ary orthogonal signalling, Ref: Chapter 8), and to generate the transmission signal per user as the concatenation of his address code of length L and one of the 2^k sequences over the period kT_b, as shown in Fig. 5.62. In the receiver, the signal is despread and maximum likelihood detection is used for decoding the

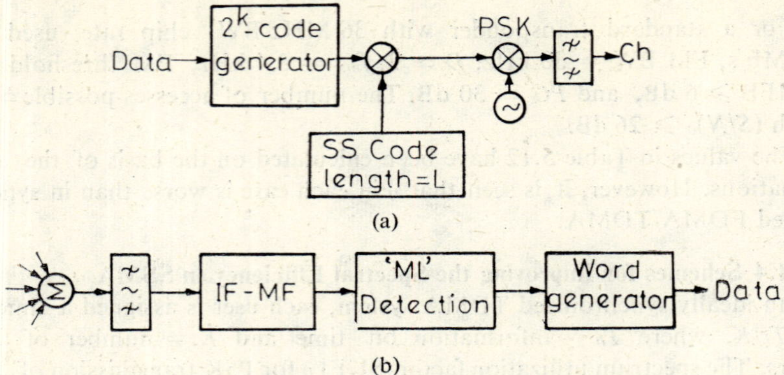

Fig. 5.62 CDMA using multi-address SS codes

transmitted codewords. Using known results of E_b/n_0 requirements in ML-detection [60], the SUF in Haber's scheme is now obtained as:

$$\text{Predetection } \gamma = \frac{kE_b}{n_0} = \frac{2^{k+1} \cdot L}{K-1}$$

and

$$\text{BWE} = \frac{2^k \cdot L}{k} \tag{5.52}$$

Therefore,

$$\eta = \frac{K}{\text{BWE}} \simeq 2n_0/E_b$$

and the minimum channel BW of the system, using PSK, is $1/\tau_c = 2^k \cdot L/kT_b$, where τ_c = chip time. For $k = 5$, and $P_e = 10^{-5}$; $E_b/n_0 = 6$ dB and $\eta \simeq 0.5$. With $k = 20$, $P_e = 10^{-5}$; $E_b/n_0 = 3$ dB and $\eta \simeq 1$; but then the signal processing would really be complex and the flexibility of the SS-techniques would be lost.

A simple alternative to the above scheme is to form k-bit words per user as above and assign 2^k orthogonal addresses to each user. This would require the provision of $2^k \cdot K$ orthogonal addresses for the total system (it is possible to generate the codes using Gold codes of length kL). An user will now transmit one of the assigned 2^k codes of length kL in time kT_b. In the receiver, the outputs of the 2^k-MF's (corresponding to the 2^k assigned addresses of the user) are processed through a 'Greatest of' decision circuit and the largest output is taken as the correct transmitted word. This is equivalent to ML-detection and the SUF is now given by:

$$\text{Predetection } \gamma = \frac{kE_b}{n_0} = \frac{2kL}{K-1},$$

and

$$\eta = \frac{K}{\text{BWE}} \simeq \frac{2n_0}{E_b}, \tag{2.52a}$$

where BWE $= L$.

For $k = 5$, $P_e = 10^{-5}$; $E_b/n_0 = 6$ dB, and $\eta = 0.5$ as in Haber's scheme. The only difference is that BWE $= L$ (and not $2^k \cdot L/k$) and $2^k \cdot K$ orthogonal addresses are to be generated. Equation (2.52a) indicates that η is the same as in Haber's scheme for large k, but with lesser bandwidth expansion.

5.11.5 Synchronization Problems

In SS-systems, using, DS/FH/TH to spread the signal, the ratio of information bandwidth to signal bandwidth is small and, consequently, the PN-modulation techniques are characterized by a very low repetition-rate-to-bandwidth ratio. Synchronization of such a receiver to the PN-code is a major problem in SS-systems. Synchronization is generally achieved in two steps: first the code is acquired and then the code is tracked for the detection of correct information bits. An overall block diagram of acquisition and tracking is shown in Fig. 5.63, where the incoming signal is first locked

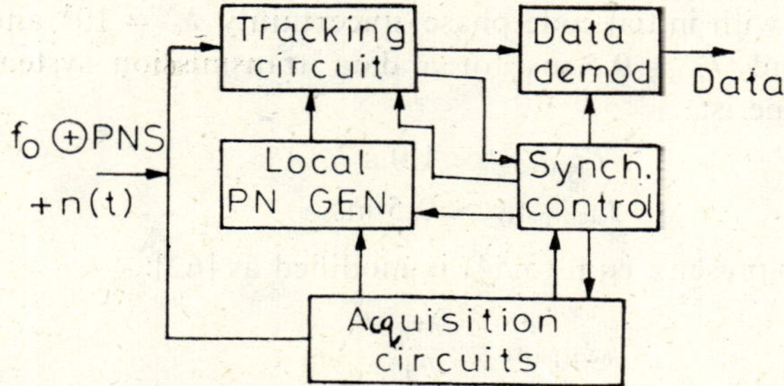

Fig. 5.63 Block diagram of acquisition and tracking loops

into the local PNS generator using the acquisition circuit and then is kept in synchronism using the tracking circuit, and finally data are demodulated.

Normally, some form of correlation is used for code acquisition, and this can be done either serially or in parallel. The sliding correlator, shown in Fig. 5.64, is a serial circuit, where the PNS generator output is shifted one

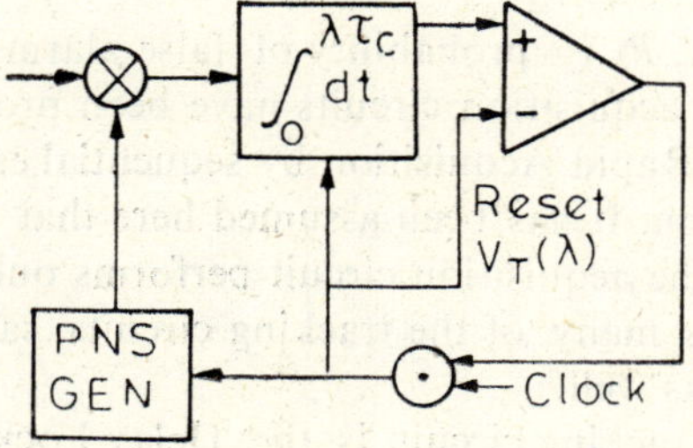

Fig. 5.64 Sliding correlator for acquisition

bit at a time through the inhibit gate and a partial correlation for λ chips is calculated through the integrator. If the output is above the threshold, then the PNS control is shifted to the tracking circuit; otherwise subsequent attempts are made to search for the correct timing of the PNS so that the threshold is exceeded. If the region of uncertainty of the code phase is λ_c, then the worst-case acquisition time is given by

$$T_{acq(ser)} = 2\lambda\lambda_c\tau_c \tag{5.53}$$

For parallel search, one needs to use $2\lambda_c$ (in the worst case $2L$) correlators in parallel whose outputs are compared and the largest value is used to lock the PNS generator. For better detectability P_d, T_{acq} increases as λ increases and a compromise is made between P_d and T_{acq}. In the FH system, similarly, the outputs of the different bandpass filters, $f_1, f_2, f, \ldots, f_L$ are squared, delayed and summed to form the correlated output. If the output crosses the threshold, then the PNS is locked at that instant. The time of acquisition is now reduced to [61]:

$$T_{acq(parl)} = \lambda\tau_c \tag{5.54}$$

For example, with initial code-phase uncertainty $\lambda_c = 10^3$ and $\lambda = T_b/\tau_c$, $\tau_c = 10^{-6}$s and $T_b = 0.5$ ms for a data transmission system, maximum acquisition time is:

$$T_{acq(ser)} \simeq 1.0 \text{ s}$$
$$T_{acq(parl)} \simeq 0.5 \text{ ms}$$

When noise is present, eqn. (5.54) is modified as [62]:

$$\overline{T}_{acq(parl)} = \frac{\lambda\tau_c}{P_d}, \tag{5.55}$$

where P_d = probability of detection. In case of the serial search circuit, eqn. (5.53) is now modified as:

$$\begin{aligned}\overline{T}_{acq(ser)} &= \left[L\left(\lambda + \frac{1}{2}\right)\tau_c + \frac{\lambda\tau_c P_F}{(1-P_F)^2}\right] \\ &\quad + \left(\frac{1-P_d}{P_d}\right)\left[2L\left(\lambda + \frac{1}{2}\right)\tau_c + \frac{\lambda\tau_c P_F}{(1-P_F)^2}\right] \\ &\simeq LT_b(3 - 2P_d)\end{aligned} \tag{5.56}$$

where $\lambda \to L$, $P_d \to 1$, P_F (=probability of false alarm) $\to 0$ and $L\tau_c = T_b$.

More sophisticated acquisition circuits have been proposed and the one by Ward [63], called Rapid Acquisition by sequential estimation, minimizes the time for acquisition. It has been assumed here that the carrier has been acquired earlier and the acquisition circuit performs only in the baseband. In practice, however, many of the tracking circuits, say, in DLL, are used for code acquisition as well.

The most popular tracking circuit is the Delay-Lock Loop (DLL) [64], where incoherent tracking of the code is achieved. The circuit, as shown in Fig. 5.65(a), works on the principles of PLL and the difference of the two correlator outputs, delayed by $\Delta = 2\tau_c$ (or τ_c) from each other, produces a saw-tooth shaped $R(\tau)$. This controls the VCO and normally the VCO operates at the null of $R(\tau)$, as shown in Fig. 5.65(b). When the tracking

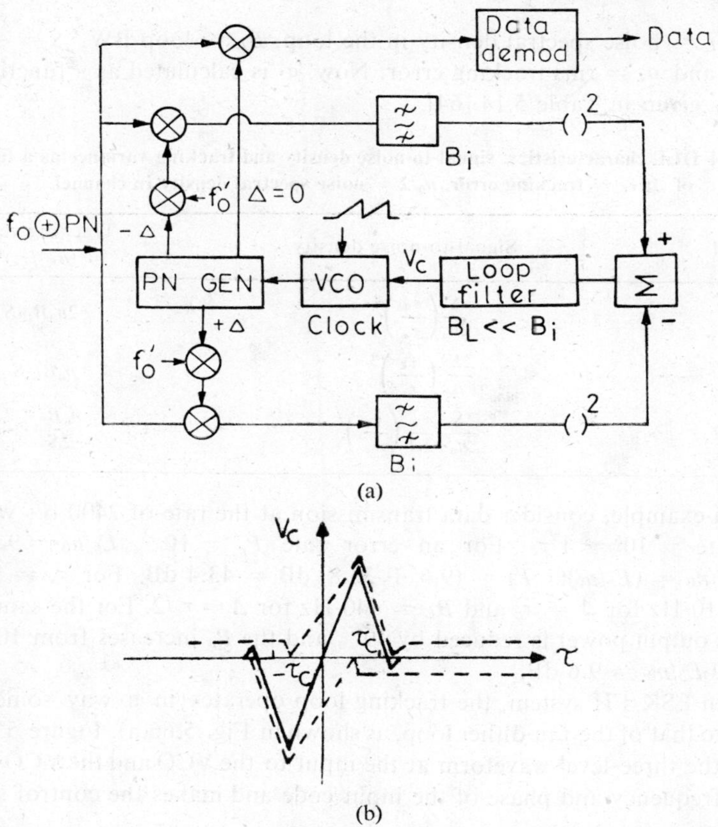

Fig. 5.65 (a) DLL for tracking DS-signal and (b) variation of control voltage V_c to VCO

error $\tau_\epsilon = 0$, the PNS with zero shift is correlated with the input to produce the despread PSK-modulated data signal, and the data are recovered through the usual coherent detector and PLL. The BPF-BW is normally kept minimum and equal to $2/T$ so that the loop SNR is maximum. The same circuit may be used for code acquisition by providing initially a saw-tooth voltage to sweep the VCO through a small range, and a parallel acquisition detector circuit, when triggered by the correlator output, switches the DLL to the tracking mode. For large carrier shifts, an auxiliary AFC circuit may also be used to aid the acquisition process [65]. A simpler DLL with a single arm uses a tau-dither loop [66] and minimizes the effect of unbalance in the two arms of the DLL of Fig. 6.65. However, in terms of loop SNR, the circuit is 3 dB worse than the two-arm loop, but is sometimes used because of its simplicity. The loop bandwidth of the DLL is an important parameter and it can be calculated, once the allowed tracking error is given. From the theory of PLL, it is known that:

$$\frac{n_L B_L}{S} = \left(\frac{\sigma_\epsilon}{\tau_c}\right)^2 \tag{5.57}$$

where n_L = noise spectral density in the loop, B_L = loop BW, S = signal power, and σ_ϵ = rms tracking error. Now, σ_ϵ is calculated as a function of Δ and is given in Table 5.14 [64].

Table 5.14 DLL characteristics: signal-to-noise density and tracking variance as a function of Δ; τ_ϵ = tracking error, $n_0/2$ = noise spectral density in channel

Δ	Signal-to-noise density	Tracking variance $(\sigma_\epsilon/\tau_c)^2$
$\Delta = \tau_c$	$\frac{S}{2n_0}\left(\frac{\tau_\epsilon}{\tau_c}\right)^2$	$2n_0B_L/S$
$\Delta_c > \Delta > \frac{\tau_c}{2}$	$\frac{2S}{n_0}\left(\frac{\tau_\epsilon}{\tau_c}\right)^2$	n_0B_L/S
$\Delta < \tau_c/2$	$\frac{2S}{n_0}\,\frac{\tau_c}{2\Delta}\left(\frac{\tau_\epsilon}{\tau_c}\right)^2$	$\frac{n_0B_L}{2S}\cdot\frac{2\Delta}{\tau_c}$

As an example, consider data transmission at the rate of 2400 b/s with a code rate $= 10^6 = 1/\tau_c$. For an error rate $P_e = 10^{-5}$, $E_b/n_0 = 9.6$ dB. Then, $S/n_0 = (E_b/n_0)(1/T) = (9.6 + 33.8)$ dB $= 43.4$ dB. For $\tau_\epsilon = \tau_c/10$, $B_L = 110$ Hz for $\Delta = \tau_c$ and $B_L = 440$ Hz for $\Delta = \tau_c/2$. For the same τ_ϵ, the data output power is reduced by 19% and the P_e increases from 10^{-5} to 10^{-4} for $E_b/n_0 = 9.6$ dB.

For an FSK/FH system, the tracking loop operates in a way somewhat similar to that of the tau-dither loop, as shown in Fig. 5.66(a). Figure 5.66(b) shows the three-level waveform at the input to the VCO and the VCO pulls to the frequency and phase of the input code and makes the control signal minimum.

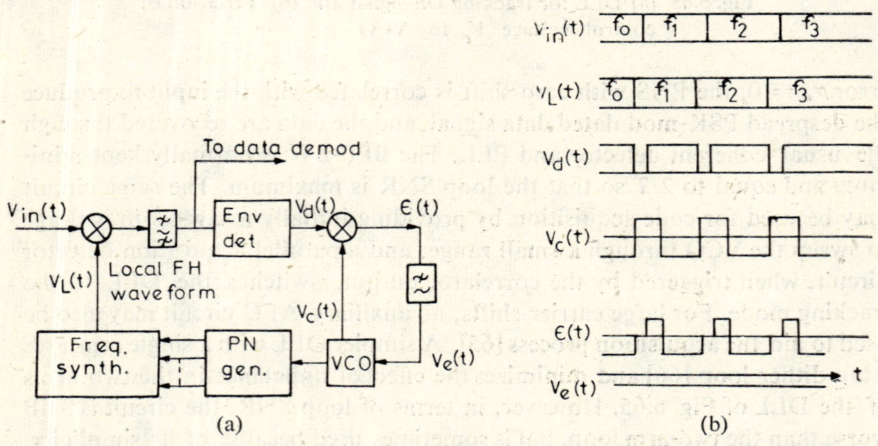

Fig. 5.66 (a) Tracking loop for FH signals and (b) waveforms of Fig. 5.66(a)

Synchronization Using SAW-TDL [67]

In contrast with the above SS techniques using active correlators for both synchronization and data detection, the concept of passive MF correlators doing the same functions in the SS-receiver has been given by Cahn *et al.* [68]. Baier *et al.* [67] have recently described in detail the operation of a SAW-TDL-MF for both acquisition and tracking. The block diagram of the system is shown in Fig. 5.67, where the periodic peaks at the SAW-TDL

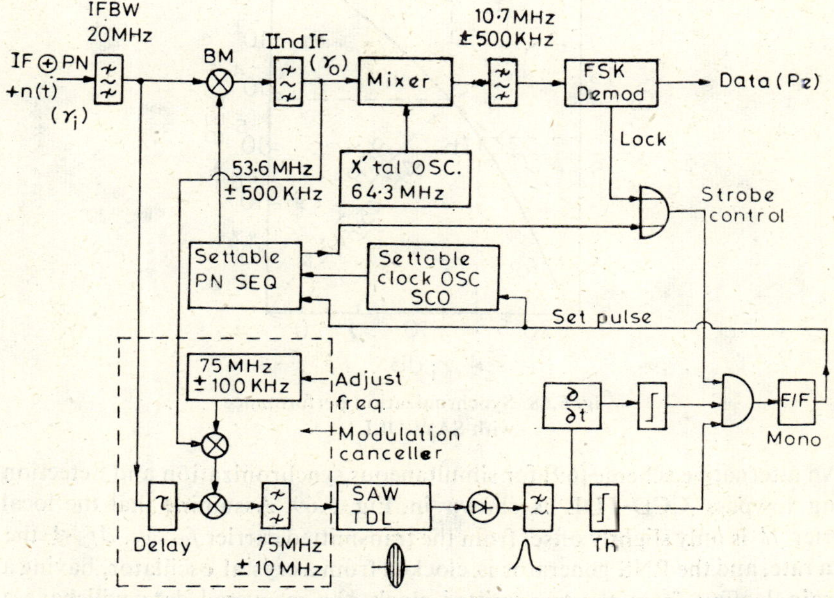

Fig. 5.67 Synchronization circuit using SAW-TDL (after Baier *et al.* [67])

output repeatedly correct the epoch of the local PNS clock phase and the code generator initial condition to the correct values. As the correlation impulses are distorted and attenuated due to the effects of message modulation, Doppler shifts, and unavoidable interference, a modulation canceller and a differentiator are employed to improve the synchronization performance. A hardware implementation of the system has also been described.

The experimental system specifications are:

Modulation	FM/FSK-DS
Clock	10 MHz
Code length, L	1278
Message bit rate	200 Kb/s
IF	75/53.6 MHz
SAW-TDL	255 taps, 75 MHz IF
Modulation canceller	Balanced modulator and adjustable PLL osc.

The system performance is shown in Fig. 5.68, where γ_i vs. γ_0 curve gives SNR gain with 80 kHz IF BW and no modulation of the IF ⊕ PNS. PG is now 24 dB and the curve shows a threshold at $\gamma_i \simeq -15$ dB, because of the τ_ϵ at low γ_i. The BER of the data at 200 Kb/s with $\Delta f = 45$ kHz and PG = 13 dB is also shown in the figure, and it is seen that BER = 10^{-4}

at $\gamma_i = -3$ dB as expected. The system seems to have great promise in solving the detection and synchronization problems at the same time.

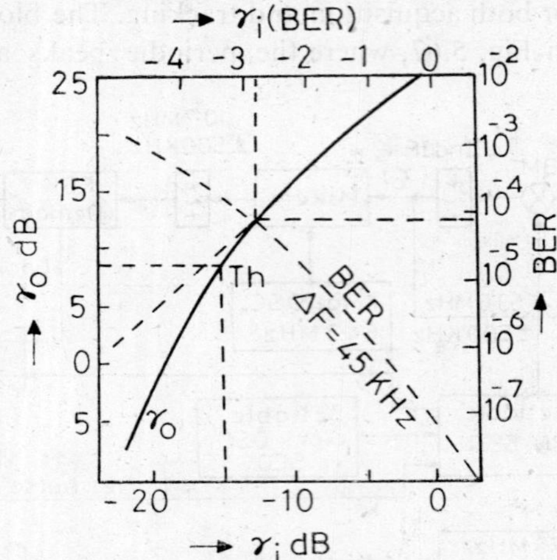

Fig. 5.68 Synchronization performance with SAW-DLL

An alternative scheme [69] for simultaneous synchronization and detection using lowpass CCD-TDL is shown in Fig. 5.69. Assuming that the local carrier f_0' is only slightly offset from the transmitted carrier f_0, say, $\Delta f \ll$ the data rate, and the PNS generator is clocked from a crystal oscillator, having a marginal offset from the transmitted clock, the recovered data will have a superimposed low-frequency modulation at the rate Δf as shown in Fig. 5.69(a).

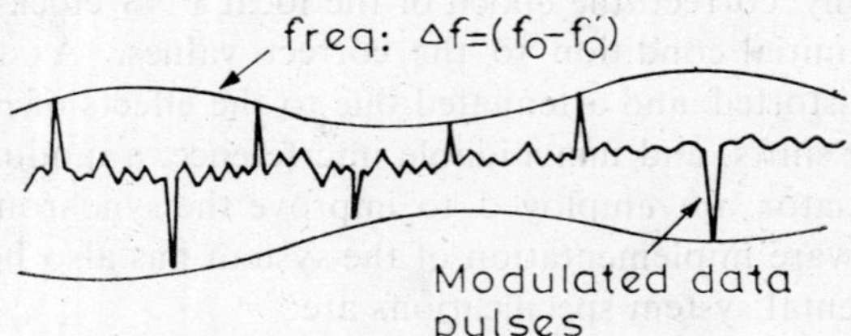

Fig. 5.69(a) Shape of data pulses before carrier synchronization

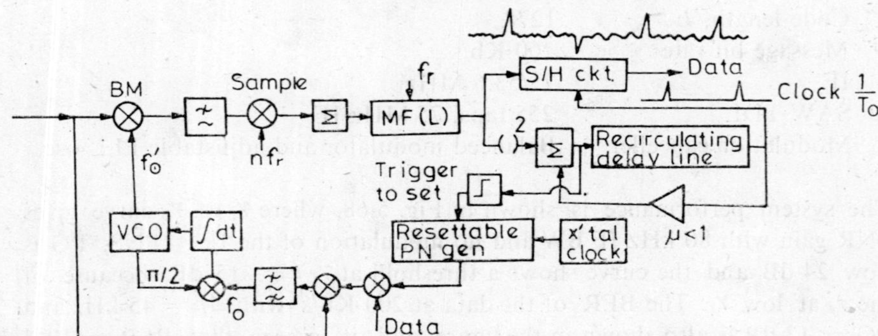

Fig. 5.69(b) An alternative synchronization scheme using TDL-MF [69]

The recirculating delay line, discussed in Sec. 5.5, has an approximate gain of 10–15 dB and clocks the MF output at the correct phase. During the peaks of the modulation cycle of the recovered data, the PNS will be reset at the correct phase, and the carrier control loop will automatically lock f_0' with f_0. The difference in the PNS clocks, if within crystal accnracy, will hardly change the MF output. If required, a secondary loop for the clock frequency may also be used. Thus, the T_{acq} will be minimum and only the tracking of f_0 will be required.

It has been shown in Chapter 2, that for saturated repeaters, multisampled MF's are the most efficient and this may easily be accommodated in the system. Moreover, large length MF's can be realized by using concatenation of smaller MF's, and as discussed in Appendix B, the MF's could be DMF as well, if the multisampling technique is utilized.

5.11.6 Some Typical Applications

Space Probe

Ward [70] has discussed the design of an integrated deep-space tracking/ communication system using a Voyager type satellite (in 3000 kg class) suitable as a Mars/Venus space probe. SS techniques have been used for command, communication and ranging, and DDL for acquisition and tracking. The system specifications briefly are:

Maximum range	2.10^8 Km
Maximum range rate	10^4 m/s
Maximum acceleration	10 m/s^2
Earth antenna diameter	210 ft; gain = 61 dB
Frequencies	Uplink—2.113 GHz, Downlink—2.295 GHz
Earth station Tx. power	100 KW
Satellite Tx. power	25 W
Satellite antenna diameter	10 ft, gain = 34 dB
Command antenna gain	3 dB
Earth receiver noise temp.	40 °K
Satellite receiver temp.	870 °K

Modulation

Code clock = 1.328 MHz, τ_c = 0.752 μS, $L = 10^6$, code period = 0.79 s, command bit rate = 1.26 bits/s, data rate = 6480 b/s.

It has been shown that the system is capable of giving the following performances:

Range ambiguity	1.18×10^5 Km
Range error	1.12 m or 1 in 10^8 whichever is larger
Range rate error	1 in 10^6
Angular error	0.05 mil rms

Acquisition time with directional antenna	4.5 min.
With omnidirectional antenna	31.2 min. average
	62.4 min. maxm.
Command BER	$< 10^{-5}$
Data BER	$< 3.10^{-3}$

The system, thus, has unique capabilities, provided due to the use of SS-techniques.

Co-channel Transmission of Data and TV Signal

Coll and Zervos [71] have shown that it is possible to transmit data at 100 Kb/s rate with a TV signal at a level of $S/N = -40$ dB and recover the data with a BER $< 2\cdot 10^{-5}$. In the transmitter, 100 Kb/s data are spread with 1 Mb/s PNS and transmitted as PSK at a carrier frequency $\simeq$ 2.5 MHz, where there is a minimum in the spectrum of the colour TV signal. Data are recovered through a BPF (with $f_0 \simeq$ 2.5 MHz) and despread with the local PNS. The overall spectrum of the signal and the experimental BER vs. data rate is shown in Figs. 5.70(a) and (b). Distortion in the TV signal is hardly noticeable.

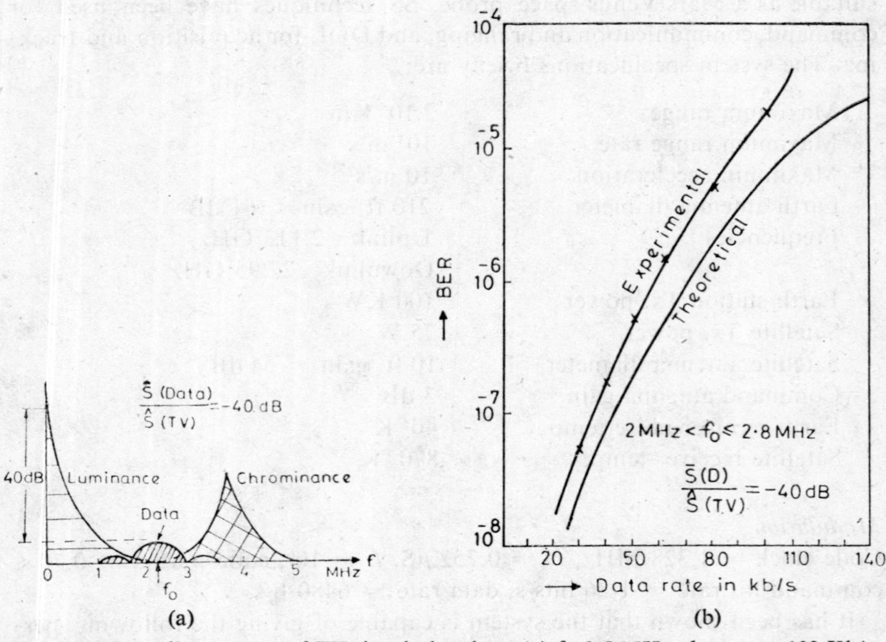

Fig. 5.70 Overall spectrum of TV signal plus data: (a) f_0: 2.5 MHz; data rate: 100 Kb/s, $\tau_c = 1\ \mu$s and (b) BER performance for co-channel transmission of data and TV signal (after Coll and Zervos [71])

Land Mobile Communication [72]

Urban mobile communication and similar defence networks always face the problem of frequency allocation and multipath distortion. The channel assignment problem with frequency pooling in the conventional FM systems has been discussed in Sec. 5.11.1. The employment of SS-techniques in land mobile cellular communication system is expected to result in substantially improved capabilities as well as considerably greater convenience in operations when compared with the use of conventional FM in the same application. Since a very large number of separately distinguishable addresses may be generated in an SS-system, this will be convenient for a mobile radio network. The same radio spectrum will be used by all cells and the handover problem as a mobile travels from cell to cell will be simpler. The protection against intercell interferences is obtained from the orthogonal addresses and not by buffering the cell's spectra as used in conventional FM systems. The protection against fading and impulsive interference is also obtained due to the spreading of the spectrum and the network can operate at sufficiently low power densities such that mutual interference between SS-systems and conventional systems in the same band can be kept within acceptable limits. Specific methods of implementation of SS will be to use FH or CDM techniques.

For the national requirements of, say, the USA, some 16 to 20 million radios including CB are now in use. Twenty million distinct addresses could be provided using a total bandwidth of about 60 MHz which is rather wasteful. However, a detailed address plan may be worked out similar to that used by the telephone network and thus the total requirement of the addresses can be minimized on a national basis. It has been shown that the number of simultaneous users possible in an SS cellular system far exceeds the number possible in an FM system as shown in Fig. 5.71. The number of simultaneous users per MHz/km² increases as the radius of the cell is decreased, as a result of geographic reuse and is common to both SS and FM

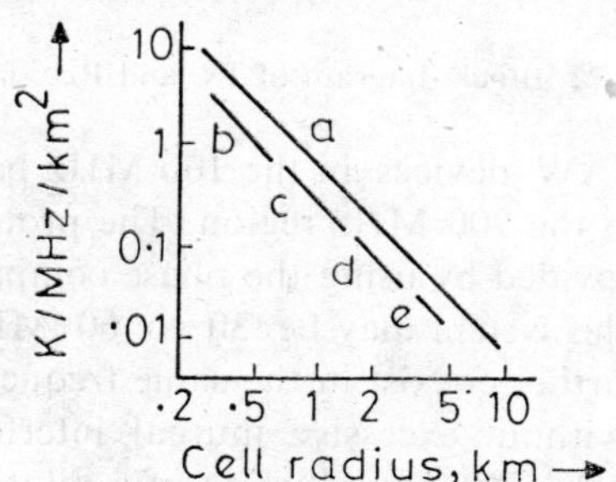

Fig. 5.71 Comparison of user density for SS and FM techniques: (a) SSMA, BT = 32, bit rate = 50 Kb/s, $(S/N)_0$ = 30 dB, (average received power $\propto 1/D^2$, where D = distances; (b) FM, 25 kHz spacing, $(S/N)_0$ = 30 dB, 63-cell pattern; (c) FM, 25 kHz spacing, $(S/N)_0$ = 30 dB, 71-cell pattern; (d) FM, 25 kHz spacing, $(S/N)_0$ = 30 dB, 57-cell pattern; (e) FM, 25 kHz spacing, $(S/N)_0$ = 30 dB, 52-cell pattern

systems. Cooper and Nettleton [73] have shown that the cellular FM systems now under consideration are not appropriate for catering to the requirements of large densities of simultaneous users and that the densities projected for the future in large cities cannot be accommodated in such systems. The SS-technique offers the capability for intensive utilization of the geographic resource in areas of high population density. Further, there is no hard limit to the number of simultaneous users in the SS-systems and when the design value is exceeded, a graceful degradation of the services occurs. The SS-system requires a fade margin of less than 3 dB under most circumstances, whereas the fade margin for FM is 10 to 20 dB. This means that the SS-system can be operated at approximately the same power level in each cell, whereas FM power requirements are much more dependent upon propagation factors which vary from cell to cell.

Although the cost of SS system is at present higher than that of FM, and FM systems can be of different complexities depending upon their use in a small or large cell, the high volume market foreseen for mobile services will cut down the cost of the SS-transreceivers. It is foreseen that SS modulators and demodulators, as shown in Fig. 5.72, will be economically

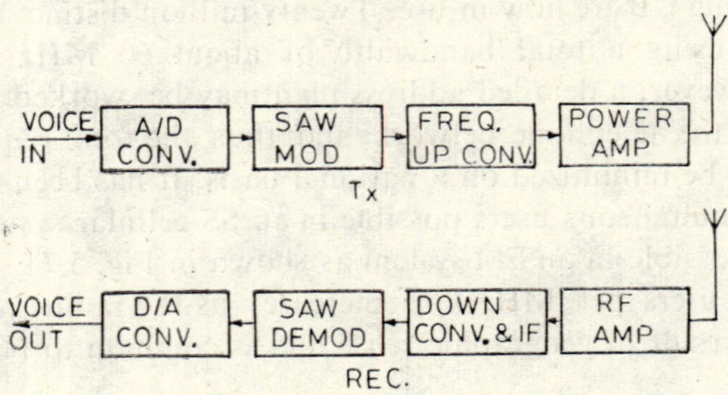

Fig. 5.72 Block diagram of Tx. and Rec. for SSMA

manufactured using SAW devices in the 100 MHz band. The actual band of operation may be in the 900 MHz region. The protection against multi-path effects may be provided by using the phase comparison technique. The overall bandwidth of the system may be 30 to 60 MHz. Such an experimental system may further coexist in the same frequency band as the conventional FM system without excessive mutual interference. This feature, quite evidently, facilitates the introduction of a pilot system. On the basis of the above, the SS-techniques are seen to be substantially superior to the conventional FM for mobile radio communication.

REFERENCES

1. Panter, P.F., *Communication Systems Design: LOS and Troposcatter Systems*, McGraw-Hill, N.Y., 1972.
2. Schwartz, M., Bennett, W.R. and Stein, S., *Communication Systems and Techniques*, McGraw-Hill, N.Y., 1966.
3. Pasupathy, S., 'Minimum Shift Keying: A Spectrally Efficient Modulation', *IEEE Comm. Soc. Mag.*, vol. 17, pp. 14–22, July 1979.

4. de Jager, F. and Dekker, C.B., 'Tamed Frequency Modulation', *IEEE Trans.*, vol. Com-26, pp. 534–542, May 1978.
5. Lucky, R.W., Salz, J. and Weldon, E.J., *Principles of Data Communication*, McGraw-Hill, 1968.
6. Dupnis, P. *et al.*, '16-QAM Modulation for High-Capacity Digital Radio System', *IEEE Trans.* Com-27, p. 1771, 1979.
7. Weber, W. III, Stanton, P.H. and Sumida, J.T., 'A Bandwidth Compressive Modulation System Using MAMSK', *IEEE Trans.*, Com-26, p. 543, 1978.
8. Oeitting, J., 'A Comparison of Modulation Techniques for Digital Radio', *IEEE Trans.*, Com-27, p. 1752, 1979.
9. Special Issue on 'Digital Radio', *IEEE Trans. on Comm.*, vol. Com-27, No. 12, Dec. 1979.
10. Barber, S. and Anderson, C.W., 'Modulation Consideration for the RD-3 91 Mb/s Digital Radio', *IEEE Trans.*, Com-26, p. 523, May 1978.
11. Horikawa, I. *et al.*, 'Design and Performance of a 200 Mbits/s 16-QAM Digital Radio System', *IEEE Trans.*, Com-27, p. 1953, Dec. 1979.
12. Miyauchi, K. *et al.*, 'A New Technique for Generating and Detecting Multilevel Signal Formats', *IEEE Trans.*, Com-24, p. 263, 1976.
13. Lender, A., Rogers, A. and Olszanski, H., '4 Bits/Hz Correlative Single Sideband Digital Radio at 2 GHz', *IEEE Int. Comm. Convention*, Boston, June 1979.
14. Monsen, P., 'Theoretical and Measured Performance of a DFE Modem on a Fading Multipath Channel', *IEEE Trans.*, Com-25, p. 1144, Oct. 1977.
15. Ehrman, L. and Monsen, P., 'Troposcatter Test Results for a High-Speed DFE Modem', *IEEE Trans.*, Com-25, p. 1500, Dec. 1977.
16. Barnett, W.T., 'Multipath Fading Effects on Digital Radio', *IEEE Trans.*, Com-27, p. 1842, Dec. 1979.
17. Anderson, C.W., Barber, S.G. and Patel, R.N., 'The Effect of Selective Fading on Digital Radio', *IEEE Trans.*, Com-27, p. 1870, Dec. 1979.
18. Komaki, S. *et al.*, 'Characteristics of a High-Capacity 16-QAM Digital Radio System in Multipath Fading', *IEEE Trans.*, Com-27, p. 1854, Dec. 1979.
19. Pieper, J.F. *et al.*, 'Design of Efficient Coding and Modulation for a Rayleigh Fading Channel', *IEEE Trans.*, IT-24. p. 457, July 1978.
20. Rogers, J.D., 'Introduction to Digital Tropo for Military Communication', *Comm. and Broadcasting*, vol. 6, no. 3, p. 3, 1981.
21. Price, R. and Green, P.E. (Jr), 'A Communication Technique for Multipath Channels', *Proc. IRE*, vol. 46, p. 555, 1958.
22. Morgan, D.R., 'Adaptive Multipath Cancellation for Digital Data Communication', *IEEE Trans.*, Com-26, p. 1380, 1978.
23. Kailath, T., 'Correlation Detection of Signals Perturbed by a Random Channel', *IRE Trans.*, IT-6, p. 361, 1960.
24. Sussman, S.M., 'A Matched Filter Communication System for Multipath Channels', *IRE Trans.*, IT-6, p. 367, 1960.
25. Kaye, A.R. and George, D.A., 'Transmission of Multiplexed PAM Signals over Multiple Channel and Diversity Systems', *IEEE Trans.*, Com.-18, p. 520, 1970.
26. Turin, G.L., 'Introduction to SS Antimultipath Techniques and Their Application to Urban Digital Radio', *Proc. IEEE*, vol. 68, p. 328, 1980.
27. Satellite Communication Issue, *IEEE Communication Magazine*, vol. 18, no. 5, 1980.
28. Pritchard, W.L., 'Satellite Communication—An Overview of the Problems and Programs', *Proc. IEEE*, vol. 65, pp. 294–307, 1977.
29. Spilker, J.J. (Jr), *Digital Communications Satellites*, Prentice-Hall, N.J., 1977.
30. Miya, C. (Ed.), *Satellite Communication Engineering*, Lattice Co., Tokyo, 1982.
31. Filipowski, R.F. and Muehldrop, E.I., *Space Communication Systems*, Englewood Cliffs, N.J., 1965.

32. Martin, J., *Communication Satellite Systems*, Prentice-Hall, N.J., 1978.
33. Van Trees, H.L. (Ed.), *IEEE Press Issue on Satellite Communications*, N.Y., 1979.
34. Pritchard, W.L., Lecture notes on 'Satellite Communication Systems Engineering', Workshop on Physics of Communications, ICTP, Trieste, Nov. 1983.
35. Northrop, G.M., 'Aids for the Gross Design for Satellite Communication Systems, *IEEE Trans. Com. Tech.*, vol. Com-11, p. 46, 1966.
36. Dicks, J.L., Schultze, P H. and Schmitt, C.H., 'INTELSAT IV Communication System—Systems Planning', *COMSAT Tech. Rev.*, vol. 2, pp. 452-469, 1972.
37. Edelson, B.I. and Werth, A.M., 'SPADE System Progress and Application', *COMSAT Tech. Rev.*, vol. 2, no. 1, pp. 221–242, 1972.
38. Ferguson, M.E., 'Design of FM SCPC Systems', *IEEE Int. Conf. on Comm.*, vol. 1, pp. 12–16, June 1975.
39. Schmidt, W.G., *et al.*, 'MAT-1: INTELSAT's Experimental 700-channel TDMA/DA System', *INTELSAT-IEE Int. Conf. Digital Satellite Comm.*, Paper No. 428, Nov. 1969.
40. Nosaka, K. *et al.*, 'TTT System (50 Mbits/s PCM-TDMA System with Time Preassignment and TASI) and Its Satellite Test Results', *IEEE Trans. Comm.* vol. COM-20, pp. 820–825, 1972.
41. Muratani, T., 'Satellite-switched Time Division Multiple Access' *IEEE Electronics and Aerospace Syst. Convention* (EASCON), pp. 189–196, Oct. 1974.
42. Koga, K., Muratani, T., and Ogawa, A., 'On-Board Regenerative Repeater Applied to Digital Satellite Communication', *Proc. IEEE,* vol, 65, pp. 401–410, 1971.
43. Das, J., 'Review of RADA Techniques', *IETE Students Jour.*, vol. 19, no. 1, p. 15, 1978.
44. Fugono, N., Yoshimura, Y. and Hayashi, R., 'Japan's Millimeter Wave Satellite Communication Program', *IEEE Trans. on Comm.*, vol. Com-27, pp. 1381–1391, 1979.
45. Carrier, L.M. and Pope, W.S., 'An Overview of the Space Shuttle Orbiter Communications and Tracking System', *IEEE Trans. on Comm.*, vol. Com-26, pp. 1494–1505, 1978.
46. Minoli, D. and Schneider, K.S., 'A Technique for Estimating the Minimum Number of Frequencies Required for Urban Mobile Radio Communication', *IEEE Trans.*, Com-25, p. 1054, 1977.
47. Haber, F., 'On the Capacity of Randomly Used Shared Channels', *IEEE Trans.*, EMC-12, p. 146, 1970.
48. Holmes, J.K., *Coherent Spread Spectrum Systems*, John Wiley & Sons, N.Y., 1982.
49. Special Issue on Spread Spectrum Communications, *IEEE Trans.*, Com-25, Aug. 1977.
50. Special Issue on Spread Spectrum Communication, *IEEE Trans.*, Com-30, May 1982.
51. Glenn, A.B., 'Code Division Multiplex Systems', *IEEE Int. Conv. Records*, pt. 6, p. 53, 1964.
52. Chesler, D., 'Performance of a Multiaddress RADA System', *IEEE Trans.*, Com Tech-14, p. 369, 1966.
53. M. Yokoyama *et al.*, 'SSRA Communication Experiment via ATS-1', *Jour. Radio Res. Lab.* (Japan), vol. 21, p. 93, 1974.
54. Cohen, A.R., Heller, J.A. and Viterbi, A.J., 'A New Coding Technique for Asynchronous Multiple Access Communication', *IEEE Trans.*, Com-19, pp. 849-855, 1971.
55. INTELSAT/IEE International Conference on 'Digital Satellite Communication', Nov. 1969.
56. Kurimura, T., 'Satellite Communications in Japan', *IEEE Trans.*, Com-20, p. 730, 1972.

57. Von Scherkel, K.D., 'Proposal of an Integrated Digital Communication System with Multiple Access for an Arbitrary Branched Wideband Transmission Network', *AEU* (Electronics and Comm.), vol. 27, p. 168, 1973.
58. Costas, J.P., 'Poisson, Shannon and the Radio Amateur', *Proc. IRE*, vol. 47, p. 2058, 1959.
59. Haber, F., 'Spread Spectrum Signals and Bandwidth Utilization', in *Communication Systems and Random Process Theory*, Ed. Skwirzynski, J.K., Sijthoff and Noordhoff, pp. 55–64, 1978.
60. Viterbi, A.J., 'On Coded Phase-Coherent Communication', *IRE Trans. on Space Electronics and Technology*, SET-7, pp. 3–14, 1961.
61. Nelson, P.T., 'On the Acquisition behaviour of Delay-lock Loops', *IEEE Trans.*, vol. AES-12, p. 415, 1976.
62. Pickholtz, R.L., Schilling, D.L. and Milstein, L.B., 'Theory of Spread Spectrum Communications—A Tutorial', *IEEE Trans.*, Com-30, p. 855, 1982.
63. Ward, R.B., 'Acquisition of Pseudonoise Signals by Sequential Estimation', *IEEE Trans., Com. Tech.*, vol. Com-13, p. 474, 1965.
64. Spilker, J.J., Jr. 'Delay-lock Tracking of Binary Signals', *IEEE Trans. Space Electron Telem.*, vol. SET-9, p. 1, 1963.
65. Cahn, C.R., *et al.*, 'Software Implementation of PN-SS Receiver to Accommodate Dynamics', *IEEE Trans.* Com-25, p. 832, 1977.
66. Hartman, H.P., 'Analysis of the Dithering Loop for PN Code Tracking', *IEEE Trans.*, vol. AES-10, p. 2, 1974.
67. Baier, P.W., Dostert, K. and Pandit M., 'A Novel SS Receiver Synchronization Scheme Using a SAW-tapped Delay Line', *IEEE Trans.*, Com-30, p. 1037, May 1982.
68. Cahn, C.R. *et al.*, 'AGARD Lecture Series No. 58', *Spread Spectrum Communication*, 1973.
69. Das, J., Maskara, S.L. and Karak, A.K., 'Studies and Development of SS Techniques', *Internal Research Report*, Dept. of Electronics and Communication Engg., Indian Institute of Technology, Kharagpur, 1982.
70. Ward, R.B., 'Application of delay-lock Radar Techniques to Deep-Space Tasks', *IEEE Trans.*, vol. SET-10, p. 49, 1964.
71. Coll, D.C. and Zervos, N., 'The Use of Spread Spectrum Modulation for the Cochannel Transmission of Data and TV', National Telecommunication Conference, USA, Paper 15.4, 1979.
72. Eckert, R.P. and Kelley, P.M., 'Implementing SS Technology in Land-mobile Radio Network', *IEEE Trans.*, Com-25, p. 867, 1977.
73. Cooper, G.R. and Nettleton, R.W., 'A Spread Spectrum Technique for High-Capacity Mobile Communication', *Proc. 27th Annual Vehicular Tech. Conf.*, Orlando, Florida, 1977.

CHAPTER 6

Computer Communication Networks

Introduction [1–7]

In the last three decades, computer architecture has gone through an evolution and today's distributed processing system has developed through (a) point-oriented computers in the 50's, (b) the family series in the 60's, (c) super computers in the early 70's and then area-oriented horizontally distributed types into late 70's. Instead of a single ultra-large computer, the new systems apply horizontal integration of several smaller computers to increase their capabilities. One of the major system advances of the early 1960's was the development of multiaccess time-sharing systems in which computer system resources were made available to a large population of users, each of whom had relatively small demands, but who collectively presented a total demand profile which was relatively smooth and of medium-to-high utilization. Thus, an expensive computing resource could be shared by many users, specially among a collection of high peak-to-average (i.e., bursty) users. This requirement led to the development of computer communication networks, as shown in Fig. 6.1, where three kinds of resources are available, viz. (a) Host computers, (b) Intelligent terminals and IMP's and (c) the Communication Subnetwork.

The communication subnetwork is mostly digital and the terminals and computers thus interconnected constitute an integral system, known as a Computer Network. The present-day computer networks serve users, both interactive, e.g., terminal dialogue, technical enquiry and process control, and non-interactive e.g., telemetry, computer program and bulk data transfer etc. The speed of data transmission required in these applications depends on the length of messages and their response/delay times required. The range of message lengths etc. are given in Table 6.1, and it is seen that data speeds of 50 to 200 bps is sufficient for 50 percent of applications, 600 to 2400 bps is satisfactory for most, while 4800 bps can cover almost all applications. For transmission of large files and for processor-to-processor communication, the higher speed of 50 Kbps to 1 Mbps may be required [8].

The available network options to the administration for data transmission is shown in Table 6.2. Although the existing telephone network can support

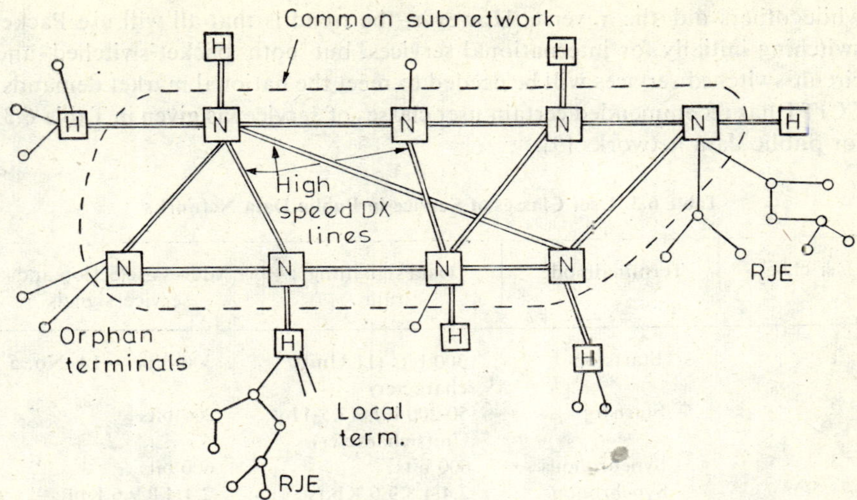

Fig. 6.1 Board outline of a computer communicaiton network; H : host computer; N: node/IMP/CC; IMP: Intermediate message processor; CC: Communication computer; RJE: Remote job eatry terminals

Table 6.1 Message Sizes and Response/Delay Times

Usage	Message size in bits	Response/delay time in seconds
Dialogue	10 to 3.10^4	1 to 10
Enquiry	10^2 to 10^5	3-100
Process control	$10\text{-}10^4$	1-500
Telemetry	$10\text{-}10^3$	1-10
Computer program	$10^3\text{-}3.10^6$	$30\text{-}3.10^3$
Bulk data	$10^5\text{-}10^8$	$3.10^3\text{-}10^5$

Table 6.2 Data Transmission Network Options

(a) Telegraph network	Private circuits 200 bauds, Circuit-switched (telex).
(b) Telephone network (analog)	Private circuits—speed upto 9.6 Kb/s, Circuit-switched (PSTN) 200-1200 bps asynchronous, Packet-switched.
(c) Telephone network (digital)	Private circuts, Circuit-switched, speed upto 48 Kbps synchronous, Packet-switched.
(d) Data network (digital)	Private circuits, Circuits-switched, speed: 200/600/1200/2400/4800/9600/48,000 bps synchronous, Packet-switched.

both switched and leased-line data services, many administrations have implemented public data networks for better efficiency. Some initially implemented circuit-switched service followed by packet-switched services, while others did the reverse. However, the trend is that all will use Packet switching initially for international services, but both packet-switched and circuit-switched services will be needed to meet the national market demands. CCITT has recommended certain user classes of service, as given in Table 6.3, for public data networks [9].

Table 6.3 User Classes of Service in Public Data Networks

User class	Terminalmode	Data signalling rate	Address selection and service signals	
1	Startstop	300 bits (11 Units/ character)	300 bits — 1A No. 5	
2	Startstop	50-200 bits (7.5-11 Units/character)	200 bits	"
3	Synchronous	600 bits	600 bits	"
4/5/6	Synchronous	2.4/4.8/9.6 Kbits	2.4/4.8/9.6 Kbits	"
7	Synchronous	48 Kbits	48 Kbits	"
8/9/10/11	Packet	2.4/4.8/9.6/48 Kbits	2.4/4.8/9.6/ 48 Kbits	*X*.25

Digital Data networks use two different types of switching. These are (a) Circuit switching and (b) Packet switching. In circuit switching a direct connection is set-up between the calling and the called subscriber and this is maintained during the whole data transmission phase, as in the case of telephone switching. This requires call set up and release times and is generally advantageous for transmission of longer messages, e.g., RJE service and file transfer. In packet switching, store-and-forward technique is employed to achieve statistical multiplexing and no direct connection exists between the subscribers. In this, each packet may experience different transmission delays and is generally considered suitable for relatively short message transmissions in such interactive applications as enquiry and on-line file access. The relative areas of application of digital data services is shown in Fig. 6.2

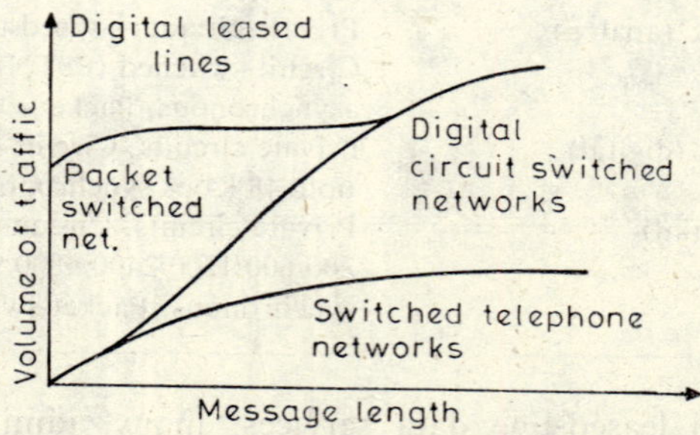

Fig. 6.2 Conceptual areas of application for digital data service

Message-switched services, somewhat similar to packet switching, provide for user information to be transmitted to a within-network storage device, from which the message can be forwarded later to some other user or users, with a copy generally retained at the message switching and storage centre. It thus differs from packet switching with regard to storage, broadcast and delayed delivery facilities. This service is not yet provided by public data networks, but many private networks provide the facilities for users. It has been forecasted that in future, message switching will also utilize new public data networks and eventually the ISDN.

6.1 EXISTING PUBLIC DATA NETWORKS

Some of the well-known data networks are the following: ARPA network for Advanced Research Project Agency, USA; SITA network for world-wide airline reservations; Tymnet and GE networks as commercial time-sharing networks, Datel Services of UK; Datapack Services, Canada; Trans-pack, France; Telenet, USA; SAENZ of Spain; DDN of Japan; EURONET for European countries and SATNET—a packet radio network via satellite in the Atlantic region.

The ARPA network [10] has probably generated more interest and excitement in the field of computer networks than any other network in use. It resulted in a tremendous amount of research in such important areas as: computer-to-computer protocol, interconnection of dissimilar networks, line protocol, communication processor hardware and software design, network topological design, reliability, adaptive routine and flow control, packet switching concepts and so on. This has also led to the development of a host of world-wide large scale computer-communication networks. The communication network of the ARPANET is shown in Fig. 6.3(a), (as of April 1972) and it initially consisted of 24 nodes and 28 circuits. This has now grown to more than 100 nodes connecting large computers, as well as smaller terminals as shown in Fig. 6.3(b). ARPANET is a packet/message-switched network and contains no mass storage and as little buffering in the nodes as is necessary to utilize the full capacity of the communication circuits. But, one or more Hosts on the net with low cost bulk-storage could provide or be dedicated to providing long term storage of messages with subsequent automatic retransmission, thus, leading to message switching. The packet lengths in the net is 1000—8095 bits and the average delay is 0.2 or less using a data speed of 50 Kbps. Dynamic routing techniques have been developed and a central controller provides the routing information to all processors or alternatively, the processors could collaborate in computing the routing information directly. As a result, the delay remains almost that of an unloaded net until the capacity of one or more 'cutsets' begins to saturate.

ARPANET has demonstrated that packet/message switching is a highly reliable and error-free method of computer-resource sharing and interactive switched communication, while providing at the same time separate/private data networks. EURONET, [11] the computer-communication network for

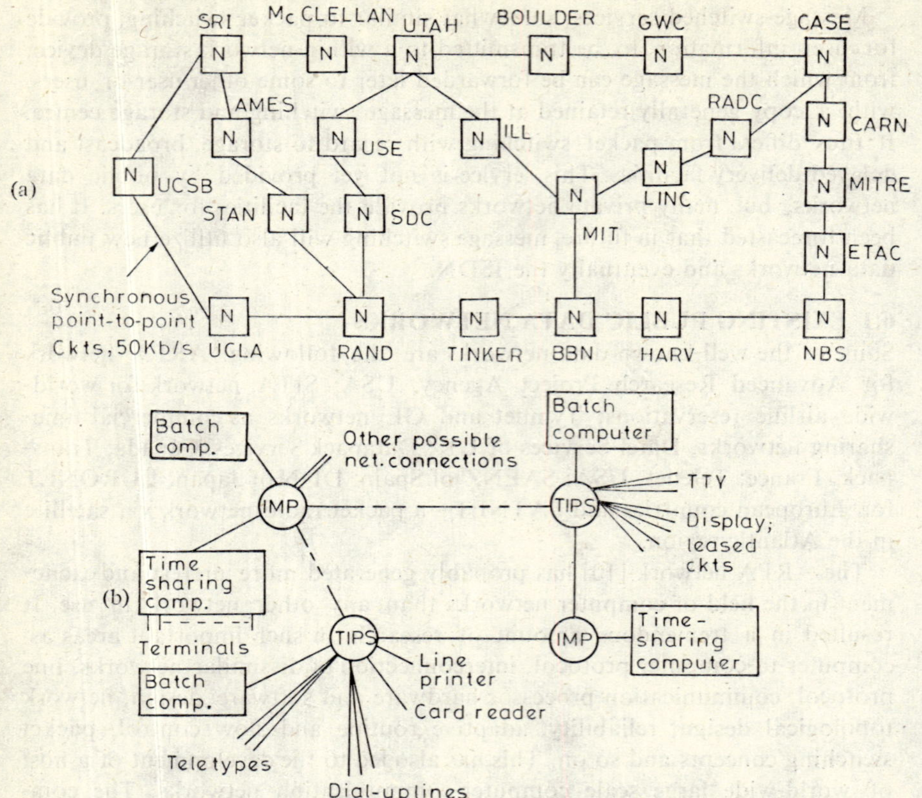

Fig. 6.3 (a) ARPANET as in 1972 (b) Various configuration of terminals and computers connected to IMP and TIPS of ARPANET. TIPS: terminal IMP.

the EEC countries, was established in the late 70's and the telecommunication network serves data bases of various countries now known collectively as DIANE (Direct Information Access Network-Europe). In the initial phase, the network connects some 20 Host computers containing over 100 data-bases, the important ones being the Space Documentation System of the European Space Agency, Italy, the German Medical Documentation and Information System at Cologne, the British Library Information System in London, Infoline and EEC's data bases in Luxemburg. Access to DIANE by terminals located in non-EEC countries and access by EEC terminals to data bases in other countries are also being provided. The network configuration currently being used is shown in Fig. 6.4, where four Packet-Switched Exchanges (PSE) have been established in Frankfurt, London, Paris and Rome with remote access facilities. The remote access facilities will enable Hosts and low-speed data terminals to gain access to the PSE's through multiplexers and the national PSTN of each country. A Central Network Management Centre (NMC) is located at London. The network has provision for virtual calls and for permanent virtual circuits between the user and the host. Fixed routing, instead of adaptive routing, is adopted in Euronet because of the problems of transit accounting arrangements and the consequent network management overheads.

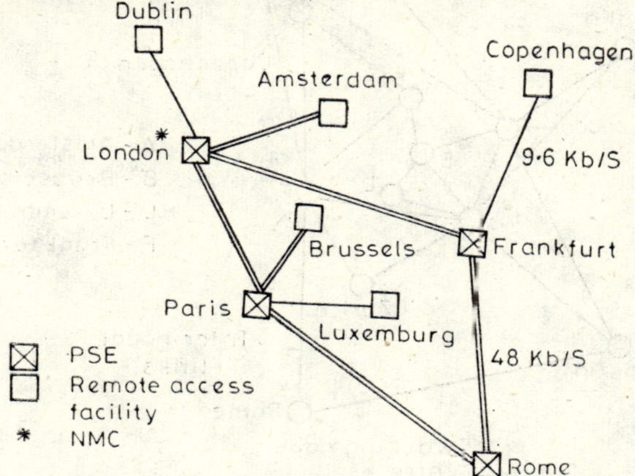

Fig. 6.4 Current configuration of EURONET

In Euronet, the Host computers are directly connected to the four PSE's over direct lines at speeds of 2.4/4.8/9.6 Kbps, but the low speed terminals (numbering approximately 480 of which 240 are synchronous packet) will generally access the network through the respective national PSTN at speed of 300 – 1200 bps or through the national data networks. The Euronet telecommunication network is based on an adaptation of TRANSPAC of France and differs only in the provision of network user identification (NUI) and facility for international accounting. It uses custom-designed hardware and the PSE's are based on the C50 processor to perform many repetitive functions involved in packet switching and on Mitra 125 general purpose minicomputer used as the command unit, both of which are of French manufacture. The performances of all PSE's are monitored and international accounting is done by the NMC, which is also based on Mitra 125 processor. Most of the processor equipment is duplicated for reliability, The Euronet will be expanded with possible interconnections as shown in Fig. 6.5, to meet the future demands. The European Informatics Network (EIN), with its centres in Paris, Ispra, Milan, Zurich and London will be connected to Euronet soon. Regarding the types of accesses to be provided, one method would be to use Euronet as the international transit data network, linking national networks in various countries.

An alternative approach, which is gaining considerable support in Europe, is to connect national packet-switched data networks to Euronet by means of *X*.75, the protocol for network interworking. Thus, the present form of Euronet may gradually disappear and give rise to an European Data Network (EDN). This will enable the EDN to have access to data networks in USA, Japan and other countries through International Packet-Switched Services (IPSS).

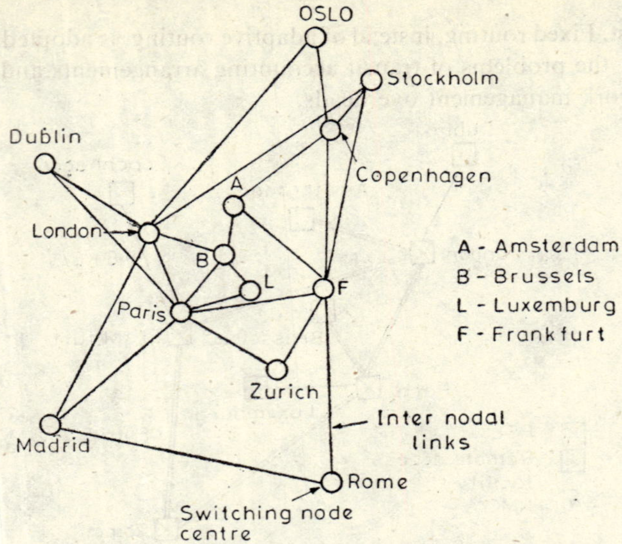

Fig. 6.5 Future expansion of EURONET

For the purpose of evolving a Global Data Network, the Conference of European Post and Telecommunication Administrations (CEPT) has carried out a survey of the plans of various countries and a study is being made as to how the various national networks could be interconnected. The current situation in Europe and the rest of the world is shown in Table 6.4. Practically each national network intends to include the facilities given in Table 6.5 [2].

Table 6.4 Current Status and Plans for Public Packet-Switched Data Services

Country	Date of Service	Network
Belgium	1979 (Expt)	—
	1980 (Public Service)	
Canada (TCTS)	1977	Datapac
(CN/CP)	1978	Infoswitch
Denmark	1980	—
FRG	1979	—
France	1978	TRANSPAC
Italy	1981	
Japan	1979 (Expt)	DDN
	1979 (Public Service)	
Netherlands	1980	DNI
Norway	1980	—
Spain	1973	RETD
Sweden	1981	—
Switzerland	1979	—
UK	1977 (Expt)	EPSS
	1979 (Public Service)	PSS
U.S.A.	1975	TELENET
	1975	TYMNET
	1978	COMPAC

Table 6.5 List of Facilities in Public P-S Networks

Initially:

- Virtual calls.
- Permanent virtual calls.
- Direct connection of packet mode terminals of data rates 2.4/4.8/9.6/48 Kbps.
- Direct connection of start-stop asynchronous character mode terminals at rates of 110/200/300 bps.
- Dialup access via PSTN at signalling rates up to 1200 bps.
- Closed user group facility, Reverse charging facility (for national facility only).

Later:

- Support of Frame mode DTE's (when defined).
- Datagram service and/or Fast select facility.
- Interworking with circuit-switched data services.

Keeping in view the recent trends in automated (electronic) offices, wired households, electronic mail, teletex, viewdata and electronic funds transfer services (EFTs), CCITT has recommended Integrated Digital Network Services (ISDN) and various standards and protocols are being evolved for this purpose. Assuming that national and international data networks have been introduced in most countries, the ISDN may take the form of a composite Telephone/Data switching centre, as shown in Fig. 6.6. This will be a total digital network including digital switching and digital transmission.

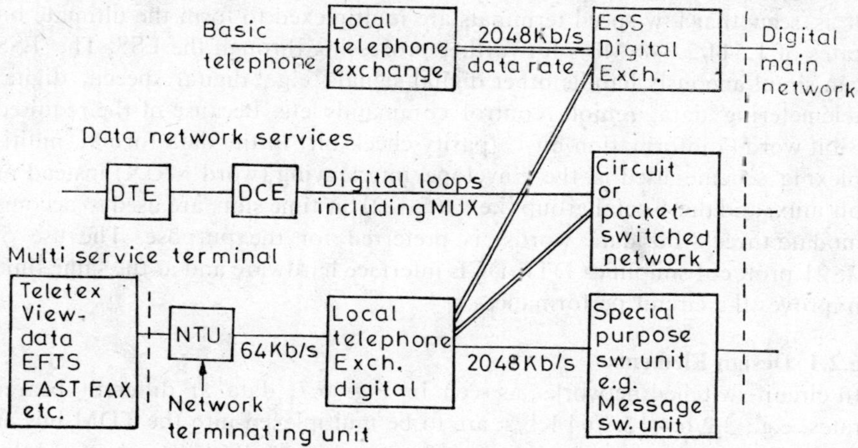

Fig. 6.6 Possible configuration of ISDN

6.2 Digital Circuit Switching Techniques [7]

With the development of Digital PCM-TDM exchanges, digital circuit switching for data has become more efficient. In a circuit-switched network, the source and the destination are connected by a dedicated communication path that is established at the beginning of the connection and broken at the end, as is done in switched telephony. To establish a connection, the subscriber provides the local central office (ESS) with an address which is used in setting up the call and the central office generates different signalling codes required for the purpose. Routing selection is performed using a set of prespecified paths, based on the first few dialled digits. Call set up time has been reduced from a few seconds to a fraction of a second in ESS. A block diagram of a circuit-switched data network is shown in Fig. 6.7, where

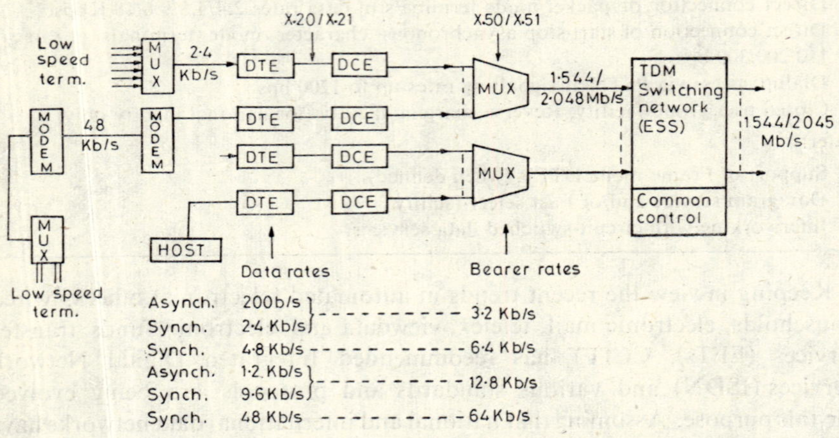

Fig. 6.7 Digital circuit switching system and its data/bearer rates. DTE: Data terminal eqpt.; DCE: Data circuit terminating eqpt.

it is seen that low speed terminals are multiplexed to form the ultimate bit rates of 1.544/2.048 Mb/s for further processing through the ESS. The ESS can simultaneously handle other digital signals, e.g., digital speech, digital telemetering data, remote control commands etc. Because of the required 8-bit word (7 information-bit + parity-check bit) in the data format, multiplexing scheme used is the Envelope interleaving (word MUX) instead of bit mux, and the 4-octal group (i.e., 4(6 + 2) bit time slots are used to accommodate three 8-bit data words) is preferred for the purpose. The use of $X \cdot 21$ protocol simplifies DTE-DCE interface hardware and at the same time improves the circuit performances.

6.2.1 Design Elements

In circuit-switched networks, as seen in Fig. 6.7, data at different bearer rates, e.g., 3.2/6.4/12.8/64 Kb/s, are to be multiplexed into the TDM bus at 64 Kb/s. Asynchronous data are synchronised by multiple sampling and the data thus synchronised along with other synchronous data are assembled into a (6 + 2) envelope format so that the bearer rates are 8/6 times higher than the corresponding data rate. The data frame is 20 times longer than a PCM frame of 125 μs in order to accommodate the lowest bearer rates of 3.2 Kb/s as shown in Fig. 6.8. PCM framing bits are used for framing and house-keeping information in the new frame (refer to Ch. 3). There are two possibilities in assignment of time slots to different bearer rates—(a) time slots in a data frame may be exclusively assigned to each of the speed classes, known as a separated system and (b) rearrangement of the slots for different

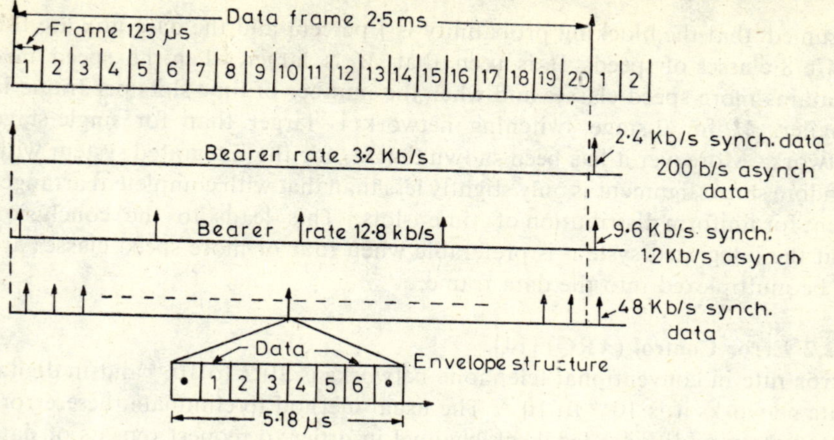

Fig. 6.8 Data multiplexing format

speed classes when a request for connection of a higher speed data is made, resulting into an integrated system with better efficiency. However, there is the finite probability of blocking while rearranging the slots in an optimum way. Assuming that the time slots are assigned randomly, resulting into non-uniformly spaced higher-speed data bits within a data frame and defining the Merit of Integration as:

$$M = \frac{A_I - A_S}{A_I} \times 100 \text{ (percent)}, \tag{6.1}$$

where A_I = maximum applicable traffic to an integrated system and A_S = maximum applicable traffic in a separated system, calculations have been made for M with number of time slots per data frame. The results of evaluation of M for single-stage and 2-stage switching networks and composite speed classes within the data frame, are shown in Table 6.6. It is

Table 6.6 Merit of Integration with Non-uniformly Spaced Time Slot Assignment for Single-Stage and Two-Stage Switching Networks

Tpye of speed Mix.	No. of slots per data frame	M for single net. in percent	M for 2-stage net. in percent
2	48	20	45
2	96	15	30
4	48	270	400
	96	150	170
	192	70	80
	384	40	45
	768	20	25
6	96	700	1000
	192	200	300
	384	70	140
	768	40	60

assumed that the blocking probability is 1 percent and the data mix are for 2/4/6/8 classes of speeds. It is seen that M is larger when the speed mix contains more speed classes and when the number of time slots per frame is smaller. M for 2-stage switching network is larger than for single-stage network. Moreover it has been shown that A_I for the integrated system with random slot assignment is only slightly less than that with complete rearrangement for uniform distribution of time slots. This leads to the conclusion that the integrated system is preferable when four or more speed classes are to be multiplexed into the data frame.

6.2.2 Error Control (ARQ) [14]

Error rate in conventional telephone networks is 10^{-5} to 10^{-7} and in digital data networks it is 10^{-6} to 10^{-8}. The usual method to eliminate these errors is to use an additional feedback channel in order to request repeats of data blocks found to be in error (ARQ). This ARQ method makes use of forward-error detection, as distinct from Forward-Error Correction (FEC). Of course, FEC may be used to improve the raw forward channel error-rate, and it is a must in cases where the total path delay is large, as in space channels. The major elements of an ARQ system are shown in Fig. 6.9,

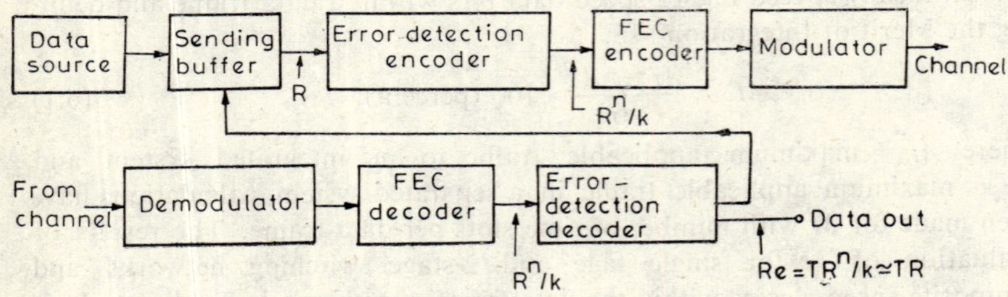

Fig. 6.9 Block diagram of an ARQ system

where a temporary buffer is required to store the data, till confirmation is received from the destination. ARQ systems are of two types: (a) stop-and-wait system, where the transmitter sends one block of encoded data at a time and waits for an acknowledgement (ACK) from the receiver before sending the next block; and (b) continuous systems, where the transmitter continuously sends the coded data and when an error is indicated by the NACK signal via the feedback channel, the transmitter repeats the block in error and subsequent blocks. The number of blocks retransmitted now depends on the round-trip delay of the channel and is usually more than one. Although more efficient, the continuous method is more complex and, therefore, difficult to organize and it also requires more storage.

Many types of error-detection codes have been used in ARQ, and the simplest of them is the odd/even parity check of a data block which is applicable only with short blocks. The error-detecting capability is improved by adding the binary counts of all marks in each block in addition to the parity check and the characters thus attached are called the block check sequence (BCS). Another variation of the error-detection code is to use two-dimensional block coding, where columns and rows are each coded through parity checks. Various subclasses of cyclic codes are also used. For example, an (n, k) BCH code having distance d is capable of detecting any combination of $(d - 1)$ or fewer errors, or any burst of $(n - k)$ or less. [Refer Ch. 8]

A particularly popular block check method, known as Cyclic Redundancy Check (CRC), uses the generator polynomial

$$G(X) = X^{16} + X^{12} + X^5 + 1 \tag{6.2}$$

or its modified version as a Frame check sequence (CRC-16). If the data block of k bits is represented as:

$$D(X) = \sum_{i=0}^{k-1} a_i X_i, \quad a_i = 0/1,$$

then multiplying $D(X)$ by X^{n-k}, $(n - k)$ = check bits,

$$X^{n-k} \cdot D(X) = Q(X)\, G(X) + R(X) \tag{6.3}$$

where $R(X)$ = remainder of the operation mod 2 sum; and $R(X)$ is added to the data block as check bits.

In the receiver, the received data $[X^{n-k}D(X) + R(X)]$ is divided (mod. 2) by $G(X)$ and for no error there is no remainder, since by equation (6.3).

$$[X^{n-k}D(X) + R(X)]/G(X) = Q(X)$$

If there is a remainder, the block is in error. By means of CRC, error bursts of length $\leqslant (n - k)$ are detected with $P_d = 1$ and bursts of length $= (n - k + 1)$ with $P_d = 1 - 2^{-(n-k-1)}$; and bursts of length $> (n - k + 1)$ with $P_d = 1 - 2^{-(n-k)}$, (P_d = probability of detection).

The effective Rate of transmission R_e in an ARQ system is lower than the Information rate R out of the buffer due to, (a) the round-trip delay time and (b) slight increase of the transmission rate (chip rate) by factor n/k. The transmission efficiency or the Throughput S is now given by:

$$S = \frac{R_e}{R \cdot n/k} = \text{(Av. no. of information bits accepted by receiver/unit time)/(Modem chip rate in the forward channel)} \tag{6.4}$$

Assuming that the CRC detects all errors and the feedback channel is error-free, the Throughput can be written as:

$$S\ (\text{stop and wait}) = \frac{k[1 - P(n)]}{n + D}$$

and

$$S\ (\text{continuous}) = \frac{k[1 - P(n)]}{n + P(n) \cdot D} \tag{6.5}$$

where $P(n)$ = probability of a block of n chips to be in error,

n = total number of channel bits per forward block

k = number of information bits per block

D = round-trip delay in chips in receiving the ACK message.

Since for practical values of $P(n)$, the influence of $n > k$ is small, one can write:

$$R_e = SR\frac{n}{k} \simeq SR \tag{6.6}$$

In practical circuits, with 1200 b/s $< R <$ 4800 b/s, the Throughput S has been achieved in the range of 0.80 to 0.98.

6.3 PACKET SWITCHING TECHNIQUE [7, 12]

A packet switching system (PSS) accepts, transmits and delivers discrete entities called packets. In this, no physical path is set up between the source and the destination, and resources (e.g., capacity, buffer storage etc.) are not allocated to its transmission in advance. Rather the source includes a destination address at the beginning of each message and the PSS uses this address to guide the message through the network to its destination, provides error control and notifies the sender of its receipt. The basic objectives of PSS are, (a) to accommodate a wide range of often incompatible terminals on the single network, (b) to increase transmission and switching reliability, (c) to achieve transmission plant economy, (d) to adopt techniques based on an extension of computer addressing and (e) to permit the adoption of a tarriff structure based on the volume of data transmitted.

In a public packet-switched data network (PSDN), three distinctive services are possible. These are (a) Virtual Call (VC) service, (b) Permanent Virtual Circuit and (c) Datagram (DG) service. In VC service, a call is established between two terminals by the exchange of call set up and acceptance packets. But the transmission facility is only utilized when packets are transmitted, thus achieving better utilization of the interconnecting facility and enabling calls to be charged on a volume basis. Calls can be established for any appropriate duration and can be cleared down whenever required. The VC service is the most popular mode of operation in a PSDN. A permanent virtual circuit may be established for an agreed period of time between any two designated terminals but is utilized only when packets are transmitted. It is somewhat similar to a leased line in character, but the charging structure is volume oriented. It is used in many international circuits. The third alternative is the Datagram Service, where the self-contained entity of data carries sufficient information for routing through the network without the need for a 'call' to be established. In DG the packets arrive with delays independent of the transmit ordering, whereas in VC the packets arrive in the same sequence as in the original message. Protocol $X{\cdot}25$ supports VC services in public PSDN, but some private networks have adopted DG principles for packet transmission. However, studies are being made by ISO to establish DG services in international PSDN if they are economically viable.

A typical PSDN is shown in Fig. 6.10, where the messages are temporarily stored and passed on from node to node on a link by link basis. Each packet of a typical length of 1000 to 8000 bits (128 bytes for international working)

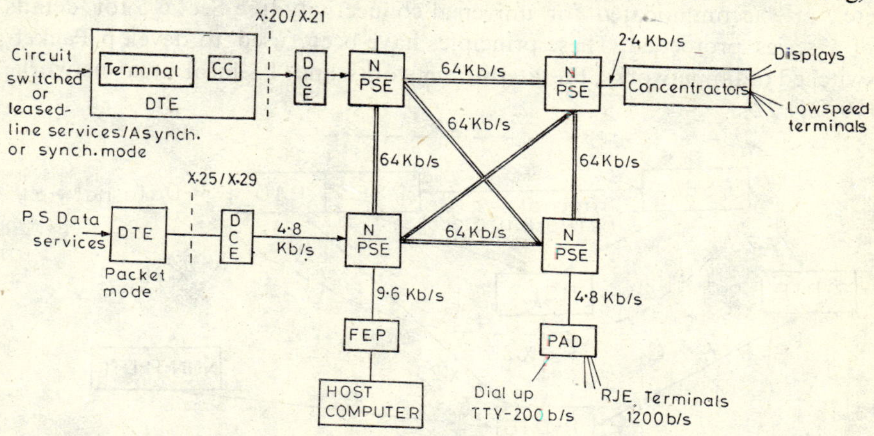

Fig. 6.10 Block schematic of a PSDN; PSE: Packet switching exchange; N: Node/PSE; CC: Communication controller; FEP: Front-end processor/IMP; PAD: Packet assembly-dissassembly facility

carries address and control information and is followed by an error-check field. According to CCITT recommendations, each packet is put inside a high-level data link control (HDLC) frame with check bytes added on a frame rather than a packet basis, as shown in Fig. 6.11. Each packet is transmitted individually but it time-shares the links with other packet-based

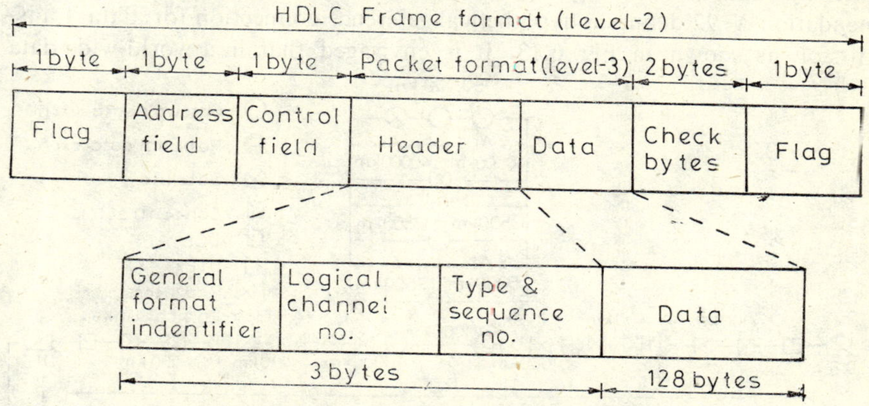

Fig. 6.11 Frame and Packet structure

traffic. The network establishes either a virtual call or a permanent virtual circuit between the terminals and the buffering of messages and packets at the nodes permit simple speed conversions required for various data terminals. By providing PAD (packet assembly and disassembly facility) the nodes can handle non-packet mode terminals as well and the interconnection of DTE/

DCE is governed by the $X \cdot 20/X \cdot 21$ and $X \cdot 25/X \cdot 29$ protocols. The overall protocol structure for PSDN is shown in Fig. 6.12, where all types of terminals, synchronous, asynchronous, packet-mode and non-packet-mode etc., are accommodated for universal connectivity [see Sec. 6.5 for details of *X*-series protocols]. These principles have been used to develop Packet-switched radio networks, PS-satellite networks and PS-local area networks [19, 21, 22].

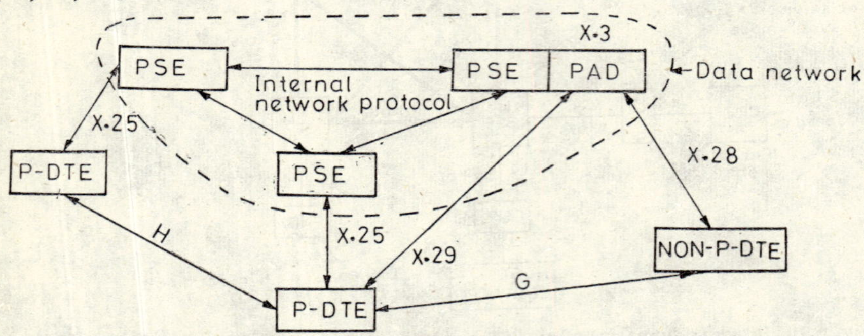

Fig. 6.12 Protocols in PSDN; P-DTE: Packet mode DTE; Non-P-DTE: non-Packet DTE; H, G: application protocols which are network independent

With the establishment of PSDN in many countries, it will be necessary to interconnect these services on an international basis. This will require international agreement on the following areas: (a) a hypothetical reference connection for data transmission, (b) a world-wide numbering plan, (c) an international signalling procedure, (d) the establishment of international gateway PSE and (e) an agreed international tarriff structure. CCITT recommendation $X \cdot 92$ defines a hypothetical reference connection for data transmission as shown in Fig. 6.13. It is envisaged that in a world-wide data

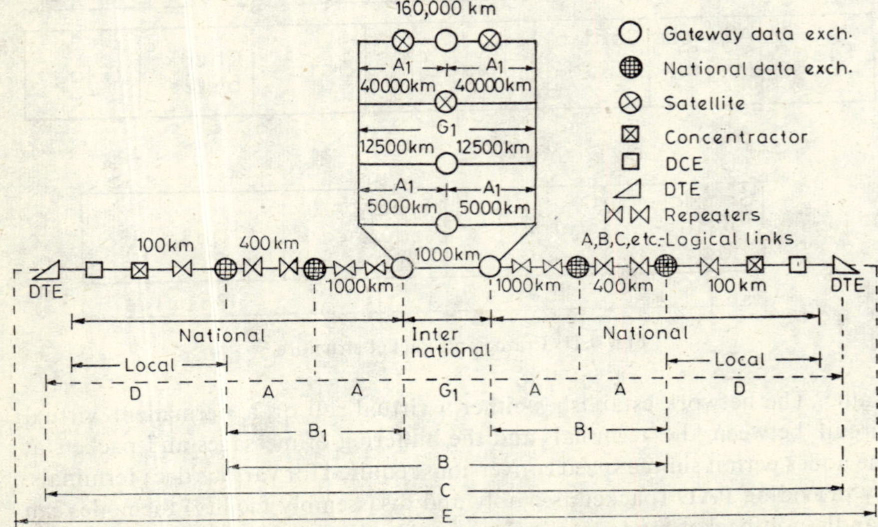

Fig. 6.13 Hypothetical reference connection for Global data network

plan, some countries will be permanently interconnected as an upper tier for intercommunication. Other countries will be connected to those forming the upper tier and use the major links in a transit mode. As seen in Fig. 6.13, the international links are provided between gateway PSE's and these could form a part of the national exchange in some countries. But the establishment of a special PSE dedicated to interfacing the national networks to overseas networks and to providing, where necessary, transit switching between two other countries has considerable advantages and is currently being adopted. CCITT recommendation $X\cdot 75$ defines the protocols necessary for the international gateway PSE's. This will require an international numbering plan acceptable on a global basis.

6.3.1 Delay and Throughput in PSDN [15]

A circuit-switched data network has an hierarchical structure and the major network performance measure is the network blocking. But in PSDN, the network is of a distributed structure using a number of store-and-forward switching nodes and the major performance measure is the Network Delay, which depends on topological design, route assignment and flow control in the network. In determining the waiting-time and service-time delay of a packet at a node, one uses the results of Queueing theory. Considering a somewhat idealized situation of M/M/1 queues, Poisson arrivals of the packets at a node and exponential service-time distribution at the node, (where M = arrival distribution, M = service discipline, and 1 = single node or server used), one obtains, the following relations:

(1) Probability of K arrivals in Ts is

$$P(K) = \frac{(\lambda T)^K \cdot e^{-\lambda T}}{K!} \tag{6.7}$$

and the average number of arrivals in Ts is

$$E(K) = \Sigma K \cdot P(K) = \lambda T$$

where λ = average arrival rate. Then, the time τ between arrivals is:

$$f(\tau) = \lambda e^{-\lambda\tau} \text{ and the average } E(\tau) = 1/\lambda \tag{6.8}$$

(2) If the messages are exponentially distributed in lengths of r units, then

$$f(r) = \mu e^{-\mu r} \tag{6.9}$$

and $E(r) = 1/\mu$ data units/message (= average message length). Then, the service-time distribution is given by

$$f(t) = \mu C e^{-\mu C t}, \tag{6.10}$$

and

$$E(t) = 1/\mu C,$$

where C = capacity of the server/outgoing trunk in data units/sec. and $1/\mu C$ = average time to transmit a message. This results in Poisson distribution of the number of message departures (service completions) in T sec. (eqn. (6.7) with λ replaced by μC).

For an M/M/1 queue, the probability of n messages present in the buffer at time t, is given by:

$$p(n) = \rho^n p(0) = (1 - \rho)\rho^n \quad \text{and} \quad \sum_{n=0}^{\infty} p(n) = 1 \tag{6.11}$$

where $\rho = \lambda/\mu C$ = traffic intensity parameter.
For finite buffer size, $\rho < 1$ to avoid blocking of messages. The average queue size is then,

$$E(n) = \sum_{n=0}^{\infty} np(n) = \frac{\rho}{1-\rho} = \frac{\lambda}{(\mu C - \lambda)} \tag{6.12}$$

and the average queue occupancy increases beyond bound as $\rho \to 1$.

[for $\rho = 0.5$, $E(n) = 1$; $\rho = 0.75$, $E(n) = 3$; $\mu = 0.9$, $E(n) = 9$ and so on].

When a message arrives, with n messages waiting ahead of it in the buffer, the time taken to serve this (i.e. to retransmit to next node) to completion = the total delay time T_d, given by:

$$T_d = \frac{1}{\mu C} + \frac{n}{\mu C} \text{ sec.}$$

Using $E(n)$,

$$E(T_d) = \frac{1}{\mu C} + \frac{E(n)}{\mu C} = \frac{1}{(\mu C - \lambda)} \tag{6.13}$$

for a single link and single node. For a broad class of problems, it has been shown that

$$\lambda \cdot E(T_d) = E(n) \tag{6.14}$$

Equation (6.14) may be used to calculate the average time delay in a large network, using M/M/1/∞/∞ type of system. Considering the network to be made of many nodes connected by links, let the outgoing buffer associated with the ith link have an average arrival rate λ_i, average number of messages either waiting or in service $E(n_i)$ and an average delay time $E(T_{di})$; then,

$$E(T_{di}) \cdot \lambda_i = E(n_i) \tag{6.15}$$

Considering the network to be a single composite unit, with the total arrival rate to the network as γ, the total packet delay = T_0 in the network of m nodes and links is given by

$$E(T_0) = (1/\gamma) \sum_{i=1}^{m} \lambda_i E(T_{di}) \tag{6.16}$$

where

$$\gamma = \sum_{i=1}^{m} \lambda_i$$

This equation is widely used as a measure for network evaluation and design- However, the waiting-time indicated in eqn. (6.13) is an overestimation by a factor of two approximately, because of the exponential service-time assumption. Actually, the length of a packet is almost constant. Hence the service-time is also almost constant and the length of the buffer is finite; thus, the independence-assumption is not valid. For a two-link tandem queue with constant service time, the mean waiting time is now modified as:

$$T(W) = \frac{\lambda_2/\mu_2 C_2}{2[\mu_2 C_2 - (\lambda_1 + \lambda_2)]}, \tag{6.17}$$

with $\lambda_1 \ll \lambda_2$, and $\mu_2 C_2 > \mu_1 C_1$. Simulation studies also confirmed the validity of eqn. (6.17).

Throughput

The calculation of buffer overflow may be done by using

$$P(n > N) = \sum_{n=N+1}^{\infty} p(n) = [(1 - \rho) \cdot \sum_{n=N+1}^{\infty} \rho^n] = \rho^{N+1} \tag{6.18}$$

From equation (6.12), $E(n) = 1.5$ for $\rho = 0.6$, and from (6.18), $P(n > 10) < 10^{-3}$. Thus, a buffer with capacity of holding 15 messages would effectively appear as infinite for $\rho = 0.6$.

For the finite buffer case, $p(n) = \dfrac{(1 - \rho)^n}{(1 - \rho^{N+1})}$ and

$$p(N) = \frac{(1 - \rho)\rho^N}{(1 - \rho^{N+1})} \tag{6.19}$$

where N is the buffer size. The probability that the buffer is filled and the messages are blocked is simply $p(N)$ = the blocking probability.

Assuming that the average arrival rate at the input to the queue is λ and the buffer size is N, the net arrival rate to the server is $\lambda(1 - p_B)$, where p_B is the blocking probability. For equilibrium to exist, this must be equal to the average message departure rate which is just the system Throughput. The average departure rate is $[1 - p(0)]\mu C$, where $[1 - p(0)]$ is the probability that the queue is non-empty and a message is waiting in the queue. Thus,

$$\lambda(1 - p_B) = [1 - p(0)]\mu C, \tag{6.20}$$

where

$$p(0) = \frac{(1 - \rho)}{1 - \rho^{N+1}} \qquad \text{and} \qquad \rho = \lambda/\mu C$$

Thus, $P(N)$ of eqn. (6.19) $= p_B =$ blocking probability from eqn. (6.20). The average Throughput $= \lambda(1 - p_B)$ for finite buffer size, but for infinite buffer size, $N \to \infty$, $p_B \to 0$, and the Throughput is simply ρ = traffic intensity parameter.

The overall design of the distributed PSDN is quite involved as it requires the solution of complex routing and capacity allocation problems, and analysis of the relation between packet delay and line-buffer utilization. Apart from the delay and Throughput briefly discussed above, one has to consider (a) route assignment problem, (b) topological design and (c) flow control. For the minimization of total packet delay or the maximization of network utilization, the dynamic route assignment is the best solution. However, to avoid complexity, quasi-static routing techniques are generally used. For the topological design, heuristic techniques, e.g., the branch exchange method and the cut-saturation method, are used. In addition to these factors, various protocols and traffic handling schemes also affect the delay and Throughput of a network. Adequate traffic flow control over the whole network is therefore necessary to avoid the problem of congestion and local clogging of data in a distributed PSDN [7].

6.4 PACKET RADIO NETWORK [16]

Packet radio communication provides an effective means to interconnect fixed and mobile computer resources over a wide geographic area. It is also an effective alternative to circuit switching in providing error-free wideband communication networks, say for, digital speech. Historically, the successful development of the ALOHA system at the University of Howaii, where a set of terminals were linked directly to the computer centre by UHF packet radio, has led to the present developments in personal terminals, multistation RADIONET, packet-net through Cable TV, mobile data networks, and General purpose Packet Satellite Networks (GPSN). The general system structure of a Radionet is shown in Fig. 6.14, which consists of a distributed

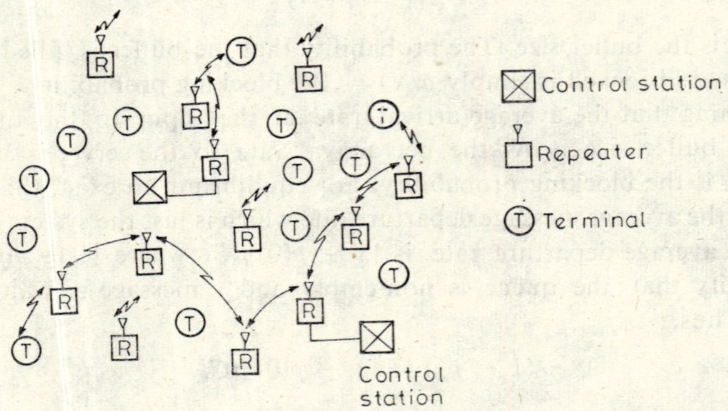

Fig. 6.14 System structure of a packet radio network

array of packet radio repeaters, each of which is able to receive and then transmit a sequence of packets, thereby serving as a relay. For line of sight operation, the repeater spacings are approximately 30–40 Km, and the total range (say up to 200 Km) is extended through multiple repeats of the data packets. Normally, a higher data rate (e.g., 400 Kb/s) is used for repeater-to-repeater communication than for terminal-to-repeater communication (say at 100 Kb/s). For fixed and mobile operation, the L-band, 1710-1850 MHz, with 20MHz bandwidth is used.

The network consists of the terminals, repeaters and control stations. The terminals transmit the packets originated by them and receive only the packets addressed to them. The users' input/output devices may be in the form of voice or keyboard input, and display, print or voice output, and the logical functions of packet processing are handled by microprocessors. Most of the system control functions are handled by control stations, distinct from repeaters, and a minicomputer provides the necessary memory and data processing facilities. The control station also serves as the gateway to other networks, and unless the destination user or Host resource is also on the same packet Radionet, all traffic must pass through it. A user's packet radio makes initial contact with the station by sending a search packet and all repeaters in the listening range forward it to the station. The station then selects one of the repeaters and transmits the address of this repeater to the user's packet radio. As a mobile user moves about within the boundaries of the Radionet, the station detects when a hand-off from one repeater to another is desirable and advices the user's packet radio accordingly. All routing computations are performed by the station and the complete routing information for each packet is inserted in it by the source and is carried along with the packet as it moves through the NET. The packet radio may be used as an $m \times n$ crossbar switch with $m + n$ radio units. It may also be used in bucket sorting for rapid storage and retrieval of data from a small table.

The main features of packet radio systems are the following:

(a) Distributed control of network management functions among multiple stations and the use of a netted array of possibly redundant repeaters for area coverage as well as reliability.
(b) The use of spread spectrum signalling for coexistence with other, possibly different, systems in the same band and for antijam and multipath protection. SAW technology is a viable current choice for MF's in the receiver [discussed in detail in Appendix B].
(c) The provision of authentication and antispoof (encryption) mechanisms.
(d) The use of system protocols that include network mapping to locate and label repeaters, route determination and resource allocation, remote debugging, and distributed network functions.
(e) The use of various implementation techniques to provide efficient operational equipment, such as, repeater power shutdown except while processing packets.

Thus, the present day packet Radionet, although a logical extension of the ALOHA system, provides for much more efficient service in a multi-user environment. This also facilitates interconnection of a large number of microprocessors through packet broadcasting, rapid deployment of an emergency network and graceful transition from inefficiently used single-user systems to more efficiently used multiple access systems through efficient spectrum management.

6.4.1 Throughput and Channel Capacity in Packet Radionet [17, 18]

In a pure ALOHA system, a common radio channel is randomly shared by a large number of users, resulting in packet collisions and transmission errors. With ARQ, the terminals retransmit the packets at randomized intervals. Assuming Poisson's process in arrival of packets, the applied traffic or channel utilization is: $S = \lambda T_p$, where λ = mean arrival rate/s, and T_p = packet duration in s.

Under steady state conditions, S is also the channel Throughput rate. Due to collision and retransmission, the actual arrival rate is higher, say λ' and the mean offered traffic is

$$G = \lambda' T_p > S \tag{6.21}$$

The possibility that no collision occurs, i.e., no packet arrives during $2T$, is given as: $\exp(-2\lambda' T_p)$. Then, the rate of retransmission is,

$$R = 1 - \exp(-2\lambda' T_p) \tag{6.22}$$

Since retransmission may be repeated as a result of successive collisions, the mean number of transmissions of a packet is given by

$$N = 1 + R + R^2 + \ldots = \frac{1}{1 - R} = \exp(2\lambda' T_p) \tag{6.23}$$

The actual arrival rate is now

$$\lambda' = \lambda N = \lambda \cdot \exp(2\lambda' T_p)$$

Therefore, the channel utilization factor (throughput) is

$$S = \lambda T = \frac{\lambda' T_p}{\exp(2\lambda' T_p)} = G \cdot e^{-2G} \tag{6.24}$$

This leads to the channel capacity $C = S$ (max) $= 1/2e = 0.184$ with $G = 1/2$ for pure ALOHA.

In pure ALOHA, collisions can take place when packets overlap partially or fully, i.e., if two packets start within $2T$ intervals. Partial overlaps are avoided in slotted ALOHA, where the time frame is divided into uniform slots and a transmission is allowed only at the start of a slot through network synchronization. Thus, the collision occurs if a second packet arrives during time T_p only, and the probability of no collision is $\exp(-\lambda' T_p)$. The channel utilization factor (throughput) now is

$$S = G \cdot e^{-}G, \tag{6.25}$$

leading to the channel capacity of 0.368, at $G = 1$.

In systems, where the propagation delay τ is much smaller than the packet transmission time T_p, as in land-mobile data networks, it is possible to avoid collision by sensing the carrier, whether present or not. Transmissions are started only when the channel is idle, either immediately or after a randomized delay. The technique may be used for both Pure ALOHA and Slotted ALOHA and is known as Carrier Sensed Multiaccess Technique (CSMA). When the transmission starts immediately after the channel is sensed to be idle, the system is known as Persistent CSMA and otherwise, the system is called non-persistent CSMA. A hybrid scheme for the slotted

ALOHA is to transmit a packet immediately after carrier sensing with probability p and to use the following slot with probability $(1 - p)$, and this is known as p-persistent CSMA [17]. The channel Throughput for the non-persistent systems are:

$$S\,(\text{pure}-\text{ALOHA, CSMA}) = \frac{G \cdot e^{-dG}}{G(1 + 2d) + e^{-dG}}$$

and

$$S\,(\text{slotted ALOHA, CSMA}) = \frac{dG \cdot e^{-dG}}{(1 - e^{-dG}) + d} \tag{6.26}$$

where $d = \dfrac{\text{packet propagation time } \tau}{\text{packet transmission time } T_p} \ll 1.$

With $d \rightarrow 0$, both equations reduce to the following (for non-persistent CSMA):

$$S_{\text{lim}\cdot d \rightarrow 0} = \frac{G}{1 + G} \tag{6.27}$$

For a given offered traffic G and a given value of p, the equation for S in the p-persistent CSMA is rather involved. However, for $d \rightarrow 0$,

$$S(G, p, d \rightarrow 0) = \frac{G[\pi_0 + (1 - \pi_0)P_s]}{G + \pi_0} \tag{6.28}$$

where

$$P_s = \sum_{n=1}^{\infty} \frac{npq^{n-1}}{1 - q^n}\left(\frac{\pi_n}{1 - \pi_0}\right), \qquad q = (1 - p).$$

and

$$\pi_n = \frac{S^n}{n!}e^{-G}$$

With $p = 1$, i.e., in 1-persistent CSMA, where all transmissions are attempted as soon as the carrier-sensing is done, the Throughput is modified as the following:

$$S(G, p = 1, d = 0) = \frac{G(1 + G)e^{-G}}{G + e^{-G}} \tag{6.29}$$

with $$P_s = \frac{G \cdot e^{-G}}{1 - e^{-G}}$$

If, however, $p \rightarrow 0$, $P_s \rightarrow 1$, and in the limit, Throughput is

$$S(G, p \rightarrow 0, d = 0) \rightarrow \frac{G}{G + e^{-G}} \tag{6.30}$$

which shows that a channel capacity of 1 can be achieved when $G \rightarrow \infty$.

Based on above values of S for various access modes, Fig. 6.15 shows the Throughput S vs. G curves, with $d = 0.01$. While the capacity of ALOHA channels does not depend on the propagation delay τ, the capacity of a

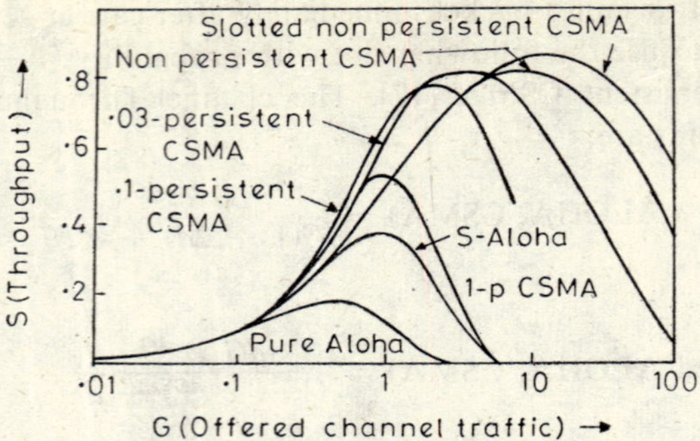

Fig. 6.15 Throughput S vs. G for various access modes

CSMA channel decreases with $d = \tau/T$, as shown in Fig. 6.16. This is because, as d increases the decisions in CMSA are made with 'older' channel informations, and with larger d, CSMA systems become inefficient. The relative channel capacities for the various protocols with $d = 0.01$, are given in Table 6.7

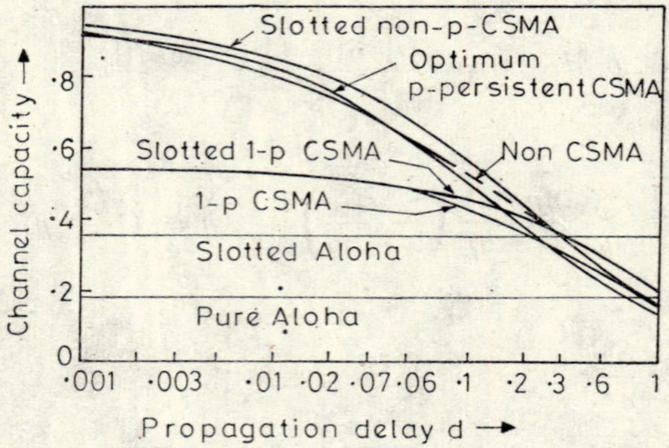

Fig. 6.17 Channel capacity vs. delay d

Table 6.7 Relative Capacities for Various Protocols ($d = 0.01$)

Protocol	Capacity C
Pure ALOHA	0.184
Slotted ALOHA	0.368
1-Persistent CSMA	0.529
Slotted 1-persistent CSMA	0.531
0.1-p CSMA	0.791
Non-persistent CSMA	0.815
0.03-p CSMA	0.827
Slotted non-persistent CSMA	0.857
Perfect scheduling	1.000

Various access modes are also compared in terms of average number of retransmissions (or average number of schedulings) = G/S, and the variation of G/S vs. S is shown in Fig. 6.17. It is seen that CSMA modes are superior in that they provide lower values for G/S than the ALOHA modes.

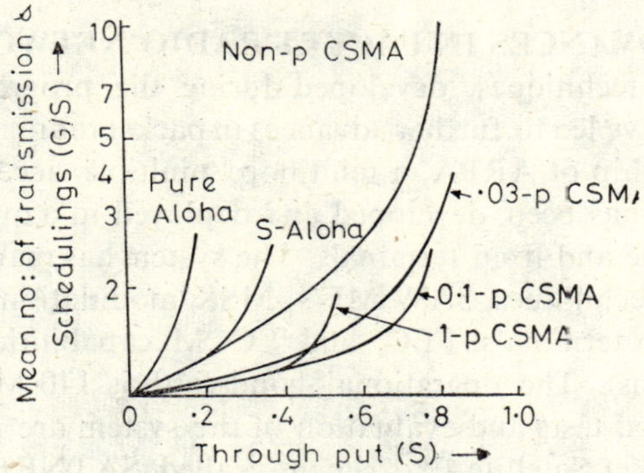

Fig. 6.17 G/S vs. S for $d = 0.01$

The delay model of ALOHA and CSMA systems may be obtained with certain simplifying assumptions, such as, (a) the ACK information is always correctly received, (b) the processing time at the destination before sending ACK information is negligible and (c) the probability of a packet's success is simply

$$P_S = S/G = \frac{\text{Throughput}}{\text{Offered traffic}}$$

The normalised delay is given by the following equation:

$$D = (G/S - 1)\, D_1 + 1 + d, \tag{6.31}$$

where D_1 includes the round trip propagation delay, transmission times of the packet and ACK signals, and the average retransmission delay δ. It is shown that $G = G(S, \delta)$ is a decreasing function of δ, and for each S, a minimum delay can be achieved by choosing an optimal δ. Simulation results of this optimization problem are given in Fig. 6.18, where the trade-

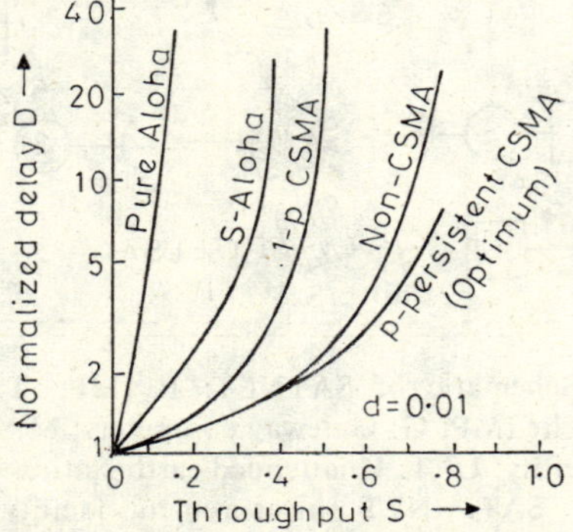

Fig. 6.18 Throughput-delay tradeoffs from simulation

off between the delay and Throughput is shown for different access modes. This is the basic performance curve and it is concluded that the optimum *p*-persistent CSMA provides the best performance, and that of the slotted non-persistent CSMA is quite comparable to it.

6.2 FURTHER ADVANCES IN PACKET RADIO NETWORK

The concepts and techniques, developed during the progress of ALOHA and ARPANET, have led to further advances in packet radio technology [19]. Under the sponsorship of ARPA, a multihop, multiple access packet radio network (PRNET) has been developed and deployed in a quasi-operational network with mobile and fixed terminals. The system has many new features such as, use of SS techniques, SAW MF's, MSK modulation, micro-processor-controlled radio terminals, FEC, and ECCM capabilities (for tactical military applications). The operational bandwidth is 140 MHz (i.e., 1710-1850 MHz). Detailed tests and evaluation of the system are in progress.

The Atlantic Packet Satellite Experiment, called SATNET, is an attempt to use satellite links as part of an integrated network to provide data communication as well as point-to-point and conference speech services [22]. Because of long propagation time to a geostationary satellite, CSMA techniques are not applicable and special protocol is required to service the class of volatile periodic traffic known as stream traffic, e.g., for voice. SATNET is composed of four earth stations which are Etam in USA, Goonhilly Downs in UK, Tanum in Sweden and Comsat Labs in Clarksburg, USA, as shown in Fig. 6.19.

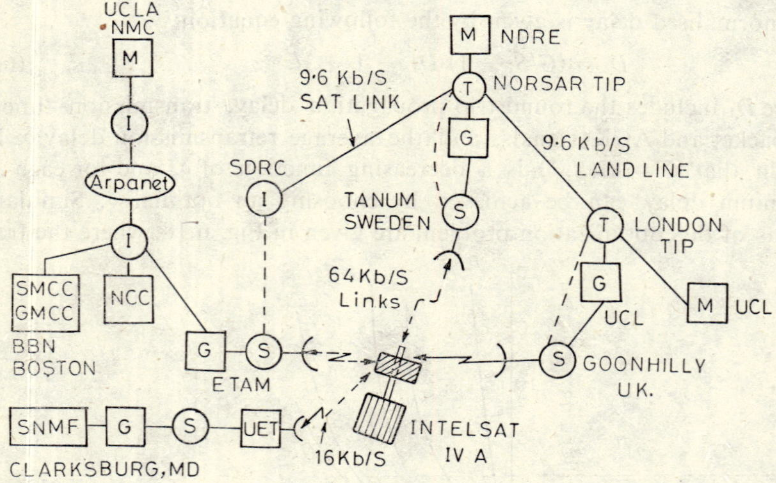

Fig. 6.19(a) Block Schematic of SATNET; I: IMP; T: Terminal IMP; S: Satellite IMP; G: Gateway computers; M: Network measurement facility; UET: Unattended earth station $G/T = 29.7$ dB/°K; SNMF: SAT. NET measurement facility; NCC: Network control centre; — — — Debugging lines.

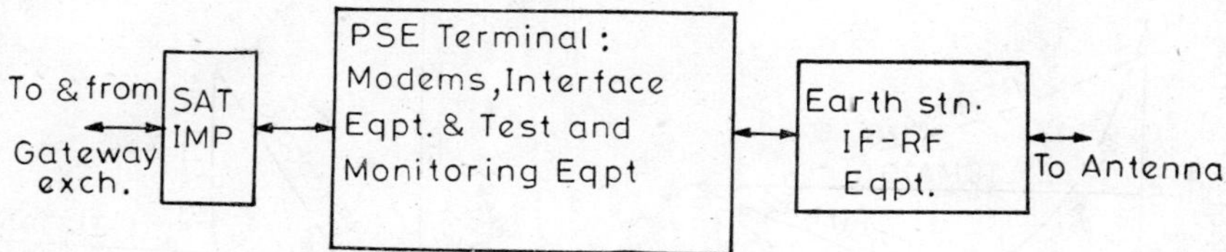

Fig. 6.19(b) Earth station equipment for SATNET

All stations are of standard *A* with $G/T = 40.7$ dB/°K, except one at Comsat Labs which has $G/T = 29.7$ dB/°K. The network uses one 38-kHz SCPC channel from the SPADE transponder of INTELSAT IV-A Satellite and the data rate is 64 Kb/s using 4ϕ-PSK modulation with BER $= 10^{-6}$ to 10^{-7}. However, for using FEC, data rate may be reduced to 16 and 32 Kb/s and normally convolutional codes are used along with Viterbi decoders. The network protocols are implemented in the satellite IMP through minicomputers and this is interfaced to the PSP terminal through a high-speed half duplex data path. SATNET is connected to the ARPANET through a separate satellite link in order to provide interworking facilities and to support certain experimental applications. The functions of the gateways, shown in the figure, are realized through PDP-II minicomputers and external data sources are connected to SATNET through these gateways only.

To improve the Throughput and delay performances of the system, various Demand Assignment approaches, e.g., pure ALOHA, S-ALOHA, and Reservation-ALOHA [20], have been investigated. One of the methods for explicit reservations, known as Reservation-TDMA (R-TDMA) has been found to be the most efficient one. In this method, a separate reservation sub-frame is used and each station is permanently assigned a slot for sending reservations. The remaining frame time is divided into data packet slots and each station is permanently assigned one data slot per data frame, with the idle station slots in each data frame dynamically allocated according to the table of outstanding reservations. Experimental results on Throughput and delay for various protocols in a two-station measurements are shown in Fig. 6.20. These curves generally show that a reservation approach (R-TDMA) exhibits better performance than fixed TDMA and S-ALOHA over a significant range of traffic values. Based on these approaches, a general purpose packet satellite network (GPSN) is being developed as a part of the futuristicglobal PSDN. This will function simultaneously as a transit network with attached inter-network gateways and as a local network supporting directly attached Hosts and terminals of differing capabilities and requirements. It is claimed that all these developments in PSDN will lead to a powerful personal data network technology which will have a revolutionary impact on both the communications and computing fields.

6.5 CCITT RECOMMENDATIONS ON DATA COMMUNICATIONS [8]

CCITT has made specific recommendations for establishing data networks

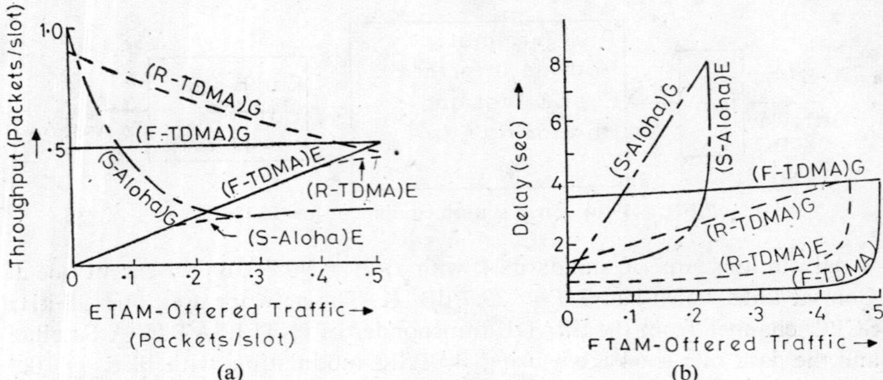

Fig. 6.20 Throughput and delay performances of different demand-assignment protocols. Subscripts E and G indicate the station where the traffic originated. (a) Throughput (b) Delay

over telephone/telex circuits (*V*–series) and over public data circuits (*X*–series). A summary of the important recommendations is given now:

The *V*–series recommendations specify:

V·2—The power levels of data signals $\leqslant$ 1 mW

V·3 and *V*·4—international alphabet No. 5

V·5 and *V*·6—synchronous data transmission rates : 600/1200/2400/4800 bps on switched telephone circuits and 600–9600 bps on leased lines.

V·16/19/20/21/23/26/27/29/35/36—these deal with specifications of different modems having speed ranges from 200–72,000 bps.

V·24—this deals with the specification for the interface between DTE and DCE and various control signals.

V·50—this gives the standards for the transmission quality.

V·51 and *V*·53—these deal with the maintenance of the data networks.

V·52/55/57—these deal with the specifications of various test equipment for data networks.

V·56—this gives the procedure for evaluating a modem.

The *X*–series recommendations for public data networks specify:

X·1—specifies the user classes of service.

X·2—specifies the user facilities in public data networks.

X·3—describes the packet assembly/disassembly facility (PAD) for a character oriented start-stop mode DTE.

X·4—gives the general structure of signals of International Alphabet No. 5 code for data transmission over public data networks.

X·20—specifies the compatible interface between DTE and DCE for start-stop transmission services.

X·21—specifies the general purpose interface between DTE and DCE for synchronous operation.

X·24—includes the list of definitions of interchange circuits for use at DTE/DCE interfaces. Also gives the functions of interchange circuits for the transfer of data, call control signals and timing signals.

X·25—specifies the interface between DTE and DCE for terminals operating in the packet mode.

X·26 and *X*·27—describes the delectrical characteristics for unbalanced and balanced-double-current interchange circuit for general use with integrated circuit equipment in the field of data communications.

X·28—describes the DTE/DCE interface for a start-stop mode DTE.

X·29—specifies the procedures for packet mode DTE to access the PAD.

X·30—gives the specifications of basic model page-printing machine in accordance with International Alphabet No. 5.

X·31/32/33—these discuss DTE/DCE, answer-back units, and an international text along with the Alphabet No. 5 for 200 baud start-stop machines.

X·40—specifies the FSK/FDM schemes for data channels.

X·50—enumerates the fundamental parameters of a multiplexing scheme for the synchronous data networks.

X·51—enumerates the fundamental parameters of the multiplexing scheme for the synchronous data networks using 10-bit envelope structure.

X·60—specifies the common channel signalling for synchronous data applications.

X·70 and *X*·71—describes the terminal and transit control signalling system for start-stop services on international circuits between asynchronous and synchronous data networks.

X·92—describes the hypothetical reference connections for public synchronous data networks.

X·95—gives the network parameters in public data networks.

X·96—describes the call progress signals in public data networks.

Some of the *X*-series recommendations have been further explained in the following pages.

Recommendation X·20

This applies to the interface between DTEs and DCEs for start-stop transmission services on public data networks, including circuit-switched service and point-to-point or multipoint leased-circuit service. It specifies electrical characteristics which at the DCE side of the interface comply with *X*·26, and at the DTE side may be applied according *X*·26, *X*·27 or *V*·28. This flexibility promotes an orderly transition from the existing equipment to the newly developed equipment which have new interface standards. For switched data network service, *X*.20 also specifies call control procedures for start-stop transmission in user classes of services 1 and 2 (data rates of 50-300 bits/sec) as given in Recommendation *X*·1. This also includes detailed state and timing sequence diagrams together with a table of interchange circuits.

Recommendation X·21

This describes a general purpose interface between DTE and DCE for synchronous operation on public data networks. Operation of *X*·21 is for user classes of service 3-7 as specified in $X \cdot 1$ and applies for synchronous data transmission at rates of 600, 2400, 4800, 9600 and 48,000 bits/sec. The electrical characteristics of the interface utilise $X \cdot 26$ and $X \cdot 27$ with different arrangements for data rate above and below 9600 bits/sec.

Annexes to this recommendation include a table of applicable interchange circuits, interface signalling state diagrams and interface signalling diagrams.

In addition to specifying procedures for call control data transfer, this recommendation includes the use of loopback testing capabilities. Two loopback levels are specified: one to enable the DTE to test itself through a loopback in the DCE to the DCE/DTE interface and another in the DCE to permit network maintenance.

Recommendation X·25

This deals with the DTE/DCE interface for terminals operating in the packet mode over public data networks. It defines a set of rules governing the way in which packet mode DTEs format control information and user data into packets, establish, maintain and clear calls, and manage the transmission and flow of data to and from the packet switching data network. It specifies three distinct and independent levels of control which are the following:

Physical interface level
Frame level
Packet level

The physical interface level is required to be in accordance with Recommendation $X \cdot 21$ or, for an interim period, $X \cdot 21$ bis. This specifies a full-duplex, point-to-point synchronous circuit between the DTE and the network.

The frame level controls the exchange of data between the DTE and the network. It specifies the use of the principle and terminology of the high-level data link control (HDLC) procedure specified by the International Organization for Standardization. The HDLC class of procedure, defined as two-way simultaneous asynchronous response mode (ARM), is specified where both the DTE and the network contain a primary and secondary function. The HDLC frame format is illustrated in Fig. 6.11.

The packet level is the biggest level of $X \cdot 25$. It supports the concurrent operation of several calls over a single access circuit where packets are contained within the link level information field, with only one packet permitted per HDLC frame. The mechanism used to indicate the call to which a packet relates is that of logical channel identifiers. At call set up time, a logical identifier is assigned to identify all packets associated with that call until it is cleared; but this association is a local matter between the DTE and the network at each end of virtual calls.

The call request packet includes the logical channel identifier chosen by the DTE as well as the network address of the called party. The called DTE is notified that a circuit is being established by an incoming call packet. The called DTE answers by returning a call-accepted packet. Subsequently the calling DTE is notified that the circuit is established by a call-connected packet. Exchange of data packets may then proceed. The call can be cleared upon completion if either the calling or the called party sends a clear indication packet to the opposite party. The acknowledgement of receipt confirms the termination of the call to the party initiating the terminating action.

Each logical channel operates independently of the other (up to 4096 on a single access link) and may be used for virtual calls or permanent virtual circuits. The maximum data field length of data packets is 128 octets. However, some administrations may support maximum data field lengths of 16, 32, 64, 256, 512, 1024 or exceptionally 2048.

Recommendation X·75

This recommendation defines the characteristics and operation of an inter-exchange signalling system for international packet-switched data transmission service. The signalling system is intended to be used for the transfer of information between two signalling terminals each within a packet-mode data network and directly connected by an international link. As was the case with *X*·25, *X*·75 is also structured in three layers of control procedures.

The physical level specifies the use of one or several high-speed full duplex synchronous circuits which are used to interconnect two international data switching exchanges. The data rate is typically 48 kbps and the circuit may be terrestrial or satellite.

The frame level specifies the use of a data link control procedure compatible with the high-level data link control (HDLC), extending numbering and balanced class of procedures. The frame level ensures that call control information and user data contained in packets are accurately exchanged across the international link. The definition of a multi-circuit data link control procedure is an item for further study within CCITT.

The packet level specifies the manner in which calls are established, maintained and cleared across the *X*·75 interface. The use of *X*·25 procedures and packet formats has been retained as much as possible in *X*·75 to maintain compatibility with virtual circuits as viewed by an *X*.25 DTE.

One notable deviation of *X*·75 from *X*·25 is the addition of a network utility field. This is a network administrative signalling mechanism which complements the user facility field and serves to separate user service-signalling from network administrative signalling. Recommendation *X*·75 also defines the manner in which call progress signals are transferred to clear and reset packets across the international link.

Recommendations X·3, X·28 and X·29

The urgent need to allow interworking between a start-stop mode DTE on a public-switched telephone network, public data network or a leased circuit, and a packet mode DTE on a packet switching network, led to the development of these three recommendations. They cover, respectively, the packet assembly/disassembly facility (PAD), the DTE/network interface and procedures for the exchange of control information and user data between a packet mode DTE and a PAD for start-stop mode DTEs.

The basic functions of the PAD ($X \cdot 3$) include the following:

(a) Assembly of characters into packets destined for the $X \cdot 25$ DTE.
(b) Disassembly of the user data field of packets destined for the start-stop mode DTE.
(c) Handling of virtual call set-up and clearing, resetting and interrupt procedures.
(d) Generation of service signals.
(e) A mechanism for forwarding packets when the proper conditions exist.
(f) A mechanism for transmitting data characters including start, stop and parity elements as appropriate to the start-stop mode DTE.
(g) A mechanism for handling a 'break signal' from the start-stop mode DTE.

User selectable functions which may be provided by the PAD include management of the assembly and disassembly of packets, and management of the procedure between the start-stop DTE and the PAD.

The DTE/network interface for start-stop mode DTEs ($X \cdot 25$) specifies the following:

(a) Procedure for the establishment of an access information path between a start-stop mode DTE and a PAD.
(b) Procedures for character interchange and service initialization between a start-stop mode DTE and a PAD.
(c) Procedures for the exchange of control information between a start-stop mode DET and a PAD.
(d) Procedures for the exchange of user data between a start-stop mode DTE and a PAD.

The procedures for the exchange of control information and user data in $X \cdot 29$ are supplemental to $X \cdot 25$ for the purpose of controlling and exchanging data with the PAD. The data qualifier-indicator in the header of data packets is used to identify whether the data field contains user data to be passed to the start-stop mode DTE or control information for the PAD. At present provision exists in the recommendations only for virtual calls initiated via a PAD by start-stop mode DTE's. A number of additional items not covered by $X \cdot 29$ are identified for further study. Among them is the possibility of a packet mode DTE establishing a virtual call to a non-packet mode DTE.

REFERENCES

1. N. Abraham and F. Kuo (Eds), *Computer Communication Network*, Prentice-Hall (1973).
2. M. Schwartz, *Computer Communication Network Design and Analysis*, Prentice-Hall (1977).
3. (a) *Proceedings of 2nd International Conference on Computer Communication*, Stockholm, Aug. 1974.

3. (b) *Proceedings of 3rd International Conference on Computer Communication* Toronto, Aug. 1976.
4. *Proceedings of 4th International Conference on Computer Communication*, Kyoto, Sept. 1978.
5. Special Issue on Computer Communication, *IEEE Trans. on Comm.* vol. Com-25, No. 1. Jan. 1977.
6. Special Issue on Packet Communication Networks, *Proc. IEEE*, vol. 66, No. 11. Nov. 1978.
7. H. Inose, *An Introduction to Digital Integrated Communication Systems*, Univ. of Tokyo Press, 1981.
8. *Report of the Panel on Computer Communication Networking*, *Electronics Information and Planning, India*, vol. 6, No. 4, p 363, 1979.
9. P.T.F. Kelley, *New Non-voice Customer Services and Terminals Using Digital Techniques*, BPO Publication, London, 1980.
10. R.E. Kahn, Resource Sharing Computer Communications Network, *Proc. IEEE*, Vol. 60, p 1397, 1972.
11. P.T.F. Kelley, the Euronet Telecommunication and Information Network, *the Radio and Elect. Engg.*, vol. 49, p 564, Nov. 1979.
12. P.T.F. Keley, Public Packet Switched Data Networks. International Plans and Standards *Proc. IEEE*, vol. 66, p 1539, 1978.
13. Issue on No. 5 ESS, *Bell Labs. Record*, Dec. 1981.
14. M.P. Ristenbart, Alternatives in Digital Communication, *Proc. IEEE*, vol. 61, p 703, June 1973.
15. L. Kleinrock, *Queueing Systems*, vol. 2, Computer Applications, John Wiley, 1976.
16. R.E. Kahn, The Organisation of Computer Resources into a Packet Radio Network, *IEEE Trans. on Comm.* vol. Com-25, p 169, 1977.
17. L. Kelinrock and F.A. Tobagi, Packet Switching in Radio Channels: Pt I—Carrier Sense Multiple Access Modes and their Throughput—Delay Characteristics, *IEEE Trans.*, Com-23, p 1400, 1975.
18. F.A. Tabagi and L. Kelinrock, Packet Switching in Radio Channels Pt. III: Polling and (Dynamic) Split–Channel Reservation Multiple Access, *IEEE Trans.* vol. Com-24, p 832, 1976.
19. R.E. Kahn, *et al.*, Advances in Packet Radio Technology, *Proc. IEEE*, vol. 66, p 1468, 1978.
20. P. Murthy, A Reservation Aloha Scheme, *The Aloha System Tech. Report*, B75-24, Univ. of Howaii, July 1975.
21. D.D. Clark, K.T. Pogran and D.P. Reed, An Introduction to Local Area Networks, *Proc. IEEE*, vol. 66, p 1497, 1978.
22. I.M. Jacobs, R. Binder and E.V. Hoversten, General Purpose Packet Satellite Networks, *Proc. IEEE*, vol. 66, p 1448, 1978.

CHAPTER 7

Electronic Switching Systems

The need to have large-scale connectivity among the telephone subscribers has led to a hierarchical structure of the national telephone network in all countries. The network is generally divided into regional, district and toll centers, which are interconnected as a star or an incomplete mesh, as shown in Fig. 7.1. Traditionally, telephone exchanges developed through manual switchboards, strowger exchanges, crossbar systems and then, 'stored-program control' (SPC) exchanges. However, in these exchanges, all circuit switching networks establish continuous and separate two-way transmission paths for a successful call. The common control SPC technique was a major breakthrough in developing the efficient large exchanges by initially using crossbar switches and later by using time-division analog switching with pulse-amplitude samples.

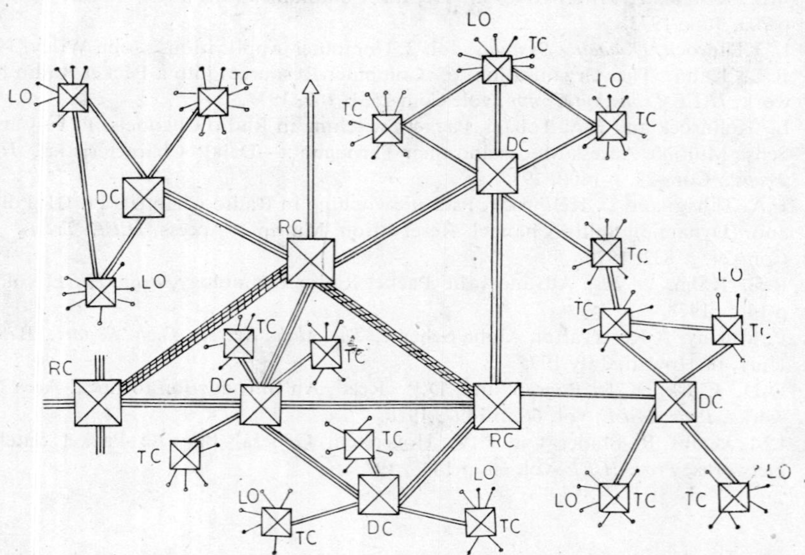

Fig 7.1 A typical telephone network; RC: Regional centre; DC: District centre; TC: Toll centre; LO: Local office

Till 1950, the switching, as well as the transmission systems in long-distance communication, were analog using FDM links and Space-division (SD) switches, e.g., Strowger/Crossbar exchanges. Even during the early 50's, only analog time-division switching was being experimented with. With the successful designing and the engineering of PCM systems in the 1950's, the trunks in the multi-exchange area were first digitized, and the hybrid systems using PCM transmission and SD-exchanges became quite efficient and popular. E. Vaughan of Bell Telephone Laboratories first proposed his experimental ESSEX system in 1959 and the concept, using both digital transmission and digital switching, led to the development of the first integrated telecommunication system [4]. This also encouraged various telecommunication administrations, notably in USA, France, UK and Japan, to initiate developmental efforts in digital switching techniques. The major techniques developed were: (a) Time-multiplex switching (TMS) using space-division switches (electronic cross-points) and (b) Time slot interchange (TSI) that switches speech samples from one time slot (of a multiplexed digital carrier) to another through temporary memories. However, digital (PCM) switching needs a large amount of memories with the result that the commercial exploitation of these techniques had to wait till the economic development of MSI's and LSI's had taken place. Because of the availability of cheap MSI/LSI in the late 1960's, digital exchanges were installed from the early 1970's, first for toll and tendem switching, and then for local telephone switching.

France and UK started developing digital switching systems in the 1960's, and installed experimental systems from 1968–71. The Bell system in USA developed a large toll switching system, No. 4 ESS and this became operative in 1976. Hitachi, NEC and others in Japan worked in collaboration and developed some large systems as well. A summary of the systems developed is given in Table 7.1, which also shows the size and type of switches used. In most cases, two methods of utilizing digital switching were evolved: (a) to start with trunk switching and then provide an overlay network (as in USA), (b) to start in the local area and then extend digital facilities to remote areas (as in France). During the mid-70's, a rapid adaptation of digital switching occurred in the field of PBX switching. Later on, the technique of digital switching was extended from small to large systems using the same network architecture (as in system-12 of ITT). This has also facilitated the introduction of switching data and non-voice services through the common network, leading to the futuristic ISDN (Integrated Service Digital Network). The principles and applications of the digital switching networks and their subsystems shall be discussed in this chapter.

7.1 PRINCIPLES OF DIGITAL SWITCHING [1, 2, 5]

A digital switching system uses speech signals in a digital form, say, 8-bit PCM, and switches the digital signal to the destination either by a space-division (SD) or by a time-division (TD) switch, under the control of a SPC processor. The different switching techniques used currently are as follows:

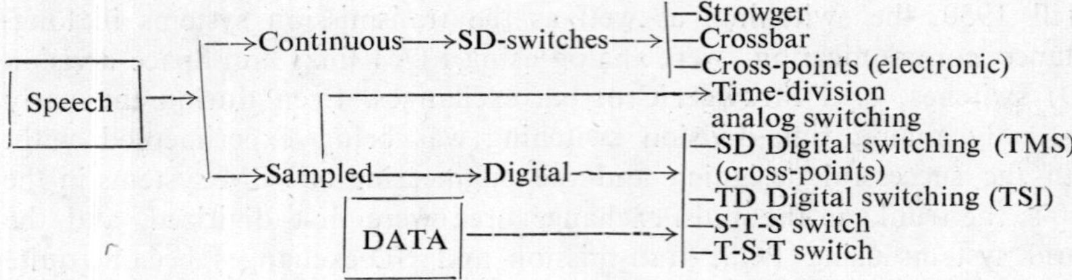

Table 7.1 A Summary of PCM Exchanges in Service [14]

System	Network	Control	Capacity	Country	Year
Empress	STS	WL	192	UK	1968
IST	TSTST	SPC	192	Australia	1971
Moorgate	SSTSS	WL	384	UK	1971
E10	TT	AT	7.2 K	France	1975
No. 4 ESS	TSSSST	SPC	107 K	USA	1976
SINTEL3	SSTSS	SPC	14.4 K	Italy	1976
PDX	TST	SPC	23 K	Netherlands	1976
AXE	TST	SPC	32 K	Sweden	1976
No. 3 EAX	SSTSS	SPC	60 K	USA	1978
DMS200	TSSSST	SPC	60 K	USA	1977
System-X	—	SPC	32 K	UK	1978
MT-20	TSST	SPC	65 K	France	1979
FETEX100T	TSSSST	SPC	60 K	Japan	1979
HTX	SSTSS	SPC	60 K	Japan	1978
NEAX61	TSST	SPC	60 K	Japan	1979
System-12	TSTSTSTS	Distributed	100 K	ITT	1981

WL: Wired logic; AT: Centralized service computer
SPC: Stored program control.

In general, digitized speech signals are time-multiplexed forming TDM bit (or word) streams. In conformity with PCM transmission standards, the first level of digital carrier known as the primary digital carrier (PDC), is either the 24-ch. PCM (DS-1, 1.544 Mb/s, 125 μS Frame, D3 signalling) or the 32-ch. PCM (CEPT, 2.048 Mb/s with signalling in the 16th channel). In large electronic exchanges, e.g., in No. 4ESS, a higher-order multiplexed (MUX) stream, e.g.,* DS-120 highway (120 ch. PCM), is used as the input to the switching network. A switching center has to serve several PDC buses simultaneously, as shown in Fig. 7.2, and it is required to switch the signals

*In DS-120 highway, each frame of 125 μs contains 128 time slots of which 120 slots are for voice (120 ch. PCM) and 8 slots are used for maintenance functions. Each slot contains 16 bits, with a basic clock rate of 16.384 MHz; and of 16 bits, 8 bits are for 8 bit-PCM data and the rest are for control and channel information signals. Similarly in DS-480, there are 512 slots in the frame of which 480 slots are for voice/data and 32 slots for miscellaneous functions.

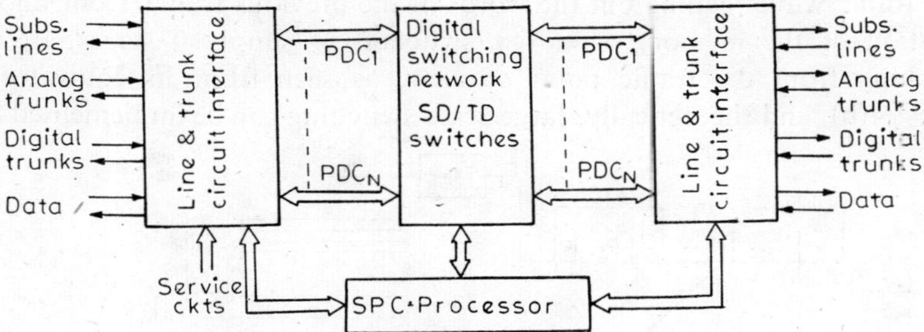

Fig 7.2 Simplified block diagram of a digital exchange

from individual time slots on one PDC bus to the same or different time slots on other PDC buses. In a small exchange this could be done by using the technique of time-slot-interchange (TSI) in conjunction with a simple SD-switch. To accommodate the different path delays of the PDC buses, time buffering (delay pad) of the PDC lines at the input to the switch is required. This ensures that all PDC's entering the exchange are in strict time slot and pulse synchronism.

7.1.1 Time-Division Switching

The switching of PCM words between the time slots of a PDC bus (i.e., TSI) can be accomplished by using variable length delay lines, where the PCM words are stored serially, and then read out during the appropriate time slots, as shown in Fig. 7.3. This was popular during the early days of

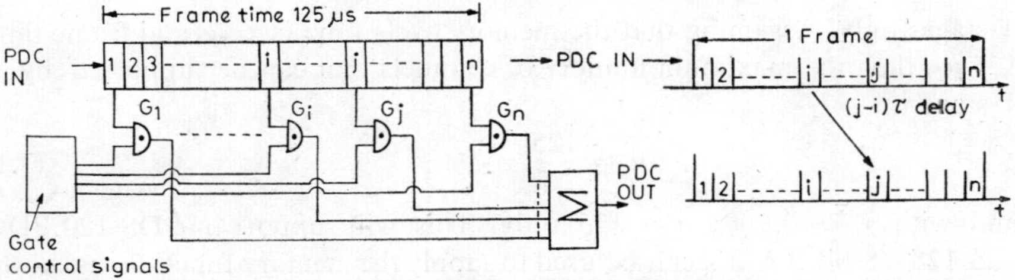

Fig 7.3 Time slot interchange using delay lines. The ith. word is shifted to the jth slot with a delay of $(j - i)\ \tau$, where τ is the chip time of the line.

digital switching; at present, however, TSI is achieved more economically by using IC digital memories with parallel access. Figure 7.4(a) shows a simple TSI circuit, where the PDC words are written in parallel in the memory locations of a RAM, under the control of a memory-write access circuit and then read out during the allotted time slot under the command of the read-access circuit. Thus, the PDC is now restructured as per the routing instructions, after TSI and parallel-to-serial (P/S) conversion; and the PCM word of the i-slot is shifted to, say, the j-slot, as shown in Fig. 7.4(a). The maximum delay introduced in the code words by the TSI may be as large as two frames, since it is necessary to store the words for one frame while reading out the words of the previous frame from another location of the memory. The TSI switches, or simply T-stages, are non-blocking from the traffic point of view, as seen from its equivalent in Fig. 7.4(b), and theoretically, large scale switching can be implemented with

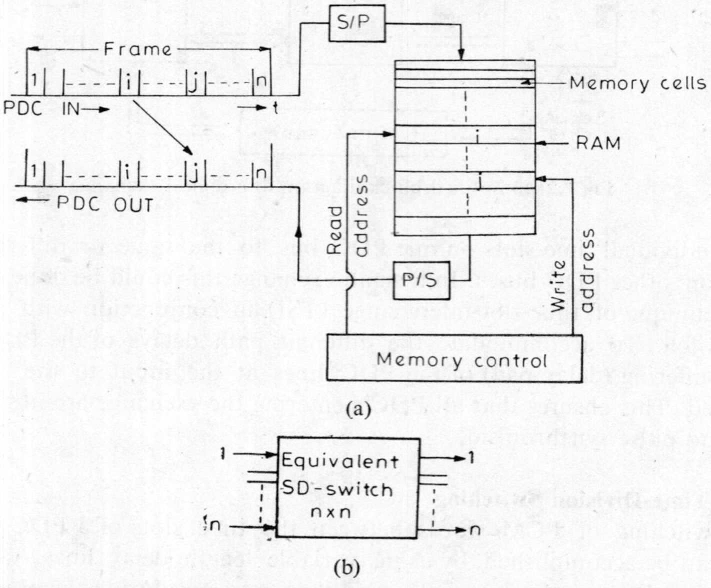

Fig 7.4 (a) TSI using a RAM, where the *i*th word is shifted to the *j*th slot
(b) Its equivalent SD-switch

T-stages only. Assuming that the memory cycle time is τ_c μs and frame time 125 μs, then the maximum number of channels that can be supported by a T-stage is:

$$n = \frac{125}{2\tau_c} \tag{7.1}$$

and with $\tau_c \simeq 0.5$ μs, $n = 125$ only. This will support one DS-120 PDC and 128×8 bit RAM's can be used to supply the memory functions as shown in Fig. 7.4. A few additional IC's will be required to provide the control functions, and the overall switch is economical and not very complex. To accommodate multiple PDC buses, say N, in one TSI-switch, it is economical to use a centralized memory with a capacity of nN words, where each PDC has n time slots. The PCM words are stored in the memory and then read out according to the routing instructions, as explained before; thus, the requirement of switching time slots between different PDC buses is easily fulfilled. Since nN-PCM words have to be written in and read out in one frame period, the read-write cycle time of memory becomes the limiting factor of its capacity. Assuming a read-write cycle of 50 ns for a large memory, the memory capacity is 2500 words; and the time switch can now support approximately 20 DS-120 PDC's. Since this capacity is equivalent to 2500 subscribers only, the scheme is useful only for small exchanges. For large exchanges, SD-switching (i.e., S-stages) in tendem with T-stages has to be used, leading to S-T-S or T-S-T configurations.

7.1.2 Space-division Switching

Traditional electromechanical switches, e.g., strowger and crossbar, are based on space-division techniques. A two-dimensional array of miniature switches, either electromechanical or solid-state, is a convenient switching matrix with N input buses and N/M output buses, as shown in Fig. 7.5(a). The cross-points of the matrix may have PNPN switches or gates which operate at the command of the control signals ($CS_1 \ldots CS_{MN}$); and these are generated by the routing instructions from the SPC-processor, as shown in Fig. 7.5(b).

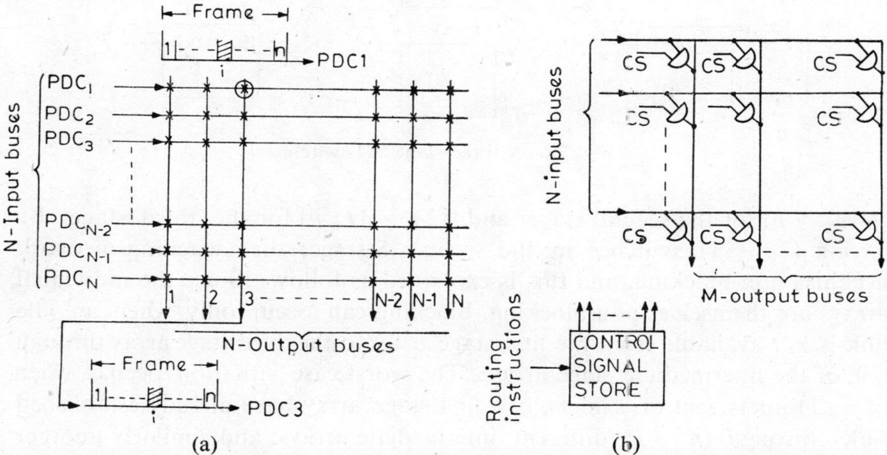

Fig 7.5 Space-division (SD) switch using cross-points. (a) $N \times N$ switch, (b) $N \times M$ switch with associated controls ($CS_1, CS_2, \ldots, CS_{MN}$)

Unlike SD-switching for analog signals where connections are maintained continuously for each call, the SD-switches for PDC-buses operate once in every time frame for the required time slot period to switch the individual PCM words to the desired output buses. While switching the words, the respective time slot positions in the PDC are maintained, but the PDC-bus allocations are interchanged according to the routing instructions (as indicated by the ⊗ cross-point in Fig. 7.5(a)). The technique has become known as 'time-multiplex switching' (TMS), or simply as S-stage of switching (in contrast with T-stages).

The $N \times N$ array of Fig. 7.5(a) is non-blocking, but the number of cross-points required is N^2, which is rather uneconomical. The realistic SD-switching networks are formed by grouping the input and output buses, say in N/n groups, and the groups are connected to N/n first stage/last stage switches. The first and last S-stages are now interconnected by an intermediate S-stage having $(2n - 1)$ array switches, as shown in Fig. 7.6. To satisfy multiple connectivity (i.e., full availability) between input and output buses, the array configurations are: $[n \times (2n - 1)]$ for the first S-stage;

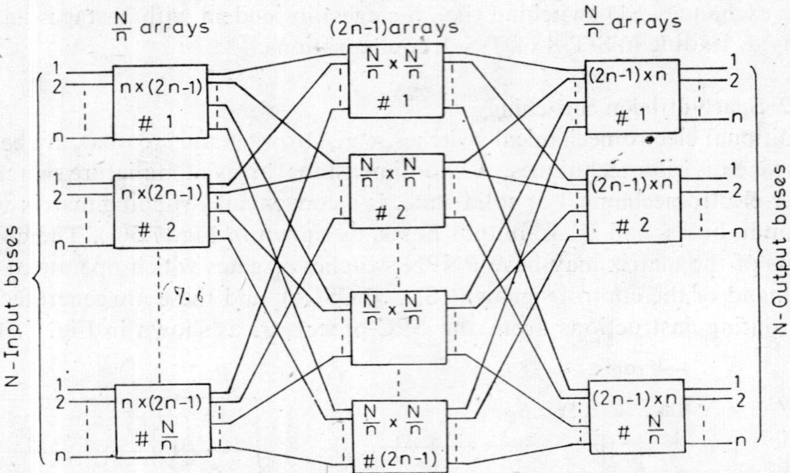

Fig. 7.6 A three-stage SD-switch

$[N/n \times N/n]$ for the second stage; and $[(2n - 1) \times n]$ for the third stage. By having $(2n - 1)$ switches in the second S-stage, the switching network becomes non-blocking, and this is explained as follows. Since the individual arrays are themselves non-blocking, blocking can occur only when an idle link is not available from the first stage array to the third stage array through one of the intermediate stage arrays. The worst case situation occurs when $(n - 1)$ inlets, out of n inlets, of a first stage array have already established links through $(n - 1)$ different intermediate arrays, and similarly another $(n - 1)$ intermediate stage arrays have been used by the $(n - 1)$ outlets of a third stage array. The only way a new link between the nth inlet of the first stage array and the nth outlet of the third stage array may be established is to have one extra intermediate array, making the total of the intermediate arrays equal to $[2(n - 1) + 1] = (2n - 1)$. Thus, the composite switch is now non-blocking, and the number of cross-points N_c required is given by

$$N_c = 2N(2n - 1) + (2n - 1)\left(\frac{N}{n}\right)^2 \tag{7.2}$$

Since N_c is dependent on the subgroup size n, the optimum value of n for a minimum of N_c is obtained by setting $dN_c/dn = 0$; then $n = \sqrt{N/2}$. Using this value of n, the minimum value of N_c is:

$$N_c\,(\min) = 4N(\sqrt{2N} - 1) \tag{7.3}$$

As an example, with $N = 128$ input buses, $N_c = 7{,}680$ for a three-stage switch and $N_c = 16{,}256$ for a single stage switch. The economy is substantial in three-stage switches and at the same time, the switching structure provides the 4-wire capabilities required in telephone circuits. Further economy in the number of cross-points may be achieved by using a five-stage non-blocking SD-switch, as an extension of Fig. 7.6. In this case, N_c is minimum if $n = 3\sqrt{N}$, and N_c is given by:

$$N_c\,(\text{min}) = 16nN - 14N + 3n^2 \tag{7.4}$$

For a large value of N, say 10^3, the number of cross-points required for the single, three- and five-stage non-blocking SD-switch is given by 10^6, 17.6×10^4 and 14.57×10^4 respectively. It is now seen that the saving in cross-points saturates as the number of S-stages increases, but significant reduction in the number of cross-points occurs when a small amount of internal blocking is allowed. This is customary in large exchanges.

Because of the limitations of both T- and S-stages of switching, a more effective approach in designing a large switching network is to use two S-stages separated by a T-stage, or two T-stages separated by a S-stage, leading to S-T-S or T-S-T networks. The present trend is to use T-S-T stages for large networks. This is discussed in the next section.

7.1.3 The Overall Switching Network

The general block diagram of a digital exchange has been shown in Fig. 7.2. In contrast to the conventional analog SD-exchanges, where the cost and performance of the system is mainly dependent on the switching network, the digital exchange switching network is the least expensive and least critical unit. However, the line and trunk interface is the costliest and most critical unit. It has been estimated that for large exchanges, the cost of the line/trunk interface is as much as 80 per cent of the total system cost, and the rest 20 per cent accounts for both the switching network and the central control. This has led to considerable architectural change in the design of digital exchanges vis-a-vis the earlier SD-exchanges. Further the line/trunk interfaces may now be located at remote places, as only the few PDC buses (along with signalling and internal control commands from the central processors) need be connected to the main exchange. This enlarges the service area of an exchange mainfold and brings the digital interface nearer to the subscribers. Thus, in a general layout, the lightly loaded subscriber lines are connected to a remote concentrator using SD/TD switching (and including line and interface circuits) and the resultant PDC buses carrying heavier traffic are connected directly at the digital trunk inputs of the main exchange. Since the trunks from other exchanges already carry high traffic load, they are also connected directly to the main exchange. The switching network of the main exchange is then designed to carry this traffic load. This does not pose a serious problem as the digital switch is almost non-blocking.

The building blocks, T- and S-stages of a digital switch, have been described in the last sections. The composite T-S-T and S-T-S networks are shown in Figs. 7.7 and 7.8 respectively. To increase the overall capacity of the network, several frame-aligned incoming PDC's are multiplexed onto a higher level highway in the multiplexer M, e.g., DS-1 to DS-120 or DS-1 to DS-480. The transmission over this highway is usually in a parallel format, i.e., eight lines in parallel for 8-bit words, to keep its speed within reasonable limits. The T-S-T network is the favoured configuration for larger exchanges and it provides multiple-choice routes through the S-stage. In Fig. 7.7, the

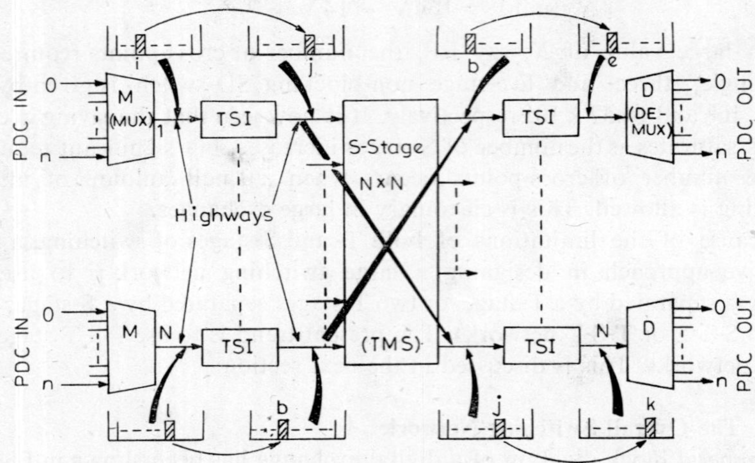

Fig 7.7 T-S-T switch. Flow of information is shown through the time-slots: $i(1) \rightarrow j(1) \rightarrow j(N) \rightarrow k(N)$ and $1(N) \rightarrow b(N) \rightarrow b(1) \rightarrow e(1)$

flow of information through the network has been indicated and it is seen that the information in the ith slot of the 1st highway may be transferred to the kth slot of the Nth highway through many alternative time-space routes. It has been seen earlier that by having sufficient number of cross-points in the S-stage, the network may be made non-blocking; however, under heavy traffic conditions, it is possible that the same time slot, say, the jth in Fig. 7.7, is not available simultaneously on both highways which interconnect the two specific TSI's. The problem is solved either by expanding the number of time slots on the highways in the S-stage or, by deloading the network ports. For ideally non-blocking condition, the time expansion required is almost 100 per cent (as in the case of three-stage SD-switch of Fig. 7.6), but with an expansion of only 10 per cent the probability of blocking is reduced to 10^{-5} for traffic densities as high as 0.9 Erlangs per network inlet. The S-T-S switch is shown in Fig. 7.8, where the flow of information

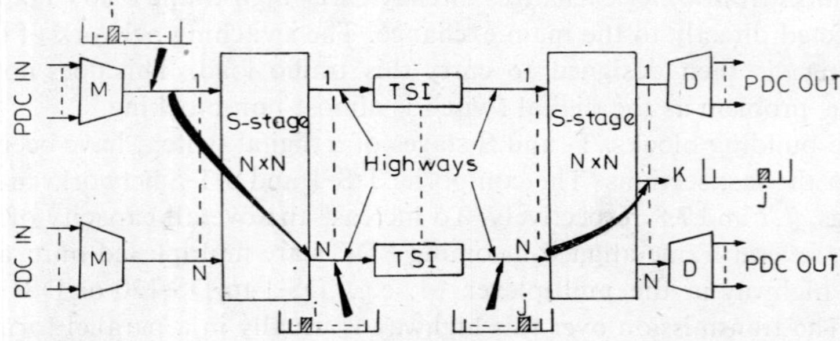

Fig 7.8 S-T-S switch. Flow of information is shown as: $i(1) \rightarrow i(N) \rightarrow j(N) \rightarrow j(k)$

has been indicated. Here also, it is necessary to increase the number of TSI units by 100 per cent to make the switch fully non-blocking. The S-T-S switch is generally efficient for smaller exchanges only and the popularity of T-S-T switch for large networks is due to the following reasons:

(a) T-S-T is more economical because of the availability of low-cost RAM's which constitute the heart of the TSI.
(b) The number of available multiple-choice routes between two ports in T-S-T networks are generally much higher than in S-T-S networks.
(c) Path search algorithm in T-S-T networks is simpler.
(d) The implementation complexities of S-T-S switches are higher for larger traffic densities.

As such, large digital exchanges, e.g., No. 4 ESS, have been designed with T-S-T switches.

For small to medium-sized exchanges, a modular structure of the T-S-T switch, as shown in Fig. 7.9, is sometimes preferred. In this configuration,

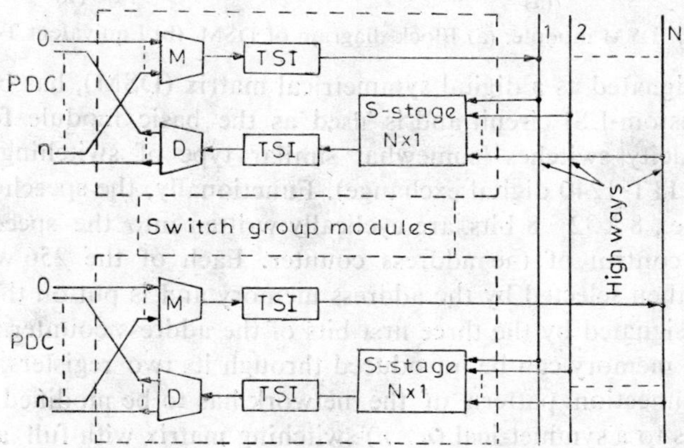

Fig. 7.9 T-S-T switch—4 wire modular design.

the incoming and outgoing TSI's along with their MUX/DEMUX units (for 4-wire connection to the line/trunk interface) and a $N \times 1$ S-stage switch are grouped together in a module, called a switch group. Such switch groups are interconnected through a number of highways (1, 2, . . . , N), which are implemented in the back plane wiring or cabling. Each switch-group transmits to its own dedicated highway and receives from all the highways via the S-stage, thus forming a somewhat independent module. The network can now be expanded with as many switch groups as required (up to 16 groups or so) and is restricted only by the number of highways already available. For large networks, however, many highways have to be distributed over several racks which results in transmission and delay problems.

DSM

A modified modular design of a large switch has been given by Charransol *et al.* [6]. The concept is somewhat similar to the use of a centralized memory in a large capacity TSI switch, as discussed in Sec. 7.1.1. Consider that 8 PDC's, each having 32 time slots of 8 bits, (equivalent to CEPT 2 Mb/s stream) are multiplexed and written in parallel into a TSI memory, and the memory is read at the appropriate time slot also in parallel, as shown in Fig. 7.10(a). The T-switch now behaves as a T-S switch for the 256 words contained in the incoming 8 PDC buses, as shown in Fig. 7.10(b). This

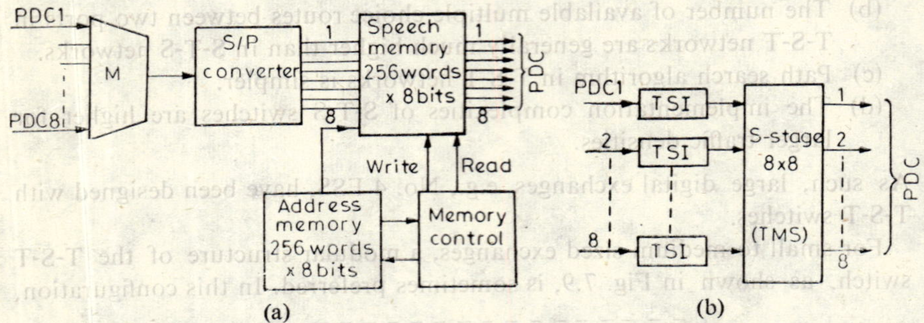

Fig. 7.10 DSM module: (a) Block diagram of DSM, (b) Equivalent T-S switch

switch, designated as a digital symmetrical matrix (DSM), has been developed as a custom-LSI circuit and is used as the basic module for building higher capacity switches (somewhat similar type of switching modules is used in the ITT-1240 digital exchange). Functionally, the speech samples of 8 PDC's, i.e., $8 \times 32 \times 8$ bits, are cyclically written into the speech memory under the control of the address counter. Each of the 256 words in the memory is then selected by the address memory and is put on the PDC bus which is designated by the three first bits of the address counter. Changes in the address memory can be introduced through its two registers, each time the interconnection pattern of the network has to be modified. The DSM corresponds to a symmetrical $(n \times n)$ switching matrix with full access from inputs to outputs and is very compact with multipurpose capability. For example, a concentrator with n inputs and $n/2$ outputs is simply organized by removing $n/2$ output links, and the ratio 2 can be easily changed without any hardware changes.

To organize larger switches using the above 8×8 DSM, it is simply required to arrange the modules in a series/parallel configuration, as shown in Fig. 7.11, for a 64×64 switch. The blocking probability of the switch is

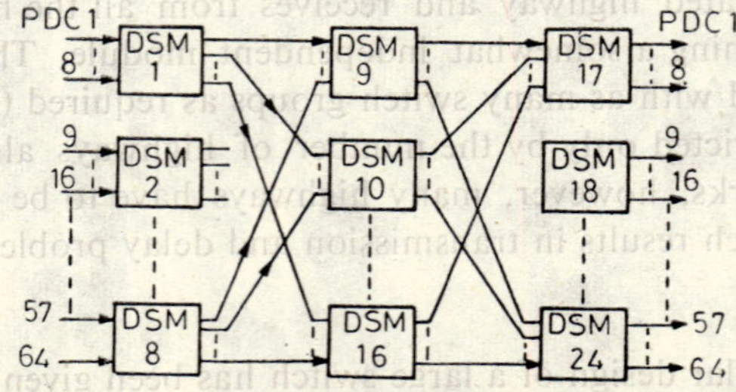

Fig 7.11. A 64×64 switch using 24 DSM modules

almost zero because of the time switching allowed in each module. The exact blocking probability is 4.10^{-23} for a traffic load of 0.8 Erlangs/link. If one of the intermediate module fails, then the increased blocking probability $\simeq 10^{-7}$, which is fully satisfactory. A higher order switch, say 512×512, is now built using the 64×64 switches as the intermediate stages, as shown in Fig. 7.12. Similarly, a switching network for 2048 PDC's may be built with

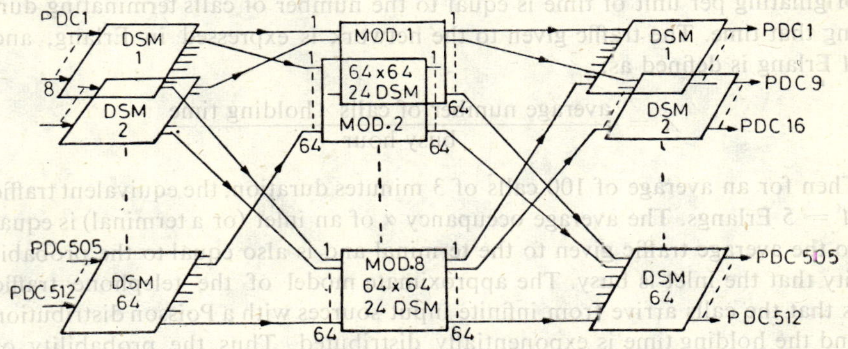

Fig. 7.12 A 512×512 switch using 320 DSM modules

256×256 modules in the intermediate stages and this requires a total of 1792 basic 8×8 DSM modules. This is equivalent to a T-S-S-S-T network having a traffic capacity of 20,000 Erlangs. But the T-S-S-S-T network will require 2.10^4 cross-points and 10^6 bits of memory, whereas the modular network requires only 1792 DSM modules. Thus, it is seen that the modular structure of the network using LSI-DSM is very economical and has the ability to implement a wide range of switch sizes. Other advantages are simplified manufacturing, maintenance and test; reduced redundancy costs and spare parts inventories, and more graceful expansion capabilities. However, the modular design introduces potentially long delays through the switch. As for example, a 5-stage DSM network of Fig. 7.12, has a maximum delay of 625 μs, which is tolerable in voice traffic, but may sometimes lead to singing conditions in local connections.

The above discussions clearly indicate the flexibility, versatility and economic realization of the digital switching networks over a broad range of sizes. Local exchanges of a few hundred to a few tens of thousands lines utilizing the same network structure are now being manufactured in quantities. In the toll switching area, digital exchanges are offered that cover a broad range from a few thousand trunks to 10^5 trunk capacities. Further, the relatively low cost of the switches have eased the problem of reliability through redundancy; and parts of the network, or even the entire network, are duplicated for greater reliability, easy maintenance and expansion. Although in modular networks, as in Figs. 7.9 and 7.12, a partial redundancy satisfies the requirements of reliability, yet the general trend is to duplicate the total switching network as it resolves several potential problems, e.g., system performance during faults, repair and expansion. Otherwise, there is the possibility of a complete breakdown of the highly multiplexed network during a hardware/software malfunction.

7.1.4 Traffic Analysis and Switch Design [2, 7]

For the purpose of traffic analysis, it is assumed that the traffic through the switching network is in statistical equilibrium, such that the number of calls originating per unit of time is equal to the number of calls terminating during that time. The traffic given to the network is expressed in Erlang, and A Erlang is defined as:

$$A = \frac{\text{average number of calls} \times \text{holding time}}{\text{busy hour}}$$

Then for an average of 100 calls of 3 minutes duration, the equivalent traffic $A = 5$ Erlangs. The average occupancy α of an inlet (or a terminal) is equal to the average traffic given to the terminal and is also equal to the probability that the inlet is busy. The approximate model of the telephone traffic is that the calls arrive from infinite input sources with a Poisson distribution and the holding time is exponentially distributed. Thus the probability of arrival of k calls during an interval t is:

$$p(k, t) = \frac{(\lambda t)^k}{k!} e^{-\lambda t} \tag{7.5}$$

where λ is the mean arrival rate. The probability that the duration of the call h is greater than t is:

$$p(h > t) = e^{-\mu t} \tag{7.6}$$

where μ is the mean departure rate (or service rate), and $h = 1/\mu$. Then, the ratio $\lambda/\mu = \lambda h = A$ Erlangs. Because of the steady-state assumption, $p(k, t) = p(k)$; and the probability of k out of N independent inlets being busy is given by (for $N \rightarrow$ large)

$$P(k) = \frac{A^k}{k!} e^{-A} \tag{7.7}$$

Consider now an $N \times M$ rectangular switching matrix, as shown in Fig. 7.5, where the traffic of A Erlangs is given to N inlets and there are M outlets (trunks). The probability that all trunks are busy (i.e., time congestion $\simeq$ the proportion of calls arising that do not find a free outlet) is given by (for $N \rightarrow$ large):

$$B_T = \sum_{x=M}^{\infty} \frac{A^x}{x!} e^{-A} \tag{7.8}$$

B_T is now the blocking probability in the network and is based on the assumption that the blocked calls are held and continue to demand service for a period of average holding time. On the other hand, Erlang and others assumed that the blocked calls are lost (with zero holding time), and this led to the well-known distribution,

$$P(k) = \frac{A^k/k!}{\sum_{x=0}^{M} A^x/x!} \qquad (7.9)^*$$

The blocking of the network occurs when all M trunks are busy, and the probability of blocking (also known is Grade of Service) is given by the well-known Erlang-B equation:

$$B = P(M) \frac{A^M/M!}{\sum_{x=0}^{M} A^x/x!} \qquad (7.10)$$

The lost traffic is then equal to AB Erlangs. The eqn. (7.10) gives the minimum number of trunks M that will be required to give a required grade of service B. In Table 7.2, some sample values of M are given for the offered traffic A and the objective value of B.

Table 7.2 Number of Trunks M Required to Give a Grade of Service B for an Offered Traffic A Erlangs

	Required values of B							
A	0.1	0.05	0.02	0.01	0.005	0.002	10^{-3}	10^{-4}
1	3	3	4	5	5	6	6	7
2	4	5	6	6	7	8	8	10
3	5	6	7	8	9	10	10	12
5	7	9	10	11	12	13	13	15
7	9	11	12	14	15	16	17	19
10	12	14	16	17	18	20	21	24
12	14	16	18	20	21	23	24	27
15	17	20	22	23	25	26	28	31

The eqn. (7.10) has been obtained with assumption that the number of inlets N originating the traffic is large; if however, N is not large and the blocked calls are lost, then the modified blocking probability B is given by Engset equation:

$$B = P(M) = \frac{\binom{N}{M} \alpha_m^M}{\sum_{x=0}^{M} \binom{N}{x} \alpha_m^x}, \qquad (7.11)$$

*The eqn. (7.9) is on the basis of the equilibrium of state of the network, where it may be shown that (refer eqn. (7.7)):

$$P(k) = \frac{A^k}{k!} P(0)$$

Since $\sum_{k=0}^{M} P(k) = \sum_{k=0}^{M} \frac{A^x}{x!} P(0) = 1$; then $P(0) = 1 \Big/ \sum_{x=0}^{M} \frac{A^x}{x!}$

where α_m is the probability that a new call arises from a free inlet, i.e.,

$$\alpha_m = \frac{A}{N - A} = \frac{\alpha}{1 - \alpha},$$

where α is the average traffic per inlet. For $N \to$ large, eqn. (7.11) tends to eqn. (7.10). If $N = M$, then eqn. (7.11) becomes the Binomial distribution.

From eqn. (7.10) and Table 7.2, the average traffic β carried by each trunk for a given B as well as the effect of overload on the grade of service may be calculated. It will then be seen that a large group of trunks is needed to obtain a trunk efficiency of, say, 0.8 Erlang per trunk. For example, for $B = 0.01$, $M = 100$ with $\beta = 0.84$; and $M = 200$ with $\beta = 0.9$. Further assume that 200 inlets, generating 10 Erlangs of traffic, are connected to an $N \times M$ cross-point switch ($N = 200$). Then for $B = 0.01$, $M = 17$ and the resulting matrix size is 200×17, i.e., 17 cross-points per inlet. This is rather uneconomical, as has been already discussed in connection with eqns. (7.2) and (7.3). The number of cross-points can be reduced only by splitting the inlets into smaller groups and interconnecting the groups in the form of 2-stage or 3-stage switches. The blocking probability B has to be now recalculated as follows.

Consider the three-stage SD switch, shown in Fig. 7.6, where individual switch groups are non-blocking (this is so in all digital switches, unless a switch is used as a concentrator). If the number of intermediate arrays $k < (2n - 1)$, then there is internal blocking (also known as the mismatch loss) in the network. It is generally assumed that conditional selection of the interstage links is adopted (meaning thereby that an attempt is made to select an interstage link which has access to a free outlet in the third switching stage) and busy conditions on the input and output links of the middle S-stage are independent of each other. If $P_1(x)$ is the probability that x (out of k) output links are idle and $P_2(x)$ is the probability that the particular x input links are busy, then the probability of blocking is given by

$$B = \sum_{x=0}^{k} P_1(x) \cdot P_2(x), \tag{7.12}$$

since blocking occurs when k idle output links find no idle input links corresponding to them. $P_1(x)$ is given by eqn. 7.9 and is written as:

$$P_1(x) = \frac{A^{k-x}/(k-x)!}{\sum_{y=0}^{k} A^y/y!}$$

$P_2(x)$ is obtained through the Bernoulli distribution and is written as:

$$P_2(x) = \frac{\alpha^x n!(k-x)!}{k!(n-x)!}$$

where α is the probability that an inlet is busy, and n is the number of inlets in each array. Combining these, the blocking probability is calculated as:

$$B_3(A, \alpha, k, n) = \left(\frac{A^k/k!}{\sum_{y=0}^{k} (A^y/y!)} \right) \Bigg/ \left(\frac{(A/\alpha)^n/n!}{\sum_{y=0}^{n} (A/\alpha)^y/y!} \right)$$

$$= B_1(A, k)/B_1(A/\alpha, n) \tag{7.13}$$

where $B_3(\ )$ indicates the blocking probability of a 3-stage switch and $B_1(\ )$ is from eqn. (7.10). The calculations of B for higher-order switches using the above technique become extremely complicated and the alternative approach using probability graphs gives a better intuitive understanding of complicated link systems.

Consider again the 3-stage SD-switch, shown in Fig. 7.6. The probability linear-graph of the network, as suggested by Lee [7], is shown in Fig. 7.13, where the link occupancy is β and inlet occupancy is α. Then, with the traffic applied to the network as A Erlangs,

$$\beta = \frac{n\alpha}{k} = \frac{nA}{kN} \tag{7.14}^*$$

Since two links are necessary to complete a connection in Fig. 7.13, the blocking probability of the network is now seen to be,

$$B_3 = [1 - (1 - \beta)^2]^k \tag{7.15}$$

where k is the number of switches in the second S-stage. The factor $\gamma = k/n$ is known as the space expansion factor; and $\gamma > 1$ when α is large, say $\alpha > 0.5$, for incoming trunks connected to tandem or toll exchanges, but $\gamma < 1$ for ordinary subscribers connected to PBX or end-office switches. For concentrators, $\gamma < 1$, leading to a large saving of hardware in the switch.

Equation (7.15) is based on the assumption that the blocking probabilities $= [1 - (1 - \beta)^2]$ of individual paths are independent of each other, but this is not strictly true, specially with $\gamma > 1$, (as it is known that with

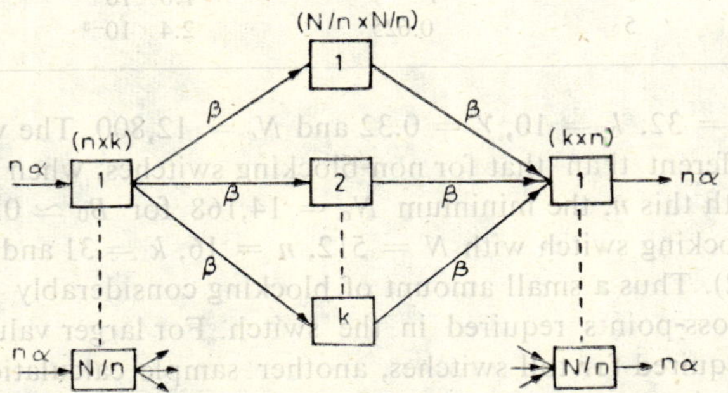

Fig. 7.13 Probability graph of a 3-stage SD-switch

*Using a PDC as an input to the network, the traffic handled by the switch at any time is the traffic offered in a particular slot, and α is the occupancy of this slot.

$k = 2n - 1$, $B_3 = 0$, but eqn. (7.15) gives a finite value of B_3). A more accurate equation for the blocking probability is given by [7],

$$B_3 = \frac{(n!)^2}{k!(2n - k)!} \alpha^k (2 - \alpha)^{2n-k} \tag{7.16}$$

Comparing eqns. (7.15) and (7.16), it will be seen that for $\gamma \leqslant 1$, the two results are almost the same, but for $\gamma > 1$, the eqn. (7.15) gives larger values of B. However, we shall use eqn. (7.15) for further discussions.

The optimum design of a switch requires that the minimum number of cross-points N_c given by,

$$N_c = 2kN + k(N/n)^2 \tag{7.17}$$

has to be obtained, subject to the constraint that the blocking specification B_0 is maintained. This may be easily done by using the method of Lagrangian multiplier and the required values of n, k, γ and N_c calculated. A sample calculation for a 3-stage switch with $N = 512$, $\alpha = 0.1$ and $A = 51$ Erlangs, is shown in Table 7.3, where it is seen that for $B_0 \simeq 0.002$, the optimum

Table 7.3 Values of k, γ, B and N_c for a Three-Stage SD-Switch with $N = 512$, $\alpha = 0.1$, $A = 51$ Erlangs

n	k	γ	B	N_c
64	16	0.25	7.9×10^{-4}	17408
64	12	0.19	0.05	13056
32	32	1	8.3×10^{-24}	40960
32	16	0,5	7.95×10^{-8}	20480
32	10	0.32	2.1×10^{-3}	12800
32	8	0.25	0.028	10240
16	16	1	2.9×10^{-12}	32384
16	7	0.44	1.9×10^{-3}	14168
16	6	0.375	0.01	12144
8	8	1	1.6×10^{-8}	40960
8	5	0.625	2.4×10^{-3}	25600

values are: $n = 32$, $k = 10$, $\gamma = 0.32$ and $N_c = 12{,}800$. The value of n is somewhat different than that for non-blocking switches, when $n = \sqrt{N/2} = 16$; and with this n, the minimum $N_c = 14{,}168$ for $B_0 \simeq 0.002$. For a purely non-blocking switch with $N = 512$, $n = 16$, $k = 31$ and $N_c = 63{,}488$ from eqn. (7.2). Thus a small amount of blocking considerably reduces the number of cross-points required in the switch. For larger values of α, say $\alpha = 0.8$, as required for toll switches, another sample calculation is shown in Table 7.4. It is now seen that for $n = 16$, $k = 24$, $\gamma = 1.5$, $B_0 = 2.6 \times 10^{-3}$, $N_c = 49{,}152$ gives the optimum value. For larger switches, the values of B and N_c may be calculated as shown above and the optimum parameters chosen. For example, for $N = 32{,}768$, $n = 128$, $B_0 = 0.002$, the values of $k = 24$ and $N_c = 3.1 \times 10^6$ for $\alpha = 0.1$; and for $\alpha = 0.7$, $k = 116$, and

$N_c = 15.2 \times 10^6$. Thus the number of cross-points required in a large exchange using 3-stage SD switches is prohibitively large. This can be minimized only by using higher-order switches, e.g., 5-stage and 7-stage switches. However, composite T-S-T and STS switches are more economical as discussed further.

Table 7.4 Values of k, γ, B and N_c for a Three-Stage SD-Switch with $N = 512$, $\alpha = 0.8$, $A = 408$ Erlangs

n	k	γ	B	N_c
32	32	1	0.27	40960
32	41	1.28	1.96×10^{-3}	52480
32	48	1.50	6.6×10^{-6}	61440
16	20	1.25	0.06	40960
16	24	1.50	2.6×10^{-3}	49152
16	32	2	6.3×10^{-7}	65536
8	12	1.50	0.05	61440
8	16	2	7.9×10^{-4}	81920

Consider now a three-stage S-T-S switch as shown in Fig. 7.8. Assume that the SD-switches are single stage and all switches including the TSI are non-blocking. Then referring to Fig. 7.4(b), the S-T-S switch is equivalent to a three-stage SD-switch as shown in Fig. 7.14(a), and its probability

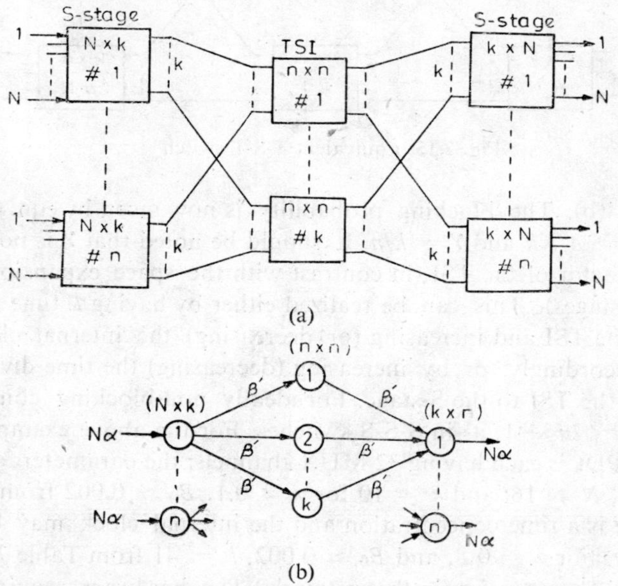

Fig. 7.14 Equivalent S-T-S switch and its probability graph (a) An S-S-S switch equivalent to the S-T-S switch of Fig. 7.8. (b) Probability graph of (a)

linear-graph is shown in Fig. 7.14(b). The eqns. (7.14) and (7.15) now hold for the S-T-S network with the modification

$$\beta' = \frac{N\alpha}{k} = \frac{A}{nk}; \quad \gamma = k/N;$$

where, n = number of time slots in PDC and N = number of PDC's as input. The blocking probability of the network is (in any slot duration):

$$B_3' = [1 - (1 - \beta')^2]^k \tag{7.18}$$

The number of TSI's required is now given by k for a desired B_3'. As an example, for an S-T-S network with an input of 16 PDC's, each having 32 multiplexed channels, the value of $k = 7$ from Table 7.3, for $\alpha = 0.1$ and $B_0 \simeq 0.002$. This gives the number of TSI = 7, number of memory cells = 224 and the number of cross-points = 224 only. For a larger value of α, say 0.8, the $k = 24$ for $B_0 = 0.002$ from Table 7.4, and for an ideally non-blocking network, $k = 31$.

The T-S-T network of Fig. 7.7 may also be represented as an S-S-S network as shown in Fig. 7.15, and its probability linear-graph is the same as

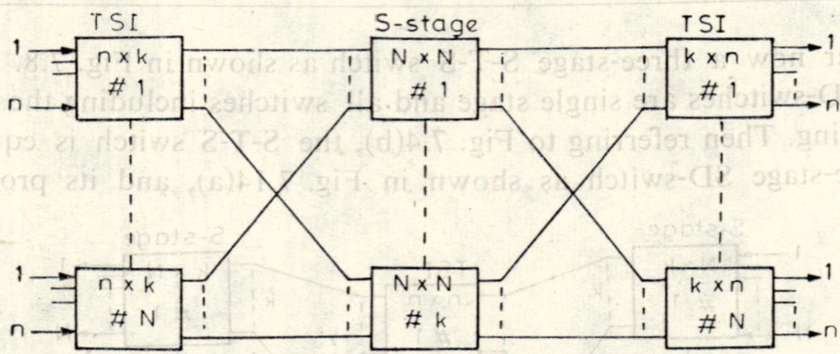

Fig. 7.15 Equivalent T-S-T switch

in Fig. 7.14(b). The blocking probability is now given by eqn. (7.18) with $\beta' = n\alpha/k = A/Nk$ and $\gamma = k/n$. It should be noted that γ is now the time expansion factor of the TSI (in contrast with the space expansion factor γ used in S-stages). This can be realized either by having k time slots at the output of the TSI and increasing (or decreasing) the internal clock of the network accordingly; or by increasing (decreasing) the time-division buses connecting the TSI to the S-stage. For ideally non-blocking condition, the value of $k = 2n - 1$, as in S-S-S switches. For the above example, with an inputof 16 PDC's each having 32 MUX channels, the parameters of Fig. 7.15 are $n = 32$, $N = 16$; and $k = 10$ for $\alpha = 0.1$, $B_0 \simeq 0.002$ from Table 7.3. Thus, there is a time concentration and the internal clock may be reduced accordingly. For $\alpha = 0.8$, and $B_0 \simeq 0.002$, $k = 41$ from Table 7.4 and this requires a time expansion in the network. The hardware required now is: No. of TSI = 32, No. of memory cells = $2Nk$ = 1312 and No. of cross-points = 256.

For larger networks, the S-stage of Fig. 7.15 may be expanded to three or four S-stages as shown in Fig. 7.6. Such a five-stage TSSST switch is shown in Fig. 7.16(a), where N PDC buses are interconnected through a 3-stage SD-switch. The sizes of the matrices, given by k_1, k_2, may be calculated on the basis of Tables 7.3 and 7.4. The probability graph of the switch is shown in Fig. 7.16(b) where,

$$\beta_1 = \frac{n\alpha}{k} = \frac{A}{Nk}$$

$$\beta_2 = \frac{\beta_1 k_1}{k_2} = \frac{Ak_1}{Nkk_2} \tag{7.19}$$

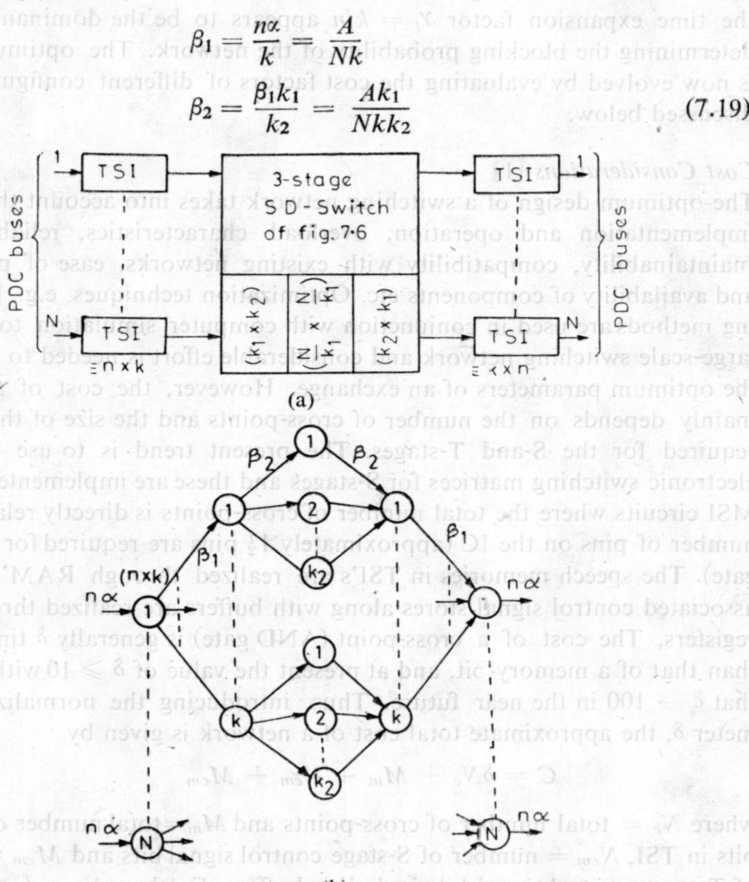

Fig. 7.16 A 5-stage TSSST switch and its probability graph:
(a) TSSST switch, (b) Probability graph of (a)

and k/n = time expansion factor γ_t; k_2/k_1 = space expansion factor γ_s. The graph is also functionally identical to the probability graph of a five-stage SD-switch. The blocking probability of the network is now calculated from the probability graph and is given by:

$$B_5 = \{1 - (1 - \beta_1)^2[1 - \{1 - (1 - \beta_2)^2\}^{k_2}\}^k \tag{7.20}$$

For an approximate design of the switch, it may be assumed that the TSI is non-blocking with $\gamma_t > 1$, and $k_1 \simeq \sqrt{N/2}$ as used in eqn. (7.3). Then the value of k_2 (= number of center stage arrays) is calculated using eqn. (7.20) for the required B_5. As for example, for an input of 512 PDC's each having 32 MUX channels and $\alpha = 0.1$, $n = 32$ and $k \simeq 16$ gives $B_3 \simeq 10^{-7}$ (from Table 7.3). Thus, the blocking probability B_5 is now mainly given by the three S-stages in the network. Using $k = 16$, $k_1 = 16$ and $k_2 = 8$, $B_5 = 1.7\times10^{-7}$. But for $\alpha = 0.8$, using $k = 40$, $k_1 = 16$ and $k_2 = 24$, $B_5 \simeq 4\times10^{-3}$; and using $k = 48$, $k_1 = 16$, $k_2 = 24$, $B_5 \simeq 5\times10^{-6}$. Thus, the time expansion factor $\gamma_t = k/n$ appears to be the dominant factor in determining the blocking probability of the network. The optimum design is now evolved by evaluating the cost factors of different configurations as discussed below.

Cost Considerations [1]

The optimum design of a switching network takes into account the cost of implementation and operation, overload characteristics, reliability and maintainability, compatibility with existing networks, ease of production and availability of components etc. Optimization techniques, e.g., hill-climbing methods are used in conjunction with computer simulation to design a large-scale switching network and considerable effort is needed to determine the optimum parameters of an exchange. However, the cost of a network mainly depends on the number of cross-points and the size of the memory required for the S-and T-stages. The present trend is to use solid-state electronic switching matrices for S-stages and these are implemented through MSI circuits where the total number of cross-points is directly related to the number of pins on the IC (approximately $1\frac{1}{2}$ pins are required for one AND gate). The speech memories in TSI's are realized through RAM's and the associated control signal stores along with buffers are realized through shift registers. The cost of a cross-point (AND gate) is generally δ times higher than that of a memory bit, and at present the value of $\delta \geqslant 10$ with the trend that $\delta \to 100$ in the near future. Thus, introducing the normalizing parameter δ, the approximate total cost of a network is given by

$$C = \delta N_c + M_m + N_{cm} + M_{cm} \tag{7.21}$$

where N_c = total number of cross-points and M_m=total number of memory bits in TSI, N_{cm} = number of S-stage control signal bits and M_{cm} = number of T-stage control signal bits including buffers. Further, $N_c = (n\cdot k)$ for an $(n\times k)$ matrix; $M_m = 2N\cdot n\cdot b.$, where N = number of PDC's, n = number of slots in a frame and b = number of bits (normally 8) in a PCM word, $N_{cm} = N(n+1)\cdot\log_2(N+1)$; $M_{cm} = (n+1)\cdot N\log_2(n+1)$. While designing a switching network, the cost of implementation has to be minimized, keeping in view that the required blocking probability B_0 is satisfied. The typical requirements of B_0 are: for a local network with $0.1 < \alpha < 0.4$, $B_0 \leqslant 0.01$; for a toll switching network with $\alpha \leqslant 0.7$, $B_0 \leqslant 0.005$; and for $\alpha \simeq 0.9$, $B_0 \leqslant 0.1$. With this formulation of the cost function C, some simple networks discussed earlier, are now re-evaluated.

Consider the problem of designing an SD-switch with the number of incoming trunks $N_t = 512$, $\alpha = 0.7$, $A = 358.4$ Erlangs, and $B_0 \leqslant 5.10^{-3}$.

For a 3-stage SD switch of Fig. 7.6, eqns. (7.15) and (7.17) give the network parameters as:

$$n = 16, k = 22 \quad \text{and} \quad N_c = 2kN_t + k(N_t/n)^2 = 45{,}056$$

Thus, the cost of the switch

$$C = (45{,}056)\delta. \tag{7.22}$$

To accommodate the same 512 trunks in a digital exchange, 16 PDC's may be used giving $n = 32$, $N = 16$ and $N_t = nN$. The switch may now be designed either as an S-T-S or a T-S-T network. For the S-T-S network of Fig. 7.8. eqn. (7.18) gives the parameter values as: number of TSI $= k = 22$ for $B_0 \leqslant 5.10^{-3}$; and $N_c = 2Nk = 704$, $M_m = 2knb = 11{,}264$, $N_{cm} = 2N(n+1)\log_2(N+1) = 5280$, $M_{cm} = k(n+1)\log_2(n+1) = 4356$. Thus the cost is:

$$C = (704)\delta + 20{,}900 \tag{7.23}$$

For the equivalent T-S-T network of Fig. 7.7, $k = 36$ for $B_0 \leqslant 5.10^{-3}$; and $N_c = N^2 = 256$, $M_m = 2Nnb + 2Nkb = 17{,}408$, $N_{cm} = N(k+1)\log_2(N+1) = 2960$, $M_{cm} = N(n+1)\log_2(n+1) + N(k+1)\log_2(k+1) = 6720$. Then the cost is:

$$C = (256)\delta + 27{,}088 \tag{7.24}$$

For the same input traffic, a higher order PDC may be used, say, $n = 128$, and $N = 4$, with $N_t = 512$, $\alpha = 0.7$. Then for the T-S-T network, $k = 128$ gives $B_0 = 6.10^{-6}$, and k may be further reduced. However, with $\gamma_t = 1$, $k = 128$, the network parameters are:

$$N_c = N^2 = 16, M_m = 4Nnb = 16{,}384$$

$$N_{cm} = N(n+1)\log_2(N+1) = 1548,$$

$$M_{cm} = 2N(n+1)\log_2(n+1) = 8192,$$

Then the cost is:

$$C = (16)\delta + 26{,}124 \tag{7.25}$$

Comparing eqns. (7.22) to (7.25), it is observed that a T-S-T or an S-T-S network is more economical than the purely SD network. For the T-S-T networks, which are the favoured configurations for large switches, a higher-order PDC gives further economy in the network and this is more evident for larger switches.

As an example, consider the design of a T-S-T network for $N_t = 16{,}384$, $\alpha = 0.7$, and $B_0 \leqslant 10^{-3}$. Then each PDC may have $n = 32$, 128 or 512, and the corresponding $N = 512$, 128 or 32. Using these combinations, the required number of cross-points and memories are:

(a) For $n = 32$, $N = 512$, $k = 36$; $N_c = 262{,}144$, $M_m = 557{,}056$; $N_{cm} = 189{,}440$, $M_{cm} = 215{,}040$;

and the cost $C = (262{,}144)\delta + 961{,}536 \simeq 3.6 \times 10^6$ for $\delta = 10$.

(b) For $n = 128$, $N = 128$, $k = 128$ giving $B_0 = 6.10^{-6}$; then
$N_c = 16{,}384$, $M_m = 524{,}288$, $N_{cm} = 132{,}096$, $M_{cm} = 264{,}192$;
and $C = (16{,}384)\delta + 920{,}576 \simeq 1.08 \times 10^6$ for $\delta = 10$

(c) For $n = 512$, $N = 32$, $k = 512$ giving $B_0 = 10^{-21}$; then,
$N_c = 1024$, $M_m = 524{,}288$, $N_{cm} = 98{,}496$, $M_{cm} = 328{,}320$;
and $C = (1024)\delta + 951.104 \simeq 0.96 \times 10^6$ for $\delta = 10$.

It is now seen that the cost decreases considerably with higher order PDC's in a T-S-T switch, and the only limitation for cost reduction is the available speed in the TSI memories. For still larger networks, the cost is reduced by approximately a factor of 4 when n in a PDC increases from 32 to 512. For very large networks, the TSSST or TSSSST configurations are more economical.

7.1.5 SPC Processor [8, 9]

The control system of an automatic exchange has the following tasks to fulfil:

(a) Scanning of input lines/trunks to determine their status, i.e., whether a call is initiated, is in progress or is in the process of termination.
(b) Detection of the changes in the status of lines/trunks.
(c) Signal processing; to send dial tone, to receive dialled pulses, and transmit them to the main controller through a register/sender circuit.
(d) Translation of the dialled pulses to a suitable routing signal and map a suitable call path.
(e) Controlling the switching network through a network controller; to test the called line, send the ringing tone and ring-back signals, and then to set-up the call path.
(f) Overall coordination.
(g) Maintenance and administration.

In the earlier electromechanical exchanges, like the Crossbar systems, the preliminary tasks, such as (a), (b) and (c), were performed by a line circuit and the major tasks, (d) to (g), by a common control unit called a marker. The inputs to the controller and the consequent responses are mostly of a routine nature which may be organized through a properly designed sequential machine, which in turn, may be replaced by a special-purpose computer.

Consider the simple problem of scanning a subscriber's line and returning the dial tone when the line is calling and then connecting a register/sender for storing the dialled pulses. This may be executed through a combinatorial switching circuit whose input is the status of the line and output is the command to send the dial tone and at the same time to access a register/sender. By changing the internal state of the switching circuit, a command may also be generated to further process the dialled digits. The switching circuit now behaves as a sequential machine and it will have input/output interfaces, an internal switching circuit, read/write memories and buffers. If the machine is synchronous and fast enough using electronic components, then the decision-making logic (i.e., the combinational switching circuit) may be time-shared between many inputs/outputs, as shown in Fig. 7.17. It is possible to include more input/output functions in this sequential machine, leading to the common control technique which is popularly used in crossbar exchanges.

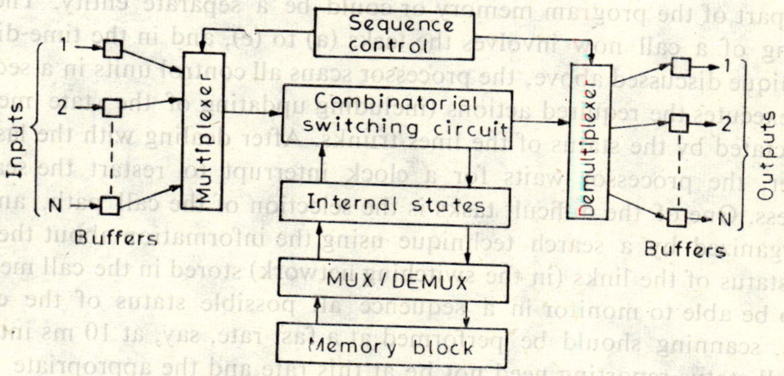

Fig. 7.17 A time-shared controller

The decision-making logic of the above common controller can also be realized by a general-purpose computer using a lookup table containing entries indexed by the new input and the old state. Using only ROM's, containing the lookup table, new decisions may be obtained whenever a new input arrives. Further, in most cases, the input/output entries in the lookup table are governed by some set of rules which may be used to calculate the new output and new state for the input and old state. This is easily organized in a computer and the lookup table along with the special-purpose decision logic need not be used. A computer may now be programmed to test the conditions of the inputs and the old states and to decide on new outputs and new states. This is the basic principle for the stored-program control (SPC) of switching systems, as shown in Fig. 7.18.

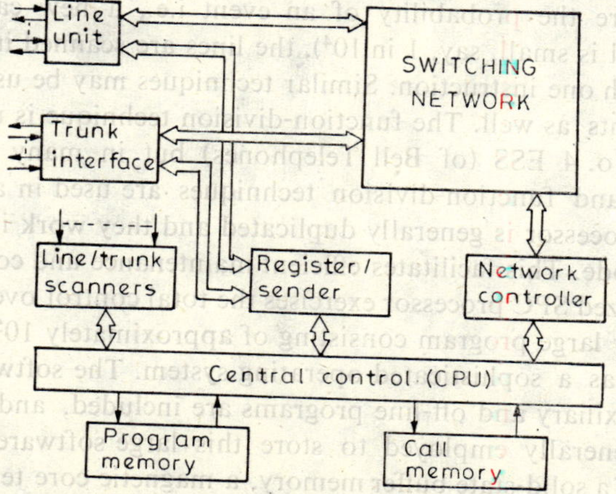

Fig. 7.18 Stored program control of an exchange

All the relevant tasks, (a) to (g), are now controlled by the central computer which functions as a time-shared decision maker and the decisions are expressed as programs. The computer memory is mainly divided into two areas: (a) program memory (for the sequence of instructions or lookup tables) and (b) call memory (for transient data, network map etc.). The actual dialled pulses are processed and stored in a translation memory, which may be a part of the program memory or could be a separate entity. The processing of a call now involves the tasks (a) to (e), and in the time-division technique discussed above, the processor scans all control units in a sequence and executes the required actions (including updating of the state memory) as dictated by the status of the lines/trunks. After dealing with the last controller, the processor waits for a clock interrupt to restart the scanning process. One of the difficult tasks is the selection of the call path, and this is organized by a search technique using the information about the busy/free-status of the links (in the switching network) stored in the call memory.

To be able to monitor in a sequence all possible status of the control units, scanning should be performed at a fast rate, say, at 10 ms intervals. But all status-reporting need not be at this rate and the appropriate scanning rates are as follows:

Line scanning—100 ms.
Call set-up (whenever a new call is detected)—100 ms.
Initial path set-up—20 ms.
Dialling—10 ms.
Final path set-up—20 ms.
Supervision—100 ms.
Path clear down—20 ms.

It is seen now that the real-time occupancy of the processor can be improved considerably if the program is split into several modules and each process is activated at the rates given as above. The processor now operates on a Function-division basis, and the 'executive' program uses a sub-program to schedule the various processes at appropriate rates. To save time on line scanning (where the probability of an event i.e., a new call arising in a 100 ms interval is small, say, 1 in 10^4), the lines are scanned in parallel, say, 16/32 lines with one instruction. Similar techniques may be used for detecting other events as well. The function-division technique is used in the 1A processor of No. 4 ESS (of Bell Telephones) but in many systems, both time-division and function-division techniques are used in a suitable mix. The central processor is generally duplicated and they work in a synchronous duplex mode. This facilitates efficient maintenance and coordination.

The centralized SPC processor exercises the total control over the ESS and requires a very large program consisting of approximately 10^5 basic instructions as well as a sophisticated operating system. The software may reach 10^6 steps if auxiliary and off-line programs are included, and a hierarchial memory is generally employed to store this large software. The memory may consist of a solid-state buffer memory, a magnetic core temporary store, a ROM, and a mass storage employing discs, drums or tapes. The large software also presents programming difficulties, and as such, the programming language has been modified and improved. Thus, the earlier assembly language was replaced by macro-instructions and then, high-level languages along with modular/structured programming techniques were introduced. All these have led to improved software productivity, sharing of programs, modular replacement of routines and easier maintenance.

The large centralized SPC processor was made economically viable by removing as many logic functions as possible from the network-elements and concentrating them in the stored logic within the central processor. But this resulted in an increase in the complexity of the control logic and a significant percentage of processing power was needed to perform relatively simple routine functions, thereby limiting the Throughput of the processor. Further, there was a necessary interdependence between the hardware and software logic and this required extensive modification of the software when a small change in the hardware was introduced. In the meantime, the general trends in computer and semiconductor development led to successful decentralized (distributed) computer systems and this trend was adopted for the SPC processor as well. Specifically, the input-output jobs dealing with incoming and outgoing lines/trunks are separated and carried out by a front-end processor (also known as the signal processor (SP)) and the capacity of the switching system is almost doubled (as in No. 1 ESS). The other trend is the use of a number of processors, either in function-sharing or in load-sharing mode, rather than having a very large single processor. (In some ESS, e.g., in NEAX61, eleven processors in load-sharing mode are used.) This reduces the increasing overhead of very large computers and the hardware cost as the minicomputers are becoming cheaper. The other significant advantage is that in a multiprocessor configuration, a partial duplication satisfies the reliability requirements, whereas in a single processor configuration, the entire processor has to be duplicated which increases the overall cost.

The front-end processors (SP) are usually used to carry out the simple and time consuming jobs, viz., (a), (b) and (c) in the list of tasks given earlier. This reduces the complexity of the main processor software and its processing load. There have been two trends in the development of the pre-processors, viz., (a) functional pre-processing and (b) module pre-processing. In functional pre-processing, a peripheral controller scans the line/trunk modules for detection of changes in their status; a second controller performs the signalling functions (register/sender) and a third acts as a marker over the network modules. All the three controllers perform for the entire system and are directed by the central controller. In module pre-processing, a general controller is assigned to each individual module in the system, as shown in Fig. 7.19, and the controller performs all the pre-processing necessary in the module. Although a combination of both methods may be used, module pre-processing is economically attractive even for small modules and the use of standardized devices such as microprocessors and

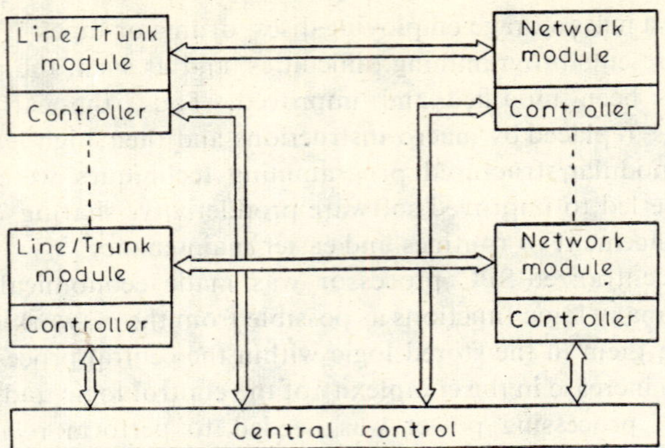

Fig. 7.19 Module pre-processing in ESS

semiconductor memories gives the modules more intelligence and self-sufficiency with clearly defined interfaces to the rest of the system. The distributed pre-processors perform most of the system's stringent real-time tasks which are logically simple, and each has the responsibility over only a limited portion of the system; thus, the pre-processor software is of relatively simple structures. At the same time, the main processor which has the responsibility of executing all major switching and control tasks, is relieved of stringent real-time tasks and fast response actions. This increases its throughput as well as the software efficiency.

The concept of module pre-processing is further extended to the distributed processing technique, where the control modules along with their software are distributed throughout the network hierarchy, including the line/trunk terminal modules (e.g., in ITT-1240 exchanges). These systems with distributed control do not use a central controller any more and neither do they use special data links between the controllers. With the availability of efficient 16-bit microcomputers, it has been possible to provide enough intelligence to the module controllers, so that the decisions for call processing can be taken at the successive modules as the call progresses through the switching network (the switching network is also modularized in this case). This results in sophistication of the control software using the concepts of virtual machines, finite message machines (FMM) and generic interfaces; and leads to the design of 'future-safe' machines, whose useful life is extended in spite of the continuous changes in module-hardware and software.

7.2 LINE AND TRUNK INTERFACES [3, 10]

The traffic carried by subscribers' lines is usually light and for loading the PDC and the network efficiently, it is necessary to concentrate the lines at the peripheral interface. This also facilitates the use of a concentrator closer to the customers and such remote concentrators considerably increase the area served by an exchange. But for heavily loaded trunk lines, the PDC is formed by multiplexing and digitization of inputs without concentration.

However, the signalling and supervisory functions to be carried out at the interface create certain design problems.

The primary functions of the line/trunk interface are (refer to Fig. 7.2):

(a) to convert analog signals to multiplexed digital PDC's
(b) to multiplex digital trunks to PDC's
(c) to provide the functions known by the acronym 'BORSCHT'.

Figure 7.20 shows a typical BORSCHT circuit in which the functions are spelt out. Since the currently used subscriber sets require high voltages for

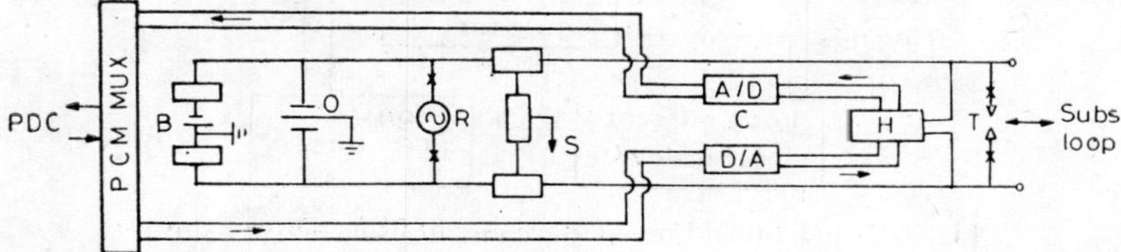

Fig. 7.20 Typical 'BORSCHT' circuit; B: battery feed; O: Overvoltage protection; R: ringing; S: Supervision; C: coding; H: hybrid; T : testing

operation of the *B*, *R* and *T* circuits, these circuits cannot be concentrated in common units through readily available low-voltage electronic cross-points and would require specially designed high-voltage semiconductor cross-points for a completely electronic concentrator. As such in the line interfaces using concentration, the functions are implemented in two parts, (a) individual line circuits (LC) with high-power and voltage requirements and (b) a common unit (CU) with low-power requirements, as shown in Fig. 7.21. The concentrator preferably uses an SD-switch and the number of

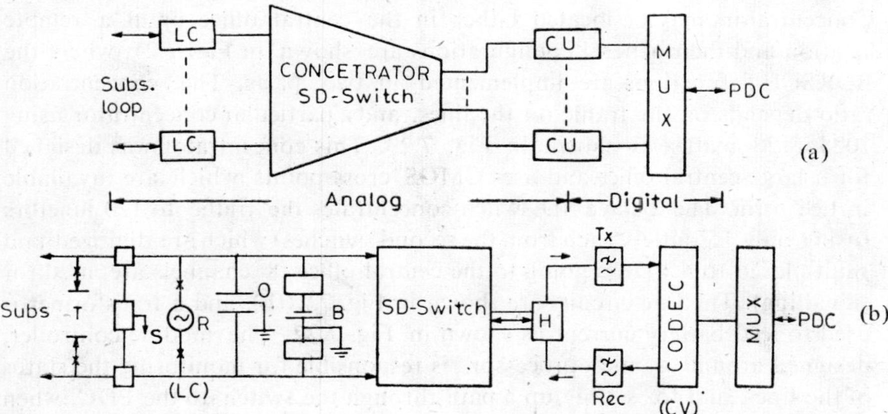

Fig. 7.21 Line interface configuration using concentrators: (a) Block diagram, (b) Division of 'BORSCHT' circuits

CU required is equal to the number of time slots in the PDC. Although this hybrid interface is cost effective, the general trend is to have an independent LC for each line input and to meet the cost objectives through the use of custom-made LSI chips, e.g., active hybrids, single-chip codecs and filters. A complete LC interface is shown in Fig. 7.22, where the electronic

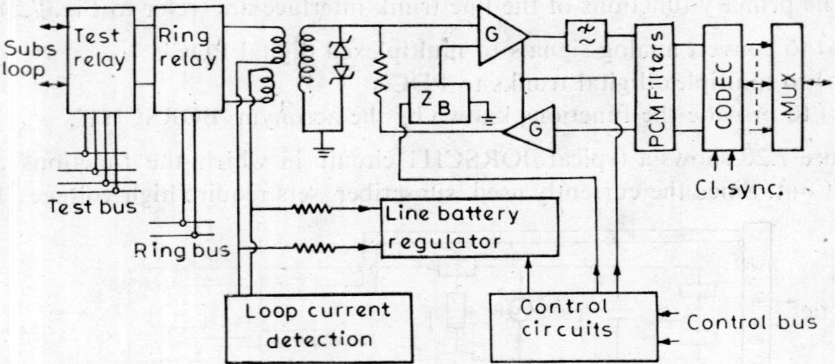

Fig. 7.22 Individual line interface circuit using active hybrid

hybrid is formed by the two amplifiers G and the balance network Z_B. A balanced transformer is used for line feed and isolation, and the battery feed is provided through a programmable DC/DC converter (to provide different DC feed for on-hook, off-hook and permanent signal conditions). The controller also sets the gain of G and selects a particular value of Z_B to match the subscriber's line [9]. Since the cost of peripheral interfaces constitutes 70 to 80 per cent of the total system cost, it is essential that the designing of the LC's should be economical. It is expected that production in large volumes will reduce the cost rapidly.

7.2.1 Concentrators [11]

Concentrators may be located either in the central office or at a remote location and their general configurations are shown in Fig. 7.21, where the BORSCHT functions are implemented in two parts. The concentration ratio depends on the traffic on the lines, and a particular concentrator using 1024×128 matrix is shown in Fig. 7.23. This concentrator was designed for a large central office and uses CMOS cross-points which are available in LSI form. The 2-stage SD-switch concentrates the traffic to 120 junctors (using only 15 outlets each from the second switches) which are digitized and multiplexed to 4 PDC's going to the central office (8 channels are used for signalling). The line circuits are shown in Fig. 7.21(b), and a transformer is used to feed battery current as shown in Fig. 7.22. The module controller, designed around a microprocessor, is responsible for monitoring the status of the lines, and for setting up a path through the switch to the PDC, when called for. The central controller, however, controls the line circuit directly via appropriate signalling links. The communication link between the local and central controllers carry information at 5 Mb/s and for a local concentrator, two controllers are connected by direct cables. But for a remote concentrator, the information is multiplexed in the signalling slots of the PDC's.

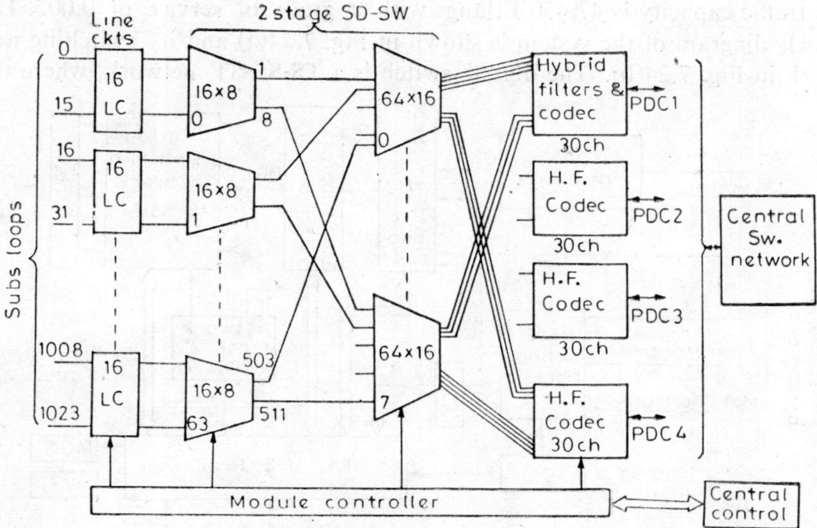

Fig. 7.23 (1024×128) line concentrator

While using a remote concentrator, it is necessary to use a Digital Carrier Interface (DCI) between the concentrator and the central office, and the DCI performs the framing and aligning of the incoming bit stream. Then the switching network makes no distinction between local and remote peripheral interfaces. In some cases, large concentrators are used to replace small exchanges and these are provided with additional features, e.g., local switching and emergency functions (when all the PDC's are disabled). There is a trend for using LSI in the speech path, but completely integrated line circuits are yet to be available in LSI form. As such, the analog concentration, as given by SD-switches, will be providing reliable service at a very low cost till LSI components become economically competitive.

7.3 LARGE SCALE ELECTRONIC SWITCHING SYSTEMS (ESS)

We have discussed so far the principles of electronic switching and SPC processors for large exchanges. Based on these principles and techniques, many large scale ESS's have been designed, installed and operated since the early 70's. Some of the major systems are: No. 4 (and No. 5) ESS (USA), System X (UK), HDX-10 (Hitachi), NEAX-61 (NEC), E-10, E-12 and MT-20 (France), and System-12 (ITT). A few of these are briefly described in the following sections.

7.3.1 No. 4 ESS [12, 13]

No. 4 ESS was developed by Bell Telephone laboratories and installed first in 1976 as a toll and tandem exchange, using a digital time-division switch and 1A SPC processor. It has a maximum capacity of terminating 107,520 analog/digital trunks and can process 550,000 peak switched calls per hour.

Its traffic capacity is 47,450 Erlangs with a grade of service of 0.005. The block diagram of the system is shown in Fig. 7.24(a) and its switching network in Fig. 7.24(b). The digital switch is a TS-SS-ST network, where the

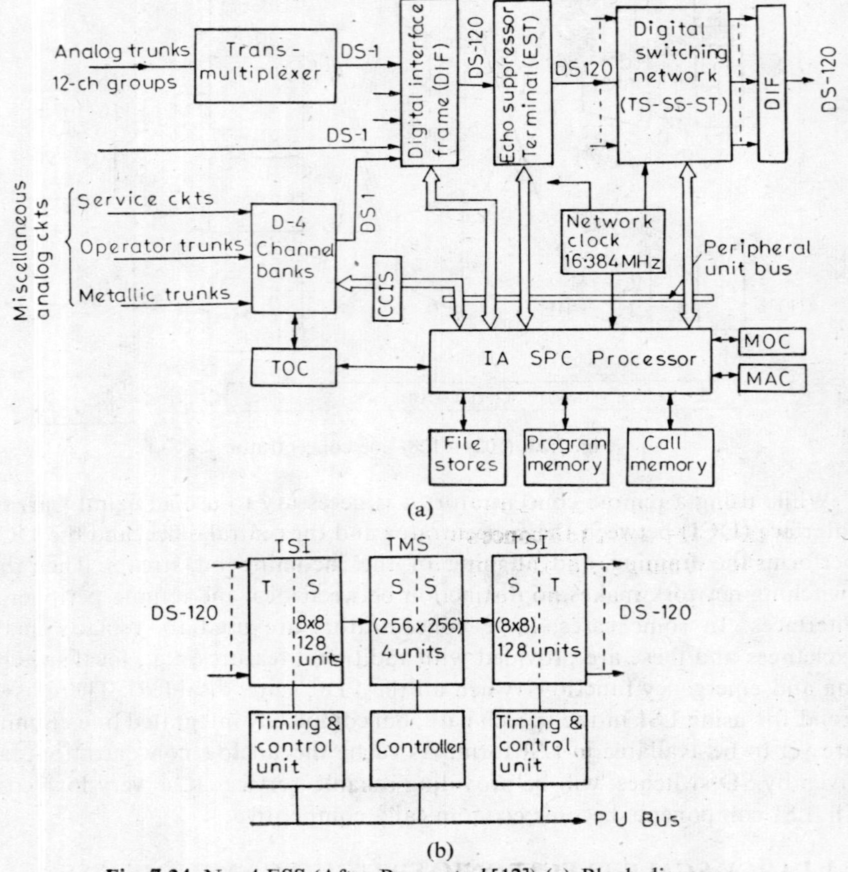

Fig. 7.24 No, 4 ESS (After Bruce *et al* [12]) (a) Block diagram;
TOC : Trunk operating centre
MOC : Machine operating centre
MAC : Machine administrative centre
CCIS : Common channel inter-office signalling
(b) Switching network of ESS

SD-stages have eight TMS frames arranged in four duplicated 2-stage (256 × 256) switch arrays. This provides 1024 unidirectional paths between the TSI-units. Since each DS-120 contains 120 information slots (8 slots are for maintenance functions), it is possible to have 112,880 network paths through the ESS. However, 12 per cent of the interface units are kept as spares, and thus, the system capacity is 107,520 trunks. The rest of the network is fully duplicated, and the TSI and TMS units are arranged in a folded, singlesided configuration.

All analog 12-ch carrier signals are converted to digital 24-channel DS-1 signals through the transmultiplexer, and the signalling information in the form of 2600-Hz tone is converted to the digital format. The central office requires 400 to 500 service circuits and operator trunks. These are digitized through D-type channel banks. All these DS-1 signals along with the digital trunks (T_1) are now terminated on the Digital Interface Frame (DIF) which converts the inputs to DS-120 signals. A fully equipped DIF contains a duplicated controller (μp-based), a duplicated peripheral unit-bus interface, thirty-two working digital interface units (DIU) and two spare DIUs. The DIF can accommodate 160 DS-1 signals from either T_1 facilities or transmultiplexers. The DS-120 signals from the DIF are now passed through the Echo Suppressor Terminal (EST), which can serve up to 1680 trunks on a time-division basis. The EST provides necessary echo suppression on long trunks and executes improved suppression algorithms that result in improved system performance. The processed DS-120 signals are now synchronized (through buffers), de-correlated, time-switched (in TSI) and space-switched, and then sent to TMS frames. After necessary switching in the TMS, the DS-120 signals are again space-switched, time-switched, re-correlated and then sent to the outgoing trunks*. The timing and control units of TSI, under the central control of the processor through the PU bus, control the sequential operation of the gates and memory circuits of the TSI. The processor then instructs the TMS controller regarding the call path to be established through the TMS, and the call is established accordingly. The crystal-controlled network clock provides a 16.384 MHz signal to all timing and control units of TSI and TMS frames.

The 1A SPC processor exercises the overall control of the switching office including maintenance and administration (through MOC and MAC) and is designed to operate at 700 ns cycle time. In the updated version, the program and call memories use 16K MOS-IC memory chips, providing 256K words of storage in a single unit. There are eight call memory units and six program memory units, giving a total of 3.5×10^6 words of memory. The updated generic program requires about 2×10^6 program words. All equipment in the ESS connected to the system buses are monitored, diagnosed and controlled by the central processor. As a result, extensive automatic reports on all aspects of system performance are provided, and the data regarding system down time, number of trunks out of service, number of calls cut-off, ineffective call attempts, interface failures etc., are suitably displayed and recorded for administrative purposes. This resulted in a system performance exceeding the earlier expectations, as shown by the following performance indicators:

---

*In case of heavy traffic or failure (in the switch) of particular multiplexed digital input, all of DS-120 signals will be lost. To avoid this, DS-120 signals are usually de-correlated (i.e., a few DS-120 links redistribute their data randomly among themselves) before the switch, and re-correlated after the switch (i.e., an inverse mapping is done). This spreads the congestion/failure throughout the remainder of the switch.

System Downtime	1.2×10^{-3} min/10^4 calls
No. of calls cutoff	$1/10^4$ calls
Trunk downtime	43 min/trunk/yr
Ineffective machine attempts	less than 0.01%
Meantime to repair (solid faults)	less than 2 hrs

With recent advances in device technology, the frames have been redesigned in the late 70's and substantial economy has been achieved. Solid-state memory devices, MSI/LSI logic and microprocessors have been incorporated wherever appropriate. The low-cost reliable and programmable μp's have replaced the earlier wired-logic peripheral unit controllers. This process of modernization is continuing and it has resulted in No. 5 ESS now introduced in service.

No. 5 ESS is a local office exchange with a subscriber capacity of 12,000 lines. It provides operational savings and ease in system growth through SPC-controlled TDM switching [14]. It has four major subsystems: (a) the central processor, (b) a message switch, (c) a time-multiplexed switch and (d) interface modules. The system uses fibre-optic links as trunk, μp-controlled interface modules and a TST switching network (using high-voltage IC switches in the SD-stages). The SPC is distributed among the central processor and interface modules. The incoming signals are converted to DS-120 formats and the information rate through the fibre-optic links is 32.768 Mb/s. The system hardware uses LSI circuitry including a digital signal processor on-a-chip, modular structures, e.g., Bell-Pac packaged units, and its software uses high-level programming languages. Because of the switching system's flexible digital design, No. 5 ESS is expected to carry forward the evolution to ISDN (Integrated Services Digital Network) in the Bell system hierarchy.

7.3.2 ITT System-12 [15, 18]

ITT has developed the concept of Network-2000 as the futuristic network to meet the requirements of ISDN incorporating new device technology and wideband media, e.g., optical fibres. The system-12 ESS is the new generation of digital switching equipment designed to fulfil the needs of Network-2000. Under System-12, four specific types of ESS, viz, ITT-1210. ITT-1220. ITT-1230 and ITT-1240 have been developed. The ITT-1210 exchange (formerly called DSS-1) is designed to meet the growing demands in USA. It is digital all the way to the subscriber's interface and uses an effective multilevel distributed control structure. The ITT-1220 is a medium-to-large-sized toll exchange and incorporates a high capacity central processor aided by distributed microprocessor control. The central processing unit deals with toll features that tend to be more centralized in nature. The ITT-1230 is designed for small rural or local exchange applications and uses widely distributed control techniques that optimize the exchange's cost effectiveness. The centralized network control processor assists the distributed call processors to perform the primary call-control functions. The ITT-1240 has an expandable digital switching network with a capacity up to 100,000 lines. It uses a modular structure and a fully distributed control, such that the microprocessor-controlled line and trunk interface units share many of the control functions of the SPC processor. While designing these exchanges (except for ITT-1210), a future-safe concept in the development of software has been adopted and this ensures maximum interchangeability of documentation and long-term availability. The concept requires that formal interfaces are defined for all software modules, particularly for those dealing with hardware devices. Further, all software, other than the individual device handler programs, deal in terms of a virtual device or virtual processor. This virtual hardware concept simplifies the introduction of new devices, features and services, since only the device handler programs need to be rewritten for the added innovations. For the purpose of administration and maintenance, a centralized operations and maintenance centre ITT-1290 has been developed. This centre can supervise more than 63 remote exchanges via data links, thus, reducing the need for skilled personnel in these exchanges.

The ITT-1220 toll/tandem exchange [15] has a capacity of 50,000 trunks with 600,000 busy hour call attempts (BHCA) and its traffic capacity is 20,000 Erlangs. The exchange has the usual configuration which includes analog/digital trunk interface, a duplicated TSST digital switching network, signalling interface modules and the central controller. The central controller uses dual ITT-3202 processors working in a call-load sharing configuration and directs the distributed control modules which are physically located in the same equipment racks as the hardware units they control. The control modules, using a common μp and a common hardware architecture, function as line signalling controller, time/space switch marker, multiregister, common channel signalling controller and as a passive check unit. Thus, the central controller carries out the basic telephonic decisions and system resource management at the centralized level; and all secondary control and execution functions are delegated to the distributed control modules. This results in a simple hierarchial control structure as well as in significant savings in central processor time occupancy per call.

The ITT-1230 digital exchange [15] is designed for economical implementation of small and medium size offices, having capacities of 240 to 6000 lines, expandable up to 20,000 lines. It is completely modular in structure and suitable for local, transit, and combined local/transit applications. The smallest local exchange configuration (also known as the line/trunk module), consists of subscriber (line)/analog trunk/digital trunk interfaces, signalling unit (e.g., multifrequency code senders and receivers), a time switch (TSI) and the control processor, as shown in Fig. 7.25(a). The line interface circuit (LC) is similar to that of Fig. 7.22 and its output is a 32-ch. PDC (CEPT). Analog and digital trunk interfaces also generate 32-ch. PDC's at 2 Mb/s. The control processor for the module consists of a pair of 16-bit microprocessors operating in an active/standby mode, the associated memories and interface logics. Control access of the processors to the submodules is

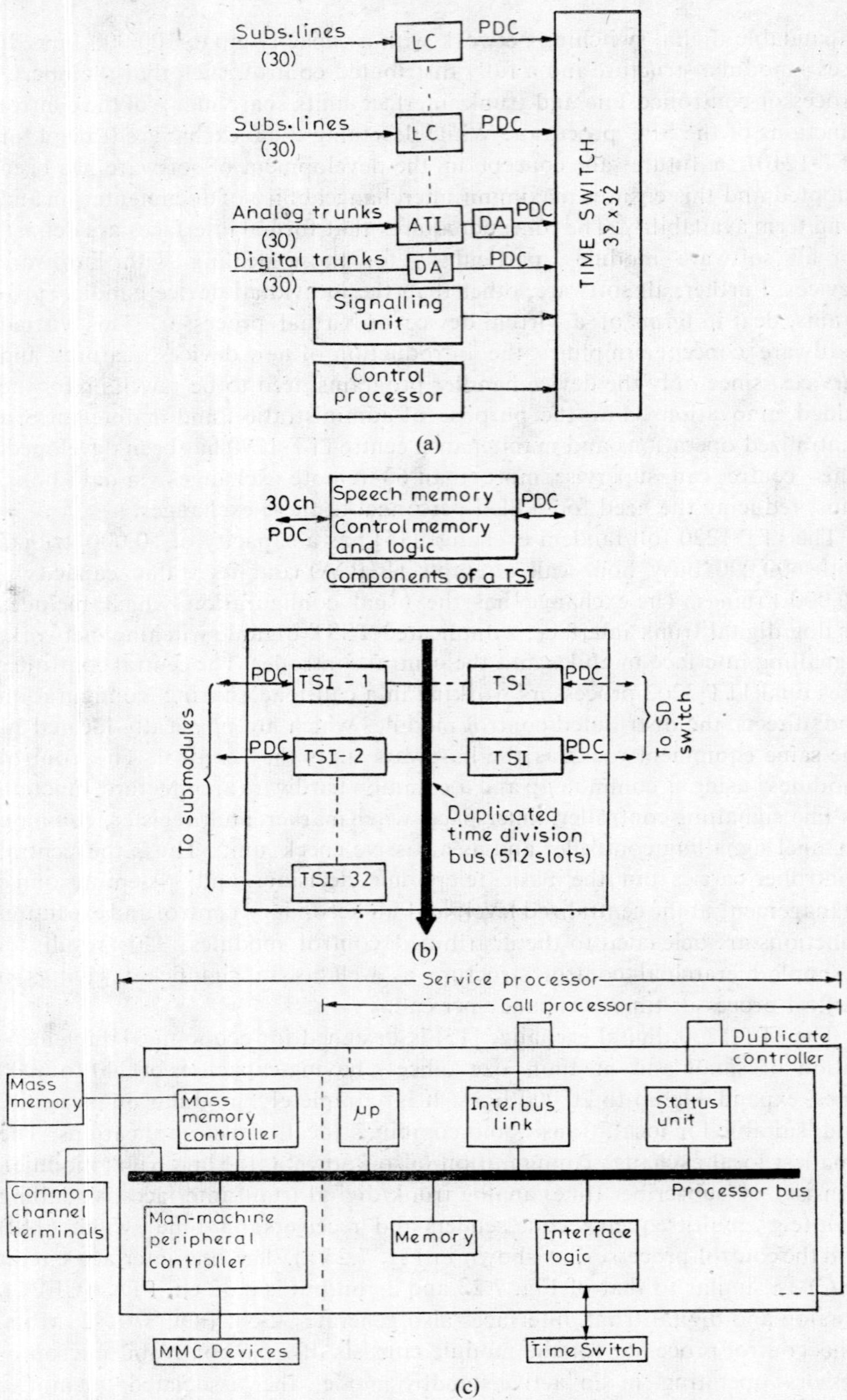

Fig. 7.25 (a) Small exchange configuration of ITT-1230 (line/trunk module). ATI: analog trunk interface; DA: Digital access (b) Time switch in a modular structure (c) Control processor of ITT-1230

via the switching network so that no additional links or interfaces with the submodules are required. Channel 16 of the PDC carries the control information for a submodule, test orders and alarms to the controller through the time switch. The time switch is modular in structure as shown in Fig. 7.25(b), where all TSI's are connected to a time-division bus with 512 time slots and all input channels have access to all of the slots on the bus. A TSI consists of a speech memory, an address memory and the control logic which responds to the marking information from the call processor. Each TSI is implemented as a custom-made LSI and larger exchanges are also configured around these LSI's. The above small exchange configuration has a capacity of up to 960 subscribers and can carry a traffic of 240 Erlangs. For very small clusters of subscribers, say less than 200, a small digital concentrator is most effective; otherwise for larger groups, autonomous digital exchanges with remote supervision are more economical.

For configuring larger exchanges, a number of line/trunk modules, shown in Fig. 7.25(a), are used as building blocks. This is shown in Fig. 7.26, where

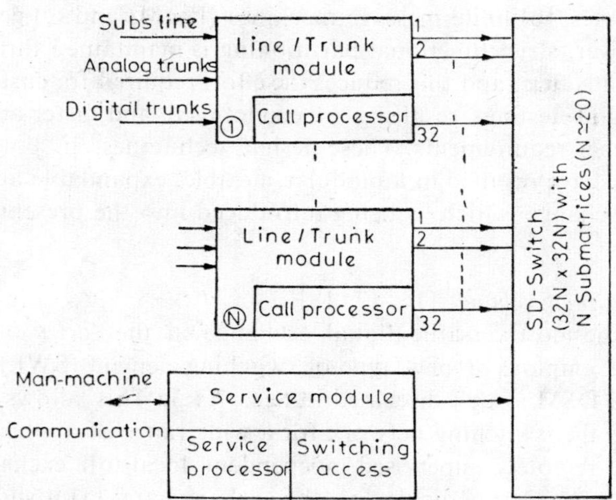

Fig. 7.26 ITT-1230 configuration for medium-size exchange

the PDC's from the time switch are interconnected through a single-stage SD-switch. The SD-switch is implemented as *N* submatrices/planes, and each of the *N* line/trunk modules is linked to one of the planes through an outgoing PDC, as shown by dotted lines in Fig. 7.25(b). This simplifies traffic engineering and extension procedures, and the new modules can be added by plugging into the SD-switch without any rearrangement of the existing cables and equipment. Approximately 20 modules (of which a few may act as concentrators) can be connected to the SD-switch, giving a line capacity of 20,000. Some of the modules may function as traffic concentrators, there by increasing the capacity of the exchange.

In addition to the line/trunk modules, a service module is connected to the SD-switch. This provides some of the call-control and service functions which are not time-critical, e.g., common channel signalling, resource management, man-machine communication and mass memory storage. The call processors in the line/trunk modules provide control access to all terminals and handles time-critical functions (e.g., subscriber and trunk signalling, parts of call-control, and some test functions) for up to 1000 lines. Cotrol information for the submodules is inserted by the call processor into dedicated fixed time slots on the bus, and these are then switched by TSI onto the channel 16 to the PDC to the submodule. The control processors for both the call/service processors are of the same type, viz., 16-bit micro-programmed μp, and both have the same structure and operation mode, as shown in Fig. 7.25(c). Communication between all processors takes place over the switching network and one plane of the SD-switch is dedicated to this function. The future-safe concept and modularity have been adopted in designing the software for ITT-1230. Software architecture is mainly characterized by the following techniques: (a) virtual machine concept in several hierarchial levels, (b) finite message machines (FMM) and (c) generic interfaces. Moreover, strict functional partitioning is maintained throughout the software architecture, and this reduces the effort required for customer design engineering and testing, resulting in economical and faster adaptation to special customer requirements. These design techniques, in both hardware and software, have resulted in a modular, flexible, expandable and economical digital exchange which is being introduced into the present-day analog networks.

ITT-1240 Digital Exchange [16, 17, 18]

ITT-1240 is the most versatile digital exchange in the series of System-12 exchanges. It employs a new type of switching element (SWE), somewhat similar to the DSM switch discussed in Sec. 7.1.3. This allows a modular expansion of the switching network for a wide range of applications, from small to large remotely supervised/independent local/toll exchanges of 10^5 line-capacity. The design goals for the exchange are: (a) it should accommodate a wider range of exchange sizes than allowed by the TST or STS network structures, (b) the network should have fully distributed control with more intelligence at the end-points of the network, and the network be controlled over the speech paths by messages generated by processors in the terminal control elements. These considerations led to the following design characteristics:

(a) The switching network should have folded single structure topology. This allows the number of switching stages to vary with the number of terminals to be connected.
(b) The switching element SWE should be a combination of T-S switching and the same SWE module is used throughout the network. The resultant network is 30 per cent more efficient than the TST network.

(c) The switching network is expandable without reconfiguration and the paths penetrate only as deeply as necessary.

(d) All connections—line-to-line, line-to-trunk, trunk-to-line and trunk-to-trunk, are switched identically; this simplifies control, and the network becomes insensitive to the type of traffic being switched.

The architecture of the ITT-1240 exchange is remarkably simple, as shown in Fig. 7.27. The overall configuration consists of a number of terminal

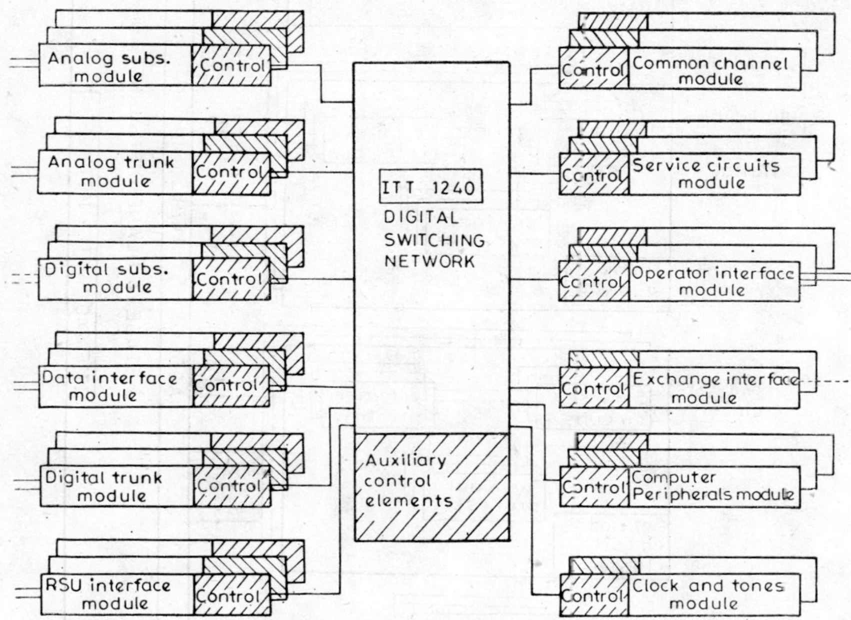

Fig. 7.27 Architecture of ITT-1240 exchange

modules, each with its own autonomous microprocessor, connected to the digital switching network. The figure also illustrates the full range of the possibilities of connecting futuristic services, e.g., digital subscribers, remote subscribers and data, directly to the switch by using identical interfaces. One of the major features of the exchange is the extensive use of the same unit for different functions, e.g., the same SWE's are used in the access switches as well as in group-switch planes. The other important feature which distinguishes ITT-1240 from similar other ESS's, is the shifting of the processing power to the terminals from the usual central processor.

The block diagram of IIT-1240 is shown in Fig. 7.28, where the terminal modules connected to the switch serve different types of subscribers/trunks. Each module serves either 60 analog subscribers or 30 trunks/service circuits. It consists of two submodules, viz., terminal and terminal control element (TCE), as shown in Fig. 7.29(a). The terminals for different types of services are different, but the TCE's are of similar design. The terminal for analog subscribers serves the same purpose as the line interface circuit of Fig. 7.22

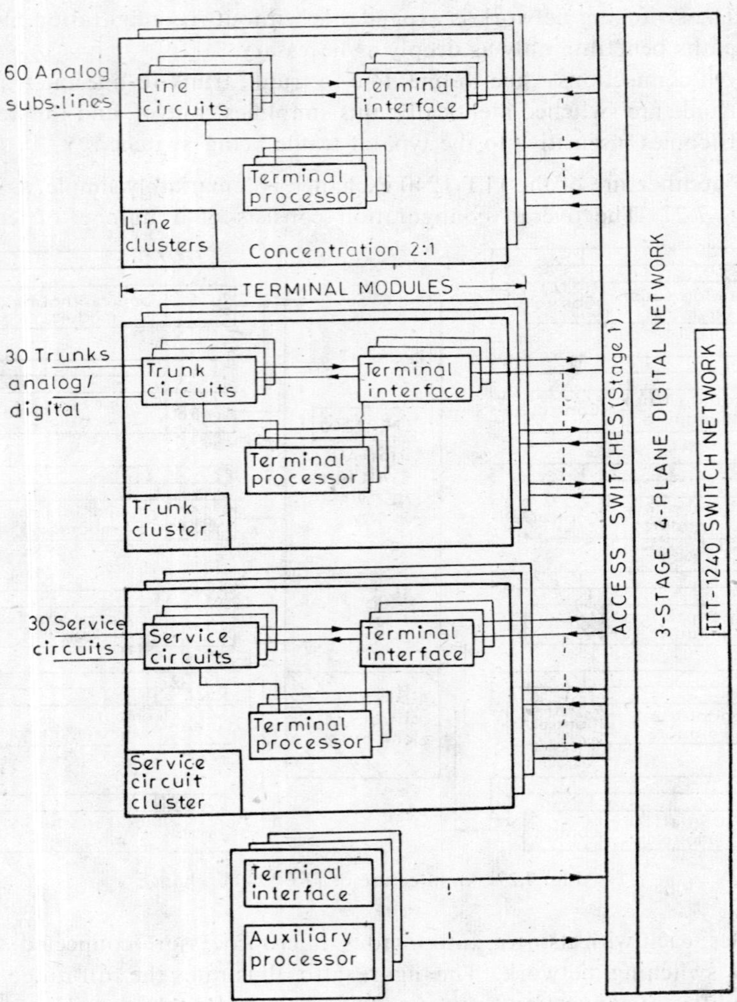

Fig. 7.28 Block schematic of ITT-1240 exchange (after Ref. [17])

and interfaces with the TCE through two 32-channel 2.048 Mb/s PDC's under the control of the microprocessor in the TCE. The terminal for analog trunks is similar to that for analog subscribers except for the ringing circuit. It performs the functions of A/D conversion, filtering, secondary protection, software-controlled pad switching, signalling and loop testing. For the digital trunks, the terminal provides HDB3/AMI conversion, clock generation and retiming, detection and supervision of alarms, and interface with the TCE through PDC's. Other terminals used are for service circuits, clock and tones, and for computer peripherals. The terminal control element consists of the terminal interface (TI), microprocessor, memory and buses, as shown in Fig. 7.29(a). The TI, consisting of receive/transmit ports and one port interfacing with μp, as shown in Fig. 7.29(b), serves as the transmission interface between the terminal and the digital switch through the 30-ch-4.096 Mb/s PDC's and also performs the TSI function. All its operations are controlled by the μp which performs basically the repetitive input/output functions, and this control can be exercised on a channel basis. Each μp exchanges messages with other control elements through the terminal interface and via the digital switch to perform call processing, maintenance and administrative functions. Thus, no special communication link is needed for transferring messages between these control elements.

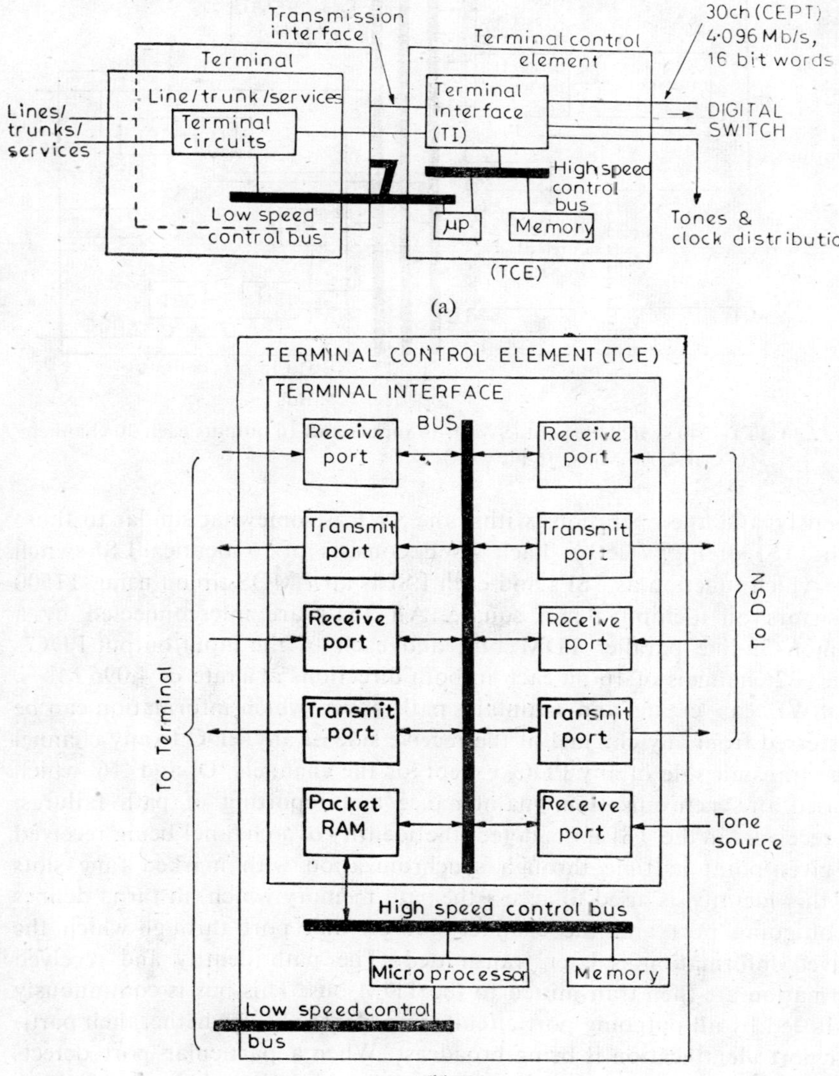

Fig. 7.29 (a) Schematic of a terminal module. Each TCE serves 60 lines or 30 trunks
(b) Schematic of a terminal interface

The digital switching network (DSN) of ITT-1240 exchange is configured using the same switching element (SWE) all through. The SWE, as shown in Fig. 7.30, provides a combination of time and space switching simul-

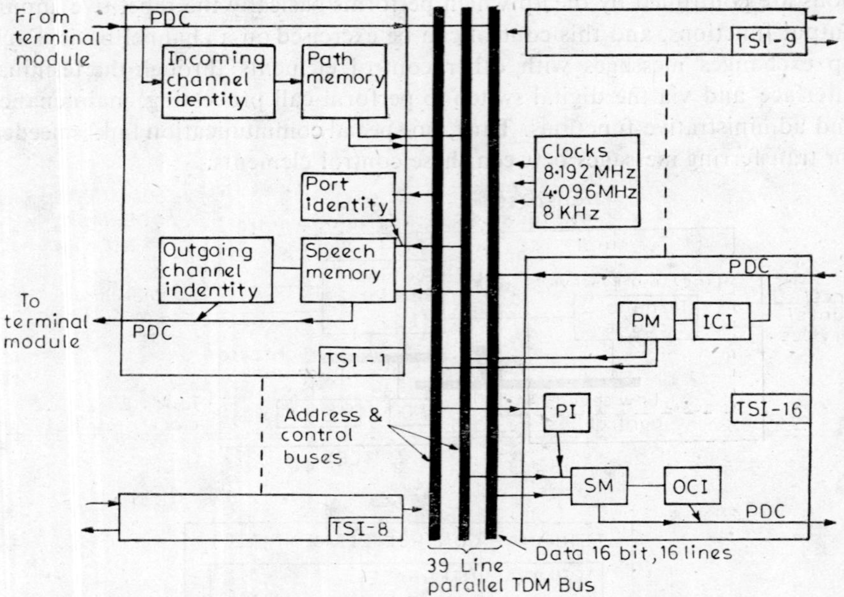

Fig. 7.30 ITT-1240 switch element (SWE); 16 inputs and 16 outputs each 30 channel MUX at 4.096 Mb/s, 16-bit words

taneously and the operations within the SWE is somewhat similar to those of the TSI of Fig. 7.25(b). Each SWE consists of 16 identical LSI switch ports which function as TSI's and each LSI is an *n*-MOS circuit using 11500 transistors on a chip 5.9 mm square. All TSI's are interconnected by a common 39-line parallel TDM bus, and each of the input/output PDC's carries 32 channels of 16 bit each in both directions at a rate of 4.096 Mb/s. The SWE can create and maintain paths over which information can be transferred from any channel of the receive side of any PDC to any channel of the transmit side of any PDC, except for the channels 'O' and '16' which are used for synchronization, maintenance and reporting of path failures. The receivers in the TSI units detect the identity of a channel being received at a given point in time through synchronization with marked time slots and this identity is used to access the path memory which, in turn, defines the outgoing port and the channel slot on that port through which the received information is later transmitted. The path identity and received information are then transmitted to the TDM bus. This bus is continuously monitored by all outgoing port circuits in order to check whether their particular port identification is being broadcast. When a particular port detects its address on the bus, its associated circuit uses its channel identity to store the corresponding information/data in the particular location of the speech memory. Subsequently, this data is transferred by the outgoing PDC to the next stage of switching. The path set-up is initiated by a request (a control code) on the incoming channel and the code is used to provide control information for the necessary search of an idle slot in a selected outgoing PDC. Once this slot is selected (i.e., a path is set-up within the SWE), the next-stage-path set-up message is transmitted to next SWE to advance the path through the network. The path may be disconnected by any SWE by transmitting the 'idle code' through the network, indicating that the path is no longer required. Thus, the TSI in the form of a custom-LSI provides not only the switching functions, but also the control functions to establish and clear down paths and to maintain network integrity without the aid of a central processor (as is needed in other ESS's, say, in No. 4 ESS).

The overall DSN is configured by interconnecting a number of SWE's, in a matrix (plane) form, as shown in Fig. 7.32, where the main switching is provided by the four group-switch planes (O-3) interconnected to the TCE's through the access switch pairs (stage-1). The group-switch plane consists of a maximum of three stages of SWE's connected in a folded matrix, as shown in Fig. 7.31. Each group in a stage consists of 8 SWE's, equivalent to a

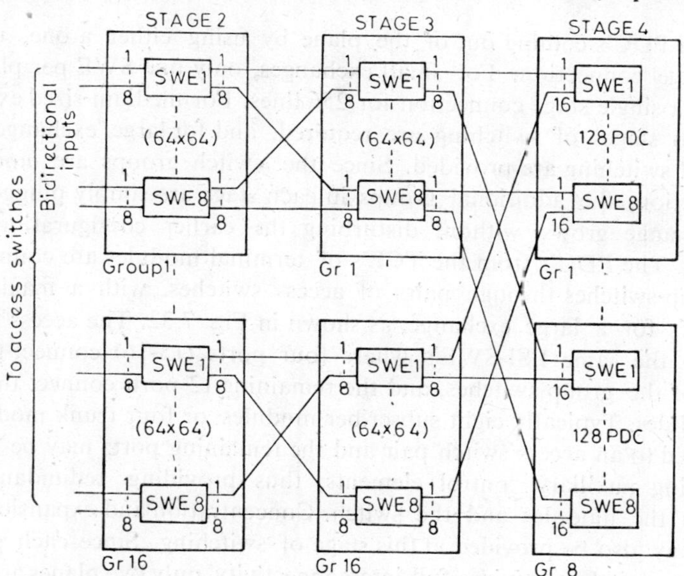

Fig. 7.31 One plane (matrix) of ITT-1240 group switches using folded matrix. Each group is (64×64) matrix. 1024 bi-directional inputs (PDC's). 30,720 line inputs for one plane

(64×64) space switch, and 16 such groups are available at each stage of switching. The switch matrix per stage is now equivalent to a (1024×1024) space switch. The last stage (stage 4) has only eight group-switches, and each SWE now has 16 inputs available; thus, making the total of 1024 inputs. It is seen that each of the incoming PDC's has full access to all of the possible

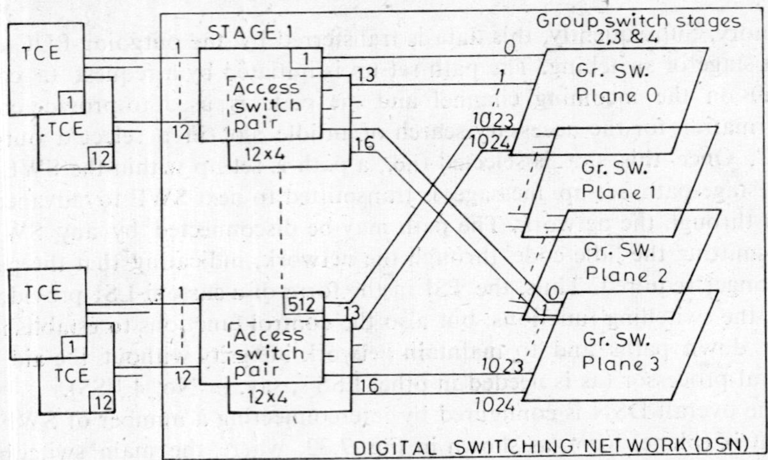

Fig. 7.32 Structure of the digital switching network of ITT-1240, showing four planes of group-switches of Fig. 7.31. Switch is connected to 512 access switch pairs having (512×12) inlets for connecting TCE's. TCE: Terminal control element

outgoing PDC's coming out of the plane by using either a one, two, or, three-stage connection. For small exchanges, only one SWE per plane will provide a single-stage connection for 256 lines. For medium-sized exchanges only two stages of switching are required, and for large exchanges, three stages of switching are provided. Since the switch groups are modular in construction, the additional groups in each stage are simply plugged in as the exchange grows, without disturbing the earlier configuration of the switches. The PDC's from the TCE's of terminal modules are connected to the group-switches through pairs of access switches, with a maximum of 512 pairs for a large exchange, as shown in Fig. 7.32. The access switches are also the same LSI-SWE's where four ports (13-16) connect the four planes of the group-switches, and the remaining 12 ports connect the terminal modules. Typically eight subscriber modules or four trunk modules are connected to an access switch pair and the remaining ports may be used for connecting auxiliary control elements, thus providing redundant paths between the modules and the switch. Concentration and expansion of the traffic may also be provided at this stage of switching. Since each plane of the group-switches provides full interconnectivity, only two planes are required for reliability of service. However, will all four planes provided, the network provides nearly non-blocking service with an average traffic of 0.25 Erlang per terminal and maximum traffic per trunk. With a line traffic concentration of 2 : 1 and trunk traffic expansion of 1 : 2, each switch plane has a blocking probability of less than 1 in 10^4 at 0.56 Erlang per terminal link. The two-stage switch can handle up to 1920 equivalent lines; a three-stage switch can handle up to 15,360 lines, and a four-stage switch handles over 100,000 lines, with nearly no blocking for an internal link occupancy of 0.5 Erlang.

It is now evident that the total intelligence of the exchange is fully distributed and the terminal modules (specifically TCE's) are assigned with more intelligence. As such, for the purpose of call processing, e.g., to set up a connection between two terminals, the originating TCE in response to the subscriber's call and the destination code, generates a string of 'select' commands (through its μp and packet RAM) and each command establishes a path through one stage of the DSN. A path is set up by one, three, five or seven commands, according to the number of stages required to establish the connection. At each stage, the SWE sets up a connection either to a port specified in the command or to an outgoing port selected by the SWE itself. The paths are initially set up on a simplex basis. However, for duplex speech paths, two simplex paths are set up. Initially a message packet is sent over the forward path, set up as above, to the destination module (TCE), and in response to this message, a return path is set up by the terminating TCE (responsible for the called subscriber); thus the two end-TCE's generate the path-control information from the knowledge of the called and the calling subscribers' code numbers. A blocked-path attempt is signalled automatically by a negative acknowledgement signal (NACK) returned to the originating SWE via the channel 16 of the PDC. The large number of possible paths for any given connection results in a virtually non-blocking switch. It should be noted that no central call-processing element is involved in the above operation, and as such, there is practically no possibility of the entire exchange going out of service at any instant in time. Moreover, the traffic overload degrades the performance gradually and does not result in a catastrophic collapse which can happen in a central processing exchange.

The ITT-1240 exchange has software-controlled operations and maintenance facilities that provide for day-to-day supervision, automatic fault location and recovery, and exchange extension and reconfiguration. A provision is also made for protecting the system before fault detection and correction, and if necessary, a diagnostic routine may be run to detect any difficult fault. A wide range of measurement facilities are provided to evaluate performance and to help in network management; as a result many corrective actions, e.g., re-routing of traffic, are taken automatically. A centralized operation and maintenance centre ITT-1290 is also available for supervision and maintenance of a group of exchanges including ITT-1240.

In conformity with the distributed control of the ITT-1240 exchange, its software is also distributed throughout the network hierarchy. The basic objectives of the software are: (a) the use of the future-safe concept, i.e., a rugged architecture that can tolerate the expected hardware changes (including the new developments in devices, specially the commercial μp's) and have sufficient flexibility to economically adapt to new markets (by addition or substitution of modules), (b) the use of high-level languages, e.g., CHILL, the CCITT high-level language designed for telephone switching, for portability and ease of use, (c) a data base management system to separate the problems of data semantics and data implementation. To achieve these objectives, the software architecture of ITT-1230 has been further refined by using the concepts of virtual machines, finite message machines (FMM) and generic interfaces. The virtual machine concept, as illustrated in Fig. 7.33,

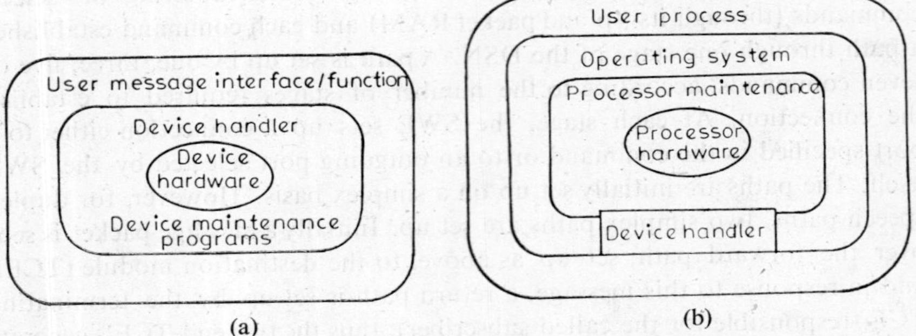

Fig. 7.33 (a) Concept of virtual hardware device, (b) Virtual processor of ITT-1240 exchange

is of particular significance and has all the functional characteristics of the actual hardware. But the specific electrical implementation is hidden by a software package called a device handler. The device handler provides both control and maintenance access, and acts as the interface with the user process. This approach is used for all modules in the exchange, e.g., line circuits, DSN and the processors. The virtual processor is shown in Fig. 7.33(b), where each ring around the actual hardware illustrates the hierarchy of program interfaces that isolate the main software from the physical hardware. As a result, a new processor hardware may be introduced by merely revising its device handler (the operating system and maintenance programs) and recompiling the high-level language programs to execute on the new processor. Thus, a substantial portion of the software becomes hard, and remains useful inspite of the hardware changes brought about due to the availability of new devices.

The distribution of software throughout the network is implemented by partitioning the software into finite message machines (FMM), i.e., the almost independent software modules, which communicate via well-defined message interfaces. They are supported by an operating system which provides concurrent processing and message communication facilities, and by a data base control system which provides data retrieval and update facilities. The FMM's reinforce the modularity of the software and these modules are essential for exchange functions. To be able to interwork with different signalling systems and to cater for futuristic applications without modifying the existing modules, the interfaces are designed as generic and not specific. All these sophisticated techniques require a high-level of processing power, which only the distributed control structure of ITT-1240 can provide. Thus, a futuristic 'future-safe' exchange has been evolved in ITT-1240, which can be easily adapted to the Integrated Digital Networks of the future.

7.3.3 Other Major Systems

The NEAX-61 ESS, developed by NEC, Japan, uses a 4-stage TSST switching network built as a modular structure. The system software is also modular. The capacity of exchange is [19]:

100,000 subscribers including 13,000 trunks for local switching;
60,000 trunks for toll switching;
30,000 international circuits for international switching;

It has a traffic capacity of 22,000 Erlang and a call handling capacity of 700,000 BHCA, giving a blocking probability of $\simeq 10^{-23}$ at 70 per cent traffic loading.

The block diagram of the ESS is shown in Fig. 7.34, where the switching network consists of 22 duplicated network modules (of which only one is shown in the figure) controlled by independent call processors. The network

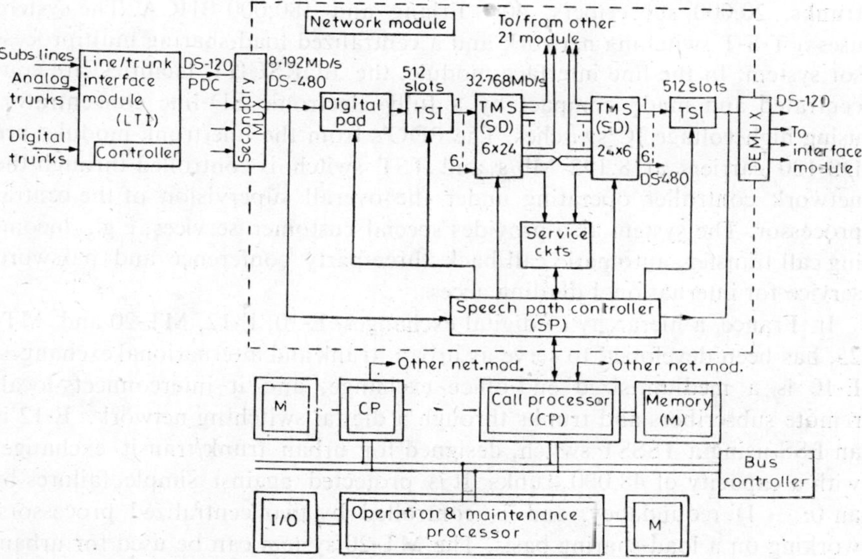

Fig. 7.34 Block diagram of NEAX-61 system (after Sueyoshi *et al.* [19])

is fully folded (although shown unfolded in the figure) and each module accesses 6 DS-480 MUX channels through individual TSI's. The SD-stage is (6 × 24) TMS connected to 21 other modules, and two junctors are used for communication across the module. Data rate across the network is 32.768 Mb/s and this provides the necessary traffic expansion and control signals over the same digital path. The line/trunk interface (LTI) module consists of terminal circuits, coder/decoder concentrator, primary multiplexer/demultiplexer and a controller, as required for different accessing lines/trunks. The input to the TSI is a DS-480 carrier and each TMS handles six such TSI outputs. The speech path (SP) controller provides communication (through data packets) between an LTI and the associated call processor

(CP), and establishes paths through the network module under the control of the CP.

The call processing is fully decentralized and is shared by several CP's, with each CP independently controlling a portion of the switching system. Each CP has its own memory containing the call processing program, network busy/idle map and translation data. The maintenance and administration subsystem is highly automated and centralized. An extensive software and hardware alarm reporting system displays alarms on the maintenance frame or at a supervisory test desk. The NEAX-61 ESS, because of its modular structure, provides a wide variety of switching services economically over a wide capacity range. Further, the system allows the introduction of technological innovations in hardware and software design as well as in network architecture.

Hitachi has developed the HDX-10 ESS with capacities as [20]: 10,000 trunks, 20,000 subscribers, 4000 Erlang and 180,000 BHCA. The system uses a T-S-T switching network and a centralized load-sharing multiprocessor system. In the line interface module, the BORSCHT facilities are concentrated and made compact by a fully electronic SD-line concentrator, using high-voltage IC switches. The PDC's from the line/trunk modules are DS-120 carriers at 8.192 Mb/s and TST switch is controlled through the network controller operating under the overall supervision of the central processor. The system also provides special customer services, e.g., incoming call transfer, automatic call back, three-party conference and password service for international dialling access.

In France, a hierarchy of digital exchanges, E-10, E-12, MT-20 and MT-25, has been developed to serve as urban, trunk and international exchanges. E-10 is a medium-sized end-office exchange, and it interconnects local/remote subscribers and trunks through a digital switching network. E-12 is an ESS using a TSSST switch, designed for urban trunk/transit exchanges with a capacity of 48,000 trunks. It is protected against simple failures by an $(n + 1)$ redundancy, and is controlled by two centralized processors, working on a load-sharing basis. The MT-20 system can be used for urban, trunk or international exchanges, with a capacity of 64,000 trunks and it uses a TSST switch with a capacity of 2048 PCM (CEPT) links. This is also controlled by two load-sharing central processors. As a purely local exchange, the system may be modified to MT-25 with a capacity of 65,000 lines. In this, the low-traffic subscribers are concentrated in the ratio of 1000 to 4 CEPT PDC's, and the line interface configurations are as shown in Fig. 7.21. The introduction of digital exchanges in the French telecommunication network has resulted in new services as well as substatial savings [21].

7.4 DIGITAL SUBSCRIBER LOOPS [23]

Traditionally, subscriber loops have been analog and the analog signals are converted to a digital PDC in the line interface module of the DSN, as shown in Fig. 7.21. However, if the subscriber loops are converted to digital working, then the subscribers will have access to the wide range of facilities and digitally-based services, e.g., packetized voice/data, facsimile/still picture communication, teletex, view data etc., which would be available in the near future. Moreover, there will be considerable improvement in the quality of transmission for regional, national and international connections that use digital techniques throughout. It is expected that, in the future, multifunctional intelligent terminals will be used by major customers and this will require end-to-end digital connectivity throughout the communication network. This will necessarily lead to the use of optical fibres in the the local area network. However, at present, the existing low-grade cables will have to be optimally adapted to serve as the digital links between the DSN and the subscribers' premises.

The average investment in the local area cable plants is approximately 40 per cent of the total network cost and, as such, the plants have to be utilized for their full life even with digital signalling. The plants consist of mostly twisted pair lines of 0.4 or 0.6 mm diameter copper wire. The average line length is 2 to 3 Km, and 90% or more of the subscribers have lines of 4 Km or less. The cable characteristics are given in Table 7.5, where it is seen

Table 7.5 Loss-Frequency Characteristics of Local Cables

Frequency in kHz	Loss in dB for 0.4 mm cable of length			Loss in dB for 0.6 mm cable of length		
	1 Km	2 Km	4 Km	1 Km	2 Km	4 Km
10	6.5	10	23	4	6.5	13
20	7.5	12.5	28	5	8	16
50	9	16	33	5.5	10	20
100	11	20	38	6	12	24
200	14	25	47	7.5	16	32
500	20	36	68	13	24	50

that the 0.4 mm cable of length 4 Km can support a data stream up to 250 Kb/s (with a loss of 40 dB at 125 kHz) with proper line codes and equalization [22]. The path delay in cables is about 5 μs/Km which also limits the range of the subscriber loop. The most disturbing noise in multipair cables is due to the near-end (NEXT) and far-end (FEXT) cross-talk, and the effect of the NEXT may be eliminated by using the time separation instead of the hybrid separation technique (for 2W to 4W conversion at the subscriber end). Extensive measurements on cable cross-talk have been made and acceptable margins have been calculated from theory and experiments. It has been found that the cross-talk margins are satisfactory for 0.6 mm cable of lengths up to 4 Km (giving a range of 10 to 26 dB), but for 0.4 mm cable the margins are unacceptable for difference of lengths less than 2 Km. For these cases, a systematic choice of pairs has to be made.

In a digital subscriber loop, two major problems have to be solved, (a) the attenuation-frequency characteristics and (b) the implementation of 2W/4W conversion. The first is partially solved by restricting the signalling rate to 80 Kb/s (or 96 Kb/s) at present, with the possibility of achieving a world standard of 144 Kb/s in future. The 80 Kb/s data rate is composed of 64 Kb/s for speech and 16 Kb/s for signalling and synchronization with a 10-bit word per frame. By using suitable line codes, e.g., AMI, Howells-Woodman, the peak of the signal spectrum may be restricted to half the bit rate transmitted. Further, the eye pattern is optimized by using adaptive equalization at the receiver input.

The implementation of 2W/4W conversion may be done by (a) hybrid separation, (b) time separation, or (c) frequency/code separation, as shown in Fig. 7.35. The hybrid separation technique is an extension of the similar

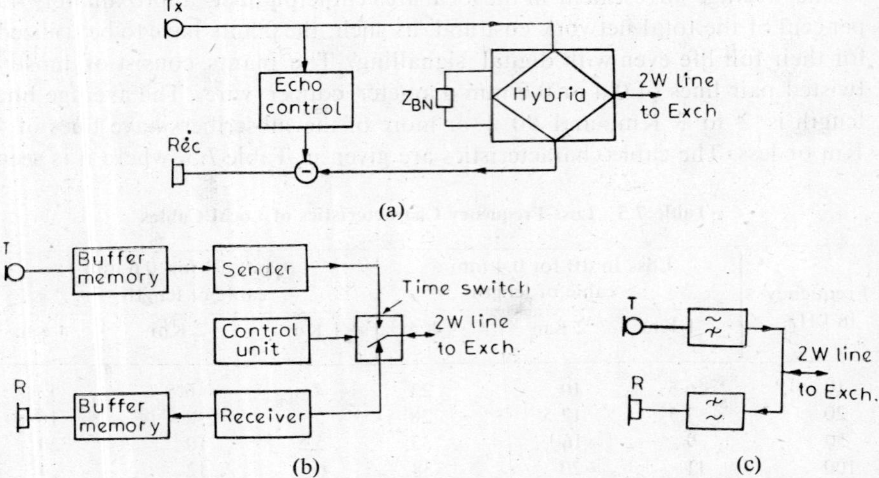

Fig. 7.35 Implementation of 2W/4W conversion (a) Hybrid-separation, (b) Time-separation, (c) Frequency/code-separation

technique presently used in analog subscriber loops, and is the only technique capable of bidirectional transmission on a 2-W cable without increasing the transmission bandwidth. A compromise balance network Z_{BN}, as used in analog loops, is however not suitable for different loops (because of the mismatch) and considerable echo may be generated due to reflections. The problem is solved by using an adaptive balancing network and an echo control as shown in the figure. The transmission rate is usually 80/96 Kb/s, and this also accommodates non-voice data at rates of 8/16/64 Kb/s.

In the time separation technique, burst inter-leaving is employed, as shown in Fig. 7.36(a), where only one burst of signal bits, either from the subscriber's set or from the exchange, will be present on the line at any instant of time, thereby, partially eliminating the echo problem. This require that the sender at each end has to wait for the burst from the other end to arrive, before the next burst is sent out. As a result, the transmission rate is increased, say to 250 Kb/s (instead of 80 Kb/s in the hybrid separation), giving a burst duration of 40 μs for a 10-bit word. This allows about 20 μs of transmission time in each direction, as seen in Fig. 7.36(a); but this time may be increased by transmitting more words in a burst, i.e., having a larger frame time. Considering the allowable path delay in duplex speech circuits, a compromise value of 2 words at a bit rate of 250 Kb/s with a frame of 250 μs is usually taken. Using AMI line codes, the peak of the signal spectrum is at 120 kHz and this gives a line loss of 40 dB, equivalent to 4 Km of 0.4 mm cable. Data buffering and transmit/receive control are obtained through the buffer memories and the control unit as shown in Fig. 7.35(b). The third method of frequency/code separation is equivalent to the usual FDM/CDM technique; however, this requires bandwidth expansion and is not a favoured technique.

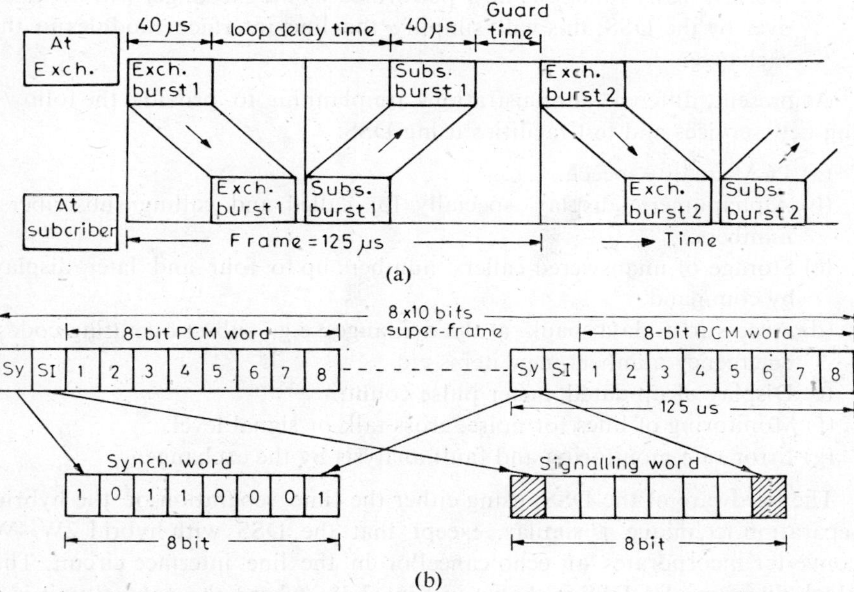

Fig. 7.36 Time-separation technique: (a) Burst interleaving in time-separation method, (b) Super-frame structure of subscriber transmission envelope

7.4.1 Digital Subscriber Set (DSS) [22, 24]

The basic functions required in a digital subscriber set (DSS) is analogous to those required in conventional analog subsets, except that the speech signals have to be digitized and a different signalling scheme has to be used, as shown in Fig. 7.37. Moreover, it is desirable to supply power for basic functions from the exchange in order to avoid service disruption in the event of a local power failure. The shifting of the A/D codec from the exchange to DSS results in several advantages. These are the following:

(a) Several services may be provided simultaneously on a single 2W-line, leading to a saving in lines.

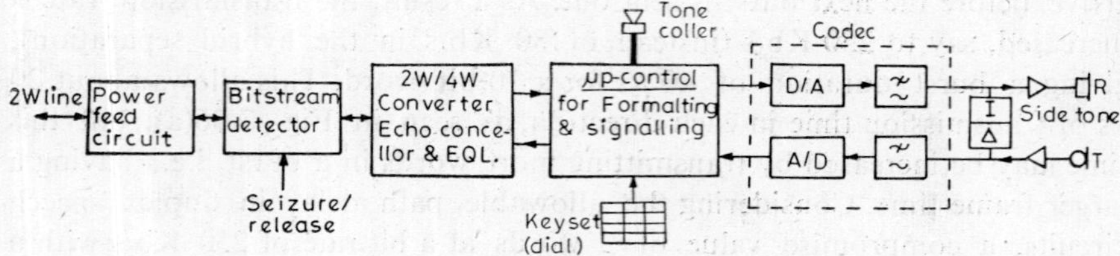

Fig. 7.37 Basic functions of a digital subset

(b) The line interfaces in the exchange will be only digital.
(c) In a future ISDN, the DSS may be easily converted to a multifunctional/intelligent terminal, providing various voice/data services.
(d) Although DSS will be hardware-wise more complex than the analog subsets, many functions now performed by the exchange, will be taken over by the DSS; this will simplify the line interface modules in the exchange.

At present, different administrations are planning to provide the following new services and test facilities using DSS:

(a) PCM quality speech.
(b) Alphanumeric display, specially for called and calling subscribers' numbers.
(c) Storage of unanswered callers' numbers up to four and later display by command.
(d) Access to a data bank at the exchange, e.g., subset operating codes, emergency numbers, new items etc.
(e) Display of up-dated meter pulse count.
(f) Monitoring of lines for noise, cross-talk or signal level.
(g) Error rate monitoring and fault analysis by the exchange.

The hardware of the DSS, using either the time separation or the hybrid separation technique, is similar, except that the DSS with hybrid 2W/4W converter incorporates an echo cancellor in the line interface circuit. The block diagram of a DSS is shown in Fig. 7.38, where the control unit is a CMOS microprocessor and the line interface unit (LIU) provides 2W/4W conversion, equalization and echo cancellation. The control unit not only handles the keyboard, display and tone caller, but also supervises synchronization, signalling and de-scrambling. The synchronization is based on the master-slave (exchange-subscriber) principle and the 8-bit SYNCH word is transmitted as the first bit of 10-bit words in eight frames, as shown in Fig. 7.36(b)*. Similarly, the signalling word is transmitted through the

*An alternate scheme for utilization of 80 Kb/s transmission rate is to have: a main channel of 64 Kb/s for PCM speech/high-rate data, an auxiliary data channel of 8 Kb/s wherein telex, view data and teletex data may be multiplexed, and a supplementary channel for signalling (at 6 Kb/s) and synchronization (at 2 Kb/s).

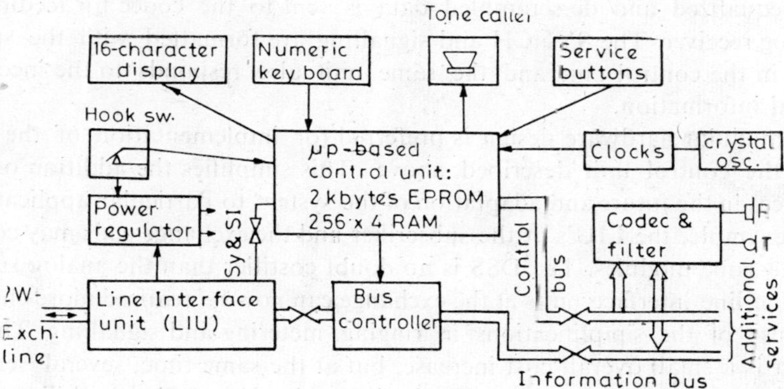

Fig. 7.38 Block diagram of a DSS. SY: synchronizing signal SI: signalling.

second bit of 10-bit words. Signalling is based on event signalling with acknowledgement and the complete format is normally transmitted in 4 ms, or in the event of an error, in 12 ms. Usually signalling for telephony has priority over other services. The subsets are line-powered for telephonic use with a 50 mA constant current source and they draw less than 0.5 mA at 10 V in the idle state. With a 60 V exchange battery and using 0.4/0.6 mm cable, the operating distance is 4 Km or more.

The major function of the LIU as shown in Fig. 7.39, is to control the line echoes and to adaptively equalize the line signal. In the transmit direction, the binary data from codec is scrambled, coded with AMI/HDB3 code

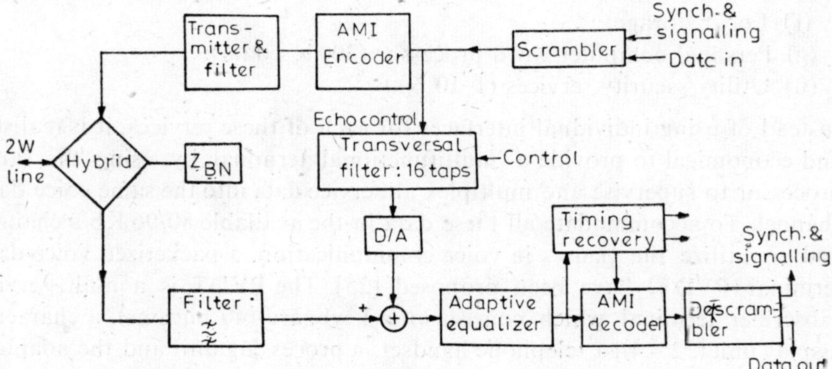

Fig. 7.39 Line interface for the DSS using hybrid separation.

and fed to the hybrid circuit. The line echo is simulated through a 16-step adaptive transversal filter using a simple algorithm with error sign as the error criterion, and the simulated echo is continuously subtracted from the received signal. The line signal is equalized through an automatic/adaptive equalizer and the signals may sometimes be pre-equalized. The clock recovery circuit is based on a conventional PLL with a crystal-controlled VCO.

The equalized and de-scrambled data is sent to the codec for feeding the analog receiver. The SYNCH and signalling are formatted with the speech data in the control unit and the same unit also responds to the incoming signal information.

A modular hardware design is preferred for implementation of the LIU and the control unit described above. This simplifies the addition of new services in the future and adaptation of the system to particular applications. For example, the LIU's at the subscriber and the exchange ends may consist of the same modules. The DSS is no doubt costilier than the analog subset, but the line interface units at the exchange can now be realized more cheaply because of the simplifications in ringing, metering and signalling. The net result is a small overall cost increase, but at the same time, several services are available simultaneously, thereby increasing the traffic capability of the digital network.

7.4.2 Multifunctional Terminals [25, 26]

With the introduction of the processor unit (control unit) in the DSS it is possible to provide some more intelligence in the subscriber's terminals and make the terminal multifunctional. The new facilities that may be provided are:

(a) High-speed data (up to 48 Kb/s).
(b) Facsimile (FAX)/still picture communication.
(c) Low-speed data.
(d) Teletex/videotex.
(e) Data retrieval.
(f) Local storage.
(g) Personal computer/word processor (300 b/s data).
(h) Utility/security services (1–10 b/s).

Instead of using individual interfaces for each of these services, it is realistic and economical to provide a multifunctional terminal by using the same processor to supervise and multiplex all service data into the same voice/data channel. To accommodate all these data in the available 80/96 Kb/s channel and to utilize the pauses in voice communication, a packetized voice/data terminal (PVDT) have been proposed [25]. The PVDT is a multi-service subscriber terminal which consists of a keyboard (40 buttons), a character display unit (32×4), a telephone handset, a processing unit and the adapters for connecting other service units, as shown in Fig. 7.40. PCM-coded voice data at 64 Kb/s are stored in the memory initially and then they are packetized according to the format shown in Fig. 7.41. Data from the keyboard or the service terminals are multiplexed and then packetized in the same format of 64 bytes with a header of 10 bytes. Usually, data packets are sent out during the intervals of voice packets. For high-speed data (up to 48 Kb/s), voice packets carrying low-level voice segments (indicated by 3-bit MSB in VL) are discarded in order to insert data packets. Packets are then transmitted through the digital subscriber loop using HDLC (high-level data link

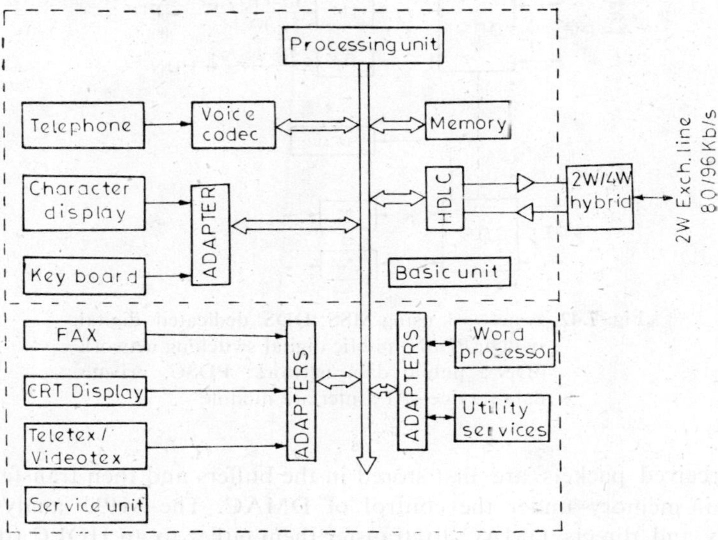

Fig. 7.40 Block diagram of the PVDT.

F	A	C	RK	DA	OA	VL	VN	DATA----------	FCS	F

Fig. 7.41 Packet frame format; F: flag ; A: address; C: control; RK: record kind; DA: Destination address; OA: originating address; VL: voice level; VN: voice packet number; FCS: frame check sequence

control) framing. In the receiver chain, the packet is processed according to its record kind (RK); and if voice, the packet is stored in the buffer, and then D/A converted to voice signal. For data packets, the transmission error is checked and recovered by the data link control procedure. For other services, the service unit in the PVDT acts as a 'data modem' and end-to-end transperancy is maintained. The PVDT hardware is built around available μp and LSI's, and further economy may be achieved by using custom-built LSI's.

To provide access to the public switching network and interconnection between PVDT's, it is necessary to process the PVDT signals by a dedicated digital switch and the associated interface, which may be housed either in the local office or in a remote location. The PVDT's are connected to the multiservice switch (MSS) which consists of a dedicated digital switch (DDS), and interface modules (IM) connected to either the PDSN or a public data network (PDN) as shown in Fig. 7.42. The block diagram of the MSS is shown in Fig. 7.43, where it is seen that two or more DDS may be interconnected through the bus controller and MSS acts as a load-shared switching network.

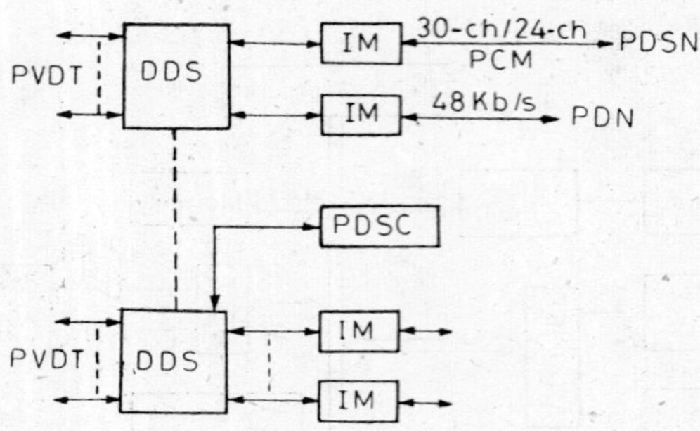

Fig. 7.42 A network using MSS; DDS: dedicated digital switch; PDSN: public digital switching network PDN : public data network; PDSC: private data service; IM : interface module

The received packets are first stored in the buffers and then transferred to the main memory under the control of DMAC. The MPU analyses the packets and directs DMAC to transfer them either to an HDLC (through buffers) or to bus interface and IM for transmission to PDSN/PDN. Bus controller polls each bus interface and transfers the packets according to their destination. The IM performs the following functions:

(a) Conversion of packet switching mode into circuit switching mode.
(b) Conversion of signalling and information formats.
(c) Echo control.

Voice links between PVDT's or between a PVDT and a conventional telephone is set up by 'virtual circuit' in the MSS. Once the virtual circuit is set up, the voice path route is fixed, resulting in minimum variation in network transit time. To avoid queuing delay in processing voice packets, preference is given to them over other data packets, and each voice packet is directly processed before the next packet from the same PVDT arrives. But the data packets are processed on a first-come first-served basis. The overall delay in processing of voice packets is less than the duration of two packets, i.e., approximately 32 msec, and a single DDS can handle more than 2000 voice packets per second. The multiaccessing scheme is: 16 PVDT's per DDS and 16 DDS per MSS, resulting in a total access of 256 subscribers to the MSS. Considering the transit time of voice packets and the load factor of the MPU, the optimum packet length is 64 bytes with a packet header of 10 bytes and a delay time of 30 msec, giving a packet efficiency of 85 per cent. Considering the minimum dynamic range required for voice signals, it was possible to discard packets with levels below −35 dBm and this resulted in a possible discard of about 60 per cent of packets in two-way voice circuits. This leads to the possibility of data transmission capacity of 48 Kb/s using 96 Kb/s line transmission.

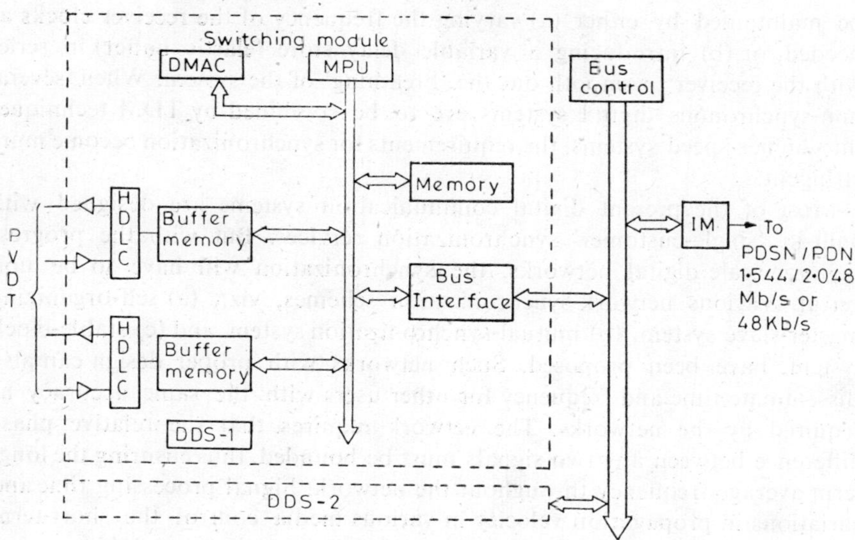

Fig. 7.43 Block diagram of the MSS using multiple DSS
DMAC: direct memory access controller
MPU : module processing unit.

The use of intelligent terminals at subscribers' premises simplifies the functions of the line interfaces at the exchange. At the same time, it is possible to use the DSS/PVDT as a multiaccess terminal for a few subscribers (in an office). Upgrading subscriber's terminal further by using three or more microprocessors as load-sharing controllers, the terminal may be used as a PABX, as well as a multifunctional terminal [26].

7.5 MULTI-EXCHANGE NETWORK: SYNCHRONIZATION [1, 27, 29]

A typical multi-exchange network has been shown in Fig. 7.1, where some of the exchanges may be analog and others digital. The transmission link between them could also be analog, i.e., multichannel FDM or digital i.e., multichannel TDM. The hierarchy of digital TDM systems has been shown in Fig. 4.24, where the channel bit-rate may be as high as 140/800 Mb/s. In the analog FDM system also, the channel bandwidth may be up to 18/60 MHz, multiplexing up to 10,000 speech circuits. In both the systems, it is essential to maintain accurate synchronization of transmit/receive/carrier/ clocks within close tolerances. Typical requirements in both synchronization and accuracy for the FDM and TDM systems range from several parts in 10^6 to a few parts in 10^{10}. But the synchronization requirements for the TDM systems are quite different from those for FDM transmission. In TDM, the coding/decoding terminals should maintain the clock phase coherence so that the information bits are not lost. Since the delay time of the transmission media is variable, due to both the propagation delay variations of the line and delay variations in repeaters, the phase coherence can be maintained by either (a) varying the frequency of the receiver clocks as needed, or (b) introducing a variable delay store (elastic buffer) in series with the receiver to smooth out the 'breathing' of the system. When several non-synchronous digital systems are to be combined by TDM techniques into higher speed systems, the requirements for synchronization become more stringent.

Most of the present digital communication systems are designed with built-in 'single-customer' synchronization services. But with the progress in large-scale digital networks, the synchronization will have to be universal. Various network-synchronization schemes, viz., (a) self-organizing master-slave system. (b) mutual-synchronization system, and (c) stable-clock system, have been proposed. Such networks with proper design can also disseminate time and frequency for other users with the same accuracy as required by the networks. The network requires that the relative phase difference between any two signals must be bounded, thus ensuring the long-term average frequency throughout the network. Signal processing time and variations in propagation velocity in various media control the short-term accuracy. A review of these problems and techniques is presented now.

In short-haul communication networks using FDM, double-sideband transmission is used. The frequency accuracy of the DSB carriers need be sufficient only to maintain information bandwidth within the passband of the channel filters, and as such, accuracy of about ± 25 Hz is adequate. But for long-haul communication networks, SSB-suppressed carrier AM techniques along with highly selective channel filters are used and a large number of channels are multiplexed. For good quality voice transmission in such wideband systems, it is necessary that the frequency of the local carrier should have an accuracy of ± 2 Hz with reference to the transmitter carrier. Considering the number of modulation steps involved in such wideband systems, it is found that the relative synchronization of all the carriers used in the system must be about 2×10^{-8} for satisfactory quality of the signal. In such systems, carrier synchronization is achieved by using crystal oscillators in the multiplexing equipment, and these oscillators are phase-locked to pilots transmitted over the medium. The system master-control is derived from the redundant highly reliable frequency generator, consisting of a number of precision crystal oscillators, whose average frequency is maintained within $\pm 1 \times 10^{-10}$ of the national standard. The short-term stability of these oscillators is sufficient to the extent that weekly comparison is adequate to maintain this degree of accuracy. To eliminate transients due to pilot-tone failures, a frequency setting system having memory (integral-type phase-locked loops) is utilized and the loss of pilot does not initiate any frequency transient in the system.

The digital network hierarchy for accommodating 120/480/1440/11,520 voice channels has been shown in Fig. 4.24, and the futuristic Integrated Digital Networks will be discussed in Chapter 10. The switching speed in such networks will be in the range of Gb/s. For point-to-point digital communication, the TDM pulse stream must be received synchronously at the receiver so that proper demultiplexing can take place. The synchronism at the receiver is obtained by the master-slave system of synchronization, where the receiver clock is recovered from the received pulse stream through a suitable PLL, and the frame synchronization is obtained from the structure of the code words used in the format. It has been estimated that for point-to-point digital transmission, the clock for each transmitting terminal can be independent with about 10^{-6} accuracy and each receiving terminal is slaved to the incoming signal. For a digital network, the problem of synchronization is much more complex and all signals should have exactly the same long-term average frequency and their phase errors have to be kept within certain limits. Otherwise, some of the bits will be lost or duplicated and the frame synchronization will be disrupted. The short-term accuracy requirement is governed by the amount of buffer storage available. With 100 bits of storage, the 7-day requirement at 1.5 Mb/s is 10^{-10}. For this purpose, some of the international and national time-frequency standard, such as, VLF radio transmissions, may be used. However, an integrated digital network, both national and international, should develop its own internal synchronization scheme, because such a common communication carrier should not have its reliability influenced by another organization not responsible for this task. It is also not very safe to tie-up such a large system to one external standard clock, as it is possible that the network can be orphaned if that clock became unavailable.

7.5.1 Synchronization Technique [27]

The various approaches that can provide a reliable clock system for digital communication have been studied and the different methods of synchronization of a digital network have been classified into three main categories. These are (i) Asynchronous/Stable Clock System: In this method, each exchange has its own timing control with a specified tolerance of the nominal frequency, so as to make negligible the inevitable information loss which occurs when a faster signal arrives at a slower exchange. (ii) Homochronous/Master-Slave System: In this method, one exchange in the network has a master timing pulse generator which determines the timing of all switching connections at the exchange and where all the other local timing pulse generators of other exchanges are locked to the master timing pulse generator. (iii) Isochronous/Mutual Synchronization System: In this method, each timing pulse generator of the exchange in a network is phase-locked to one incoming signal or the average of several incoming signals.

We now discuss these systems in greater detail.

Asynchronous Stable Clock System

If it is possible to generate stable enough frequencies at each node of the communication network, then there will be no need for a clock dissemination scheme. No clock can achieve the requirement that the phase error be held within bounds. However, the state-of-the-art precision crystal oscillators are capable of remaining within about 1×10^{-10} for a week or more and the atomic clocks (cesium beam) are capable of accuracy of the order of 10^{-13}. Thus, the clock variation between the systems may be kept within 10^{-9} or so; the relative phases will drift very slowly and the lost or repeated bits will occur infrequently. At 1.5 Mbits/s, loss of one bit per day requires a frequency accuracy of about one part in 10^{11}. The entire network can be divided into regions with stable clocks functioning as master in each region and they can in turn be loosely tied together by mutual synchronization methods.

For multiplexing asynchronous data, pulse-stuffing technique [28] is sometimes used, specially for point-to-point communication, as shown in Fig. 7.44.

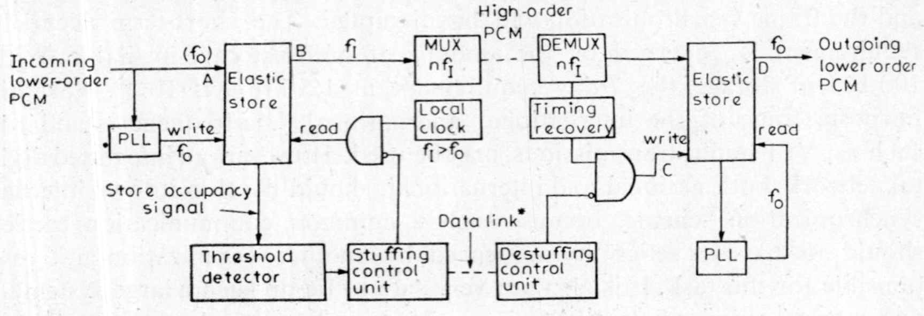

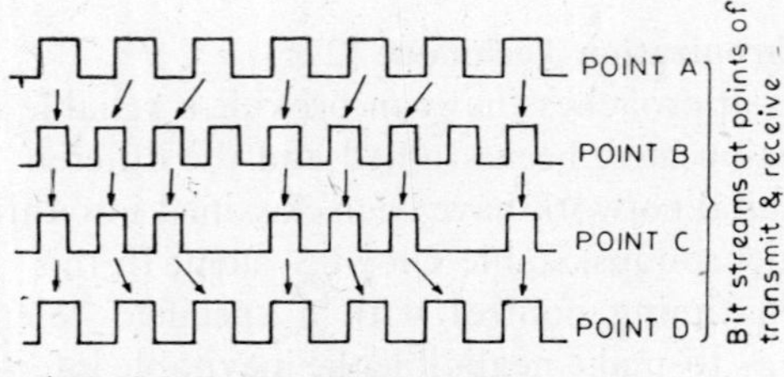

* MULTIPLEX ON TO PCM TRAIN

Fig. 7.44 Pulse-stuffing synchronization

In the circuit, the received clock f_0, extracted through the PLL, is used to store the incoming lower-order PCM data in the elastic buffer, and the buffer is read out at a faster rate f_I by the local clock to feed the higher-order TDM stream. Since $f_I > f_0$, the stored pulses in the elastic store are gradually reduced and this information is given to the threshold detector. The threshold detector supervises this and instructs the stuffing control unit to inhibit reading (the elastic store) according to a predetermined rule. As a result, null pulses, called stuff pulses, are inserted in the PCM train to make up for the difference between clock pulses. The location of the stuff pulses is also transmitted to the receiver through a data link (multiplexed onto the PCM train). At the receiver, the stuff pulses are removed by the de-stuffing control unit and the PCM data is read out of the elastic store at f_0, obtained through the local PLL. The lower-order PCM is now recovered from the higher-order MUX signals at a data rate of f_0, as required. The pulse-stuffing technique is quite efficient because the initial frequency stability allows the pulse to be stuffed at the rate of one per several hundred and the location of stuffed pulses is signalled through a separately multiplexed channel. Pulse-stuffing, of course, does not result in a synchronous network and, thus, cannot be used for frequency dissemination. This technique also, has the advantage of simplicity over synchronous network operation because it requires the least amount of buffer storage. For example, a 1000-Km coaxial cable carrying 3×10^8 pulses per second will have about 1 million pulses in transit. An increase of 0.01 percent in propagation velocity, as a result of the decrease in temperature by a few degrees, will result in 100 fewer pulses in the cable. In a synchronous network, this must be absorbed by buffer storage. With pulse-stuffing, a surplus or a deficit of a single pulse is immediately compensated. Each pulse-stuffing multiplexer can have an independent clock with an accuracy of 10^{-6}, thus avoiding the clock reliability problems. The disadvantage of this technique is the amount of processing required.

Homochronous/Master-Slave System

The homochronous technique may be used for both point-to-point and multi-point digital networks. A typical master-slave synchronization scheme, as used for transmission links between a digital exchange and its TDM channel banks, is shown in Fig. 7.45(a). In the system, the digital switch

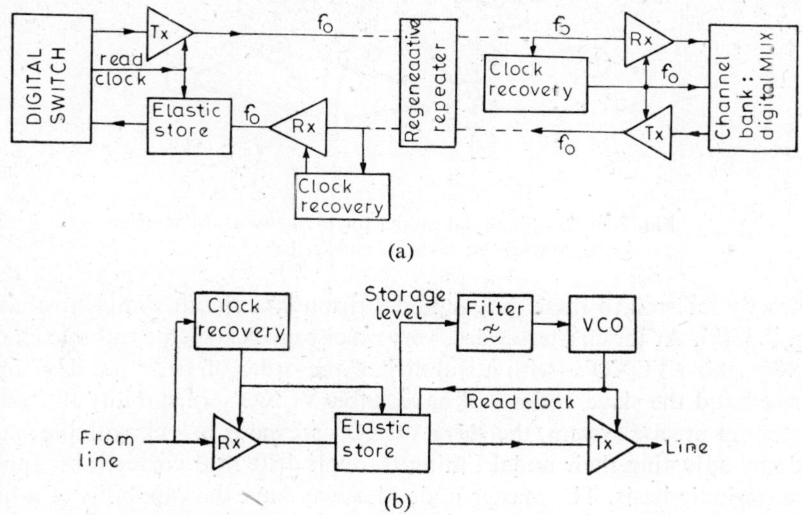

Fig. 7.45 (a) Master-slave synchronization scheme between a digital exchange and a TDM channel bank. (b) Regenerative repeater with jitter-free clock

provides the master clock for all outgoing TDM links and the distant channel bank derives its clock from the incoming master clock sent from the exchange. In the return path, the clock instabilities due to timing jitter, noise, changes in path length etc., are absorbed in the elastic store as shown in the figure. It is now seen that this loop timing maintains a constant integral number of clock intervals between the inlet and outlet of the switch. Thus, from the point of view of timing, the inlets and outlets operate as though directly connected to a common source. In the case of the regenerative repeater used along the transmission path, the transmit timing is directly established from the received bit stream, as shown in Fig. 7.45(b). The elastic store absorbs the short-term instabilities, but the long-term stability of the clock is maintained through a PLL with a large time constant. In practice, the large time constant is derived from a certain average level of storage in the elastic buffer, and considerable memory is required for sufficient stability of the clock. Such elastic buffers are also used at the input trunks to the digital switch.

Usually a digital communication network is composed of many widely dispersed nodes that operate over a variety of different media, as seen in the simple nodal model of Fig. 7.46. One of the approaches used for synchronizing such a network is to designate some of the nodes (usually the toll switches) but not all, as 'master' nodes. All other nodes will be directly or

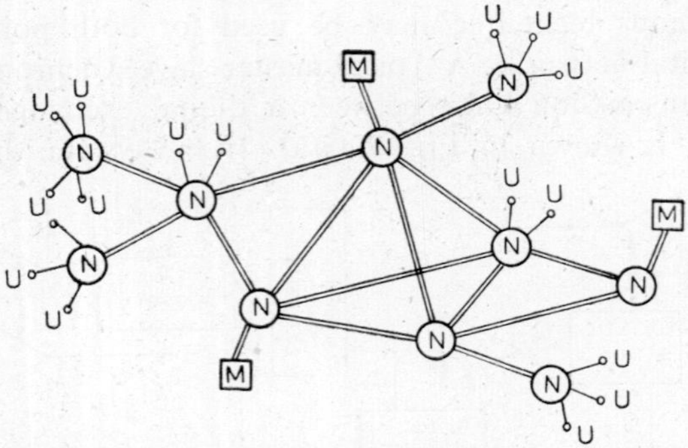

Fig. 7.46 Simple nodal model for the network M: master control; N: node; U: subscriber

indirectly referred to these master nodes through a circuit similar to that of Fig. 7.45(a). At the master nodes, very precise clocks (such as atomic clocks) or very stable VCXO's (with a stability of the order of 10^{-10} per day) could be used and the slave nodes will have stable VCXO's (of stability 10^{-5} only). In a more precise system, the slave VCXO's are made to lock with the master node by adjusting their nodal timing through drift rate corrections applied on a periodic basis. The master node also may have the capability of adjusting their own timing through similar correction. This periodic adjustment is based upon a method of weighting signals proportional to the contents of the buffers and is used to minimize the effect of phase delay variations. This technique, periodic in nature, is known as 'Discrete Control Correction' [29] which absorbs large delay variations provided that they are not rapid, by increasing the number of periodic adjustments to the node clocks. The block diagram of a multi-input synchronization system using this technique is shown in Fig. 7.47, where the timing recovery and data regeneration operations essentially prepare the received data streams S_j to be read into the buffers under the control of the recovered timing signal f_j. The clock controller provides the necessary controls to weigh the buffer contents and to derive a periodic correction for the local VCXO. To avoid undesirable transients, the differential buffer increment with buffer states is fed to the controller to provide the error signal for node clock adjustment. The system stability with such techniques has been found to be within practical limits.

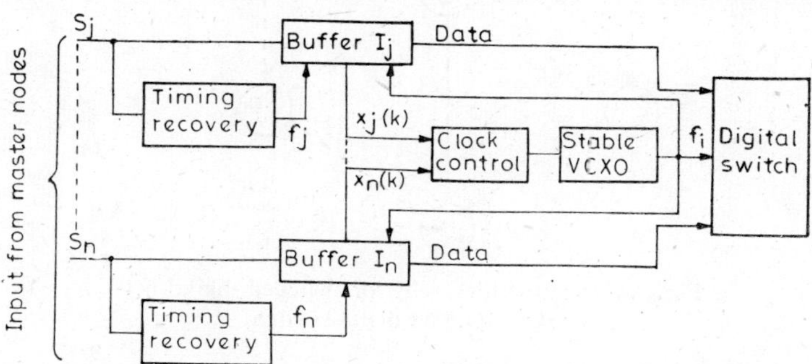

Fig. 7.47 Functional block diagram of discrete control correction. S_j: received data; f_j: received timing data; f_i: local frequency; $x_j(k)$: level of buffer j at an instant k

To achieve reliability in the distribution of master clocks, a self-organizing master-slave system has been proposed. In this system, the master clock and all intermediate distribution clocks are rank-ordered and redundant communication paths are provided to maintain the connectivity of the network in the event of failures in some part of the network. The rank-ordering is done by sending a 3-integer signature with the clock signal. The slave node then can recognise a higher-order clock and lock to it. This ensures that if the network is connected, the highest ranked node will always be the master and if the network becomes disconnected, a unique new master is chosen for each part of the network. The long-term accuracy of the network is determined by the master clock while the short-term accuracy is determined by propagation factors.

The master-slave system of synchronization is used in the Bell System's digital switching network [30]. The timing hierarchy of this network is shown in Fig. 7.48, where the master clock (of the highest order) is supplied by the Bell System Reference Frequency. This is derived from three cesium-beam frequency standards with an accuracy of one part in 10^{11} and their outputs are jointly compared to identify any malfunction. The output frequencies of these standards are 20.48 MHz and 2.048 MHz, and these are used to slave the 16.384 MHz crystal oscillators used in No. 4 ESS toll exchanges. The primary standard frequencies are transmitted to the toll exchanges via the existing broadband analog transmission facilities. The toll

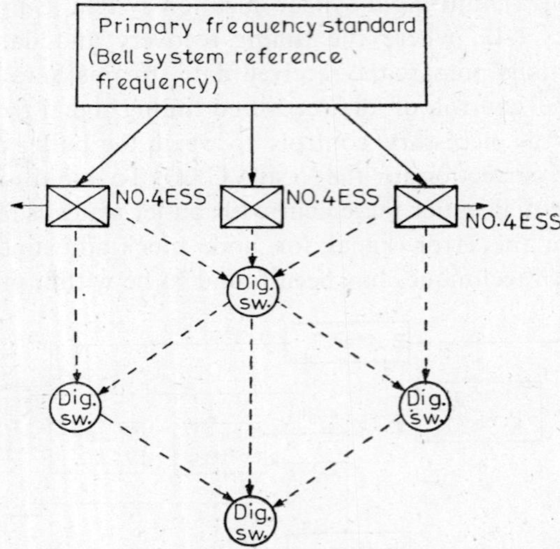

Fig. 7.48 Timing hierarchy for switched digital network; DIG.SW: digital switch

switches now form the second tier of the master-slave hierarchy. Clocks in these selected digital switches then supply frequency reference to other PCM-TDM switches over the existing digital transmission facilities. The digital MUX systems are also slaved to the same timing hierarchy.

Isochronous/Mutual Synchronisation System

The master-slave system is suitable for applications in simple star networks but it has serious problems with a generalized network and also with the master-clock failure. A more organic method of mutual synchronization is to interconnect the nodes with many redundant paths and the node clock is locked to the average reference phase contributed by all incoming clock phases plus that of the locally generated clock. The average clock reference is in turn distributed to other nodes to be used as a part of each of their references. A simple three-input scheme for mutual synchronization is shown in Fig. 7.49, where the local VCO is controlled by the sum of the outputs of three-phase comparators, which in turn compare the phases of the local clock and the incoming clocks. Thus, the VCO operates at the mean clock frequency. The frame phase synchronization is obtained by using variable delay pads (similar to the elastic stores) at the input to each trunk. This allows the digital switch to operate synchronously with the signals on the digital trunks.

A mutual synchronization technique, called 'Equational Timing System', has been proposed and investigated in detail [31]. In this system, each node sends out to the other nodes the information on the phase of incoming signals, arriving at this node from their respective nodes, with reference to the phase of the clock of the receiving node. Each clock generator is

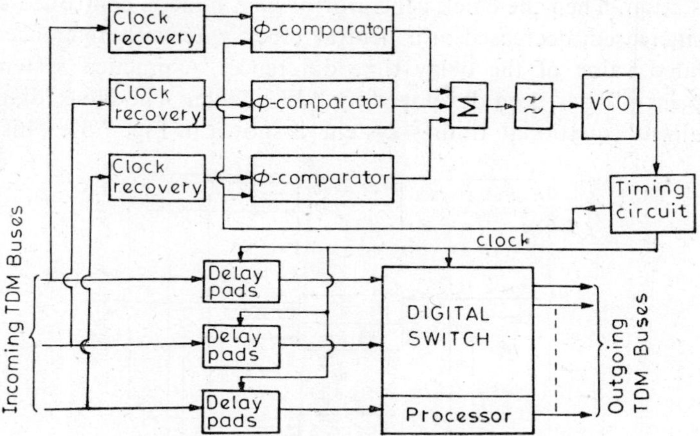

Fig. 7.49 A 3-input mutual synchronization system

controlled so as to have a common final clock rate among the nodes that is independent of the propagation delay of the transmission line. A scheme for only two nodes is shown in Fig. 7.50, where τ_{ab} and τ_{ba} are the propagation delays of paths from A to B and B to A respectively. These two paths include variable delay pads τ_a and τ_b, and the bit streams from the delay pads are read out in synchronism with the timing of respective clocks in each node. If the total loop delay $(\tau_{ab} + \tau_{ba} + \tau_a + \tau_b)$ is kept an even multiple of a frame repetition period, then the digital signals can be controlled under a common clock pulse generator in respective nodes. When the propagation delay of the paths varies due to climatic or geographical condition of the transmission cable, τ_a and τ_b are corrected to satisfy the above loop delay condition and the respective clock rates are controlled so that $\tau_a = \tau_b$. Thus, a common clock rate and a frame phase relationship exist at both the nodes.

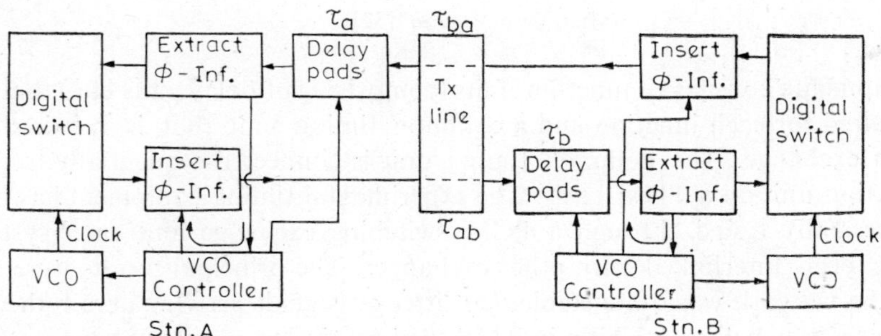

Fig. 7.50 A two-node scheme for Equational timing system

In an integrated network composed of many switching nodes that are interlinked by several digital junctions, it should be arranged that the delay time differences of the delay pads of the respective junctions are accumulated in each station. Then the clock generator of each node is controlled so as to have an increased, decreased or a neutral clock rate with reference to the accumulated value of the delay time differences. A practical system based on this principle has been developed by NEC, Japan. The block diagram of a generalized equational timing system is shown in Fig. 7.51. This timing

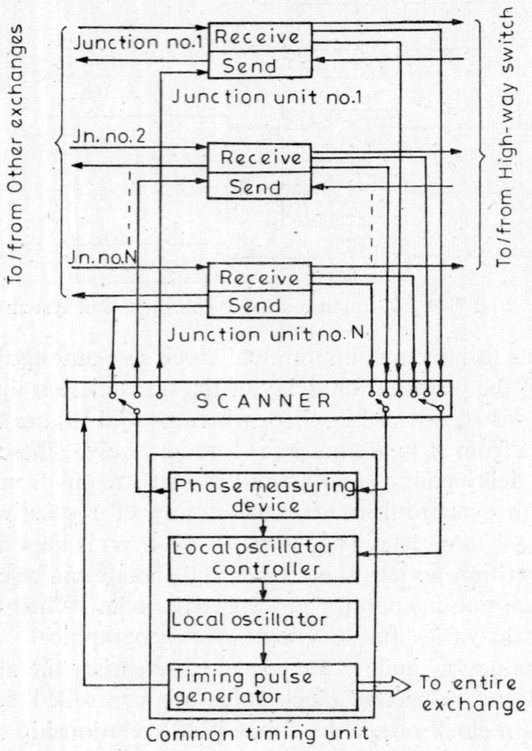

Fig. 7.51 A generalized Equational timing system (after Yamato *et al.* [32])

equipments consists of junction units (comprising of delay pads etc.) that are required for each junction and a common timing unit that is required for each exchange. The common timing unit is connected sequentially to each junction unit on a TDM basis. The experimental timing equipment has been successfully tested through a PCM switching exchange and such systems have been interlinked with other exchanges. The principal results were that (a) the network was quite stable, (b) jitter on signals arriving at all the exchanges was below ± 3 bit/s in RMS, and (c) timing accuracy between four generators was within ± 0.1 μs. Further theoretical analysis of the dynamic behaviour of this system has been done with mesh, ring and chain networks, and also for an hypothetical world-wide network [32]. The computer simulation of the control system has shown that a chain network has better stabilizing characteristic than a ring network. In the case of the world-wide network, it was possible to control the accuracy of the clock pulse generator within 1/50 bit. This requires that the clock generator frequency accuracy should be of the order of 10^{-7} for 1.5 Mb/s PCM signal. Thus, the stable world-wide integrated network is quite possible in principle.

REFERENCES

1. Inose, H., *An Introduction to Digital Integrated Communications Systems,* University of Tokyo Press, 1981.
2. Bellamy, J., *Digital Telephony*, John Wiley, 1982.
3. Special Issue on Digital Switching, *IEEE Trans. Comm.*, vol. Com-27, July 1979.
4. Joel, A.E., Jr., 'Digital switching—How it was developed', *IEEE Trans. Comm.*, vol. Com-27, pp 948-959, 1979.
5. Skaperda, N.J., 'Some Architectural Alternatives in the Design of a Digital Switch', *IEEE Trans., Comm.*, vol. Com-27, pp 961-972, 1979.
6. Charransol, P., *et al.*, 'Development of a Time division switching network usable in a very large range of capacities', *IEEE Trans. Comm.*, vol. Com-27, pp 982-987, 1979.
7. Lee, C.Y., '*Analysis of Switching Networks'*, BSTJ, pp 1287-1315, 1953.
8. Hills, M.T., '*Telecommunications switching principles*', George Allen and Urwin, London, 1979.
9. Hills, M.T., and Kano, S. '*Programming electronic switching systems*', Peter Peregrinus, 1976.
10. Cotton, J.M., 'Hardware for a wide range of digital exchange sizes', *Electrical Communication*, vol. 54, No. 3, pp 215-224, 1979.
11. Fritz, P., 'Subscriber line connection units—Present realizations and future trends', *IEEE Trans. Comm.*, vol. Com-27, pp 973-978, 1979.
12. (a) Bruce, R.A., Giloth, P.K. and Siegel, E.H. Jr, 'No. 4 ESS—Evolution of a digital switching system', *IEEE Trans. Comm.*, vol. Com-27, pp. 1001-1011, 1979.
    (b) Special Issue on No. 4 ESS, *BSTJ*, vol. 56, No. 7, Sept. 1977.
13. (a) Joel, A.E. Jr, (Ed), Electronic switching, central office systems of the world, *IEEE Press* (1976).
    (b) Special Issue on Telecommunication Circuit Switching, *Proc. IEEE*, vol. 65, No. 9, Sept. 1977.
14. Special Issue on No. 5 ESS, *Bell Labs. Record*, vol. 59, No. 9, Nov. 1981.
15. Special Issue on ITT System-12, *Electrical Communication*, vol. 54, No. 3, 1979.
16. ITT-1240 Digital Exchange: A Reprint of Articles, *Electrical Communication*, Supplement to vol. 55, No. 2, 1980.
17. Special Issue on ITT-1240 Digital Exchange, *Electrical Communication*, vol. 56, No. 2/3, 1981.
18. Special Issue on System-12 Digital Exchanges, *Electrical Communication*, vol. 59, No. 1/2, 1985.
19. Sueyoshi, H., *et al.*, 'System design of digital telephone switching system—NEAX 61', *IEEE Trans. Comm.*, vol. Com-27, pp 993-1000, 1979.
20. Yoshiaki, M. and Funakoshi, M., 'HDX10 Digital switching system', *Hitachi Review*, vol. 28, pp 279-284, 1979.
21. Andre, G. *et al.*, 'Digital switching in France', *IEEE Trans. Comm.*, vol. Com-27, pp 1047-1055, 1979.
22. Meyer, J., *et al.*, 'A digital subscriber set', *IEEE Trans. Comm.*, vol. Com-27, pp 1096-1103, 1979.
23. Kaderali, F. and Weston, J.D., 'Digital subscriber loops', *Electrical Communication*, vol. 56, pp 71-79, 1981.
24. Gasser, L., *et al.*, 'Digital subset characteristics for a future digital-telephone network', *Electrical Communication*, vol. 56, pp 103-108, 1981.

25. Tsuda, T., *et al.*, 'An approach to multi-service subscriber loop system using packetized voice/data terminals', *IEEE Trans. Comm.*, vol. Com-27, pp 1112-1117, 1979.
26. Okada, K., *et al.*, 'Microprocessor application to telephone terminal switching system architecture', *IEEE Trans. Comm.*, vol. Com-27, pp 1088-1095, 1979.
27. Pan, J.W., 'Synchronizing and multiplexing in a digital communication network', *Proc. IEEE*, vol. 60, p 594, 1972.
28. Mayo, J.S., 'Experimental 224 Mb/s PCM Terminals', *BSTJ*, vol. 44, p 1813, 1965.
29. Bittel, R.H., *et al.*, 'Clock synchronisation through discrete control correction', *IEEE, Trans.* vol., Com-22, p 836, 1974.
30. Cooper, C.A., 'Synchronisation for telecommunications in a switched digital network', *IEEE Trans. Comm.*, vol. Com-27, pp 1028-1033, 1979.
31. Yamato, J. *et al.*, 'Synchronization of a PCM Integrated telephone network', *IEEE Trans.*, vol. COM-16, p 1, 1968.
32. Yamato, J., *et al.*, 'Dynamic behaviour of a synchronization control system for integrated telephone network', *IEEE Trans.* vol. Com-22, p 839, 1974.

CHAPTER 8

Information Theory and Coding

The basic elements of a digital communication system have been shown in Fig. 2.1. Some of the important subblocks of the system, e.g., A/D conversion, modulation schemes, multiplexer etc., have been discussed in earlier chapters. However, to obtain maximum efficiency in the systems, it is necessary to incorporate source encoding and error-correcting techniques as well. The error performances of digital modulation schemes, e.g., CPSK, QPSK, FSK etc., show a threshold (for $P_b = 10^{-5}$) at the input SNR in the range of 9–15 dB; and further improvements in this threshold are only possible through the use of forward error-correcting (FEC) codes. Shannon's classical channel capacity theorem [2], on the other hand, promises a coding gain of 11 dB with an infinite channel bandwidth and a large coding delay. Since the publication of these results in 1948, theoreticians and designers have been constantly trying to discover new methods of coding information to approach the potential coding gain. Recent developments in coding techniques and optimum signal design are the result of these efforts [1].

The communication channels are broadly classified as: (a) discrete noiseless channel, (b) discrete channel with discrete noise, (c) discrete channel with continuous noise, and (d) continuous channel with continuous noise. For each of these types, a specific coding technique is applicable, e.g., source encoding for noiseless channels, FEC codes for discrete channels and wideband modulation schemes for continuous channels.

The FEC codes, as discussed in this chapter, are of two classes, viz., Block codes and Convolutional codes. Each of these codes performs optimally for certain practical channels and concatenation of certain codes results in further improvement in their error performances. An alternative approach to improve channel efficiency is to use Orthogonal codes, but these require larger bandwidths as compared to FEC codes. It has been shown that in both systems, an exponential bound on the error probability is obtained and $P_b \to 0$, in the limit of the codelength $n \to \infty$ [1, 7].

The 'Code capacity' and 'Channel capacity' theorems in Information Theory specify the limits of the maximum information that may be generated by a set of probabilistic symbols and the information that may be carried by a physical noisy channel. Shannon's theorems do not provide the methods of achieving the ideal channel capacity, given the statistics of the channel, but only indicate a benchmark, with which all practical communication systems can be tested to evaluate their efficiency. However, the coding theorists have achieved the goal partially. As a result, in wideband systems, a coding gain of 6.5–7.5 dB is achievable with realizable hardware by using either long-length orthogonal codes or concatenated codes [10].

In view of the above, we discuss information measure, language capacity, code capacity and channel capacity in this chapter. Techniques for noiseless coding as well as for coding in noisy channels (Orthogonal and FEC codes) have been discussed. Based on the bounds of performances developed, many of the useful practical communication systems have been evaluated and compared.

8.1 ENTROPY OF DISCRETE SOURCES [1, 2]

Based on the work of Gabor, Shannon and others [2], the Entropy $H(X)$ of a zero-memory discrete source S is defined as the average amount of information associated with the source symbols $\{X\} = \{x_1, x_2, \ldots, x_i, \ldots x_m\}$, and

$$H(X) = -\sum_{i=1}^{m} p_i \log p_i, \qquad \Sigma p_i = 1 \tag{8.1}$$

where the symbol probabilities $p\{X\} = \{p_1, p_2, \ldots, p_i, \ldots, p_m\}$. The Eq. (8.1) is based on the intuitive definition of the information associated with an event as

$$\begin{aligned} I(x) = \log \frac{1}{p(x)} &= -\log_2 p(x) \text{ bits} \\ &= -\log_e p(x) \text{ nats} \\ &= -\log_{10} p(x) \text{ Hartleys} \end{aligned} \tag{8.2}$$

and $I(x)$ of an event increases with its uncertainty. Further, the total information of m independent events is the sum of the informations associated with individual events. The Entropy, as defined by Eq. (8.1), is associated with a source, whereas information is associated with a message. $H(X)$ is also referred to as marginal entropy and $p(x_i)$ as marginal probabilities. It follows from Eq. (8.1) that:

(i) $H(X) = 0$, if all $p(i)$'s, except one, are zero;

(ii) $H(X)$ is maximum and $H(X) = -\sum_m \frac{1}{m} \log \frac{1}{m} = \log m$,

if $p_1 = p_2 \ldots = p_m = 1/m$.

For a binary source with alphabet $\{0, 1\}$ and probabilities $\{p, 1 - p\}$,

$$\begin{aligned} H(X) &= p \log \frac{1}{p} + (1 - p) \log \frac{1}{1 - p} \\ &\triangleq H(p) \end{aligned} \tag{8.3}$$

$H(p)$ then increases with p for $0 < p < 0.5$ and decreases for $p > 0.5$, with a maximum $H(P) = 1$ bit for $p = 0.5$.

Since binary transmission is efficient in terms of hardware, it is usually convenient to encode multi-alphabet source outputs into words using binary (or lesser) alphabets; and this is known as the extension of the coder alphabets. Consider that 8 messages are to be coded with a binary (0/1) alphabet, then the binary codewords are: {000, 001, 011, 010, 100, 101, 110, 111}, leading to the third extension of the source coder. In general, for m messages, with $m = 2^n$, an nth extension of the source is required. Thus, if a zero-memory source S has the source alphabets $\{x_1, x_2, \ldots, x_m\}$, then the nth extension of S, called S^n, is a zero-memory source with m^n symbols $\{y(1), y(2), \ldots, y(m^n)\}$ as its higher order alphabet. It may then be shown that the entropy of the extended source S^n is given by:

$$H(S^n) = nH(S) \tag{8.4}$$

Example 1

Consider a source with alphabets $\{x_1, x_2, x_3, x_4\}$ and apriori probability $p\{x_i\} = \{1/2, 1/4, 1/8, 1/8\}$. Thus, a typical message will be:

$$\{\ldots x_1x_1x_2x_1x_3x_1x_2x_4x_2x_1x_3x_4x_1x_2x_1x_1 \ldots\},$$

satisfying $p\{x_i\}$. The entropy of the source $H(X)$ is then calculated as $H(S) = H(X) = 1.75$ bits/ symbol. Suppose the source is now extended to deliver messages with two symbols only, then the new alphabet $\{y(j)\}$ are:

$$\{x_1x_1,\ x_1x_2,\ x_1x_3,\ x_1x_4,\ x_2x_1,\ x_2x_2,\ x_2x_3,\ x_2x_4,\ x_3x_1,\ x_3x_2,\ x_3x_3,$$
$$x_3x_4,\ x_4x_1,\ x_4x_2,\ x_4x_3,\ x_4x_4\}$$

Using the corresponding values of $p\{y(j)\}$, the entropy of the extended source is calculated as:

$$H(S^2) = 3.5 \text{ bits/symbol pair} = 2H(S)$$

8.1.1 Source Encoding

In practical communication systems, it is necessary to transform the m-ary source alphabets $\{x_1, x_2, \ldots, x_m\}$ to a convenient form, say, binary or D-ary digits to match the channel. For efficient transmission of information, it is desirable that the transformed code alphabets $\{C_1, C_2, \ldots, C_D\}$ have equal probability of occurrence, hence the technique is also known as Entropy coding. To avoid ambiguity in deciphering the codes as well as to minimize the time of transmission of the message, the codes should have the following properties:

(a) Codes should be 'separable' or uniquely decodable. They should also be 'comma-free'. This restricts the selection of the codes in such a way that no shorter code can be a prefix of a longer code. Such codes are also called Instantaneous codes.

(b) The average length of the codewords

$$\bar{L}_i = \Sigma p_i L_i \tag{8.5}$$

where, L_i = length of the ith codeword, should be minimum and $\bar{L}$ should approach the value $[H(X)/\log D]$, with the condition $\bar{L} \geqslant H(X)/\log D$. Since $\Sigma p_i L_i$ is to be minimum, the codewords with larger L_i should have smaller p_i, i.e.,

$$L_i \propto \log (1/p_i)$$

(c) The code efficiency η is defined as:

$$\eta = \frac{H(X)}{\bar{L} \log D}; \tag{8.6}$$

and $\eta \rightarrow 1$ for the optimum (also called compact) codes.

Further, Redundancy of codes is defined as: $R = (1 - \eta)$. The optimum codelength L_i, according to Shannon [3], has the bound,

$$\log \frac{1}{p(x_i)} \leqslant L_i \leqslant \log \frac{1}{p(x_i)} + 1 \tag{8.7}$$

Black has suggested that $L_i \approx \log 1/p(x_i)$, may be taken as the codelength. The Shannon-Fano technique of generating the codewords from the given symbol probabilities is illustrated in the following example.

Example 2

Consider an 8-alphabet source with $p(x_i)$ as given in Table 8.1. The symbols are first arranged according to their decreasing probability, and then, they are divided into two groups, such as $\Sigma\, p(x_i)$ in each group ≈ 0.5. The

Table 8.1

Symbols	$p(x_j)$	Code Selection						Codes	log $1/p(x_i)$
A	0.30	$\Sigma = 0.5$	1	1				11	1.7
B	0.20		1	0				10	2.3
C	0.15		0	1	1			011	2.7
D	0.12		0	1	0			010	3.05
E	0.10	$\Sigma = 0.5$	0	0	1			001	3.3
F	0.07		0	0	0	1		0001	3.8
G	0.04		0	0	0	0	1	00001	4.6
H	0.02		0	0	0	0	0	00000	5.6

upper group is assigned '1' as the first code bit and the lower group is assigned '0'. Then each group is further divided into two subgroups, such that $\Sigma p(x_i)$ in each subgroup ≈ 0.25, and the next bit is assigned as '1' to the upper subgroup and '0' to the lower one. This subdivision according to equal $\Sigma p(x_i)$ is continued till all alphabets have been assigned distinct codes. This assignment of the codes is shown in Table 8.1. In the Table, the values of $\log 1/p(x_i)$ are also shown, and it is seen that $L_i \approx \log 1/p(x_i)$, as suggested by Black. However, L_i's given by Eq. (8.7) is larger than the actual values in the Table. Further,

$$H(X) = \Sigma p(x_i) \log 1/p(x_i) = 2.663 \text{ bits/symbol}$$

and

$$\overline{L} = \Sigma p_i L_i = 2.69$$

Then,

$$\eta = \frac{2.663}{2.69} = 99\%;$$

and at the coder output $p(0) = 0.53$ and $p(1) = 0.47$.

Huffman [5] has suggested a more efficient method of constructing separable codes with minimum redundancy (i.e., minimum $\overline{L}$). The rules of code construction are:

(a) The symbols are arranged according to their decreasing probability.
(b) For minimum $\overline{L}$, $L(i)$ should be greater or equal to $L(i-1)$. Only in case of last two symbols, $L(m) = L(m-1)$, and they should differ only in the last digit.
(c) Select the last two symbols x_{m-1} and x_m, and assign 0/1 to them. Form a composite symbol with probability $= p(x_{m-1}) + p(x_m)$, and construct a reduced source with $(m-1)$ alphabets and order them according to their decreasing probability.
(d) Assign 0/1 to the last two symbols of the new column and again construct a second reduced source with the ordering as above.
(e) Continue this till all symbols are assigned 0/1 codes. Example 3 shows the coding technique.

Example 3

Consider a source with 8 alphabets with $p(x_i)$ as given in Table 8.2. The reduced sources are formed as indicated above and 0/1 codes are assigned accordingly. By tracing the code path backwards, the assigned code is defined as shown. The efficiency of the code is calculated as:

$$H(X) = 2.75 \text{ bits}; \qquad \overline{L} = \Sigma p_i L_i = 2.8$$

and,

$$\eta = 2.75/2.8 = 98.2\%$$

Shannon-Fano method also gives a similar code with $\overline{L} = 2.8$.

Consider now the encoding of an extended source S^n and it may be shown that the code efficiency is improved considerably by encoding the extended source-outputs, instead of coding the direct source outputs.

Table 8.2

Assigned code	Symbol	$p(x_i)$	Reduced Sources					
10	A	0.22	0.22	0.22	→0.25	→0.33	→0.42	→0.58(0)
11	B	0.20	0.20	0.20	0.22	0.25	0.33(0)	0.42(1)
000	C	0.18	0.18	0.18	0.20	0.22(0)	0.25(1)	
001	D	0.15	0.15	0.15	0.18(0)	0.20(1)		
011	E	0.10	0.10	→0.15(0)	0.15(1)			
0100	F	0.08	0.08(0)	0.10(1)				
01010	G	0.05(0)	→0.0.7(1)					
01011	H	0.02(1)						

Example 4

A two-alphabet source with $p(x_1) = 0.25$ and $p(x_2) = 0.75$, has $H(X) = 0.815$ bits, and the best binary code as : $x_1 = 0$, $x_2 = 1$; giving $\eta = 0.815$. However, extending the source as S^3 with the alphabets $\{y_1, y_2, \ldots, y_8\} = \{000, 001, \ldots, 111\}$, and the corresponding $p\{y_j\}$, a new set of codes may be developed (using Huffman's technique) as given in Table 8.3. We now have $H(S^3) = 2.445$ bits and $\bar{L} = \Sigma p_i L_i = 2.47$ bits, giving $\eta = 99\%$, thus showing remarkable improvement over the coding without extension. Further extension of the source gives better η, but the rate of improvement is slower.

8.1.2 Code Capacity Theorem [1, 2, 3]

It has been seen in these examples that by efficient (noiseless) encoding, the average length $\bar{L}$ of the codewords at the coder output approaches $H(X)$/symbol of the source, particularly with the extended source S^n. Since the average information per source symbol is $H(X)$ bits, a message of length n symbols gives $nH(X)$ bits of information. Equivalently, the source generates a number of messages M_s given by,

$$M_s = 2^{nH(X)} = 2^{H(X)T/t_0} \tag{8.8}$$

where, T = total duration of n-length message, t_0 = duration of each symbol and $n = T/t_0$. At the coder output, the average information per codeword is $\bar{L} \log D$ bits ($D \geqslant 2$), and in T sec, the number M_c of code messages generated is,

$$M_c = 2^{n\bar{L} \log D} = 2^{(T/t_0)\bar{L} \log D} \tag{8.9}$$

where $n = T/t_0$ and t_0 = average duration of codewords. It is intuitively evident that M_c has to be greater or equal to M_s, if the overflow of the buffer memory at the coder input is to be avoided. Thus, we have the bound:

$$M_c = 2^{n\bar{L} \log D} \geqslant M_s = 2^{nH(X)}$$

Table 8.3

Assigned code	Symbols	$p(y_j)$	Reduced Sources					
1	y_8	27/64	27/64	27/64	27/64	27/64	27/64	→37/64(0)
001	y_7	9/64	9/64	9/64	→10/64	→18/64	→19/64(0)	27/64(1)
010	y_6	9/64	9/64	9/64	9/64	10/64(0)	18/64(1)	
011	y_4	9/64	9/64	9/64	9/64(0)	9/64(1)		
00000	y_5	3/64	→4/64	→6/64(0)	9/64(1)			
00001	y_3	3/64	3/64(0)	4/64(1)				
00010	y_2	3/64(0)	3/64(1)					
00011	y_1	1/64(1)						

Therefore, $$\bar{L} \geqslant H(X)/\log D \tag{8.10}$$

It now remains to be shown that with $n \to \infty$, $\bar{L} \to H(X)/\log D$. Shannon has formalized these results in the following (Ist. Fundamental theorem):

'Suppose that there are a zero-memory source S with entropy $H(X)$/symbol and a noiseless channel with a capacity C bits/message. Then it is possible to encode the output of S (with the encoder delivering messages to the channel) so that all the information generated in S is transmitted through the channel without loss if, and only if, $C \geqslant H(X)$. If $C < H(X)$, then all messages generated by S in T sec, cannot be transmitted through the channel in T sec, without buffer memory overflow.' [Here the channel capacity C bits/message is equivalent to the code capacity of the coder $= \bar{L} \log D$ bits per codeword, since the coder has to be matched to the channel].

It is observed from the encoding examples of the earlier section that $\bar{L} = H(X)$, and $L_i = \log 1/p(x_i)$, if $\log 1/p(x_i)$ is an integer; and in a general case, the Eq. (8.7) holds. Thus with D-ary encoding, $D > 2$, we have the bound:

$$\log 1/p(x_i) \leqslant L_i \log D \leqslant \log 1/p(x_i) + 1 \tag{8.11}$$

Multiplying the above by $p(x_i)$ and summing over all i's, we get

$$\sum_{i=1}^{m} p(x_i) \log 1/p(x_i) \leqslant \sum_{i=1}^{m} p_i L_i \log D$$

$$\leqslant \sum_{i=1}^{m} p(x_i) \log 1/p(x_i) + \sum_{i=1}^{m} p(x_i)$$

or

$$\frac{H(X)}{\log D} \leqslant \bar{L} \leqslant \frac{H(X) + 1}{\log D} \tag{8.12}$$

To obtain better efficiency, one uses the nth extension of S, giving $\bar{L}_n$ as the new codelength. Then using $H(S^n) = nH(S)$, Eq. (8.12) is modified as:

$$\frac{nH(X)}{\log D} \leqslant \bar{L}_n \leqslant \frac{nH(X) + 1}{\log D}$$

or

$$\frac{H(X)}{\log D} \leqslant \bar{L}_n/n \leqslant \frac{H(X)}{\log D} + \frac{1}{n \log D}$$

and with $n \to \infty$, $$\lim_{n\to\infty} \bar{L}_n/n = \frac{H(X)}{\log D} \tag{8.13}$$

Here $\bar{L}_n/n$ is the average number of code symbols used per single symbol of S, using the extended source S^n, and in general, $\bar{L}_n/n \leqslant \bar{L}$, where $\bar{L}$ is for the original source S. The code capacity is now $(\bar{L}_n/n) \log D = C$ bits/message of the channel, and for successful transmission of messages through the channel, we have,

$$H(X) \leqslant (\bar{L}_n/n) \log D = C \text{ bits/message} \tag{8.14}$$

The theorem is also valid for sources with finite memory, i.e., the Markov sources, as discussed in the next section. Moreover, all code symbols may not be of equal duration t_0 as assumed above; even then, the theorem remains valid with minor modifications.

8.2 MARKOV SOURCES: CONDITIONAL ENTROPY [4]

We have so far considered zero-memory sources, i.e., sources with no intersymbol correlation. In real-life sources, however, there are intersymbol influence (i.e., probabilistic constraints) present such that the occurrence of x_i in the zeroth position s_0 of the message depends on the previous q symbols $\{s_1, s_2, \ldots, s_q\}$. Such a source is known as a qth order Markov source S_q, (English or any other language is an excellent example of such sources), and these are generally specified by the set of conditional probabilities:

$$p(x_i/s_1, s_2, \ldots, s_q) \tag{8.15}$$

where, x_i is the symbol in the s_0 position and each s has an m-symbol alphabet $\{x_1, x_2, \ldots, x_m\}$.

To define conditional entropies and other properties of Markov sources, consider first two correlated sources S_1 and S_2 delivering symbols $\{x_i\}$ and $\{y_j\}$. Their joint and conditional entropies are defined as:

Joint entropy

$$H(X, Y) = -\sum_{i=1}^{m}\sum_{j=1}^{n} p(x_i, y_j) \log p(x_i, y_j) \tag{8.16}$$

Conditional Entropy

$$H(Y/X) = -\sum_{i=1}^{m}\sum_{j=1}^{n} p(x_i, y_j) \log p(y_j/x_i) \tag{8.17}$$

where $\quad p(x_i, y_j) = p(x_i)p(y_j/x_i)$.

If $\{X\}$ and $\{Y\}$ are independent, then,

$$H(X, Y) = H(X) + H(Y)$$

and

$$H(Y/X) = H(Y) \tag{8.18}$$

Further, it may be shown that:

(i) $H(X, Y) = H(X) + H(Y/X)$
$\quad = H(Y) + H(X/Y)$

(ii) $H(X) \geqslant H(X/Y)$
$H(Y) \geqslant H(Y/X)$

(iii) $H(X, Y) \leqslant H(X) + H(Y)$ (8.19)

These relations for two variables may be extended to three and more variables, giving the relations:

$$\begin{aligned} H(X, Y, Z) &= H(X, Y) + H(Z/X, Y) \\ H(X, Y) &\geqslant H(X, Y/Z) \end{aligned} \tag{8.20}$$

and for q-variables,

$$H(S_1, S_2, \ldots, S_q) = H(S_1, S_2, \ldots, S_{q-1}) + H(S_q/(S_1, S_2, \ldots S_{q-1})) \tag{8.21}$$

where, $H(S_1, \ldots, S_q)$ is the joint entropy of all sources $\{S_1, S_2, \ldots S_q\}$.

Assuming that the Markov source is ergodic, where the intersymbol influence lasts only for a finite number of earlier q symbols (i.e., the memory of the source is finite) and its distribution of 'states' is stationary, the joint probability and different entropies may now be calculated from the given state transition probabilities by extending the results of Eqs. (8.16) and (8.20). Thus, we have,

$$p(s_1, s_2, \ldots, s_q, x_i) = p(s_1, s_2, \ldots, s_q) \cdot p(x_i/s_1, s_2, \ldots s_q)$$

and

$$H(s_1, s_2, \ldots, s_q, x_i) = H(s_1, s_2, \ldots, s_q) + H(x_i/s_1, s_2, \ldots, s_q) \tag{8.22}$$

where the conditional entropy $H(x_i/s_1, s_2, \ldots, s_q)$ is given by

$$\begin{aligned} H(S_q) &= -\sum_{s_{q+1}} p(s_1, s_2, \ldots s_q, x_i) \log p(x_i/s_1, s_2, \ldots, s_q) \\ &= H(s_1, s_2, \ldots, s_q, x_i) - H(s_1, s_2, \ldots s_q) \end{aligned} \tag{8.23}$$

We can now calculate the conditional entropy of a Markov source by using Eq. (8.23).

In similarity with the extended zero-memory source S^n, a qth-order Markov source may also be extended to S_q^n-source, with m^n symbols and reduced intersymbol influence. Thus, the nth extension of S_q is a μth-order Markov source where $\mu = q/n$, and a qth-order Markov source may be converted into a 1st order Markov source by making $n = q$. This means that the intersymbol influence is between two adjacent n-symbol words only, and as such, the coding problem is considerably simplified. In similarity with eq. (8.4), the entropy of the extended first order Markov source S_1^n is:

$$H(S_1^n) = nH(S_1) \tag{8.24}$$

Further, for an extended qth-order Markov source, S_q^n, the entropies are related as:

$$H(\overline{S_q^n}) = nH(S_q) + \epsilon, \qquad \epsilon > 0$$

and for $n \to \infty$,

$$\lim_{n\to\infty} \frac{H(\overline{S_q^n})}{n} = H(S_q) \tag{8.25}^*$$

*Given a qth-order Markov source S_q with the source alphabet $\{x_1, x_2, \ldots, x_m\}$, an Adjoint, $\overline{S_q}$, is defined as a zero memory source with the same source alphabet and with the 1st order symbol probabilities. $\{p_1, p_2, \ldots, p_m\} = \{p(x_1), p(x_2), \ldots, p(x_m)\}$. Then the two sources S_q and $\overline{S}_q$ have identical 1st order probabilities, but because of the intersymbol constraints in S_q,

$$H(S_q) \leqslant H(\overline{S_q}) \tag{8.25a}$$

The Code Capacity theorem, given by Eq. (8.13), may now be shown to hold for Markov sources as well. The result is [1, 2]:

$$\lim_{n\to\infty} \frac{\overline{L}_n}{n} = \frac{H(S_q)}{\log D} \tag{8.26}$$

8.2.1 Entropy of English Language [6]

In English language, the probabilities of occurrence of the 27 symbols (26 letters + space) are not equal and they are tabulated in various texts [4], e.g., $p(E) = 0.107\ldots$ and $p(Z) = 0.0006$. From these results,

$$H(X) = -\sum_{1}^{27} p_i \log p_i = 4.065 \text{ bits/symbol.}$$

For equal probabilities of all letters,

$$H(X)_{\max} = -\sum_{1}^{27} \frac{1}{27} \log 1/27 = 4.76 \text{ bits/symbol.}$$

This means that the length of messages could be shortened by the ratio 4.065/4.76 = 0.854, if equal probability random coding is used in forming English words. In other words, approximately 17 letters ($\text{Antilog}_2\, 4.065$) would be sufficient to express English messages, if the choice of the letters are ideally random. (However, this would mean that English could not be distinguished from French or German languages.)

The actual entropy of English is much less than 4.065 bits/symbol, because there are many constraints in forming words. These are in the form of some combinations, e.g., *th*, *qu*, etc., and the grammatical and syntactic constraints, which make the prediction of the following letter after initial few letters quite easy. Thus the English language source, and all other language sources, are Markov sources with finite memory, since

$$p(x_i/s_1, s_2, \ldots s_q)$$

for various letters are well established. It is then possible to calculate the digram, trigram and q-gram entropies for the language by using Eq. (8.23).

Consider initially that the language source is only a first order Markov source, i.e., the intersymbol influence extends only to the earlier letter in the words. Then the digram entropy $H(S_1) = H(x_i/s_1) = H(s_1, x_i) - H(s_1)$. Statistics of digrams, trigrams etc. for English are available, and from tabulated values, $H(s_1, x_i) = 7.70$ bits, $H(s_1, s_2, x_i) = 11.0$ bits. Thus,

$$H(S_1) = 7.7 - 4.065 = 3.635 \text{ bits/letter} \tag{8.27}$$

and the diagram structure has reduced the information value of the language to 3.635/4.76 = 76% of the maximum possible. Considering the intersymbol influence to extend to earlier two letters, the source is now a 2nd order Markov and the trigram entropy is:

$$\begin{aligned} H(S_2) &= H(x_i/s_1, s_2) = H(s_1, s_2, x_i) - H(s_1, s_2) \\ &= 11.0 - 7.7 = 3.3 \text{ bits/letter} \end{aligned} \tag{8.28}$$

The q-gram entropy is now calculated by using Eq. (8.23), but all such statistics are not easily available. It has been shown that the 8-gram entropy, $H(s_7)$, is approximately 2.3 bits/letter.

Shannon [6] has shown that the entropy per word is approximately:

$$H/\text{word} = 11.82 \text{ bits/word} \tag{8.29}$$

Taking each word $\simeq$ 5.5 letters, H/letter $\simeq$ 2.14 bits. Finally if one considers all possible intersymbol influences in English, then H/letter $\simeq$ 1.5 bits only. Thus the English language has an efficiency = 1.5/4.76 = 30% only and a redundancy of 70%.

8.2.2 Generalised Code Capacity Theorem [1, 2]

Equations (8.13) and (8.14) were proved in Sec. 8.1.2 assuming a zero-memory source and equal duration of source/code symbols. Relaxing the condition of zero intersymbol influence, and using Eq. (8.26), the number of messages generated by a qth-order Markov source in time T sec. is given by:

$$M_s(S_q) = \exp_2 [(T/t_o)H(S_q)] \tag{8.30}$$

Considering the general case of long sequences, β symbols long and of duration T, made up of $\beta/q = \gamma$ groups of q-symbol words, it may be shown that the number of messages generated in T sec. grows exponentially with time and is given by:

$$M_T(S_q) = A_m R_m^T, \tag{8.31}$$

where A_m and R_m are functions of q, t_0 and γ. The asymptotic source (language) capacity may now be defined as:

$$\lim_{T\to\infty} \left[\frac{\log M_T(S_q)}{T}\right] = \frac{\log A_m + T \log R_m}{T} \to \log R_m \tag{8.32}$$

On the other hand, by relaxing the condition of equal duration of code symbols, the number of possible code messages $N(T)$ generated in T sec is given as:

$$M_c(T) = N(T) = A_R R_o^T \quad \text{for } T \to \infty \tag{8.33}$$

The asymptotic code capacity is now defined as:

$$\lim_{T\to\infty} \left[\frac{\log N(T)}{T}\right] = \frac{\log A_R + T \log R_o}{T} \to \log R_o \tag{8.34}$$

It may now be concluded that for successful transmission of all information generated by the source in time T, the condition to be satisfied is:

$$M_c(T) = N(T) \geqslant M_T(S_q)$$

and for $T \to \infty$,

$$R_o \geqslant R_m \tag{8.35}$$

Shannon's first fundamental theorem may now be restated for general sources and code channels as:

'If the code capacity of a code channel and encoder exceeds the average information per unit time generated in the information source, i.e. iff $R_o \geqslant R_m$, then there is a method of encoding all the information into the channel (even if one has to wait for a long time T). There is no method of encoding all the information if it is generated at a rate exceeding the code capacity, i.e., if $R_0 < R_m$.'

The proper encoding, of course, will require a large buffer memory at the input to the coder and a second memory at the output if the codes of unequal length and duration are to be transmitted through the channel in a synchronous mode.

8.3 DISCRETE NOISY CHANNELS

In an actual transmission channel, the statistics of the transmitted symbols $\{x_i\}$ is changed due to the noise in the channel, and the received symbols $\{y_j\}$ have to be optimally interpreted from the inverse probability matrix $p(X/Y)$. It is usual to describe the channel characteristics (channel matrix) by the conditional probability matrix $p(Y/X)$ as shown below. The matrix $p(Y/X)$ is also known as the transition/noise matrix and is graphically shown in Fig. 8.1.

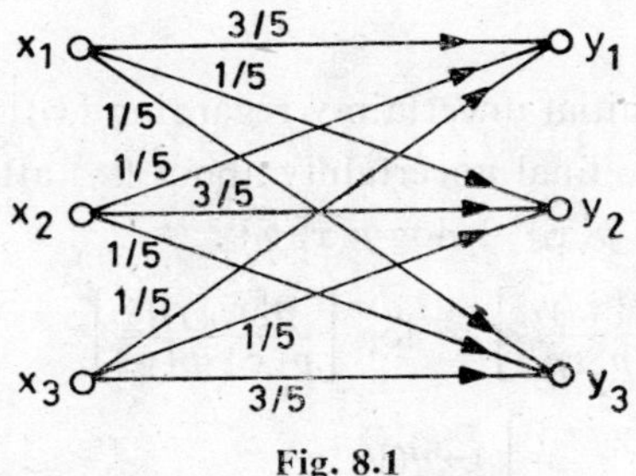

Fig. 8.1

Example 5

Consider the channel matrix $p(Y/X)$ as given below and also shown graphically in Fig. 8.1.

		y_1	y_2	y_3
	x_1	3/5	1/5	1/5
$p(Y/X) \rightarrow$	x_2	1/5	3/5	1/5
	x_3	1/5	1/5	3/5

and $p(X) = \{1/8, 1/4, 5/8\}$

Then, we may easily calculate the matrix $p\{X, Y\}$ and then, the matrix $p(X/Y)$ along with $\{p(y_j)\}$. The results are:

$$\{p(y_j)\} = \{1/4, 3/10, 9/20\},$$

and $p(X/Y) \rightarrow$

	y_1	y_2	y_3
x_1	3/10	1/12	1/18
x_2	1/5	1/2	1/9
x_3	1/2	5/12	5/6

Finally the different entropies are calculated as:

$$H(X) = 1.3 \text{ bits/symbol},$$
$$H(Y) = 1.54 \text{ bits/symbol},$$
$$H(Y/X) = 1.37 \text{ bits/symbol},$$
$$H(X/Y) = 1.13 \text{ bits/symbol},$$

and
$$H(X,Y) = 2.67 \text{ bits/symbol} < H(X) + H(Y).$$

8.3.1 Mutual Information

Assuming that the symbols $\{x_i\}$ are transmitted and $\{y_j\}$ are received in a channel, where $\{p(x_i)\} \neq \{p(y_j)\}$ due to noise (or due to probability constraints between $\{x_i\}$ and $\{y_j\}$), the Information gain through the channel (also known as Mutual information or Transinformation of the channel) is defined as:

$$\begin{aligned} I(x_i, y_j) &= [\text{the initial uncertainty regarding } \{x_i\}] \\ &\quad - [\text{the final uncertainty about } \{x_i\} \text{ after receiving } \{y_j\}] \\ &= -\log p\{x_i\} + \log p\{x_i/y_j\} \\ &= \log\left[\frac{p\{x_i/y_j\}}{p\{x_i\}}\right] = \log\left[\frac{p\{x_i, y_j\}}{p\{x_i\}\cdot p\{y_j\}}\right] \\ &= \log\left[\frac{\{y_j/x_j\}}{p\{y_j\}}\right] = I(y_j, x_i) \end{aligned} \tag{8.36*}$$

Averaging over all possible values of x_i and y_j, we obtain the Average Information gained by the receiver and is given as,

$$\begin{aligned} I(X, Y) &= E[I(x_i, y_j)] \\ &= H(X) - H(X/Y) \\ &= H(Y) - H(Y/X) \\ &= H(X) + H(Y) - H(X, Y) \text{ bits/symbol} \end{aligned} \tag{8.37}$$

where $H(X/Y)$ is specially called 'Equivocation', meaning thereby the loss of information due to channel noise.

---

*For noiseless channel, $p(x_i/y_j) = 1$, and $I(x_i, y_j) = -\log p(x_i)$ and this is known as: Self Information $= I(x_i, x_i) = -\log p(x)$ of Eq. (8.2). It then follows that

$$H(X) = \sum_i^m p_i I(x_i, x_i),$$

which is the same as Eq. (8.1)

Further, for a noiseless channel, $H(Y/X) = H(X/Y) = 0$, and

$$I(X, Y) = H(X) = H(Y) = H(X, Y) \tag{8.38}$$

For a channel, where $\{x_i\}$ and $\{y_j\}$ are independent and have no correlation between them,

$$H(X/Y) = H(X) \quad \text{and} \quad H(Y/X) = H(Y)$$

and
$$I(X, Y) = H(X) - H(X/Y) = 0; \tag{8.39}$$

thereby no information is transmitted through the channel.

As an example, the mutual information in Example 5 is calculated by using Eq. (8.37) as:

$$I(X, Y) = 1.54 - 1.37 = 0.17 \text{ bits/symbol}$$
$$= 1.30 - 1.13 = 0.17 \text{ bits/symbol.}$$

In digital communication, binary or multilevel (D-ary) signals are normally transmitted. and the channel noise produces errors at the output. The channel characteristic is normally defined by the error rate in the channel or equivalently by the transmission/noise matrix. Mutual information is defined by Eq. (8.37). $I(X, Y)$ is the rate of transmission in bits/symbol and the rate of transmission in bits per second is defined as:

$$R(X, Y) = I(X, Y)/t_0 \tag{8.40}$$

where t_o = duration of each symbol.

In general, the channels may have m inputs and n outputs, and all the entropies and mutual information are calculated using Eqs. (8.1), (8.17) and (8.37). However, in systems that are practically useful, $m = n$, and among these, the Binary Symmetric Channel (BSC) and the Binarp Erase Channel (BEC) are the most popular ones. From Eq. (8.37), it is evident that: given the transition matrix $p(Y/X)$, the $I(X, Y)$ may be maximized by maximizing $H(Y)$, and the resultant $H(X)$ gives the values of $p\{X\}$. This maximum value of mutual information is defined as the Channel Capacity C; thus,

$$C = I(X, Y)_{\max} \text{ bits/symbol}$$
$$= R(X, Y)_{\max} \text{ bits/sec.} \tag{8.41}$$

The transmission efficiency of the channel η is then defined as:

$$\eta = I(X, Y)/C \tag{8.42}$$

and the Redundancy of the channel $= (1 - \eta)$

$$= \frac{C - I(X, Y)}{C} \tag{8.43}$$

8.3.2 Binary Channels [1, 4]

One of the most common and widely used channel is the Binary Symmetric Channel (BSC), whose transition matrix is given by:

$$p(Y/X) = \begin{array}{c|cc} & 0 & 1 \\ \hline 0 & q & p \\ 1 & p & q \end{array} \tag{8.44}$$

If $p(x_1 = 0) = p_1$ and $p(x_2 = 1) = p_2$, then for a given error probability p,

$$p(y = 0) = p_1 q + p_2 p,$$
$$p(y = 1) = p_1 p + p_2 q,$$

Thus, the rate of transmission $R = H(Y) - H(Y/X)$

$$= H(Y) - H(p), \tag{8.45}$$

where $H(p) = -[p \log p + q \log q]$, given by Eq. (8.3). The maximum R is now obtained by making $p_1 = p_2 = 0.5$, giving $H(X) = H(Y) = 1$ bit; and the channel capacity of BSC, whose channel diagram is shown in Fig. 8.2, is given by:

$$C = R_{\max} = 1 - H(p)$$
$$= 1 + p \log p + q \log q \tag{8.46}$$

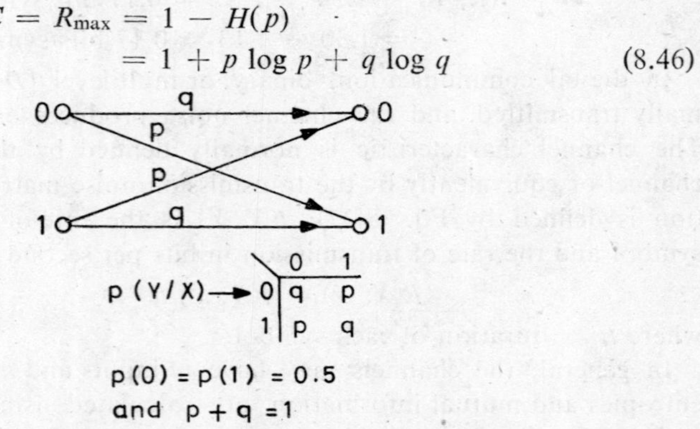

Fig. 8.2 A BSC and its channel matrix

Example 6

Consider that a source is transmitting equiprobable I/O symbols at the rate of 10^3 bits/sec, and the probability of error in the channel is $p_e = 1/16$. Then the rate of transmission and channel capacity are:

$$C = R/\text{sec} = 10^3[1 - H(p_e)]$$
$$= 663 \text{ bits/sec.}$$

If, however, $p_1 \neq p_2$, and $p_1 = 1/4$, $p_2 = 3/4$, then R/sec $= 523$ bits/sec. which is less than C.

It is interesting to note that if p and q are interchanged in the channel matrix, then also R remains the same. But with $P_e \to 0.5$, R decreases and $R = 0$ for $P_e = 0.5$.

Binary Erase Channel (*BEC*)

In data communication, it is a common practice to use ARQ (automatic request for retransmission) techniques for 100% correct data recovery. Such an ARQ channel is an equivalent BEC, whose transition matrix is shown in Fig. 8.3. Assuming $p_1 = p_2 = 0.5$, the mutual information is given by,

$$I(X, Y) = H(X) - H(X/Y)$$
$$= 1 - p = q \text{ bits/symbol} \tag{8.47}$$

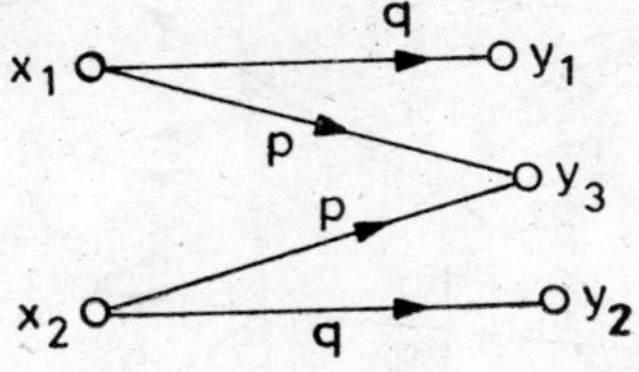

Fig. 8.3 A BEC

In this particular case, use of Eq. $I(X, Y) = H(Y) - H(Y/X)$ will not be correct as the information due to y_3 is rejected at the receiver.

We have discussed nth extension of the sources, giving $H(S^n) = nH(S)$, in Sec. 8.1. Similarly the nth extension of the channel results in a greater mutual information, giving,

$$I(X^n, Y^n) = nI(X, Y) \tag{8.48}$$

As an example, the second extension of a BSC gives the channel matrix as:

	$X \downarrow$ \ $Y \rightarrow$	00	01	10	11
	00	q^2	pq	pq	p^2
$p(Y/X) \rightarrow$	01	pq	q^2	p^2	pq
	10	pq	p^2	q^2	pq
	11	p^2	pq	pq	q^2

For $p(0) = p(1) = 0.5$, $p(x_1) = p(x_2) = p(x_3) = p(x_4) = 0.25$. Similarly, $p(y_1) = p(y_2) = p(y_3) = p(y_4) = 0.25$. Therefore,

$$H(Y^2) = 2 \text{ bits},$$

$$H(Y^2/X^2) = -2[q \log q + p \log p] \text{ bits}$$

and
$$I(X^2, Y^2) = 2[1 + q \log q + p \log p] = 2I(X, Y) \tag{8.49}$$

Thus, for the nth extension, $H(Y^n) = nH(y)$, and because of the symmetry of the channel, $H(Y^n/X^n) = nH(Y/X)$, giving Eq. (8.48).

To improve the channel efficiency (equivalently to reduce the error rate), a useful technique is to repeat the signals $\{X\}$ at the channel input and to detect only the repeated signals as $\{Y\}$, as shown in Fig. 8.4. Consider the BSC of Fig. 8.2, with inputs $\{X\} = \{00, 11\}$, and acceptable $\{Y\} = \{00, 11\}$, thereby discarding the outputs $\{01, 10\}$, as in a BEC. Then evaluating the new entropies, the mutual information is shown to be:

$$I(X, Y)_2 = (p^2 + q^2)[1 - H(p^2/(p^2 + q^2))] \tag{8.49}$$

It is seen that the channel is now equivalent to a BSC with error probability $p' = p^2/(p^2 + q^2) \simeq {}^2$ for $q \gg p$. Since $p' < p$, $I(X, Y)_2$ is now greater

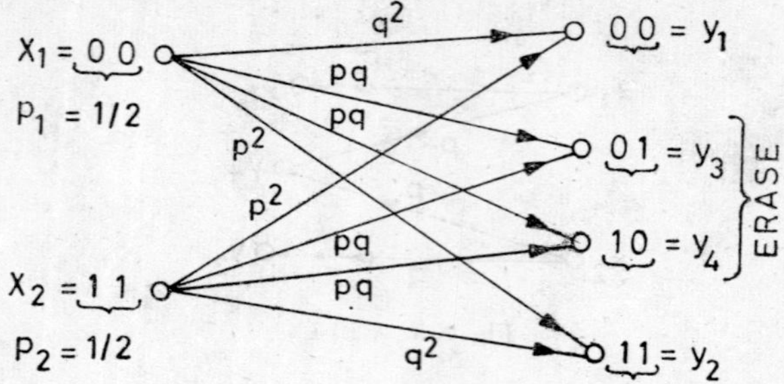

Fig. 8.4 A repetitive BSC

than the original value of $[1 - H(p)]$ of a BSC. It may be shown that with more repetitions of the input, the mutual information is further increased [4].

Sometimes, the channels are cascaded for operational reasons, and if both channels are noisy, there is an information loss in the final outcome. For two cascaded noisy BSC's with inputs and outputs as $\{X\}$, $\{Y\}$, $\{Z\}$, the overall mutual information is shown to be:

$$I(X, Z) = 1 - H(2pq) \leqslant I(X, Y) = 1 - H(p) \tag{8.50}$$

For a cascade of three BCS's with final output $\{U\}$, the mutual information is:

$$\begin{aligned} I(X, U) &= 1 - H(3pq^2 + p^3) \\ &\leqslant I(X, Z) \end{aligned} \tag{8.51}$$

Thus, for a cascade of noisy BSC's, the equivalent channel has a large error probability and there is increased information loss.

8.3.3 Capacity of Generalized Discrete Channels

For general channels where the channel transition matrix is not symmetric, the maximization of mutual information $I(X, Y)$ involves some variational techniques so that $H(Y)$ and $H(Y/X)$ are simultaneously optimized, and at the same time, a realizable $p\{X\}$ is obtained. $p\{X\}$ is restricted by the constraints: $p(x_i) \geqslant 0$ and $\sum_i p(x_i) = 1$. Thus maximization can be easily done only in restricted channels, e.g., in the symmetric/uniform channels. Consider a $[m \times m]$ channel matrix $p(Y/X) = [p_{ij}]$ with the restrictions as:

(i) $\sum_{i=1}^{m} p(y_j/x_i) = \sum_{i=1}^{m} p_{ij}$ is independent of j, i.e., the sums of all columns of $[p_{ij}]$ are the same. Then all $p\{Y\} = \{p_1', p_2', \ldots, p_m'\}$ will be equal if all $p\{X\} = \{p_1, p_2, \ldots, p_m\}$ are equal.

(ii) Further, $$\begin{aligned} H(Y/x_i) &= -\sum_{j=1}^{m} p(y_j/x_i) \log p(y_j/x_i) \\ &= k \text{ (constant) for all } i\text{'s}. \end{aligned}$$

With these restrictions, $$\begin{aligned} H(Y/X) &= \sum_{i=1}^{m} p(x_i) \cdot H(Y/x_i) \\ &= k \sum_i p(x_i) = k; \end{aligned}$$

and the mutual information $I(X, Y) = H(Y) - k.$ (8.52)
Maximum $H(Y) = \log m$, and the channel capacity

$$C = I(X, Y)_{\max} = \log m - k \tag{8.53}$$

Such a restricted channel is called a symmetric/uniform channel, and the restrictions are satisfied, if the entries in every row and every column of the matrix $[p_{ij}]$ consist of an arbitrary permutation of the entries in the first row (e.g., the matrix of Examples 5). Such symmetric $[p_{ij}]$ occurs in case of multilevel PAM (m-ary) or in orthogonal code transmission, where the error is equally distributed among all $(m - 1)$ levels (codes) not transmitted. This gives,.

$$p_{ij} = \begin{cases} \dfrac{p}{(m-1)}, & i \neq j \\ q\ , & i = j \end{cases}$$

and $\quad p + q = 1.$

The channel capacity C is now given by:

$$C(\text{PAM}) = \log m - p \log (m - 1) - H(P) \tag{8.54}$$

For channels with square transition matrix $[p_{ij}]$, but without symmetry, the channel capacity is calculated by using the method suggested by Muroga [1]. Introducing a set of new variables $\{Q_1, Q_2, \ldots, Q_m\}$, we can write a set of equations given by:

$$\begin{bmatrix} p_{11} & p_{12} & \cdots & p_{1m} \\ \vdots & & & \vdots \\ p_{i1} & p_{i2} & \cdots & p_{im} \\ \vdots & & & \vdots \\ p_{m1} & p_{m2} & \cdots & p_{mm} \end{bmatrix} \begin{bmatrix} Q_1 \\ \vdots \\ Q_i \\ \vdots \\ Q_m \end{bmatrix} = \begin{bmatrix} \sum_{j=1}^{m} p_{1j} \log p_{1j} \\ \vdots \\ \sum_{j=1}^{m} p_{ij} \log p_{ij} \\ \vdots \\ \sum_{j=1}^{m} p_{mj} \log p_{mj} \end{bmatrix} \tag{8.55}$$

Solving this Eq. (8.55), Q_i is obtained. Then the rate of transmission is shown to be:

$$R = I(X, Y) = -\sum_{j=1}^{m} p'_j \log p'_j + \sum_{i,j=1}^{m}\sum^{m} p'_j Q_i,$$

where $p'_j = p(y_j)$. Maximizing the value of R leads to:

$$C = R_{\max} = Q_i - \log p'_j, \qquad i = j$$

$$= \log \sum_{i=1}^{m} 2^{Q_i} \text{ bit/symbol} \tag{8.56}$$

Further, $p'_j = \exp(Q_1 - C), \qquad i = j$

and $[p_i] = [p'_j][p_{ij}]^{-1}, \qquad p(i) = p(x_i).$ (8.57)

The value of C may not necessarily lead to realizable values of $\{p_i\}$, and also the solution of Eq. (8.55) may not exist. However, within certain limits of the values in the $[p_{ij}]$ matrix, the method will provide a satisfactory solution.

Example 7

Consider the channel matrix given by:

$$p(Y/X) \rightarrow \begin{bmatrix} 0.4 & 0.6 & 0 \\ 0.5 & 0.0 & 0.5 \\ 0.0 & 0.6 & 0.4 \end{bmatrix}$$

Using Eq. (8.55), $Q_1 = -1$, $Q_2 = -0.95$, and $Q_3 = -1$,

and $C = \log[2^{-1} + 2^{-0.95} + 2^{-1}] = \log_2(1.5) = 0.58$ bit;

$p'_1 = 0.35, \qquad p'_2 = 0.3, \qquad p'_3 = 0.35$

$p_1 = 0.25, \qquad p_2 = 0.5, \qquad p_3 = 0.25$ (Using Eq. (8.57).

Based on the above discussions and the results of coding theory (discussed in the next few sections), Shannon's second fundamental theorem on coding for memoryless noisy channels has been formulated. The theorem is discussed in Sec. 8.6.2.

8.4 ERROR CORRECTING CODES [1, 7, 8]

The optimum detection techniques for signals corrupted by channel noise have been discussed in Chapter 2; however, the rate of transmission of information through a noisy channel requires to be improved through proper signal design and coding. Shannon's channel capacity theorem (discussed in Section 8.7) promises a potential coding gain of 11 dB as compared to PSK signalling with no coding. Basically two approaches have been used to improve the rate of transmission in noisy channels: (a) one approach has been to use error-detecting and correcting codes, e.g., block codes and convolutional codes, where the redundant codes are non-orthogonal, and zero-error rate in the limit is achieved by closely packing the waveforms within a finite bandwidth, (b) the other approach has been to expand the information words into larger and larger sets of orthogonal or bi-orthogonal codes and to use optimum detection techniques at the receiver by using matched filters/correlators, as discussed in Sec. 8.9. In a practical communication system, the improvement in the rate of transmission may be obtained either by using Forward-error-control techniques (FEC) or by using a feedback channel along with error detection and retransmission technique (ARQ), as discussed in Chapter 6.

For discussion in this section, the channels are modelled as either a BSC or a BEC, as shown in Figs. 8.2 and 8.3. FEC is generally used in a BSC and ARQ in a BEC. Errors occur in the channel either in a random manner or in bursts, e.g., in telephone channels and HF radio channels. In general, the error correcting/detecting codes are of two types: (a) Block codes and (b) Convolutional codes. Among the block codes, Hamming, Golay and BCH codes have been widely used. But convolutional codes with Viterbi/sequential decoders show better performances in real channels. Shannon's coding theorem states that for a fixed transmitter power and a transmission rate $R < C$, the error probability P_e in the channel is given by [1, 2]:

$$P_e \leqslant \exp[-nE(R)], \quad R < C \tag{8.58}$$

where $E(R)$ is a function of R and n = number of bits in a codeword. Then $P_e \to 0$ for $n \to \infty$, which requires a long delay for coding and decoding. In practical codes, a gain of 5 dB in E_b/n_0 has been obtained using low-rate convolutional codes and soft decision Viterbi decoding. In this and the following sections, we shall discuss some of the important FEC codes and also their performances.

8.4.1 Block Codes [7, 8]

For the purpose of encoding, the continuous message is broken into blocks of k bits and $(n - k)$ parity bits are added according to certain rules of coding. The n-bit codewords known as (n, k) codes are then transmitted through the noisy channel. In the receiver-decoder, the 2^k distinct messages are selected out of the possible 2^n words (any of which may be received due to the addition of noise bits in the channel) using suitable decoding algorithms. The set of 2^k words is called a block code.

A simple scheme of error detection/correction is mechanized by arranging the serial data bits in a $[m \times n]$ matrix form and by adding parity checks (either even or odd) for every row/column in the $(n + 1)$th column and the $(m + 1)$th row respectively, as shown below. If a single error occurs in a row or column, then the corresponding row and column checks will fail in the decoder, indicating the location of the error, which may now be corrected. Thus, a total of m or n errors (whichever is less) may be corrected in a block of $[(m + 1)(n + 1)]$ bits transmitted through the channel. If, however, even number of errors occur in a row or a column, then the parity check over that row/column will not fail, but the presence of the errors will be indicated only due to the check failures in corresponding column/row. This parity-check matrix technique may be extended to a third dimension by stacking the $[(m + 1)(n + 1)]$ matrices, one above the other, in the vertical (z) axis. Checks are then introduced along the z-axis, say, over a block of $(r - 1)$ bits, and two errors along any axis may now be corrected if the errors along other axes are single. This technique is known as Elias's iteration technique [14], and may be made more robust by adding multiple check bits along any of the axes using Hamming or other error-correcting algorithm.

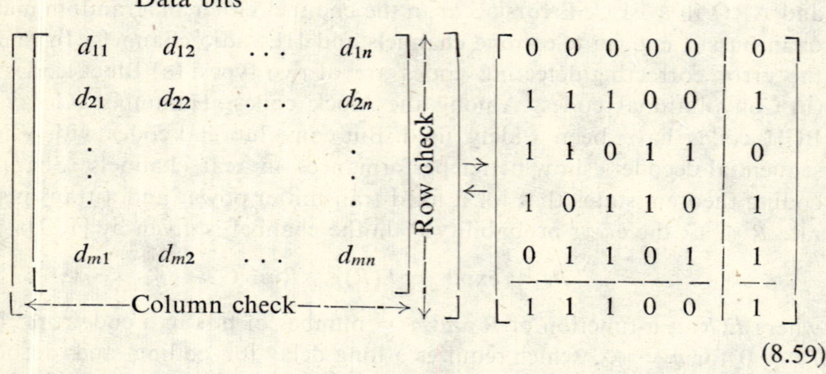

(8.59)

The process of encoding and decoding data blocks may be looked upon as a problem of mapping the source outputs to match the channel characteristics, as shown in Fig. 8.5. The encoder generates 2^k distinct codewords

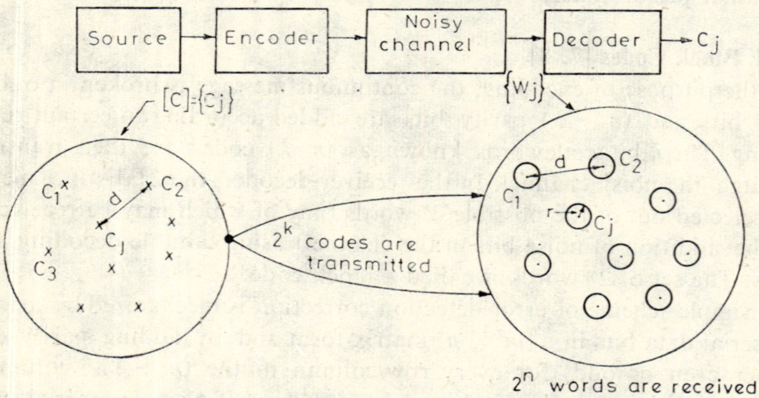

Fig. 8.5 Code mapping for SEC-codes, $d \geqslant 3$, $r = 1$

$\{C_j\}$, which are corrupted by noise in the channel. The decoder now observes the received words $\{W_j\} = \{C_j + e_j\}$, where $\{e_j\}$ is the error vector; and decides which particular codeword was most likely transmitted. This amounts to optimum partitioning of the n-dimensional space enclosing 2^n points, each a word, into 2^k distinct spheres, each containing a codeword. If now the received word W_j falls within the sphere around C_j, i.e., if W_j is one of the words clustered around C_j, then the decoder decides that the transmitted code was C_j. This means that the correct decision is possible only when the received W_j differs from C_j in only a few bit positions and this is bounded by Hamming distance defined now.

R.W. Hamming [12] first developed the concept of single-error correcting (SEC) codes and the Hamming Distance (HD). HD between two codewords is defined as the number of positions in which the codewords differ. Then a single error results in a HD = 1, between $\{C_j\}$ and $\{W_j\}$. Thus, for detecting error patterns of t or fewer bist, it is necessary and sufficient that

$$\text{HD} \geqslant (t + 1) \tag{8.60}$$

For correction of t or fewer errors, $\text{HD} \geqslant (2t + 1)$ (8.61)
and for combined error-correction of t_1 errors and error-detection of t_2 errors,

$$\text{HD} \geqslant (t_1 + t_2 + 1), \qquad t_2 \geqslant t_1 \tag{8.62}$$

For n-bit words, one may imagine that the codewords are placed at the vertices of an n-dimensional cube with the required HD $(= d)$. Then the number of codewords N with $d \geqslant (2t + 1)$, is given by:

$$2^k = N(n, d) \geqslant \frac{2^n}{\sum_{i=0}^{t} \binom{n}{i}} \tag{8.63}$$

Based on this equation, $N = 16$ for a single-error correcting code using $n = 7$, $k = 4$, and the code is known (7, 4) SEC code. For SEC, (7, 4) (15, 11), (31, 26), (63, 57) etc. are valid codes and for double-error correction, (15, 7), (31, 21), (63, 51) etc. are valid codes.

The Hamming codes [12] are (n, k) SEC block codes, whose parameters are:

$$\begin{aligned}
&\text{Codelength } n = 2^r - 1, \\
&\text{Number of parity check bits } r = (n - k) \\
&\text{HD} = 3 \\
&\text{Number of information bits } k = (2^r - r - 1) \\
&\text{Error correcting capability } t = 1
\end{aligned} \tag{8.64}$$

and for any integer r, a Hamming code is a SEC code. For a (7, 4) code, the parity check bits P_1, P_2, P_3 for the message bits X_1, X_2, X_3, X_4 are generated by the parity check equations (other equations are also possible):

$$\begin{aligned}
P_1 &= X_2 + X_3 + X_4 \\
P_2 &= X_1 + X_3 + X_4 \\
P_3 &= X_1 + X_2 + X_4
\end{aligned} \qquad (8.65)^*$$

*All '+' signs in Secs. 8.4–8.6 indicate modulo sum $\oplus$.

Equation (8.65) may be written in the matrix form as:

$$\underline{P_1 P_2 P_3} = \underline{X} \cdot [P],$$

where,

$$[P] = \begin{bmatrix} 0 & 1 & 1 \\ 1 & 0 & 1 \\ 1 & 1 & 0 \\ 1 & 1 & 1 \end{bmatrix} \tag{8.66}$$

For a systematic code, where the message bits are transmitted at the beginning or at the end of the code C_j, a code generating matrix may now be specified as:

$$[G] = [I_k P], \tag{8.67}$$

where $[I_k]$ is a unity matrix of order k. The resulting code vector C_j will now be of the form $\{X_1 X_2 X_3 X_4 \; \vdots \; P_1 P_2 P_3\}$. In the above (7, 4) code, the generating matrix is then given by:

$$[G] = \begin{bmatrix} 1 & 0 & 0 & 0 & 0 & 1 & 1 \\ 0 & 1 & 0 & 0 & 1 & 0 & 1 \\ 0 & 0 & 1 & 0 & 1 & 1 & 0 \\ 0 & 0 & 0 & 1 & 1 & 1 & 1 \end{bmatrix} \tag{8.68}$$

and the code vectors $C_j = \underline{m_j} \cdot [G] \cdot$ All the code vectors are obtained from

$$[C] = [m] \cdot [G] \tag{8.69}$$

For the sixteen message vectors, the corresponding code vectors may be easily tabulated.

Decoding

Assuming that the error vector of the channel is $[E]$, the received vector $[W]$ is given by:

$$[W] = [C] + [E]$$
$$= [mG] + [E]$$

The parity check matrix (used for decoding) is written as:

$$[H] = [P^T \; \vdots \; I_{n-k}] \tag{8.70}$$*

such that $[G] \cdot [H^T] = 0$. In the decoder, the syndrome vector $[S]$ is obtained as:

*It is equally valid to have $[G] = [P \; \vdots \; I_k]$ and $[H] = [I_{n-k} \; \vdots \; P^T]$. Further, $[G] \cdot [H^T] = [I_k \; \vdots P] \cdot \begin{bmatrix} P \\ \cdots \\ I_{n-k} \end{bmatrix} = [P + P] = 0.$

$$
\begin{aligned}
[S] &= [W]\cdot[H^T] \\
&= [C\cdot H^T] + [E\cdot H^T] \\
&= [E\cdot H^T] \to 0, \quad \text{if } [E] = 0 \qquad (8.71)
\end{aligned}
$$

Further, the code vector is recovered from Eq. (8.71) as:

$$[E] = [S]\cdot[H^T]^{-1}$$

and

$$[W + E] = [C] \qquad (8.72)$$

Example 8

Consider the $[G]$ matrix of Eq. (8.68). For a message $\{X_1, X_2, X_3, X_4\} = \{0\ 0\ 1\ 0\}$, the code $\{C_j\} = \{0\ 0\ 1\ 0\ 1\ 1\ 0\}$. If the received vector $\{W_j\} = \{0\ 0\ 1\ 0\ 0\ 1\ 0\}$, then the syndrome is generated as:

$$[S] = [W]\cdot[H^T]$$

i.e.,

$$[S] = \lfloor 0010010 \rfloor \begin{bmatrix} 0 & 1 & 1 \\ 1 & 0 & 1 \\ 1 & 1 & 0 \\ 1 & 1 & 1 \\ 1 & 0 & 0 \\ 0 & 1 & 0 \\ 0 & 0 & 1 \end{bmatrix} = \lfloor 100 \rfloor \qquad (8.73)$$

By inspection of H^T, the error is seen to be in the 5th position, giving $E_j = \{0000100\}$, and the correct $C_j = \{0010110\}$ corresponding to the message $\{0010\}$. If it is required that the syndrome pattern should give the error location in a binary form, then $[P]$ matrix has to be modified accordingly.

8.4.2 Cyclic Code [7,8]

An important class of block codes is the cyclic code, where the cyclic shifts of the codes are also code vectors of $[C]$. Thus in a polynomial form, if $C_1(X)$ is a code vector given by,

$$C_1(X) = C_0 + C_1X + C_2X^2 + \ldots + C_{n-1}X^{n-1}, \qquad C\text{'s} = 0/1,$$

then the shifted vector $C_2(X) = C_{n-1} + C_0X + C_1X^2 + \ldots + C_{n-2}X^{n-1}$ is also a code vector. The generator polynomial of an (n, k) cyclic code is a factor of $(1 + X^n)$, e.g., $(1 + X^n) = g(X)\cdot h(X)$. If $g(X)$ is a polynomial of degree $(n - k)$, then $g(X)$ generates an (n, k) cyclic code. The code polynomial $C(X)$ can be expressed as:

$$
\begin{aligned}
C(X) &= m(X)\cdot g(X) \\
&= m(X)\cdot[1 + g_1X + g_2X^2 \ldots + X^{n-k}] \\
&= (m_0 + m_1X + m_2X^2 + \ldots + m_{k-1}X^{k-1})\cdot g(X) \qquad (8.74)
\end{aligned}
$$

where g's and m's are only 0/1; and there are 2^k code vectors in a (n, k) code. However, this method generates non-systematic codes.

To generate systematic codes of the form $[m \, \vdots \, P]$, as given by Eq. (8.68), or equivalently of the form $[P \, \vdots \, m]$, a shift register encoder may be used. The encoding algorithm is:

Multiply $m(X)$ by X^{n-k} (to shift $m(X)$ to the rightmost entries in the rows), and dividing $[X^{n-k} \cdot m(X)]$ by $g(X)$, we have,

$$X^{n-k} \cdot m(X) = Q(X) \cdot g(X) + R(X)$$

or

$$C(X) = R(X) + X^{n-k} \cdot m(X) = Q(X) \cdot g(X), \tag{8.75}$$

where $Q(X)$ is the quotient and $R(X)$ is the remainder polynomial. The above equation indicates that $[X^{n-k} \cdot m(X) + R(X)]$ is a polynomial of degree $(n - 1)$ or less and is a multiple of $g(X)$ as well, as required by Eq. (8.74). Hence $C_j = [X^{n-k} \cdot m_j(X) + R_j(X)]$ is a valid code corresponding to the message $m_j(X)$. The codes are generated by divider circuits using shift registers (SR) as shown in Figs. 8.6(a) and (b).

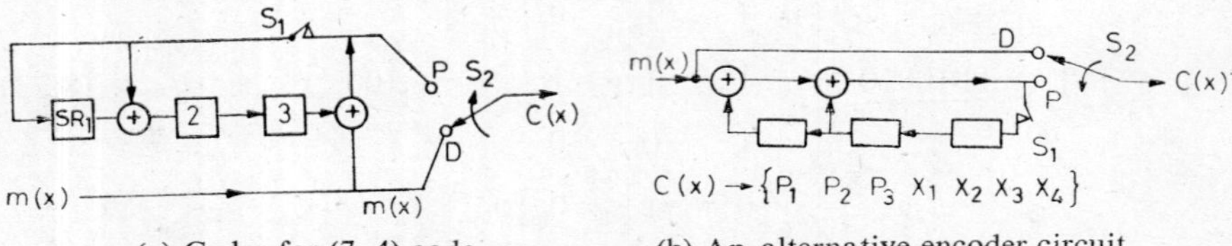

(a) Coder for (7, 4) code (b) An alternative encoder circuit

Fig. 8.6 $(n - k)$ stage encoder for SEC code

Example 9

To generate (7, 4) SEC code, $m(X) = m_0 + m_1X + m_2X^2 + m_3X^3$, and $1 + X^n = 1 + X^7 = (1 + X + X^3)(1 + X + X^2 + X^4)$. Thus using $g(X) = 1 + X + X^3$,

$$R(X) = X^3 \cdot m(X) + Q(X) \cdot g(X) \tag{8.76}$$

$R(X)$ is now obtained by the divider circuits of Fig. 8.6. In Fig. 8.6(a), initially S_1 is closed and S_2 at D. The 4-bit message is sent out in the line and at the same time $m(X)$ is loaded in the SR's producing $Q(X)$ and $R(X)$ simultaneously. Connecting $m(X)$ at the rightmost point of the divider circuit is equivalent to multiplying $m(X)$ with X^3. After $m(X)$ is loaded in the SR's, and the division is complete, the content of the SR's is $R(X)$. S_1 now opens and S_2 moves to P. During the later shifts, the parity bits are sent out in the line. It may be shown that for $m(X) = \{1011\}$, the SR content $= [P] = \{100\}$, giving $C_j = \{1001011\}$. The alternative coder of Fig. 8.6(b) gives the same result.

For the (15, 11) SEC code, $g(X)$ is obtained from, $(1 + X^{15}) = (1 + X)(1 + X + X^2)(1 + X + X^4)(1 + X^3 + X^4)(1 + X + X^2 + X^3 + X^4)$. We may choose either $(1 + X + X^4)$ or $(1 + X^3 + X^4)$ as the $g(X)$. The coder using $g(X) = 1 + X + X^4$ is shown in Fig. 8.7, where S_1 opens on the last 4 bits and S_2 is down for first 11 bits, and a total of 15 shifts are required for each codeword.

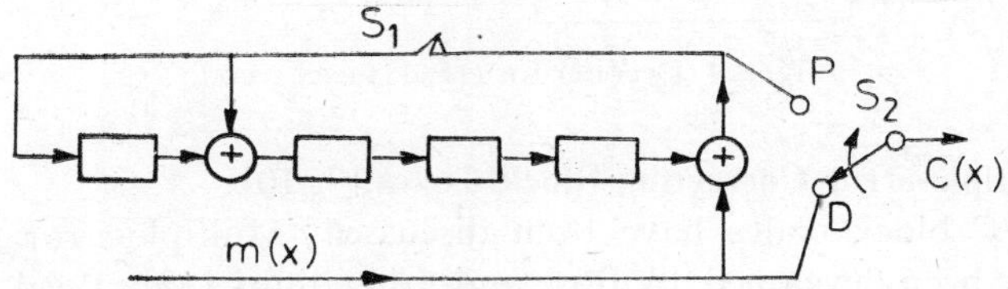

Fig. 8.7 Encoder for (15, 11) code with $g(X) = 1 + X + X^4$.

The decoding of the cyclic codes is a complementary operation. Since $W(X) = C(X) + E(X)$, the syndrome is given by:

$$S(X) = \text{Rem. } [W(X)/g(X)] = \text{Rem. } [E(X)/g(X)] \tag{8.77}$$

Thus, the syndrome of any n-tuple is its residue modulo $g(X)$, and all legitimate codewords $C(X)$ have all-zero syndromes. Since the syndrome of $W(X)$ contains the information about $E(X)$, this information is used for error detection and correction. A simple decoder for (7,4) cyclic code, using a divider and a buffer is shown in Fig 8.8, where the entire received vector W_j is entered into the SR's bit by bit and at the same time W_j is stored in

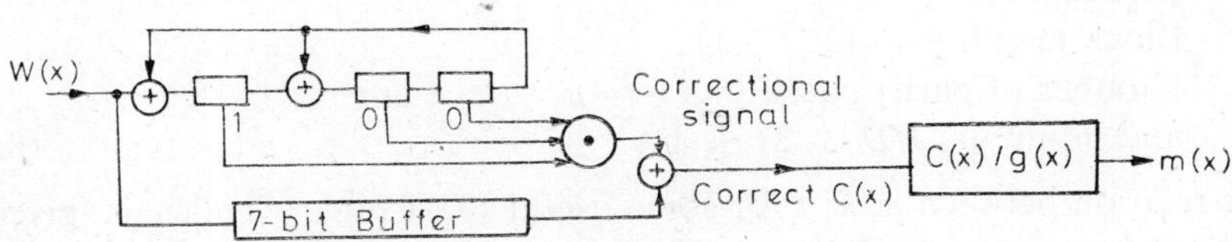

Fig. 8.8 The Meggitt decoder for (7, 4) cyclic code

the buffer memory. After the 3rd shift, the division process starts and after the 7th shift, the syndrome will be stored in the SR's. If $S(X) = \{000\}$, then $E(X) = 0$, and $\{W_j\}$ is read out of the buffer. If, however, $E(X) \neq 0$, then, $S(X) \neq 0$ and the SR-content will change with successive shifts till it becomes $\{100\}$ indicating that the bit coming out of the buffer at that instant is in error. The 'AND' circuit gives a correction signal and the bit in error is corrected accordingly. As for example. with $E(X) = \{0000100\}$, the SR-content at the 7th shift will be $S(X) = \{011\}$, and at the 10th shift, the SR-content will be $\{100\}$ indicating the location of the error bit. The correction signal will now correct the X^4 bit of W_j. Corresponding to the coder for $\{15, 11\}$ SEC cyclic code, shown in Fig. 8.7, the decoder is shown in Fig. 8.9, which operates in the same way as the decoder of Fig. 8.8. In this case, the SR-content will be $\{1000\}$ to indicate the error bit at the buffer output and this is corrected as explained above.

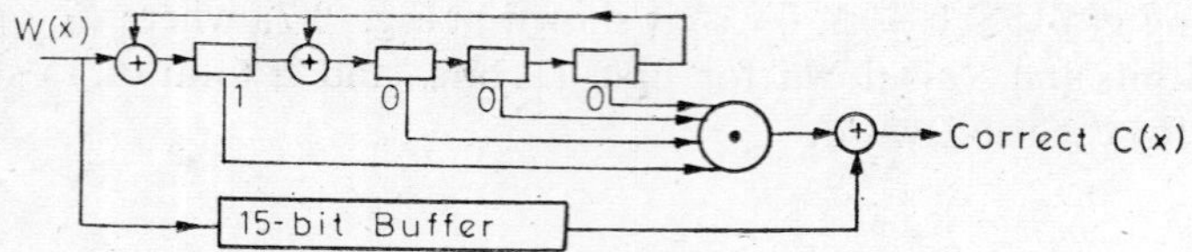

Fig. 8.9 Decoder for (15, 11) cyclic code

8.4.3 Multiple-error Correcting Block Codes [7, 10]

So for SEC block codes have been discussed. Multiple-error correcting codes have been developed by Bose and Chaudhuri [13], Reed and Solomon [16], Golay and others; and most of them are cyclic codes. They are decoded by Meggitt or Error-trapping decoders. Decoding of BCH and RS codes is rather complex, requiring solution of simultaneous equations and matrix inversion. Fortunately, some of these codes are majority-logic decodable and the decoders for them are very simple.

Bose-Chaudhuri-Hocquenghem (BCH) codes [13] are a class of cyclic codes with powerful error-correcting properties and well-known implementation algorithms. BCH codes have the following properties:

(a) For a t-error correcting (n, k) binary group code, an $(n \times r) = [H^T]$ matrix exists, such that $2t$ row vectors of the matrix are mutually independent, $(r = n - k)$.

(b) For any positive integer m and t $(t < 2^{m-1})$, a BCH code has the parameters;
Block length $n = (2^m - 1)$
Number of parity check bits $r = n - k \leqslant mt$
and minimum $HD \geqslant 2t + 1$ (8.78)

The relation between n, k, t for some useful BCH binary codes is given in Table 8.4.

Table 8.4 Parameters of BCH Codes

n	7	15	15	15	31	31	31	31	31	63	63	63	63	63	63	63	63	63
k	4	11	7	5	26	21	16	11	6	57	51	45	39	36	30	24	18	7
t	1	1	2	3	1	2	3	5	7	1	2	3	4	5	6	7	10	15

The alphabet of a BCH code for $n = (2^m - 1)$ may be represented as the set of elements of an appropriate Galois Field, GF(2^m), whose permitive element is α. Then the generator polynomial of the t-error correcting BCH code is the least common multiple (LCM) of

$$M_1(X), M_3(X), \ldots, M_{2t-1}(X),$$

$$\text{i.e., } g(X) = \text{LCM}\,[M_1(X), M_3(X), \ldots, M_{2t-1}(X)] \tag{8.79}$$

where $M_i(X)$ is the minimum polynomial of α^i, $i = 1, 2, \ldots, 2t$, as given in Appendix C, for $m = 4$. Since the degree of each polynomial is m or less, the degree of $g(X)$ is at most mt, and $r = n - k \leqslant mt$. For many of the BCH codes of Table 8.4, the minimum HD $= 2t + 1$.

From Eq. (8.78), $m = 4$ for $n = 15$, $m = 5$ for $n = 31$ etc. Thus, for a (15,7) double-error correcting BCH code over GF(2^4), the generating polynomial $g(X)$ is:

$$\begin{aligned} g(X) &= \text{LCM}\,[M_1(X), M_3(X)] \\ &= M_1(X) \cdot M_3(X) \\ &= (1 + X + X^4)(1 + X + X^2 + X^3 + X^4) \\ &= 1 + X^4 + X^6 + X^7 + X^8 \end{aligned} \tag{8.80}$$

and α, α^2, α^3 and α^4 are the roots of $g(X)$. The code is cyclic with HD $= 5$, since $g(X)$ has weight 5. Similarly, for the tripple-error correcting (15, 5) BCH code, $g(X)$ is:

$$\begin{aligned} g(X) &= M_1(X) \cdot M_3(X) \cdot M_5(X) \\ &= 1 + X + X^2 + X^4 + X^5 + X^8 + X^{10} \end{aligned} \tag{8.81}$$

This is also a cyclic code with HD $= 7$.

With $m = 5$, $n = 31$ and $g(X)$ of different codes over GF(2^5) are:

$$\begin{aligned} &\text{(i) } (31, 26),\ t \leqslant 1,\ g(X) = 1 + X^2 + X^5 \\ &\text{(ii) } (31, 21),\ t \leqslant 2,\ g(X) = 1 + X^3 + X^5 + X^6 + X^8 + X^9 + X^{10} \\ &\text{(iii) } (31, 16),\ t \leqslant 3,\ g(X) = 1 + X + X^2 + X^3 + X^5 + X^7 + X^8 \\ &\qquad + X^9 + X^{10} + X^{11} + X^{15} \end{aligned} \tag{8.82}$$

With $m = 6$, $n = 63$ and $g(X)$ of different codes over GF(2^6) are:

$$\begin{aligned} &\text{(i) } (63, 57),\ t \leqslant 1,\ g(X) = 1 + X + X^6 \\ &\text{(ii) } (63, 51),\ t \leqslant 2,\ g(X) = (1 + X + X^6)(1 + X + X^2 + X^4 + X^6) \\ &\text{(iii) } (63, 45),\ t \leqslant 3,\ g(X) = (1 + X + X^6)(1 + X + X^2 + X^4 + X^6) \\ &\qquad (1 + X + X^2 + X^5 + X^6) \\ &\text{(iv) } (63, 39),\ t \leqslant 4,\ g(X) = (1 + X + X^6)(1 + X + X^2 + X^4 + X^6) \\ &\qquad (1 + X + X^2 + X^5 + X^6)(1 + X^3 + X^6) \end{aligned} \tag{8.83}$$

The equivalent generating matrix $[G]$ and the party check matrix $[H]$ of the above codes can be formulated by using certain rules. The coder circuits are simple shift register circuits similar to that of Fig. 8.7. For decoder algorithms, the reader is referred to Ref. [1, 7].

Reed-Solomon (RS) codes [16] are a subclass of BCH codes, where the code symbols are from GF(q) ($q \neq 2^m$ in general, but usually q is taken as 2^m). A t-error correcting RS code has the following parameters (where the symbols are in terms of m-tuples):

Block length $n = (q - 1)$ symbols
Number of parity check symbols $r = n - k = 2t$
Minimum HD $= d = 2t + 1$ symbols (8.84)

Consider $q = 2^m$, α as primitive, then the generator polynomial for the RS-code is given by:

$$g(X) = (X + \alpha)(X + \alpha^2) \ldots (X + \alpha^{2t}) \tag{8.85}$$

The polynomial has a degree of $2t$ and is a maximum distance separable (MDS) code, i.e., RS code has maximum HD for given values of n and r.

Consider a (15, 11) double-error correcting RS-code with $n = 15$, $r = 4$. Then $q = 2^4$; and using Eq. (8.85) along with the Table of $G(2^4)$ (see Appendix C),

$$\begin{aligned} g(X) &= (X + \alpha)(X + \alpha^2)(X + \alpha^3)(X + \alpha^4) \\ &= \alpha^{10} + \alpha^3 X + \alpha^6 X^2 + \alpha^{13} X^3 + X^4 \end{aligned} \tag{8.86}$$

Comparing the binary double-error correcting BCH code of Eq. (8.80), it is seen that the degree of the $g(X)$ of BCH code is 8, but the degree of $g(X)$ for the RS-code is 4 only. As such, RS code is a (15, 11) code, whereas the binary BCH is a (15, 7) code. For a message vector $(1 + X)$, the transmitted code $C_1(X)$ for binary BCH is:

$$\begin{aligned} C_1(X) &= 1 + X + X^4 + X^5 + X^6 + X^9 \\ &= \{110011100100000\} \end{aligned}$$

The RS-code for the same message vector is:

$$\begin{aligned} C_2(X) &= (1 + X)(\alpha^{10} + \alpha^3 X + \alpha^6 X^2 + \alpha^{13} X^3 + X^4) \\ &= \alpha^{10} + x^{12} X + \alpha^2 X^2 + X^3 + \alpha^6 X^4 + X^5 \\ &= \{\alpha^{10}, \alpha^{12}, \alpha^2, 1, \alpha^6, 1, 0, 0, 0, 0, 0, 0, 0, 0, 0\} \end{aligned}$$

The equivalent C_2 in the binary form is then given by:

$$\begin{aligned} \{C_2\} = \{&0111, 1111, 0100, 0001, 1100, 0001, 0000, 0000, 0000, 0000, \\ &0000, 0000, 0000, 0000, 0000\} \end{aligned}$$

In terms of binary digits, the RS code here is a (60, 44) code, where the encoder accepts 11 binary 4-tuples of data and forms 15 binary 4-tuples of coded symbols. The Binary BCH code, on the other hand, accepts 7 bits of data and converts these into a 15-bit code. Both codes have a minimum distance of 5, but in RS code, the distance is measured in terms of 4-tuple symbols. Thus, the rate of transmission in RS code is much higher for the same error correcting capability. Further, the RS codes are very efficient in multiple-burst correction. For channel transmission purposes, one can transmit RS codes either in the binary-word form or as 16-level symbols. For decoding the RS codes with errors, the steps are somewhat similar to those used in decoding BCH codes.

Binary Golay code [1, 8] is a (23, 12) perfect binary cyclic code, since it satisfies Eq. (8.63) with equality and corrects all errors $t \leqslant 3$. The code has been used in several practical systems. The generator polynomial of the code is obtained from

$$(1 + X^{23}) = (1 + X)\cdot g_1(X)\cdot g_2(X)$$

where

$$g_1(X) = 1 + X^2 + X^4 + X^5 + X^6 + X^{10} + X^{11}$$

$$g_2(X) = 1 + X + X^5 + X^6 + X^7 + X^9 + X^{11} \tag{8.87}$$

One can use the SR-implementation of the encoder using either g_1(X) or g_2(X) as the divider polynomial. The minimum HD of the code is 7, and hence $t = 3$. Besides the binary Golay code, there is also a perfect ternary (11,6) Golay code with $d = 5$.

For the purpose of decoding, one may use a table-look up decoder, an error-trapping decoder or a systematic search decoder. It is observed that the syndrome polynomial $S(X)$ is given by Eq. (8.77), and as such, the degree of $S(X) \leqslant (n - k - 1) = 10$. If now the error $E(X)$ is confined to $(n - k)$ parity check positions $(1, X, \ldots, X^{n-k-1})$ of $W(X)$, then $E(X) = S(X)$, since the degree of $E(X)$ is less than that of $g(X)$. Thus, correction can be carried out by simply adding $S(X)$ to $W(X)$. Even if $E(X)$ is not confined to the $(n - k)$ parity check positions of $W(X)$, suppose that $E(X)$ has non-zero values clustered together in a span of 11 bit-positions, then also the $S(X)$ will exhibit an exact copy of the error pattern after some cyclic shifts of $E(X)$. For each error pattern, the syndrome content $S(X)$ (after the required shifts) is subtracted from the appropriately shifted $W(X)$, and the corrected $W(X)$ recovered. If, however, $E(X)$ has a length greater than 11, then the error pattern is not trapped by cyclically shifting $S(X)$. In this case it is shown that one of the three error bits must have at least five zeros on one side of it and at least six zeros on the other side. This property is utilized in a modified error-trapping decoder as shown in Fig. 8.10.

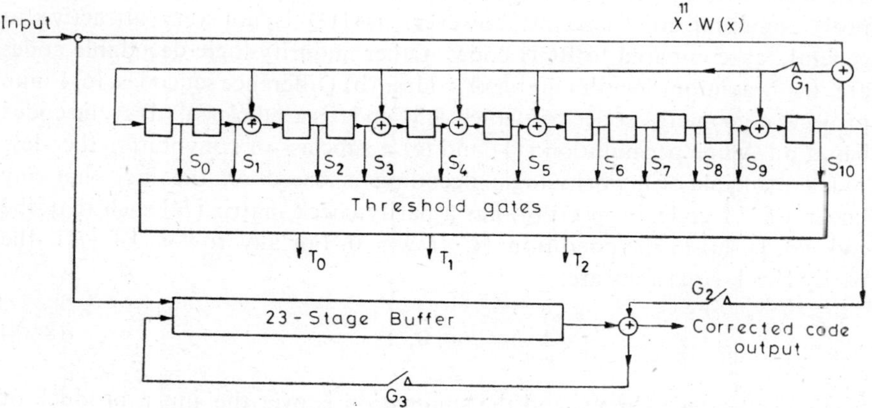

Fig. 8.10 Error-trapping decoder for Golay code (23, 12)

An alternative method of decoding is based on the assumption that when the block of 23 bits with $t \leqslant 3$ is cyclically shifted, it will be observed that 'at most one error will lie outside the 11 high-order bits of $W(X)$' at some shift. This fact isutilized in a Systematic search decoder, where the first test to be performed is 'whether the three errors are confined to $\{X^{12}, X^{13}, \ldots, X^{22}\}$ of $W(X)$, i.e., if $S(X)$ matches the $E(X)$ pattern and the weight (wt) of $S(X) \leqslant 3$'. This is checked by a threshold gate in Fig. 8.10 and the gate output T_0 switches G_2 on and G_1 off. $W(X)$ is now received from the buffer and corrected by syndrome bits (as they are clocked bit by bit) through the mod-adder circuit. If the first test is unsuccessful, then the 1st bit of $W(X)$ is inverted and the check is made if wt of $S(X) \leqslant 2$. If this test is successful, then the non-zero positions of $S(X)$ give the two error locations and the other error is in the 1st position. If this also fails, then the syndrome content is shifted cyclically, each time testing for wt of $S(X) \leqslant 3$; if not, invert 2nd, 3rd, . . . , and 12th bit of $W(X)$ successively and test for wt of $S(X) \leqslant 2$. Since not all errors are in the parity check section, an error must be detected in one of the shifts. Once one error is located and corrected, the other two errors can easily be located and corrected by the above test. Sometimes, this systematic search coder is simpler in hardware, than the error-trapping decoder, but the latter is faster in operation. The systematic search decoder can be generalized for other multiple-error correcting cyclic codes also.

8.4.4 Majority-Logic Decodable Codes [17]

Majority-logic decoding techniques are quite useful in decoding many of the cyclic and extended cyclic codes. In general, it has been shown that $(2^m - 1, 2^m - m - 1)$ Hamming codes, m any integer, are completely orthogonalizable in $(m - 1)$ steps. Among the BCH codes, the completely orthogonalizable codes are: (a) (15, 7) code with $t \leqslant 2$ is 1-step orthogonalizable, (b) the subclass of $(2^m - 1, m + 1)$ codes with $t \leqslant (2^{m-2} - 1)$, $m \geqslant 3$ are 2-step orthogonalizable, (c) (31, 16) code with $t \leqslant 3$ is 3-step orthogonalizable. The Reed-Muller (RM) code is L-step decodable, and has been used in space probe communication systems; however, its HD is not very attractive in general, as compared to BCH codes. Other majority-logic decodable codes are, (a) Maximum length (simplex) codes, (b) Difference set codes, (c) Finite projective geometry codes (of which RM code is a subclass), (d) cyclic codes based on Affine permutation [11] and (e) a subclass of convolutional codes.

The principle of majority-logic decoding is based on the fact that any linear (n, k) code over GF(q) has a parity check matrix $[H]$ such that the codewords satisfy the condition: $[C \cdot H^T] = 0$. For any row $\underline{|h_i|}$ of $[H]$, the parity check equations are:

$$\sum_{j=1}^{n-1} h_{ij} C_j = 0, \tag{8.88}$$

where j indicates columns and the summation is over the inner product of the row vector $\underline{|h_i|}$ and the code vector. Any linear combinations of the rows of $[H]$ also forms another parity check equation and there are a total of q^{n-k} equations that can be formed in this way.

Example 10

Consider the (7, 3) cyclic simplex code, which is the dual of the (7, 4) Hamming code. This has HD $= 4$ and its $g(X) = 1 + X^2 + X^3 + X^4$. The syndrome at the decoder may be generated by using a divider circuit, as shown in Fig. 8.11. Based on the above $g(X)$, the parity matrix is:

$$[H] = \begin{bmatrix} 1 & 0 & 0 & 0 & 1 & 1 & 0 \\ 0 & 1 & 0 & 0 & 0 & 1 & 1 \\ 0 & 0 & 1 & 0 & 1 & 1 & 1 \\ 0 & 0 & 0 & 1 & 1 & 0 & 1 \end{bmatrix} \quad (8.89)$$

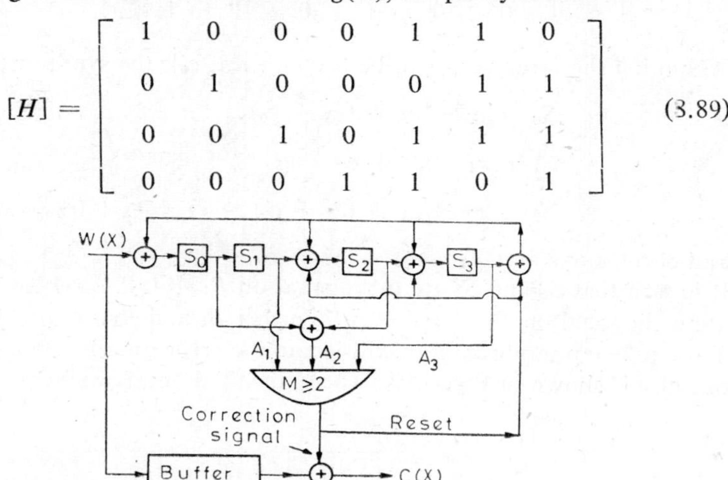

Fig. 8.11 A majority-logic decoder for (7, 3) simplex code

Since $[CH^T] = 0$, the syndromes are calculated as (with the error vector $= \{e_0, e_1, \ldots e_6\}$):

$$S_0 = e_0 + e_4 + e_5; \qquad S_1 = e_1 - e_5 + e_6;$$
$$S_2 = e_2 + e_4 + e_5 + e_6; \qquad S_3 = e_3 + e_4 + e_6.$$

Further, forming the parity-check sums as:

$$A_1 = S_1 = e_1 + e_5 + \underline{e_6}$$
$$A_2 = S_0 + S_2 = e_0 + e_2 + \underline{e_6}$$
$$A_3 = S_3 = e_0 + e_4 + \underline{e_6} \quad (8.90)$$

it is observed that all of the check sums A_1, A_2, A_3 check the error bit e_6 and no other error bit is checked by more than one check sum. Then a majority decision can be taken that $e_6 = 1$, if two or more A_j's are non-zero. If $e_6 = 0$, and any other bit is in error, then only one of the A_j's will be non-zero. It is said that the check sum A_j's are orthogonal on the error bit e_6. Thus, by using the majority logic circuit $M \geqslant 2$, as shown, the error-bit e_6 is corrected. The successive bits are then checked for a single error in the block. The feedback signal is necessary, if the decoder is to correct some two error patterns as well, otherwise this may be omitted.

Example 11

Consider another example of the (7, 4) Hamming code generated by $g(X) = 1 + X + X^3$ where the code is 2-step orthogonalizable. Its parity-check matrix $[H]$ is:

$$[H] = \begin{bmatrix} 1 & 0 & 0 & 1 & 0 & 1 & 1 \\ 0 & 1 & 0 & 1 & 1 & 1 & 0 \\ 0 & 0 & 1 & 0 & 1 & 1 & 1 \end{bmatrix} \tag{8.91}$$

Assuming the error vector to be $\{e_0, e_1, \ldots, e_6\}$, the syndromes are:

$$\begin{aligned} S_0 &= e_0 + e_3 + (e_5 + e_6) \\ S_1 &= e_1 + e_3 + e_4 + e_5 \\ S_2 &= e_2 + e_4 + (e_5 + e_6) = e_2 + e_5 + (e_4 + e_6) \end{aligned} \tag{8.92}$$

and check sums are: $A_1 = S_0 + S_1 = e_0 + e_1 + (e_4 + e_6)$.
It is seen that S_0 and S_2 are orthogonal on $B_1 = (e_5 + e_6)$; and A_1 and S_2 are orthogonal on $B_2 = (e_4 + e_6)$. Further B_1 and B_2 are orthogonal on e_6. Thus, a 2-step mojority vote will locate the error on e_6. The corresponding decoder is shown in Fig. 8.12, where the 2nd level majority logic circuit

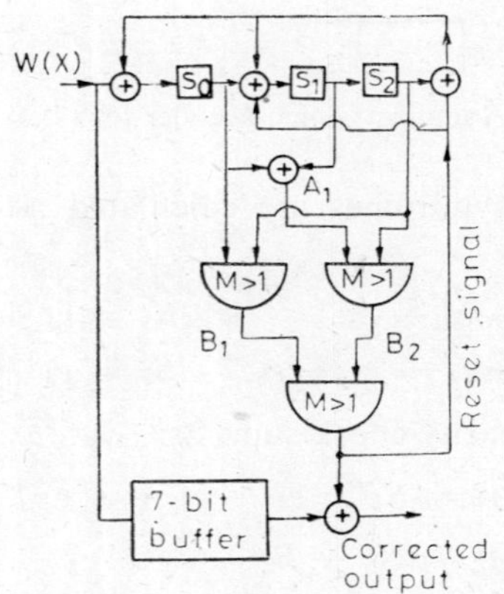

Fig. 8.12 2-step majority-logic decoder for (7, 4) Hamming code

gives the correction signal and the stored $W(X)$ is corrected as the bits are read out from the buffer. Correct decoding is achieved if $t \leqslant j/2 = 1$ error. Comparing this decoder with that of Fig. 8.8, it is seen that the majority logic gives a '1' signal when the syndrome state is {101}, where as {100} was used as for error correction in Fig. 8.8. But the basic principles of the two decoders are the same.

The results of these examples are generalized in the following:

*(a) A set of parity-check sums $A_1, A_2, \ldots, A_j$ is said to be orthogonal on the error bit e_l, if e_l is checked by all check sums A_j in the set and no other error bit is checked by more than one check sum. The error bit e_l is now detected by a majority vote of the check sums A_j.

(b) Successive bits are tested at each cyclic shift of the syndrome and e_l located. $W(X)$ is corrected as the bits are shifted out of the buffer.

(c) If from the syndrome equations, the number of check sums formed $j = (d - 1)$, where d = minimum HD of the code, then the cyclic code is said to be completely orthogonalizable in one step. The errors can be corrected if $t \leqslant j/2$.

(d) A set of j parity-check sums is said to be orthogonal on a group of error bits $\{e_{il}\}$, if and only if (i) every error-bit in $\{e_{il}\}$ is checked by all check sums A_j and (ii) no other error bit is checked by more than one check sum.

(e) A code is said to be L-step orthogonalizable (or L-step majority-logic decodable) if L steps of orthogonalization are required to make a decoding decision on the error bit e_l. If $j = (d - 1)$, then the code is completely L-step orthogonalizable; and the decoder requires L levels of majority gates for decision making.

Using these principles, the (15, 7) BCH code, $t \leqslant 2$, is decoded by a majority-logic decoder similar to that of Fig. 8.11. Assuming an error vector to be $\{e_0, e_1, \ldots, e_{14}\}$, the eight syndrome equations are calculated from $[E \cdot H^T]$ and four check sums orthogonal on e_{14} are obtained as:

$$\begin{aligned} A_1 &= S_3 = e_3 + e_{11} + e_{12} + \underline{e_{14}} \\ A_2 &= S_1 + S_5 = e_1 + e_5 + e_{13} + \underline{e_{14}} \\ A_3 &= S_0 + S_2 + S_6 = e_0 + e_2 + e_6 + \underline{e_{14}} \\ A_4 &= S_7 \ldots = e_7 + e_8 + e_{10} + \underline{e_{14}} \end{aligned} \tag{8.93}$$

It is now seen that the check sums A_1, A_2, A_3 and A_4 are orthogonal on e_{14} and this is corrected by a majority vote, $M \geqslant 3$. If e_{14} and any other bit are in error, then at least three sums will be '1' and a correct decision will take place with $t \leqslant 2$. But the decoder will fail with $t > 2$ except for few special combination of error bits. Since the code is cyclic, it is possible to form another set of 4 check sums orthogonal on any other bit e_l, and the decoder will operate in the usual way.

The (15, 5) BCH code $t \leqslant 3$, is a 2-step orthogonalizable code and its decoder is similar to that of Fig. 8.12. (for details, see Ref. [1]). Other orthogonalizable codes are simplex (maximum length) and RM codes. Although HD of RM codes for a given n, are slightly less than those of

*The term 'orthogonal' as used here directly conflicts with its usage in the study of vector spaces, and as such, some authors [11] prefer the word 'concurring'. So it may be stated that 'A set of check sums $A_1, \ldots, A_j$ is concurring on the coordinate l, if e_l is checked by all check sums in the set'. However, here the term orthogonal is used as is common in many references.

comparable BCH codes, RM codes for moderate n are attractive because of the simple decoding process. For large n, however, RM codes are definitely inferior to BCH codes.

8.5 CONVOLUTIONAL CODES

We have considered the block codes, where k-bit words are coded into n-bit words by adding $(n - k)$ parity check bits, and decoding is done word by word. In contrast with this technique, a coder for convolutional codes (also called Recurrent codes and a subclass of Tree codes) transforms k bits of data into n-bit codes, but the coder output depends on the earlier blocks of data that have been processed by the coder; thus, the coder memory couples the currently processed data with a few earlier data blocks. The size of the coder memory is indicated by the constraint length K; and the code is designated by (n, k, K) along with the code rate $R = k/n$. Thus, a convolutional code designated as (3,2), $K = 3$, means $n = 3$, $k = 2$, $R = 2/3$ and the memory length is 3. Again in contrast with block codes, k, n and K are generally small integers for convolutional codes.*

Consider a simple convolutional encoder and a decoder, shown in Fig. 8.13, where the parameters are: $R = 1/2$, $n = 2$, $k = 1$, $K = 2$. The parity bit P is generated by the mod-2 sum of the SR outputs as:

$$P = (X + 1), \quad \text{or} \quad g(1, 1) = (11) \tag{8.94}$$

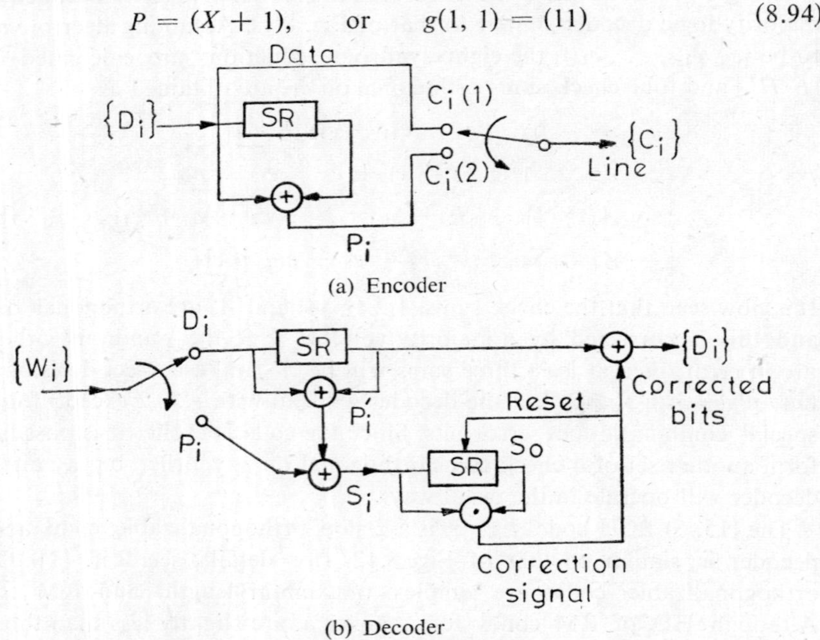

(a) Encoder

(b) Decoder

Fig. 8.13 Encoder and decoder for $g(1, 1) = (11)$

*Because of the constraint length K, a convolutional code is designated, by some authors, as: (nK, kK) code, i.e., a (3, 2, 3) code is also called as a (9, 6) convolutional code.

Conventionally, the subgenerator for P is also written as $g(1, 1)$ and the generator sequence $g(1) = (\underline{11}\ \underline{01})$, which gives the coder output for a data input $\{100 \ldots 0\}$. For an arbitrary data input $\{D_i\}$: $\{1011001100 \ldots 0\}$, the coder output $\{C_i\}$ is: $\{\underline{11}\ \underline{01}\ \underline{11}\ \underline{10}\ \underline{01}\ \underline{00}\ \underline{11}\ \underline{10}\ \underline{01}\ \underline{00} \ldots\}$. This is a systematic code as the first bit of each pair in C_j is the data bit.

The code corrects a single error in a span of 4 bits of the received data plus parity bits, $\{W_i\}$ (i.e., one error in two bits of information). Decoding is done through the calculation of syndrome bits by comparing the received parity bits P_i with the locally generated parity bits P_i' in the decoder, as shown in Fig. 8.13(b). It is easily checked that for an error vector $\{e_i\} = \{100 \ldots 0\}$, with the error in the data bit, the syndrome content will be: $\{S_1 S_0\} = \{11\}$, and the AND gate will give a correction signal when the corresponding data bit is shifted out of the decoder. Simultaneously, the syndrome register will be reset to zero. If the error is in the parity bit, then $\{S_1 S_0\}$ is either $\{01\}$ or $\{10\}$, and the AND gate will give zero output. Thus, the error patterns of the form $\{0100010\}$ will be corrected, but not the patterns of the form $\{0100100\}$, $\{0101000\}$, $\{011000\}$ etc.*

The coder/decoder of Fig. 8.13 may be generalized by noting that for systematic convolutional codes, the coded outputs consist of Kk bits of data and $K(n - k)$ bits of parity check per block of nK bits of $\{C_j\}$. Moreover, the number of subgenerators is equal to k and the number of SR's required in the encoder is $(K - 1)$ per data bit (there is an alternate configuration where the number of SR's is $(K - 1)$ only). Based on these principles, a generalized encoder structure for systematic codes in shown in Fig. 8.14,

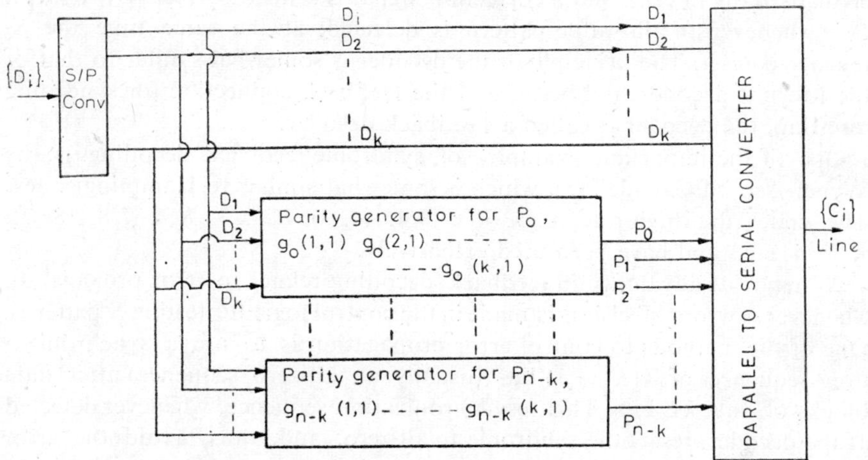

Fig. 8.14 Encoder (Type II) for systematic codes, having k-bit data and $(n - k)$-bit parity

*It will be seen that the ordering of the data and error vectors in this section is from the leftmost bit onwards, i.e., the leftmost bit is first processed and transmitted. This convention is in contrast with the convention used for cyclic codes. The reason for this is, that, in convolutional codes, the starting point in coding and decoding is the first bit transmitted and the coder memory uses this information for subsequent coding/decoding. The polynomial form of the generators, however, is unchanged.

where parity bits are generated through SR-delay units and mod-2 sums, as per the given subgenerator polynomials. The outputs of the subgenerators and data bits of the same block are multiplied to give $\{C_j\}$. The complementary generalized decoder is shown in Fig. 8.15, where the auxiliary parity

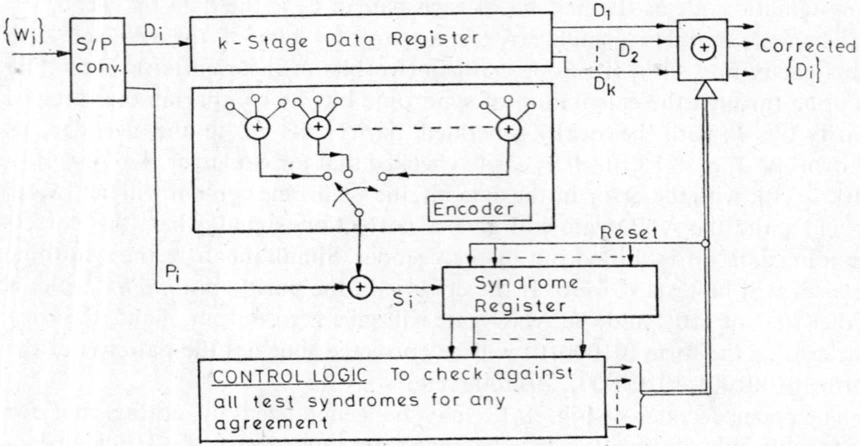

Fig. 8.15 A generalized feedback decoder

bits P_i' are generated through a local encoder and the syndrome register stores the syndrome bits $\{S_i\}$, given by $S_i = (P_i + P_i')$. The $\{S\}$ vectors are checked against the allowable test patterns precalculated for allowable error patterns in $\{D_i\}$, and a correction signal is sent to correct $\{D_1, D_2, \ldots, D_k\}$, whenever an allowable pattern is detected; at the same time the S-register is reset. The principle of the decoder is somewhat similar to that of the Meggitt decoder, and because of the feedback connection for syndrome resetting, the decoder is called a Feedback decoder.

One of the important examples of syndrome/feedback decoding is the Wyner-Ash SEC code [18], which is somewhat similar to Hamming codes. The codes are high-rate, e.g., $R = 3/4$, (4, 3), $K = 3$; $R = 7/8$, (8, 7), $K = 4$, etc., and have been used extensively.

A major disadvantage in feedback decoding relates to error propagation whenever a wrong decision is made in the control logic for testing S-patterns. One of the methods to control error propagation is to use a synchronization sequence of $k(K - 1)$ bits (usually an all-zero sequence) after data blocks of, say, kL bits. This resynchronization sequence, whenever detected at the decoder, resets the syndrome to all-zero, and, thus, avoid the error propagation in future data bits. The transmission rate, of course, decreases due to the addition of the resynchronization sequence and the new rate is:

$$R' = \frac{LR}{L + K - 1}, \quad \text{where} \quad R = k/n \tag{8.95}$$

Thus, it is necessary to make $L \gg K$, to maintain $R' \simeq R$. In general, however other types of decoders, e.g., majority logic decoder, Viterbi decoder, etc. are preferred for large-K codes. However, for simple codes, feedback decoders are easily implemented.

8.5.1 Majority-Logic Decoding of Convolutional Codes

Majority-logic decoding techniques for block codes have been discussed in Sec. 8.4.4. Some of the convolutional codes are also majority-logic decodable, and the theorems, established for block codes, are also valid for convolutional codes. The majority-logic decoder circuits for convolutional codes are similar to that of Fig. 8.11 for block codes, where the check sums are calculated and the majority vote is taken.

Example 12

Consider the $R = 1/2$, (2, 1), $K = 4$ convolutional code, with

$$g(1, 1) = (1011) \rightarrow (X^3 + X + 1) \tag{8.96}$$

The generator sequence is $g(1) = \{\underline{11}\ \underline{00}\ \underline{01}\ \underline{01}\}$, and the encoder is an extension of Fig. 8.13. The corresponding syndrome decoder is based on Fig. 8.15. However, the code is also majority-logic decodable and the corresponding decoder is shown in Fig. 8.16. Assume that an error vector

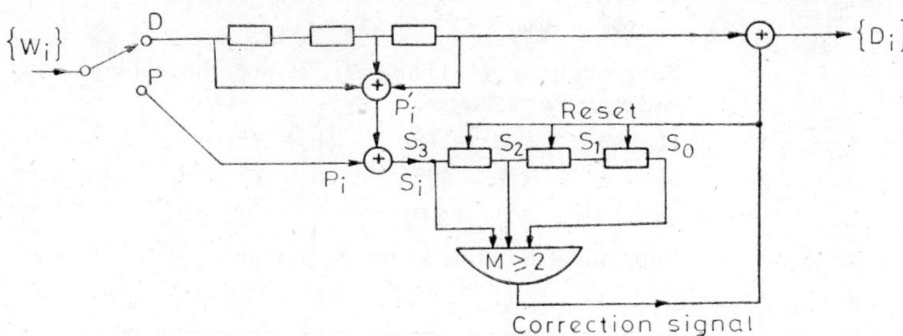

Fig. 8.16 Majority logic decode for (2, 1), $K = 4$, code

$\{e_d^0, e_d^1, e_d^2 \ldots\}$ is present in the received data vector $\{D_i'\}$ and an error vector $\{e_p^0, e_p^1, e_p^2 \ldots\}$ is present in the received parity vector $\{P_i\}$, where $\{W_i\} = \{D_1'P_1, D_2'P_2, \ldots D_l'P_l \ldots\}$. Then from the given subgenerator polynomial, the syndrome equations are:

$$\begin{aligned} S_0 &= e_d^0 + e_p^0 \\ S_1 &= e_d^1 + e_p^1 \\ S_2 &= e_d^0 + e_d^2 + e_p^2 \\ S_3 &= e_d^0 + e_d^1 + e_d^3 + e_p^3 \end{aligned} \tag{8.97}$$

It is seen now that S_0, S_2, S_3 check on e_d^0 and no other error bit appears in more than one equation. Thus, the error bit e_d^0 can be identified by a majority vote in the decoder shown in Fig. 8.16. It may be noted further that the syndrome taps for majority vote appear to form a matched-filter (MF) for the coder (since both coder and decoder may be considered to be digital filters) and the syndrome bits are processed through an MF to give the optimum output. The code is now known as Self-orthogonal. Some of the majority-logic decodable convolutional codes are self-orthogonal and others are L-step orthogonalizable, as discussed in Sec. 8.4.4. Table 8.5 gives a short-list of self-orthogonal convolutional codes and Table 8.6 gives a short list of Orthogonalizable codes, obtained through computer search [17].

Table 8.5 A Short-List of Self-Orthogonal Convolutional Codes [17]

(n, k)	k	t	Subgenerator polynomials*
(2, 1)	3	1	$(1 + X^2)$
,,	7	2	$(1 + X^2 + X^5 + X^6)$
,,	18	3	$(1 + X^2 + X^7 + X^{13} + X^{16} + X^{17})$
,,	36	4	$(1 + X^7 + X^{10} + X^{16} + X^{18} + X^{30} + X^{31} + X^{35})$
,,	56	5	$(1 + X^2 + X^{14} + X^{21} + X^{29} + X^{32} + X^{45} + X^{49} + X^{55}$
(3, 2)	3	1	$(1 + X)$; $(1 + X^2)$
,,	14	2	$(1 + X^8 + X^9 + X^{12})$; $(1 + X^6 + X^{11} + X^{13})$
,,	41	3	$(1 + X^3 + X^{15} + X^{28} + X^{35} + X^{36})$; $(1 + X^2 + X^6 + X^{24}$ $+ X^{29} + X^{40})$
(3, 1)	—	—	Subgenerators $g(1, 1)$ and $g(1, 2)$ are the same as for (3, 2) codes; t is now larger.
4, (3)	4	1	$(1 + X)$; $(1 + X^2)$; $(1 + X^3)$
,,	20	2	$(1 + X^3 + X^{15} + X^{19})$; $(1 + X^8 + {}^{17} + X^{18})$, $(1 + X^6 + X^{11} + X^{13})$
(4, 1)	—	—	Subgenerators same as for (4, 3) codes
(5, 4)	5	1	$(1 + X)$; $(1 + X^2)$; $(1 + X^3)$; $(1 + X^4)$

*In the above codes, the complements of the polynomials $f_n(X)$ are equally valid subgenerator polynomials, where $\overline{f_n(X)} = X^n[f_n(X^{-1})]$. As for example, $X^6(1 + X^{-2} + X^{-5} + X^{-6}) = (X^6 + X^4 + X + 1)$ is an equally good subgenerator polynomial for (2, 1), $K = 7$ code.

The equivalent subgenerator in binary form is written as (cf. Eq. (8.96): For $R = 1/2$, $K = 18$, $t \leqslant 3$ code, $g(1, 1) = (110010000010000101) = 1+X^2+X^7+X^{13}+X^{16}+X^{17}$.

In the orthogonalizable convolutional codes, which have better error-correcting capabilities, the syndrome equations and check sums are formed as in Eq. (8.93). The majority vote on these check sums $(A_1, A_2, \ldots, A_j)$ is then taken to identify the error bit and the correction signal is generated accordingly.

Example 13

Consider a (3, 1), $K = 4$, code with subgenerators given by:

$$g(1, 1) = (1011)$$

$$g(1, 2) = (1101) \tag{8.98}$$

and

$$g(1) = (\underline{111}\ \underline{001}\ \underline{010}\ \underline{011})$$

Table 8.6 A Short List of Orthogonalizable Convolutional Codes [17]

(n, K)	K	t	Subgenerator Polynomials*
(2, 1)	6	2	$(1 + X + X^2 + X^5)$
,,	12	3	$(1 + X + X^2 + X^4 + X^5 + X^{11})$
,,	22	4	$(1 + X + X^2 + X^4 + X^5 + X^8 + X^{10} + X^{21})$
(3, 1)	3	2	$(1 + X^2)$; $(X + X^2)$
,,	5	3	$(1 + X + X^2 + X^4)$; $(X^3 \times X^4)$
,,	8	4	$(1 + X^6 + X^7)$; $+ (X + X^3 + X^4 + X^5 + X^7)$
,,	11	5	$(X + X^9 + X^{10})$; $(X + X^2 + X^5 + X^7 + X^8 + X^9 + X^{10})$
,,	18	6	$(X + X^2 + X^3 + X^{16} + X^{17})$; $(X + X^4 + X^5 + X^8 + X^{10} + X^{11} + X^{12} + X^{13} + X^{17})$
(5, 1)	2	3	$(1 + X)$; $(1 + X)$; (X); (X)
,,	3	4	$(1 + X + X^2)$; $(X + X^2)$; $(1 + X^2)$; (X^2)
,,	4	5	$(1 + X + X^2 + X^3)$; $(X^2 + X^3)$; $(X + X^3)$; $(1 + X^3)$
,,	6	6	$(X + X^2 + X^3 + X^4 + X^5)$; $(X^4 + X^5)$; $(1 + X^3 + X^5)$; $(1 + X^2 + X^5)$
$(2^m, 1)$	$(m + 1)$	$[(m + 2)\cdot 2^{m-2} - 1]$	Uniform code; dual of Wyner–Ash codes. Subgenerators same as Wyner–Ash codes

*Contrary to the polynomials of Table 8.5, the complements of these polynomials are not valid as subgenerator polynomials.

The two sets of syndrome equations are:

$$\left.\begin{aligned} \nearrow S_0(1) &= e_d^0 + e_P(1) \\ S_1(1) &= e_d^1 + e_p^1(1) \\ \nearrow S_2(1) &= e_d^0 + e_d^2 + e_p^2(1) \\ S_3(1) &= e_d^0 + e_d^1 + e_d^3 + e_p^3(1) \end{aligned}\right\} \quad \begin{aligned} \nearrow S_0(2) &= e_d^0 + e_p^0(2) \\ \nearrow S_1(2) &= e_d^0 + e_d^1 + e_p^1(2) \\ S_2(2) &= e_d^1 + e_d^2 + e_p^2(2) \\ S_3(3) &= e_d^0 + e_d^2 + e_d^3 + e_p^3(2) \end{aligned}$$

and $\quad \nearrow A_1 = S_1(1) + S_3(1) = e_d^0 + e_d^3 + e_p^1(1) + e_p^3(1).$

It is now seen that $S_0(1)$, $S_2(1)$, $S_0(2)$, $S_1(2)$ and A_1 are all orthogonal on e_d^0; then the majority vote locates the error in the data bit. Since $j = 5$, the code corrects two errors in a span of $12W_i$ bits. The decoder is schematically shown in Fig. 8.17, and the code is classified as orthogonalizable.

Uniform codes are low-rate orthogonalizable codes and are dual to Wyner-Ash codes. Their parameters are: $R = 1/2^m$, $m > 0$, $n = 2^m$, $k = 1$, $K = (m + 1)$; and the code is designated as $(2^m, 1)$ code. The $(2^m - 1)$ subgenerators for the codes are the same as those for the Wyner-Ash codes. It has been shown that the codes are orthogonalizable with $j = [(m + 2)2^{m-1} - 1]$ parity check sums. Thus, the code corrects $t \leqslant [(m + 2)2^{m-2} - 1]$ errors in a span of $[(m + 2)\cdot 2^m - 1]$ channel bits.

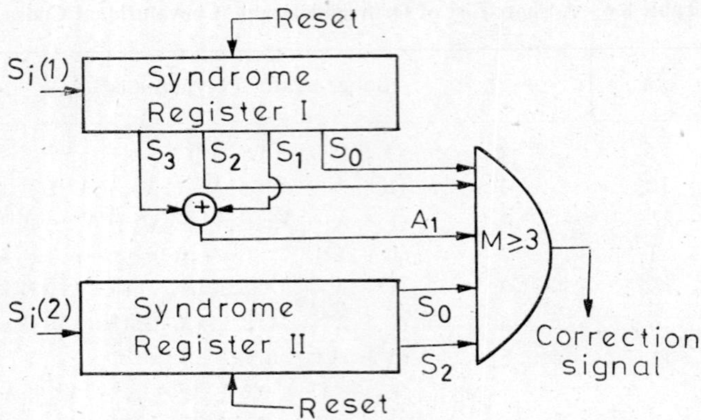

Fig.8.17 Decoder for (3, 1), $K = 4$ code

The problem of error propagation persists in majority-logic decoders, as in feedback decoders. It has been shown that in self-orthogonal convolutional codes, the decoder automatically recovers from a decoding error over a short span of received bits; but it is not so in orthogonalizable codes. The error propagation may be eliminated if the decoder operates in the Definite decoding mode, i.e., with the feedback connection removed from the syndrome resetting points. This, however, reduces the error-correcting capability of the self-orthogonal codes by a factor of two approximately; and the reduction is considerable in orthogonalizable codes. However, the technique of resynchronization after kL bits, $L \gg K$, can be used whenever required.

8.5.2 Non-Systematic Code: Viterbi Decoding [10, 19, 20]

We have discussed the systematic convolutional codes along with some simple decoding techniques. However, the non-systematic codes have been shown to be more powerful, but they can be decoded only through probabilistic techniques, *viz.*, (a) Viterbi decoding algorithm and (b) Sequential decoding algorithm. Viterbi's maximum-likelihood deconding scheme is very effective for codes with short constraint length, say, $K \leqslant 7$, and has several advantages over the sequential decoding scheme. But the complexity of it's decoding makes it impractical for larger K where the coding efficiency is much better. However, the sequential decoding scheme, using computer-like hardware, can be implemented for large K at a modest cost. Table 8.7 gives a short list of non-systematic convolutional codes obtained through computer search [11], and it is seen that their error-correcting capabilities are better as compared to equivalent systematic codes.

For probabilistic decoding, it is necessary to represent a convolutional code in the form of special graphs, known as State diagram, Tree-graph and Trellis diagram. Consider a $R = 1/3$, (3, 1), $K = 3$, systematic code with subgenerators given by:

$$g(1, 1) = (101); \qquad g(1, 2) = (110)$$

Table 8.7 A Short List of Non-Systematic Convolutional Codes [11]

(n, k)	K	$d_{\min}$	Subgenerator polynomials*
(2, 1)	3	5	$(1 + X^2)$; $(1 + X + X^2)$
,,	4	6	$(1 + X + X^3)$; $(1 + X + X^2 + X^3)$
,,	5	7	$(1 + X^3 + X^4)$; $(1 + X + X^2 + X^4)$
,,	7	10	$(1 + X^2 + X^3 + X^6)$; $(1 + X + X^2 + X^3 + X^6)$
,,	10	12	$(1 + X^3 + X^4 + X^5 + X^7 + X^8 + X^9)$; $(1 + X + X^3 + X^4 + X^7 + X^9)$
,,	12	15	$(1 + X^4 + X^5 + X^7 + X^8 + X^9 + X^{11})$; $(1 + X^2 + X^3 + X^4 + X^5 + X^7 + X^{10} + X^{11})$
(3, 1)	3	8	$(1 + X^2)$; $[1 + X + X^2)$; $(1 + X + X^2)$
,,	4	10	$(1 + X^2 + X^3)$; $(1 + X + X^3)$; $(1 + X + X^2 + X^3)$
,,	5	12	$(1 + X^2 + X^4)$; $(1 + X + X^3 + X^4)$; $(1 + X + X^2 + X^3 + X^4)$
,,	7	15	$(1 + X^2 + X^3 + X^5 + X^6)$; $(1 + X + X^4 + X^6)$; $(1 + X + X^2 + X^3 + X^4 + X^6)$
,,	10	20	$(1 + X^3 + X^6 + X^7 + X^8 + X^9)$; $(1 + X + X^3 + X^4 + X^5 + X^7 + X^9)$; $(1 + X + X^2 + X^5 + X^6 + X^8 + X^9)$
(4, 1)	3	10	$(1 + X^2)$; $(1 + X + X^2)$; $(1 + X + X^2)$; $(1 + X + X^2)$
,,	5	16	$(1 + X^2 + X^4)$; $(1 + X + X^3 + X^4)$; $(1 + X + X^2 + X^4)$; $(1 + X^2 + X^2 + X^3 + X^4)$
,,	7	20	$(1 + X^2 + X^3 + X^4 + X^6)$; $(1 + X^2) + X^3 + X^4 + X^6)$; $(1 + X + X^4 + X^5 + X^6)$; $(1 + X + X^2 + X^5 + X^6)$

*Complements of the polynomials $f_n(X)$ are also valid subgenerator polynomials as in the case of the polynomials in Table 8.5.

and $$g(1) = (111\ 001\ 010) \tag{8.99}$$

The coder is shown in Fig. 8.18, where for an input $\{D_i\} = \{10110\ldots\}$, the output is:

$$\{C_i\} = \{111\ 001\ 101\ 110\ 011\ 010\ 000\ldots\}.$$

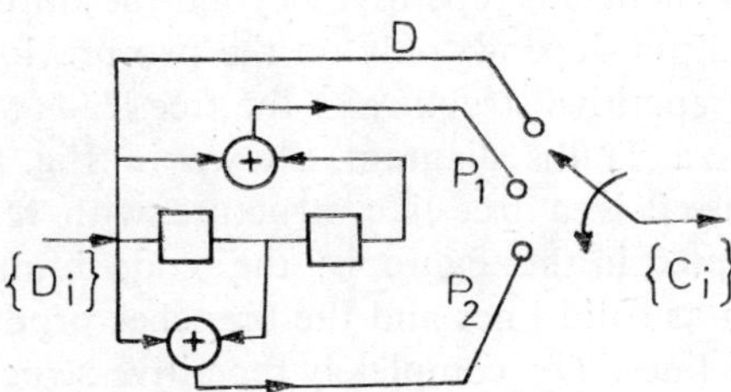

Fig. 8.18 Coder for (3, 1) code

The code is majority-logic decodable and corrects two errors in a block of nine W_i-bits.

The coder can be represented by the state diagram of Fig. 8.19, where the states of the SR's {00, 01, 10, 11} are designated as the machine states. As the data bits move into the coder, the state transitions and the coder outputs are obtained from the state diagram. The state diagram can be redrawn as a Tree graph, representing the coder, as shown in Fig. 8.20. Here also,

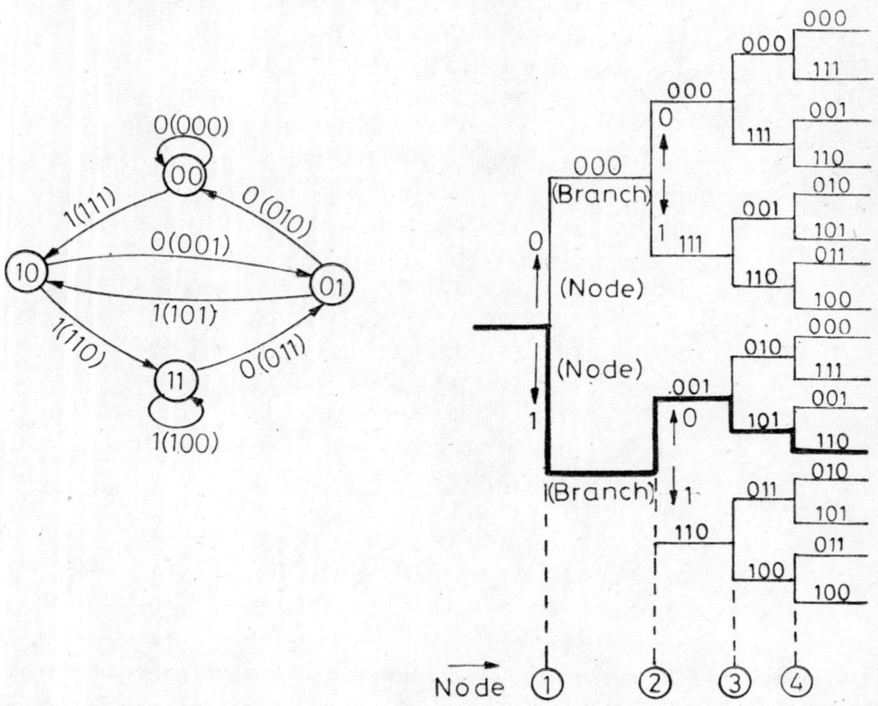

Fig. 8.19 State diagram of the coder of Fig. 8.18

Fig. 8.20 Tree graph of the coder

the coder output is easily traced through the tree paths and the particular path traversed for $\{D_i\} = \{10110 \ldots\}$ of the above example is shown by a heavy line. The coder output also checks with this example. It is further seen that the tree structure is repetitive beyond the third node level; this is because the coder output depends only on the two previous inputs and the current input. The repetitive structure of the tree leads to the redrawing of the tree diagram as a Trellis diagram, shown in Fig. 8.21. The trellis is called as such because it is a tree-like structure with remerging branches. The convention adopted in the figure is: the code branches produced by '0' inputs are shown as solid lines and the branches produced by '1' inputs are shown as dotted lines. The completely repetitive structure of the trellis diagram suggests a further reduction in the representation of the coder

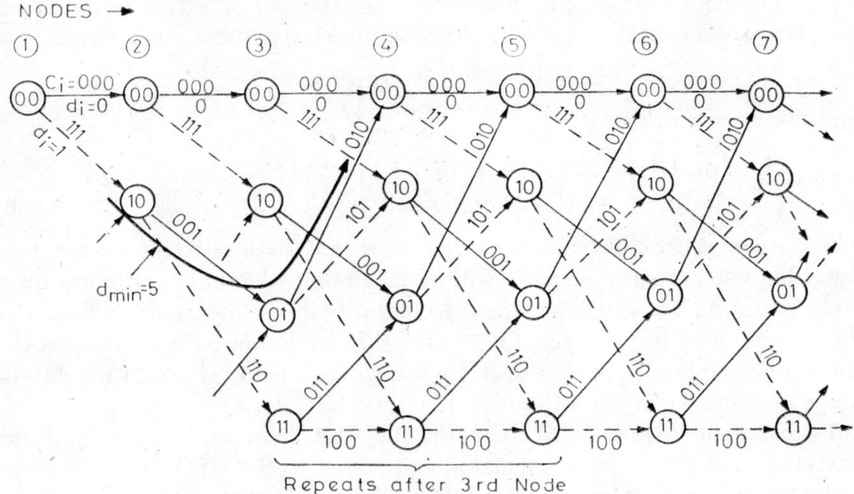

Fig. 8.21 Trellis diagram of the coder of Fig. 8.18

and this results in the state diagram of Fig. 8.19. Thus, the tree, trellis and the state diagrams, all represent the input-output relation of the coder and are used for decoding coded data bits by tracing the most likely path that was traversed while the code was generated. Specifically, the trellis graph is most useful for probabilistic decoding.

As in block codes, the distance property of convolutional codes is also important in evaluating their error-correcting capabilities. Referring to the trellis diagram of Fig. 8.21, consider all paths that start from the first (00) node and merge with all-zero path at some later node, say, j. Then it is seen that:

(a) there is one path with $d = 5$ (for the code word {00100 . . .}) from the all-zero path.
(b) there is one path with $d = 8$ (for the code word {001100}) which diverged at 4 nodes earlier.
(c) there is another path with $d = 8$ (for the code word {0010100 . . .}) which diverged at 5 nodes earlier, and so on.

The minimum distance among all paths is, then, $d_{\min} = 5$, and this is sometimes called the minimum 'free' distance. With $d_{\min} = 5$, the corresponding $t \leqslant 2$.

Viterbi Decoding [10]

The Viterbi decoding is based on maximizing the likelihood function, defined as the conditional probability pr. $(W_i/C_i^{(N)})$, where $\{W_i\}$ is the received sequence with errors and $C_i^{(N)}$ is one of the possible transmitted sequences (refer to Sec. 8.2.2). Such a maximum likelihood decoder gives the minimum error in decoding for memoryless noisy channels including BSC. Considering a BSC with $p_e < 0.5$, the likelihood function for 'd' errors in $\{W_i\}$ (i.e., the HD between $\{W_i\}$ and $\{C_i\} = d$ for N bit long sequence) is given by:

$$\text{pr.}\ (W_i/C_i^{(N)}) = p^d(1-p)^{N-d}$$

and taking logarithm,

$$\log\ [\text{pr.}\ (W_i/C_i^{(N)})] = -d \cdot \log\ [(1-p)/p] + N \log\ (1-p)$$
$$= -Ad + B \qquad (8.100)$$

where A and B are constants. It is clear now that the log-likelihood function (also known as 'path metric') will be maximum when d is minimum for a given p and N. Thus, it is sufficient to find a path in the trellis which has the minimum d for a given $\{W_i\}$, and this minimum path is taken as the transmitted code $\{C_i\}$. Such a decoder using '$d_{\min}$' as the criterion for decoding is also known as the Minimum Distance Decoder.

To calculate the most likely path, the HD between W_i-bits and the branches of the trellis, are calculated and added to the HD's of the previous nodes in order to compute the HD of the nodes at the next stage. Those paths are said to survive which result in the minimum HD at the nodes. These paths (and their history) and the HD of the corresponding nodes are stored and similar computations are repeated for other nodes. It can be shown that the paths which have survived at any stage of computation emerge from one old branch at a distance of $3K$ or $4K$ bits (decoding window) in the past, and thus, the oldest bit of all paths is expected to be the same. The oldest bit is then declared to be the decoded data bit. This process is continued until all received W_i-bits are decoded. For small blocks of coded data, the computations may be recorded in a tabular form, but for large lengths/coutinuous data, a computer algorithm is necessary for the decoding process.

Consider now a non-systematic convolutional code: $(2, 1), K = 4$, $d_{\min} = 6$

and $g(1, 1) = (1101)$; $g(2, 1) = (1111)$ (8.101)

The coder and its state diagram are shown in Fig. 8.22(a) and (b). It is now possible to write $\{C_i\}$ for a given $\{D_i\}$ from the state diagram. Since $d_{\min} = 6$, the code corrects $t \leqslant 2$ in a span of 8 W_i-bits. Consider that $\{C_i\} = \{00 \ldots 00\}$ as the transmitted vector and the received $\{W_i\} = \{$ 00 01 10 00 00 01 00 01 00 00 $\}$. By following the different probable paths through the state diagram, the different probable $\{C_i\}$ sequences along with their HD with reference to $\{W_i\}$ are given below:

(a) {00 11 11 01 11 00 00 00 00 00 } : HD = 7

(b) {00 11 00 01 01 10 11 00 00 00} : HD = 9

(c) {00 11 11 01 00 00 01 01 10 11} : HD = 8

(d) {00 11 00 10 10 11 00 00 00 00} : HD = 6

(e) {00 11 11 10 00 01 11 00 00 00} : HD = 6

(f) {00 00 00 00 00 00 00 00 00 00} : HD = 4

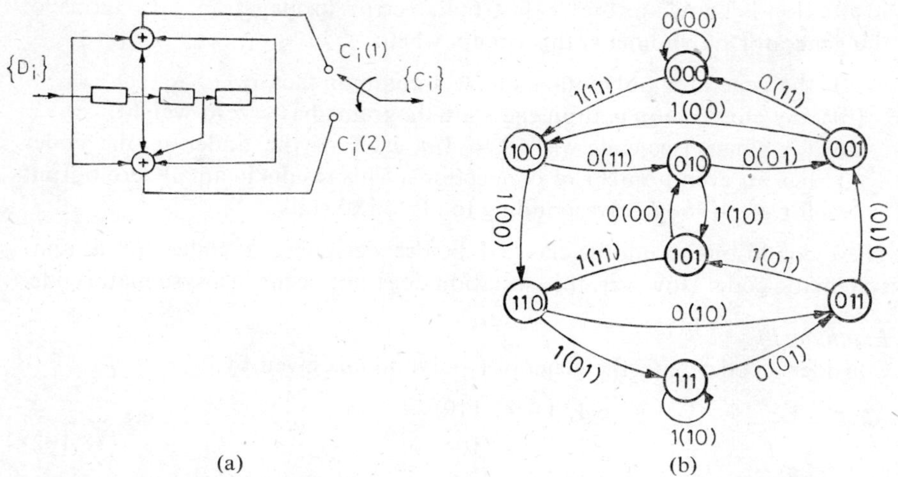

Fig. 8.22 (a) Coder for the (2. 1) $K = 4$ (b) State diagram of coder in (a) codes

It is observed that the sequence with $d_{min} = 4$ is the all-zero sequence, and the decoder has corrected 4 errors in 16 W_i-bits.*

It is possible to have ambiguities with a particular error pattern and the solution in these cases is to arbitrarily discard the lower trellis branch. It is, however, convenient to truncate the sequence at certain intervals by introducing 000 in D_i after blocks of kL bits. This ensures that the final state of the code must be 000, and as such, the ultimate survivor at node (000). The corresponding trellis, therefore, is truncated in the same way as it was begun.

The performance of a convolutional code can be improved by increasing the constraint length K, but the Viterbi decoder becomes impractical for large K, say $K > 10$. With $K = 10$, the decoder has 1024 nodes and must store 1024 surviving paths; at the same time, the trellis must be searched over approximately 100 levels to make a decision. On the other hand, Viterbi decoders can be very fast with K small, and for $K = 7$, a decoder can be built to iterate 10^7 times per second with 128 nodes at each level. A typical hardware description for a Viterbi decoder is given in Ref. [21]. For decoding codes with large K, sequential decoding is used, and the reader is referred to Wozencraft and Reiffen [25].

Catastrophic Error Propagation

With certain convolutional codes, it is possible to have too many ambiguities, leading to a large number of errors in decoding a long sequence. Such

*In this example, if $\{Wi\} = \{$ 00 01 10 00 00 01 01 00 00 00 $\}$, then it may be checked that HD for a probable sequence (*e*) is also 4. Thus, there is a tie with the all-zero sequence. Since the all-zero sequence is at the top of the trellis, this sequence is accepted as the transmitted $\{Ci\}$.

a situation is known as the "catastrophic error propagation." In terms of the generator polynomials, this occurs when:

(a) the generator polynomials have a common factor,
(b) any closed loop path in the state diagrams has a zero weight,
(b) for binary tree code with $R = 1/n$, each of the adder in the coder has an even number of connections. This results in an all-zero output for a self-loop corresponding to all 'ones' state.

These conditions should be checked before designing a coder for a non-systematic code. However, this situation does not occur in a systematic code.

Example 14
Consider a coder with the generator polynomials given by:

$$g(1, 1) = (110)$$
$$g(2, 1) = (101) \tag{8.102}$$

The coder and its state diagram are shown in Fig. 8.23. For an input data $\{D_i\} = \{110000\ldots\}$, the coder output is:

$$\{C_i\} = \{\underline{11}\ \underline{01}\ \underline{11}\ \underline{01}\ \underline{00}\ \underline{00}\ \ldots\}$$

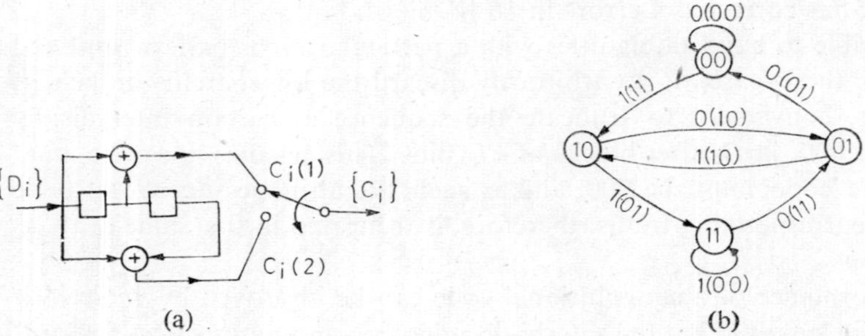

Fig. 8.23 Example of catastrophic error propagation: (a) Coder for Eq. (8.102) (b) State diagram for the coder

If the received sequence is $\{W_i\} = \{\underline{11}\ \underline{01}\ \underline{01}\ \underline{11}\ \underline{00}\ \underline{00}\ldots\}$, the decoder may follow either of the paths:

(a) $\{\underline{11}\ \underline{01}\ \underline{11}\ \underline{01}\ \underline{00}\ \underline{00}\ \ldots\}$ with HD = 2
(b) $\{\underline{11}\ \underline{01}\ \underline{00}\ \underline{11}\ \underline{01}\ \underline{00}\ldots\}$ with HD = 2

Since each path has HD = 2, there is an ambigutiy and the decoder may follow the incorrect path (b). This will lead to a large number of errors, leading to catastrophic error propagation. It can now be seen that the code violates all the three conditions mentioned before and hence, it is an improper one.

Soft Decision Decoding
The path metric $= -Ad + B$, defined by Eq. (8.100), was based on the hard decision of the demodulator output. The optimum decoder, however,

maximizes the log-likelihood function (or the 'path metric') given by (refer to Eq. (2.17)):

$$\log [\text{pr}\cdot(W_i'/C_i^{(N)})] = -\sum_{j=1}^{N} (W_{ij}' - C_{ij}^{(N)})^2/n_0 + N \log (1/\pi n_0) \qquad (8.103)$$

where, W_{ij}' are the soft decision (multilevel) values for the j-components of the received vector and $C_{ij}^{(N)}$ are the j-components (1/0) for the possible transmitted sequence. It is seen now that the expression (8.103) is maximized when

$$d_m^2 = \sum_{j=1}^{N} (W_{ij}' - C_{ij}^{(N)})^2 \qquad (8.104)$$

is minimized. The summation is simply the Euclidian distance between the possible transmitted sequence and the received multilevel signal. The quantity d_m is defined as the soft decision path metric and it is only necessary to add the values of d_m to evaluate the total path difference Σd_m, (since Eq. (8.104) is monotonically increasing with d_m, it is not necessary to calculate Σd_m^2) as one moves along the trellis in search of the most likely path. It has been shown that to approximate the idealized results with infinitely quantized samples, it is necessary to quantize the samples with 8-levels only, and the loss in coding gain is negligible. Further, the approximate coding gain with soft decision decoders is 2 dB, which is indeed the SNR loss in hard limiters [10]. The technique is applicable for decoding both block codes and convolutional codes.

In the Viterbi decoding technique, the soft decision decoding is implemented by simply replacing the Hamming distances by the soft decision metric d_m; and all other decoding operations remain the same. However, to economize on the total memory required, modified branch-metric assignments are used in fast decoders. The following example shows the improvement in the decisions obtained through 8-level quantization of the received signal samples.

Example 15

(a) Consider that the decoder has to decide between two possible 8-bit codes (00 . . . 00) and (11 . . . 11), and the received samples give the quantized values (0-7 units) as:

$$\{W_{ij}'\} = \{2, 5, 1, 6, 5, 0, 4, 3\}$$

The metric sum $= 26 < 28$, where 28 is the decision threshold, and the decision will be in favour of an all-zero vector. However, if the decision was taken bit by bit, then the result is {0 1 0 1 1 0 1 0}, which gives an ambiguity in the decision.

(b) Consider now the decoder for the non-systematic convolutional code given by Eq. (8.101). It was observed that if the received (hard decision) vector $\{W_i\} = \{\underline{00}\ \underline{01}\ \underline{10}\ \underline{00}\ \underline{00}\ \underline{01}\ \underline{01}\ \underline{00}\ \underline{00}\ \underline{00}\}$, then there was an ambiguity in decision, i.e., either the sequence (e) or the all-zero sequence (f) could be the transmitted code. Assume now that the soft decision received symbols are (this gives the hard-decision vector $\{W_i\}$ as above):

$$\{W'_{ij}\} = \{(2, 3), (1, 5), (5, 2), (2, 1), (1, 2), (1, 4), (2, 6), (1, 2), (2, 1), (2, 3)\}$$

Then the path metric Σd_m with all-zero vector $= 48$ and the Σd_m with the sequence $(e) = 50$. Thus, the proper decision is that the transmitted vector is the all-zero sequence.

8.5.3 Burst-Error Correcting Codes [23]

So far, the FEC codes have been discussed on the assumption that the channel is memoryless and the errors occur independently with stationary statistics. With the exception, perhaps, of the space channel, all real channels tend to have error bursts. An error burst of length 'b' is defined as a sequence of b error symbols, the first and last of which are non-zero, e.g., an error vector (0010110011000) is a burst of $b = 8$. For proper decoding, the 'b' symbol bursts have to be separated by a guard space of 'g' symbols. Various specific block and convolutional codes, e.g., Burton, Fire, RS, Berlekamp-Preparata-Massey, Iwadare and Adaptive Gallager codes, have been designed for correcting bursts. However, Interleaved cyclic or convolutional codes offer the advantage of simultaneous correction of random and burst errors, and are almost insensitive to channel statistics.

A simple interleaver is a block interleaver, in which the (n, k) coded words are placed in the rows of a $(b \times n)$ matrix, and read out from the columns. In the receiver, the codewords are reassembled in a complementary way, thereby breaking the burst of errors into t correctable errors per coded word. A somewhat simpler and more effective type of interleaver is a periodic (or convolutional) interleaver, shown in Fig. 8.24, where the successive symbols

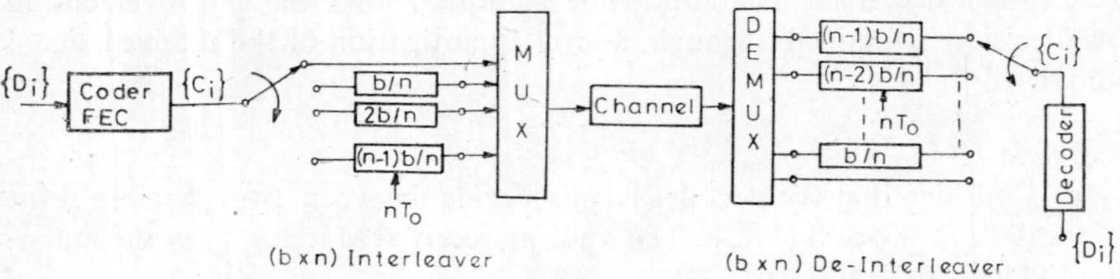

Fig. 8.24 $(b \times n)$ periodic interleaver and de-interlever; T_0 = symbol duration

of an (n, k) codeword are delayed by $\{0, b, 2b, \ldots (n - 1)b\}$ symbol units respectively. As a result, the symbols of one codeword are placed at distances of b-symbol units in the channel stream, and a burst of length 'b' separated by a guard space of $(n - 1)b$ units affect only one symbol per codeword. In the receiver, the codewords are reassembled through complementary delay units and decoded to correct the single errors so generated. If the burst length $l > b$ but $l \leqslant 2b$, then the (n, k) code should be capable of correcting two errors per codeword. To economize on the number of SR's used, one could use b/n, $2b/n$ etc. number of SR's as shown in Fig. 8.24, and clock them at a period of nT_0, where T_0 = symbol time, thereby producing the required delay of $\{b, 2b, \ldots, (n-1)b\}$ symbol units. Thus, if ones use a (15, 7) BCH code with $t \leqslant 2$ for FEC, then a burst of length $\leqslant 2b$ can be corrected with a guard space of $13b$. This 13-to-2 guard-space-to-burst ratio is rather too large, and the codes with small values of n are preferable. Among all the block codes, cyclic codes can be easily implemented for burst correction and the 'Fire' code has been widely used [1, 7]. However, the correctable burst length b for cyclic codes is small even for large (n, k), and as such, interleaved convolutional codes are preferred for long burst correction.

A simple and efficient scheme of correcting large-length bursts with a rate 1/2 code is shown in Fig. 8.25, where the $(b \times n)$ interleaver degenerates into

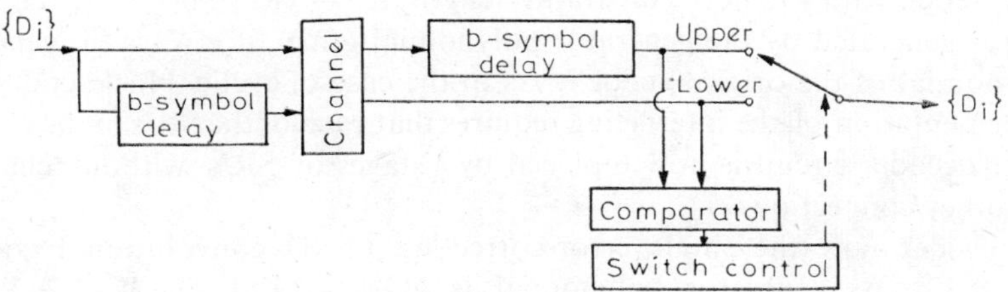

Fig. 8.25 A time-diversity scheme

a simple time diversity of order 'b'. In the coder, the data is repeated through a delay line of length 'b' and at the decoder the upper and lower channels are compared continuously to detect the start of the burst whenever it occurs. Initially, when the channel is quiet, data can be received through one of the channels, but as soon as a burst it detected, the receiver switches to the upper channel (as the burst will be received first in the lower channel) and remain there for $(b - a)$ symbol units, where a = number of symbols required to detect the burst. Then the receiver switches to the lower channel and receives data through the lower channel for b units, and the burst will pass out of the upper delay line by this time. The receiver now reverts to the quiet state. The scheme corrects bursts of length $(b - 2a - 1)$ channel symbols and requires a guard space of $(b + 2a + 1)$ symbols. The probability of a decoding failure is 2^{-a}, and $a \ll b$, but $a \geqslant 15$ (say). This scheme has been successfully used for data transmission in telephone channels. The technique may be further sophisticated by using erasure-correcting (n, k) RS code with $(b \times n)$ interleaver, and burst lengths $\leqslant (n - k)b$ are corrected with a guard space of kb symbols only.

Convolutional Codes for Correcting Bursts

As in the case of block codes, the burst correcting capability of convolutional codes may also be extended by interleaving, and interleaving could be by either bit or block interlacing. The bit interlacing is obtained through the interleaver circuit given in Fig. 8.24, and the resultant code can correct a combination of bursts and random errors.

Example 16
Consider form Table 8.6, the (2, 1), $K = 6$, $t \leqslant 2$, orthogonalizable code with $g(X) = (1 + X + X^2 + X^5)$. By using bit interleaving of the degree $\lambda = 6$ with the circuit of Fig. 8.24 the (12, 6) code, so generated, will correct any burst of length $b = 12$ within a span of 72 channel bits. At the same time, it will correct many random error patterns. However, for only burst-correction, other techniques are more efficient, as the guard space required ($g = 60$) in this case is too large.

Block interleaving can be implemented by noting that "an interleaved (n, k, K) burst-correcting convolutional code generates another convolutional code with the new constraint length $K' = (K - 1)\lambda + 1$. The new code is generated by the generator polynomial $g(X^\lambda)$, if $g(X)$ is the generator polynomial of the original code". As in the case of cyclic block codes, the implementation of the interlacing requires that each of the SR's in the original coder/decoder circuit is now replaced by λ stages of SR's without changing any other connection.

Consider now the single error-correcting (2, 1) convolutional code of Fig. 8.13. Its generator polynomial is $g(X) = (X + 1)$, $K = 2$. With a simple modification, the coder-decoder can be made to correct a burst of $b = 2$ in a span of 8 bits. The modified $g(X) = X^{-1}(X + 1)$, $K = 3$, and the coder and decoder are shown in Fig. 8.26. Now to obtain a coder/deco-

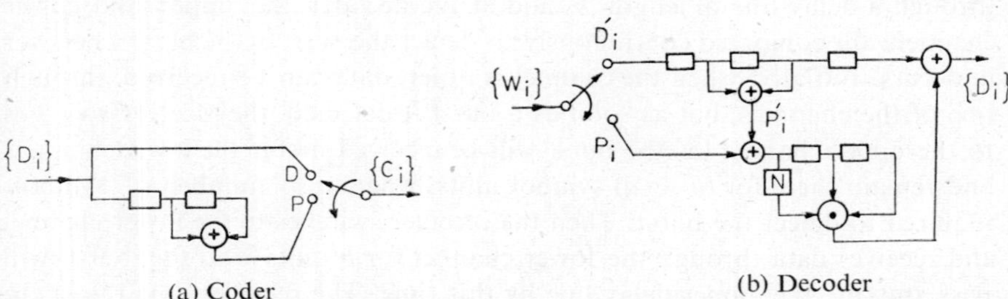

Fig. 8.26 Coder/decoder for (2, 1) code, $K = 3$, $b = 2$, $g = 7$, $N =$ NOT circuit

der with $b = 8$, the above coder may be interleaved to a degree $\lambda = 4$, and the new $g(X^\lambda) = X^{-4}(X^4 + 1)$, as shown in Fig. 8.27, where the required $g = 25$.

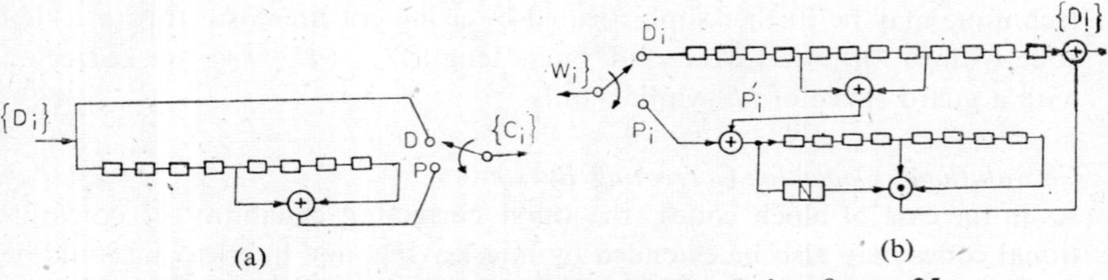

Fig. 8.27 Coder/decoder for (2, 1) code, $K = 9$, $b = 8$, $g = 25$

It is observed that the function of the 4 SR's outside the mod-2 circuit in the parity bit path of the coder is to delay the P-bits in such a way that in the decoder, after the compensating delay of the data bits, the errors in the P_i bits (maximum of 4 P-bits) are shifted out, and the P'-bits are added to only the correct P_i-bits. As an example, with all-zero transmitted code, let the erroneous $\{W_i\} = \{00001101011100 0\}$. Then $\{D'\} = \{0010010\}$ and $\{P_i\} = \{0011110\}$, and the resultant syndrome is $\{(\overleftarrow{001111})\ (10011001)\ 0000\}$. This will give the correction signal to $\{D'\}$ after 4 bit delay and the corrected $\{D_i\}$ is obtained at the output of the decoder. The total time required for decoding the 4 data errors is now equal to 12 bits, equivalent to 24 W_i-bits; thus, the maximum guard space required (with the possibility of g starting with a parity bit) : $g = 25$ bits. For all rate 1/2 codes, these principles may be extended for any value of 'b', and Table 8.8 gives a short list of such codes.

Table 8.8 Parameters of Burst-Correcting Codes

(n, k)	(2, 1), Ref. Fig. 8.26 and 8.27						(3.1), Ref. Fig. 8.28 and Eq. (8.108)			(3.2) Ref. Fig. 8.29 and Eq. (8.112)			(4,3) Ref. Eq. (8.112)		
K	3	5	7	9	11	13	5	7	11	13	18	23	20	27	34
b	2	4	6	8	10	12	7	10	16	6	9	12	8	12	16
g	7	13	11	25	31	37	14	20	32	38	53	68	79	107	135

For low-rate codes, e.g., rate 1/3, 1/4 etc., a simple and efficient burst-correcting technique is possible by using interleaving with delayed repeat codes. Consider a rate 1/3 code with $K = 3$ and generator polynomials given by:

$$g(1, 1) = (010)$$
$$g(1, 2) = (001) \tag{8.105}$$

The coder, shown in Fig. 8.28(a), simply rearranges the data bits and the corresponding parity bits with delay of one or two units, as shown in matrix (8.106). It is seen that the matrix on the left with data and corresponding parity bits in the same column is converted to the matrix on the right with shifted parity bits. The coded bits are now transmitted column by column. In the decoder of Fig. 8.28(b), the data and parity bits are regrouped as in the matrix on the left,

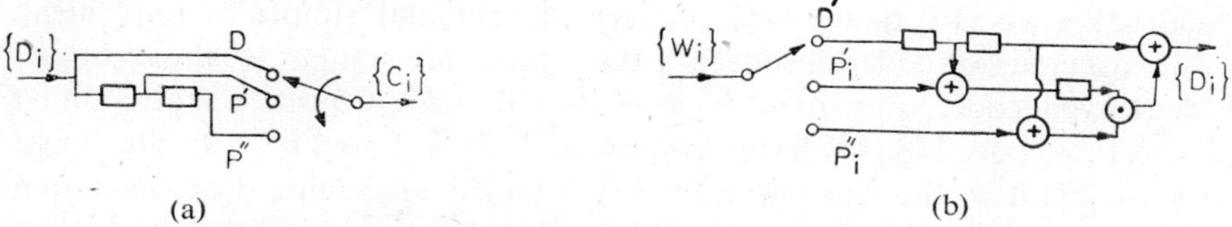

Fig. 8.28 Coder/decoder for rate 1/3, $K = 3$ code, $b = 4$, $g = 8$
(a) Coder, (b) decoder

$$\begin{bmatrix} D_0 & D_1 & D_2 & D_3 \\ P_0' & P_1' & P_2' & P_3' \\ P_0'' & P_1'' & P_2'' & P_3'' \end{bmatrix} \rightarrow \begin{bmatrix} D_0 & D_1 & D_2 & D_3 & \vdots & D_4 & D_5 & \cdots \\ P_{-1}' & P_0' & P_1' & P_2' & \vdots & P_3' & P_4' & \cdots \\ P_{-2}'' & P_{-1}'' & P_0'' & P_1'' & \cdot & P_2'' & P_3'' & \cdots \end{bmatrix} \tag{8.106}$$

and as such the errors in parity bits are shifted out before the corresponding data bits are decoded. The decoding is done by a simple coincidence of the syndrome bits as shown in the figure. As an example, if the data D_0 is wrong, then P_0' and P_0'' locally generated are also wrong and the AND gate will give a correction signal. If, however, D_0 is correct and one of the received P_0' or P_0'' is wrong, then the AND gate output is zero. From the matrix above, it is further observed that with a burst length $b = 4$ covering any four contiguous bits in the 1st and 2nd columns, the possible errors in D_0 and D_1 bits are corrected with a gap length $g = 8$ bits. As a result, the coder-decoder corrects one 4-bit burst in a span of 12 channel bits.

For any value of n, $n \geqslant 3$ and rate $1/n$ of the $(n, 1)$ code, the parameters are now generalized as:

With $K = n = 3$, $$b = \frac{nK - 1}{2} = \frac{(3n - 1)}{2}$$

and $$g = (nK - 1) \tag{8.107}$$

Interleaving these codes to a degree λ, the resultant codes have the parameters:

$$K_I = (n - 1)\lambda + 1; \quad b = \frac{(nK_I - 1)}{2} \tag{8.108}$$

$$g = (nK_I - 1)$$

and the corresponding generator polynomials for $n = 3$, $\lambda = 2$, $K_I = 5$, $b = 7$ and $g = 14$ are;

$$g(1, 1) = (00100)$$
$$g(1, 2) = (00001) \tag{8.109}$$

The operation of the coder/decoder can be easily checked by formulating the corresponding matrices similar to that in (8.106). It has been shown that error propagation does not occur with these codes and the codes are optimal burst correctors.

Iwadare Codes [8]

For high-rate codes, e.g., rate 2/3, 3/4, etc., a class of burst correcting codes, known as Iwadare codes, is very efficient and simple to implement. The principles of coder/decoder for these codes are similar to those of the interleaved codes, shown in Figs. 8.26 and 8.27. Consider that a burst correcting codec is to be designed for $n = 3$, $k = 2$, and $b = 9$. The burst error of 9 bits will disturb at most 4 sub-blocks of 3 bits, i.e., maximum of 3 errors in data bits is possible. The code subgenerators should be distinct for the data bits D_1 and D_2 of the same sub-block, and at the same time, should be capable of generating syndrome bits which will correct 3 errors. Following the method of Fig. 8.27, the subgenerator polynomials may be:

$$g(1, 1) = (10001)$$
$$g(2, 1) = (1001) \tag{8.110}$$

The delay required in the parity bits corresponding to, say, D_2 is 3 bits, but at the same time the syndrome register of (3 + 1) bits have to be cleared of earlier P-errors, thus making the total of 7 bits of delay required for the parity bits of D_2. Since the parity bits of D_1 now have to be distinct from those corresponding to D_2, the delay required for the D_1-parity bits is equal to (7 + 6) bits (6 bits for clearing the syndrome generator of D_2-parity bit errors). The subgenerator are now given by:

$$g(1, 1) = (000000000000010001)$$
$$g(2, 1) = (000000010010000000) \tag{8.111}$$

The last 7-zeros in $g(2,1)$ are to bring D_1 and D_2 in the same time frame at the decoder output. The decoder, of which the local coder is the same as the one in the transmitter, is shown in Fig. 8.29, where D_2-bits are first corrected, and then D_1-bits. (But the order may be reversed). It can now be calculated that for $b = 9$, the guard space required in the decoder is 53 bits.

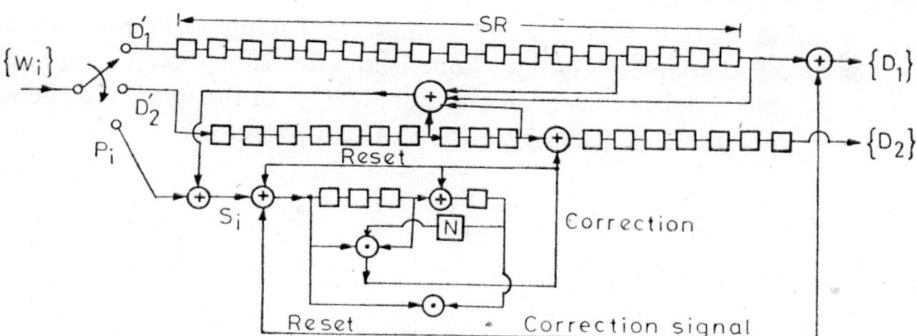

Fig. 8.29 Decoder for Iwadare code (3, 2), $b = 9$, $g = 53$

The parameters of the $(n, n - 1)$ Iwadare codes can now be generalized as:

For any $\lambda, \lambda > 0,$

$$k = \lambda(2n - 1) + (1/2)\, n(n - 1)$$
$$b = \lambda n$$
$$g = (nK - 1) \tag{8.112}$$

and the subgenerator polynomials are given by:

$$g(i, 1) = (\underbrace{00 \ldots 00}_{a_0}1\underbrace{000 \ldots}_{b_0}1\underbrace{000 \ldots 00}_{c_0}) \tag{8.113}$$

where the number of zeros a_0, b_0 and c_0 are specified by:

$$a_0 = \left(\frac{n - i}{2}\right)(4\lambda + n - i - 3) + (n - 1)$$
$$b_0 = (n + \lambda - i - 2) \tag{8.114}$$

and $$c_0 = (K - a_0 - b_0 - 2)$$

A short list of burst correcting codes and their parameters are given in Table 8.8.

Example 17

Consider that a (4, 3) code with $b = 12$ is required. Then, from Eq. (8.112), $\lambda = 3$, $K = 27$ and $g = 107$. The subgenerators are specified by Eq. (8.113) with the values of a_0, b_0 and c_0 as given by:

$$a_0(1, 1) = 21, \quad b_0(1, 1) = 4, \quad c_0(1, 1) = 0$$
$$a_0(2, 1) = 14, \quad b_0(2, 1) = 3, \quad c_0(2, 1) = 8$$
$$a_0(3, 1) = 8, \quad b_0(3, 1) = 2, \quad c_0(3, 1) = 15$$

Then, the subgenerator polynomials are:

$$g(1, 1) = (000000000000000000000100001)$$
$$g(2, 1) = (000000000000001000100000000)$$
$$g(3, 1) = (000000001001000000000000000)$$

The coder and decoder may now be drawn as based on Fig. 8.29.

Adaptive Gallager Code [26]

Many communication channels are composite having both random and burst errors. The adaptive Gallager code is an effective technique for simultaneously correcting both random and burst errors in the channel. The rate 1/2 coder, shown in Fig. 8.30(a), consists of two parts: (a) an inner coder

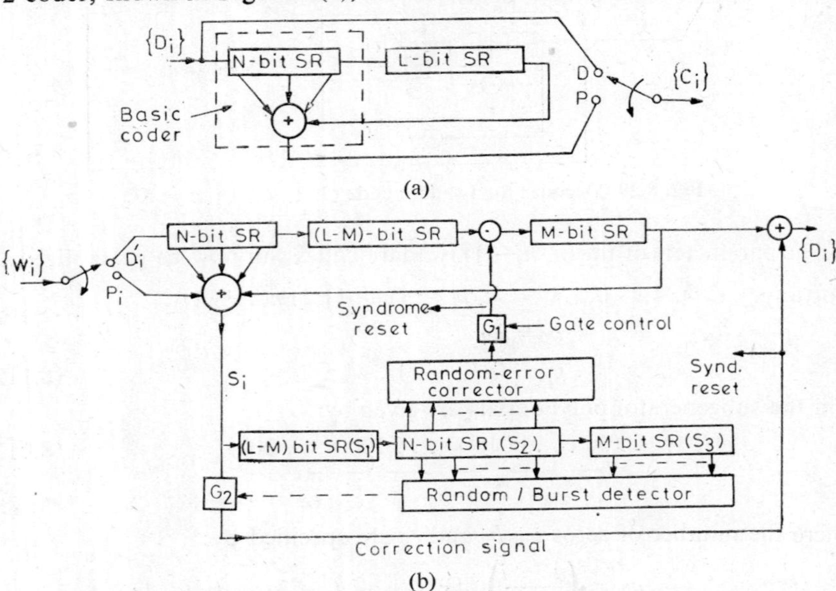

Fig. 8.30 Coder and decoder for $R = 1/2$ adaptive code: $b \leqslant 2L$, $g = 2L + N$ (a) coder, (b) decoder

which uses a random-error correcting code, say, an orthogonalizable convolutional code, for a design value of t, and (b) the outer coder which uses the basic principles of burst correction by the technique of Fig. 8.27. The decoding scheme, shown in Fig. 8.30(b) is an adaptive one, where the decoder is usually in the random-correcting mode with gate G_1 open, and t-random errors are corrected through the N-bit-S_2 syndrome register and the associated logic circuit, say, the majority logic. When a large burst occurs in the channel, the S_2 and S_3 registers with the associated burst detector sense the arrival of the burst and switch the decoder to the burst mode, i.e., gate G_2 is opened and G_1 closed. Since the burst is sensed after $(L + N)$ syndrome bits, $N \leqslant L$, the erroneous data bits of length $(L + N)$ are stored simultaneously in the upper information registers of length $[N + (L - M) + M]$ bits. If the burst length $b \leqslant 2L$, then incoming $\{W_i\}$ is now error free, since all erroneous L-bit data and L-bit parity are by this time shifted into different registers. If the gaurd space is little more than $2L$, then erroneous bits shifting out of the upper M-bit register will now be corrected by the syndrome bits through G_2. At the end of the burst, few all-zero syndrome bits in S_2 and S_3 will restore the decoder to the random mode. The lengths of the registers, i.e., L, M, N and t are chosen to suit the requirements of the channel, with the restriction that $N \ll L$, and $M < L$. The guard space required in the scheme is: $g = 2L + N$, and the ratio $b/g \approx 1$, showing that the code is optimal. However, all bursts of $b \leqslant 2L$ are not corrected in the decoder, as some burst-error patterns may not be detected by the burst detector, but that is compensated by the code's capability of correcting bursts of length b with guard space of $g \simeq b$. The above rate 1/2 adaptive coding scheme may be generalized for other code rates as well.

8.5.5 Two-Dimensional Codes [14, 27]

In Section 8.4.1, it was shown that an array of data vectors may be simultaneously checked by simple parity checks along the rows and columns, and the resultant coded array can be processed in the decoder to correct single or double errors. This iteration technique of Elias [14] can be extended to more powerful codes, e.g., cyclic codes including BCH, RS and convolutional codes. Such codes using an array of data vectors are generally known as Product codes. Alternatively, two codes, one called the inner code and the other the outer, may be concatenated as suggested by Forney [27], to form a very powerful code correcting both random and burst errors. Thus, the product code and the concatenated codes are in a class of two-dimensional codes, where step-by-step decoding is performed. The techniques can be extended to three and higher-dimensional codes as well.

In developing product codes, the information symbols $k = k_1k_2$ are arranged in rows and columns, and the parity checks of a systematic code (either block or convolutional) are added to these rows and columns, as shown in Fig. 8.31. Thus, the product code $C = C_1 \otimes C_2$, where the rows are codewords in C_1 and the columns are codewards in C_2. If C_1 and C_2 are

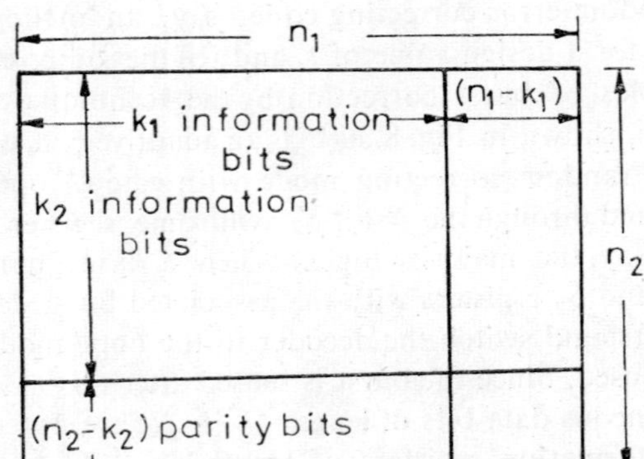

Fig. 8.31 Structure of a systematic product code

systematic, then C is also a systematic code. As is evident from Fig. 8.31, the k_1 symbol in each row is coded by the encoder for C_1 and the k_2 symbol in each column is coded by the encoder for C_2. The resultant code array is then transmitted serially either row by row or column by column. Such two-dimensional codes have the following properties:

(a) if the minimum distances of the codes C_1 and C_2 are d_1 and d_2 respectively, then the d_{min} for the Code C is (d_1d_2).

(b) the code C will correct all error patterns of weight

$$t \leqslant (d_1d_2 - 1)/2 = (2t_1t_2 + t_1 + t_2).$$

(c) the code C also corrects all bursts of length up to $b = \max(n_1t_2, n_2t_1)$, where t_1 and t_2 are the error correcting capabilities of C_1 and C_2 respectively.

(d) for a BSC, the error probability in the receiver using the array code C is better or equal to $f_2[f_1(p)]$, if $f_1(p)$ and $f_2(p)$ are the error probabilities obtained through the individual codes C_1 and C_2, and p = transition probability in the BSC.

The properties (a) and (b) are evident from the array organization. The property (c) is obtained from the consideration that the array is equivalent to the interleaving of the coded words as discussed in Sec. 8.5.3 and illustrated in Fig. 8.24. With $t = t_1$ for C_1 and $t = t_2$ for C_2, the length of the bursts that will be corrected by the interleaved array is either n_1t_2 or n_2t_1, which ever is the larger. It is then appropriate to use the code C_2 with parameter (n_2, k_2) as a burst correcting code (assuming that the array is transmitted column by column) and the code C_1 for (n_1, k_1) as a random-error correcting code. The residual burst plus the random errors will now be corrected by C_1; however, the decoders now will have to be more intelligent than the single decoders discussed so far. Such composite decoders have been discussed in [Ref. [11], Chapter 10].

The overall error probability for a BSC using two dimensional array codes is: $p_1 \leqslant Np^2, p < 1/N$ and $N = m \times n$; and this conforms to the property (d). More specifically, if SEC Hamming codes with an extra parity for double-error detection is used as C_1, then the residual probability of error p_1 per codeword is

$$p_1 \leqslant n_1 p^2 \tag{8.115}$$

If Elias's iteration technique is used now for $n_2, n_3, \ldots, n_r$ blocks, and the blocks are coded as $C_2, C_3, \ldots, C_r$, then, it can be shown that the residual error probability after each iteration is given by:

$$\begin{aligned} p_2 &\leqslant n_2 p_1^2 < n_2 n_1^2 p^4 \\ p_3 &\leqslant n_3 p_2^2 < n_3 n_2^2 p_1^4 < n_3 n_2^2 n_1^4 p^8 \end{aligned} \tag{8.116}$$

and

$$p_r \leqslant (n_r n_{r-1}^2 \, n_{r-2}^4 \ldots n_1^{2(r-1)}) p^{2^r}$$

Then, $p_r \to 0$ as $r \to \infty$, for $n_r > n_{r-1} \ldots > n_1$ and $n_1 p < 1/2$. At the same time, the information rate in the channel decreases according to:

$$R_r = \left(1 - \frac{k_1}{n_1}\right)\left(1 - \frac{k_2}{n_2}\right) \ldots \left(1 - \frac{k_r}{n_r}\right) \tag{8.117}$$

but $R_r > 0$.

Thus, the technique provides an efficient method of coding and decoding for binary symmetric channels.

For the concatenated codes, block diagram of the scheme is shown in Fig. 8.32, where the inner code is generally a random-error correcting code, either block or convolutional, and the outer code is a burst-correcting code, say, an RS-code using symbols from GF(2^k). The coding scheme is as follows:

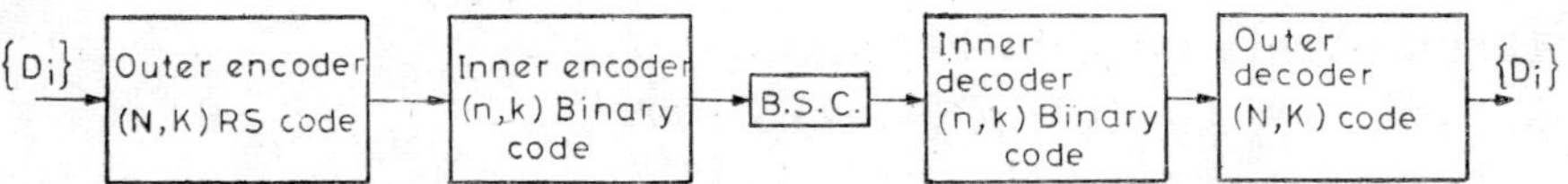

Fig. 8.32 A scheme for a concatenated code

The input data $\{D_i\}$ is divided into K bytes of k bits, and each byte of k bits is considered to be a symbol in GF(2^k). The K bytes are then coded into an N-byte codeword in the outer encoder. Each of the N bytes of k bits is now considered as a k-bit word for encoding in the inner encoder, which encodes the k-bit words into n-bit codewords. The set of N codewords of the (n, k) inner code is then a codeword of the concatenated (Nn, Kk) code.

As an example, one may use the (15, 7) RS code with symbols from GF(2^4) as the outer code, which will correct t errors, $t \leqslant 4$ (i.e., 4 bytes, each of 4 bits, out of 15 bytes in the codeword will be corrected). The inner code now has $k = 4$, and the (7, 4) binary Hamming code may be used.

Thus, 28 bits of $\{D_i\}$ are taken as 7 bytes of 4 bits each, and converted into the (105, 28) code, whose rate $\approx$ 1/4. For decoding, the inner code is decoded first delivering 4-bit words (= one byte) to the outer decoder. Single-errors in 7-bit codewords will be corrected in the inner coder, and the residual errors in the output will be in the form of erroneous bytes delivered to the outer coder. If the number of wrong bytes $\leqslant 4$ out of 15 bytes, then all the errors will be corrected by the outer decoder. If there are bursts of length $b \leqslant 28$ bits in 105 channel bits, then this will give a 4-byte errors at the input to the outer decoder, and hence, will be corrected.

Alternatively, a combination of an inner convolutional code with a simple decoder and an outer RS code is also very effective. The inner code will correct random errors and the burst errors will be corrected by the outer RS-code. The outer code can also be any other efficient burst-correcting code, say, the Iwadare code. It has been estimated that the coding gain of a concatenated RS-cum-convolutional (with Viterbi decoding) code is 6.5-7.5 dB at $P_b = 10^{-5}$ at the output [10]. Thus, a very powerful code is obtained through the concatenation of simpler codes specially for Gaussian-noise channels.

8.6 PERFORMANCE OF CODES

The methods of generation of codewords and their decoding techniques have been discussed so far. In this section, the error-correcting capabilities and their bounds will be briefly discussed. Consider the simple single-error correcting block codes, say, Hamming code, where the probability of correct decoding of a codeword is (assuming binary symmetric channels with pr. (1) = pr. (0) = 0.5, pr. (1/1) = pr. (0/0) = q and pr. (1/0) = pr. (0/1) $= p = (1 - q)$,):

$$p_c = \text{pr. (correct decoding)} = q^n + \binom{n}{1} p \cdot q^{n-1}$$

Then the word-error probability $P_w = 1 - P_c$

$$= 1 - (1 - p)^n - np\,(1 - p)^{n-1}$$
$$\simeq n(n - 1)p^2, \qquad p < 1 \tag{8.118}$$

The bit-error probability $P_b = p_w/k$ and with no error-correction $P_w \approx kp$, where k-bit words are coded into n-bit words. A figure of merit F may now be defined as:

$$F = \frac{P_b(\text{uncoded})}{P_b(\text{coded})} = \frac{kp}{n(n - 1)p^2} = \frac{k}{n(n - 1)p} \quad \text{for SEC.} \tag{8.119}$$

Based on Eq. (8.119), the values of F for (n, k) SEC block codes are given in Table 8.9.

Table 8.9 Valuer of F and P_b for SEC Block Codes

(n, k)	(3, 1)	(7, 4)	(15, 11)	(31, 26)	(63, 57)
F	$(1/6)p$	$(2/21)p$	$(11/210)p$	$(13/465)p$	$(19/1302)p$
P_b	$6p^2$	$42p^2$	$210p^2$	$930p^2$	$3906p^2$

For multi-error correcting codes with $t > 1$, the probability of correct decoding is given by:

$$P_c = q^n + \binom{n}{1} pq^{n-1} + \binom{n}{2} p^2 q^{n-2} + \ldots + \binom{n}{1} p^t q^{(n-t)} \quad (8.120)$$

and

$$P_w = (1 - P_c) \geqslant \binom{n}{t+1} p^{(t+1)} q^{(n-t-1)}, \quad t < n; p \ll 1$$

$$\simeq n^{(t+1)} p^{(t+1)} \quad (8.121)$$

For systematic convolutional codes, similar equations are derived based on the equivalent codeword length $= nK$. As for example, for $R = 1/2$, $K=4$, systematic convolutional code, 2 errors in 8 bits are corrected. Then P_w for the code is:

$$P_w = 1 - \left[q^n + \binom{n}{1} pq^{(n-1)} + \binom{n}{2} p^2 q^{(n-2)} \right], \quad \text{with } n = 8.$$

$$\simeq \frac{n^2 p^2}{2} + \frac{n^3 p^3}{2} \simeq 3 \times 10^{-3} \quad \text{for } p_e = 10^{-2} \text{ in the channel}$$

$$(8.122)$$

But for $p_e = 10^{-4}$, $P_w \simeq 3 \times 10^{-7}$ and $P_b \simeq (3/8) \times 10^{-7}$.

For non-systematic codes, however, such simple relations for P_w are not available, and most of the error performances of such codes are obtained through computer simulation. A short table of the figure of merit F for some typical codes is given in Table 8.10, where the complexity of the decoder is

Table 8.10 Values of F for Some FEC Codes [28]

Codes	Rate	No. of SR's required	F for P_b (uncoded) =		
			10^{-2}	10^{-3}	10^{-4}
Block code $t = 1$	26/31	93	1/10	1	10
BCH, $t = 2$	7/15	87	10	10^3	10^5
BCH, $t = 6$	30/63	—	10^3	10^6	—
Golay code $t = 3$	12/24	53	10^2	10^5	10^7
Convolutional Codes:					
Wyner-Ash, $t = 1$, Interleaved	1/2	23	10	10^2	10^3
CC-W-A, $t = 1$, Interleaved	3/4	43	5	25	2×10^2
CC-W-A, $t = 1$, Interleaved	7/8	87	1.2	10	10^2
Self-Orthogonal CC, Interleaved, $t = 2$	1/2	51	10^2	10^3	$>10^5$
As above	3/4	209	10	5×10^2	$>10^4$
As above	7/8	873	1	50	$>10^3$
Convolutional code: Viterbi decoding, hard decision	1/2	—	5×10^2	10^5	10^7
Convolutiona code: $K = 7$, Viterbi decoding, soft decision	1/2	—	10^6	10^9	—

also indicated in terms of the shift registers (SR) required [28]. It is seen that low-rate codes give better values of F and F is larger for lesser values of channel errors. For low-rate, $R=1/2, 1/4, \ldots 1/16$, convolutional codes, the improvement at the decoder output is indeed spectacular, as is shown in Fig. 8.33. It is possible to recover data with $P_b \leqslant 10^{-5}$, even with a

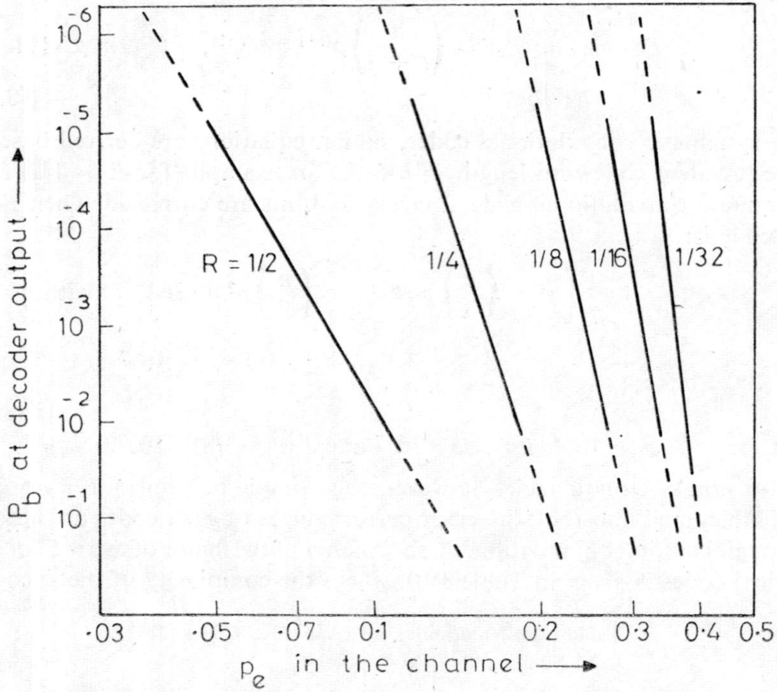

Fig. 8.33 P_b vs. P_e (ch) for low-rate convolutional code (after Shaft [29])

channel error-rate $p_e \simeq 0.3$, (but $p_e < 0.5$), if one uses a composite code with $R = 1/32$. Generally, $R = 1/4$ codecs are built as a single unit and the tandem connections of such codecs produce the low rate codec as desired. The simulation results for P_b vs. E_b/n_0 for various cascaded codecs ($n = 1, 2, 4$; $R = 1/4, 1/8$, etc.) are shown in Fig. 8.34, where it is seen

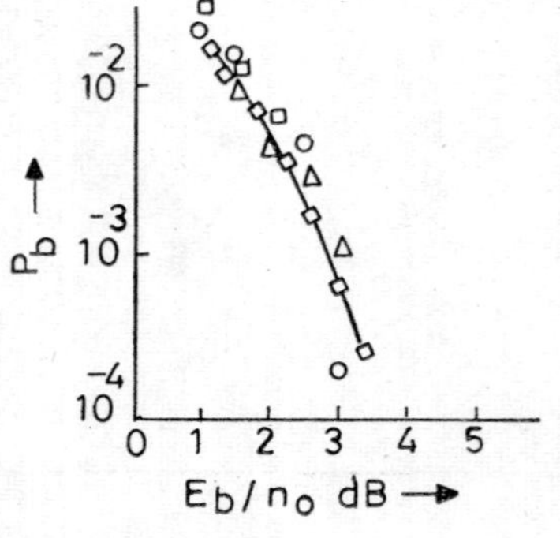

Fig. 8.34 Comparison of FEC code rates in Gaussian noise (after Shaft [29]):
$\Delta R = 1/4$, $n = 1$; ○$R = 1/8$, $n = 2$; □$R = 1/16$, $n = 4$; ◇$R = 1/4$, Laylands results; n = number of coders in cascade

that the results are invariant with decreasing code rates, as is predicted theoretically [29]. An alternative evaluation of error performance of the codecs is obtained by noting the coding gain in terms of E_b/n_0 at a given P_b. As for example, P_e (uncoded PSK modulation) $= 10^{-4}$ at $E_b/n_0 = 8.5$ dB. Referring to Table 8.10, the figure of merit F for $R = 1/2$, convolutional code with Viterbi decoding is 10^7, i.e., P_b at the decoder output is 10^{-11}, but the new E_b/n_0 (coded) $= 11.5$ dB, and the value of P_b at E_b/n_0 (coded) $= 8.5$ dB is calculated to be 10^{-7}. The coding gain at $P_b = 10^{-7}$ is seen to be 2.5 dB. The coding gains of different codes may be obtained from the curves of Fig. 8.35 [10], which shows the variation of P_b (coded) vs. E_b/n_0

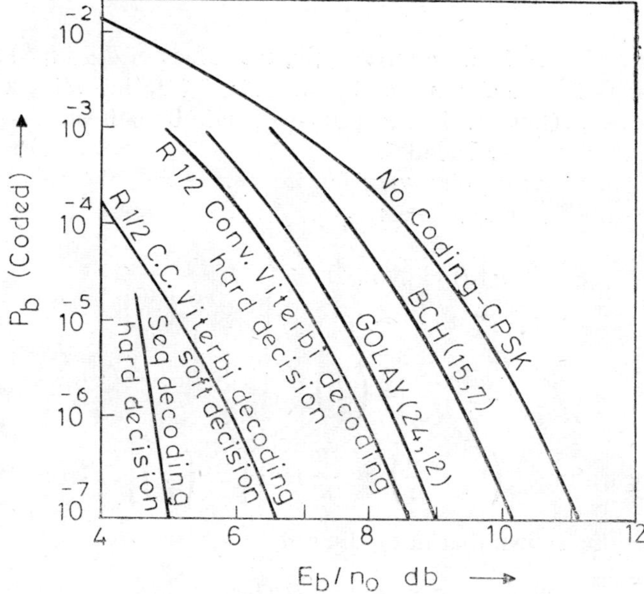

Fig. 8.35 P_b vs E_b/n_0 for various codes

(the results obtained through computer simulation). For Gaussian cum burst noise in the channel, low-rate convolutional codes offer considerable advantage. As for example, with a $R = 1/32$ codec, the performance is acceptable even with a 70 percent interference environment, as shown in Fig. 8.36 [29]. It is seen that, in general, the convolutional codes with Viterbi/sequential decoding perform much better than the block codes. However, for non-Gaussian channels, interleaved or adaptive codes have to be used and their performances have to be individually calculated. Further application of FEC codes in fading channels and SSMA systems will be discussed in Chapter 9.

8.6.1 Bounds on Code Performances [1, 7]

Among the linear (group) codes, the best code is the one which gives the maximum HD $= d_{max}$, as d is directly related to the error-correcting

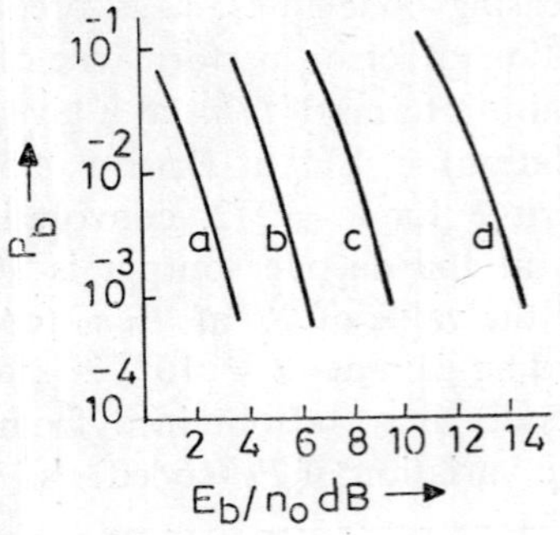

Fig. 8.36 Error rate with $R = 1/32$ FEC (CPSK) in presence of Gaussian and burst noise, (after Shaft [29]); E_b/n_0 for Gaussian noise, Burst length = 1 symbol, and burst amplitude distribution unifrom: (a) only Gaussian noise (b) Gaussian plus 25% burst (c) Gaussian plus 50% brust (d) Gaussian plus 75% burst

capability $t \leqslant (d - 1)/2$. Unfortunately, the values of $d_{\max}(n, k)$ are known only for $k \leqslant 5$, and for the codes with $d_{\max} \leqslant 3$, including a few other isolated cases. Thus, it is necessary to obtain bounds on $d_{\max}$ and consequently on P_w for (n, k) codes.

For the (n, k) group (binary) codes, Eq. (8.63) gives the number of codewords N_c with $d \geqslant (2t + 1)$ as:

$$2^k = N(n, d) = N(n, 2t + 1) \leqslant \frac{2^n}{\sum_{i=0}^{t} \binom{n}{i}} \tag{8.63}$$

or

$$2^{n-k} \geqslant \sum_{i=0}^{t} \binom{n}{i}$$

Taking logarithm,

$$\left(1 - \frac{k}{n}\right) \geqslant \frac{1}{n} \log_2 \left[\sum_{i=0}^{t} \binom{n}{i}\right] \triangleq B_H(n) \tag{8.123}$$

It may then be shown that in the limit $n \to \infty$,

$$\lim_{n \to \infty} B_H(n) = H(\lambda) = H(t/n), \tag{8.124}$$

where, $H(\lambda)$ = entropy function $= -\lambda \log \lambda - (1 - \lambda) \log (1 - \lambda)$. The above limiting value of $B_H(n)$ is known as the sphere-packing bound or Hamming upper bound for the relation between n, k and t.

Other bounds for k/n (for binary codes only) are:

(a) Plotkin's upper bound on d:

$$(n - k) \geqslant (2d - 1) - 1 - \log_2 d \tag{8.125}$$

and for $n \to \infty$, $(1 - k/n) \geqslant 2d/n$.

(b) Elias's upper bound on d :

$$\lim_{n \to \infty} d_{\min} \leqslant 2t_h \left(1 - \frac{t_h}{n}\right), \tag{8.126}$$

where t_h is given by: $(1 - k/n) = H(t_h/n)$

These bounds indicate that the d_{min} of linear (n, k) codes cannot be greater than the upper bound. The lower bound on d is given by Varsharmov-Gilbert bound as:

$$\text{(c)} \quad (1 - k/n) \geqslant 1 + \binom{d-2}{n-1} \to \lim_{n\to\infty} H(d/n) \tag{8.127}$$

From these equations, it is seen that Elias's bound is tigher than others, and the rate $k/n \to 0$ for $d/2n \geqslant 0.25$ (i.e., for $t/n \geqslant 0.25$). For large rates, $t/n \ll 1$ only will satisfy the bounds [1].

For convolutional codes, the bound for d_{min} is approximately the same for large n. When n is not large, d_{min} for convolutional codes is slightly larger than the d_{min} for best block codes. Thus, for moderate values of n, convolutional codes are slightly superior to block codes.

For obtaining bounds on error performances, it is to be noted that for perfect and quasi-perfect (n, k) codes, the probability of correct decoding of a codeword (using a BSC) is given by:

$$P_c \geqslant \sum_{i=0}^{t} \binom{n}{i} p^i \cdot q^{n-i} \tag{8.128}$$

The word error probability $P_w = 1 - P_c$, where P_c is the upper bound for correct decoding, and $P_b = P_w/k$. To show the asymptotic behaviour of P_w, define a quantity E, called Reliability, as:

$$E = \lim_{n\to\infty} (-1/n) \log_2 P_w \tag{8.129}$$

Denoting E_s as the sphere-packing bound on reliability, one obtains

$$\begin{aligned} E \leqslant E_s &= \lim_{n\to\infty} (-1/n) \log_2 (1 - P_c) \\ &= \lim_{n\to\infty} (-1/n) \log_2 \left[\sum_{i=t+1}^{n} \binom{n}{i} p^i q^{n-i}\right] \\ &= E(t/n, p) \\ &= E(\lambda, p) > 0, \qquad \lambda = t/n \end{aligned} \tag{8.130}$$

where $E(\lambda, p)$ is defined as:

$$E(\lambda, p) = H(p) + (\lambda - p)H'(p) - H(\lambda)$$

and $$H'(p) = \frac{dH(p)}{dp} = \log\left(\frac{1-p}{p}\right)$$

Therefore, $$\begin{aligned} P_w &= \exp_2[-nE(\lambda, p)] \\ &= \delta \to 0 \text{ in the limit } n \to \infty \end{aligned} \tag{8.131}$$

This is the exponential bound of error probability that may be obtained for best linear codes (cf. Eq. 8.58). From Eq. (8.123),

$$\frac{1}{n} \log_2 \left[\sum_{i=0}^{t} \binom{n}{i}\right] \leqslant (1 - k/n) \leqslant \frac{1}{n} \log_2 \left[\sum_{i=0}^{t+1} \binom{n}{i}\right]$$

and in the limit $n \to \infty$, $(1 - k/n) = H(\lambda) = H(t/n)$ (8.132)

This value of t has to be used in Eq. (8.130) for evaluating E_S.

These bounds for d_{min} and the reliability function E are hardly achieved by any practical codes. However, the convolutional codes, specially with soft decision Viterbi decoding and sequential decoding, perform much better than the block codes, as is shown in Fig. 8.35. The bounds have been evaluated on the assumption of a BSC, and Gaussian noise statistics. The performance of the codes is non-Gaussian channels, e.g., in channels with memory, can be evaluated only by computer simulation. For burst-error correcting codes, the requirements of parity checks have been discussed in Sec. 8.5.3. In general, the burst-error correcting capability of convolutional codes is superior to that of block codes, for large (n, k).

8.6.2 Channel Capacity Theorem (Discrete Case) [1, 4]

The code capacity theorem (Sec. 8.1.2) was proved for a noiseless channel; but for a noisy channel, the relation between the source transmission rate R and the amount of reliable information received by the sink, is controlled by the channel capacity C as discussed below. The formulation of error-correcting codes, as discussed earlier, is one of the best known techniques for reliable transmission of information. For an (n, k) code with $R = k/n$, the total number of messages (code words: $C_1, C_2, \ldots, C_M$) generated, $M = 2^k = 2^{nR}$; and these are corrupted by the channel noise to generate a maximum of 2^n words, which are then presented to the receiver (decoder) for correct decoding. With reference to Fig. 8.5, it is seen that the n-dimensional space generated by the 2^n words (with errors) is partitioned into 2^k distinct spheres $S_j(np)$, each corresponding to a code C_j and with radius np $(= e_j)$, p = error probability in the BSC. For correct decoding, the errors in the channel have to be less or equal to np (since in a block of n bits, the average number of errors $\simeq np$) and the received vector W_j lies within $S_j(np)$. There is, of course, a finite probability that $W_j(i \neq j)$ may fall within the sphere around C_i, if the original distance between the codes $d(C_j, C_i)$ is reduced due to noise in the channel. Thus, there are two types of possible errors: (a) P_m, the probability of miss, where $d(W_j, C_j)$ increases due to noise and W_j falls outside the sphere $S_j(np)$, and (b) P_f, the probability of false alarm, where $d(W_j, C_i)$, $i \neq j$ is reduced and W_j falls within $S_i(np)$. The overall probability of error P_e is now written as:

$$\begin{aligned} P_e &= \text{pr}\cdot\{W_j \notin S_j(np)\} + \text{pr.}\{W_j \in S_i(np)\}, \quad i \neq j \\ &= P_m + P_f \end{aligned} \tag{8.133}$$

where $\in$ is read as 'is contained in' and $\notin$ the inverse.

The exponential bound, given by Eq. (8.131), gives the value of P_m. The value of P_f is obtained from Shannon's random coding argument, and $P_f = (M-1)\overline{P_f} \leqslant M\overline{P_f}$, where the average $\overline{P_f}$ is given by (using Eq. (8.123)):

$$\begin{aligned} \overline{P_f} &= \overline{\text{pr.}\{W_j \in S_i(np)\}}, \qquad i \neq j \\ &= \text{number of sequences in } S_i(np)/(2^n) \\ &= \sum_{r=0}^{np} \binom{n}{r} \Big/ 2^n \\ &\leqslant 2^{nH(P)}/2^n = 2^{-n[1-H[P)]} \end{aligned} \tag{8.134}$$

By combining Eqs. (8.131) and (8.134),

$$P_e \leqslant \delta + M \cdot 2^{-n[1-H(p)]} = \delta + M \cdot 2^{-nC}, \text{ for } C = [1 - H(p)] \text{ in BSC.}$$

Assuming that at $R = (C - \epsilon)$, $M = 2^{nR} = 2^{n(C-\epsilon)}$, and the average overall error rate $\bar{P}_e$ is given by,

$$\bar{P_e} \leqslant \delta + 2^{-n\epsilon} = \exp_2 [-nE(\lambda, p)] + 2^{-n\epsilon} \tag{8.135}$$

For $E(\lambda, p) > 0$ and $\epsilon > 0$ (but $\epsilon \to 0$), $\bar{P}_e \to 0$ with limit $n \to \infty$, and this is true only for $R \leqslant C$, and $M \leqslant 2^{nC}$. The ultimate channel capacity is then given by:

$$\lim_{n \to \infty} \frac{\log M}{n} \to C \qquad \text{with } \bar{P}_e \to 0. \tag{8.136}$$

If, however, $R > C$, say $R = C + \epsilon$, then $\bar{P}_e \leqslant \delta + 2^{n\epsilon}$ which tends to ∞ with $n \to \infty$, and error-free transmission is not possible. Thus, Shannon's second Fundamental Theorem on Coding for memoryless noisy channels may be stated as:

"Given that a source has a transmission rate of R bits/symbol and the capacity of the discrete noisy channel without memory is C bits/symbol, then it is possible to transmit the messages through the channel with as small a probability of error (or equivocation) as desired, only if $R < C$ and the coded message length $n \to \infty$."

Thus, for $n \to \infty$, $R = (C - \epsilon)$, $\epsilon > 0$, we have $P_e \to 0$. Conversely, if $R > C$, then $P_e \to 1$ and reliable transmission is not possible.

P. Elias and others [15] have obtained the specific values of $\bar{P}_e$ for different codelengths n for BSC and BEC using (n, k) block codes. Their results are: For a BSC with $p = 0.05$, $C = 0.7136$, $R = 1/2$, $550 < n < 650$; $\bar{P}_e = 10^{-10}$. Thus, for $R \to C$; and $P_e \to 0$, the codelength n will be very large indeed.

8.7 INFORMATION IN CONTINUOUS SYSTEMS [1, 2, 30]

In conformity with the definition of entropy for discrete signals (Eq. (8.1)), the entropy of continuous variable is defined as:

$$H(X) = -\int_{-\infty}^{\infty} p(x) \log p(x)\, dx \tag{8.137*}$$

and for m variables,

$$H(m) = -\int \cdots \int p(x_1, x_2, \ldots, x_m) \log p(x_1, x_2, \ldots, x_m) \cdot dx_1\, dx_2 \ldots dx_m$$

where $$\int_{-\infty}^{\infty} \int_{-\infty}^{\infty} p(x_1, x_2, \ldots, x_m)\, dx_1\, dx_2 \ldots dx_m = 1.$$

*In the continuous case, $H(X)$ is simply the entropy function, and the concept of self-information associated with $H(X)$ in the discrete case is no longer valid. However, when the difference of entropies is taken, as in the case of mutual information, the quantities are meaningful.

The other related entropies may now be defined as:

$$H(X, Y) = -\int_{-\infty}^{\infty}\int_{-\infty}^{\infty} p(x, y) \log p(x, y)\, dx\, dy$$

$$H(X/Y) = \int_{-\infty}^{\infty}\int_{-\infty}^{\infty} p(x, y) \log p(x/y)\, dx\, dy \tag{8.138}$$

In the discrete case, all entropies were positive, since $p(x_i)$ was bounded as: $0 < p(x_i) < 1$. In the continuous case, the only conditions to be satisfied are that

$$\int_{-\infty}^{\infty} p(x)\, dx = 1$$

and

$$\int_{-\infty}^{\infty}\int_{-\infty}^{\infty} p(x, y)\, dx\, dy = 1, \tag{8.139}$$

and then, for certain distributions, the entropy may turn out to be negative. However, mutual information, as discussed later, is never negative.

In case of discrete signals, it was shown in Sec. 8.1, that the entropy $H(X) = -\sum_m p_i \log p_i$ is maximum when all p_i's are equal. In case of continuous signals, this is equivalent to a rectangular distribution leading to the entropy,

$$H(X) = -\frac{1}{A} \log \frac{1}{A} \int_0^A dx = \log A, \tag{8.140}$$

where, $p(x) = 1/A$, $0 \leqslant x \leqslant A$. In practical systems, the sources, e.g., in radio transmitters, are either average-power or peak-power limited, and the entropy has to maximized under such restrictions. Using the calculus of variations, the optimum distributions and corresponding entropies are calculated and they are:

(a) Symmetric distribution with peak power limitation, e.g., in AM, FM and Pulse modulation systems:

$$p(x) = \frac{1}{2A}, \qquad -A \leqslant x \leqslant A$$

$$H(X) \text{ per sample} = \frac{1}{2} \log (4\hat{S}), \quad \text{where } \hat{S} = A^2, \tag{8.141}$$

and

$$H/\text{sec.} = W \log (4\hat{S}) = W \log (12\bar{S}),$$

where average power $\bar{S} = \hat{S}/3$.

(b) Symmetrical distribution with average power limitation, as applicable for random noise with given variance σ^2, audio frequency telephony and similar systems.

$$p(x) = \frac{1}{\sqrt{2\pi\sigma^2}} \cdot \exp(-x^2/2\sigma^2), \quad \overline{x^2} = \sigma^2,$$

$$H(X)/\text{sample} = 1/2 \log (2\pi e\sigma^2), \tag{8.142}$$

and $$H/\text{sec} = W \log (2\pi e \sigma^2)$$

It is seen that the optimum $p(x)$ is Gaussian and this has the maximum entropy for a given power $\sigma^2 = \overline{N}$, among all realistic *pdf*'s for continuous signals.

Mutual Information

As in the discrete case, the mutual information $I(X, Y)$ of a continuous noisy channel is also defined by Eq. (8.37) and this remains invariant under all linear transformations at the input and output of the channel.

If the noise in the channel is additive (as with AWGN) and statistically independent of the transmitted signal, then $p(Y/X)$ depends on $(Y - X)$ only. Thus,

$$H(Y/X) = -\int_{-\infty}^{\infty} \int_{-\infty}^{\infty} p(x)\, dx\, p(n) \log p(n)\, dn$$

$$= H(n), \qquad \text{since} \int_{-\infty}^{\infty} p(x)\, dx = 1$$

and the Rate of transmission/sample = Mutual Information

$$= H(Y) - H(n) \tag{8.143}$$

8.7.1 Channel Capacity Theorem [1, 2]

The channel capacity theorem for discrete channels has been discussed in Sec. 8.6.2. For the continuous channel, the channel capacity is defined as R max, giving,

$$C = R_{\max} = [H(Y) - H(n)]_{\max} \tag{8.144}$$

A case of particular interest is the power limited source and AWGN in the channel. For such a source, the optimum *pdf* and $H(X)$ are given by the Eq. (8.142), which shows that the signal *pdf* has to be Gaussian. With the Gaussian noise added to the channel, the statistics of the received y's is also Gaussian with average power $S + N$, where $N = \sigma^2 =$ noise power. The entropy $H(Y)$ is now:

$$H(Y) = 1/2 \log [2\pi e(S + N)] \text{ per sample}$$

and $$H(n) = 1/2 \log (2\pi e N)/\text{sample} \tag{8.145}$$

Therefore,

$$C/\text{sample} = 1/2 \log (1 + S/N) \text{ bits}$$

and C per second in a bandwidth W

$$= W \log_2 (1 + S/N) \text{ bits}$$

The overall capacity with a signal duration of T sec, and BW = W, is:

$$C = WT \log_2 (1 + S/N) \text{ bits} \tag{8.146}$$

This is the well-known Shannon's Channel Capacity theorem, which gives an ideal limit for a given SNR in the channel. This limit, however, is achieved only by using large-length orthogonal codes and in the limit $W \to \infty$, as will be discussed later.

The maximum entropy per sample for an ensemble with peak power limitation at sample points has been given in Eq. (8.141). But it is possible that while reconstructing the continuous signal by an ideal lowpass filter, the signal peaks may occasionally exceed the peaks of the samples, and the entropy due to larger peaks will have to be subtracted from that given by Eq. (8.141). Shannon [1] has shown that the loss of entropy due to peak limitation through a triangular filter is $2W$ bits per sec. Thus, the entropy of a truly peak-limited signal is:

$$\begin{aligned} H/\text{sec} &\leqslant W \log_e (4\hat{S}) - 2W \\ &= W \log_e (4\hat{S}/e^2) \end{aligned} \tag{8.147}$$

Accordingly, the channel capacity of a peak-limited source is bounded by:

$$W \log_e \left(\frac{4\hat{S}}{2\pi e^3 N}\right) \leqslant C \leqslant W \log_e \left(1 + \frac{4\hat{S}}{2\pi e N}\right) \tag{8.148}$$

However, using a modified peak-limiting amplifier after the reconstructed signal, a tighter bound is obtained as [1]:

$$W \log_e \left(1 + \frac{4\hat{S}}{2\pi e^2 N}\right) \leqslant C \leqslant W \log_e \left(1 + \frac{4\hat{S}}{2\pi e N}\right) \tag{8.149}$$

For practical purposes, with $S/N \gg 1$, C may be approximated as:

$$C \simeq W \log_e [4\hat{S}/(2\pi e^2 N)] \tag{8.150}$$

Some Useful Properties

A discussion of the Channel capacity theorem of Eq. (8.146) leads to the following useful properties.

(a) Equations (8.146) may be written as:

$$(W/C) \cdot \log_e (1 + S/N) = 1 \tag{8.151}$$

This shows that for small S/N, the same capacity may be obtained by increasing W. However, this exchange of bandwidth for S/N is advantageous for $S/N \leqslant 10$; and for $S/N > 10$, the bandwidth reduction for larger S/N is small. Use of larger bandwidth for smaller S/N is generally known as coding upwards and use of smaller BW for larger S/N is called coding downwards. SNR improvements in *FM*, *PM*, and *PCM* systems using larger bandwidths are examples of coding upwards.

(b) In practical channels, the noise spectral density n_0 is generally constant, and $N = n_o W$. Then the capacity per Hz, C/W, is an indication of the efficiency η of the system. Writing C/W as,

$$\eta = C/W = \log_e (1 + S/n_o W) \text{ nats,}$$

and differentiating,

$$\frac{\partial \eta}{\partial W} = -\frac{1}{W}\frac{S/n_o W}{(1 + S/n_o W)} \tag{8.152}$$

This indicates that η decreases with the increase in W, as is seen in FM and PM systems.

Further, writing $S = n_0 W_0$, i.e., W_0 is the bandwidth for which $N = S$, then,

$$C/W_0 = \frac{W}{W_0} \log_e (1 + W_0/W)$$

and for $W \gg W_0$,

$$C/W_0 \simeq 1, \quad \text{and} \quad C = S/n_0. \tag{8.153}$$

Thus, the limiting value of C/W_0 is one nat (= 1.433 bits) per second. It may appear somewhat anomalous that communication is possible, even when $S/N = W_0/W \ll 1$. But this is made practically realizable by the use of correlation detection and integration of signals over a large T, thereby improving the receiver SNR considerably, e.g., in SSMA and pulse-compression radar.

(c) The most significant result of Eq. (8.146) is the limiting value of E_b/n_0 required for reliable communication. Rewriting the equation in the form,

$$\begin{aligned} C/W &= \log_2 [1 + (S/n_0C)(C/W)] \\ &= \log_2 [1 + (E_b/n_0)(C/W)] \end{aligned} \tag{8.154}$$

where $S/n_0C = E_b/n_0$, a curve showing the relation between E_b/n_0 and C/W is plotted in Fig. 8.37. It is seen now that for a fixed S/n_0, more efficient

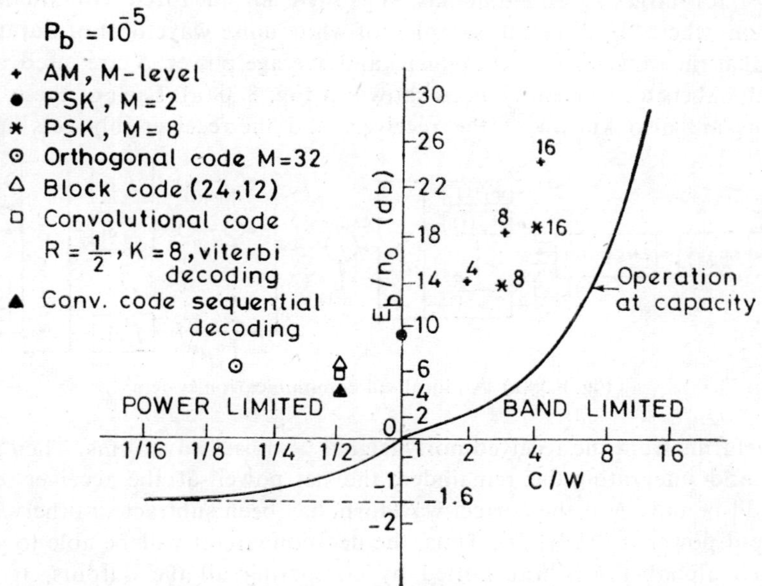

Fig. 8.37 C/W vs E_b/n_0 (after Ristenbatt [36])

communication is possible as $W \to \infty$. In the limit $W \to \infty$,

$$C/W = (\log_2 e) \log_e [1 + (E_b/n_0)(C/W)]$$
$$\to \log_2 e \cdot \left(\frac{E_b}{n_0} \cdot \frac{C}{W}\right), \quad \text{since} \quad \left(\frac{E_b}{n_0} \cdot \frac{C}{W}\right) \ll 1.$$

Therefore, reliable communication is possible for (with $W \to \infty$)

$$\lim_{W \to \infty} \frac{E_b}{n_0} \to \ln 2 = 0.69 = -1.6 \text{ dB} \tag{8.165}$$

This value of $E_b/n_0 = -1.6$ dB is the so-called Shannon's limit for transmission at capacity and the communication fails if $E_b/n_0 < 0.69$. Although the E_b/n_0 limit is reached only at $W \to \infty$, even at 1/2 bit/Hz (or for a code rate of 1/4 with binary PSK), we are practically there. Since ideal PSK signalling with no coding requires $E_b/n_0 = 9.6$ dB for $P_b = 10^{-5}$, it is seen that the capacity theorem promises a potential coding gain of 11 dB, as it should be possible to obtain $P_b \to 0$ with $E_b/n_0 = -1.6$ dB only with ideal coding and $W \to \infty$. However, no practical system has been designed as yet to reach this limit, as is indicated by the results of E_b/n_0 required in practical coded systems, plotted in Fig. 8.37. The best results obtained so far have been for the Rate 1/2 convolutional coding using Viterbi soft decision or sequential decoding techniques. In the bandwidth limited case, i.e., for large C/W, coding no longer offers such dramatic gains, as is shown in the figure for M-level AM or PSK signalling [36].

8.7.2 Ideal Communication System [2]

Based on the assumption of random coding and a large set of alphabets $M(T)$ for large T, Shannon has suggested an idealized communication system where M different samples of white noise waveform of duration T (so that the samples are orthogonal) and average power S are used as the M alphabets for transmission, as shown in Fig. 8.38(a). Each of these waveforms are also known at the receiver, and the receiver subtracts these M

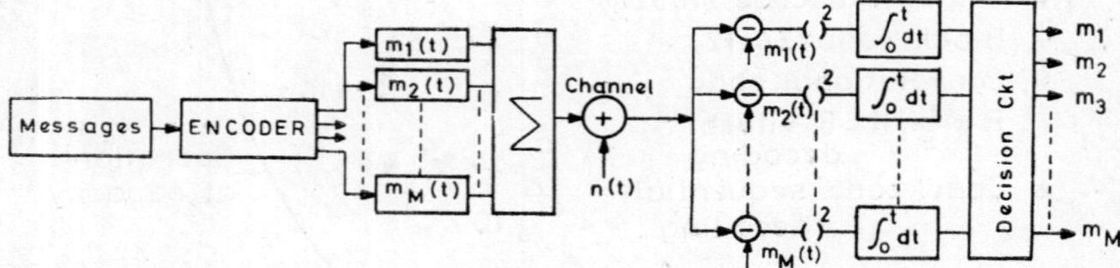

Fig. 8.38(a) An idealised communication system

waveforms from the received noisy (signal + noise) waveforms. Then squaring and integrating the remainder, the net power at the receiver output would be only N if the correct waveform has been subtracted; otherwise the output power is $(2S + N)$. Thus, the decision circuit will be able to decide which alphabet was transmitted by comparing all the outputs. If all M messages are equiprobable, then

$$H(X) = \frac{\log M}{T} \text{ bits/sec.}$$

To maintain $H(X)$ constant (i.e., the rate of transmission constant), $M(T)$ has to grow exponentially with T (as discussed in Sec. 8.2.2). It has been shown that for $P_b = 10^{-5}$ with $S/N = 10$, the number of symbols required to achieve a rate $R \simeq 0.96C$, is equal to 10^{12}, indeed a very large number.

M-ary Orthogonal Systems [1, 31]

The nearest approximation to the above ideal communication system is obtained by using M-orthogonal (digital) alphabets as the transmitted words, along with correlation/MF detection and maximum likelihood decision. The equal-energy digital orthogonal signals are generated through Hadamard matrices or Walsh functions (refer to Appex. C), and BPSK transmission is generally used. The optimum receiver for such an orthogonal set of signals is shown in Fig. 8.38(b), where M cross-correlators or M-MF's

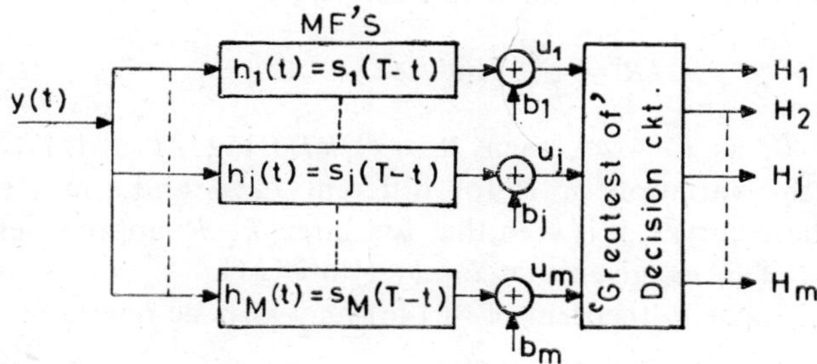

Fig. 8.38(b) Optimum receiver for M-ary othogonal signals

are used to generate the maximum u_j at the desired output and the 'Greatest Of' decision circuit decides on the transmitted word. The word error probability in such a receiver is [1]:

$$P_w(M) = 1 - \int_{-\infty}^{\infty} \frac{\exp(-y^2/2)}{\sqrt{2\pi}} \left[erf\left(y + \sqrt{\frac{2E_w}{n_o}} \right) \right]^{M-1} dy \tag{8.156}$$

where $M = 2^k$, and $E_w = kE_b$ (k bit messages are converted to orthogonal code sets). It may further be shown that the limiting value of P_e with $k \to \infty$, $M \to \infty$, is given by:

$$P_w(M) \to 0 \quad \text{for } E_b/n_0 > \ln 2 = -1.6 \text{ dB}$$

and

$$P_w(M) \to 1 \quad \text{for } E_b/n_0 < \ln 2 \tag{8.157}$$

Thus, Shannon's limit, given by Eq. (8.155), based on the Channel capacity theorem, is satisfied only by the above M-ary orthogonal system of communication. For practical values of k and M, the error rate $P_w(M)$ is bounded as:

$$1/2 \exp\left(-\frac{E_w}{2n_o}\right) \leqslant P_w(M) \leqslant \frac{(M-1)}{2} \exp\left(-\frac{E_w}{2n_0}\right) \tag{8.158}$$

Some typical values of $P_w(M)$ are given in Table 8.11, where it is shown that with $k = 20$, $P_w \simeq 10^{-5}$ for $E_b/n_0 = 3$ dB; and $P_w < 10^{-10}$ for $E_b/n_0 = 5$ dB, a result better than that obtained by using convolutional codes (refer to Fig. 8.35).

Such M-ary orthogonal systems are also equivalent to M-ary symmetrical channels, (discussed in Sec. 8.3.3) where

$$p_{ij} = \begin{cases} p/(M-1) & \ldots \quad i \neq j; \quad p = P_w \\ q & \ldots \quad i = j; \quad q = (1 - P_w) \end{cases} \tag{8.159}$$

and the channel capacity C is given by Eq. (8.54) as:

$$\begin{aligned} C(M) &= \log M - p \log (M-1) - H(p) \\ &= \log M - P_w \log (M-1) + P_w \log P_w \\ &\quad + (1 - P_w) \log (1 - P_w) \end{aligned} \tag{8.160}$$

If the data transmission rate is $1/T_b$ bits/sec, then the word transmission rate is $1/kT_b$ symbols/sec, and R is given by:

$$R = \frac{C}{kT_b} \text{ bits/sec.} \tag{8.161}$$

Since $P_w \to 0$, as $k \to \infty$, then $R \to (1/kT_b) \log M = 1/T_b$. Using these equations, the variation of R for different E_b/n_0 and k may be obtained; and from these curves, it is seen that for large k, R approaches $1/T_b$ with small values of E_b/n_0, given that $E_b/n_0 \geqslant \ln 2$ [31].

The channel capacity theorem of (8.146) may now be rewritten as:

$$\begin{aligned} C &= W \log (1 + S/N) \\ &= \frac{2^k}{kT_b} \log_2 \left[1 + \frac{k}{2^k}(E_b/n_o)\right] \text{ bits/sec.} \end{aligned} \tag{8.162}$$

and in the limit $k \to \infty$,

$$C \simeq \frac{1}{T_b \ln 2}(E_b/n_0) \text{ bits/sec,} \tag{8.163}$$

since $\log_e (1 + x) \simeq x$, $x \ll 1$ and $(k/2^k)(E_b/n_0) \to 0$ for large k. From the variation of C with E_b/n_0, using Eq. (8.162), it may be seen that the limiting value of C is reached with $k \geqslant 10$ [31].

Although the above technique of M-ary signalling leads to the ideal requirements of $P_w \to 0$, $R \to 1/T_b$ and $C \propto E_b/n_0$, these are attainable only with the condition that $E_b/n_0 > \ln 2$, and $k \to \infty$, $W \to \infty$. For practical systems, however, the basic results are: for $P_w = 10^{-5}$, E_b/n_0 required are 7.0, 4.7 and 3 dB respectively with $k = 5$, 10 and 20 [1]. Thus, with a code length of 10^6 ($\simeq 2^{20}$) and an equal number of MF's in the M–L detector, the coding gain is only 7 dB. To obtain the coding gain of 11 dB, one would require a formidable length of codes and an equally formidable number of MF's in the receiver. To improve the coding gain of M-ary orthogonal systems, a modified scheme has been suggested [32], and this is discussed now.

Consider orthogonal signalling with an alphabet size of $M = 2^k = m^2$ (k even). It is now possible to replace M orthogonal codes with two sets of m-length codes $\{C_1, C_2, \ldots, C_i \ldots C_m\}$ and $\{D_1, D_2, \ldots, D_j \ldots D_m\}$ and each k-bit word w_l will be designated by a concentrated sequence $\{C_i \cdot D_j\}$, thus forming a DS-PSK system as shown in Fig. 8.39. The chip rate for

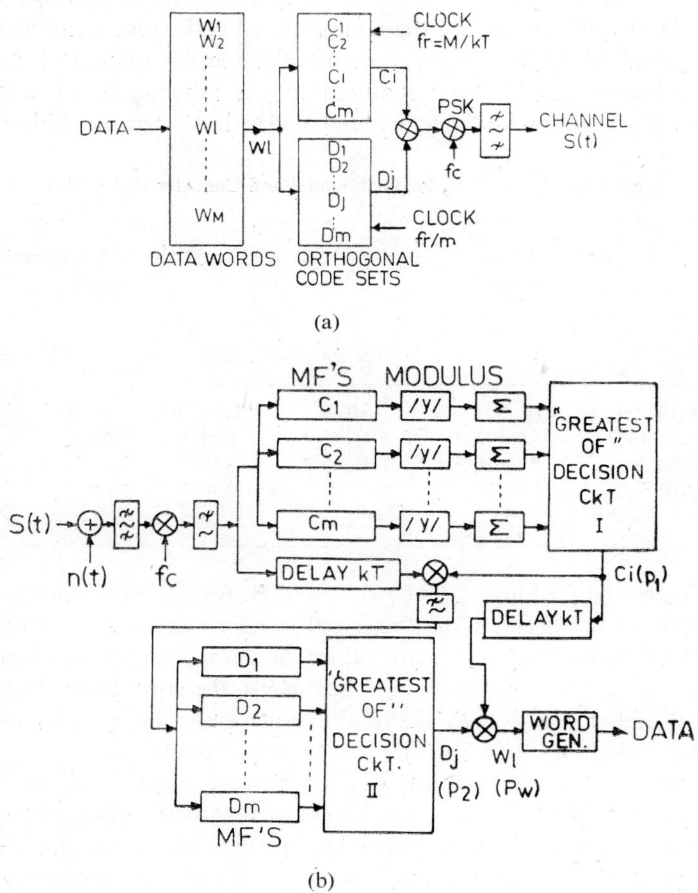

Fig. 8.39 A DS—PSK system using concatenated codes; (a) Coder-transmitter (b) Receiver-decoder (after Das [32]).

code C_i is $M/kT = f_r$, and the rate for code D_j is m/kT; thus the code D_j forms the outer code in the concatenation. In the decoder, the code C_i is passed through a bank of MF's of length m and then summed over the period kT after taking the modulus only. The peak SNR at the inputs to the decision circuit I (for determining 'the Greatest of inputs') is then $(2E_w/n_0)$ and the word error rate in C_i estimate is now bounded by Eq. (8.158). Once C_i has been decided upon, the delayed input is multiplied by C_i and the remaining code D_j is decided upon through a second bank of MF's and decision circuits. Finally the coincidence of C_i and D_j gives the desired word w_l.

The overall errorrate P_w is now approximately doubled and the upper bound of P_w is given by:

$$P_w \leqslant (m - 1) \exp(-E_w/2n_0), \quad \text{for } m \gg 1 \quad \text{and} \quad E_w/n_0 \gg 1, \tag{8.164}$$

and $P_w \to 0$, in the limit $m \to \infty$, if $E_b/n_0 > \ln 2$. Comparing Eqs. (8.158) and (8.164), it is observed that the concatenated codes give an improvement of $m/2$ approximately for large k, because of the lesser number of decisions $[2(m - 1)$ instead of $(M - 1)]$ required in the scheme. Some calculated results for P_w (max) with concatenated codes is shown in Table 8.11 for

Table 8.11 P_w vs E_b/n_0 for Orthogonal and Concatenated Codes

E_b/n_0	P_w (max) for concatenated codes with . . .			P_w or M-length codes [1] with . . .		
	$k_1 = k_2 = 5$	$k_1 = k_2 = 10$	$k_1 = k_2 = 20$	$k = 5$	$k = 10$	$k = 20$
0	6.10^{-2}	10^{-2}	3.10^{-4}	1.5×10^{-1}	1.5×10^{-1}	10^{-1}
1	1.4×10^{-2}	5.6×10^{-4}	1.4×10^{-6}	10^{-1}	6.10^{-2}	4.10^{-2}
2	2.10^{-3}	1.8×10^{-5}	1.4×10^{-9}	5.10^{-2}	2.10^{-2}	5.10^{-3}
3	3.10^{-4}	3.10^{-7}	4.10^{-13}	2.10^{-2}	2.10^{-3}	2.10^{-5}
5	1.2×10^{-6}	10^{-11}	—	2.10^{-3}	10^{-5}	$< 10^{-10}$

different E_b/n_0 and k along with the values of P_w for M-length codes as given in Ref. [1]. It is observed that reasonable error rates are obtainable with concatenated codes even at small values of E_b/n_0 and practical lengths of MF's. For a specified length $m = 2^{k/2}$ of MF's, the concatenated code provides 3 dB higher effective E_w's and the same error rates in w_l estimates now occur with E_b/n_0, 2 to 3 dB less, as seen in Table 8.11.

The process of concatenation may be further augmented by making $M = m^q$, $m = 2^{k/q}$, and q sets of m-length orthogonal codes concatenated successively to form an M-length code $\{C_i \cdot D_j \cdot E_k \cdot F_l \ldots \text{etc.}\}$. The q-sets of m-length MFs in the receiver will now decode the desired word w_l through q decision circuits and the final coincidence of $\{C_i, D_j, E_k, \ldots,$ etc$\}$. The hardware will be an extension of Fig. 8.39, with multiple banks of MF's and multiple 'Greatest Of' decision circuits. Following the reasonings of Eq. (8.164), the overall error-rate will be q times that of individual estimates, and the error bound will now be given by:

$$P_w \leqslant (q/2)(m - 1) \exp\cdot(-E_w/2n_0) \ldots, m = 2^{k/q} \tag{8.165}$$

It is seen now that the improvement factor in P_w is m^{q-1}/q approximately. As an example, with

$$k = 20, q = 4, \quad M = 2^k = m^4, \quad m = 2^5 = 32;$$

$$P_w \simeq 5.10^{-4} \quad \text{for} \quad E_b/n_0 = 0 \text{ dB},$$

indicating more than two orders of improvement in the error rate. For a given length m of MF's and a specified error rate, the higher order concatenation will reduce the E_b/n_0 requirement by (10 log q) dB approximately, at least theoretically. This would reduce the required length of MF's and a large design value of M may be used with reasonable hardware, say SAW-MF's with $BT \leqslant 1000$.

It is observed from Table 8.11 that for a given error rate $\simeq 10^{-5}$, the MF length required is only 2^{10} ($\simeq 10^3$) with $E_b/n_o \simeq 2$ dB, for concatenated codes, (thus giving a coding gain of 8 dB), whereas for M-length codes, the MF length $> 2^{40}$ ($\simeq 10^{12}$) with $E_b/n_0 \simeq 2$ dB. Thus, the bandwidth expansion required is 10^6 times more in the later case than in the system using concatenated codes. The requirement of E_b/n_0 and band expansion in the q-concatenated system will be still more conservative, as explained earlier.

A second concatenation with a burst-correcting FEC as the outer code as suggested by BerleKamp [33], will improve the system efficiency still further. Even with a simple FEC using say $R = 2/3$ convolutional code and a feedback decoder, one may operate the inner orthogonal code at $P_w = 10^{-2}$, and improve the overall P_e to 10^{-5}, thus obtaining a very robust system at an E_b/n_0 much less than that required by a straightforward M-ary system (although the 3 dB advantage due to the first concatenation is reduced by 2 dB.)

8.8 PRACTICAL COMMUNICATION SYSTEMS

Basic theorems on channel capacity have been discussed in the last section, and these form the basis of the comparison of the existing and evolving communication systems in terms of the rate of transmission and channel efficiency. Using sampling theorem, a continuous signal may be discretized both in time and amplitude; and the information contents of the analog signal and its digital equivalent are approximately the same. Thus, a continuous signal of bandwidth W is represented by $2W$ pulses per second, each having an amplitude range of 0-A, divided into L levels,

$$(0, 1, 2, \ldots, L - 1).$$

Then in a noiseless channel, the rate of transmission is:

$$R = 2W \log_2 L \text{ bits/sec.} \tag{8.166}$$

Since each quantum in A is $\Delta A = A/(L - 1)$, the PAM pulses will be received without error, if the disturbing noise $n(t)$ in the channel is peak-limited to $\pm \Delta A/2$. This half a quantum step is known as the noise threshold for error-free reception, and may be defined as:

$$I_{th} = 20 \log \frac{A}{\hat{n}} = 20 \log \frac{A}{\Delta A} = 20 \log (L - 1) \text{ dB} \tag{8.167}$$

where $\Delta A \geqslant \hat{n}$, $\hat{n}$ = peak-to-peak noise voltage at the instants of sampling. Equation (8.166) is now rewritten as:

$$R = 2W \cdot \log_2 L = 2W \log \left(1 + \frac{A}{\hat{n}}\right) \text{ bits/sec.} \tag{8.168}$$

For small error-rates, say $P_e \leqslant 10^{-4}$, the crest factor (= peak to *rms* value) for Gaussian noise $\simeq 4$. Then using $\hat{n} \geqslant 8\sigma$, one gets the approximate R in the channel. This equation also leads to an appropriate exchange between bandwidth and the noise threshold. If L-level PAM pulses are converted to m-bit PCM words, then

$L = 2^m$; $I_{th} = 0$ dB, and the channel bandwidth required $= mW$ Hz.

Example 18

Consider that speech with $W = 4$ kHz is sampled at a rate of 8 kHz/sec. and each sample is quantized into one of 64 possible levels, i.e., $L = 64$. Then for the equal step 64-ary codes,

$$I_{th} = 20 \log_{10} (64 - 1) = 36 \text{ dB}$$

and the required BW = 4 kHz. If, however, 64 levels are converted to 6-bit PCM words, then,

$$I_{th} = 0 \text{ dB, and the BW required} = 6 \times 4 = 24 \text{ kHz.}$$

In case the bandwidth reduction is called for, then two 64-level PAM pulses may be combined into one pulse having,

$$L = (64)^2 = 4096 \text{ levels, and BW required} = 2 \text{ kHz;}$$

but $\quad I_{th} = 20 \log_{10} (4096 - 1) = 72.2$ dB.

In channels disturbed by random noise, the condition that $\Delta A \geqslant \hat{n}$, cannot be maintained even with large signal power A^2 and it is necessary to consider the finite error probability in the channel to calculate the channel capacity of the link. For a given acceptable error probability P_e, the separation ΔA between the levels of PAM pulses has to sufficiently large as compared to the *rms* noise voltage σ. Assuming $\Delta A = K\sigma$, $K > 1$, the probability of symbol error in a PAM channel is given by (refer to Eq. (2.40)):

$$P_e = 2\left(1 - \frac{1}{L}\right) \text{erfc} \left[\left(\frac{3}{L^2} \cdot \frac{S}{N}\right)^{1/2}\right] \tag{8.169}$$

where $N = \sigma^2$, $S = \dfrac{A^2}{12}\left(\dfrac{L+1}{L-1}\right) \simeq A^2/12$ for $L \gg 1$ and the amplitude is in the range of $-A/2$ to $A/2$. The bit error rate is:

$$P_b \simeq P_e/\log_2 L.$$

For small P_e, $\quad N = \dfrac{A^2}{K^2(L-1)^2} \simeq \dfrac{A^2}{K^2L^2}$, and $S = \dfrac{A^2}{12}\left(\dfrac{L+1}{L-1}\right)$; and

this gives $\quad S/N = \dfrac{K^2(L^2-1)}{12} \simeq K^2L^2/12$, $L \gg 1$. The rate of message transmission is now given by (neglecting equivocation):

$$R(M) \leqslant W \log L^2 = W \log \left[1 + \frac{12}{K^2} S/N)\right] \tag{8.170}$$

It is seen that the actual rate of transmission is much lesser than the channel capacity C of Eq. (8.146), if no coding is used. To reach the transmission at capacity, the signal power has to be increased by a factor of $K^2/12$ (K defined for acceptable P_e), in case of binary and multilevel PCM systems. $(K^2/12)$ is known as the excess power required to reach the ideal capacity.

In a binary system, to achieve $P_e = 10^{-5}$, $K = 8.4$ (from Tables) and excess power required $= 10 \log_{10} (K^2/12) = 7.8$ dB; and for $P_e = 10^{-6}$, $K = 10$, excess power required $= 9$ dB. It has been calculated that for L-level PCM systems with $P_b = 10^{-5}$, the excess power required is approximately 8 dB higher than that given by Eq. (8.146).

In considering the baseband, AM and FM transmission systems, the above considerations along with those given in Sec. 8.7.1 have to be borne in mind. It has been observed that in practical channels, the rate of transmission of information $R(\text{Ch})$ is always less than the channel capacity C. Further, the source/message transmission rate is less than capacity. Based on these observations, it is convenient to define the two efficiencies as:

$$\text{Message efficiency: } \eta\,(M) = \frac{R(M)}{C(M)}$$

$$\text{Channel (transmission) efficiency: } \eta(\text{Ch}) = \frac{R(\text{Ch})}{C(\text{Ch})} \tag{8.171}$$

A few illustrative examples are discussed now.

Example 19: Speech Signals

Consider the baseband transmission of speech signals. The S/N required in a toll-quality transmission is better than 32 dB, with a dynamic range $\geqslant 40$ dB. The nominal BW $= 4$ kHz. Thus, the channel capacity C_M at the baseband is (assuming Gaussian *pdf* of the signal):

$$C_M = \log_2 (1 + S/N) = 4 \times 10^3 \log_2 (1.6 \times 10^3)$$
$$= 4.25 \times 10^4 \text{ bits/sec.}$$

Assuming, $P_e = 10^{-4}$, $12/K^2 \simeq 1/8.5$; and from Eq. (8.170),

$$R(M) \leqslant W \log_2 \left[1 + \frac{12}{K^2}\,(S/N)\right]$$
$$\simeq 30.16 \times 10^3 \text{ bits/sec.} \tag{8.172}$$

This result is more realistic since multichannel ADM systems operate at 32 Kb/s only, as discussed in Chapter 3. The more realistic *pdf* for speech is the Laplacian density function (refer to Appx. A), and C_M is modified accordingly to give:

$$C_M = W \log_2 \left(1 + \frac{S}{\pi e N}\right) \tag{8.173}$$

and the rate of transmission is now further reduced by 3 bits/Hz/sec. The effective information content in speech signals is then approximately 16 Kb/s, and the intersymbol constraints further reduce this rate to approximately 10 Kb/s or so.

From the above calculated values, the message efficiency is obtained as:

$$\eta(W) = \frac{16}{40} = 40\% \tag{8.174}$$

Similarly for TV signals, $W = 5$ MHz and $S/N \geqslant 40$ dB, giving $C_M \simeq 66.5$ Mb/s, whereas excellent TV pictures can be obtained at a rate of 24-32 Mb/s (refer to Ch. 3) Thus,

$$\eta(M) \leqslant \frac{32}{66.5} = 48\% \text{ only} \tag{8.175}$$

Example 20: FM and PPM Signals

In an FM system, the system parameters are:

$$W_{(Ch)} = 2(D+1)W_M, \quad W_M = \text{message bandwidth}$$

and $$(S_o/N_o)_M = 3D^3(S_i/N_i)_{ch},$$

where D = deviation ration, $N_i = 2(D+1)n_0 W_M$ and $(S_0/N_0)_M$, $(S_i/N_i)_{ch}$ refer to the average SNR's in the message and the channel respectively (assuming S_i/N_i to be above threshold). Considering that the message signal has similar statistics as the speech signals, it has been shown in Ex. 19, that $R_{(M)}$ is given by:

$$R_{(M)} \simeq W_{(M)} \log_2 \left[1 + \frac{12S_0}{K^2 N_0}\right]$$

$$\simeq 30 \text{ Kb/s for speech signals.}$$

$R_{(M)}$ may also be expressed as:

$$R_{(M)} \leqslant W_M \log_2 \left[1 + \frac{3D^3}{(8.5)} (S_i/N_i)_{ch}\right]$$

$$\simeq W_M \log_2 \left[\frac{3D^2}{8.5} \frac{S_i}{2n_0 W_M}\right] \tag{8.176}$$

The ideal capacity of the RF channel is:

$$C_{(ch)} = 2(D+1)W_M \log_2 (1 + S_i/N_i)$$

and when the signal in the channel is considered to be peaklimited, then,

$$C_{(ch)} \simeq W_{(ch)} \log_2 \left[1 + \frac{4S_i}{2\pi e^2 N_i}\right]$$

Thus, the channel efficiency may be written as:

$$\eta_{(Ch)} = \frac{R_{(ch)}}{C_{(ch)}} = \frac{R_{(M)}}{C_{(ch)}}$$

$$\simeq \frac{1}{2(D+1)} \cdot \frac{\log_2 \left[\frac{3D^2}{8.5} S_i/2n_0 W_M\right]}{\log_2 \left[\frac{S_i}{2(D+1)n_0 W_M}\right]}, \quad D > 1 \tag{8.177}$$

A plot of $\eta_{(ch)}$ vs. $(S_i/n_0 W_M)$ for various values of D, shows that η decreases with the increase of D, i.e., with the increase of $W_{(Ch)}$, as was shown in Eq. (8.152). It is also seen from Eq. (8.176), that $R_{(M)}$ increases with W only as 2 log D, whereas $C_{(Ch)}$ increases directly as $W_{(Ch)}$. In addition, the information rate suffers a penalty of 9 dB due to the factor $(12/K^2)$ but this is partly offset by 4.75 dB (= 10 log 3) due to the gain through the triangular noise spectrum in FM detector output. With $S_i/N_i \gg 1$,

$$\eta(\text{ch}) \to \frac{1}{2D}$$

In PPM systems, the improvement in SNR is given by [35]:

$$(S_0/N_0)_M = \frac{2\Delta d_m^2 W^3{}_{(\text{ch})}}{W_M} (S_i/N_i)_{\text{ch}} \tag{8.178}$$

where, Δd_m =emaximum time shift of the transmitted pulses due to modulation. The channel efficiency is now given as:

$$\eta(\text{ch}) = \frac{R_{(M)}}{C_{(\text{Ch})}} = \frac{W_M \log_2\left[1 + \frac{1}{8.5}(S_0/N_0)_M\right]}{W_{(\text{Ch})} \log_2[1 + S_i/N_i)]} \tag{8.179}$$

where $(S_0/N_0)_M$ is given by Eq. (8.178). It will be again seen that with increasing $W_{(ch)}$, $\eta_{(Ch)}$ will decrease, as in case of *FM* systems.

Example 21: Multilevel PCM

Channel capacity and transmission rate of multilevel PAM/PCM have been discussed earlier in the section. If the number of levels L in PAM is large, then the signal may be considered to be almost continuous with peaks of $\pm A/2$ volts. Then using Eq. (8.141),

$$\begin{aligned} H(X) &= W \log_2 (4\hat{S}) \\ &= W \log_2 (12\bar{S}), \end{aligned}$$

where $\hat{S} = A^2/4$, and $\bar{S} = \frac{A^2}{12} = \hat{S}/3$.

For Gaussian noise,

$$H(n) = W \log_2 (2\pi e N).$$

Then the rate of transmission $R \leqslant W \log_2 \left(1 + \frac{12\bar{S}}{2\pi e N}\right)$

$$\leqslant W \log_2 \left[\frac{6}{\pi e}(\bar{S}/N)\right], \qquad \bar{S}/N \gg 1 \tag{8.180}$$

Sanders [34] has given a similar education, for R in continuous PAM channels, and the equation is:

$$R \simeq W \log_2 (S/N) + 2W(0.75)\sqrt{N/\bar{S}} - 0.51W \tag{8.181}$$

Equation (8.181) gives approximately the same result as Eq. (8.180). The channel efficiency is now given as (assuming that $W_{(ch)}$ is the same as W_M,

$$\eta(\mathrm{Ch}) = \frac{\log_2\left[\frac{6}{\pi e}(\bar{S}/N)\right]}{\log_2[\bar{S}/N]} \tag{8.182}$$

Comparing Eqs. (8.177) and (8.182), it is obtained that the channel efficiency of multilevel PCM is much better than that of FM.

8.8.1 Exchange of Bandwidth for SNR

The channel capacity theorem indicates that in an ideal receiver (assuming no information loss), the input-output information rate should be the same, and thus,

$$W_{(Ch)} \log\left(1 + \frac{S_i}{N_i}\right) \leftrightarrows W_M \log(1 + S_0/N_0) \tag{8.183}$$

where $W_{(Ch)}$, S_i/N_i refer to the input and W_M, S_0/N_0 refer to the output of the receiver. For $S_i/N_i \gg 1$ and $S_0/N_0 \gg 1$, the output SNR is given by:

$$S_0/N_0 \simeq (S_i/N_i)^{W_{(ch)}/W_M} \tag{8.184}$$

This indicates the maximum possible exchange between input and output SNR's for an ideal receiver. In practical modulation schemes, the ideal is never achieved and is approached only by coded communication systems [more realistic Eq. (8.170) for R may also be considered]. From known results, the ideal and actual SNR values for some systems are given in Table 8.12. For PCM, it is normally assumed that the effective noise at the

Table 8.12

System	BW expansion	Ideal S_0/N_0	Actual S_0/N_0
SSB	1	S_i/N_i	S_i/N_i
AM-DSB-SC	2	$(S_i/N_i)^2$	$2S_i/N_i$
WBFM	$2(D+1)$	$(S_i/N_i)^{2(D+1)}$	$3D^3(S_i/N_i)$
PPM	$W_{(ch)}/W_M$	$\left(\frac{S_i}{N_i}\right)^{(W_{(ch)}/W_M)}$	$KW^3_{(ch)}(S_i/N_i)$
PCM-PSK	m	$(S_i/N_i)^m$	2^{2m}

output due to quantization and that due to the channel errors are equal, i.e., $\sigma_q = \sigma_N$, thus the output SNR is given as:

$$\frac{S_0}{N_0} = 2^{2m} = 6m \text{ dB}$$

and the corresponding BW-expansion for PCM-PSK is $2m$ (m = number of bits in the PCM-word). It is seen from the Table that PCM gives the best result amongst the usual modulation schemes. However, the coded systems give better results as discussed in Sec. 8.7.2.

Example 22

Assume that 64-level PAM pulses are converted into 6 bit PCM words. The $S_i/N_i \simeq 10$ for binary transmission (for $P_e \simeq 10^{-4}$). For the bandwidth expansion of 6, the ideal $S_0/N_0 = (10)^6$, whereas the actual $S_0/N_0 = 2^{12} = 36$ dB. If the PAM pulses were transmitted as a multilevel signal, then $S_i/N_i \geqslant 4 \times 10^4$ ($= 46$ dB); thus m-bit PCM has the advantage of requiring lower S_i/N_i, as discussed in Sec. 8.8, although the ideal improvement is not obtained. The channel η for the binary channel is:

$$\eta_{\text{ch(binary)}} = \frac{R_{(M)}}{C_{(\text{ch})}} = \frac{12}{6 \log_2 (1 + 10)} \simeq 0.6,$$

and for PAM,

$$\eta_{\text{ch}}(\text{PAM}) = \frac{12}{\log_2 (4.10^4)} \simeq 0.8$$

8.8.2 Communication Efficiency [34, 35]

We have discussed above the message efficiency and channel efficiency of some of the commonly used communication systems. The overall system efficiency is related to the values of S_i/N_i and the channel BW, $W_{(\text{ch})}$, for a given bit error rate P_b or the output SNR S_0/N_0. Accordingly, Sanders [34] has defined a composite efficiency parameter β, called here as the Communication efficiency, and β is given as:

For a given P_b or S_0/N_0,

$$\beta = \frac{E_{b(\min)}}{n_0} = \frac{S_{(\min)}}{n_0 R} = \frac{S_i}{N_i}\frac{W_{(\text{ch})}}{R} = \frac{S_i}{N_i}(\alpha) \tag{8.185}$$

where $E_{b(\min)}$ is the minimum signal energy required per message bit, n_0 = noise spectral density in the channel, R = actual information rate of the received digital/analog message and $\alpha = W/R$ = bandwidth expansion due to the particular modulation scheme. A lower bound on β can be obtained from the channel capacity theorem as:

$$R = W \log_2 \left(1 + \frac{S_{(\min)}}{n_0 W}\right)$$

or

$$R/W = \log_2 e \cdot \log_e \left[1 + \frac{\beta_{\min}}{\alpha}\right] \tag{8.186}$$

and in the limit $\alpha \to \infty$, $\quad \frac{1}{\alpha} = \log_2 e \cdot \frac{\beta_{\min}}{\alpha},$

which leads to the condition that:

$$\beta_{\min} \geqslant \log_e 2 = 0.69,$$

as has been shown in Eq. (8.155).

For digital modulation schemes, e.g., in PCM-PSK, PCM-FSK and PCM-ASK, β for a given P_e is calculated from the results of Chapter 2. As an example, for Binary-coherent PSK, $\alpha = W/R = 1$ and S_i/N_i for $P_e = 10^{-5}$ is 10 dB, then $\beta = 10$. In Binary-incoherent ASK, S_i/N_i for $P_e = 10^{-5}$ is 17 dB, $\alpha = 1$, then $\beta = 50$. Similarly, in narrow-band FSK with $\alpha = 1$, S_i/N_i for $P_e = 10^{-5}$ is 13 dB; and $\beta = 20$. For orthogonal M-ary communication systems, E_b/n_0 and α for various P_e's are given in Ref. [1], and the β values are directly obtained from these results. Using the results of Sanders, β-efficiency for ASK, FSK*, PSK and orthogonal M-ary systems with $P_b = 10^{-6}$ are shown in Fig. 8.40.

Along with the results of the practical systems, Shannon's limit of β value, given by Eq. (8.186), is also shown in Fig. 8.40. It is observed that for BW-expansion $\alpha > 1$, M-ary orthogonal systems (also the FEC-coded digital systems) are the only modulation schemes which give decreasing values of β with increase of α, as predicted by Shannon's Channel capacity theorem. FM/FSK with $\alpha > 1$, on the other hand, has increasing β, thus showing lesser channel efficiency, as discussed in Example 20.

For analog modulation systems, a similar comparison can be made by using the relation,

$$\beta = \frac{S_i}{N_i}(\alpha) \text{ for a given } S_o/N_o \tag{8.187}$$

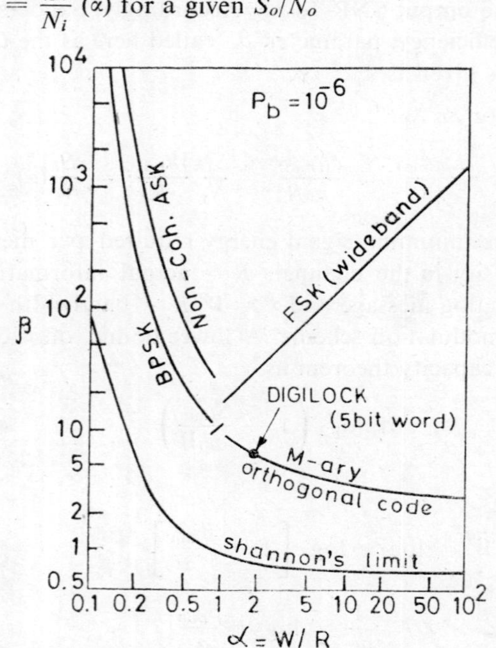

Fig. 8.40 β-efficiencies vs. α (after Sanders [34]).

*For FSK with $\alpha > 1$, the threshold of S_i/N_i increases proportionately with the deviation ratio D and the receiver has to operate above threshold only. Thus β increases proportionately with α. At threshold, β is given by: $\beta = 16\alpha$, and this value is plotted in Fig. 8.40.

and the corresponding ideal R may be calculated from

$$R = W_M \log_2 \left(1 + \frac{S_o}{N_o}\right) = 40.7 \text{ Kb/s}$$

for a complex baseband signal with $W_M = 3.5$ kHz and $S_0/N_0 = 35$ dB.

The digital systems, e.g., PCM and DM, may also be compared for a given S_0/N_0 in the receiver output, using the equivalence between P_b and S_0/N_0, as discussed in Chapter 3. The relations are:

$$\begin{aligned}(S_0/N_0)_{PCM} &= 1/4P_b \\ (S_0/N_0)_{DM} &= (f_s/f_m)\cdot 10^{-2}/P_b \end{aligned} \qquad (8.188)$$

where f_s = bit rate and f_m = maximum message frequency.

Using now the known results of S_0/N_0 vs. S_i/N_i and the corresponding α for different systems [1], β-efficiency may be calculated [35]. These results are shown in Fig. 8.41. It is seen now that among the analog systems, SSB

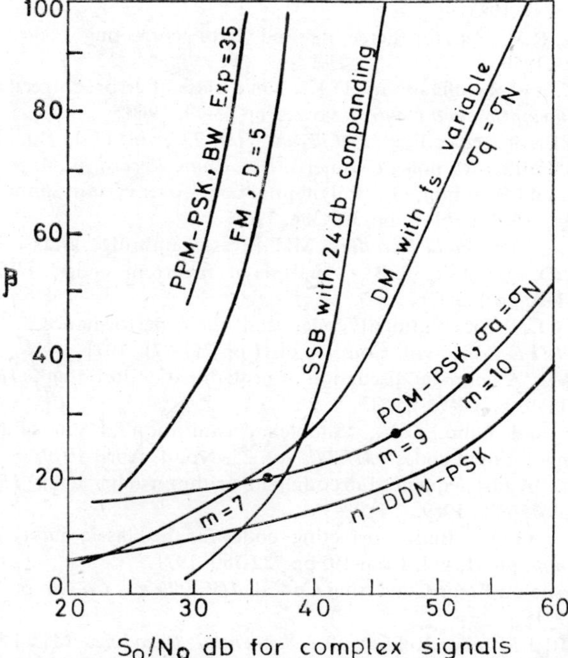

Fig. 8.41 β-efficiencies for different modulation schemes (after Das [35]).

with 24 dB companding requires the minimum β, and among the digital systems, n-stage DDM has the best β-efficiency. As was observed in the earlier sections, the channel efficiencies of digital systems are better than those of analog systems; similarly, β-efficiencies are also seen to be better in case of digital systems. With FEC coding, β-efficiencies improve further.

REFERENCES

1. Das, J., Mullick, S.K. and Chatterjee, P.K., *Principles of Digital Communication*, Wiley Eastern Publishers, New Delhi, 1986.
2. Shannon, C.E., 'A Mathematical Theory of Communication', *BSTJ*, vol. 27, pp 379-423, July 1948 and pp 623-654, Oct. 1948.
3. Shannon, C.E., and Weaver, W., *The Mathematical Theory of Communication*, Univ of Illinois Press, 1949.
4. Abramson, N., *Information Theory and Coding*, McGraw-Hill, N.Y., 1961.
5. Huffman, D.A., 'A method of construction of minimum redundancy codes', *Proc. IRE*, vol. 40, pp. 1098-1101, Sept. 1952.
6. Shannon, C.E., 'Prediction and Entropy of Printed English', *BSTJ*, vol. 30, pp 50-64, Jan. 1951.
7. Peterson, W.W. and Weldon Jr., E.J., *Error-Correcting Codes,* 2nd Edition, MIT Press, Combridge, Mass. 1971.
8. Lin, S., *An Introduction to Error-Correcting Codes*, Prentice Hall, N.J., 1970.
9. BerleKamp, E.R., *Algebric Coding Theory*, McGraw-Hill, N.Y., 1968.
10. Clark, Jr., G.C. and Cain, J.B., *Error-Correcting Coding for Digital Communica-*tions, Plenum Press, NY, 1981.
11. Blahut, R.E., *Theory and Practice of Error Control Codes*, Addison-Wesley Publishing Co., Mass. 1983.
12. Hamming, R.W., 'Error detecting and error correcting codes', *BSTJ*, vol. 29, pp 147-160, 1950.
13. Bose, R.C. and Raychaudhuri, D.K., 'On a class of error correcting binary group codes', *Information and Control*, vol. 3, pp 68-79, 1960.
14. Elias, P., 'Error-free coding', *IRE Trans., Inf. Th.*, vol. IT-1, No. 4, pp 29-37, 1954.
15. Elias, P., 'Coding for noisy channels', *IRE Conv. Record*, pt. 4, pp 37-47, 1955.
16. Reed, I.S. and Solomon, G., 'Polynomial codes over certain finite fields', *J. Soc. Indust. App. Math.*, vol. 8, pp 300-304, 1960.
17. Massey, J.L., *Threshold Decoding*, MIT Press, Cambridge, Mass. 1963.
18. Wyner, A.D. and Ash, R.B., 'Analysis of recurrent codes', *IEEE Trans. on Inf. Th.* vol. IT-9, pp 143-156, 1963.
19. Viterbi, A.J., 'Convolutional codes and their performance in communication systems,' *IEEE Trans.* vol. Com 19, pt II pp 751-771, 1971.
20. Fano, R.M., 'A heuristic discussion of probabilistic decoding', *IEEE Trans. Inf.* Th., vol. IT-9, pp 64-74, 1963.
21. Sinha, V. and Babu, G.V., 'Simplex organisation for Viterbi decoding of short binary convolutional codes', *JIETE*, vol. 25, No. 4, April 1979.
22. Jelinek, F., 'A fast sequential decoding algorithm using a stack' *IBM J. Res. Dev.*, vol. 13, pp 675-685, 1969.
23. Forney Jr., G.D., 'Burst correcting codes for the classic bursty channels', *IEEE Trans. Comm.* pt. II, vol. Com-19, pp 722-780, 1971.
24. 'Special Issue on Error Correcting Codes', *IEEE Trans. Comm.* pt. II, vol. Com-19, October 1971.
25. Wozencraft, J.M. and Reiffen, B., *Sequential Decoding*, MIT Press, Cambridge, Mass. 1961.
26. Gallager, R.G., *Information Theory and Reliable Communication*, McGraw-Hill, NY, 1968.
27. Forney, Jr., G.D., *Concatenated Codes*, MIT Press, Cambridge, Mass. 1967.
28. Matsushita, I., et al., 'Applications of error-correcting codes to TDMA Satellite Communication', *Fujitsu Sc. and Tech. Journal*, vol. 9 (3), pp 87-110, 1973.
29. Shaft, P.D., 'Low rate convolutional code applications in SS Communications,' *IEEE Trans*, vol. Com-25, p. 876, 1977.
30. Goldman, S., *Information Theory*, Prentice-Hall, 1953.

31. Viterbi, A.J., *Principles of Coherent Communication*, McGraw-Hill, N.Y., 1966.
32. Das, J., 'A Technique for Improving the Efficiency of *M*-ary Signalling', *IEEE Trans.*, vol. Com-32, pp 199-201, 1984.
33. BerleKamp, E.R., 'The Technology of Error-correcting codes', *Proc. IEEE*, vol. 68, pp 564-593, 1980.
34. Sanders, R.W., 'Communication Efficiency Comparison of Several Communication Systems', *Proc. IRE*, vol. 48, pp 575-588, 1960.
35. Das, J., 'Optimum detection and communication in presence of noise, Pt V', *IETE Student's Journal*, vol. 14, pp 222-237, 1973.
36. Ristenbatt, M.P., 'Alternatives in digital communication', *Proc. IEEE*, vol. 61, pp 703-731, 1973.

CHAPTER 9

Survival of Communication

INTRODUCTION

The demand for widespread telecommunication applications is increasing rapidly, leading to congestion in the available RF spectrum—MF, HF, VHF and MW; thus causing serious RFI, EMI and EMC Problems. The classical solution to the RFI problems was in the form of shielding, grounding and filtering, optimized in a 'trial and error' method. This is being replaced by damping and filtering using newly developed distributed-parameter and integrated dielectric-magnetic devices, along with predictive EMC planning and control. These techniques, however, deal mainly with the intra-system and subsystem RFI problems. For inter-system EMI, narrowband signalling and frequency reuse with necessary power control and frequency-management are being used as a partial solution in low density and mobile communication environment. The multiplexed wideband channels use FDM and TDM for heavy traffic densities and some protection against Gaussian and impulsive noise is provided through FEC.

Shannon's channel capacity theorem promises a coding gain of 11 dB with ideal wideband coding and infinite channel bandwidth. This may be approached only through large-length orthogonal codes and convolutional FEC, which are now hardware-realizable because of the progress in LSI technology and SAW devices. The modern trend, therefore, is to use larger and larger bandwidth for communication and radar, and their coexistence and survival are ensured through suitable coding.

Till recently, the coding theorists were mainly concerned with Gaussian and impulsive noise corrupting a dedicated channel; however, the problems of jamming, multipath phenomena, rapid time-varying interference, network synchronization and the uncontrolled behaviour of media and channels are engaging the attention of the modern theorists and designers. The solution of consequent EMI and EMC problems are being attempted through the use of EMC planning, pulse compression techniques and interference blankers in radar and by using orthogonal codes, FEC and SSMA techniques in communication.

Yet another dimension in the survival of communication is the security of voice and data information, as the communication networks are becoming more widespread resulting in the possibility of eavesdropping, message falsification and lack of authentication. Although the art of ciphers was known and used by military and diplomatic communities for centuries, the problem of data security has been appreciated only recently by a much broader cross-section of users as a result of the popularization of computer-to-computer communication for such important transactions as banking and credit verification. The use of cryptography for authentication of the message source as well as for prevention of eavesdropping is now a must for most of the computer-communication networks. Fortunately, the progress in LSI and computer techniques has enabled the specialists to design and operate almost foolproof cryptographic systems, and in some countries like the USA, Data Encryption Standards (DES) have been introduced for application in data processing and digital communication networks.

This chapter presents a review of the available signal processing techniques to solve some of the EMI and EMC problems in Radar and Communication [3]. The review discusses signal resolution, orthogonality, RFI suppression in general, and ambiguity diagram, Interference blankers and quasi-orthogonal FM signals in a radar environment. In a communication environment, the channel assignment problem, channel capacity, error-control techniques and finally the application of SS-techniques in a multi-access mode have been reviewed. It is shown that for the coexistence of different modes of communication, both classical and modern, the FEC and SSMA play a vital role. The performance of SSMA is further improved by using low-rate codes in tandem and it is now possible to receive reliable data using a sufficiently large bandwidth and efficient coding, even when the channel is disturbed by burst interference for 50-70 per cent of time. Similarly, in remote area communication through satellites, it is now possible to operate with small earth stations having $G/T \simeq$ OdB, at least for telegraph signals through SSMA. Finally, voice and data encryption techniques, including DES and public-key cryptography, are discussed from the point of view of secrecy and authentication.

9.1 SIGNAL RESOLUTION AND ORTHOGONALITY [1, 2]

The central problem in a communication or in a radar environmental is to be able to detect signals with minimum distortion even in the presence of co-channel and adjacent-channel interference. This brings in the problem of signal resolution and orthogonality. Consider the problem of detecting two interfering signals mixed with Gaussian noise, given as:

$$y(t) = h_1 s_1(t) + h_2 s_2(t) + n(t) \tag{9.1}$$

The maximum-likelihood detection is now optimum and the estimates of h_1 and h_2 are given by

$$\hat{h}_1 = \frac{1}{(1 - R_{12}^2)} \int_0^T [s_1(t) - R_{12} s_2(t)] \cdot y(t)\, dt \tag{9.2}$$

$$\hat{h}_2 = \frac{1}{(1 - R_{12}^2)} \int_0^T [s_2(t) - R_{12}s_1(t)] \cdot y(t)\, dt \tag{9.3}$$

where R_{12} = cross correlation = $\int_0^T s_1(t)s_2(t)\, dt.$

The probability of detection P_d of h_1 is now given by

$$P_d = p\{\,|\,\hat{h}_1\,| > h_0\}_{h_1 \neq 0} = 1 - \text{erfc}\, z_2 + \text{erfc}\, z_1 \tag{9.4}$$

where, $z_1 = h_1\sqrt{(1 - R_{12}^2)/n_0} + h_0\sqrt{(1 - R_{12}^2)/n_0}$

$z_2 = h_1\sqrt{(1 - R_{12}^2)/n_0} - h_0\sqrt{(1 - R_{12}^2)/n_0}$

h_0 = the threshold for detection of a given false alarm probability P_f

n_0 = one-sided noise spectral density for $n(t)$.

It is seen that P_d increases for smaller values of R_{12}. As the signals overlap more and more, R_{12} is nearer to one and a larger SNR is required to attain a given P_d for either of the signals. R_{12} is also called a 'Resolution parameter' in the Radar signal theory.

In the presence of multiple signals, the resolution parameter

$$R_{mn}(t) = \int_0^T S_m(t)S_n(t)\, d(t)\, dt \ll 1, \qquad m \neq n \tag{9.5}$$

for efficient detection. In the ideal situation, $R_{mn}(t) = 0$, $m \neq n$, and the signals are called Orthogonal. In the conventional multiple-signal systems, the orthogonality is ensured by Frequency-division or Time-division multiplexing. However, because of the waveform 'Uncertainty principle', given by [3],

$$\alpha\beta \geqslant \pi$$

where α = time spread of the signal

β = frequency spread of the signal,

the time-limited signals are not bandlimited and vice versa. Thus, strict orthogonality among signals is not achieved in practice. A looser definition of orthogonality is then,

$$R_{mn}(t) = \int_0^T S_m(t)S_n(t)\, dt \leqslant \epsilon, \qquad m \neq n \tag{9.6}$$

where ϵ is a measure of some acceptable level of cross-interference between the signals. In multi-alphabet digital communication, $R_{mn}(t)$ is made as small as possible by using orthogonal and semi-orthogonal codes, e.g., in SSMA. For radar signals, the waveforms are so designed that $R_{mn}(t)$ is sufficiently small even for large T and Doppler shift, e.g., in Pulse compression FM signals. In such situations, the strict definition of orthogonality does not hold, since the signals occupy the same frequency bands and a certain degradation in the capability of the receiver to correctly detect the signal of interest has to be accepted.

9.1.1 Matched Filter [2, 4]

One of the most important hardware for orthogonal signalling is the matched filter (MF) or the equivalent correlation receiver. The ideal MF, given by the impulse response $h(t) = k \cdot s \cdot (\tau_0 - t)$, gives a peak SNR $= 2E/n_0 = 2BT(S_i/N_i)$, where E is the signal energy, $N_i = n_0 B$, and BT = time-bandwidth product of the signal. In case the interference is band-limited and is of fixed total power, as in the case of jamming, the output SNR of the MF is: (S/N) peak $= 2BT \cdot (S_i/N_i)$. Clearly then, output SNR may be improved by spreading the signal energy more and more, and the BT product of the signal and its MF are very important parameters. Since the signal may be represented by 2BT independent samples, the corresponding MF requires no more than 2BT elements or parameters to be specified for its synthesis. In the FM pulse compression technique used in radars, a large BT signal is compressed in the receiver to a narrow pulse of approximate width = 1/B and the corresponding ambiguity surface is given by [2]

$$\chi(\tau, \Delta B) = \begin{cases} \left(1 - \dfrac{|\tau|}{x}\right)\left|\dfrac{\sin x}{x}\right|, & |\tau| < T \\ 0 \quad , \quad . \quad . \quad . \quad . & |\tau| > T \end{cases} \tag{9.7}$$

In the case of digital signalling using large-length codes, both for radar and communication, the MF may be designed with digital hardware or with the linear tapped-delay lines. Again the resultant width of the correlation pulse is approximately 1/B.

The advent of SAW and CCD devices and LSI chips has completely revolutionized the MF hardware. It is now possible to fabricate, with high replicability, SAW-MF's with $BT \leqslant 1000$, up to a bandwidth of 100 MHz. The DMF's with LSI technology may have $BT \leqslant 500$, fabricated on one or a few chips. For a bandwidth of less than 1 MHz, CCD's and IC's have better performance because they can handle delays of tens of msecs, needed for MF's with lesser bandwidth [5, 6]. Both linear and digital MF's have been discussed in detail in Chapter 2 and Appendix B.

9.2 RFI SUPPRESSION

As distinct from the inter-system interference discussed before, intra-system interference i.e., the classical RFI suppression problem, has so far been dealt in an empirical way: by trial and error, and the brute-force method of shielding, grounding and filtering. Mayer [7] has given an excellent review of the modern trends in RFI suppression by using the newly developed distributed-parameter and integrated dielectric-magnetic devices, along with predictive EMC planning and control. The classical technique of shielding, grounding and filtering is being replaced by damping and filtering, where a part of RFI is converted to heat and absorbed by equivalent resistive effects.

Basically RFI suppression circuits use L, C and R which may be voltage-sensitive, frequency-sensitive or fixed elements. The principle of damping is used to make filter characteristics non-resonant, so that the common-mode interference, ground-loop currents and oscillatory phenomena in switched circuits are minimized. The trends in RFI inductors are: Lossy L, $Q \simeq 1$, special magnetic materials, e.g., high saturation lossy ferrites, magnetic composites and the dielectromagnetic approach in the design of L. Similarly, in case of C's, the trends are: ceramic, film and paper, aluminium-electrolytes, tantalum-electrolytic capacitors.

The classical technique for filtering of power lines while connecting equipment and sub-systems has been to use 4-pole LC filters, but sometimes the insertion of a filter is worse than having no filter. These are being replaced by RFI feed-through filters, electrolytic RFI filters, distributed filters and active RFI filters. Damped RFI suppression ignition cables are made of magnetic conductors coiled around a flexible magnetic absorptive core using ferrite mixture; and damped RFI suppression power cables are made of coaxial structures with absorptive magnetic dielectric between the centre conductor and outer braid. The distributed dielectromagnetic RFI filters use closely spaced insulated copper wire wound on lossy magnetic cores and have an outer grounded conductive sheath with a dielectromagnetic layer in between. These have insertion loss characteristic of 25-30 dB/oct. up to 50 MHz. Active RFI filters are being used now for isolating unwanted frequencies other than power supply frequencies. Thus, adapted RFI suppression components will take into account the intrinsic advantage of damping over shielding and grounding. Such damped components will follow modern technology trends—sophisticated structures, and partition-cum-integration in the general concept of dielectromagnetics.

9.3 RADAR ENVIRONMENT [3]

Because of the RF spectrum becoming more and more crowded, more efficient techniques of spectrum utilization have to be implemented. In Radar signal design, the techniques of coding, bandsharing etc. are becoming more common and ultimately the waveform design is matched to the Radar environment (e.g., ground and sea clutter, weather and other interference signals). These problems of EMI and Compatibility are sought to be solved by using radar waveforms having a thumbtack ambiguity diagram and approximately orthogonal waveforms.

9.3.1 Ambiguity Diagram [1]

The technique of using overlays of ambiguity diagram of a radar waveform upon the environmental diagram along with the desired target trajectories, provide a pictorial view for approximate EMC analysis and spectrum usage, as suggested by Aasen [8]. On the same diagram, interfering RF emissions and the different clutters are shown in Fig. 9.1. Knowing the clutter and interference locations in the diagram, one can superimpose the ambiguity diagram of the radar waveform and then the signal-to-interference (S/I) and the signal-to-clutter (S/C) ratios can be determined. In a complex situation

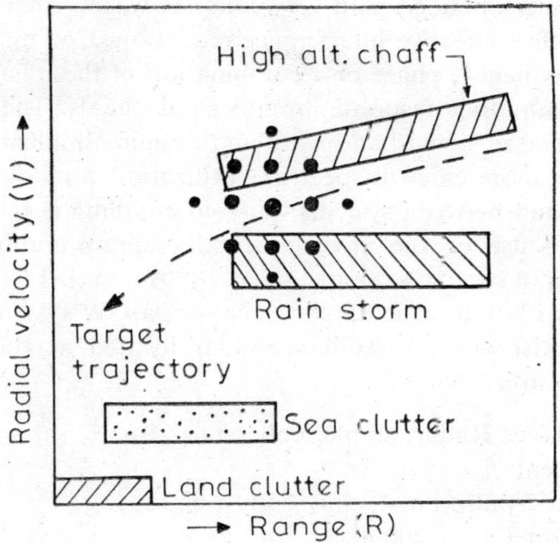

Fig. 9.1 Basic environment at diagram showing Ambiguity diagram overlay (R-V)

as shown in Fig. 9.2, one has to consider the unstable and asynchronous pulse signal, low stability CW signals, stable and synchronous pulse signals and different types of clutters. It then becomes necessary to avoid ambiguity functions which have a two-dimensional spread as given by waveforms having a single pulse or a uniform pulse burst. Although the ideal thumbtack ambi-

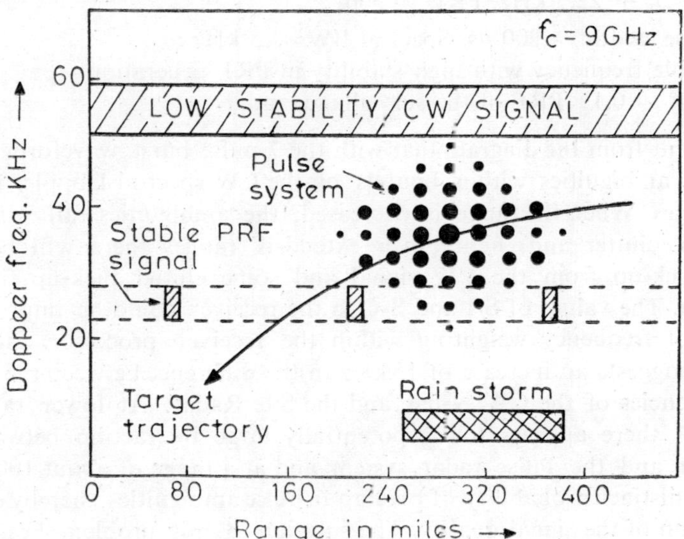

Fig. 9.2 Environment diagram for a Surveil ance radar with pulse burst ambiguities in presence of other CW and pules systems (after Aasen [8])

guity function is only obtained with very complex waveforms, the technique of signal coding offers effective interference reduction. Coding of the waveforms in time, frequency, phase or a combination of these has the effect of decreasing or eliminating response to undesired signals. The coding, however, requires increased signal bandwidth and design complexity, but on the other hand, allows more efficient spectrum utilization and the requirement of a wide guard band between systems to avoid coupling is relaxed.

To illustrate the use of the environmental diagram combined with an ambiguity diagram, a specific example given by Aasen [8] is presented in Fig. 9.2, where within a general site location, a CW system and a pulse systemhave to coexist with a surveillance radar located at the origin. The system parameters are:

(a) Site Surveillance Radar: stable, coherent pulse.
Doppler system: $f_c = 9$ GHz.
Waveform: a 7-pulse burst, uniformly weighted.
Burst duration $T_d = 1200\ \mu$s.
The interpulse period $T_i = 197\ \mu$s.
System integration time = 1.4 ms.

(b) C.W. System: C.W. emission:

$f_I = f_c + 52$ kHz.

Short term stability = 1.1 ppm in 1.4 ms.
Effective BW = 10 kHz.

(c) Pulse system:

$f_I = f_c + 22.5$ kHz, PRF = 500.

Pulse width = 200 μs, Spectral BW = 5 kHz.
Stable frequency with high stability in PRF generation;
PRF = $0.1 \times$ PRF of the surveillance radar.

It is evident from the diagram that with the 7-pulse burst waveforms, there are distal ambiguities within bounds of the CW spectral Doppler band at large ranges. When the range is decreased, the ambiguities fall within the rainstorm clutter and, hence, over extended ranges there will be some energy pick-up from the CW signal and some clutter pick-up from the rainstorm. The values of S/I and S/C in the receiver could be improved by the use of frequency weighting within the receiver processor. Also the diagram suggests an increase of 15 kHz in the difference between the operating frequencies of the CW system and the Site Radar. At lower ranges of the target, there appears to be potentially large interaction between the Site radar and the Pulse radar system and at a range of about 100 miles, there is a distinct probability of overlap in pulse ambiguities, thereby causing degradation of the signal quality. The solution to this problem lies in the selection of a long duration random binary phase-coded waveform, which will give an approximate thumbtack ambiguity diagram. Using a waveform with the following parameters:

Burst duration = 600 μs,
Waveform = 10 bits, 60 μs per bit, phase-coded,
BW = 16.6 kHz.

The thumbtack ambiguity diagram has been overlaid on the environmental diagram in Fig. 9.3. This waveform provides good S/I and S/C ratio at larger

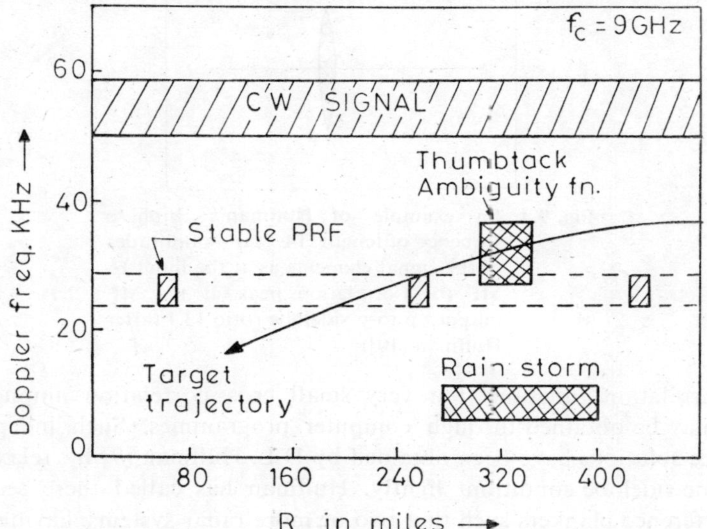

Fig. 9.3 Environment diagram with thumbtack ambiguity diagram in presence of other CW and pulse systems and clutter (after Aasen [8])

ranges, and only at closer ranges an overlap with the pulse signal is possible. Since the floor level of the thumbtack diagram is approximately one percent of the peak, this interaction will be very small. Selection of wider signal bandwidth and more bits or longer pulse duration will further improve the S/I ratio.

9.3.2 Interference Blankers [9]

It is apparent from the environmental diagrams of radar systems that the coexistence of two or more radar systems in the same site requires proper signal design. Almost orthogonal and quasi-orthogonal signalling has been proposed by D.L. Huffman [9], Cook [11] and others. D.A. Huffman [10] originally proposed polyphase codes which have idealized autocorrelation function having a large peak-to-sidelobe ratio. An example of Huffman's sequence of length 14 which has only real coefficients is given by

$$\{A_i, \phi_i\} \rightarrow \{A_i\}\cdot \exp\{jo, j\pi\} = \{b_i\} \tag{9.8}$$

and $\{b_i\} \rightarrow \{-0.57, 0.27, -0.56, 0.55, -0.14, -0.28, -0.31, -1, -0.43, 0.5, 0.037, -0.34, 0.22, 0.43\}$.

Its auto-correlation function is shown in Fig. 9.4. Such sequences have been useful in designing pulse compression radars. Further studies in the technique of designing such sequences reveal that alternative sequences giving similar

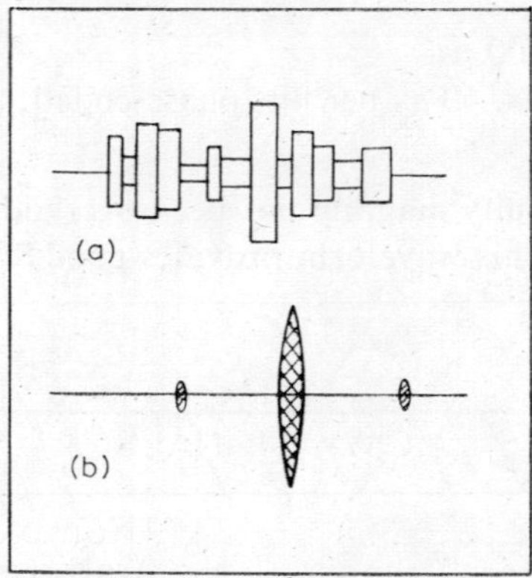

Fig. 9.4 An example of Huffman's Biphase sequence of length 14. (a) Amplitudes of the signal elements as at the input to MF (b) Correlation peak at the MF output: p-to-p sidelobe rotio 13:1 (after Huffman [10])

auto-correlation functions but very small cross-correlation among themselves may be obtained through computer programmes. Such independent sequence solutions have been obtained by D.L. Huffman [9] by relaxing the zero time-sidelobe condition slightly. Huffman has called these sequences as Interference blankers such that two or more radar systems having quasi-orthogonal signalling waveforms may coexist. A modified Huffman code of length 13 (as in Table 9.1) has its auto- and cross-correlation values (with another similar pulse train) as shown in Fig. 9.5. Huffman has made exhaustive studies of different quasi-orthogonal sequences of length 13 and has

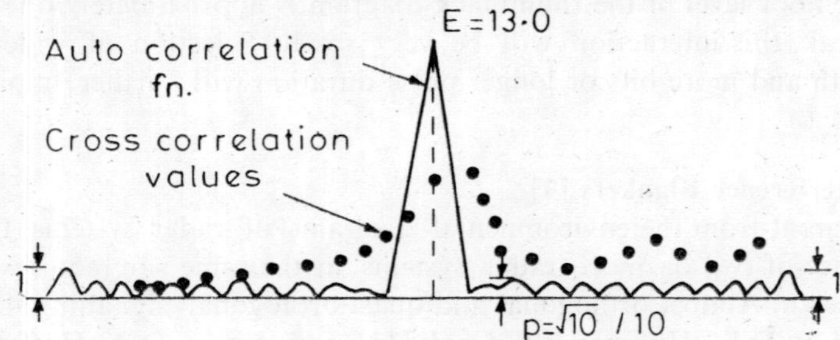

Fig. 9.5 MF response to the modified Huffman sequence with n = 13. Peak-to-sidelobe ratio: 32 dB; Uncorrelation value between 1st and 2nd sequences: 16.8 dB (after Huffman [9])

given ten such sequences that are most favourable, a few of which (given by $\{A_i, \phi_i\}$) are shown in Table 9.1. The uncorrelation values, defined as the DB difference between the ratios of impulse energy to time-sidelobe energy that are measured in the auto-correlation function and in the cross-correlation function, have also been given by him and some typical values for sequences (m, n) are shown in Table 9.2, using the sequences of Table 9.1.

Table 9.1 Excitation Values for Ten Most Favourable Modified Huffman Sequences: Length 13, $E = 13.0$, $p = \sqrt{10}/10$

Peak-to-sidelobe ratio = 32 dB (After Huffman [9])

1st Sequence			2nd Sequence			3rd Sequence		
i	$A(i)$	$\phi(i)$	i	$A(i)$	$\phi(i)$	i	$A(i)$	$\phi(i)$
1	1.000	0	1	1.244	0	1	0.997	0
2	1.153	0	2	1.393	−16.4	2	1.400	−32.2
3	0.823	0	3	0.939	−26.8	3	1.173	−34.4
4	0.964	0	4	0.930	−46.7	4	0.253	−42.4
5	0.932	0	5	0.434	111.7	5	0.334	155.4
6	0.957	180	6	1.398	104.7	6	1.079	89.0
7	1.248	180	7	0.976	−86.1	7	0.524	−34.0
8	0.957	0	8	0.864	−104.0	8	1.118	−148.2
9	0.932	0	9	1.303	57.3	9	0.431	35.5
10	0.964	180	10	0.617	−134.6	10	0.823	−40.4
11	0.823	0	11	0.622	25.8	11	1.135	80.0
12	1.153	180	12	0.900	−163.6	12	1.409	−147.0
13	1.000	0	13	0.804	0	13	1.003	0

4th Sequence			5th Sequence			6th Sequence		
i	$A(i)$	$\phi(i)$	i	$A(i)$	$\phi(i)$	i	$A(t)$	$\phi(i)$
1	1.215	0	1	0.957	0	1	1.000	0
2	1.246	−38.3	2	0.216	−31.2	2	0.802	−17.0
3	1.374	−26.7	3	1.363	27.4	3	0.461	−23.0
4	0.546	−58.2	4	0.683	64.2	4	1.415	59.3
5	0.132	142.0	5	1.412	19.7	5	1.248	42.3
6	1.410	95.5	6	0.996	168.7	6	0.986	−43.7
7	1.147	−62.9	7	0.931	37.4	7	0.477	−180.0
8	1.115	167.1	8	1.005	−70.6	8	0.986	−136.3
9	0.847	−50.9	9	1.308	−99.4	9	1.248	−42.3
10	0.893	29.8	10	0.723	67.0	10	1.415	120.7
11	0.562	131.1	11	1.211	142.9	11	0.461	23.0
12	0.844	−141.7	12	0.236	−148.8	12	0.802	−163.0
13	0.823	0	13	1.045	0	13	1.000	0

7th Sequence			8th Sequence			9th Sequence		
i	$A(i)$	$\phi(i)$	i	$A(i)$	$\phi(i)$	i	$A(i)$	$\phi(i)$
1	1.215	0	1	2.973	0	1	0.977	0
2	1.433	−47.0	2	0.792	−31.8	2	1.211	−26.1
3	1.065	−16.4	3	1.442	14.7	3	1.465	−24.1
4	1.215	−61.1	4	1.331	−36.6	4	0.826	−68.3
5	1.065	128.1	5	0.522	94.7	5	0.607	87.1
6	0.967	39.1	6	0.416	−32.1	6	0.276	116.6
7	0.521	−145.4	7	1.223	179.3	7	0.659	94.0
8	0.467	−124.1	8	0.306	99.9	8	1.172	−84.8
9	0.144	174.5	9	0.551	−4.3	9	0.766	169.0
10	1.234	−13.0	10	1.332	−36.8	10	1.363	1.5
11	1.067	118.5	11	1.307	130.4	11	0.628	100.6
12	0.970	−133.0	12	0.835	−143.2	12	1.269	−153.9
13	0.823	0	13	1.027	0	13	1.024	0

(Contd.)

i	10th sequence $A(i)$	$\phi(i)$
1	0.973	0
2	0.701	−16.3
3	1.467	3.4
4	0.893	−9.3
5	0.439	−34.7
6	0.728	112.6
7	1.050	−114.2
8	1.396	127.2
9	1.381	−34.4
10	0.767	−0.6
11	0.861	153.9
12	0.739	−163.7
13	1.027	0

Table 9.2 Uncorrelation Values Between Ten Most Favourable Modified Huffman Sequence Solutions: Length 13, $E = 13.0$, $p = \sqrt{10}/10$

Time sidelobe level: −32 dB. (After Huffman [9])

m	n	Uncorrelation value in dB
1	2	16.8
1	3	20.4
1	4	21.8
1	5	22.4
1	6	20.6
1	7	21.0
1	8	18.1
1	9	20.0
1	10	18.7
2	3	13.0
2	4	14.7
2	5	23.0
2	6	23.2
2	7	15.3
2	8	20.8
2	9	16.7
2	10	20.4
3	4	10.3
3	5	18.5
3	6	21.0
3	7	10.3
3	8	16.5
3	9	11.7
3	10	16.7
4	5	17.9
4	6	21.2
4	7	11.9
4	8	17.6
4	9	14.2
4	10	14.5
5	6	18.7
5	7	19.4
5	8	19.8
5	9	18.7
5	10	19.1
6	7	21.4
6	8	21.0
6	9	22.0
6	10	20.8
7	8	14.5
7	9	11.5
7	10	17.2
8	9	16.4
8	10	11.9
9	10	17.9

Uncorrelation value between mth sequence solution and a hypothetical uncoded sequence

m	Uncorrelation value
1	25.2
2	25.2
3	25.1
4	25.1
5	25.1
6	25.1
7	25.1
8	25.2
9	25.1
10	25.2

Most uncorrelation values are seen to be quite large and in practical situations, a S/I ratio of 10 dB is good enough to differentiate between the interfering signal and the desired signal. Further, modification of the sequences is possible by a trade-off between time-sidelobe improvement and the degraded pulse energy utilization value. Such modifications have shown that the ranges of uncorrelation values from 7.5 to 20 dB is obtained for a time-sidelobe suppression level of 25 dB; from 12.6 to 26.3 dB for 30 dB; from 17.4 to 30.7 dB for 35 dB and from 21.8 to 33.7 dB for 40 dB. It follows then that a pulse compression decoder, designed for time-sidelobe suppression level of 25, 30, 35 and 40 dB, is expected to be an Interference blanker, if an effective time-sidelobe suppression level of 17.4 to 18.2 dB is obtained. Thus, it is shown that independent sequence solutions that are notably dissimilar from each other provide a desirable EMC capability, when simultaneous operation of similar pulse compression systems is desired. Sequences longer than 13 would provide many more independent solutions then given above. However, pulse suppression capabilities of a longer sequence can be developed from the product of two short sequence lengths, where each pulse component of one of the short sequences consists of the other short sequence. It is, therefore, not absolutely necessary to consider longer sequence lengths.

9.3.3 Linear FM Signals [11]

The pulse compression techniques using linear FM signals for better detectability and signal resolution have been studied in detail. Cook [11] has suggested that FM signals with different slopes may be used as quasi-orthogonal signals in the bounded time-frequency plane. He has examined the interaction among the FM indices, time-bandwidth product and cross-talk interference criterion that determine the number of different linear FM signals that may coexist. The general class of FM signals, shown in Fig. 9.6,

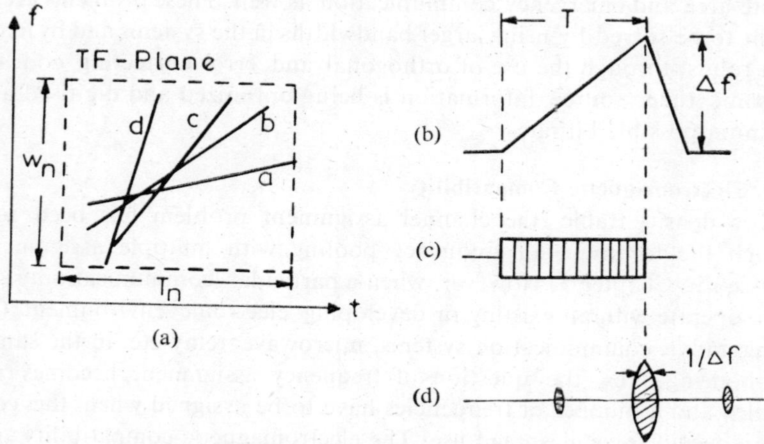

Fig. 9.6 General class of FM signals: (after Cook [11]) (a) Joint occupancy of common TF-plane by several signals, (b) A linear FM function, (c) Uncompressed FM pulse, (d) Compression filter output: $\tau \approx 1/\Delta f$

is given by

$$S_n(t) = \cos\left(\omega_n t + \frac{1}{2}\mu_n t^2\right) \tag{9.9}$$

where $\mu_n = (2\pi W_n)/T_n.$

When an FM signal with slope μ_m is introduced into a linear matched filter, matched to the slope μ_n, then the peak amplitude is given by

$$R = (\mid \gamma \mid T_n W_n)^{-1/2} \tag{9.10}$$

where $\gamma = [(\mu_m - \mu_n)/\mu_n].$

The quantity R^2 is a measure of mutual interference and depends on the mismatch factor γ and the time-bandwidth product TW. As an example, for a 20 dB $S/I(= 1/R^2)$ ratio, $\mid \gamma \mid T_n W_n = 100$, and for a 30 dB S/I, $\mid \gamma \mid T_n W_n = 1000$. One may now calculate the number N of discrete signals that may coexist with the allowable interference in the given TW-plane and N is given by,

$$N = R^2[(TW)_{\max} - (TW)_{\min}] \tag{9.11}$$

Then for an $S/I = 100$, $N = 15$, for $(TW)_{\max} = 2000$ and $(TW)_{\min} = 500$. Thus, it is possible to have approximately 16 radar systems operating in the TW plane having an area of 1500 Hz-sec., with an acceptable mutual interference. Cook has further examined the question of identification of the signals by using composite modulation on the FM signals.

9.4 COMMUNICATION ENVIRONMENT

The problems of co-channel and adjacent channel interference, random and burst noise, time-varying interference, multipath effects and compatibility are more severe in the communication environment, specially in mobile and low-density static communication. There is a trend to use satellite communication for not only the international links, but to extend the services to remote area and emergency communication as well. These problems are now sought to be solved by using larger bandwidths in the systems and by making them robust through the use of orthogonal and error-correcting codes. At the same time, source information is being optimized and digitised at the minimum possible bit rates.

9.4.1 Electromagnetic Compatibility

For low density traffic, the channel assignment problem has been solved through the technique of frequency pooling with multiple assignment as discussed in Chapter 5. However, when a particular communication system has to operate with an existing or developing electronic environment (EC), having radar, communication systems, microwave retay etc. in the same or neighbouring sites, the question of frequency assignment becomes more complex and a number of frequencies have to be assigned when the equipment is meant for widespread use. The electromagnetic compatibility analysis has now to consider [12]

(a) The frequency separation requirement between each pair of equipment in a proposed EC environment.

(b) The optimum channel assignment for each equipment, for a given set of frequency separations.

In such analysis, the signal-to-interference ratio (S_i/I_j) is dependent on the antenna-coupled interaction between systems and the selectivity characteristics of the receivers. This S/I is compared with a predetermined interference threshold level $(S/I)_{th}$, which the receiver can tolerate. In certain cases, it may be necessary to provide isolation between systems using additional terrain shielding and off-tuning between receivers. Based on the above, a $[\beta_{ij}]$ is formed, where the entries correspond to the amount of rejection in dB that is necessary between the receiver i and the signal j for compatible operation. A channel separation matrix $[\Delta f_{ij}]$, as shown in Fig. 9.7, is then

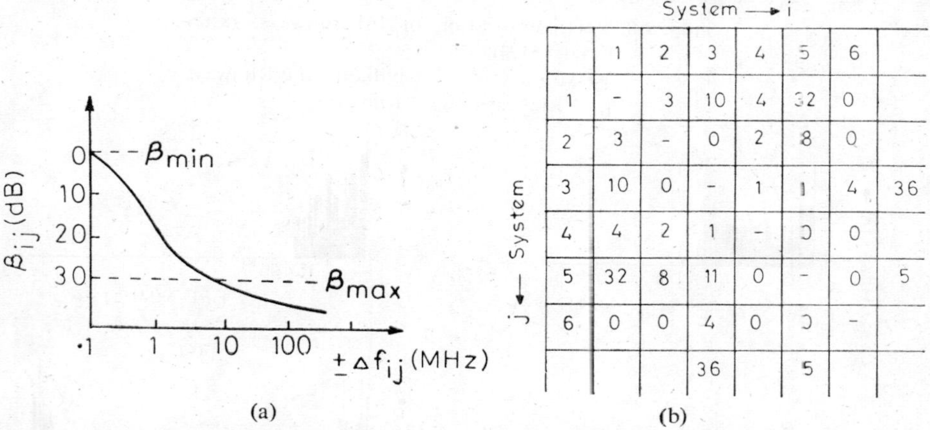

Fig. 9.7 (a) Typical symmetric off-frequency rejection (OFR) curve for interfernce from systems i to j
(b) Example of channel separation matrix $[\Delta f_{ij}]$ (after Frazier [12])

formed by calculating the bounds on the required number of discrete channels and the total channel spread. The entries in $[\Delta f_{ij}]$ is a function of $[\beta_{ij}]$, where $\beta_{ij} \neq \beta_{ji}$. Based on this data, the frequency assignment algorithm uses an assignment priority scheme that assign frequencies first to those transmitters that have the greatest constraints imposed on them by the channel separation matrix. It has been shown that the reuse of frequencies is to be emphasized throughout the assignment procedure and some sort of 'frequency packing' both at the low and at the high-end of the band minimizes the total bandwidth required.

As an example of the application of these principles, a Microwave Landing Guidance System (MLS) has been planned for US Civil/Military authorities for worldwide use in about 500 runways. The MLS uses two frequency bands: (5-5.25) GHz and (15.4-15.7) GHz with 20-30 nmi range under the most severe conditions. An assumed location of the subset of 161 runways is shown in Fig. 9.8, and the channel assignment along with the reuse factor r is shown in Fig. 9.9. With only low-end packing, the number of channels required in the (5-5.25) GHz band was 79 and with frequency sharing, the

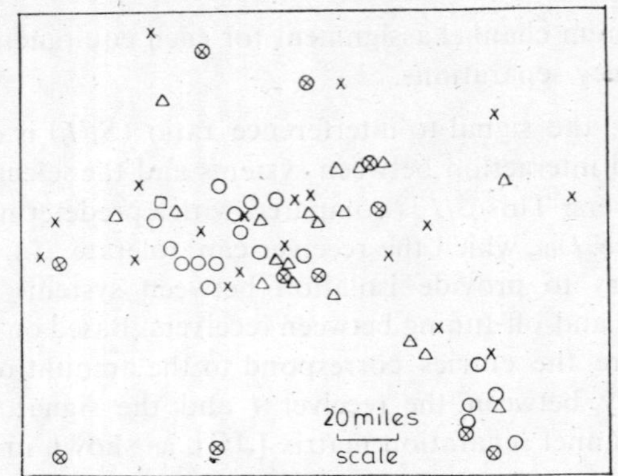

Fig. 9.8 Relative location of 161 runways; (after Frazier [12])
⊗ Military, × Civilian, △ 2 Runways
□ 3 Runways ○ 4 Rnnways

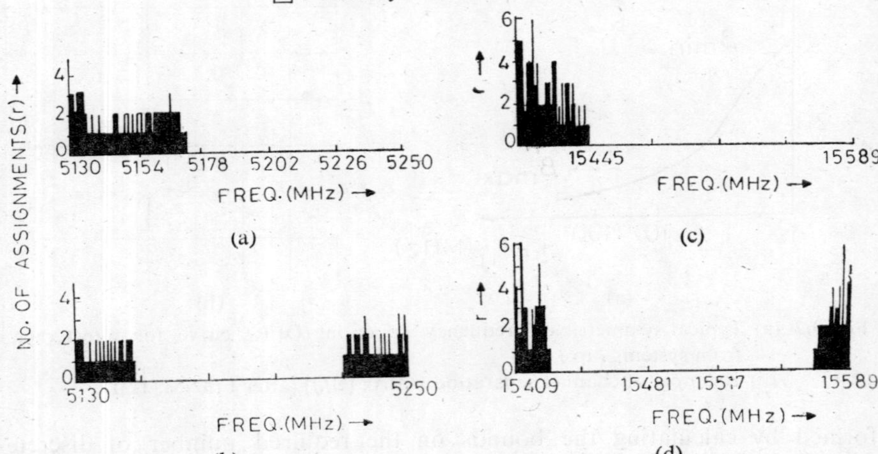

Fig. 9.9 Spectrum occupancy nomograph for MLS with 161 runways (after Frazier [12]) (a) 5-5.25 GHz band, Frequency sharing, Low end packing (b) 5-5.25 GHz band, Frequency sharing; both end packing; (c) 15.4-15.7 GHz band, Low end packing (d) 15.4-15.7 GHz band, both-end packing

number is reduced to 71. For the (15.4-15.7) GHz band, the number of channels required using frequency sharing and both-end packing, was 43 only. This is a considerable saving over the previous case, and shows that in the lower band, EMC constraints are more severe. It has been found that both-end packing is generally more economical and sometimes it is convenient to operate different systems at different portions of the available frequency band.

9.4.2 Channel Capacity and FEC [13, 14]

The classical technique, for survival of communication in presence of noise (mostly thermal noise), is the wideband FM, where an SNR improvement of 15 to 25 dB may be obtained, when the receiver operates above its threshold CNR (approximately 10 to 15 dB). However, all real channels are composite having both thermal and burst noise, but the FM technique does not protect the signal in such real-life situations. It has also been shown in Chapter 8 that the communication efficiency of FM is poor, as compared to digital signals, and the modern techniques of coding and signal processing cannot be applied to analog signals. In view of this, digital signalling has become popular and the signals are now pre-coded and protected through source encoding and forward-error correcting codes (FEC) or orthogonal codes. For analog signals, minimum digital representation is obtained through DPCM, ADM and Transform coding, and speech signals are optimally digitized by using APC or LPC codecs (Ref. Chapter 3). For channel signalling FSK, CPSK, DPSK, QPSK and QAM are used, and for the purpose of bandwidth conservation. *M*-PSK or APK are used (Ref. Chapter 2). Assuming now that all signals are available in the digital form, the current techniques of digital signal processing may be used to optimize channel transmission and reception and to combat noise, interference, jamming and multipath effects.

Shannon's capacity theorem, as discussed in Chapter 8, provides a potential coding gain of 11 dB, as it is theoretically possible to have $P_b \to 0$ with $E_b/n_o = -1.6$ dB only with ideal coding (by using large-alphabet orthogonal codes) and bandwidth $W \to \infty$ (since for PSK, the threshold $P_b = 10^{-5}$ for $E_b/n_0 = 9.6$ dB). The ideal coding gain of 11 dB, however, has not been obtained through any practical FEC codes so far, as has been shown in Fig. 8.35. For ready reference, the performances of several popular FEC codes are shown in Fig. 9.10, where it is seen that the theoretical limit for the convolutional codecs in the AWGN channel is $E_b/n_o \geqslant 1.4$ dB, exactly 3 dB above Shannon's limit [13]. This limit can be approached using PSK with low-rate codes and soft decision. The practical result with a Rate 1/2 code is: $E_b/n_o = 4.5$ dB for $P_b = 10^{-5}$. Of the two decoders for convolutional codes, the Viterbi decoder can be implemented with the present-day hardware up to a speed of 10 Mb/s, whereas the sequential decoder becomes uneconomical. Higher rater ($R > 1/2$) convolutional codes have been widely used for conventional communication systems, including satellite links operating in AWGN.

In the discussions of the channel capacity theorem and ideal code performances, it has been assumed that the channel is disturbed by AWGN only. However, in practice, no channel is ideal as all channels (except perhaps the earth station-to-earth station satellite link) are disturbed by a combination of random and burst noise. As such, it is necessary to use, in practical systems, either an interleaved/adaptive code, e.g., Gallager code, or a concatenated code, e.g., convolutional-cum-*RS* code, to protect the signal against both random and burst noise. In certain situations, e.g., for SSMA in a jamming environment, the major interference is time-varying and impulsive, rather than Gaussian noise, and this can be countered by using low

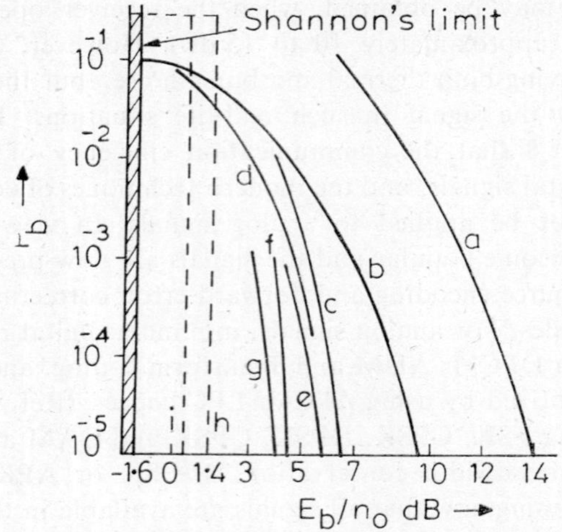

Fig. 9.10 Variation of P_b with E_b/n_o for different FEC-coded signals (after Ristenbatt [13]) (a) PSK, $M = 8$; (b) PSK, $M = 2$; (c) Block codes (24, 12); (d) Simplex code, $M = 32$; (e) $R = \frac{1}{2}$ convolutional code-Viterbi decoding, hard decision; (f) $R = 1/2$, Conv. code, sequential decoding; (g) $R = 1|2$, Conv. code, Viterbi decoding, 8-level; (h) Theoretical limit for sequential decoding ($E_b/n_o = 1.4$ dB); (i) Channel capacity with hard decisions($E_b/n_o = 0.4$ dB)

rate convolutional codes, say $R = 1/32$, which can tolerate a large symbol error-rate and still give an acceptable P_c at the output, as shown in Fig. 8.36 [15].

Many experimental and simulation results are available to show the effectiveness of FEC for HF, VHF, troposcatter and satellite channels [15]. Since these channels are mostly disturbed by burst errors, an experimental codec using heavily interleaved (15, 5) BCH code was used for 1500 Km HF path and for 1300 Km VHF path, and low data rate signals (< 100 b/s). The error improvement factor F, defined as:

$$F = \frac{P_{e(\text{uncoded})}}{P_{b(\text{coded})}} = \text{coding gain in BER} \quad (9.12)$$

was evaluated and is given in Table 9.3. It is to be noted that none of the 10-min samples examined had decoded character error, when $P_e \leqslant 10^{-3}$.

Brayer [16] has examined the performances of four FEC techniques, viz., Golay, Adaptive Golay, Massey and adaptive Gallager codes (all $R = 1/2$), over several real channels, e.g., a transcontinental HF 4800 bit/s channel, a mixed MW Tropo link of length 583 miles at a data rate of 2400 b/s, and a satellite link with necessary tail ends at a data rate of 2400 b/s. Figure 9.11(a) and (b) show the average Improvement Factor F for the three

Table 9.3 Error Improvement Using Coded Transmission

Antenna type	P_e uncoded (= channel BER)	F for VHF channels	F for HF channels
Dual Yagi	0.5	1	1
or	10^{-1}	10	6
Dual Rhombic	10^{-2}	10^4	10^3
	10^{-3}	10^6	10^5

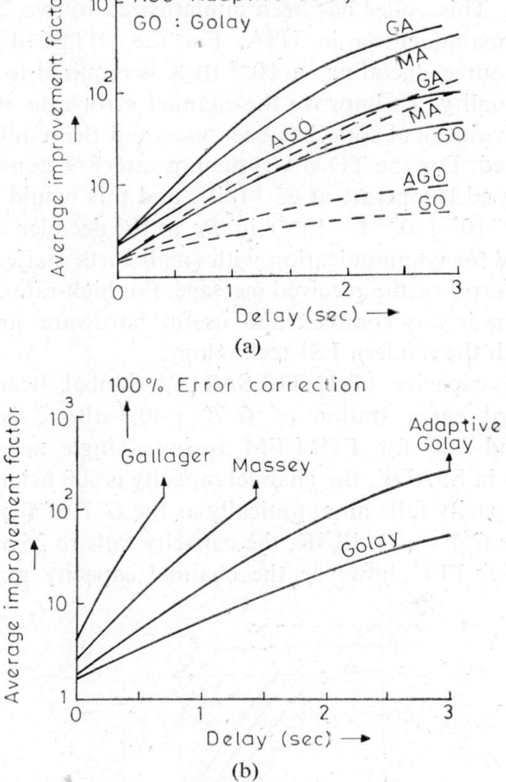

Fig. 9.11 Code performances in HF, Tropo and Satellite links (after Brayer [16])
(a) HF and Troposcatter channels—for HF ch. at 4800 b/s; · · · for Tropo-channels at 2400 b/s. (b) Satellite channel at 2400 b/s.

different channels. The value of F are shown against Delay (in sec), which is a measure of the complexity of the codec, cost and size, and also the visible transmission effects. It is seen that the adaptive Gallager code performs better than others, and obtains values of F in the range of 10^2 to 10^3, with reasonable storage requirements. These results, indeed, show that the real channels are somewhat bursty and a composite code is necessary to protect the signal adequately.

9.4.3 Satellite Applications [17]

COMSAT has made extensive studies with the convolutional codes for application in satellite communication. In the SPADE, the rate 3/4 convolutional codes have been used and an error improvement from 10^{-4} to about 10^{-9} has been obtained. The overall performance of the codec over a satellite link shows that : 10^{-2} channel error-rate can be improved to 10^{-4} at the decoder output. The E_b/n_0 performance also improves considerably and for the 4-phase PSK modem of the SPADE, a 10^{-5} error-rate is obtained with 7.5 dB of E_b/n_0. This codec has been miniaturized to two 24-pin chips by a semiconductor manufacturer in USA. For the DITEC-digital TV system using DPCM source encoding, a 10^{-8} BER is required to meet the CCIR recommended quality. To improve the channel error-rate of 10^{-4} to 10^{-8}, the rate 7/8 convolutional code has been used and the required performance has been obtained. For the TDMA system, a rate 8/9 convolutional codec has been developed to operate at 65 Mbit/s and this would give a moderate coding gain of 10^2 (10^{-4} to 10^{-6}) in P_b at the decoder output. For data transmission and for communication with small earth stations, it is essential to have error control of the received message. For high-rate codes, the coder/decoder hardware is very complex and useful hardware are being implemented only with the modern LSI technology.

The Channel capacity of INTELSAT–IV Global Beam Transponder, using a standard earth station of G/T = 40.7 dB/°K, is 1000 for PCM/PSK/TDMA and 900 for FDM/FM (using a single access). For PCM/PSK/FDMA, as in SPADE, the channel capacity is 800 using voice-activated carriers. This capacity falls monotonically as the G/T of the earth station is reduced, and for G/T = 21 dB/°K, the capacity falls to 25 only, as is shown in Fig. 9.12. Using FEC, however, the channel capacity may be improved

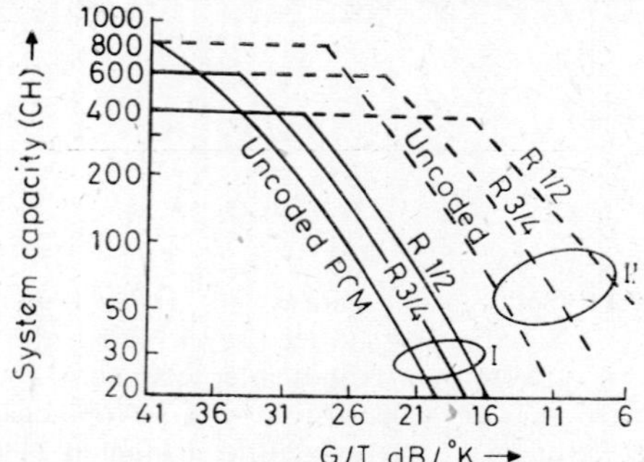

Fig. 9.12 Intelsat IV Transponder (SPADE) capacity vs. G/T of earth stations, and improvement with FEC.
I : Global beam with EIRP=23 dBW
II : Spot beam with EIRP=34.8 dBW.

considerably, e.g., with $R = 3/4$ coding the channel capacity improves from 400 to 600 for $G/T = 36$ dB/°K; with $R = 1/2$ coding channel capacity improves from 80 to 200 for $G/T = 26$ dB/°K. These improvements in channel capacity with FEC are shown in Fig. 9.12 for the global beam and spot beam applications and, thus, confirm that the channel capacity and G/T ratio can be conveniently traded for better overall efficiency.

The required G/T of a small earth station may also be reduced by using orthogonal codes. For example, a calculation was made for the G/T requirement for transmission of TDM-TP signals using Symphonic Satellite which provides SCPC channels of bandwidth = 38 kHz and capable of transmitting 32 Kbits CPSK signals. For a C/N of 11.5 dB at the detector input, G/T required was 21 dB/°K. Using lower transmission rates and coding the data to 32 Kbits by orthogonal codes, the G/T requirement decreases considerably as shown in Table 9.4. It is seen that the data at 5 Kbits can be received with $G/T = 6.7$ dB/°K only. For TP channel rate of 75 bits/s, 24 channel TDM-TP would require 1800 bit/s and this may be received with $G/T = 1.5$ dB/°K only. Use of bi-orthogonal codes will increase the rate of transmission further. Thus, to optimize the overall system, it is necessary to have minimum bit-rate A/D converters, effective error-control codes and efficient modulation processes. The ADM codec, orthogonal and low-rate convolutional codes and PSK/partial-response SSB techniques have been found to be most useful.

Table 9.4 Reduction of G/T by Using Ortho/Biorthogonal Codes. EIRP same as that of one channel in TRACT (STEP) = 8.2 dBW, Detector input $C/N \simeq 11.5$ dB as in TRACT, $BW = 38$ kHz, $P_b = 10^{-4}$. Reference $G/T = 21$ dB/°K

R in Kb/s	C/N dB	G/T required
12	2.8	12.3 dB/°K
8	0	9.5 ”
5	−2.8	6.7 ”
3	−5.6	3.9 ”
1.75	−8.3	1.2 ”

9.5 SPREAD SPECTRUM (SS) SYSTEMS—ANTI JAMMING PROPERTIES [18, 19,20]

The multiaccessing property of the SS systems was discussed in Chapter 5, but the most important application of SS techniques is to enable the communication system to operate in a hostile and jamming environment. Both DS and FH systems show tolerance to jamming, although the actual mechanism may be different in them.

9.5.1 DS-PNS System

In Chapter 5, it was shown that for the DS-system using MF/correlator receiver and under wideband noise jamming, the effective SNR at the MF output is (cf. Eqn. (5.43)):

$$\gamma_{\text{eff}} = \frac{2ST}{n_0 + J/B} = 2BT\left(\frac{C}{N_{\text{eff}}}\right)_i \simeq \frac{2BT}{K + K_1} = \frac{2G_p}{k + K_1}$$

and (9.13)

$$P_e = \frac{1}{2} \exp\left(-\frac{2BT}{K + K_1}\right)$$

where S = desired signal power, J = Jamming power, $B = BW$ of the channel, $BT = G_p$ = Processing Gain, K = No. of simultaneous talkers and $K_1 = J/S$ = No. of interfering talkers equivalent to the jamming power. For a given error-rate P_e, say, equal to 10^{-4}, $\gamma_{\text{eff}} \simeq 8$, and

$$K \simeq \left(\frac{BT}{4} - K_1\right) \tag{9.14}$$

assuming thermal noise being included in the jamming noise power J.
The trade-off between K and J for a given P_e, neglecting thermal noise, is shown in Fig. 9.13, where it is seen that J could be 15 to 35 dB above S and P_e is still better than 10^{-4}, if G_p is 30 to 50 dB [18].

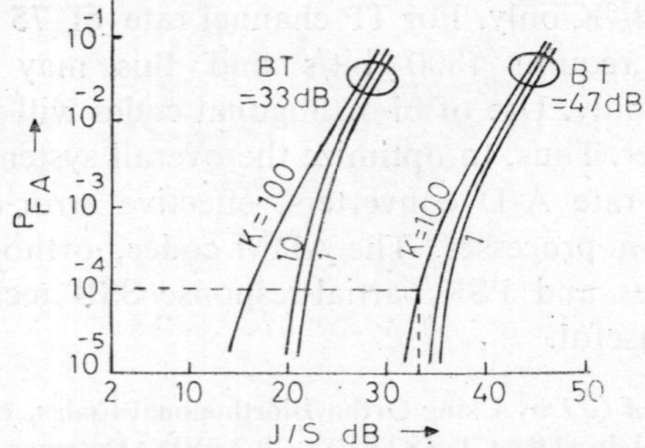

Fig. 9.13 Antijamming capability of a CDMA system; P_{FA}: false-alarm probability; K = number of simultaneous talkers; BT = Time-bandwidth product in dB.

For narrow-band/tone jamming, the frequency domain representation of DS-PNS-SS is shown in Fig. 9.14, where it is again seen that the equivalent jamming noise N_j entering the data filter (LPF) is

$$N_j = \frac{J}{BT} = J/G_p \tag{9.15}$$

and $(S/N)_0$ at MF output is now $(C/N)_i \times G_p$. Thus the number of simultaneous talkers allowed is :

$$K = \frac{G_p}{(S/N)_o\,(\text{min})} - J/S + 1 \tag{9.16}$$

$$= \frac{G_p}{(S/N)_o\,(\text{min})} - K_1,$$

where $(S/N)_o$ (min) is for the given P_e.
The worst jamming for a DS-PNS system is the Pulse jamming (partial or full-band), where the jammer transmits high-peak-power pulses, at least for T seconds. This produces burst errors and these errors are produced by an

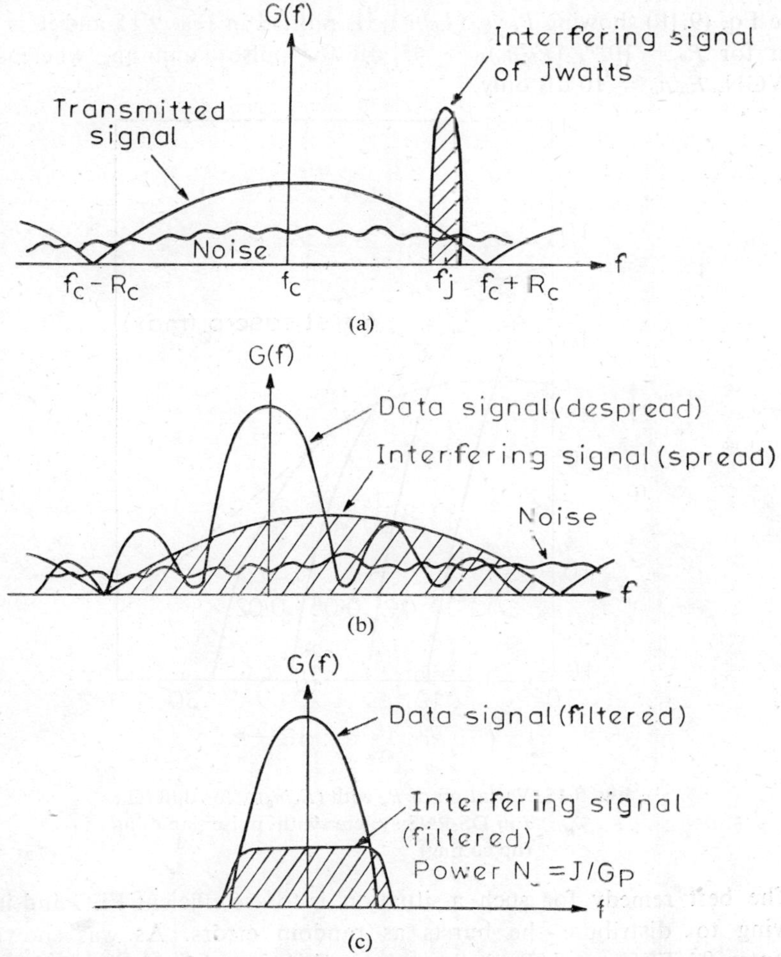

Fig. 9.14 Frequency domain representation of narrow band jamming in DS-PNS-SS systems (a) Received power spectra (precorrelation) (b) Received de-spread spectra at the detector input (c) Spectra at the output of data filter.

average J_{av}/S ratio much less than that required by 100 percent duty factor jammer. Assuming the duty ratio α for the jamming pulses, $\alpha < 1$, the effected symbols have the equivalent S/N as:

$$(E_b/n_o)_{\text{eff}} = \left(\frac{G_p S}{J_\alpha}\right), \; J_\alpha = \text{peak pulse power} = J_{av}/\alpha$$

and,
$$P_e \text{ (for effected symbols only)} = \frac{1}{2} \exp\left(-2\,\frac{G_p S}{J_\alpha}\right) \tag{9.17}$$

One can maximize P_e by varying α and in the worst case [21],

$$P_e(\max) = \frac{0.083 J_{av}}{G_p S} = \frac{0.083}{(E_b/n_o)_{av}}. \tag{9.18}$$

The Eq. (9.18) showing P_e vs. $(E_b/n_o)_{av}$ is plotted in Fig. 9.15 and it is seen that for $P_e = 10^{-5}$, $(E_b/n_o)_{av} \simeq 45$ dB for pulse jamming, whereas for AWGN, $E_b/n_o = 10$ dB only.

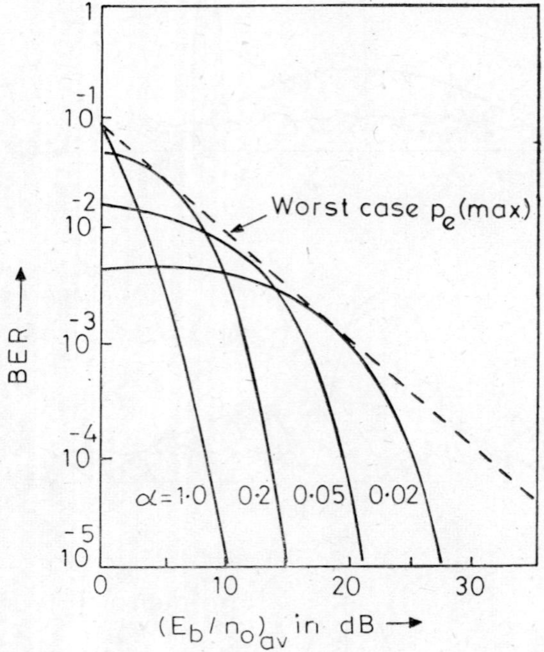

Fig. 9.15 Variation of P_e with $(E_b/n_0)_{av}$ for different α in DS-PNS system with pulse jamming (no coding)

The best remedy for such a situation is to use efficient FEC and interleaving to distribute the bursts as random errors. As was shown in Chapter 8, FEC can effectively reduce E_b/n_0 requirement to 5 dB with reasonable codec complexity and many authors have investigated the combined performance of FEC-cum-SS-systems [20]. The improvement due to FEC is shown in Fig. 9.16, where it is seen that BER performance improves with larger values of α (at BER $= 10^{-7}$), and the $(E_b/n_0)_{av}$ requirements are similar to the case of AWGN with the coding gain of approximately 5 dB.

Ketchum *et al.* [22] have investigated the performance of an adaptive narrowband prefilter to reduce the BER deterioration due to narrowband jamming in a DS-system. The adaptive filter was designed using linear predicting and conventional spectral analysis techniques. The improvement obtained by such an adaptive filter is shown in Fig. 9.17, where the dependence of BER on G_p is also seen. Li and Milstein [23] have also shown similar improvements in BER with adaptive TDL filters.

The DS-PNS system has a large-talker capture problem, specially when the repeater has a limiter at the front end. Since the large talker interference

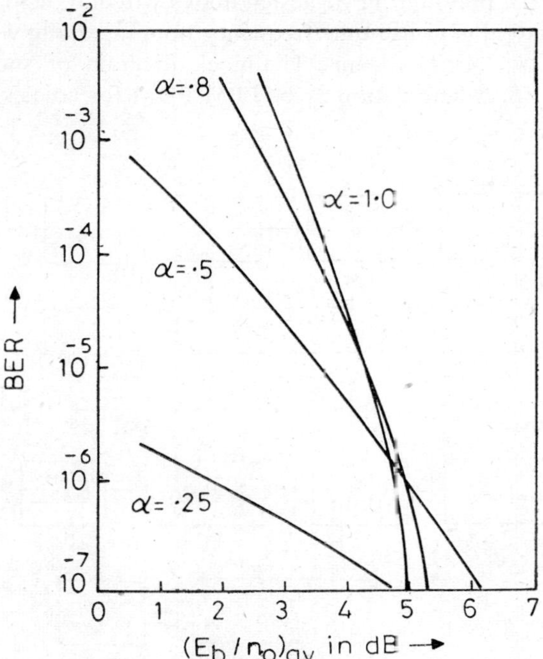

Fig. 9.16 Variation of P_e with $(E_b/n_0)_{av}$ with $R = 1/2$ coding, using (48, 24), $d_{min} = 12$, quasi-cyclic code for DS-PNS and pulse jamming (after Bhargava [21])

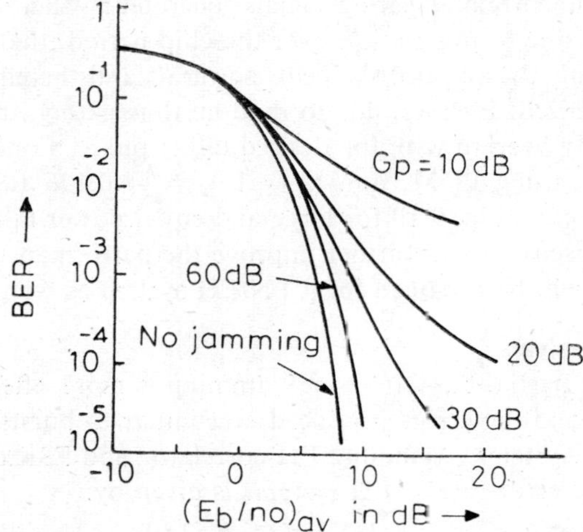

Fig. 9.17 BER under Gaussian assumption for 4-tap predictor with MF. $J/S = 20$ dB, non-dispersive channel (after Ketchum et al. [22])

is broadband, the pre-filtering suggested above will not help. The solution for such a situation is to use time/frequency hopping along with DS, resulting in PN/TH or PN/FH systems. The block diagram of such a system is shown Fig. 9.18, where data may be PPM/PDM for voice signal or digital

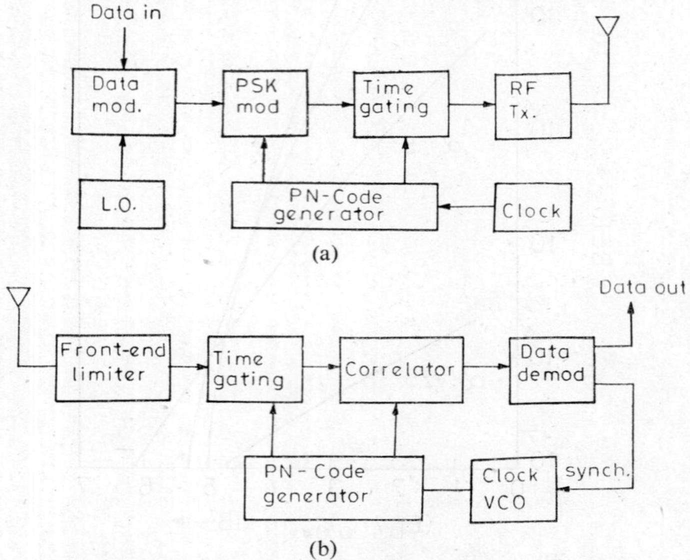

Fig. 9.18 Block schematic of PN/TH-SS system. (a) Transmitter (b) Receiver

for PCM/ADM. After PN-modulation, the signal is randomly time-gated to transmit only for a fraction of τ_c, and it is assumed that different transmitters will have uncorrelated gating signals generated by their PNS-generators. In the receiver, due to integration over the chip period, the effective jamming power is only J_{av} and not the peak power J_α and the effect of the larger talkers is minimized. Further, due to random time gating, interfering talker pulses will rarely overlap with the desired talker pulses. For a message bandwidth of 3 kHz, using PDM, and $\tau_c = 1\ \mu$s, $\alpha = 0.125$ (duty factor), the $(S/N)_0$ is shown in Fig. 9.19 for large and equal power talkers. For digital data, FEC as discussed will further improve the performance of the system. Similar results will be obtained for a PN/FH system as well [19].

9.5.2 FH-SS Systems

In FH systems, partial-band (or tone) jamming is more effective than full-band jamming and the errors produced are similar to bursts caused by pulse jamming in DS-systems. Assuming FH correlator and FSK demodulation in the receiver, the error-rate in FH systems is given by:

$$P_e = 1/2 \exp(-E_b/2n_0) \tag{9.19}$$

Assuming partial band jamming, with duty ratio α, effective E_b/n_0 is:

$$(E_b/n_0)_{\text{eff}} = \alpha G_p S/J_{av},$$

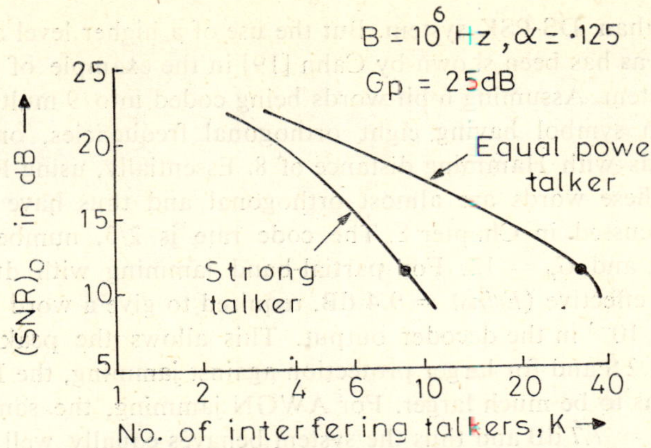

Fig. 9.19 SNR characteristic with strong and equal power talker for a PDM-SS system. PDM (S/N) improvement = 5 dB (after Cahn et al. [19])

where $G_p = B/R$, R = rate of transmission, and P_e due to jamming

$$= (\alpha/2) \exp\left(-\frac{\alpha G_p S}{2J_{av}}\right) \tag{9.20}$$

The worst case jamming by optimizing α, is obtained as:

$$P_e\,(\max) \simeq \left(\frac{J_{av}}{eG_pS}\right) \tag{9.21}$$

The variation of P_e with (G_pS/J_{av}) is shown in Fig. 9.20 for various values of α. It is seen that binary signalling in a non-coherent FH system is about

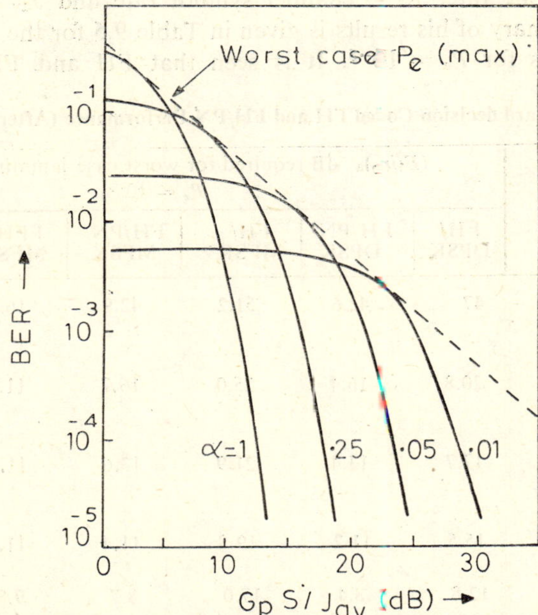

Fig. 9.20 Variation of P_e for FH-SS system in partial-band jamming (no coding)

6 dB worse than DS-PSK system. But the use of a higher level alphabet is very efficient, as has been shown by Cahn [19] in the example of an M-ary coded FH system. Assuming 6 bit words being coded into 9 multifrequency symbols, each symbol having eight orthogonal frequencies, one obtains 64 code words with Hamming distance of 8. Essentially, using Reed-Solomon codes, these words are almost orthogonal and thus have the noise immunity discussed in Chapter 8. The code rate is 2/3, number of slots required is 72, and $G_p = 12$. For partial-band jamming with duty factor $\alpha = 1/6$, the effective $(E_b/n_0) = 9.4$ dB, required to give a word error probability $P_w = 10^{-3}$ in the decoder output. This allows the peak jamming power $J_\alpha = 1.2$ S and for larger protection against jamming, the FT-matrix of the code has to be much larger. For AWGN jamming, the same system requires $E_b/n_0 = 7.7$ dB and thus the system behaves equally well for both types of jamming.

Huth [24], has shown that FEC techniques significantly improve the performance of FH-SSMA in environments containing jamming, multipath and unregulated multiple access. He has studied non-coherent FH, FFH-and FH-PN SS-modulation systems using DPSK, DQPSK and MFSK techniques. He has also determined the output bit error-rate P_e vs. $(E_b/n_0)_{av}$, defined as:

$$(E_b/n_0)_{av} = \frac{1}{R_c}(E_c/n_0)_{av}$$

and

$$(E_c/n_0)_{av} = \left(\frac{B}{R_s}\right)\cdot\left(\frac{S}{J_{av}}\right) \tag{9.22}$$

where R_c = code rate, R_s = channel symbol rate and J_{av} = interference power. A summary of his results is given in Table 9.5 for the FH and FH-PN SS-systems for $P_e = 10^{-5}$. It is seen that FH and FH-PN systems

Table 9.5 Hard decision Coded FH and FH/PN Performance (After Huth [24])

Type of codes	$(E_b/n_0)_{av}$ dB required for worst case jamming to give $P_e = 10^{-5}$					
	FH/ DPSK	FH/PN/ DPSK	FH/ MFSK	FH/PN/ MFSK	FFH/ MFSK	FFH/PN/ MFSK
No. coding	47	42.6	51.2	42.9	16. 5	12.1
$R_c = 1/2$, convolutional, feedback decoding	20.8	16.4	25.0	16.7	11.9	11.6
$R_c = 1/2$, convolutional, Viterbi decoding	17.7	13.4	21.9	13.6	11.1	11.1
$R_c = 1/3$, convolutional, Viterbi decoding	15.5	11.2	19.8	11.4	11.9	11.4
$R_c = 1/2$, sequential decoding	12.8	8.4	17.0	8.7	9.8	8.7

require large $(E_b/n_0)_{av}$ for operation without FEC; but fast FH (FFH) modulation has a much better performance. With FEC, however, the performances of FH and FH-PN improves remarkably and $(E_b/n_0)_{av} \simeq 9.0$ dB only is required to give $P_e = 10^{-5}$ for worst case jamming, as compared to (E_b/n_0) 4.5 dB for Gaussian noise. Further improvement in the antijamming performance of SSMA systems may be obtained by using low-rate convolutional codes ($R_c = 1/8, 1/16, \ldots,$ etc.) as shown by Shaft [25], and discussed in Chapter 8. It has been shown that the high redundancy of the low-rate codes makes them a natural match for SS-communication systems, particularly FH system, and provides a large antijamming capability.

9.6 SS-SYSTEMS—ANTIMULTIPATH PROPERTIES [19, 20, 26]

Fading and multipath effects on Digital radio have been discussed in Chapter 5. General solutions to the problems of error bursts were found in the use of diversity, DFE and interleaved FEC. The Rake receiver [27], based on the technique of channel measurements, has been also discussed. However, in land-mobile radio (say, urban) and high-rate packetized data transmission (say, in ARPA packet radio) networks, SS-techniques may be used to resolve the multipath and then the path components are added coherently to give a time diversity gain, somewhat similar to that obtained in a Rake receiver. The multipath resolution is illustrated in Fig. 9.21,

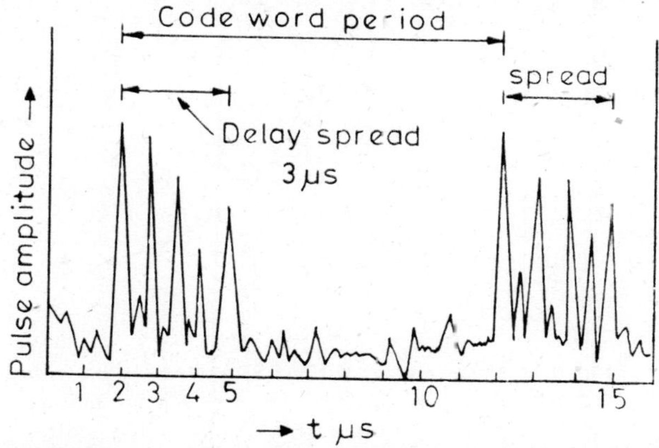

Fig. 9.21 Time resolution of multipath using wideband signalling; urban propagation at 1370 MHz.

which was obtained by using an SS-test waveform for an urban propagation path at 1.37 GHz and it is seen that significant delay components span over a 3 μs interval [28]. Thus, in this example, it will be possible to transmit data at Mb/s rate, if coherent correlation of the resolved signal peaks can be achieved. SS-techniques using both DS and FH signalling are now being widely used for high speed mobile data networks. For typical avionics applications, however, the LOS component tends to predominate over multipath and by itself must suffice for proper operation of the receiver. A PN receiver, then, would be designed to receive the LOS component and discriminate against the multipath components, by using diversity and DFE.

In non-coherent FH systems, there is no multipath discrimination, but there is a frequency diversity advantage leading to multipath tolerance. For CW transmission, multipath effects show up as frequency fading with Rician distribution for a specular plus diffuse path. With FH, the receiver effectively integrates over a number of pulses in randomly selected slots; for example, in the *M*-ary coded FH system, discussed in Section 9.5.2, an approximately 8th order diversity operates due to the combination of coding and FH. It has been shown [19] that P_e in the range of 10^{-2} to 10^{-4} is obtained for effective (E_b/n_o) of 9 to 13 dB.

Milstein and Schilling [20] have investingated the performance of a non-coherent FH-FSK system operating in the presence of partial-band tone interference and frequency selective Rician fading. Their results are shown in Fig. 9.22, where 100 interfering tone jammers were assumed giving

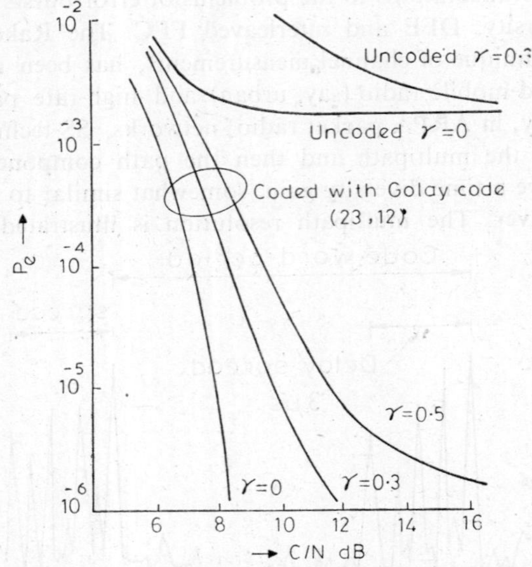

Fig. 9.22 Performunce of FH-FSK coded system with Partial-band jamming and Rician fading. No. of interfering tones = 100 (after Milstein and Schilling [29])

$J/S = 10$ to 40 dB. The worst case $P_e > 10^{-3}$, even with large C/N, where C/N is defined as the signal-to-ratio in the information bandwidth of $1/T$, and $\gamma^2 =$ (power in the scatter component)/(power in the specular component). But with rate 1/2 coding [Golay code (23, 12)], the BER improves considerably giving $P_e \simeq 10^{-6}$ with $C/N = 12$ dB and $\gamma = 0.3$. However, the system is sensitive to the value of J_{av}/S per slot (and not so much on the total J/S). Thus, the FEC-coded FH-FSK system is effective against both jamming and multipath interference.

Turin [26] has made extensive studies on SS-antimultipath techniques, specially for urban networks, and has suggested simplifying modification of the 'Channel sounding RAKE' receiver. In the DPSK-RAKE receiver, shown in Fig. 9.23(a), the MF-DPSK demodulator output is in the form of multiple responses $X_o, X_1, \ldots, X_3$, due to multipath, shown in Fig. 9.23(b) (assuming that the transmitted signal has a large BT product). The sounding receiver estimates the amplitudes and arrival times, $\{\hat{a}_k\}$, $\{\hat{t}_k\}$, shown in Fig. 9.23(c), and these estimates are used as the coefficients of the TDL correlator (with K taps). The correlated outputs, showing the large main peak in Fig. 9.23(d), are fed to the decision circuit to obtain the data out-

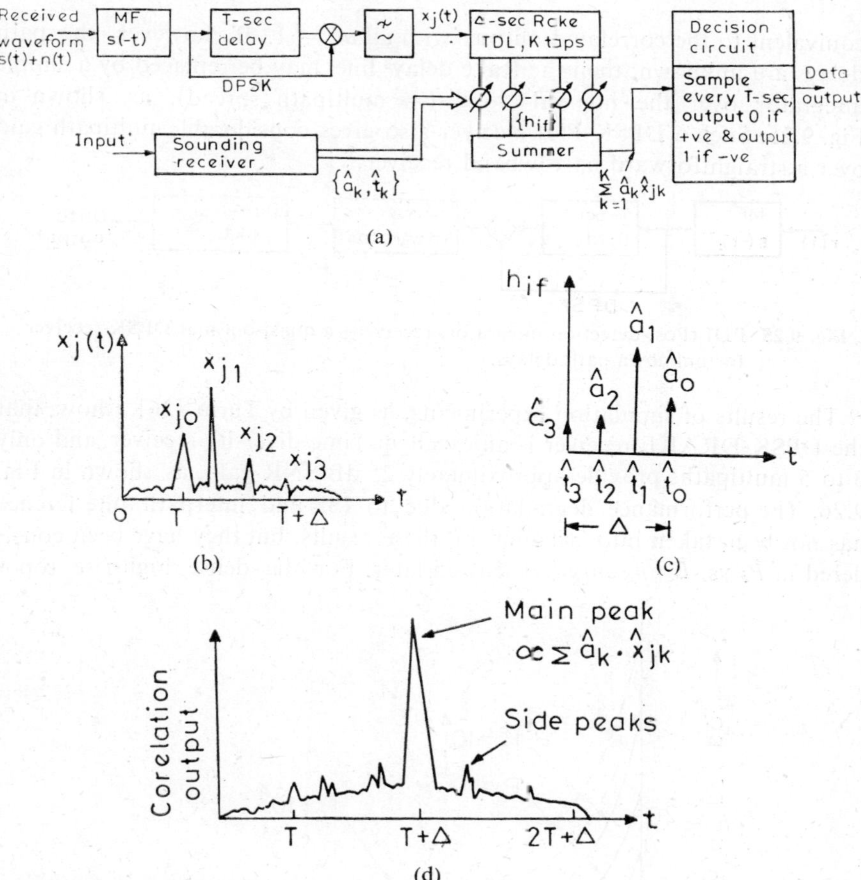

Fig. 9.23 RAKE receiver and its responses (after Turin [26]) (a) A channel-sounding DPSK receiver (RAKE); (b) Filter output; (c) $\{h_{if}\}$—weighting functions obtained from the sounding receiver; (d) correlated output

put. Because of the complexities in the sounding receiver, Turin has suggested a simplified Digital RAKE (DRAKE) receiver, where only the arrival times $\{\hat{t}_k\}$ of the multipath signals will be estimated and used to switch TDL-taps at times $\hat{t}_k$, as shown in Fig. 9.24. The summed output is now

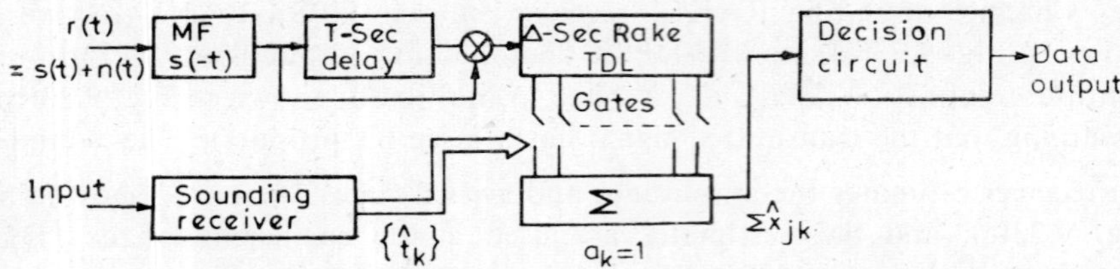

Fig. 9.24 Digital RAKE (DRAKE)—a quasi-optimal receiver when only path delays are estimated

equivalent to the correlated output with all $a_k = 1$. If, however, the path delays are unknown, then the Rake delay line may be replaced by a simple integrator over the interval Δ-sec (= multipath spread), as shown in Fig. 9.25. Such a DPSK-PDI receiver also gives considerable multipath gain over a straightforward narrowband receiver.

Fig. 9.25 PDI (Post-detection integrator) receiver—a quasi-optimal DPSK receiver for unknown path delays.

The results of simulation experiments, as given by Turin [26], show that the DPSK-DRAKE receiver is an excellent Time-diversity receiver and only 3 to 5 multipaths provide approximately 25 dB SNR gain, as shown in Fig. 9.26. The performance degradation due to ISI and interpath interference has not been taken into account in these results, but they have been considered in P_e vs. $\bar{E}_{av}/n_o$ curves discussed later. For the dense high-rise topo-

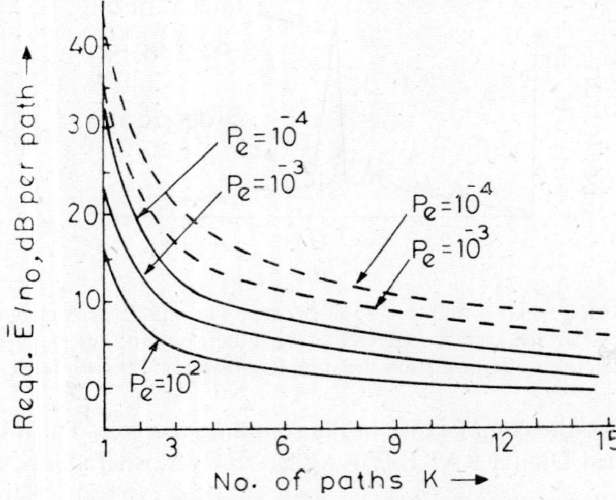

Fig. 9.26 Required $\bar{E}/n_0$ per path for DPSK/DRAKE and PDI systems, K independent Rayleigh fading paths. (After Turin [26]); – – – PDI; ——— DRAKE

graphy of the urban network used for simulation studies, the multipath spread was about 7.0μs, and as such, the information rate was limited below 100 Kb/s. The chip rate was 10 Mb/s, B = 10 MHz, BT = 127, $1/T$ = 78.7 Kb/s and Δ/T = 0.55, so that the ISI was negligible. The results of P_e measurements are shown in Fig. 9.27, for various configurations of the

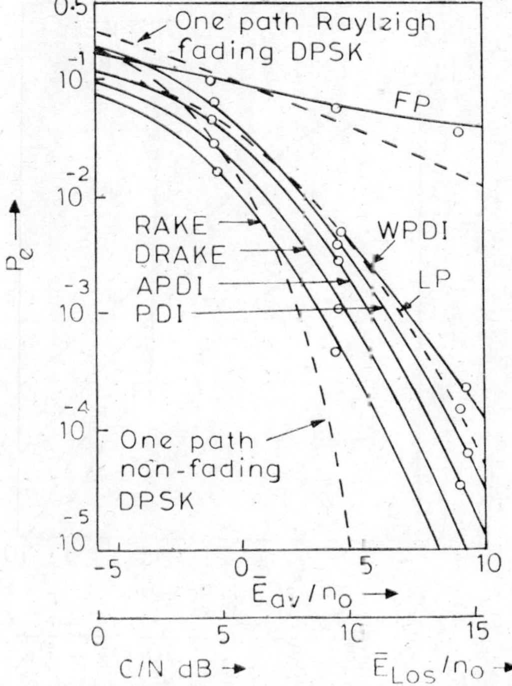

Fig. 9.27 P_e for the low-rate case, from simulation, for dense-highrise topography (after Turin [26]) BT = 127. Δ/T = 0.55, $R = 1/T$ = 78.7 Kb/s; FP: first path; LP: largest path; PDI: post detection integration; WPDI: weighted PDI; APDI: adaptive PDI; DRAKE: digital RAKE

modified RAKE, and both $\bar{E}_{av}/n_o$ and $\bar{E}_{Los}/n_o$ parameters have been used, where $\bar{E}_{av}$ = average energy received per bit per path, determined by calculating the mean square path strength of all paths in all profiles, used in the experiment, and $\bar{E}_{Los}$ = average energy received per bit only on LOS paths It. is seen from the curves that RAKE and DRAKE perform only 1 to 3 dB better than APDI and PDI receivers. This moderate gain of the more complex receivers suggests that PDI could be used in most situations, where ISI is not a serious problem. However, PDI structure is quite susceptible to ISI and as such DRAKE/RAKE have to be used at high data rates, where $\Delta/T \gg 1$. The quasi-optimal DRAKE automatically cancels the ISI as the ISI responses will fall on slots of the RAKE delay line which are not switched by the time estimates $\{\hat{t}_k\}$ of the sounding receiver. Of course with RAKE, where both $\{a_k\}$ and $\{t_k\}$ are estimated, the performance is still better. The experimental results at higher data rates, with $1/T = 787$ Kb/s, $\Delta/T = 5.5$, $BT = 127$ and effective $B = 100$ MHz, are shown in Fig. 9.28

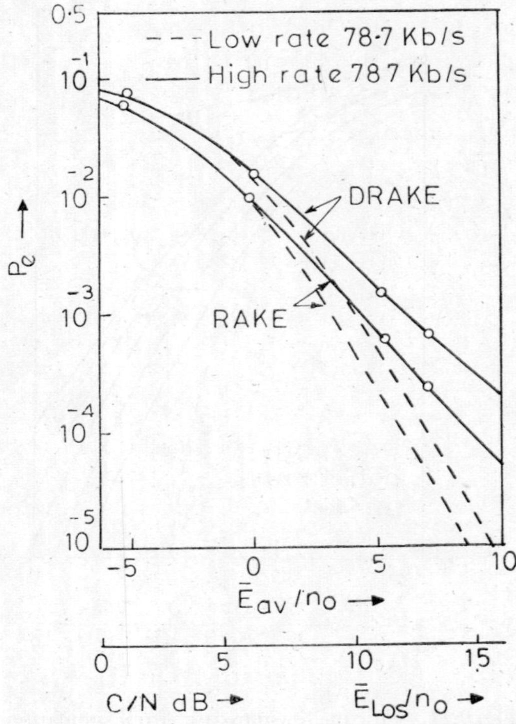

Fig. 9.28 P_e for the high-rate case; $BT = 127$: $\Delta/T = 5.5$; $R = 1/T = 787$ Kb/s; (after Turin [26])

for RAKE and DRAKE. It is seen that the performance is only slightly inferior as compared to that of the lower rate system. The degradation due to multipath induced interference is given by:

$$D = 1 + \frac{\bar{K}}{BT} \cdot \frac{\bar{E}}{n_o} \tag{9.23}$$

and with the number of taps $\bar{K} \simeq 23$, $D = 1.3$ dB for $E_{av}/n_o = 3$ dB; $D = 2$ dB for $E_{av}/n_o = 5$ dB; and $D = 2.8$ dB at $E_{av}/n_o = 7$ dB. These losses are seen in the curves. With RAKE, the improvement is about 2–3 dB over DRAKE. In both figures, it is seen that the time diversity gain in RAKE/DRAKE/PDI is approximately 6 dB (i.e., the difference of E_{LOS}/n_o and E_{av}/n_o). Moreover, an LP (Largest path) receiver performs much better than the one path receiver with Rayleigh fading. Sussman [30] has suggested an alternative MF receiver structure for performing the coherent correlation of the resolved multipath signals, as discussed in Chapter 5. This has also confirmed the efficiency of the SS-technique for pulse transmission through fading channels.

Kahn *et al.* [28] have discussed the many advantages that are obtained in Packet radio networks (e.g., in PRNET) through the use of SS-techniques. The advantages are: SNR improvement due to the processing gain, time diversity gain through coherent correlation of the resolved multipaths, lower electromagnetic profile due to band spreading, better multiaccess efficiency due to capture property of the receiver and ISI suppression by changing chip patterns from symbol to symbol. Both FH and PN techniques have distinct advantages and a combination of FH and PN can result in a waveform with desirable properties of both techniques. Thus, SS-techniques are finding wide applications in futuristic data networks as well.

9.7 SPEECH AND DATA SECURITY [32, 33, 34]

For centuries, the security of speech and data signals has been an important aspect of military and diplomatic communication; and recently with the introduction of computer networks for defence, banking and commercial purposes, the problems of security have attracted the attention of cipher theorists and practical designers. It has been shown experimentally that passive techniques of message interception in radio (including microwaves) links are quite successful. Further, it has been estimated that in typical switched speech/data network, the cost of interception is not very significant, e.g., \$ 100-1000 K for satellite links, \$ 10–100 K for MW links, and \$ 50–250 K for cable trunks. To avoid the unauthorized interception, the code makers have evolved various ciphers (encryption techniques) to protect the data/signal, and at the same time, the code breakers (cryptanalysts) have been trying to break the codes. One of the major objectives of data encryption is to make the cost of codebreaking prohibitive (in terms of computer hardware and processing time).

Historically, simple letter substitution methods (for written text), implemented by small mechanical machines or even by paper and pencil, were used first. But all these codes were broken eventually. During the two world wars, many complex mechanical and electromechanical machines were built using telegraph codes as the source information; but cryptanalysts were able to break these codes too by using sizeable computing power [32]. Shannon's 'theory of secrecy systems' [35] first gave a sound mathematical basis for cryptography, and new algorithms and techniques were then developed using number theory and computer techniques, leading to the present-day highly complex and unbreakable codes. The development of custom-made LSI chips and computer software has helped in designing security equipment of a complexity that could not be thought of in the pre-microelectronics era. At the same time, the cost of cipher systems has been drastically reduced, and the customers, other than governments and the military, are now encouraged to use the secure communication for their banking and commercial purposes. All these are leading to a more widespread use of speech/data encryption techniques, requiring more efforts by the codemakers, and the codebreakers, as well. The traditional cipher systems were mainly used for point-to-point links, but the present-day computer networks, with multi-user accesses, require an efficient key management policy. This has ultimately resulted in the development of a new type of cipher system, and the problems of public-key cryptography appear to be solved, at least with the present status of cryptanalytical capabilities [33].

9.7.1 Speech Security Systems [36]

One of the earliest techniques used for speech security is the speech frequency scrambler in the form of either a speech inverter or a speech-band scrambler, as shown schematically in Fig. 9.29. The inversion of the total speech band,

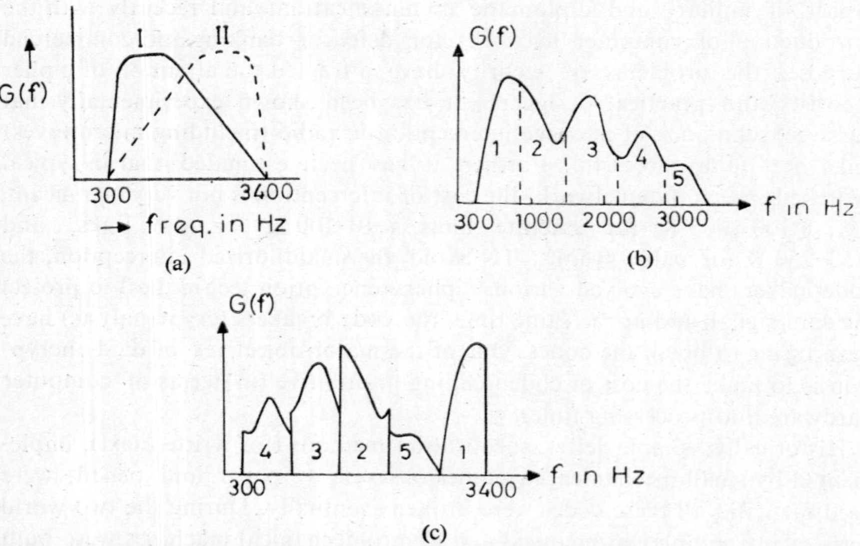

Fig. 9.29 Speech scrambling in frequency domain (a) Total band inversion; (b) Speech band divided into five subbands; (c) Five subbands are reordered as shown

say 300 to 3400 Hz, by a modulator at random intervals, gives marginal security, and can be used only as a privacy system. In the band scrambling technique, the audio spectrum is divided into n sub-bands and the locations of the sub-bands may be rearranged along with optional sub-band inversion (as in speech inverter), according to a random number generator, say through a PN-sequence (PNS) generator. Thus, the number of possible rearrangements is $(n! \times 2^n)$, of which may combinations give considerable residual intelligibility. The reason is that whenever some of the sub-bands remain in their original positions, the combinations give high intelligibility. Since the speech signal has 40% of the energy in the first format (i.e., in the band 300-1000 Hz) comprising of first 2 or 3 sub-bands, it is possible to concentrate on these only for recovering the best part of the speech. Then it is desirable to use only those rearrangements, (reordering of the sub-bands) which give minimum intelligibility. In a practical system, the useful combinations are stored in a ROM, and a PNS is used to select a new combination every few hundred milliseconds or so. The PNS may have a period of more than 10^6 bits so that the cycle of combinations does not repeat itself for days. However, with the techniques of fast spectral analysis and other related software being available, the frequency scramblers are regarded more as privacy systems than as fully secure systems.

The other possible methods for analog speech security systems are amplitude scrambling and time scrambling. In the amplitude scrambling technique, the analog signal is sampled at the Nyquist rate, say, at 6 to 8 kHz rate, and the samples are multiplied by random numbers, say 1-100, generated by a random-number generator of a large period. The resultant scrambled samples are filtered and then transmitted through the speech channel. The original speech is now recovered by multiplying the received sampled-signal by the inverse of the random numbers used in the coder. The system is, however, sensitive to the quality of the speech channel, as the delay characteristic of the channel modifies the received samples, and, thus, causes some distortion of the recovered speech.

The time scrambling technique (also known as time-division multiplexing) is somewhat similar to the time-slot interchange (TSI) technique used in ESS as discussed in Chapter 7. In this method, a frame T of the speech signal is divided into n equal segments of duration τ and stored in an analog delay line, say a TDL. The stored segments are now gated to the output in a random manner under the control of the PNS with a key as shown in Fig. 9.30. Thus, the segments {1, 2, 3, . . ., 9, 10} may be reordered in time

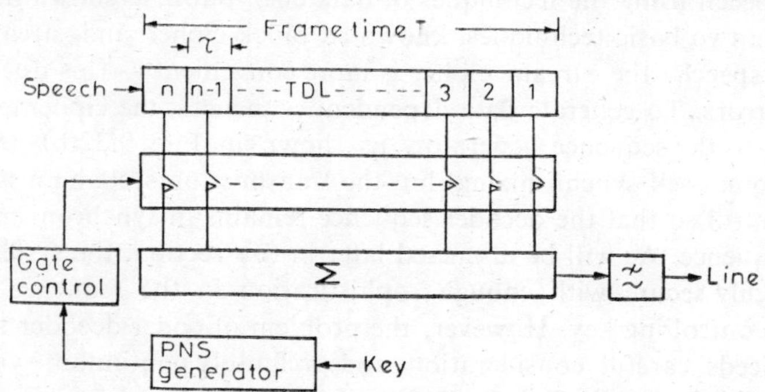

Fig. 9.30 Time-slot scrambling; τ = 20-50 msec; n = 10-20; T = 200-1000 msec.

as {8, 6, 4, 9, 2, 7, 5, 1, 10, 3}. This permutation of the segments may be changed every frame by the PNS. Usually the TDL parameters are:

$$\tau = 20\text{-}50 \text{ msec}, \quad n = 10\text{-}20 \quad \text{and} \quad T = 200\text{-}1000 \text{ msec}.$$

Since the total delay in the coder and decoder is $2T$ sec, it is not possible to increase n and the possible permutations are $n!$. Of these, many combinations give considerable residual intelligibility and to avoid this, a minimum shift for all segments has to be maintained. As in the case of speech-band scramblers, here also it is desirable to pre-select the useful permutations, say, 2^k-patterns, giving minimum intelligibility and store them in a ROM. The gate control signal is then provided by the ROM under the control of the PNS with a large period. It is more convenient to realize the system digitally, as shown in Fig. 9.31, where the digital version of the speech frame is stored in a RAM serially and then read out to the line under the control of the selected stored patterns in the ROM (the selected pattern provides the addresses in the order desired). Since the number of acceptable permutations are limited, the time-scrambling systems provide privacy only and not true security.

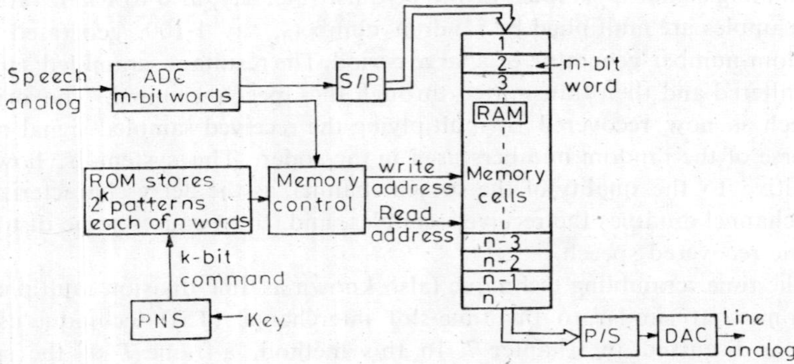

Fig. 9.31 Digital time-slot scrambler

A fully digital cipher system for speech may be realized by encryption of digital speech using the techniques of data encryption, as shown in Fig. 9.32. There are two basic techniques, known as block cipher and stream cipher; but for speech, the stream cipher is more convenient as this does not propagate errors. To generate data-dependent sequences, the cipher text may be fed back to the sequence generator, as shown in Fig. 9.32(b), (the system now becomes self-synchronizing) but the transmission system must now have error control so that the decoder sequence remains in synchronism with the coder sequence. As will be discussed later in this section, the system can be made highly secure with enough sophistication in the sequence generator and the controlling key. However, the problem of coder-decoder synchronization needs careful consideration, and preferably continuous synchronization should be used. To avoid detection of the synchronizing signal at predictable intervals, this should be sent at random intervals along with the interval information to the decoder. All this can be organized by using a PNS to determine both the synchronizing signal and the time at which it is to be sent. The other serious problem is the occurrence of the cipher bits

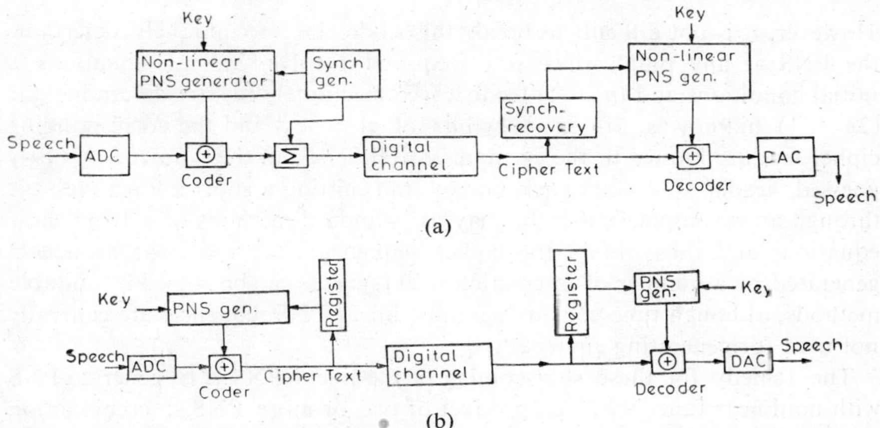

Fig. 9.32 Two forms of digital stream cipher systems (a) a stream cipher system (b) a cipher-feedback system.

only in the channel during the pauses in the speech signal. The avoid detection of the cipher during the pauses, a low-level noise is substituted in place of the signal; thus, the channel always carries the cipher mixed with some signal, either speech or noise. With the acceptance of the Data Encryption Standard (DES, as described later), the coding can now be done more effectively using the cipher code generated by DES [37].

It should be noted that the digital cipher system for speech requires a digital transmission channel capable of transmitting 16-64 Kb/s. The ADM and PCM codes can operate at these bit rates. To be able to further reduce the bandwidth requirements of the channel, one has to use some form of vocoders, preferably LPC codecs, operating at 1.2/2.4/4.8 Kb/s (Ref. Ch. 3). The cipher text can now be transmitted through telephone channels using a suitable modem. An interesting possibility of achieving security in LPC codecs is to encrypt the LPC parameters (including the pitch information) with a number of random sequences and transmit them in a multiplexed form, as in a normal LPC codec [38]. Since the LPC parameters are no longer as redundant as plain speech, the security of the system (assuming that the random sequence is adequately secure) is now much better; however, because of the same reason, the cipher text should be protected against channel errors by using proper error control in the system.

9.7.2 Nonlinear Sequences (33)

It is evident that all cipher systems require very long random number generators to effectively secure the data transmitted in the channel. Pseudo-noise sequences (PNS) are usually generated by linear feedback shift registers and new sequences may be obtained by controlling the feedback connections $\{k_1, k_2, \ldots, k_{n-1}\}$ and the initial conditions $I_v = \{z_1, z_2, \ldots, z_n\}$. Then the output sequence $\{R_n\}$ is a function of the initial vector $\underline{\mathbf{Z}}$ and the feedback connection $\underline{\mathbf{K}}$ (equivalent to the KEY of PNS), giving (Ref. Appendix C),

$$\{R_n\} = f(\underline{\mathbf{Z}}, \underline{\mathbf{K}}) \tag{9.24}$$

However, it is not difficult to break this code, i.e., to uniquely determine the PNS at any point, since it is required to solve $(2n - 1)$ equations (n initial conditions and $(n - 1)$ feedback connections) only to determine the $(2n - 1)$ unknowns. If $(2n - 1)$ bits of clear text and the corresponding cipher text are known to the cryptanalyst (either from the known or easily guessed header of a data packet or by transmitting a short known message through an accomplice), then he may use standard methods of solving these equations and thus obtain the cipher sequence $\{R_n\}$. Even the sequences generated by a number of interacting shift registers can be solved by suitable methods, although time consuming; thus, linear PNS sequences are generally not used for generating cipher texts.

The remedy for these shortcoming of the linear PNS is to generate PNS with nonlinear functions, e.g., product of two or more PNS's, combination of two or more PNS's using *J-K* flip-flops, time scrambling of a PNS etc. To be able to break such nonlinear sequences, it is necessary to solve a number of nonlinear equations, but so far, a general solution to nonlinear equations over $GF(2)$ has not been found. Thus, it seems that the introduction of nonlinearity into the cipher algorithm will considerably increase the complexity of the cryptanalyst's problem. It has been further shown that the 'time complexity function' of an algorithm (equivalent to the maximum time required to break the cipher) may be either of the polynomial form or of the exponential form (i.e., either r^2, r^3, etc. or 2^r, 3^r, etc. where r is the length of the portion of the sequence used for solving the problem). It has been estimated that for a problem with an exponential time function, say 3^r, the time required for solution will be a few years or more with $r \geqslant 30$. The cryptographer, therefore, attempts to design cipher algorithms which have a somewhat exponential time-complexity function.

A few simple techniques for the generation of nonlinear sequences are shown in Figs. 9.33-9.35. In Fig. 9.33(a), the multiple outputs from the same

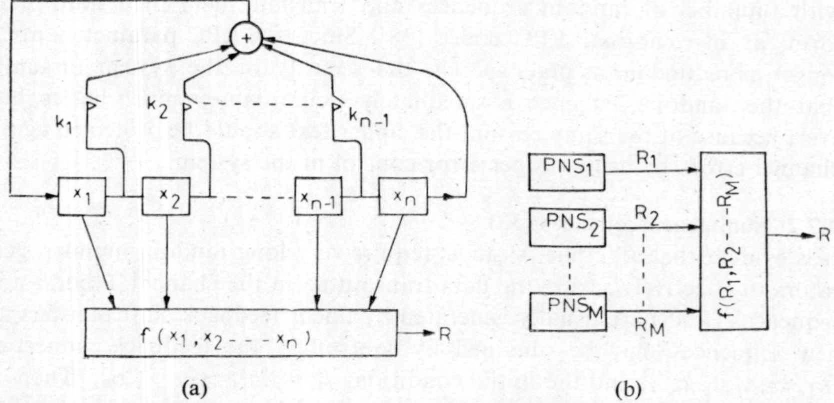

Fig. 9.33 Generation of nonlinear sequences (a) Nonlinear sequences from a single PNS; (b) Nonlinear sequences from multiple PNS's

PNS generator is mixed through some nonlinear function $f(\cdot)$, which may be of the form,

$$f(\cdot) = x_1 + x_2 \ldots + x_1x_2 + x_1x_3 \ldots + x_1x_2x_3 + \ldots \qquad (9.25)$$

Alternatively, the feedback function for the generator may be a nonlinear function of $\{x_1, x_2, \ldots, x_n\}$, and the possible number of such feedback function is $(2)^{2n}$. In Fig. 9.33(b), a possible configuration using multiple linear PNS generators are shown, where the overall period of the output sequence R is also increased. If the number of shift registers in the generators are n_1, $n_2, \ldots, n_M$, and if $(n_1 - n_2) = (n_2 - n_3) \ldots = (n_{M-1} - n_M) = 1$, then the period of the output sequence R is $(2^{n_1} - 1)(2^{n_2} - 1) \ldots (2^{n_M} - 1)$. However, in many cases, the balance property of the output sequence R is not satisfied and this should be carefully checked while using.

An useful configurarion using *J-K* flip-flops are shown in Fig. 9.34, where

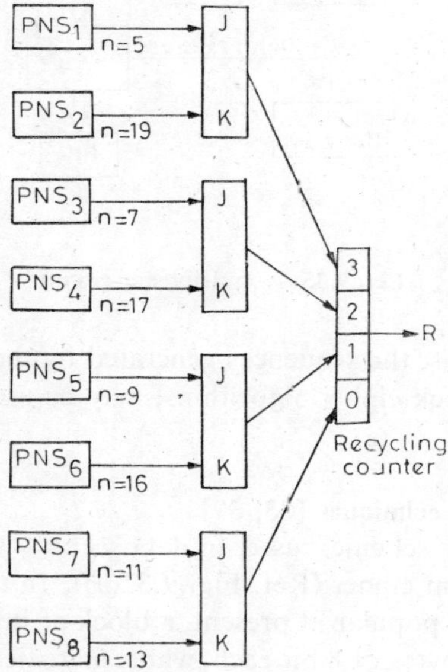

Fig. 9.34 Nonlinear sequence using JK flip-flops

the pairs of PNS's are mixed in the flip-flops and the resulting four outputs are passed through a recycling counter [33]. The final output is now a nonlinear function of the eight PN sequences, and the key for the system consists of I_v of the shift registers, the feedback vectors $\underline{K}$ and the initial states of the flip-flops. Using the number of SR's in each generator as shown in the figure, the period of the resulting sequence R is 1.52×10^{29}, and the balance property is reasonable. The number of possible keys are more than 10^{51}, and an exhaustive search for breaking the code is not possible. It has been estimated that with 167 bits of the R-sequence known, it will take more than 10^9 trials to break the code generated by the system.

The other simple method of obtaining a nonlinear sequence is the Multiplexer (or a time-slot scrambler, somewhat similar to that shown in Fig. 9.31), where the bits/words of PNS_1 is stored in a RAM and these bits/words are read out to form R randomly under the control of a second PNS_2 which generates the Read-addresses, as shown in Fig. 9.35. The multiplexed sequence R has a period less or equal to the product of the periods of PNS_1 and PNS_2, and its time-complexity function is approximately exponential for suitable choice of the parameters of the two primary PNS's. Nonlinear sequences are also generated by blockcipher algorithms which are fed with random blocks of bits from a large-period PNS and a suitable key, as

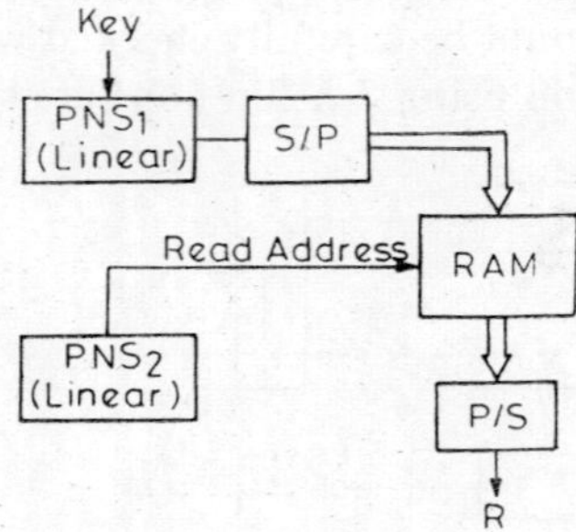

Fig. 9.35 A multiplexer circuit

discussed below. Thus, the sequences generated by the circuits of Figs. 9.34 and 9.35 and by block cipher algorithms, may be used in the stream cipher systems of Fig. 9.32.

9.7.3 Block Cipher Techniques [33, 39]

There are two basic schemes used in data security systems, viz., (a) Block cipher, and (b) Stream cipher (Ref. Fig. 9.32(a)). In the block cipher technique, which is more popular at present, a block of information bits (usually 32 bits consisting 4 bytes of 8 bit each, whose statistics are almost unknown), are transformed into a new block of cipher text by using a suitable cipher algorithm, generally using substitution and permutation as in classical cipher systems, but much more complex in implementation and for code breaking. Consider a simple example of a block cipher using a block of 3 message bits, where any of the eight messages could be substituted by any of the remaining seven messages, as shown in Fig. 9.36. The transformation (or permutation) used in the example is of the form*.

$$[C] = [A][M] \tag{9.26}$$

*A uniquely reversible transformation (or substitution) from any set into itself is called a permutation. In the example of Fig. 9.36, the key of the cipher merely selects one of the the eight permutations possible.

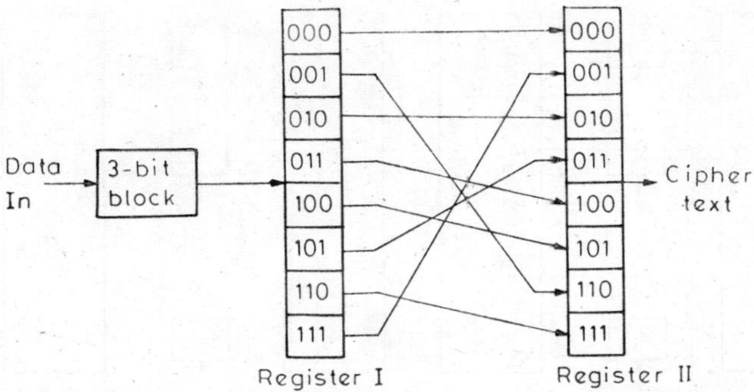

Fig. 9.36 Example of a simple substitution/permutation cipher

where $[A] = \begin{bmatrix} 101 \\ 011 \\ 100 \end{bmatrix}$, $[M]$ is the message vector $(m_1m_2m_3)$, and $[C]$ is the resultant cipher. The transformation could be slightly modified to give an Affine block cipher by having,

$$[C] = [A][M] + [t], \tag{9.27}$$

where $[A]$ is a nonsingular transformation matrix, and $[t]$ is the translation vector, say $\lfloor 111 \rfloor$, for the above example. Since $[A]$ is nonsingular, the decipherment is given by,

$$[M] = A^{-1}[C + t] \tag{9.28}$$

Since the above cipher will be easy to break, the system is made more complex by having a larger alphabet size and repeated use of such transformations with different keys (i.e., repeated superencipherment).

In practical systems, it is required that the length of data blocks should be at least 32 bits, then the permutation box of Fig. 9.36 would require 2^{32} links and a key space capable of producing any of the $(2^{32})!$ permutations. This will be an impossible system, and as such, the straight forward substitution scheme is not practically used. In a practical scheme with large complexity, the data block is subdivided into smaller blocks, say 4 bits each, and each of the subblocks is now processed in a substitution (S) box under the control of a separate key, similar to Fig. 9.36. The outputs of the parallel-operated S-boxes are then permuted bit by bit in a permutation (P) box, as shown in Fig. 9.37, and this process of substitution and premutation (SP) is repeated a number of times (rounds) to strengthen the cipher system. It may be shown that the minimum number of substitution functions generated by the scheme is given by:

$$\text{Minimum } \{[2^4)!]^{rn}, (2^{4r})!, 2^{nk}\}, \tag{9.29}$$

where r = number of S boxes per round, n = total number of rounds and n_k = total number of key bits. It is now evident that it will be almost impossible to find an inverse of the above nonlinear transformation, specially with the keys unknown.

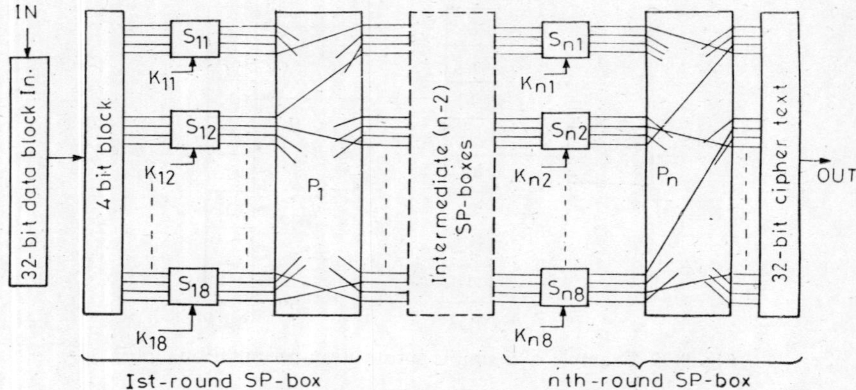

Fig. 9.37 A parallel scheme of substitution-cum-permutation cipher

To avoid realizing (in hardware) a complicated inverse operation for decoding the cipher, Feistal [40] has given a novel scheme based on the above principles, where the cipher text is decoded without using an inverse of the *SP*-transformation. The scheme is shown in Fig. 9.38, where the data

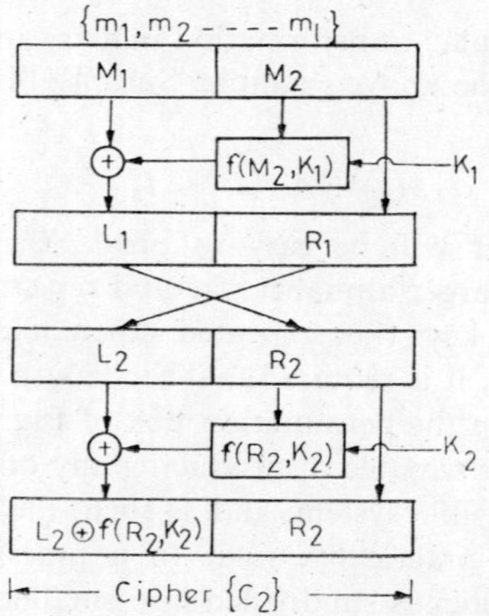

Fig. 9.38 Fiestal's scheme of Block cipher [40]

block $\{m_1, m_2, \ldots, m_l\}$ is divided into two parts M_1 and M_2 with

$$\{M_1\} = \{m_1, m_2, \ldots, m_{l1}\},$$
$$\{M_2\} = \{m_{l_1+1}, m_{l_1+2} \ldots, m_{l_1+l_2}\}, \quad \text{and} \quad l_1 = l_2.$$

Then the one half, say the right-half $\{M_2\}$, is transferred to the next register R_1 unaltered, but at the same $\{M_2\}$ is transformed by a cipher function $f(\cdot)$ in an *SP*-box under the control of a key $\{K_1\} = \{k_1, k_2, \ldots, k_i\}$. The transformed $f(M_2, K_1)$ is now added to the other half of the data and the result $[M_1 \oplus f(M_2 k_1)]$ is transferred to the register L_1. The first-level cipher $\{C_1\} = \{L_1, R_1\}$ is then interchanged blockwise to the registers L_2 and R_2, and R_2 is processed again under the control of the second key K_2. The second-

of possible key combinations are more than 7.10^{16}, and it will be almost impossible to break the code with random-search methods. No algorithm for breaking the code has been found so far. The idea of a well-publicized algorithm in cryptography is indeed revolutionary and for the first time the cryptanalysts, all over the world, have been challenged to break the algorithm. The DES has been implemented in hardware by using either LSI chips and ROM's, or by microprocessors having ROM's or microprogrammed devices using microcode for hardware level control instructions.

The basic block diagram of DES is shown in Fig. 9.40, whose important components are the initial permutation box IP, its inverse IP^{-1}, the cipher function $f(\cdot)$, and the key schedule KS. The IP-box permutes the data bits (i.e., interchanges the time slots) according to Table 9.6, and it is seen that 58th, 50th, 42nd, . . . 7th data bits are transformed as the 1st, 2nd, 3rd, . . ., 64th bits at the IP-box output. Similarly, the last transformation, given by IP^{-1}, is according to Table 9.7. The output of the IP-box is divided into two halves and they are alternately transformed using the cipher function $f(R_{i-1}, K_i) = f(L_i, K_i)$. The technique of calculating the cipher function is shown in Fig. 9.41, where it is seen that the 32-bits of R_{i-1} is converted to a 48-bit block through an extended permutation (EP) as in Table 9.8. These are mixed with the 48 bits of K_i, and the K_i' bits are now transformed into 32 bits of cipher function using a limited substitution process and the final P-box.

Table 9.6 Permutation of IP-box of Fig. 9.40 [33]

For the location of the bits at the IP-box output in the order {1, 2, 3, . . ., 64}, the input bit locations are:

{58, 50, 42, 34, 26, 18, 10, 2, 60, 52, 44, 36, 28, 20, 12, 4, 62, 54, 46, 38, 30, 22, 14, 6, 64, 56, 48, 40, 32, 24, 16, 8, 57, 49, 41, 33, 25, 17, 9, 1, 59, 51, 43, 35, 27, 19, 11, 3, 61, 53, 45, 37, 29, 21, 13, 5, 63, 55, 47, 39, 31, 23, 15, 7}.

Table 9.7 [33] Permutation of IP^{-1}-box of Fig. 9.40

For the location of the bits at the output of [IP^{-1}] in the order {1, 2, 3, . . ., 64}, the input bit locations are (Ref. Fig. 9.40):

{40, 8, 48, 16, 56, 24, 64, 32, 39, 7, 47, 15, 55, 23, 63, 31, 38, 6, 46, 14, 54, 22, 62, 30, 37, 5, 45, 13, 53, 21, 61, 29, 36, 4, 44, 12, 52, 20, 60, 28, 35, 3, 43, 11, 51, 19, 59, 27, 34, 2, 42, 10, 50, 18, 58, 26, 33, 1, 41, 9, 49, 17, 57, 25}

Table 9.8 [33] Extended Permutation of R_{i-1}-bits of Fig. 9.41

For the 48-bit output of the EP-box in the order {1, 2, 3, . . ., 48}, the input bit locations are (some of the locations are repeated to fill up the output locations):

{32, 1, 2, 3, 4, 5, 4, 5, 6, 7, 8, 9, 10, 11, 12, 13, 12, 13, 14, 15, 16, 17, 16, 17, 18, 19, 20, 21, 22, 23, 24. 25, 24, 25, 26, 27, 28, 29, 28, 29, 30, 31, 32, 1}.

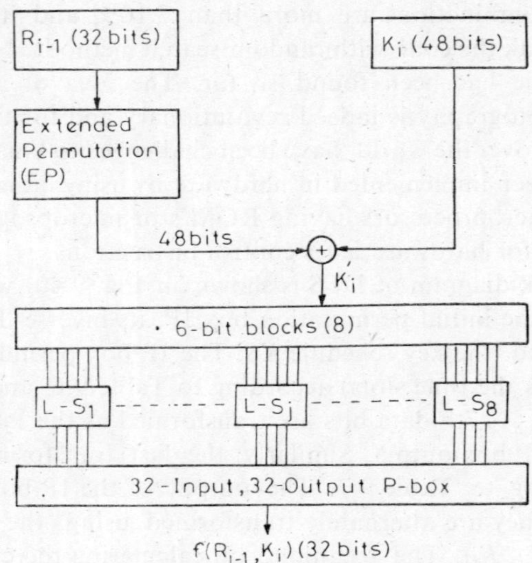

Fig. 9.41 Calculation of the cipher function $f(R_{i-1}, K_i)$

In the boxes $(L-S_j)$, six input bits are converted to 4-bit numbers using a random table and the bits are multiplexed at the input of the P-box. The permutation of the P-box is given in Table 9.9. Thus the cipher function for each round of SP-transformation is generated.

The key schedule KS for the generation of the keys $\{K_1, K_2, \ldots, K_{16}\}$ from the master key K_0 is shown in Fig. 9.42; where 64 random bits of K_0 are permuted in the selective P-box I, according to Table 9.10. The output is divided into two halves K'_{01} and K'_{02} and each half is placed in a register of 28 bits. These are now shifted in the left by one or two bits according to a predetermined table. The shifted outputs are then combined and transformed

Table 9.9 [33] Permutation of the *P*-box of Fig. 9.41

For the bit location at the output in the order {1, 2, 3, . . ., 32}, the input locations are:

{16, 7, 20, 21, 29, 12, 28, 17, 1, 15, 23, 26, 5, 18, 31, 10, 2, 8, 24, 14, 32, 27, 3, 9, 19, 13, 30, 6, 22, 11, 4,25}.

Table 9.10 Permutation of the Selective *P*-box I

The input locations of the bits to give the output bits in the order {1, 2, 3, . . ., 28}, and {29, 30, . . ., 56} are:

{57, 49, 41, 33, 25, 17, 9, 1, 58, 50, 42, 34, 26, 18, 10, 2, 59, 51, 43, 35, 27, 19, 11, 3, 60, 52, 44, 36}

and, {63, 55, 47, 39, 31, 23, 15, 7, 62, 54, 46, 38, 30, 22, 14, 6, 61, 53, 45, 37, 29, 21, 13, 5, 28, 20, 12, 4}

The first half is stored in the register K'_{01} and the second half in the register K'_{02}

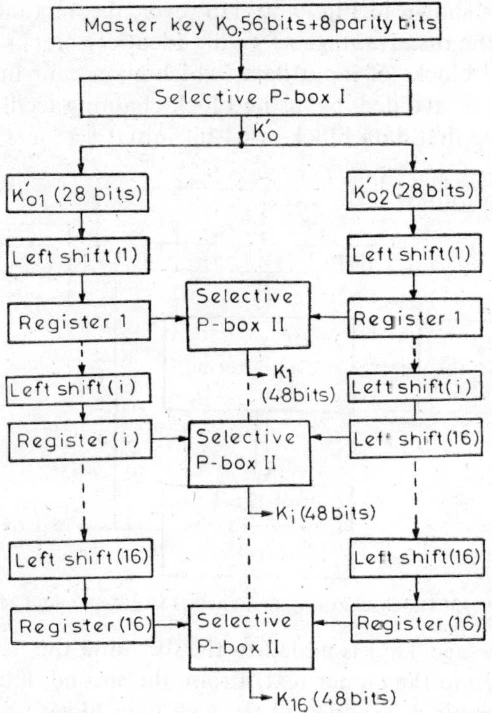

Fig. 9.42 Generation of keys $K_1, \ldots, K_{16}$ from the master key K_0

through another selective *P*-box II according to Table 9.11. It is seen that the selective *P*-box II is the same for all K_i''s, but the left shift values are different (or same in some cases) for different K_i. The DES, so designed, can be used for both block cipher and stream cipher systems, as discussed in the next section.

Table 9.11 Permutation of Selective *P*-box II (56-bit input and 48-bit output)

The input bit locations for the output bits to be in the order {1, 2, 3, . . ., 48}.

{14, 17, 11, 24, 1, 5, 3, 28, 15, 6, 21, 10, 23, 19, 12, 4, 26, 8, 16, 7, 27, 20, 13, 2, 41, 52, 31, 37, 47, 55, 30, 40, 51, 45, 33, 48, 44, 49, 39, 56, 34, 53, 46, 42, 50, 36, 29, 32}.

9.7.4 Practical Cipher Systems [33]

Two basic techniques for designing data security systems are: (a) block cipher (as in Fig. 9.38) and (b) stream cipher (as in Fig. 9.32). The DES transform is one of the elegant techniques for designing block cipher systems and the same may be used for designing stream cipher systems as well. The straightforward application of DES (or similar block cipher algorithm) as a block cipher for an *l*-bit data block with a *k*-bit key is known as the code-book method, as shown in Fig. 9.40. However, the method, although data dependent, has the disadvantage of giving identical blocks of output cipher text for identical blocks of input data (which may occur in highly formatted data), and this is avoided by using block-chaining feedback, as shown in Fig. 9.43. For the first data block, a 64-bit initial vector I_v (which may be

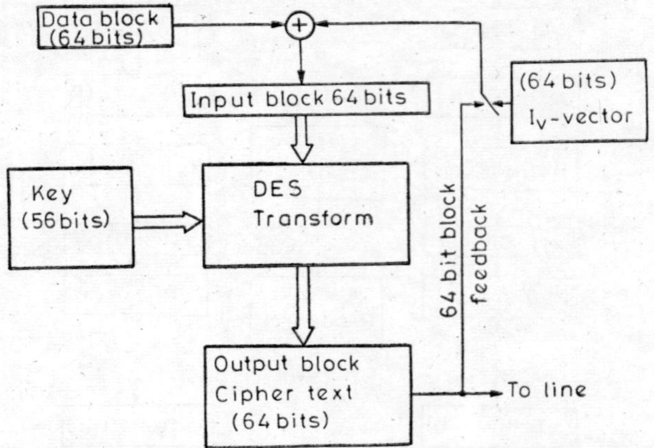

Fig. 9.43 Block-chaining feedback (Data-dependent cipher)

used as the message key) is added to the data and the new vector is fed to the DES to generate the cipher text. From the second data block onwards, the earlier cipher block is added to the next data block to form the input to the DES; and this produces different cipher texts for the identical input blocks, assuming that the preceding blocks are different. In the receiver, the first cipher block is decoded in a complementary way and the result is added to the same initial vector I_v to give the clear text. From then onwards, each cipher block is deciphered and then added to the previous cipher block to obtain the clear text. Block-chaining is particularly useful as a protection against deletions or insertions of spurious data in the original cipher text, but the system is subject to error propagation in the decoder.

To organize a stream cipher system using a block cipher algorithm, it is necessary to feed the DES with blocks of PNS bits in lieu of data blocks. Alternatively a block of I_v-bits may be used to generate the first cipher block at the output, and then onwards, the output of the DES is fed back to its input to generate a continuous stream of random bits. This is now added to the clear data stream to finally form the cipher text, as shown in Fig. 9.44(a). It is possible to add data to the random sequence bit-by-bit or in blocks of j-bits, where $1 \leqslant j \leqslant 64$. One of the advantages of the system is that there is no error propagation in the decoder, but the system requires exact synchronism between the two key streams at the coder and decoder. To make the stream cipher data dependent and self-synchronizing, the cipher-text feedback may be used as shown in Fig. 9.44(b), where the number of bits j in the feedback block is $1 \leqslant j \leqslant 64$. NBS cryptolaboratory has implemented a scheme with $j = 8$.

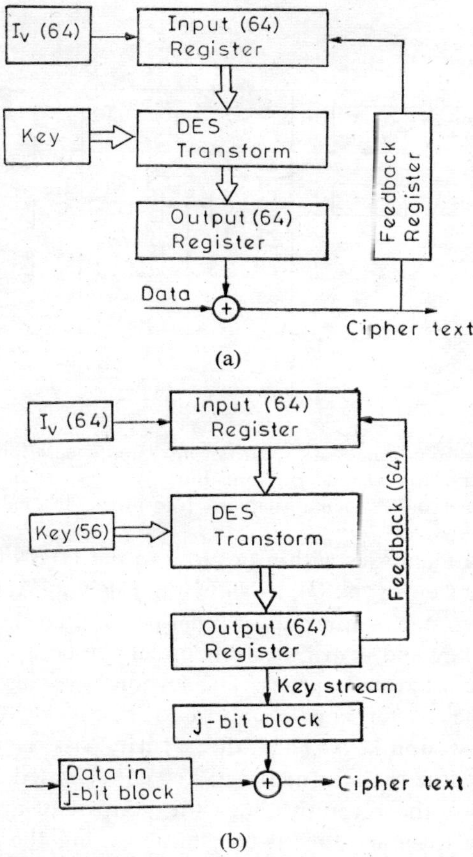

Fig. 9.44 Stream cipher system (a) Key-autokey mode; (b) With feedback

Key Management [43]

The key management scheme for data networks is based on a hierarchy of cryptographic keys, known as (a) Master (base) keys, (b) sub-master (user) keys*, and (c) Session (message) keys. To make the job of cryptoanalysis difficult, it is necessary that the cryptographic key should be somewhat similar to that of the one-time-pad system. To make the electronic cipher system behave in the similar way, the session/message keys should be generated and distributed to the receivers, each time a message is required to be sent. This is done under the control of the master key, but it is now required that a group of users should have the same master key. This creates a security risk, and to avoid this, the individual master keys for the users are held in

*In cipher systems, where algorithms other than that of DES are used, the user key (also called customer option/algorithm changing key) performs the extension of the key size of the system without increasing the work involved in changing the keys. This gives the user some confidence that he is using an algorithm unique to him. However, in systems using DES, this is not possible.

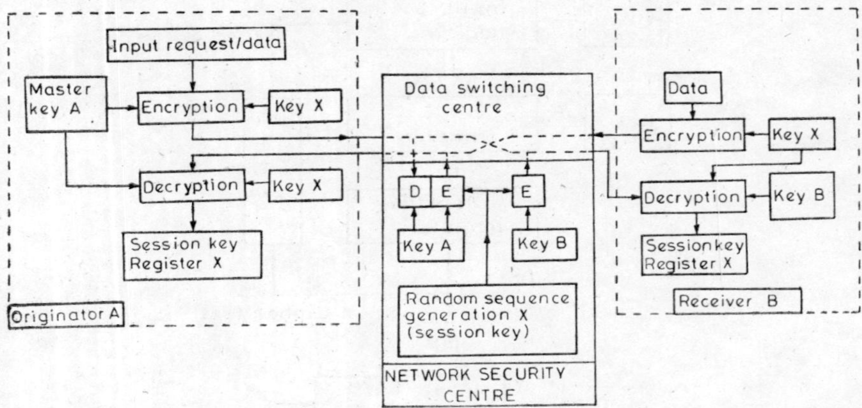

Fig. 9.45 Organisation of network security centre and generation of session keys *X*; *E*: encryption; *D*: decryption. Initially keys *A* and *B* are used at stations *A* and *B*; then all data are transmitted through key *X*.

physically secured memories within or close to the DES of the users and the Network Security Centre (NSC), as shown in Fig. 9.45. When a user requests for a connection to one or a group of other users, then the NSC generates a session/message key and distributes this under the control of the individual master key to the interested parties. The session key is used for data transmission while the master key is only used as the key-encrypting key for distributing user/session keys. Thus, the security risk is minimized. For a two-user network, the session (message) key is generated by the originator and transmitted to the receiver under the control of the common master key. Then onwards, the message is transmitted using the session key only, as shown in Fig. 9.46. In general, the session keys have a lifetime equal to

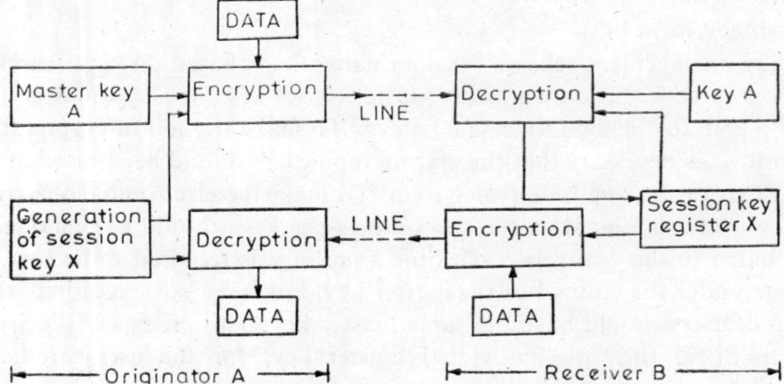

Fig. 9.46 Two-user cipher network

the length of the session, and master keys have a fixed, finite lifetime unless they have been compromised. The submaster/user keys have a lifetime selected by the user, and may be used for identification or for encrypting other keys/information.

To ensure a high level of security, it is also necessary to have full authentication of the users or nodes (in a network) and also of the data transmitted. The simple method for authentication of the users has been to transmit passwords and transaction numbers between the terminals, but an interceptor can decode these informations if used repeatedly. A better method is to use the socalled 'full handshake authentication', as described below [33]. Once the session key K_{AB} between the users A and B has been set up, the originator A sends as encrypted message m_{AB} to B, where m_{AB} includes A's identification I_A plus a unique identifier I_{AU}. The receiver B decodes the message using K_{AB}, identifies A and retransmits to A an encrypted message consisting of $f(I_{AU})$ plus a unique identifier I_{BU}. The user A then decodes the message from B and checks $f(I_{AU})$ (which is known to him) to identify B, since only B could encrypt $f(I_{AU})$ using K_{AB}. In the third stage of authentication, A returns a message including $f(I_{BU})$ and this, when decoded and checked by B, confirms that A and B are the actual communicators in the link. For better security, the unique identifiers I_{AU} and I_{BU} should be used as randomly as possible and should not be repeated between the same pair of users for a long time.

For the authentication of data (i.e., to protect data from insertion or deletion of bits by an unauthorized (active interceptor), the cyclic redundancy check (CRC) as used for error detection in data networks, has been suggested and used. But CRC does not guarantee that the data has not been tampered in some form, even though the check sum has been ciphered. A better method is to generate a cryptographic check digital field through a cryptographic algorithm and transmit this after encryption. Referring to Fig. 9.43, it is seen that, due to chain feedback, the cipher text is a function of the present input data block and the earlier cipher blocks. At the end of the message, the final DES output is a residue which is a cryptographic function of the entire message and all or part of this may be used as the check digits for the message. Any tampering with the transmitted cipher text will now result into a chain of garbles at the receiver output.

9.7.5 Public Key Cryptography [33, 44]

We discussed, in the last section, the problems of key management and a hierarchy of properly secured keys was suggested. In contrast with this, a radically different approach has been suggested for the public-key cryptosystems, where the enciphering key is made public without compromising the deciphering key. While the two keys are no doubt related and perform inverse operations, it is so arranged that the deciphering key cannot be computed easily without the knowledge of the particular algorithm (equivalent to the secret key) used in the formulation of the key-pair. At the best, the cryptanalyst would require millions of years to solve the problem, if the key consists of 200 digits or more, and the problem has an exponential time-complexity function. Since 1976, various proposals have been made for public-key and related cryptosystems. Two of the well-known techniques are

(a) RSA scheme proposed by Rivest *et al.* [45], and (b) the Merkle-Hellman scheme [46] based on the 'trap-door Knapsack' problem. These are briefly discussed now.

The RSA scheme is based on the fact that if two large prime numbers p and q, are multiplied to give $n = pq$, it is very difficult to factor n and obtain p and q independently (e.g., if $p = 15107$, $q = 34351$, $n = 518940557$, then factorizing n is a very time-consuming task). The value of n may be made public as part of the enciphering key without compromising the factor p, q which effectively constitute the deciphering key. Usually the numbers p and q are 100 digits (or more) long; the multiplication can be done easily, but the factorization is extremely difficult. The algorithm for choosing the enciphering-deciphering key pair is that:

Choose two large prime numbers p and q, such that $n = pq$; and Euler's function $\phi(n) = (p - 1)(q - 1)$. Then for any integers α, $0 < \alpha \leqslant (n - 1)$, and β,

$$\alpha^{[\beta\phi(n)+1]} = \alpha \bmod n \tag{9.32)*$$

Choose a random number E (enciphering exponent), $3 < E \leqslant [\phi(n) - 1]$, such that E has no common factors with $\phi(n)$. The deciphering exponent D is now given by:

$$D = E^{-1} \operatorname{Mod} \phi(n)$$

and (9.33)

$$ED = \beta\phi(n) + 1$$

The formation (E, n) is made public as the enciphering key (in a directory of users' keys) and is used to code the plain-text messages into the cipher text. The steps of encipherment are:

Represent the message as a sequence of integers $\{\alpha_1, \alpha_2 \ldots \alpha_i\}$, where all α_i's are in the range $0 < \alpha_i \leqslant n - 1$. For an integer α_i (equivalent message block), the corresponding cipher text is:

$$C_i = \alpha_i^E \operatorname{Mod} n \tag{9.34}$$

The information (D, n), where D is the secret key of the receiver, is now used as the deciphering key giving,

$$\alpha_i = C_i^D \operatorname{Mod} n \tag{9.35}$$

Equations (9.34) and (9.35) are inverse transformations, since from Eq. (9.33),

$$C_i^D = \alpha_i^{ED} = \alpha_i^{[\beta\phi(n)+1]} = \alpha_i \operatorname{Mod} n \tag{9.36}$$

---

*$\phi(n)$ is the number of integers between 1 and n which have no common factors with n, i.e., $\phi(n)$ excludes all pth numbers and all qth numbers between 1 and n. Equation (9.32) is illustrated by taking $p \cdot q = 3 \times 7 = 21$, and $\phi(n) = 2 \times 6 = 12$. If $\alpha = 4$, $\beta = 1$, then $\alpha^{13} = 4^{13} = 67108864 = 4 \operatorname{Mod} 21$. It should be noted that computing $\phi(n)$ is easy if p and q are known, but computing $\phi(n)$ directly from n is equally difficult as factorizing n.

This scheme is illustrated by a simple example using $p = 7$, $q = 11$, $n=77$; $\phi(n) = 6\times 10 = 60$. Then with $E = 13$, $D = 37$ (since, $ED = 13\times 37 = 481 = 1$ Mod 60). Assume that the message integer $\alpha_i = 20$, then the cipher text* is:

$$C_i = 20^{13} \text{ Mod } 77 = 69$$

and its inverse,
$$\alpha_i = 69^{37} \text{ Mod } 77$$
$$= 20 \text{ (as required)}$$

To show how difficult it is to factorize a large number n, it has been estimated that for n expressed as a b-bit number ($n = 2^b$), the best factoring algorithm requires approximately N number of machine cycles (in a large computer), where

$$N = \exp \{[\ln(n)\cdot \ln(\ln(n))]^{1/2}\} \tag{9.37}$$

Using this equation the number of operations and the time to factor a b-bit number are shown in Table 9.12, (assuming 1 μs instruction time) [44].

Table 9.12

b (bits)	100	200	300	500	1000
Number of operations	2.8×10^7	2.3×10^{11}	2.9×10^{14}	3.6×10^{19}	1.8×10^{29}
Time	30 sec.	3 days	9 year	10^6 year	6×10^{15} yr.

It is thus seen that the problem of factorizing n has an exponential time-complexity function.

Markle and Hellman's scheme makes use of the well-known Knapsack problem**. The Knapsack problem is a combinatorial problem, where a vector of n integers, $\underline{A}$, is given and an integer S which is the sum of a subset

*It is to be noted that both enciphering and deciphering involve an exponentiation in modular arithmetic and this can be done with atmost 2 ($\log_2 n$) multiplication Mod n. To evaluate $Y = a^X$, the exponent X is represented in binary form, the base a is raised to the 1st, 2nd, 4th, 8th etc. powers (each step involving only one squaring or mulipli-cation), and the appropriate sets of these are multiplied together to form Y. As for example,

$$C = 20^{13} \text{ Mod } 77 = 20^8\times 20^4\times 20^1 \text{ Mod } 77 = 71\times 71\times 71\times 20 \text{ Mod } 77$$
$$= 36\times 71\times 20 \text{ Mod } 77$$
$$= 69 \text{ Mod } 77.$$

**The classical Knapsack problem is that the Knapsack is filled with a subset $\{a_i\}$ of items $\{a_1, a_2, \ldots a_m\}$, whose weights are given as $\{w_1, w_2, \ldots w_m\}$. Given that the weight of the Knapsack is W, is it possible to calculate which of the items are included in the Knapsack? For small values of n, 2^n equations for 2^n possible subsets may be solved. But it becomes computationally infeasible when $m > 100$. However, if the set of weights $\{w_1, w_2, \ldots w_m\}$ have nice properties, then the person with the special trapdoor' information can quickly solve the problem even with $m > 100$.

of the $\{a_i\}$ is specified. The subset can be denoted by the inner (dot) product of a binary vector $\underline{X}$ and the vector $\underline{A}$, giving,

$$S = \underline{A} * \underline{X}$$

or $$S = (a_1 x_1 + a_2 x_2 + \ldots + a_i x_i \ldots a_m x_m) \tag{9.38}$$

where $x_i = 0/1$. The problem is to calculate the vector $\underline{X}$. The general solution of the problem is very difficult; however, if $\{a_i\}$ have some specific properties, then the problem can be solved. One such property is that:

$$a_{k+1} > \sum_{i=1}^{k} a_i \tag{9.39}$$

i.e., each component of $\underline{A}$ is larger than the sum of the preceding components. For example, consider the Knapsack vector $\underline{A}' = \{2, 5, 9, 18, 37\}$, where the inequality (9.39) is satisfied. If the sum of the subset of $\{a_i'\}$, $S' = 48$, then, it is easy to see that $x_5 = 1$ (since sum $(a_1 + a_2 + a_3 + a_4) = 34$ only). Subtracting a_5' from S', the solution is continued recursively and we get, $x_4 = 0$, $x_3 = 1$, $x_2 = 0$, $x_1 = 1$. Thus the binary vector

$$\underline{X} = \{10101\}.$$

The problem has been trivial so far. Consider now the vector

$$\underline{A} = \{1273, 1823, 1650, 581, 439\}$$

with $$S = 3362 \tag{9.40}$$

Since the vector $\underline{A}$ does not have the property (9.39), there is no simple solution for $\underline{X}$. However, if we multiply $\underline{A}$ by a quantity D Mod n, it is possible that we may find a new $\underline{A}'$ satisfying the property (9.39). In this particular example, assume $D = 1053$ and $n = 2719$. Then multiplying each component of $\underline{A}$ by 1053 Mod 2719, one obtains the vector

$$\underline{A}' = \{2, 5, 9, 18, 37\}$$

and $$S' = 48 \tag{9.41}$$

which are the starting Knapsack vector of this problem. The solution $\underline{X} = \{10101\}$, obtained earlier, also satisfies the larger problem of the vector $\underline{A}$ and the sum S. The procedure for obtaining the cipher text is as follows:

The vector $\underline{A} = \{a_1, a_2, \ldots, a_i \ldots, a_m\}$, similar to Eq. (9.40), is the public key. The message is expressed in blocks of m binary digits equivalent to $\underline{X}$. Then the ciphered text is the sum $S = \underline{A} * \underline{X}$, and the secured message is transmitted in the form of digits given by the successive sums. The transformation variables $\{D, n\}$ are the secret trap-door information, used by the receiver to reconstruct the sum $S' = S(D \text{ Mod } n)$, and the vector

$$\underline{A}' = \underline{A} \cdot (D \text{ Mod } n),$$

which has the special property of Eq. (9.39). The solution of $\underline{X}$ (i.e., the binary message bits) is now obtained from,

$$S' = \underline{A}' * \underline{X} \tag{9.42}$$

The two solutions of $\underline{X}$ given by Eqs. (9.38) and (9.42) are the same, provided the modulus n is greater than the sum of the $\{a_i'\}$. It is observed that without the knowledge of D, the solution of X with a large value of m is almost impossible.

The best known method of cryptanalyzing the trap-door Knapsack system requires a number of operations of the order of $2^{m/2}$, where m is the size of the vector $\underline{A}$. The computational time required to solve $\underline{X}$, without the knowledge of the secret key, is approximately : 4×10^{16} yrs for $m = 200$, 6×10^{61} yrs for $m = 500$. Thus, the system provides a high security level. Both enciphering and deciphering require less computation than that in the RSA system, and the code-breaking time is also much more than that in the RSA system.

There have been a few more proposals for public-key cryptosystems [33, 34]. One interesting scheme [47], based on the discrete exponentials and logarithm functions, uses the relations,

$$Y = a^X \text{ Mod } q$$

and

$$X = \log_a Y, \quad \text{over GF}(q) \tag{9.43}$$

where, q is a prime number, a is a primitive element and $1 \leqslant X,\ Y \leqslant (q-1)$. Each user chooses a random element (secret key) and makes the associated Y public. When the users A and B wish to establish a key for secure communication, they use,

$$K_{AB} = a^{X_A X_B} = (Y_A)^{X_B} = (Y_B)^{X_A} \tag{9.44}$$

Thus the secure key K_{AB} is easily computed from the public keys Y_A and Y_B and the user's secret keys X_A and X_B. While the discrete exponentials are easily evaluated, no general fast algorithms are known for evaluating the discrete logarithm function of Eq. (9.43). Thus, the cryptanalyst's task can be made computationally infeasible for large values of q. It has been estimated that for computation of a logarithm in modular arithmetic, it is required to carry out $2^{b/2}$ or more operations, where b is the number of bits in the binary representation of the modulus q. The computational time required for the above solution, without the knowledge of X, is then the same as that required in the Knapsack system, thus providing a high security level. The system, however, is not a true public-key cryptosystem, but is a public-key distribution system, where a key is exchanged securely over an insecure channel. Then the key is used in a conventional cryptosystem, as discussed in Sec. 9.7.4.

Authentication

User authentication (digital signature) can be conveniently provided in public-key cryptosystems by using the same public and secret keys. As an example, in the RSA system, the messages are encoded by the key E and decoded by the key D. Suppose now that user A interchanges the order of operations and uses his secret key D_A first to sign the message M. Then the recepient B uses the public key E_A of the sender to validate the signature. The successive operations are:

A signs the message M by coding M with D_A and transmits to B the cipher text,

$$C = D_A(M) \tag{9.45}$$

User B decodes the message by using A's public key E_A and obtains,

$$E_A(C) = E_A[D_A(M)] = M, \tag{9.46}$$

since from Eq. (9.36) E_A and D_A are inverse processes. User B keeps C as a proof that the message M was received from A, since none else could have generated C (as D_A is only known to A). It is also not possible for B to change the message, as any change in the bit stream will result in a garble message.

Although it is generally understood that the public keys will be stored in a directory, a higher level of security may be achieved by following the procedures (for distribution of public keys) given now [48]:

(a) At the beginning of each session, selected users or nodes in a network generate mated pairs of keys (K_E, K_D).
(b) Each selected node sends (over insecure lines) K_E to its neighbours.
(c) Each neighbour responds by randomly generating a 'session key' and sends it to the selected node, encrypted under K_E. The selected node decrypts it by using K_D.
(d) Once the 'session key' is established, all K_E and K_D are destroyed.
(e) Messages are encrypted under the current 'session key' and sent from node to node.
(f) At the end of the session, all session keys are destroyed.

In this system, there are no long-term keys, stored or otherwise, but only temporary public and session keys, giving additional security.

Ref. [48] has indicated interesting applications, e.g., Electronic fund transfer system, Securing automatic teller machines, Read only communications, where the public key and conventional techniques have been usefully combined to evolve much better secure communication. It is no doubt that cryptography is playing an important role in protecting vital information, specially in the expanding computer networks. Many commercial systems have been developed incorporating public key and digital signatures. However, a major problem of certification of the systems is yet to be solved. Hopefully, the cryptanalysts and cryptographers will join hands in evolving the future standards and specifications of cipher systems.

9.8 SUMMARY

It has been amply shown that for the coexistence of different modes of communication, radars and other electromagnetic emitters and for their survival in an interactive environment, the only solution is to go wideband and mostly digital, using orthogonal codes, FEC and SS techniques. This is also the only solution to approach Shannon's theoretical limit of the Channel Capacity. The classical solution of narrowband signalling and frequency management is no longer valid with the tremendous increase in the use of the available RF spectrum. Thus the efficient interference blankers and pulse compression techniques are necessary for multiple radar operation; at the same time, antijamming capability can be provided only through SS techniques. In VHF, UHF, Tropo and mobile communication, the problem of EMC, multipath and time-varying interference can also be solved through FEC and SS techniques.

For the remote area and emergency communication through satellite, it is now possible to use cheap portable earth station of $G/T \simeq 0$ dB, to communicate with the central station by again using FEC and SS techniques. SSMA gives a new dimension to mobile communication, where it is seen that the proposed multi-address system will be much more flexible and efficient than the classical FM systems using frequency and power management techniques (also see Ch. 5). In the ultimate limit, the personnel radio may become a reality, if millions of SS codes are generated and decoded. The progress in the design of MF hardware using LSI and SAW devices seems to be promising for such applications.

In the area of speech and data security, considerable progress has been made to safeguard the information in both speech and data channels through the use of sophisticated encryption techniques. Using number theory, coding theory and computer techniques, new algorithms have been developed, e.g., DES in USA, and the problem of key management has been solved by evolving the public-key cryptosystems. However, this area remains as a challenge to the innovative capabilities of both the cryptographers and the cryptanalysts.

REFERENCES

1. Cook, C.E. and Bernfeld, M., *Radar Signals*, Academic Press, N.Y., 1967.
2. Das, J., Mullick, S.K. and Chatterjee, P.K., *Principles of Digital Communication*, pp 22, 263, 363, 369, Wiley Eastern Ltd., New Delhi, 1986.
3. Das, J., 'Signal processing for Electromagnetic Interference and Compatibility—a Review and a Case for SS techniques', *JIETE*, vol. 26, pp 41-56, 1980.
4. G.L. Turin, 'An Introduction to matched filter', *Trans. IRE*, vol. IT-6, p. 311, 1960.
5. L.B. Milstein and P.K. Das, 'Spread spectrum receiver using surface acoustic wave technology', *Trans. IEEE*, vol. Com-25, p. 841, 1977.
6. Dutta Roy, S.C. *et al.*, 'Signal processing applications of charged coupled devices', *JIETE*, vol. 24, p. 400, 1978.
7. Mayer, F., 'R.F.I. suppression components: State of the Art: New development', *Trans. IEEE*, vol. EMC-18, p. 58, 1976.
8. Aasen, M.D., 'Improvement of EMC by applying ambiguity and environmental diagrams to the design of radar waveforms', *Trans. IEEE*, vol. EMC-18, p. 74. 1976.
9. Huffman, D.L., 'Interference blanker for dissimilar modified Huffman sequences', *Trans. IEEE*, vol. EMC-18, p. 118, 1976.
10. Huffman, D.A., 'The generation of impulse—equivalent pulse trans,' *Trans. IRE*, vol. IT-8, p. 810, 1962.

11. Cook, C.E., 'Linear FM signal formats for Beacon and communication systems', *Trans. IEEE*, vol. AES-10, p. 471, 1974.
12. Frazier, R.A., 'Compatibility and the frequency selection problem,' *Trans. IEEE*, vol. EMC-17, p. 248, 1975.
13. Ristenbatt, M.P., 'Alternatives in digital communications', *Proc. IEEE*, vol. 61, pp. 703-721, 1973.
14. Das, J., 'Advances in digital communication,' *IETE, St. Journal* vol. 19, p. 109, 151, 1978.
15. Special Issue on 'Error Correcting Codes,' *IEEE Trans. Com. Tech.*, vol. COM-19, No. 5, Oct. 1971.
16. Brayer. K., 'Error correction code performance on HF troposcatter and satellite channels', *IEEE Trans.* Com-19, p. 781, 1971.
17. Cacciamani, E.R., 'The SPADE system as applied to data communieation and small earth station operation,' *Comsat Tech. Review*, vol. I, p. 171, 1971.
18. Das. J., 'Review of the RADA technique,' *IETE St. Journal*, vol. 19, p. 15, 1978.
19. Cahn, C.R. *et al.*, AGARD Lecture Series No. 55, *Spread Spectrum Communication*, 1973.
20. Special Issue on 'Spread Spectrum Communication,' *IEEE Trans.* vol. Com-30, May 1982.
21. Bhargava, V.K., 'Error control coding for SS-systems', *Proc. International Symp. on MW and Communications*, I.I.T., Kharagpur, Dec. 1981.
22. Ketchum, J.W. and Proakis, J.G., 'Adaptive algorithms for estimating and suppressing narrowband interference in PN-SS systems,' *IEEE Trans.* vol. Com-30, p. 913, 1982.
23. Li, L.M. and Milstein, L.B., 'Rejection of narrowband interference in PN-SS systems using transversal filters', *IEEE Trans.*, vol. Com. 30, p. 925, 1982.
24. Huth, G.K., 'Optimization of coded spread spectrum system performance,' *Trans. IEEE*, vol. Com-25, p. 763, 1977.
25. Shaft, P.D., 'Low-rate convolutional code applications in spread-spectrum communications,' *Trans. IEEE*, vol. Com-25, p. 815, 1977.
26. Purin, G.L., 'Introduction to SS-antimultipath techniques and their application to urban digital radio', *Proc. IEEE*, vol. 68, p. 328, 1980.
27. Price, R. and Green, P.E. (Jr), 'A communication technique for multipath channels,' *Proc. IRE*, vol. 46, p. 555, 1958.
28. Kahn, R.E., *et al.*, 'Advances in packet radio technology,' *Proc. IEEE*, vol. 66, p. 1468, 1978.
29. Milstein, L.B. and Schilling, D.L., 'The effect of frequency selective fading on a non-coherent FH-FSK system with partial-band tone jamming,' *IEEE Trans.*, vol. Com-30, p. 904, 1982.
30. Sussman, S.M., 'A matched filter communication system for multipath channels,' *IRE, Trans.* IT-6, p. 367, 1960.
31. Raut, R.N., 'SS-signal communication over a fading channel—a simulation study', *JIETE*, vol. 26, p. 57, 1980.
32. Kahn, David., *The codebreakers; The story of secret writing*, Macmillan, N.Y., 1967.
33. Beker, H. and Piper, F., *Cipher Systems—the Protection of Communications*, John Wiley, N.Y., 1982.
34. Denning, D.E.R., *Cryptography and data security*, Addisson-Wesley, Mass., 1982.
35. Shannon, C.E., 'Communication theory of secrecy systems,' *BSTJ*, vol. 30, pp 50-64, 1949.
36. Brunner, E R., *Speech security systems today and tomorrow*, Gretag Ltd., 1980.
37. Orceyre, M.J. and Heller, R.M., 'An approach to secure voice communication based on the data encryption standard,' *IEEE Comm. Soc. Mag.*, vol. 16, No. 6, pp. 41-50, 1978.

38. Jayant, N.S., *et al.*, 'A comparison of four methods for analog speech privacy,' *IEEE Trans. in Comm.*, vol. Com-29, pp. 18-23, 1981.
39. Meyer, C.H., 'Enciphering data for secure transmission', *Comput. Design*, vol. 13, pt 4, pp. 129-134, 1974.
40. Feistal, H., 'Cryptography and computer privacy,' *Scientific American*, vol. 228, pp 15-23, May 1973.
41. National Bureau of Standards, U.S.A., 'Encryption Algorithm for Computer data protection: Requests for comments,' *Federal Register*, vol. 40, p. 2134, 1975.
42. Morris, R., 'The data encryption standard-retrospective and prospects,' *IEEE Comm. Soc. Mag.*, vol. 16, No. 6, pp. 11-14, 1978.
43. Branstad, D.K., 'Security of computer communication,' *IEEE Comm. Soc. Mag*, vol. 16, No. 6, pp. 33-40, 1978.
44. Hellman, M.E., 'An overview of public key cryptography,' *IEEE Com. Sc. Mag*, vol. 16, No. 6, pp. 24-32, 1978,
45. Rivest, R.L., Shamir, A. and Adleman, L., 'On digital signatures and public key cryptosystems,' *Comm. Ass. Comput. Mach.*, vol. 21, pp. 120-126, 1978.
46. Merkle, R.C. and Hellman, M.E., 'Hiding information and signatures in Trapdoor Knapsacks,' *IEEE Trans. Inf. Th.* vol. IT-24, pp. 525-530, 1978.
47. Diffie, W. and Hellman, M.E., New directions in cryptography,' *IEEE Trans. Inf. Th.* vol. IT-22, pp. 644-654, 1976.
48. Adleman, L.M., and Rivest, R.L., 'The use of public key cryptography in communication system design,' *IEEE Comm. Soc. Mag*, vol. 16, No. 6, pp. 20-30, 1978.

CHAPTER 10

Future Trends

We have discussed in the previous chapters, the state of the art in switching, transmission, data networking and in signal processing, as applied to telecommunications. It has been shown further that universal connectivity of global and local networks is being achieved more and more through digital techniques, and the cost of telecommunication is becoming less and less every decade. This has been possible due to the spectacular progress in digital hardware and software during the last two decades. Over the years, two types of distinct networks have evolved, one for voice communication and the other for data communication. In voice communication, the traditional analog methods (FDM) of transmission have been cheaper for a long time, but during the last decade, the availability of cheap LSI/VLSI chips for signal processing and the new media, e.g., satellites and optical fibres, have been instrumental in bringing down the cost of long distance digital transmission. At the same time, remarkable progress has been made in designing and installing of electronic (digital) switching systems (ESS). These two parallel developments have made possible the integration of switching and transmission of voice signals into an Integrated digital network (IDN), where the individual channels basically carry 64 Kb/s data, based on the PCM format.

Till recently, voice communication constituted virtually the entire telecommunication traffic and the small amount of non-voice traffic was being carried through voice networks by analog techniques. Because of the rapid growth in non-voice data traffic, separate data networks, e.g., ARPA NET, have been developed, as discussed in Chapter 6. With the availability of 64 Kb/s data channels in IDN, there is hardly any requirement for separate data networks (except for wideband data, e.g., digital video or high speed computer-to-computer data transfer) and the existing as well as the future data networks will now be integrated with the IDN. The current information explosion and new dimensions in non-voice data traffic, in the form of Teletext, Viewdata, Electronic Mail, Teleconferencing and Interactive CATV, require that these new services should also be integrated with the IDN, leading to the Integrated services digital network (ISDN). This ISDN

will provide global coverage, as in the current international telephone network, for all types of users and services; at the same time, there will be virtually no extra cost or performance penalty for voice services already carried on the IDN, and the cost to data services will be marginal. In view of these emerging technologies, we discuss, in this concluding chapter, some aspects of the futuristic ISDN.

10.1 PROGRESS IN HARDWARE AND SOFTWARE

The progress in digital hardware, specifically for LSI and VLSI, has been indicated in Fig. 1.6. LSI/VLSI are generally characterized by three important parameters, viz., (a) transistor feature size, (b) functional complexity and (c) computing power. Over the last decade, the feature size has decreased from 10 μm to 1 μm, and this will continue to decrease reaching 0.5 μm in 1990. Similarly, the number of transistors on a chip has increased to 10^6 and will perhaps reach 10^7 by 1990, as shown in Fig. 10.1. In the early

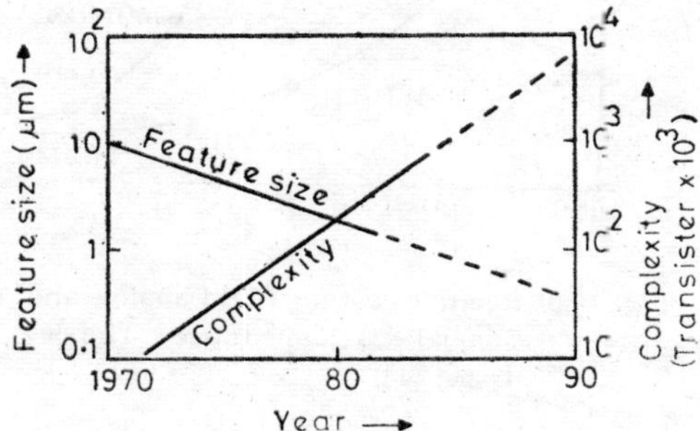

Fig. 10.1 Trends in MOS technology (after Soloway [2])

1970's, the development of the 4-bit μP was a great achievement, and today, 32-bit microcomputers with a processing power of one million instructions per second (one MIPS) are available off the shelf. The cost per chip has also decreased considerably, as shown in Fig. 10.2. However, the ultimate

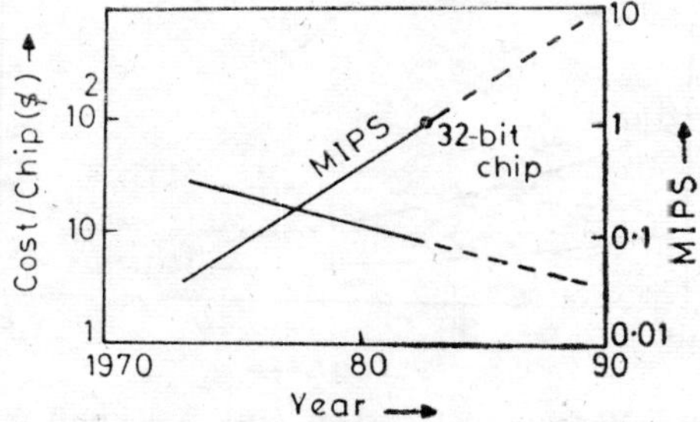

Fig. 10.2 Trends in μ-computer chip technology (after Soloway [2])

performance in VLSI depends not only on the refinement in process technology to achieve reduction in feature size, but also on the advanced system architecture that is used to provide the expected processing power.

Concurrently with the downward trends in the cost of LSI/VLSI, the capital cost of digital transmission circuits, through coaxial cables, optical fibres and satellites, has also decreased considerably over the years, as shown in Figs. 10.3-10.5 [1]. Even in the traditional coaxial systems, the introduction of digital TDM techniques reduces the capital cost at a faster rate than in analog FDM systems, as shown in Fig. 10.3. The same trend

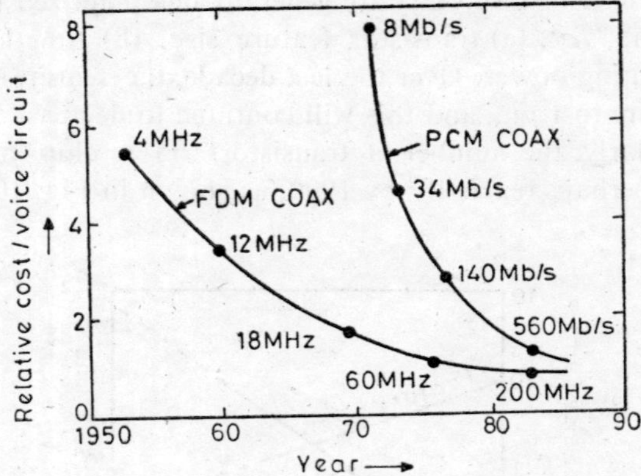

Fig. 10.3 Relative cost for FDM analog and PCM coaxial system (after Horsley and Usher [1])

is reflected in Fig. 10.4, where a cost comparison of digital systems using coaxial and fibre-optic cables is shown. The fibre-optic cables are more

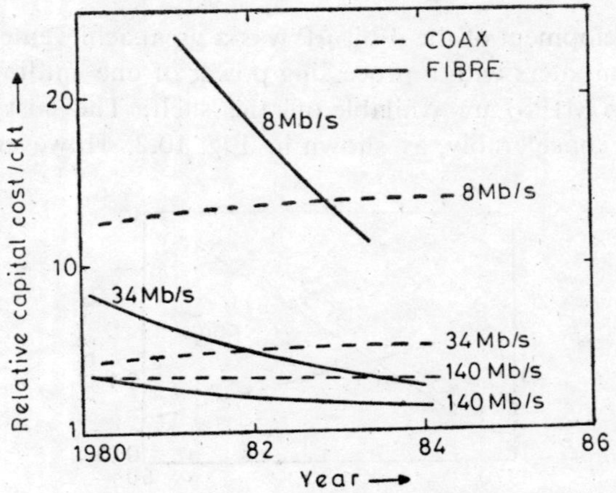

Fig. 10.4 Variation of cost/circuit-Km with time for digital systems (after Horsley and Usher [1])

economical both in capital and recurring costs (annual charges) as the system capacity increases and this is shown in Fig. 10.5. In statellite systems, the cost per circuit has come down from $20,000 to $5000 only in

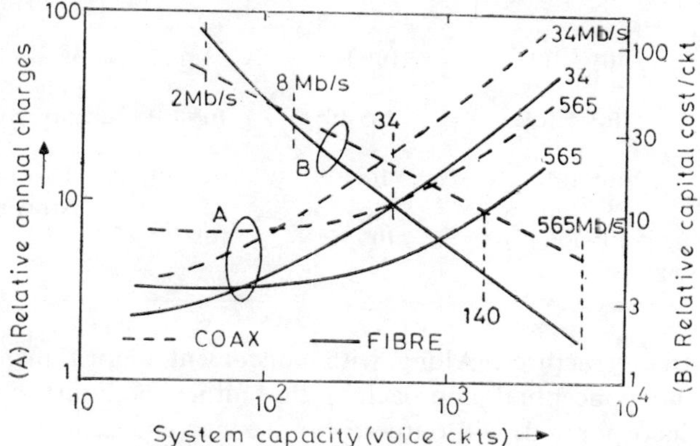

Fig. 10.5 Relative capital cost and annual charges vs. system capacity. Transmission rates are given as Mb/s. Annual charges are based on duct installation over a 100 Km route. (after Horsley and Usher [1])

recent years, and this trend is going to continue as more digital techniques, e.g., TDMA, DA and TASI, are introduced in the overall system. In ESS, only 20% of the total system cost is in the digital switches and the rest of the investment goes in the cable network and in the interfaces for lines/ trunks. As more and more digital subscribers and multifunctional terminals alongwith fibre-optic cables are introduced into the IDN, the overall cost structure of the ESS will also be more favourable than it is today. All these are happening due to the introduction of digital techniques in switching and transmission of voice and no-voice information through the evolving IDN.

10.1.1 Digital Signal Processing (DSP)

The computational requirements of signal processing tasks, e.g., speech processing and recognition, image processing etc., are large, as given in Table 10.1. For real-time processing, user-independent speech recognition requires approximately 10^3 MIPS and this will require an array processor. In a tactical radar system, about 2.10^4 MIPS are required to obtain real-time results and this is a very large task to be handled by available computers. To make this processing economically attractive, it will be necessary to improve the performance of today's supercomputers by 10 to 100 times, with corresponding cost and size reductions. Apart from the improved performance of VLSI through feature reduction and introduction of gallium arsenide technology, advanced system architecture and algorithms will have to be developed to provide the required processing power. The architectures, which are being considered now, are associative arrays, systolic

Table 10.1 Computational Requirements for DSP

Process	Sampling rate	No. of operations per sample	Required MIPS	Suitable processing (real time) computers
Speech processing	5.10^3–5.10^4	5.10^2–10^5	1–100	Mini/large computer
Speech recognition	5.10^3–5.10^4	10^5–10^6	10^2–10^3	Array processor
Image processing	10^5–10^8	10–10^4	10–10^4	Large computer/ Array processor
Tactical radar signal processing	10^8–10^9	10^3–10^5	10^4–10^5	—

arrays and tree structures. Along with concurrent algorithms, statistical modelling, and conceptual approaches, the future computers, with new architecture incorporated, will outperform the current generation of Von Neuman machines by a factor of 50 to 100. These will be uniquely suited to the computationally intensive real-time tasks and at the same time, provide built-in fault tolerance [2].

In telecommunication, the analog signal processing techniques, as have been used in traditional systems, are being replaced by digital signal processing (DSP) techniques. We have today wide application of DSP in both switching, and transmission systems, e.g., in ADC, LPC, speech detector, speech recognizer and synthesizer, TASI, modems, echo suppressor, digital filters, interfaces for line signalling, conference modules, testing modules and spectrum analyzer. Custom-built DSP chips are now available for ADC, LPC and digital filters [3]. However, many of these processes require DSPs to be used in a multiprocessing environment. To achieve this, the application requirements are partitioned so that the processing functions are distributed to several single-chip processors and/or microcomputers. ITT has developed, in their STK Research Center, a speech synthesizer based on a three-chip configuration consisting of the synthesizer, a μP and an external vocabulary ROM [4]. A three-chip speech synthesizer (consisting of TMC 0280, TMC 0350-ROM, and TMC 0270-controller) based on the LPC technique is also manufactured by Texas Instruments. The synthesizer performs the calculations required for simulating the lattice digital-filter and uses one pipeline multiplier, one adder/subtracter and delay circuits. The filter coefficients are stored in the ROM and the digital data rate is approximately 2.4 Kb/s [5].

To meet the needs of various DSP tasks as stated above, several programmable processors are now available, e.g., Intel 2920, NEC-μPD 7720, TI-TMS 320, Fujitsu-MB8764, and STC-DSP 128. The basic functions in a DSP with an accumulator-based architecture is shown in Fig. 10.6, where the functional arrangement optimizes signal processing throughput. The arithmetic unit for data, consisting of a multiplier, ALU (arithmetic and

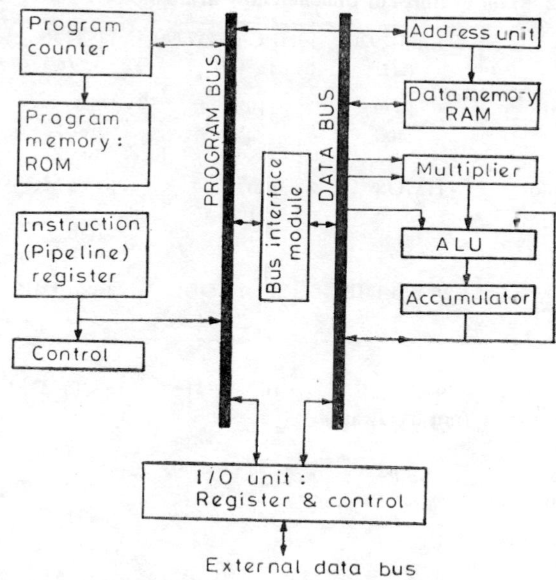

Fig. 10.6 Block diagram of a single-chip digital signal processor (DSP)

logic unit) and accumulator, performs the intensive arithmetic processing. Separate arithmetic units are provided for data address calculations and program control. The data and program memories are in separate address spaces, RAM and ROM respectively. To increase the throughput of the processor, the instructions are prefetched with the help of the instruction register associated with the program ROM. Prior to the 1970's, all these were organised using multi-chip configuration, but with the progress in device technology, single-chip general-purpose VLSI circuits are now manufactured by several industries. The single-chip DSPs reduce the number of components in a total system, and at the same time, their programmability makes them suitable for a variety of applications.

Some of the main features including architectural characteristics of commercially available DSPs are compared in Table 10.2 [6]. The processors (*A-D*) have been introduced in that order, and the procesrors *B* and *C* have been popular DSPs since their introduction. The processor *D* is the first CMOS-DSP to be introduced, and has a power dissipation about one-third of that required by others. However, because of its large size (93 mm^2), the cost and availability of the device may be adversely affected. Machine cycle times have been decreasing over the years, but the processor throughput also depends on memory size, addressing capability, amount of parallelism and pipelining, and instruction set. Processor *A* is the only chip with an on-chip ADC incorporated, but in later DSP chips, this has been excluded. The reasons are: ADCs are required to meet different specifications and the total area of the DSP would have to be doubled to provide an analog interface

Table 10.2 Main Features of Commercially available DSP (after Kneib [6])

Features	Intel-2920 (*A*)	NEC-μPD7720 (*B*)	TI-TMS320 (*C*)	Fujitsu-MB8764(*D*)
Date of availability	1979	1981	1983	1984
Machine cycle time (*nS*)	400	250	200	100
Fabrication technology	HMOS	3μm, E/D NMOS	3μm SMOS	CMOS
Maximum power dissipation (*W*)	0.8	0.9	0.9	0.3
Packaging	28-pin DIP	28-pin DIP	28-pin DIP	28-pin grid array
Arithmetic				
Multiplier:	9×25–25 (not hardware)	16×16–31	16×15–32	16×16-26
In×in-out (bits)				
Time to multiply-accumulate in an inner loop	~ 5 μs	250 ns	400 ns	100 ns
Arithmetic and logic unit	Yes	Yes	Yes	Yes
Accumulator width (bits)	28	Two 16	32	26
Saturation mode	Hardware	Software	Hardware	Hardware
Memory				
On-chip:				
Program ROM	192×24	512×23	1536×16	1024×24
Data RAM	40×25	128×16	144×16	256×16
Data ROM	24×25	510×13	In program ROM	In program ROM
Table lookup	No	Yes	Yes	Yes
Data shift (Z^{-1})	No	Yes	Yes	Yes
Direct access of external memory:				
Program	No	No	Upto 4*K*×16	Upto 1*K*×24
Data	No	No	No	Upto 1*K*×16
Program Control				
Width of instruction word	24	23	16	24
Stack depth	No	4	4	2
Loop counter	No	No	Yes	Yes
Conditional branch	No	Yes	Yes	Yes
Input/Output				
Parallel data *I/O* (bits)	4 in/8 out	8	16	16
Analog-on-chip	9 bit	No	No	No
Hardware interrupts	No	1	1	No
Serial port (rate)	No	Yes (200 kHz)	No	No
Direct memory access (block transfer)	No	Yes	No	Yes

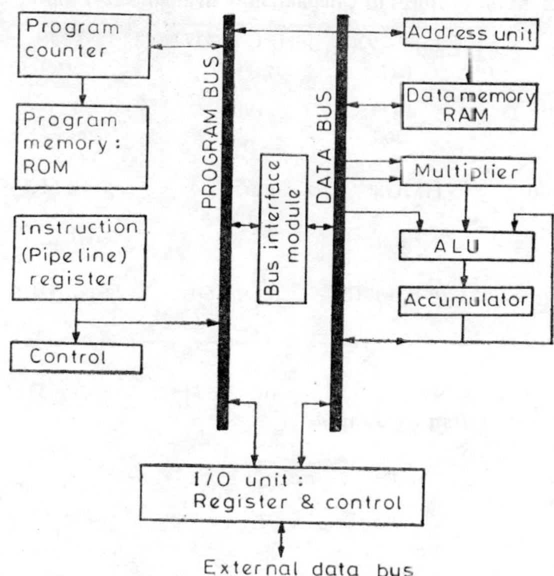

Fig. 10.6 Block diagram of a single-chip digital signal processor (DSP)

logic unit) and accumulator, performs the intensive arithmetic processing. Separate arithmetic units are provided for data address calculations and program control. The data and program memories are in separate address spaces, RAM and ROM respectively. To increase the throughput of the processor, the instructions are prefetched with the help of the instruction register associated with the program ROM. Prior to the 1970's, all these were organised using multi-chip configuration, but with the progress in device technology, single-chip general-purpose VLSI circuits are now manufactured by several industries. The single-chip DSPs reduce the number of components in a total system, and at the same time, their programmability makes them suitable for a variety of applications.

Some of the main features including architectural characteristics of commercially available DSPs are compared in Table 10.2 [6]. The processors (*A*-*D*) have been introduced in that order, and the procesrors *B* and *C* have been popular DSPs since their introduction. The processor *D* is the first CMOS-DSP to be introduced, and has a power dissipation about one-third of that required by others. However, because of its large size (93 mm²), the cost and availability of the device may be adversely affected. Machine cycle times have been decreasing over the years, but the processor throughput also depends on memory size, addressing capability, amount of parallelism and pipelining, and instruction set. Processor *A* is the only chip with an on-chip ADC incorporated, but in later DSP chips, this has been excluded. The reasons are: ADCs are required to meet different specifications and the total area of the DSP would have to be doubled to provide an analog interface

Table 10.2 Main Features of Commercially available DSP (after Kneib [6])

Features	Intel-2920 (*A*)	NEC-μPD7720 (*B*)	TI-TMS320 (*C*)	Fujitsu-MB8764(*D*)
Date of availability	1979	1981	1983	1984
Machine cycle time (*nS*)	400	250	200	100
Fabrication technology	HMOS	3μm, E/D NMOS	3μm SMOS	CMOS
Maximum power dissipation (*W*)	0.8	0.9	0.9	0.3
Packaging	28-pin DIP	28-pin DIP	28-pin DIP	28-pin grid array
Arithmetic				
Multiplier:	9×25–25 (not hardware)	16×16–31	16×15–32	16×16-26
In $\times$ in-out (bits)				
Time to multiply-accumulate in an inner loop	~ 5 μs	250 ns	400 ns	100 ns
Arithmetic and logic unit	Yes	Yes	Yes	Yes
Accumulator width (bits)	28	Two 16	32	26
Saturation mode	Hardware	Software	Hardware	Hardware
Memory				
On-chip:				
Program ROM	192×24	512×23	1536×16	1024×24
Data RAM	40×25	128×16	144×16	256×16
Data ROM	24×25	510×13	In program ROM	In program ROM
Table lookup	No	Yes	Yes	Yes
Data shift (Z^{-1})	No	Yes	Yes	Yes
Direct access of external memory:				
Program	No	No	Upto $4K\times16$	Upto $1K\times24$
Data	No	No	No	Upto $1K\times16$
Program Control				
Width of instruction word	24	23	16	24
Stack depth	No	4	4	2
Loop counter	No	No	Yes	Yes
Conditional branch	No	Yes	Yes	Yes
Input/Output				
Parallel data *I/O* (bits)	4 in/8 out	8	16	16
Analog-on-chip	9 bit	No	No	No
Hardware interrupts	No	1	1	No
Serial port (rate)	No	Yes (200 kHz)	No	No
Direct memory access (block transfer)	No	Yes	No	Yes

with capabilities that matched those of the DSP. It is, therefore, preferable to use an external ADC when required. These single-chip DSPs are based on the Harvard architecture which is an improvement over the traditional von Neumann architecture. In this, separate data and program memories are provided, along with independent buses for data, and instruction transfers and control, so that a program fetch can take place at the same time as the previous instruction is being executed. Further pipelining may be used within the operation of the DSP, allowing the processor to perform a set of operations with a result every machine cycle; thus increasing the throughput by 2 to 5 times (e.g., in processors *C* and *D*). Stages of parallelism have also been introduced in some of the DSP-chips, mainly within the arithmetic units and address generation parts, to increase the speed still further. The arithmetic unit of the processor *C* contains a full (16×16 bit) multiplier producing a 32-bit product, and this is most desirable for signal processing applications. All devices provide parallel I/O, having 8/16 bits, and wider parallel I/O handles I/O burst rates more efficiently than the smaller I/O bus.

Since many of the DSP tasks require a multiprocessor configuration, it is necessary to have a combination of higher I/O data rates together with control signals for handshaking and synchronizing multiple processors. The interprocessor capability of a DSP would require features which allow a device to function either as a master or a slave and to interface with slower devices. These facilities are not provided in the currently available DSP chips, and considerable external hardware is used to configure them as a multiprocessing system. The future trends in DSP chips are: CMOS as the processing technology; major enhancements in data addressing facilities, memory space and multiprocessing I/O; advances in basic architecture to incorporate a register file suited to array processing applications and full-precision floating point arithmetic. These improvements will expand the range of applications for single-chip DSPs.

10.2 NEW DIMENSIONS IN INFORMATION TECHNOLOGY [7]

Changing social needs and the development of new technology are bringing about the evolution of new communication capabilities, e.g., Videotex, Electronic mail, Teleconferencing and Interactive CATV. These are also leading to Automated offices and Wired houses, where most of the future information will be generated, processed and utilized for running the industries, business and the government, and for education and recreation. Communication media are basically of two types: (a) Record media e.g., books, magazines, newspapers, letters etc. and (b) real-time media, e.g., telephone, radio, TV etc. Of the new media, Teletext and Viewdata (jointly known as Videotex) may replace the record media and customers may now get the desired information on demand from a central database which has all these informations (books, newspapers, etc.) in its memory. Teleconferencing, Electronic mail and Interactive CATV will surely cater for all the needs of the real-time information, characterized now by telephone, radio and TV.

Videotex is implemented in two ways, as Viewdata and Teletext. Viewdata is a two-way, interactive system which uses a combination of video display (often a TV receiver), local processing and a remote data base, accessible through the public telephone network. Pages of information are stored in the database ready to be accessed through a tree search protocol which allows the user to page the increasing levels of detail. Viewdata public service is now available in UK (Prestel); France (Teletel), Canada (Talidon), Japan (Captain) and other countries have installed limited test systems. An agricultural information system (Green Thumb) has been developed by USA. Teletext, on the other hand, is a one-way information system which employs unused lines in the broadcast TV signal to transport data to modified TV receivers without interfering with normal program. In this system, the details available in viewdata is not possible and the service is described as an Electronic magazine. Teletext public service is again available in UK (Coefax, Oracle), in France (Antiope) and in several other countries. In UK and France, where viewdata and teletext services are offered concurrently, the format and data encoding are made compatible so as to reduce the cost of signal processing at the display.

In Electronic mail, text messages, derived from letters and telegraph, can be exchanged between many persons using terminals communicating through a large common memory under the control of a modest computer. Messages addressed to a specific recipient (or a group of recipients) are stored until the recipient logs-on, identifies himself and requests for his messages. Most systems also provide answering, forwarding and filing functions, and some of them handle voice messages as well. Commercially electronic message systems are operating in USA at present. Postal automation, on the other hand, aims at increasing the efficiency of the present system of delivering letters and telegrams to the addresses.

Teleconferencing, using videophone, provides an interactive meeting space and substitutes telecommunications for travel (resulting in energy saving). Its contents are derived from telephone, TV and data communications. The real-time aspects of the medium facilitate unique executive discussions between several levels of management in several locations in an industry or government and proves the richness of the medium for any corporate discussion or conference. In USA, several major companies maintain facilities for teleconferencing, using leased and/or dial-up connections. Most major telecommunication organizations in the world now offer similar services on a limited basis.

Cable TV (CATV), although originally planned for TV program transmission to rural subscribers, now provide a wideband, one-to-many, downstream connection together with limited bandwidth and one-to-one, upstream connections. It provides such services as Network and Premium TV, local access channels, opinion polling, alarm services, meter reading etc. CATV reflects the real-time nature of its use and may partly serve as a teleconferencing network. In USA, several systems provides security monitoring and a few afford the opportunity for answer back to the head end.

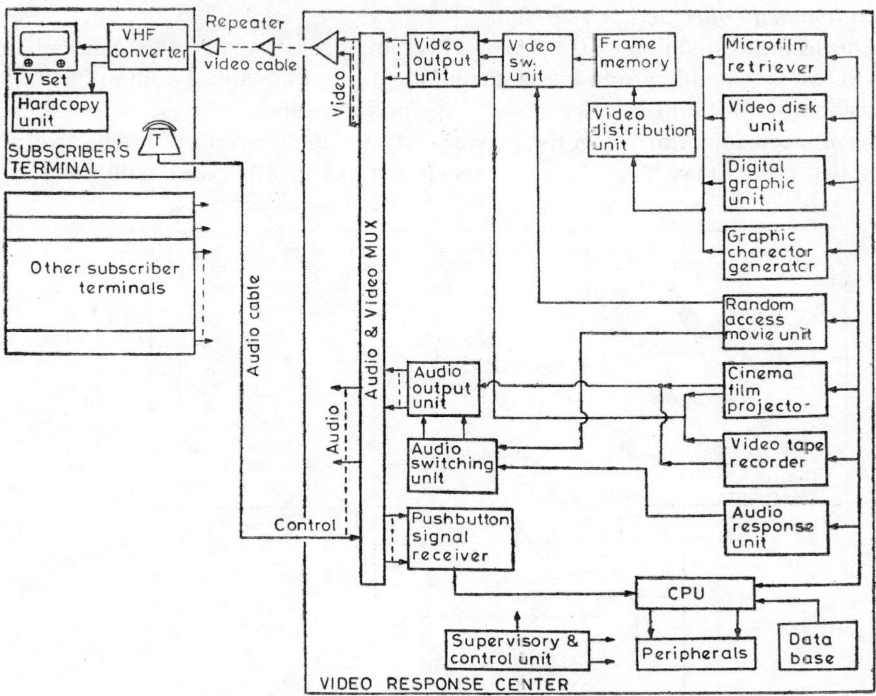

Fig. 10.7 Block diagram of a Video response system

A more sophisticated videotex service is the video response system (VRS), developed by NTT, Japan [8]. The VRS as shown in Fig. 10.7 is an advanced user-oriented video information system (somewhat similar to CAPTAIN, but with more facilities) where the subscriber can call for both still and movie pictures, graphics, and films along with its audio. The VRS centre maintains video and audio files consisting of 8000 frames of colour still picture, 15,000 frames of graphic characters, 10 reels of colour movie, 120 movie cassettes, 5000 audio messages each of 15 sec duration, and announcement tapes. The display in the subscriber's terminal is on the TV screen with audio guide, and the terminals are linked to the VRS centre through video and audio cables. Multiple terminals share the VRS facilities and individual requests are responded to by the centre through the video and audio switching units of the system. The system may be used for interactive TV, video conferencing, TV broadcast and retransmission, video telephone, telephone facsimile, video information indexing, public opinion survey, mass education, shopping, reservations, monitoring, telemetering, counselling and remote medical diagnosis. A similar voice response system has also been developed by Hitachi, Japan, and this may be used for order entry, making reservations and enquiries, providing information etc. With the growing customer demands, these VRS systems will now find wide acceptability, at least in the developed urban areas.

Automated Office and Wired Homes [9]
Due to the availability of these exotic technologies, the concept of Automated Offices is gaining ground and the application of advanced communication technology within the office as well as for inter-office purpose is going to have spectacular impact on the productivity of functions and products. Automated offices may have the elements shown in Fig. 10.8, and will incorpo-

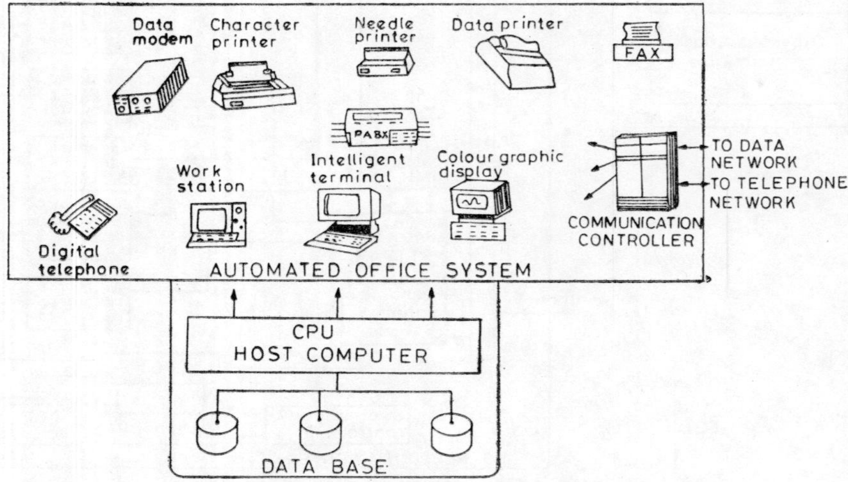

Fig. 10.8 Elements of an Automated office, offering services, such as, word processing, electronic mail, automated management, information retrieval, status reporting, audio/video/computer conferencing, security etc.

rate data processing services to achieve desk-to-desk transfer of text, data and image information, computer services to accomplish the design and drafting of complex products and to control and support manufacturing, administrative program for coordinating all activities, and communication services to provide electronic mail, information retrieval and conferencing. Through these applications, resources will be conserved, the reliability and consistency of the product improved and productivity will be increased.

As an extension of the concept of automated officers, the automated homes, or wired household, will incorporate microcomputer based entertainment and data processing functions, residential communication media such as, radio, telephone and TV, and will provide an environment in which entertainment, information and personal communications are readily available, and administrative, security and conservation functions can be performed automatically. Teletext, viewdata and interactive CATV will be used in addition to the traditional communication media. Many of the applications in wired homes will require development of specialized infrastructures (databases, networks, administrative centres etc.) which will have a social or commercial basis. Experimental systems, using cables or fibres, have been set up in Japan and Canada and surely other countries will follow them.

Thus the new communication technologies, as discussed above, will continue to evolve under the influence of consumer demands, supplier rewards and the opportunities presented by hardware and software developments. Whether the new media introduced in the early 80's will achieve as widespread a use and influence society as much as telephone, radio, TV and data communication, will necessarily depend on the vigour of public demand and the afluence of the respective societies in various countries. Media such as viewdata, electronic mail and teleconferencing, which will contribute to improved productivity, have the greatest chance of survival and growth, and they would play major roles in automated offices of the future. For home needs, interactive CATV, Teletext and viewdata will surely be popular, at least, in the afluent societies.

10.2.1 Local Area Networks

In the area of computer networking, there is a major demand for intelligent terminals and workstations along with efficient local area networks (LAN). The workstation should be a powerful single-user computer with display, keyboard, disk, pointing device and a network interface (LAN) to other workstations. It may be connected to a mainframe computer and will be a major step beyond the time-sharing terminals in use today. The workstation (or a network of workstations as shown in Fig. 10.9) will create the necessary

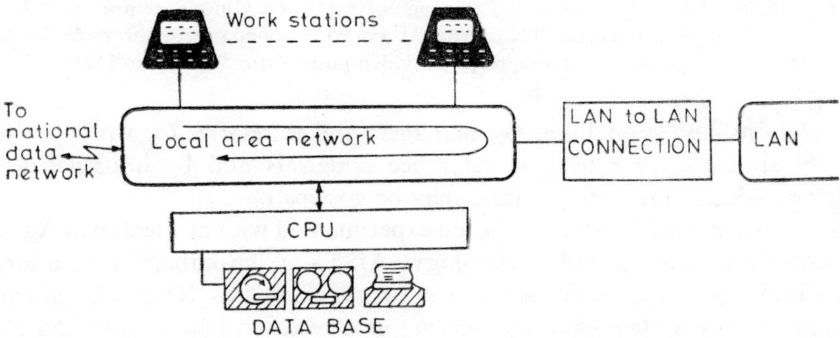

Fig. 10.9 A network of work stations

design environment of the future and provide design assistance, as well as communication and coordination facilities between the design team members. It will also be a tutor to the novice and, at times, give expert advise to the user.

These concepts have brought in a new trend in developing Personal Information Processing and Communication Systems (PICS) using opto-electronics [10]. It is thought to be an exciting challenge for information processing and communication engineers to improve the human/social environment by developing easy-to-use person-oriented systems. PICS will consist of man-machine interface, communication network and information processing facilities including computer centres and data banks as shown in Fig. 10.10. PICS may be used by professional engineers and scsentists for scientific

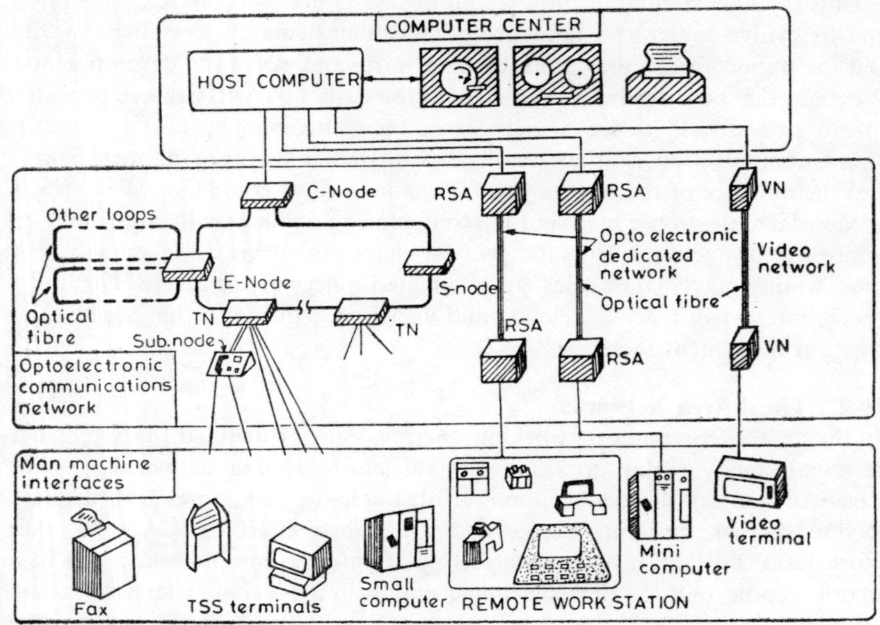

Fig. 10.10 Block schematic of PICS using optical fibres. C-node: centre node; LE: loop exchange; TN: terminal node; S-node: supervisor node; RSA: Remote station adapter; VN: video node; (after Yamamto [10])

calculations and modelling, by managers and secretaries for word processing, by students for retrieving reference materials and by housewives for storing their favourite recipes (and may be gossips!).

The concept of this PICS has been experimented with at the Japan Agency of Industrial Science and Technology (AIST) at Tuskuba, where a large-scale local opto-electronic computer network, known as Research Information Processing System (RIPS) is serving nine research laboratories and 3000 engineers/scientists and supports research activities in electronics, mechanical engineering, chemistry and microbiology. It has 3000 terminals of 50 different types connected through optical fibre highway at 48 Kbps, 64 Kbps (speech), 16.9 Mbps and 33 Mbps. It uses 360 km of fibre and 3000 fibre connections. The synthesis of first generation opto-electronic technology with computer systems has proved to be a highly effective means of personalizing laboratory and office systems. With further development of integrated opto-electronics in the 2nd and 3rd generation, PICS will take a quantum leap forward. This emphasis on personalization can be expected to help overcome differences in language, culture, religion, politics and economics as man-machine interfaces enable persons to interact with machines on a simple level, using every-day language. It is hoped that PICS will provide a richer environment for people to live in and that the system will progress along with the world community.

10.2.2 Demand Profile of New Services

The existing telephone services in most of the developed countries will be saturated in the near future. (Of course, the situation in developing countries is quite different). The telecommunication services will then evolve from the current data and voice communication to text, graphic and video communication in the 1990s. Table 10.3 shows the chronological evolution of the

Table 10.3 Evolution of Telecommunication Services

Year	Introduced services	New services
1920	Telegraph, Telephone	
1950	Telegraph, Telephone, Telex, Telepicture	
1980	Telex, Telephone, Telepicture, Colour TV, Data Services, Telefax	Viewdata, Teletex
1990	—do—	Electronic mail, Interactive Cable TV, Cable-text, Videophone

services and the trends in the 1980s and the 1990s. It is hoped that the progress in the key technologies of microelectronics, fibre optics, and digital transmission and switching will lead to a dramatic reduction in transmission costs over the next decade and videophone will replace most of the telephones by the turn of the century.

Some demand forecasts for non-voice services have been made for moderately developed countries and Table 10.4 shows a representative study [11].

Table 10.4 Anticipated Rise in Demand for Non-Voice Services by 1990

Services and data rates	Status in 1982 (No. of lines)	Forecast for 1990 (No. of lines)
Telephone (analog)	8.10^6	1.2×10^7
Telex (50 bauds)	3.10^4	5.10^4
PSDN* (2.4/4.8/9.6 Kb/s)	10^4	4.10^4
Videotex (1.2/2.4 Kb/s and 75 Kb/s) (Business)	—	3.10^4
Videotex (Home)	—	2.10^4
Teletex (2.4 Kb/s)	—	1.2×10^4
FAX (up to 48 Kb/s)	10^3	10^4

*PSDN: Packet switched data network.

Some studies have also been made to identify the potential users of the futuristic ISDN. It is generally accepted that the data services will be used in a variety of environments which include banking, insurance, education, national and local government, utilities, transportation, investment, manufacturing, law, health care, wholeselling and retailing, information services and the home. These services will operate in a number of modes which include interactive (inquiry/response, remote access/time-sharing, graphics), data collection (polling/sensing), data distribution, remote display/documentation, transactions, video coferencing and voice. These modes of operation can be categorised in terms of such characteristics as, connect time, holding time, error-rate tolerance and data rate. The study shows that all new services will fall within CCITT user classes (UC) 1 to 11 (Ref. to Table 6.3), with the exception of wideband services; thus, they will be easily accommodated within the 64 Kb/s PCM bit stream of digital exchanges. Most of the users will be satisfied with a fast connect (100 to 600 msec), medium speed ($\leqslant$ 9.6 Kb/s) transaction system, and the transmission quality should give an error-rate of 10^{-6} to 10^{-9}. For access to local and remote data bases, the users would now connect the terminals through PBXs and local exchanges, rather than through point-to-point networks [12].

At present, the number of subscribers served by the public and private data networks is only about 1% of the number with access to the telephone network. It is estimated that by the introduction of new data services in the next decade, the additional traffic would be about 7.5% of the telephone traffic (even assuming that all business mail becomes electronic and 90% of telephone subscribers use the videotex service for six minutes per day). If all data traffic is now carried by the evolving IDN, then the impact on the grade of service and service performance would be very small. As such, it will be economical to integrate the future non-voice services with the (digital) telephone service of the IDN. It has been shown that the additional investment required in an integrated network to cater for the expanding data services is much lesser than that required for augmenting the distinct data networks. But there wiil be a marginal extra cost to the telephone subscribers as well. Given the limit of performance, as indicated in Table 10.5, most of the data services can be catered by an economical ISDN. However, the high performance data services will need specialised networks, although

Table 10.5 Performances of Data Networks and ISDN Objective

Parameter	Public-switched telephone network (analog)	Public-switched data network	Private network	ISDN objective (proposed)
Bit rate	2.4 Kb/s	Up to 48 Kb/s	Up to Mb/s	Up to 64 Kb/s
Error-rate	10^{-4}	Circuit-switched: 10^{-6} Packet: 10^{-10}	10^{-6}	10^{-6}
Call set up time	3 to 20 sec	Circuit: 0.1–1 sec. Packet: 1–10 sec.	—	1–3 sec.
Information transfer time	10 sec.	Circuit: 10 msec Packet: 0.1–1 sec.	10 sec.	10 msec.
Call duration	120 sec.	Circuit: 10–3600 sec Packet: 0.01–1 sec.	Leased path	Variable mix

at a much higher cost, and interworking will have to be provided between the ISDN and the dedicated networks [13].

10.3 INTEGRATED SERVICES DIGITAL NETWORK (ISDN) [13, 14]

It is seen from these discussions that the current demand for non-voice services can be economically and efficiently met by offering universal connectivity to these users through the evolving IDN. The present PCM-based (for voice services) IDN will gradually grow into a multi-service IDN, which has end-to-end digital connectivity and through which customers are provided with a multiplicity of services. The ISDN will thus grow out of the generalization of the services and protocols of IDN. The ISDN has been conceived as a general-purpose digital network capable of supporting (and integrating) a variety of services, e.g., voice, data, text and image, using a few standard multi-purpose user-to-network interface. A key difference between the multi-service IDN and the ISDN is the provision of a unified (or integrated) digital customer access (interfaces and protocols) between the customer premises and the ISDN to support a multiplicity of services along with 'out-slot' signalling. In order to assist the orderly development of ISDN's, CCITT is studying the proposed *I*-series recommendations, dealing with services, network aspects, interfaces and protocols; and these will be finalized soon [14].

To meet the multi-purpose user needs, the ISDN, as shown in Fig. 10.11, will have capabilities for: (a) Circuit switching at 64 Kb/s (for voice and bulk information transfer), (b) Packet switching at bit rates up to 64 Kb/s

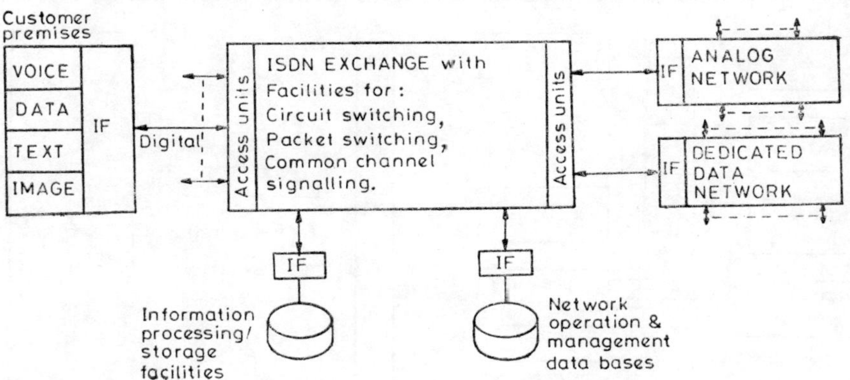

Fig. 10.11 An ISDN configuration. IF: interface

(for interactive data applications), (c) Information processing and storage facilities (for telematic services—teletex and videotex), along with the common channel signalling (CCITT No. 7 signalling system enhanced to support ISDN environment), and network operations and management data bases. The ISDN will of course be interconnected with other analog/data networks in the region. Although a variety of customer access channels may be provided using different types of cables/optical fibres, it is preferable to restrict the types at the developmental stage of ISDN, and as such, CCITT have recommended two types of channels, one for data only and the other for control and signalling information. These access channels are specified as [14]:

(a) *B*–channel at 64 Kb/s for data only;
HO–channel at 384 Kb/s (= 6 PCM channels) for data only.
(b) *D*–channels at 16 Kb/s or 64 Kb/s for signalling (*S*-type) only; *E*-channel at 64 Kb/s, where packet data (*p*-type) and telemetry data (*t*-type) may also be carried along with the signalling information in the same channel using statistical multiplexing.

CCITT has further recommended that for the user-network interface, the access will be through:

(a) Basic access interface structure comprising of two *B*-channels and one *D*-channel, i.e., the line rate capability is $(2B + D) = 144$ Kb/s.
(b) Primary rate access (or PABX access) interface structures, e.g., $23B + D$ or $23B + E$ (equivalent to T_1 carrier); $30B + E$ (equivalent to CEPT-PDC).

Initially the customer access is being provided at 80/96 Kb/s and comprises of $(B + D)$ channels, (since the existing cables can carry only that much of data). The PABX accesses may use either $(HO + E)$ channels or $(23B + E)$ channels (for large PABX). The customer access to 80 Kb/s may be realized either using fixed channel allocation or using dynamic channel allocation. In the realization with fixed channel allocation, as shown in Fig. 10.12, a

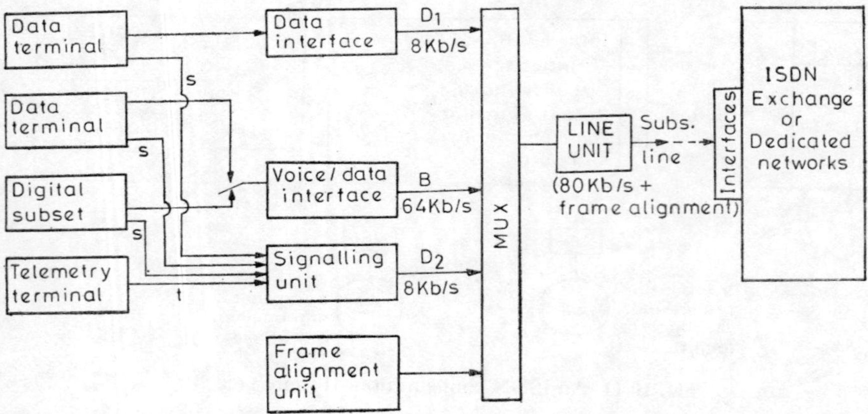

Fig. 10.12 Possible realization of customer access with fixed channel allocation. D_1, D_2: 1/2 *D*-channel

separate data channel D_1 at 8 Kb/s is provided, and *s* and *t* signals are carried by the D_2 channel using fixed frame or variable frame formats. Alternatively, the *D*-channel at 16 Kb/s may be dynamically allotted to carry *s*, *t* and some data information (upto 9.6 Kb/s) in a message-interleaved manner. With the improvement in local area cables, it will be possible to offer extended customer access at 144 Kb/s, using $(2B + D)$ channels. For a PABX access, say in an automated office (refer to Fig. 10.8), $(HO + E)$ channels may be used. For a large PABX, the accesses will probably be arranged in a multiplex scheme using $(nB + D)$ channels, $n > 2$. The most probable bit rates would be the CEPT-PDC at 2048 Kb/s or 704 Kb/s, providing 30 or 10 data channels of 64 Kb/s with an additional 'out-slot' signalling channel. Such accesses could be useful for wideband services as well.

10.3.1 Signalling and Protocols [14, 15]

An important feature of the proposed ISDN is the use of the 'out-slot' common channel signalling technique for customer-network access and inter-exchange signalling. The signalling information is carried over the *D* or *E* channel and is used to control multiple circuit-switched connections from user to user. Within the network, the CCITT Common Channel Signalling System (CCSS) No. 7 is used for inter-exchange signalling. This separation of the signalling path (*D*/*E*-channel) from the data path (*B*-channel), facilitates the specification of a flexible, universal signalling protocol, supporting multiple services. (This is in contrast with the 'in-slot' signalling technique as used in X.25 and X.21 access protocols). However, within the *D*/*E*-channel, an in-slot signalling technique (similar to that used in packet data networks) is used to establish logical signalling connections. These are then used to transfer signalling messages required by the controllers to establish multiple circuit-switched connections. Thus there is a logical separation between user information (*U*-plane) and signalling/control information (*C*-plane), as shown in Fig. 10.13. The common channel signalling technique also provides a separate signalling path between end users (which may be used for controlling the service characteristics or communication modes during an established call, e.g., alternate voic/text transmission without changing the logical or physical connection.)

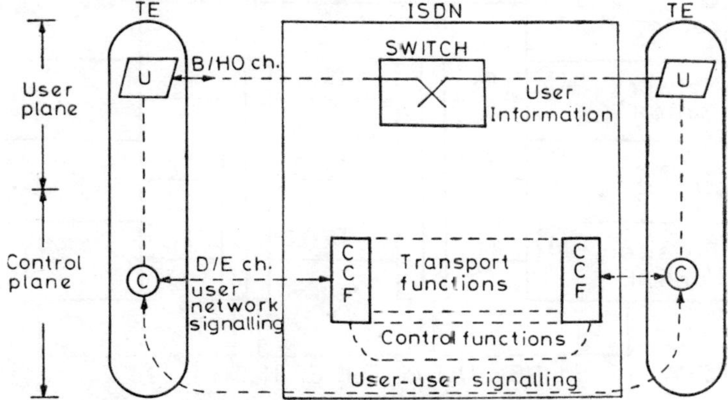

Fig. 10.13 Common channel signalling and Information flow in ISDN. TE: ISDN terminal equipment; CCF: connection control function

The ISDN Protocol Reference Model, with the objective of modelling information flows in the *U*-plane and the *C*-plane, is being finalised by CCITT. The model consists of two logical protocol planes corresponding to *U* and *C*, and each plane is divided into seven layers representing seven distinct ordered partitions. Each layer offers a specific layer service or a set of layer services to the layer above. The functions of each layer and the services provided by them are defined in general terms in the OSI (Open System Interconnection) Reference Model. The proposed set of layers in the *C*-plane comprises of three upper user service layers, and four lower transport service layers, as shown in Fig. 10.14. The user-to-network signalling is performed by three lower layers as:

Physical layer (level 1): provides the functional and procedural characteristics needed to establish, maintain and release physical connections between the user terminal and the exchange.
Link control larger (level 2): provides reliable transmission over a single data link including frame management, link flow control, and link initiation/release procedures.
Network control layer (level 3): provides the control for establishing and clearing the calls through the switching network nodes.

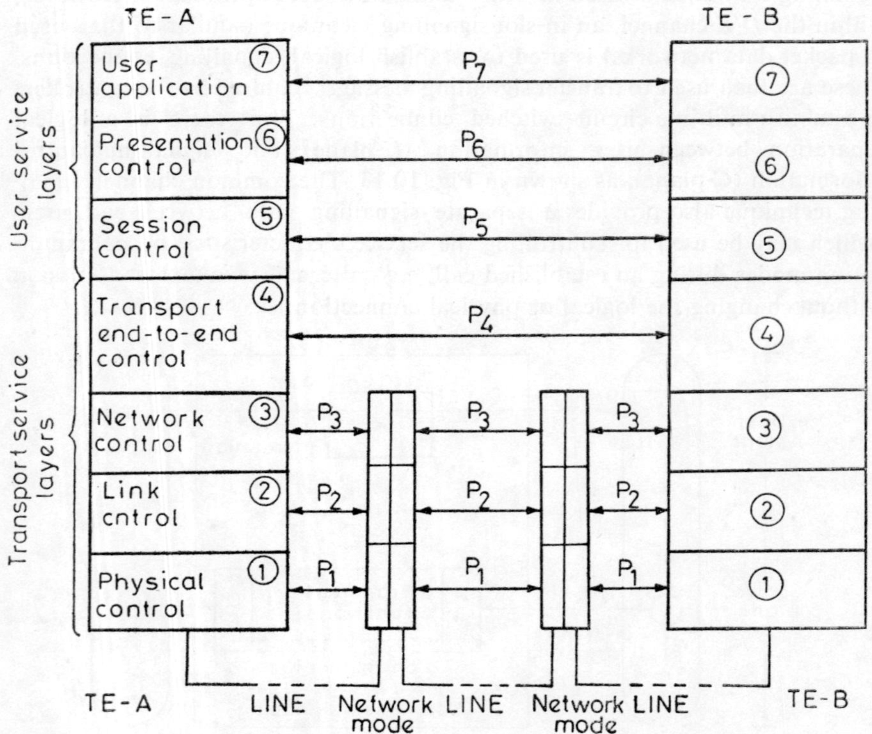

Fig. 10.14 Logical functional layers of the *C*-plane in an ISDN

The user-to-user protocols define the functions related to end-to-end communication between user terminals or between user and network resources, using the following layers:
Transport control layer (level 4): provides end-to-end (user-to-user) control signals across the network, e.g., end-to-end acknowledgement of received information.
Session control layer (level 5): establishes, maintains and terminates logical connections for transfer of data between processes. This provides, for example, dialogue control (simplex, half-duplex, duplex), message unit flow control and segmentation of message data units.
Presentation control layer (level 6): provides data formats and data information, if needed.
Application layer (level 7): is the source or sink of data including services which process data. User applications include the interactive use of a computer for reservations, banking etc.

The CCITT No. 7 common channel signalling has been developed primarily for the interchange of signalling information between exchanges and is specifically defined to handle a wide range of signalling information needed for voice and data services. In its primary role, it is carried on a 64 Kb/s channel, serves approximately 1000 terminals and carries only control information, not user data. Its frame structure, as shown in Fig. 10.15, comprises

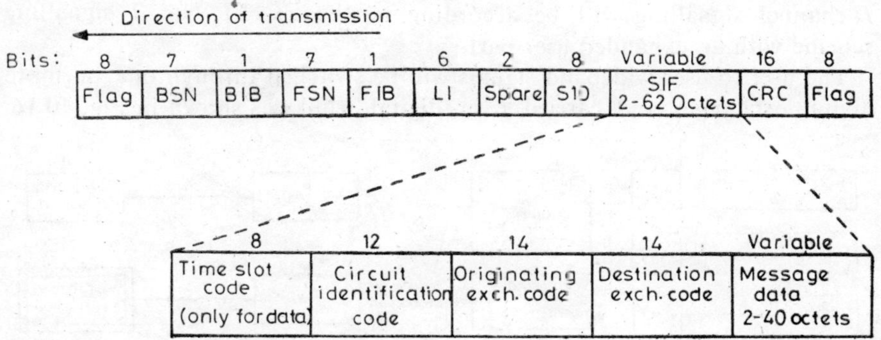

Fig. 10.15 CCITT No. 7 Common Channel Signalling frame structure; Flag: a unique word 01111110; BSN: Backward sequence number; BIB: Backward indicator bit; FSN: Forward sequence number; FIB: Forward indicator bits; LI: Length indicator; SIO: Service information octet; SIF: Signalling message content 2-62 octets; CRC: Cyclic redundancy check.

of a message transfer part, which is service independent, and a user part (SIF) which is related to the procedures needed for a particular service (e.g., a telephone user part and a data user part). In the case of packet-switched services (including packetized voice and voice-cum-data), the *X*.75 protocol has been recommended. In general, a frame mode transport protocol is being considered for future ISDN's, as this method is suitable for both circuit and packet-switched calls. However, the CCITT No. 7 signalling system is too comprehensive for use in customer access, but for a large PABX access, it is possible that the system or its subset will be used.

10.3.2 Wideband/Broadband ISDN [16, 17]

We have discussed so far the ISDN configurations satisfying user requirements within the bit rate of 64 Kb/s. But for high-resolution facsimile, videophone and TV signals, much higher bit rates are required. It is expected that videophone signals will be digitized at 1.92 Mb/s using efficient ADPCM codecs and only for live and interactive TV, higher bit rates ($\simeq$ 50-100 Mb/s) will be required. To meet these requirements, CCITT has defined H_1 channel having rates of 1.536 Mb/s (compatible with American T_1 carrier) and 1.92 Mb/s (compatible with European CEPT standard of 30 channels). Services at rates upto 1.92 Mb/s are known as Wideband and services above this rate (2 to 140 Mb/s) are classified as Broadband. The present plan is that the wideband service will be provided as a transparent circuit-switched channel at the selected bit rate ($N \times 64$ Kb/s; $2 \leqslant N \leqslant 30$) from subscriber-to-subscriber and it will be established on a per-call or continuous service basis. Both point-to-point and point-to-multipoint services will be required and the switched service will have to be provided in both local and transit exchanges. Further the wideband ISDN structure will include a D-channel in additlon to the B, HO or H_1 channels; and the D-channel signalling will be according to the CCITT No. 7 signalling scheme with an expanded user part.

The user-to-user wideband signals will be switched through one or more digital exchanges and carried over digital trunks as shown in Fig. 10.16.

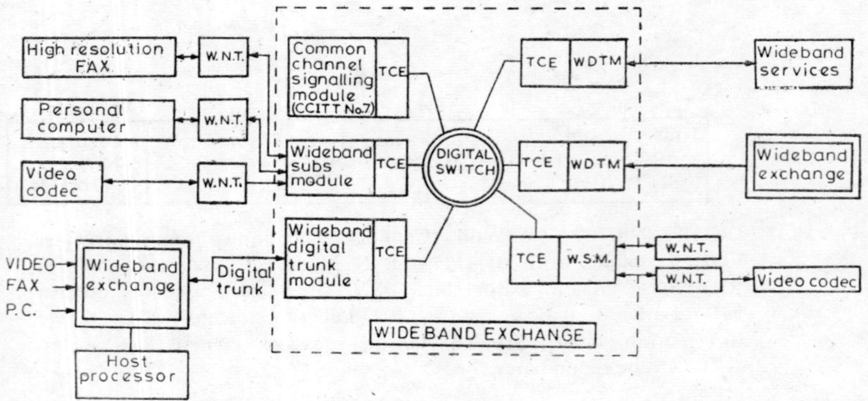

Fig. 10.16 Wideband ISDN using digital switches and trunks; W.N.T: Wideband network termination; TCE: Terminal control element; WDTM: Wideband digital trunk module; WSM Wideband subscriber module

Since the bandwidth available in digital trunks is sufficient to carry the wideband signals, there will be no problem of time skew (unequal transit delays) in the transmission media. Thus, the main-problem in providing wideband ISDN (WISDN) services through the currently available digital exchanges, is that the switching network should be able to handle H_1 channels without any time skew.* But unequal transit delays invariably occur when the associated channels (bytes), comprising the wideband word, are switched through N paths in the switching network. As a result, the consecutive bytes in a word are reassembled incorrectly at the switch output. To correct the situation, compensation must be provided for the time skew, so that the reassembled words have correctly positioned bytes. One possible method is to add a common frame identifier at the switch input to each of the N bytes (assuming that the wideband signal consists of frames of N bytes each, a frame occurring every 125 μ secs) and then use this frame identifier to resolve the time skew at the switch output. The overall wideband user-to-user connection through a digital switch is shown in Fig. 10.17,

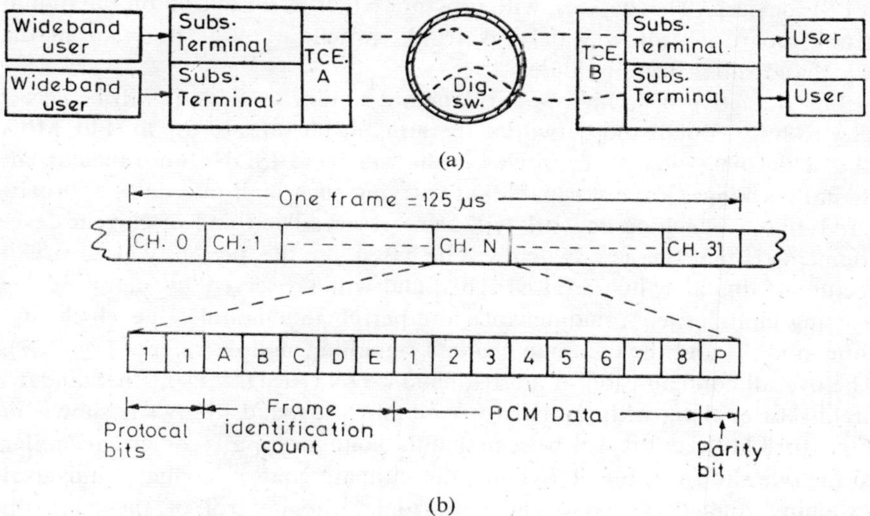

Fig. 10.17 Wideband user-to-user connection and Frame structure; (a) user-to-user connection; (b) Frame structure showing byte identification bits

where the subscriber modules consist of terminal control elements (TCE) and one or more wideband subscriber terminals. The transmit terminal adds the frame identifier signal which is used by the receive terminal to remove any skew. Multiple paths are established through the T-S-T switch between the wideband subscribers modules, thus producing unequal transit delays. It may be recalled (from Chapter 7) that the PDC at the switch input usually consists of 16 bit words, of which 8 bits are for information and the other 8 bits for control and signalling. Since in the ISDN structure, the D-channel

*From the discussion in Chapter 7, it will be seen that the data speeds up to 32.768 Mb/s (refer to Fig. 7.34) are handled by digital switching networks. However, the different words in a frame are switched through different paths and the time skew problem is not encountered.

provides 'out-slot' signalling information, the 8 bits in the bytes of the wideband words are spare and out of these, 5 positions are used to carry a mod-32 frame identification count which is indexed once per frame. The current value of this count is inserted into these positions on all N associated bytes. At the switch output, the bytes are reassembled in proper sequence by using an array of RAM's and CAM's (content addressable memory) and following the frame identification number associated with each byte.

A versatile cost-effective wideband switching module requires all 30 CAM/RAM pairs and their associated control circuitry to be built as a single custom-made VLSI circuit. The control requirement is that any of the 30 channels (bytes) can be a single channel, any can be wideband and any combination of channels can be involved simultaneously in one or more paths. This completely general scheme will be implemented in future through VLSI devices. The devices will function either as an option on the digital trunk board to allow wideband trunk switching or as interface in the wideband subscriber's module.

For iteractive TV, high speed data and similar services, it will be necessary to have Broadband networks operating at bit rates of up to 140 Mb/s and with the switching facility as in the wideband ISDN. Since the current digital exchanges are not capable of carrying such high rate data, a broadband digital switching network will be necessary. The trend of current developments is that the new switch will be based on the presently used architecture of digital switches (at 64 Kb/s) and will be served by many of the existing modules, e.g., maintenance and peripherals module, the clock and tone module, and the common channel signalling module (refer to Fig. 7.27). The overall configuration of a broadband ISDN (BISDN) using the standard architecture, along with the narrowband and wideband ISDN's is shown in Fig. 10.18 [17], and it will be seen that the configuration is somewhat similar to the one shown in Fig. 1.7. Thus, the ultimate goal of having a universal switching and transmission network under the control of the common signalling system will be achieved in a decade or so.

The progress in optical fibre technology has further enhanced the prospect of the success of BISDN, as the broadband trunks, broadband subscriber's lines as well as the interconnections within the broadband switching network, will all be using optical fibres as the interconnecting media. The development of integrated optical devices and WDM techniques will result in efficient opto-electronic broadband switches and multiplexed transmission in optical fibres. Such a broadband integrated optical fibre network (a local network) is shown in Fig. 10.19, where the users and the network switches are interconnected through optical fibres. It has been forecast that such local area networks will be vary economical and the cost of optical fibre networks will not exceed twice that of the audio cable network. For longhaul circuits using 2 Gb/s optical fibre system, the cost of video-signal transmission at 2 Mb/s rate will be less than that of the audio cable transmission. Thus, the universal ISDN of the future encompassing WISDN and BISDN, will not

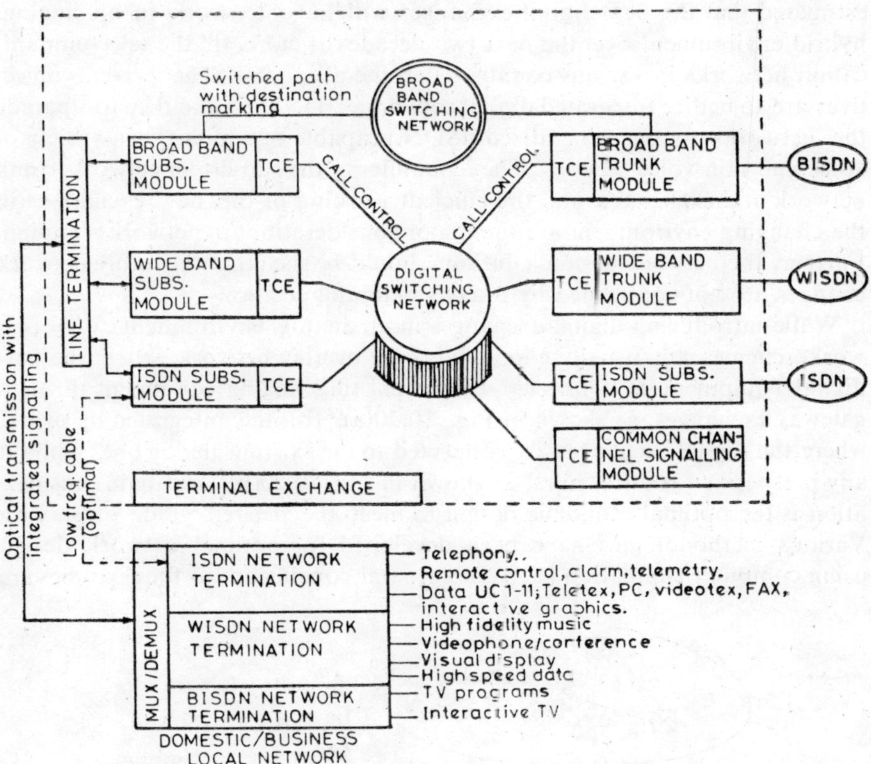

Fig. 10.18 Overall configuration of an ISDN/WISDN/BISDN universal network

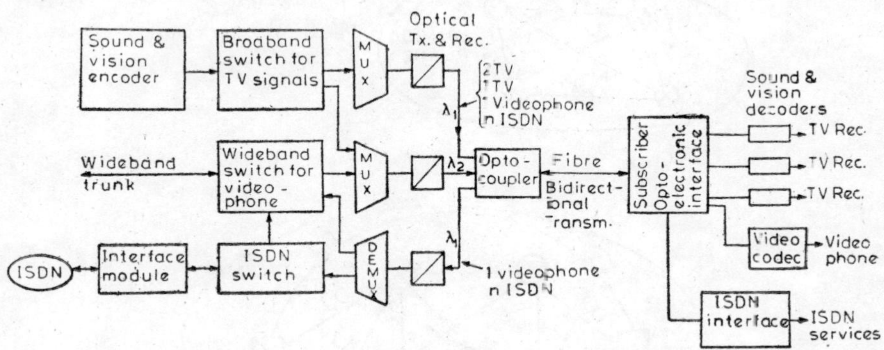

Fig. 10.19 A broadband local network using optical fibres and WDM

only provide exotic user services, but at the same time, will be cost-effective as well [18].

10.3.3 Network Planning [19, 20]

Till date, the investment in analog exchanges has been large, and as such, digital exchanges will be introduced in phases as the demand grows. It is estimated that the new digital exchanges will have to work in an analog/hybrid environment over the next two decades or more, till the telecommunication networks in various countries become all digital. The present objectives are to realize Integrated digital networks (IDN) first and then to upgrade the networks to the generalized ISDN, capable of supporting a variety of voice and non-voice services. The economics of the subscriber network, trunk network and exchanges, and the efficient working of the new exchanges in the changing environment are the major considerations in network planning. Further, the network design solutions should be such that the future network changes are not contrained by today's planning decisions.

While introducing digital exchanges in an analog environment, two network schemes are usually adopted: (a) the overlay network, where the new digital equipment is connected to the existing analog equipment through gateway exchanges, as shown in Fig. 10.20(a), (b) the integrated network, where the digital exchanges are connected to the existing analog ones without any prespecified routing rules, as shown in Fig. 10.20(b). The main consideration is the optimal economic design to meet the desired grade of service. Various methodologies have been developed for optimal network design using computer simulation [19]. The general conclusions of these studies are

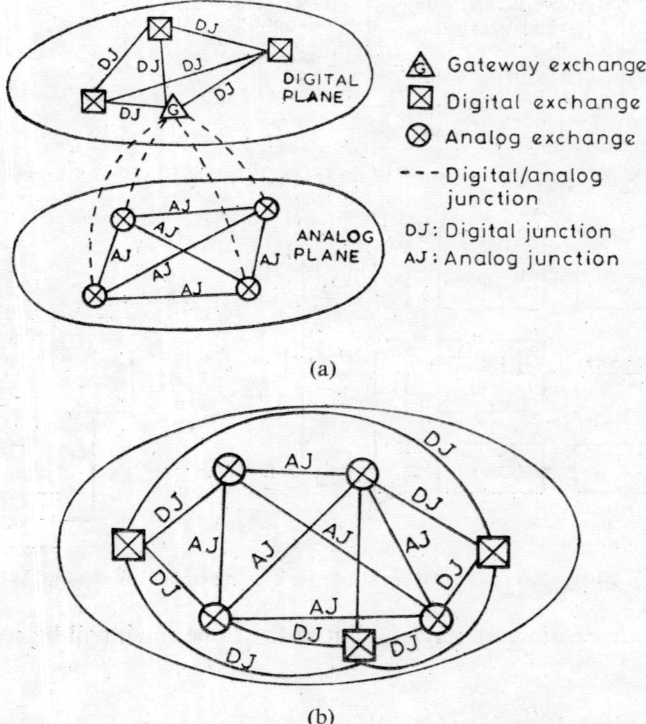

Fig. 10.20 (a) Typical overlay network configurationn; (b) Typical integrated network configuration

that: each application has its own optimal network structure and a distributed control architecture (of the digital switch) offers greater freedom in the choice of the optimal network. A typical study of a multi-exchange area, in Fig. 10.21(a), shows that the network digitization may be introduced in two successive stages, as in Fig. 10.21(b) and (c). In the intermediate stage of

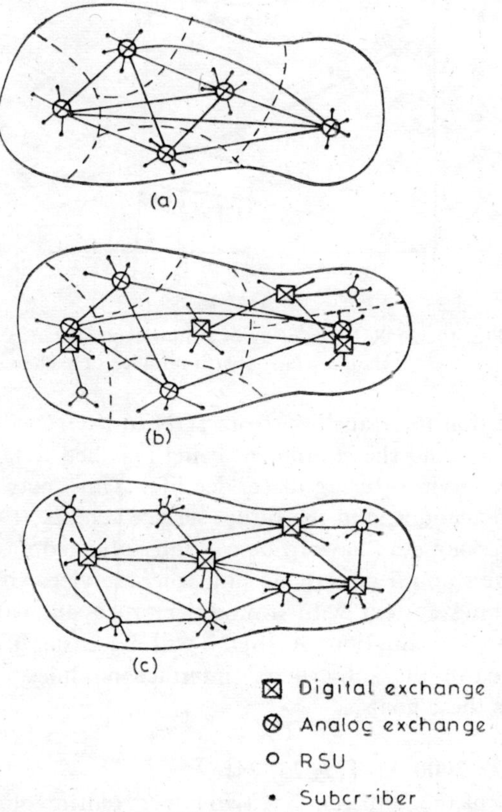

Fig. 10.21 Progressive digitization of a multiexchange area (a) Analog exchange only, (b) Partial digitization of the network, (c) complete digitization of the network

Fig. 10.21(b), the present service area borders and existing exchange buildings are retained; and only two new digital exchanges are cut over. The traffic demand is served by two more digital exchanges, co-located with analog exchanges, and two digital RSU's (remote subscriber unit). Ultimately, the optimum (ideal) solution is achieved by having only four main digital exchanges alongwith six RSU's as shown in Fig. 10.21(c). The overall economy of the total network depends on the costs of exchanges, junction network and subscriber network, and their variation with the number of exchanges is shown in Fig. 10.22. The use of small RSU's with a capacity of 100 or more reduces the cost of the subscriber plant when connecting distant subscribers. It is seen from Fig. 10.22, that there is a minimum in the network cost function and this minimum indicates the optimum number of exchanges to be installed.

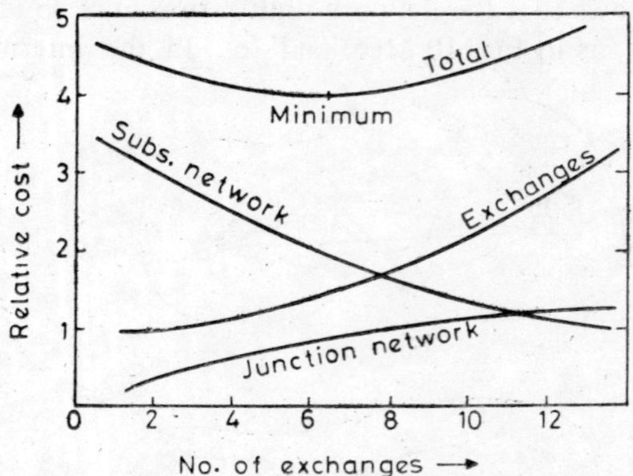

Fig. 10.22 Relative costs in a multiexchange area and their variation with number of exchanges

It is expected that the transition from IDN to ISDN will take place in the next 15 to 20 years and the equipment being installed today will be required to handle ISDN services during its service life. Thus, network planning has to take this into account, and the future service centres (for videotex, data and word processing etc.) have to be carefully planned along with the total network. For the smooth growth of non-voice services, it is important to choose a switching system with a modular processing capacity (e.g., ITT-1240). Also, the introduction of ISDN will be easier if more processing power is provided in the subscribers' interface modules [20]. The present trend is towards these goals.

10.4 TOWARDS 2000 AD [22, 23, 24]

In the later half of the 20th century, two concurrent revolutions in telecommunication technology, viz., the Information Revolution (see Sec. 10.2), and the Digital Revolution (refer to Fig. 1.8), are taking place. It is estimated that in this decade, the world's total knowledge base is doubling every two or three years, and it may double every year at the turn of the century. The key to this rapid growth of knowledge in this century lies in our ability to communicate at long distances, to process and store information in large data bases, and to bring in new dimensions in information technology in the form of videotex, high speed FAX, interactive TV and others. At the same time, the digital revolution has integrated both transmission and switching, resulting in IDN and the evolving ISDN, and leading to universal digital connectivity. Technological advances in devices indicate the possibility of packing 10^7 transistors on a chip and of increasing the processing speed of μP from 10^6 instructions to 10^7 instructions per second with a concurrent reduction in cost. Hopefully, the next century will see the realization of an entire 'system on a chip'. To achieve the expected goals, research and development efforts are being made in the following key technologies (goals are also indicated):

(a) Micro-electronics:	'System on a chip'
(b) Fibre optic technology:	$\lambda \rightarrow 5\ \mu$m, loss $\rightarrow 10^{-3}$ dB/Km, data rate $\rightarrow$ 10 Gb/s, WDM, coherent detection, integrated optics.
(c) Storage technology:	High density low cost mass storage, capacity—5 to 10 Gbytes, speed—10 Mbytes/s, access time—0.1 s, storage life—10 to 30 yrs, BER—10^{-10}.
(d) Display technology:	Liquid crystal flat-screen displays with colour capability.
(e) Speech processing:	Recognition, synthesis, voice dialling, hand-free operation of terminals, directory and assistance.
(f) Computer architecture and software development:	Distributed processing, breakthrough models and algorithms, associated and systolic arrays, automation of software development and maintenance.
(g) Human factor engineering:	More friendly man-machine interfaces.
(h) Artificial intelligence:	Expert systems.
(i) Executive computer:	Intelligent information processing and decisoin making in response to voice commands.

Many futurists think that the hardware goal of having 10^8 or so transistors on a chip will result in the complete 'system on a chip', thus delivering the exotic products, e.g., ESS-on-a-chip, office-on-a-chip, library-on-a-chip etc. However, a system is more than hardware; it is hardware, software, microcode, user interface, documentation and performance evaluation. It is an aggregate of parts and functions too complex to comprehend by a single person. The enormous complexity of tomorrow's systems requires completely new approaches to system design. It will be necessary to realize new description languages, advanced CAD and CAM techniques and sophisticated design environment using a network of work stations, so that the futuristic systems are efficiently designed and economically manufactured [23].

It is amply evident that the computer and communication industries are converging in their technologies, architecture and applications. The underlying technology that adds value is communication. Without high-rate communication facilities, the high processing power of the present day computer systems could not be fully and usefully utilized; and this leads to the evolution of C and C (computer and communication) networks with their flexibility and multiservice facilities to the users. The concept of ISDN has evolved from this desired interaction between computer and communication, and its applications have resulted not only in modern automated office and home, but also in fully integrated factory automation systems. Such integrated automation systems include computer-aided design (CAD), computer-aided manufacturing (CAM), materials management, robotics, workstations and security systems. Further, an overall factory information system, capable of integrating the exchange of data, voice, text and video between different departments, including the factory office and shop floor, is operated through an ESS. Such all-embracing design, management and information systems are only possible through the extended application of the ISDN concept [17].

As a consequence of the expected progress in ISDN (even partial), the scenarios at the office, home, factory, bank and at many other activity centres, will be completely revolutionized. Consider, for example, a futuristic personal office, shown (in sketch) in Fig. 10.23, where many functions will be automated and papers replaced by versatile flat-screen displays. Speech recognition systems will make the office 'user friendly', although keyboards and other input/output devices will also be used. With a little sophistication, it will be possible for the office to recognise the identification code of the authorized person and welcome him to function in the office (or give an alarm, if he is unauthorized). An extension of this concept leads to the 'paperless' corporate office, shown schematically in Fig. 10.24, where all

Fig. 10.23 Futuristic office with flat panel display, user friendly personal computers (with voice commands), video displays and key boards (after Wexelblat [23])

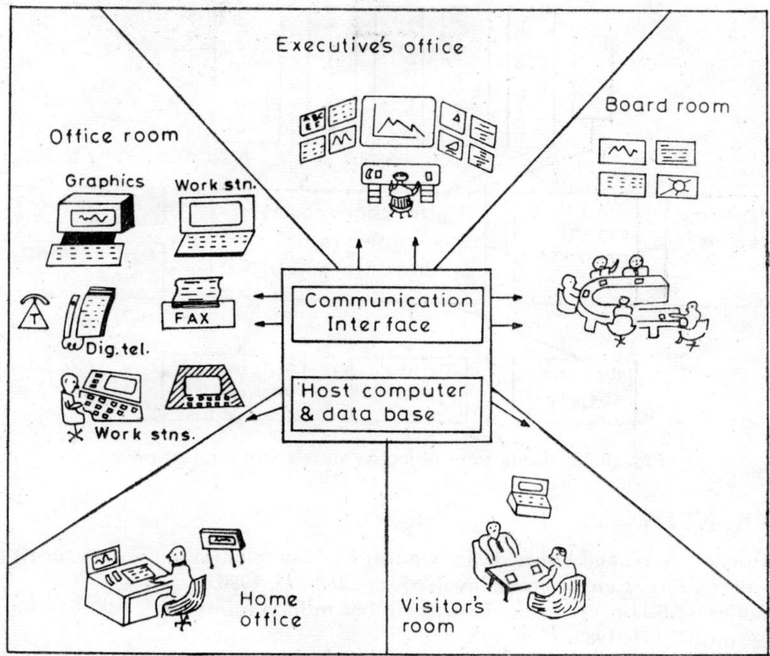

Fig. 10.24 Futuristic paperless office, using fibre optic links and ISDN (after Kobayashi [24])

transactions, discussions, conference and decision will be carried out through the *C* and *C* network interconnecting all the suboffices. The visitor will no longer wait to have an interview with the executive as he can discuss business through computer terminals. Flat-panel displays in the board room and in the executive's office will give all the information (on pressing a button) required for taking decisions. All the office files will, of course, be stored in the data base of the host computer and one may access other data bases, if necessary.

The concurrent digital revolution will not only provide the global digital connectivity, but also the required signal processing, data compression, data protection, storage and most of all, data transmission through digital techniques, as indicated in Fig. 10.25. The resultant *C* and *C* network and the evolving ISDN will not only lead to better and stimulating work environment in the form of the automated office, home and factory, but at the same time, will contribute significantly towards mutual understanding among the people of the world, leading to a better prospect for durable world peace. Hopefully, the Information Revolution and Digital Revolution will jointly succeed in creating a more open society and a global family—the long term objectives in telecommunications [24, 25].

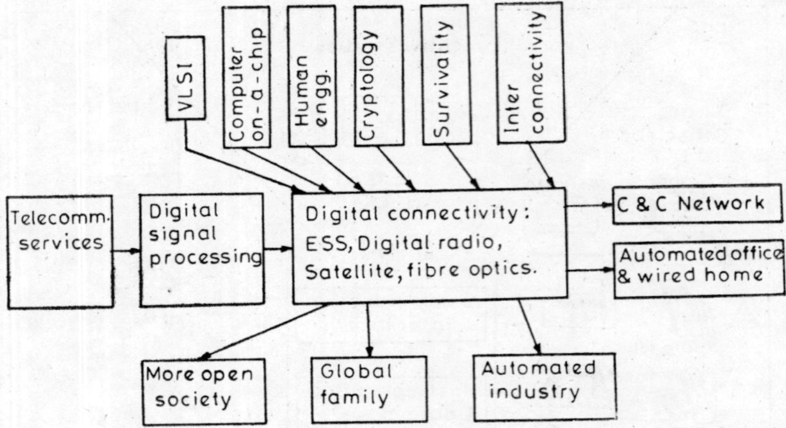

Fig. 10.25 Long term objective in telecommunications.

REFERENCES

1. Horsley, A.W. and Usher, E.S., 'Optical fibre communication systems in PTT networks', *Elec. Communication,* vol. 55, pp. 268-275, 1980.
2. Soloway, B.H., 'VLSI: the challenge of one million transistors', *Elec. Comm.*, vol. 58, pp 109-111, 1983.
3. Special issue on Signal Processing, *Elec. Communications*, vol. 59, No. 3, 1985.
4. Christensen, R., Johnsen, O., and Patovan, B., 'Speech processing in public telephone exchanges', *Elec. Comm.*, vol. 59, pp 266-272, 1985.
5. Wiggins, R. and Brantingham, L., 'Three-chip system synthesizes speech', *Electronics*, pp 109-116, Aug. 31, 1978.
6. Kneib, K.N., 'Trends in single-chip programmable digital signal processors', *Elec. Comm.*, vol. 59, pp 312-319, 1985.
7. Carne, E.B., 'New dimensions in telecommunications', *IEEE Comm. Mag.*, vol. 20, pp 17-25, Jan, 1982.
8. Nakajima, H., *et al.*, 'Recent developments in video information systems in Japan', *Hitachi Rev.*, vol. 28, pp 295-300, 1979.
9. Special issue on 'Communications in the automated office,' *IEEE Trans. on Communications*, vol. Com-30, No. 1, 1982.
10. Yamamoto, T., 'Information processsing by optoelectronics', *IEEE Comm. Mag.*, vol. 20, pp 4-12, May 1982.
11. Villar, J.E., 'Telecommunication research in the 1990s', *Elec. Comm.*, vol. 58, No. 1, pp 120-123, 1983.
12. Smith, E.A. *et al.*, 'Impact of non-voice services on network evolution', *Elec. Comm.*, vol. 56, No. 1, pp 17-29, 1981.
13. Robin, G., and Treves, S.R., 'An introduction to integrated services digital networks', *Elec. Comm.*, vol. 56, No. 1, pp 4-16, 1981.
14. Duc, N.Q., and Chew, E.K., 'ISDN protocol architecture', *IEEE Comm. Mag.*, vol. 23, No. 3, pp 15-22, 1985.
15. Carter, C.R., *et al.*, 'Transport methods for Integrated services digital networks', *Elec. Comm.*, vol. 56, No. 1, pp 57-70, 1981.
16. Treves, S.R., and Upp, D.C., 'Technique for wideband ISDN applications', *Elec. Comm.* vol. 59, No. 1/2, pp 131-136, 1985.
17. Gimpelson, L.A. and Treves, S.R., 'Telecommunication networks beyond ISDN Transport', *Elec. Comm.*, vol. 59, No. 1/2, pp 220-227, 1985.

18. Ohnsorge, H , 'From voice to video communication', *Elec. Comm.*, vol. 58 No. 1, pp 127-130, 1983.
19. Casas, F.A., and Casali, F., 'Network planning and applications', *Elec. Comm.*, vol. 56, No. 2-3, pp 302-314, 1981.
20. Caballero, P.A., *et al* , ''Digital network planning', *Elec. Comm.*, vol. 59, No. 1/2, pp 200-206, 1985.
21. Robin G, and Treves, S.R., 'Pragmatic introduction of digital switching and transmission in existing networks', *IEEE Trans. on Comm.*, vol. Com-27, pp 1071-1078, 1979.
22. Fontaine, B.J., 'Future direction', *Elec. Comm.* vol. 59, No. 1/2, pp 228-234, 1985.
23. Wexelblat, R.L., 'Designing the systems of tomorrow', *Elec. Comm.*, vol. 58, No. 1, pp 93-97, 1983.
24. Kobayashi, K., 'Telecommunication in the future', *IEEE Comm. Mag.*, vol. 18, No. 4, pp 23-27, 1980.
25. Das, J., 'The information age and the digital revolution', *JIETE*, vol. 28, pp 633-642, 1982.

APPENDIX A

Random Signals

The major emphasis in the text has been on the effect of channel noise and other imperfections in the overall system performances. While some of the imperfections, e.g., intersymbol interference and bandwidth constraints, are deterministic, the channel noise, and sometimes, the propagation characteristics, e.g., fading and multipath, are probabilistic. It is assumed that the reader has prior exposure to the theory of probability and stochastic processes. However, for ready reference, some important results of Random signal theory are briefly discussed here.

A.1 PROBABILITY FUNCTIONS [1, 2]

For a continuous random variable, say, the sample values of a noise waveform, the probability density function (*pdf*) is defined as:

$$p(x) = \lim_{\Delta x \to 0} \frac{P(x - \Delta x/2) \leqslant X \leqslant x + \Delta x/2)}{\Delta x} \tag{A.1}$$

where $P(\cdot)$ denotes the probability that X lies between $(x - \Delta x/2)$ and $(x + \Delta x/2)$. Thus, the probability that X lies between x_1 and x_2 is :

$$P(x_1 \leqslant X \leqslant x_2) = \int_{x_1}^{x_2} p(x)\, dx$$

and

$$\int_{-\infty}^{\infty} p(x)\, dx = 1 \tag{A.2}$$

On the basis of this, the probability distribution function $P(X_1)$ (also called Cumulative distribution function, (*cdf*)), is defined as:

$$P(X_1) = \int_{-\infty}^{X_1} p(x)\, dx$$

and

$$p(x) = \frac{dP(x)}{dx} \tag{A.3}$$

Two Dimensional pdf

The joint *pdf* $p(x, y)$ for variables x, y is defined such that the probability of a joint observation falling in dx and dy is $p(x, y)\, dx\, dy$. Hence,

$$P(x_1 \leqslant X \leqslant x_2; y_1 \leqslant Y \leqslant y_2) = \int_{x_1}^{x_2} \int_{y_1}^{y_2} p(x, y)\, dx\, dy$$

Then the two-dimensional *cdf* is given by:

$$P(x_1, y_1) = \int_{-\infty}^{x_1} \int_{-\infty}^{y_1} p(x, y)\, dx\, dy \tag{A.4}$$

For statistically independent variables,

$$p(x_1, y_1) = p_1(x_1) \cdot p_2(y_1)$$

or

$$p(x, y) = p_1(x) \cdot p_2(y) \tag{A.5}$$

If however the variables are not independent, then,

$$\begin{aligned} p(x, y) &= p_1(x) \cdot p_{21}(y/x) \qquad \text{(A.6)*} \\ &= p_2(y) \cdot p_{12}(x/y) \end{aligned}$$

where $p_{21}(y/x)$ is a conditional density function, i.e., the conditional probability of y, given that X lies between x and $x + dx$. $p_{12}(x/y)$ is similarly defined. Further,

$$\int_{-\infty}^{\infty} p(x, y)\, dy = p(x) \quad \text{and} \quad \int_{-\infty}^{\infty} p(x, y)\, dx = p(y) \tag{A.7}$$

A.1.1 Statistical Averages

The nth moment of a continuous variable is defined as:

$$E[x^n] = \int_{-\infty}^{\infty} x^n \cdot p(x)\, dx$$

and for a second order density function,

$$E[x^n \cdot y^m] = \int_{-\infty}^{\infty} \int_{-\infty}^{\infty} x^n \cdot y^m p(x, y)\, dx\, dy \tag{A.8}$$

The first moment $m_1 (n = 1)$ is a measure of the average $\bar{X}$ of the variable or the expected value of the random process at time t_1 for which X was defined. Thus,

$$m_1 = \bar{X} = E[X] = \int_{-\infty}^{\infty} x\, p(x)\, dx \tag{A.9}$$

Similarly, the expected value of a function $f(x)$ is defined as:

$$E[f(X)] = \int_{-\infty}^{\infty} f(x) \cdot p(x)\, dx$$

The second moment m_2 gives the mean-square value of X, i.e.,

$$m_2 = \overline{X^2} = E[X^2] = \int_{-\infty}^{\infty} x^2 p(x)\, dx \tag{A.10}$$

The variance σ_x^2 of the variable X is now given by:

$$\begin{aligned} \sigma_x^2 &= E\{[X - E(X)]^2\} \\ &= E(X^2) - [E(X)]^2 \\ &= m_2 - m_1^2 \end{aligned} \tag{A.11}$$

*The relation $p(x/y) = \dfrac{p(x, y)}{p(y)}$ is also known as Bayes' theorem on Inverse probability.

Based on this, the nth order central moment μ_n is defined as:

$$\mu_n = E[(X - m_1)^n] = \int_{-\infty}^{\infty} (x - m_1)^n \cdot p(x)\, dx \tag{A.12}$$

It is then observed that $\mu_1 = 0$, but the second order central moment μ_2 is (from Eq. (A.11)):

$$\begin{aligned} \mu_2 &= E[(X - m_1)^2] = m_2 - m_1^2 \\ &= \sigma_x^2 \end{aligned} \tag{A.13}$$

In the case of two random variables, X and Y, with joint *pdf* $p(x, y)$, the joint moment is defined as:

$$E[X^k Y^n] = \int_{-\infty}^{\infty} \int_{-\infty}^{\infty} x^k y^n\, p(x, y)\, dx\, dy$$

and the joint central moment as:

$$E[(X - \overline{X})^k \cdot (Y - \overline{Y})^n] = \int_{-\infty}^{\infty} \int_{-\infty}^{\infty} (x - m_{1x})^k (y - m_{1y})^n\, p(x, y)\, dx\, dy \tag{A.14}$$

Of particular interest are the moments with $k = n = 1$, and these joint moments are called the correlation ρ_{ij} and the covariance μ_{ij} of the random variables X and Y, respectively. These joint moments are most useful in practical applications, e.g., in predictive and adaptive algorithms (refer to Chapter III). For a set of random variables $\{X_i\}$, $i = 1, 2, \ldots, n$, with the joint *pdf* $p(x_1, x_2, \ldots, x_n)$, the values of the correlation ρ_{ij} and covariance μ_{ij} are given as:

$$\rho_{ij} = E[X_i X_j] = \int_{-\infty}^{\infty} \int_{-\infty}^{\infty} x_i x_j p(x_i, x_j)\, dx_i\, dx_j$$

and

$$\begin{aligned} \mu_{ij} &= E[(X_i - m_{1i})(X_j - m_{1j})] \\ &= E[X_i X_j] - m_{1i} m_{1j} \end{aligned} \tag{A.15}$$

The $(n \times n)$ matrix with μ_{ij} is called the covariance matrix of the variables X_i.

Two random variables are said to be uncorrelated if

$$E(X_i X_j] = E(X_i) \cdot E(X_j) = m_{1i} m_{1j}$$

and

$$\mu_{ij} = 0 \tag{A.16}$$

Thus, when X_i and X_j are statistically independent, they are also uncorrelated. But if X_i and X_j are uncorrelated, they are not necessarily statistically independent. The variables X_i and X_j are said to be orthogonal, if $E[X_i X_j] = 0$. This holds when X_i and X_j are uncorrelated and either one or both of the variables have zero mean.

Examples

(a) Rectangular or Uniform *pdf* has:

$$p(x) = 1/a, \qquad 0 \leqslant x \leqslant a$$

$$m_1 = a/2, \qquad m_2 = a^2/3$$

Therefore,

$$\sigma_x^2 = m_2 - m_1^2 = a^2/12.$$

(b) Exponential *pdf*:

$$p(x) = a \cdot e^{-b|x|}, \qquad -\infty < x < \infty, \qquad b > 0$$

The moments are: $\quad m_1 = 0;\ m_2 = \dfrac{4a}{b^3};$

Therefore, $\quad \sigma_x^2 = 4a/b^3.$

(c) Gaussian *pdf*: $\quad p(x) = \dfrac{1}{\sqrt{2\pi}\sigma} \exp\,[-(x - x_0)^2/2\sigma^2].$

The moments are $\quad m_1 = x_0; \qquad m_2 = x_0^2 + \sigma^2$

Therefore, $\quad \sigma_x^2 = \sigma^2.$

(d) Rayleigh *pdf*: $\quad p(x) = \begin{cases} \dfrac{x}{\mu^2}\, e^{-x^2/2\mu^2}, & 0 \leqslant x \leqslant \infty \\ 0 &, \ x < 0 \end{cases}$

The moments are: $\quad m_1 = \sqrt{\dfrac{\pi}{2}}\,\mu; \qquad m_2 = 2\mu^2$

Therefore, $\quad \sigma_x^2 = 0.43\mu^2.$

Bivariate Gaussian pdf

Among the many possible multivariate or multidimensional distributions, the bivariate or two-dimensional Gaussian *pdf* is an important case, where, from Eq. (A.15), the mean m_x and the covariance matrix M are:

$$m_x = \begin{bmatrix} m_{11} \\ m_{12} \end{bmatrix}; \quad [M] = \begin{bmatrix} \sigma_{x1}^2 & \mu_{12} \\ \mu_{12} & \sigma_{x2}^2 \end{bmatrix} \tag{A.17}$$

Defining normalised covariance:

$$\mu_{oij} = \frac{\mu_{ij}}{\sigma_{xi}\sigma_{xj}}, \qquad i \neq j,$$

where $0 \leqslant |\mu_{oij}| \leqslant 1$, the matrix $[M]$ is now given by (simplifying some of the notations):

$$[M] = \begin{bmatrix} \sigma_1^2 & \mu_0\sigma_1\sigma_2 \\ \mu_0\sigma_1\sigma_2 & \sigma_2^2 \end{bmatrix}$$

and its inverse is:

$$[M^{-1}] = \frac{1}{\sigma_1^2\sigma_2^2(1 - \mu_0^2)} \begin{bmatrix} \sigma_2^2 & -\mu_0\sigma_1\sigma_2 \\ -\mu_0\sigma_1\sigma_2 & \sigma_1^2 \end{bmatrix}$$

with $\quad \det|M| = \sigma_1^2\sigma_2^2(1 - \mu_0^2).$

The joint *pdf* of the correlated variables x_1 and x_2 is now given by:

$$p(x_1, x_2) = \frac{1}{2\pi\sigma_1\sigma_2\sqrt{1-\mu_0^2}} \times$$

$$\exp\left[-\frac{\sigma_2^2(x_1 - m_{11})^2 - 2\mu_0\sigma_1\sigma_2(x_1 - m_{11})(x_2 - m_{12}) + \sigma_1^2(x_2 - m_{12})^2}{2\sigma_1^2\sigma_2^2(1-\mu_0^2)}\right] \tag{A.19}$$

With $\mu_0 = 0$, $p(x_1, x_2) = p(x_1)\cdot p(x_2)$, i.e., x_1 and x_2 are statistically independent. This theorem is extended to the multivariate case in a straightforward way.

A.1.2 Sums of Random Variables

When independent random variables X and Y with distinct *pdf*'s are added to give a new variable Z, then, $p(z)$ is given by the convolution integral,

$$p(z) = \int_{-\infty}^{\infty} p(x)p(z-x)\,dx, \quad z = (x+y) \tag{A.20}$$

The theorem may be extended for successive additions of variables U, V, W etc., and the result shows that the final *pdf* tends to be Gaussian, independent of the individual *pdf*'s, given that the number of variables added are sufficiently large. This is known as the Central limit theorem.

Consider now the set of random variables $\{X_i\}$, $i = 1, 2, \ldots, n$ which are statistically independent, but identically distributed, and each having a finite mean m_{1x} and a finite variance σ_x^2. If Y is defined as the normalised sum, called the sample mean, then,

$$Y = \frac{1}{n}\sum_{i=1}^{n} X_i;$$

$$E(Y) = m_{1y} = m_{1x};$$

and the variance of Y, $$\sigma_y^2 = \frac{\sigma_x^2}{n} \tag{A.21}$$

It may then be shown that $p(y)$ approaches a Gaussian *pdf* as $n \to \infty$, and the result leads to the Central limit theorem.

A.2 RANDOM PROCESSES : ACF and CCF [2]

Random processes may be stationary or non-stationary. A stationary random process is the one whose statistical properties are invariant with time, because the underlying physical mechanism generating the process is not changing with time, e.g., shot noise and thermal noise in vacuum tubes. A class of Random processes is further defined as Ergodic, if all types of their ensemble averages are interchangeable with the corresponding time averages. Then a process is ergodic if :

$$\langle x(t) \rangle = \overline{X} = \int_{-\infty}^{\infty} x\cdot p(x)\,dx = \lim_{T\to\infty}\frac{1}{2T}\int_{-T}^{T} x(t)\,dt$$

$$\langle x^2(t) \rangle = \overline{X^2} = \int_{-\infty}^{\infty} x^2\cdot p(x)\,dx = \lim_{T\to\infty}\frac{1}{2T}\int_{-T}^{T} x^2(t)\,dt$$

and in general,

$$\langle x^n(t) \rangle = \overline{X^n} = \int_{-\infty}^{\infty} x^n \cdot p(x)\, dx = \lim_{T\to\infty} \frac{1}{2T} \int_{T}^{T} x^n(t)\, dt \qquad (A.22)$$

We have discussed so far the *pdf* of random processes, and this mainly provides the information regarding the amplitude fluctuations of the process. Further characteristics of the processes are obtained from the correlation functions and the spectral densities, as discussed now.

A.2.1 Correlation Functions

The autocorrelation function (ACF) of a random (ergodic) process is defined as:

$$R_x(t_1, t_2) = E(X_1 X_2) = \iint x_1 x_2 \cdot p(x_1, x_2)\, dx_1\, dx_2 \qquad (A.23)$$

where X_1 and X_2 are random variables at times t_1 and t_2. If X_1 and X_2 are independent variables, then,

$$E(X_1 X_2) = E(X_1) \cdot E(X_2)$$

and if $t_1 \to t_2$, $R_x(t_1, t_2) = E(X_1 X_2) = E(X_1^2) = \overline{X_1^2} = m_2$

Let $t_2 = t_1 - \tau$, then,

$$R_x(\tau) = \lim_{T\to\infty} \frac{1}{2T} \int_{-T}^{T} x_i(t_1) \cdot x_i(t_1 - \tau)\, dt_1$$

$$= \lim_{T\to\infty} \frac{1}{2T} \int_{-T}^{T} x(t) \cdot x(t - \tau)\, dt \qquad (A.24)$$

The ACF $R_x(\tau)$ (of ergodic random processes) has the following properties:

(a) $R_x(0) \geqslant | R_x(\tau) |$

(b) $R_x(\tau) = R_x(-\tau)$ (A.25)

(c) If $x(t)$ contains periodic components in addition to random noise, then $R_x(\tau)$ also contains periodic components with the same period.

(d) Although a given $x(t)$ has a unique $R_x(\tau)$, a given $R_x(\tau)$ does not correspond to a unique $x(t)$.

If a sample function $y(t)$ of an ergodic random process is such that $y(t) = S(t) + n(t)$, and $S(t)$ and $n(t)$ are uncorrelated, the ACF of $y(t)$ is:

$$R_y(\tau) = \langle y(t) \cdot y(t-\tau) \rangle = E[S(t) \cdot S(t-\tau)] + E[s(t-\tau) \cdot n(t)]$$
$$+ E[n(t-\tau) \cdot S(t)] + E[n(t) \cdot n(t-\tau)] \qquad (A.26)$$
$$= R_s(\tau) + R_n(\tau), \text{ if either } S(t) \text{ or } n(t) \text{ is zero mean.}$$

This result is often used to detect $S(t)$ in present of noise.

If in Eq. (A.23), X_1 and X_2 belong to two different random processes, X and Y, and both processes are ergodic, then their cross correlation (CCF) is defined as:

$$R_{xy}(\tau) = \langle x(t) \cdot y(t - \tau) \rangle$$

$$= \lim_{T\to\infty} \frac{1}{2T} \int_{-T}^{T} x(t) \cdot y(t-\tau)\, dt \qquad (A.27)$$

Spectral Density G(f)

$R_x(\tau)$ and $G(f)$ are related by the Fourier transform pair given by:

$$G(f) = \int_{-\infty}^{\infty} R_x(\tau)e^{-j\omega\tau}\,d\tau$$

$$R_x(\tau) = \int_{-\infty}^{\infty} G(f)e^{j\omega\tau}\,df$$

$$= \frac{1}{2\pi}\int_{-\infty}^{\infty} G(\omega)e^{j\omega\tau}\,d\omega \tag{A.28}$$

As a corollary, $R_x(0) = \int_{-\infty}^{\infty} G(f)\,df = \overline{X^2}$.

Examples

(a) Consider $x(t) = E\sin(\omega t + \theta)$, then,

$$R_x(\tau) = \frac{1}{2T}\int_{-T}^{T} E^2 \sin(\omega t + \theta)\sin[\omega(t - \tau) + \theta]\,dt$$

$$= \frac{E^2}{2}\cos\omega\tau.$$

(b) A random telegraph signal has ± 1 as the amplitude and the number of zero crossings per second is β. Then the ACF of the signal is:

$$R_x(\tau) = e^{-2\beta|\tau|}$$

and the corresponding $G(f) = \dfrac{4\beta}{\omega^2 + 4\beta^2}$

(c) Consider a random binary signal (PNS) with amplitude ± 1 and time period T. Then the ACF is:

$$R_x(\tau) = \begin{cases} 1 - |\tau|/T, & |\tau| \leq T \\ 0, & |\tau| > T \end{cases}$$

and $$G(f) = \frac{\sin^2(\omega T/2)}{(\omega T/2)^2}$$

Grussian Noise

For White noise with uniform $G(f) = G_0$, the ACF is a delta function given by:

$$R_x(\tau) = \int_{-\infty}^{\infty} G_0 e^{j\omega\tau}\,df = G_0\delta(\tau) \tag{A.29}$$

For band-limited lowpass Gaussian noise with $G(f) = G_0$, $-f_0 \leq f \leq f_0$, ACF is given by:

$$R_x(\tau) = \int_{-f_0}^{f_0} G_0 e^{j\omega\tau}\,df = 2f_0G_0\,\frac{\sin\omega_0\tau}{\omega_0\tau} \tag{A.30}$$

and $$\overline{X^2} = R(0) = 2f_0G_0.$$

For bandpass noise with $G(f) = G_0$ in the band $(f_0 \pm \Delta f/2)$, the ACF is:

$$R_x(\tau) = 2G_0\Delta f \cos \omega_0\tau \left(\frac{\sin \Delta\omega\tau/2}{\Delta\omega\tau/2}\right), \quad \omega = 2\pi\Delta f \tag{A.31}$$

A.2.2 Networks with Random Inputs

Using convolution integral, the output of a network with the impulse response $h(t)$ is given by:

$$y(t) = \int_{-\infty}^{\infty} h(\tau)x(t - \tau)\, d\tau$$

If $y(t)$ is ergodic, then,

$$R_y(\tau) = \langle\, y(t_1)y(t_2)\,\rangle\,, \qquad \text{where} \quad (t_2 - t_1) = \tau$$

Using $y(t_1)$ and $y(t_2)$ as convolution integrals, the final result is:

$$R_y(\tau) = \int_{-\infty}^{\infty} h(\tau_1) \int_{-\infty}^{\infty} h(\tau_2)R_x(\tau - \tau_2 + \tau_1)\, d\tau_2\, d\tau_1 \tag{A.32}$$

The cross-spectral density of the output of the network is now obtained by using the transform pair (A.28) and the result is:

$$G_y(f) = \int_{-\infty}^{\infty} R_y(\tau)\cdot e^{-j\omega\tau}\, d\tau,$$

and using Eq. (A.32),

$$\begin{aligned} G_y(f) &= \int_{-\infty}^{\infty} h(\tau_1)e^{j\omega\tau_1}\, d\tau_1 \cdot \int_{-\infty}^{\infty} h(\tau_2)\cdot e^{-j\omega\tau_2}\, d\tau_2 \\ &\quad \times \int_{-\infty}^{\infty} R_x(\tau - \tau_2 + \tau_1)\exp\{-\, j\omega(\tau - \tau_2 + \tau_1)\}\, d\tau \\ &= H(-j\omega)\cdot H(j\omega)\cdot G_x(f) \\ &= |\, H(j\omega)\,|^2 \cdot G_x(f), \end{aligned} \tag{A.33}$$

where $|\, H(j\omega)\,|$ is the transfer function of the network.

Example

Consider that white noise with $G(f) = G_0$ is the input to a lowpass RC filter. Then, using Eq. (A.32),

$$R_y(\tau) = (G_0/2\text{RC})\cdot e^{-\tau/\text{RC}}, \qquad \tau \geqslant 0$$

where $h(t)$ of the RC network is

$$\frac{1}{\text{RC}}e^{-t/\text{RC}}, \qquad \text{and} \qquad R_x(\tau) = G_0\delta(\tau).$$

Since $R_y(\tau)$ is symmetric,

$$R_y(\tau) = \frac{G_0}{2\text{RC}}e^{-|\tau|/\text{RC}}, \; -\infty < \tau < \infty$$

and

$$\overline{Y^2} = R(0) = G_0/2\text{RC} \tag{A.34}$$

The cross-spectral density of the output of the RC network is:

$$G_y(f) = G_x(f)\cdot \mid H(j\omega) \mid^2$$

$$= G_0 \frac{(1/\mathrm{RC})^2}{\omega^2 + (1/\mathrm{RC})^2}, \tag{A.35}$$

and
$$\overline{Y^2} = \frac{1}{2\pi}\int_{-\infty}^{\infty} G_y(\omega)\, d\omega = G_0/2\mathrm{RC}.$$

Equivalent Noise Bandwidth

The equivalent noise bandwidth W_n of a network is defined as

$$W_n = \frac{\int_0^{\infty} \mid H(j\omega) \mid^2 d\omega}{H_0^2}, \tag{A.36}$$

and
$$\overline{Y^2} = (G_0 W_n H_0^2)/\pi,$$

where H_0 is the value of $H(j\omega)$ at $W = 0$.

For the above RC lowpass filter, $W_n = \pi/2\mathrm{RC}$. If the half-power bandwidth is W_1, then, $W_n = \dfrac{\pi}{2} W_1$.

A.3 NONLINEAR TRANSFORMATIONS [2]

When a variable X is processed through a nonlinear device, the *pdf* of the output Z is given by:

$$p(z) = p(x)\cdot \left| \frac{dx}{dz} \right| \tag{A.37}$$

where, the device characteristic is given by, $Z = F(X)$. If $F(X)$ is multi-valued, then

$$p(z) = p(x_1) \left| \frac{dx_1}{dz} \right| + p(x_2)\cdot \left| \frac{dx_2}{dz} \right| + \ldots \tag{A.38}$$

Examples

(a) For a square law device given by $Z = X^2$, and an exponential *pdf* of x, $p(x) = e^{-x}$. The *pdf* of the output is:

$$p(z) = \frac{e^{-\sqrt{z}}}{2\sqrt{z}}, \qquad 0 \leqslant z \leqslant \infty$$

and
$$E(z) = \overline{Z} = 2.$$

(b) Assume a Gaussian *pdf* for the input to a linear diode. Then the *pdf* of the output is given as:

$$p(z) = \frac{1}{2}\,\delta(z) + \frac{1}{\sqrt{2\pi}\sigma} e^{-z^2/2\sigma^2}, \qquad z \geqslant 0.$$

(c) Consider that a random waveform is passed through an ideal limiter, then the output Z is a random telegraph signal. If the number of zero-crossings obey a Poisson distribution given by,

$$P(n, T) = \frac{(\beta T)^n}{n!} e^{-\beta T},$$

then, the *pdf*'s of (+1) and (−1) are:

$$p(+1) = \frac{1}{2}(1 + e^{-2\beta|\tau|})$$

and $$p(-1) = \frac{1}{2}(1 - e^{-2\beta|\tau|}).$$

The ACF and $G(f)$ of the waveform has been given in Sec. A.2.1.

A.3.1 Envelope and Phase of a Sinewave Pluse Narrowband Gaussian Noise [1]

For a Gaussian noise input $n(t)$, the output $n_0(t)$ of a narrowband bandpass filter may be represented as:

$$\begin{aligned} n_0(t) &= r(t) \cos [\omega_0 t + \theta(t)] \\ &= x(t) \cos \omega_0 t + y(t) \sin \omega_0 t \end{aligned} \tag{A.39}$$

where $r(t)$ is the envelope function; $\theta(t)$, phase function; and

$$x(t) = r(t) \cos \theta(t), \qquad y(t) = -r(t) \sin \theta(t),$$

are the inphase and quadrature components of $r(t)$. Further,

$$\langle x^2 \rangle = \langle y^2 \rangle = \langle n_0^2 \rangle.$$

Assuming the filter bandwidth $B \ll f_0$; $r(t)$, $\theta(t)$, $x(t)$, and $y(t)$ are all slowly varying components. $x(t)$ and $y(t)$ have Gaussian amplitude distributions with zero mean, and the correlation between x and y, $E(XY) = 0$; hence their joint *pdf* is:

$$p(x, y) = p(x) \cdot p(y) = \frac{1}{2\pi\sigma^2} \exp\left(-\frac{x^2 + y^2}{2\sigma^2}\right) \tag{A.40}$$

The *pdf*'s of $\theta(t)$ and $r(t)$ are given by

$$p(\theta) = \frac{1}{2\pi}, \text{ i.e., uniform over 0 to } 2\pi,$$

and $p(r)$ is the well-known Rayleigh distribution

$$p(r) = \frac{r \cdot e^{-r^2/2\sigma^2}}{\sigma^2}, \qquad r \geq 0 \tag{A.41}$$

When a sinusoidal signal $A \cos \omega_0 t$ is added to the narrowband noise $n_0(t)$, then the output of the envelope detector is modified as:

$$v(t) = R(t) \cos (\omega_0 t + \phi)$$

where the envelope and phase functions are:

$$R^2 = (x + A)^2 + y^2, \text{ } x \text{ and } y \text{ are those of Eq. (A.39);}$$

$$\phi = \tan^{-1} \frac{y}{x + A}. \tag{A.42}$$

Using Rice's results [3], the *pdf* of $R(t)$ is given by the Rice distribution:

$$p(R) = \frac{R}{\sigma^2} \exp\left[-\frac{(R^2 + A^2)}{2\sigma^2}\right] \cdot I_0\left(\frac{RA}{\sigma^2}\right) \ldots, \qquad R \geqslant 0 \tag{A.43}$$

where $I_0(z)$ is the modified Bessel function of the first kind and zeroth order with imaginary argument. Defining the SNR γ at the input to the detector as:

$$\gamma = A^2/2\sigma^2,$$

the *pdf* of the envelope is written as:

$$p(R) = \frac{R}{\sigma^2} \exp\left[-\left(\frac{R^2}{2\sigma^2} + \gamma\right)\right] \cdot I_0\left(\frac{R}{\sigma}\sqrt{2\gamma}\right) \tag{A.43}$$

For $\gamma \ll 1$, the *pdf* of $R(t)$ is the Rayleigh distribution given by Eq. (A.41). But for $\gamma \gg 1$, $p(R)$ is approximated as:

$$p(R) \simeq \sqrt{\frac{R}{2\pi A\sigma^2}} \cdot \exp\left[-\frac{(R - A)^2}{2\sigma^2}\right] \tag{A.44}$$

This distribution has a peak at $R \simeq A$ and approaches the Gaussian distribution for large γ. Thus with large γ, the output of a linear envelope detector is Gaussian with the mean value of A, as shown in Fig. A.1.

In phase-modulated systems, e.g., in M-PSK modulation, the *pdf* of the phase, $p(\phi)$, is of interest. It has been shown that $p(\phi)$ is given by:

$$p(\phi) = \frac{1}{2\pi} e^{-\gamma}[1 + \sqrt{4\pi\gamma} \cos\phi \cdot \exp(\gamma \cos^2 \phi) \cdot \text{erfc}(\sqrt{2\gamma} \cos\phi)] \tag{A.45}$$

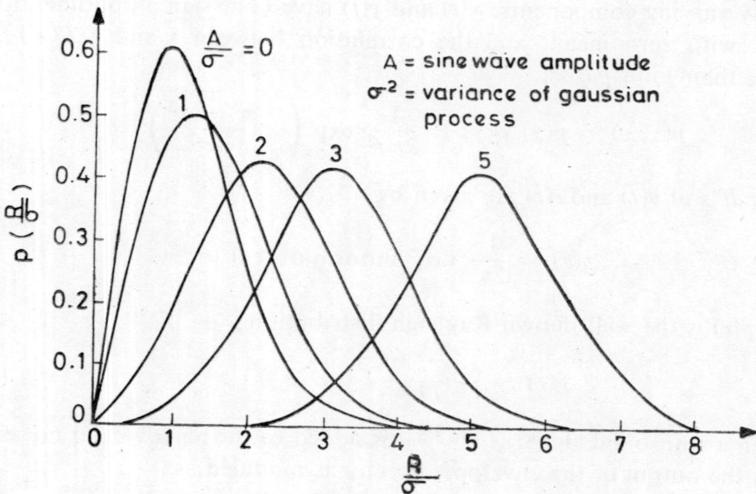

Fig. A.1 *pdf* for the envelope R of narrowband noise plus sinewave

where erf $(x) \triangleq \frac{1}{\sqrt{2\pi}} \int_{-\infty}^{x} e^{-z^2/2}\, dz$. Equation (A.45) is plotted in Fig. A.2, for various values of γ. The equation is also approximated, for small γ, as:

$$p(\phi) \simeq \frac{1}{2\pi}(1 + \sqrt{\pi\gamma}\, \cos\phi + \gamma \cos 2\phi), \qquad \gamma < 0.1; \tag{A.46}$$

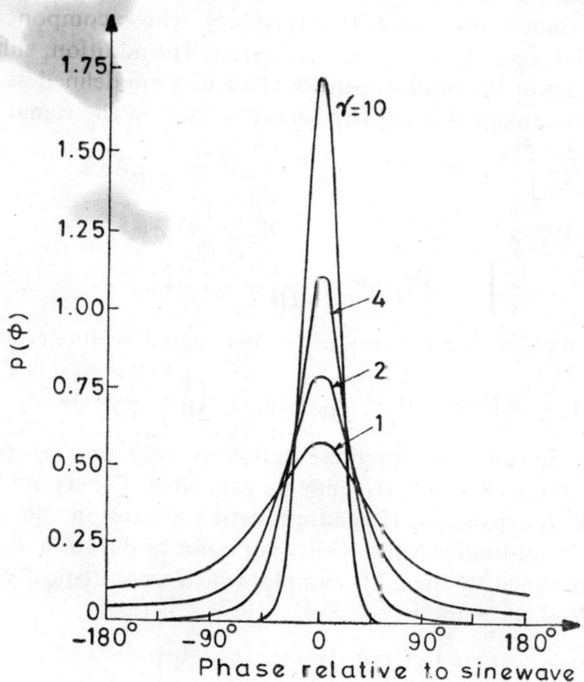

Fig. A.2 *pdf* of the phase of sinewave plus narrowband noise

But for large $\gamma (\gamma > 5)$,

$$p(\phi) \simeq \begin{cases} \sqrt{\frac{\gamma}{\pi}} \cos\phi \cdot \exp(-\gamma \sin^2\phi), & |\phi| < \pi/2 \\ 0 & , \quad |\phi| > \pi/2 \end{cases} \tag{A.47}$$

These results are used in evaluating the error-rates in M-PSK systems.

A.4 SIGNAL SPACE [1]

According to Shannon's sampling theorem, a signal $x(t)$, band-limited to 0–W Hz and time-limited to 0–T sec, is completely represented by $2TW$ samples (equivalent impulses), $x_1, x_2, \ldots, x_{2TW}$ taken at intervals of $1/2W$ sec. For $TW \gg 1$,

$$x(t) = \sum_{n=1}^{2TW} x_n \frac{\sin \pi(2Wt - n)}{\pi(2Wt - n)} \tag{A.48}$$

If the sampling pulses are of finite width τ, then the total response of the interpolation (rectangular) filter to the $2TW$ samples is:

$$2W\tau \sum_{1}^{2TW} x_n \cdot \frac{\sin \pi(2Wt - n)}{\pi(2Wt - n)} = 2W\tau \cdot x(t) \tag{A.49}$$

The signal space concept now allows the representation of $x(t)$ as a vector in the $2TW$-dimensional space (hyper space), whose components along the $2TW$ coordinates are $X = \{x_1, x_2, \ldots, x_{2TW}\}$. The addition, subtraction and the dot product of the multidimensional vectors are defined as an extension of the three-dimensional case. The total energy E in the signal is then,

$$E = \int_{-\infty}^{\infty} x^2(t)\, dt = \frac{1}{2W} \sum_{1}^{2TW} x_n^2 = \frac{1}{2W} \overline{X} \cdot \overline{X} \tag{A.50}$$

and its time-average is:

$$P = \frac{1}{T} \int_{-\infty}^{\infty} x^2(t)\, dt = \frac{1}{2WT} \sum_{1}^{2TW} x_n^2 \tag{A.51}$$

In a similar manner, the dot product of two signal vectors is represented as:

$$\int_{-\infty}^{\infty} x(t) \cdot y(t)\, dt = \frac{1}{2W} X \cdot Y = \frac{1}{2W} \sum_{1}^{2TW} x_n y_n \tag{A.52}$$

This sampling function reprepresentation is only one set of orthogonal functions in terms of which $x(t)$ may be expanded. Others are Fourier series expansion, K–L expansion, Legendre function expansion, etc.

In case of band-limited white Gaussian noise of duration T, the noise is completely specified by the $2TW$ samples which are statistically independent. The joint *pdf* of the samples $\{n_1, n_2, \ldots, n_{2TW}\}$ is then

$$\begin{aligned} p(n_1, n_2, \ldots, n_{2TW}) &= p(n_1) \cdot p(n_2), \ldots, p(n_{2TW}) \\ &= \left(\frac{1}{2\pi\sigma^2}\right)^{TW} \cdot \exp\left[-(n_1^2 + n_2^2 + \ldots + n_{2TW}^2)/2\sigma^2\right] \end{aligned} \tag{A.53}$$

This result has been used for obtaining solutions in Chapter 2.

A.5 STATISTICAL CHARACTERISTICS OF SPEECH AND IMAGE SIGNALS [1]

The digital representation of speech and image signals was obtained in Chapter 3, based on the sampling theorem and statistical properties of these signals. These statistics have been discussed in detail in Appx. A of Ref. [1]. The most important statistics for speech are the *pdf* of the speech samples $\{X_r\}$ (at Nyquist rate) and the *pdf* of the difference signals

$$\{D_r\} = \{(X_r - X_{r-1})\}.$$

It has been shown that the *pdf* of $\{X_r\}$ may be approximated by a special form of the Gamma density as [4]

$$p(X_r) = \frac{\sqrt{x}}{2\sqrt{\pi}} \cdot \frac{\exp(-k \mid X_r \mid)}{\sqrt{\mid X_r \mid}} \tag{A.54}$$

where $k = \sqrt{0.75}/\sigma_x$. A slightly poorer but simpler approximation is given by the Laplacian density as:

$$p(X_r) = \frac{\alpha}{2} \cdot \exp(-\alpha \mid X_r \mid) \tag{A.55}$$

where $\alpha = \sqrt{2}/\sigma_x$. These two approximations of $p(X_r)$ along with the measured $p(X_r)$ are shown in Fig. A.3, where the closeness of the approximations are easily seen.

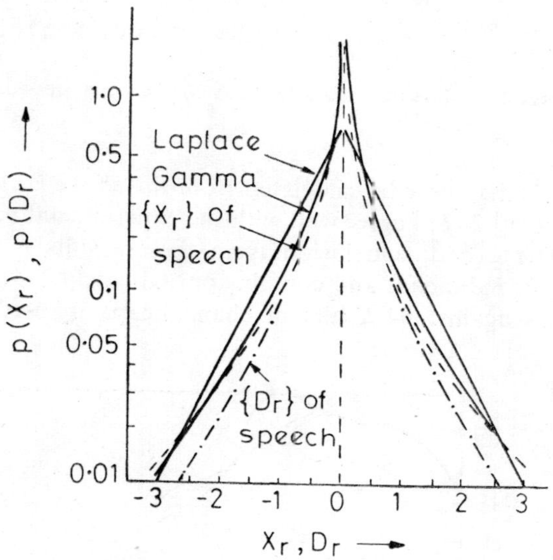

Fig. A.3 *pdf* of speech samples $\{X_r\}$ and the differential $\{D_r\}$ of samples alongwith Gamma and Laplacian *pdf*'s (after Paez and Glisson [4])

For DPCM codecs, the *pdf* of the difference signals D_r has been experimentally determined and is also shown in Fig. A.3. It is seen that $p(D_r)$ is closely approximated by the Gamma density, given by Eq. (A.54).

The successive speech samples are highly correlated and this correlation $\rho(n)$ decreases with the sample separation n. The long-time average values of $\rho(n)$, with $\rho(0) = 1$, is given by:

$$\rho(n) = \frac{1}{N} \sum_{r=0}^{N-n-1} X_r \cdot X_{r+n}, \ldots, \quad N \text{ large.} \tag{A.56}$$

These correlation values for $0 \leqslant n \leqslant 10$ have been experimentally determined and are shown in Fig. A.4, where the curve (a) is for lowpass and the curve (b) is for bandpass speech signals. The spread around the curves is due to different speakers. It is observed that $\rho(n)$ is large for $n \leqslant 3$, and this property has been used in estimating the speech samples for the DPCM codecs discussed in Chapter 3. The long-time power spectral density $F(f)$

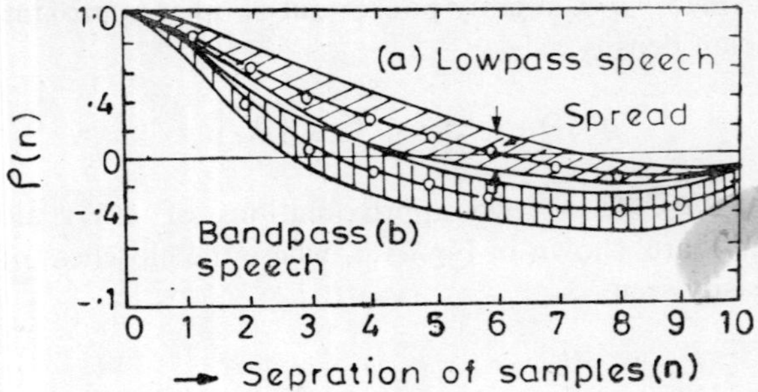

Fig. A.4 Measured correlation $\rho(n)$ between speech samples

of speech signals may now be calculated from the above $\rho(n)$ by using DFT and the calculated $|F(f)|$ agree well with the experimental values, as shown in Fig. A.5. The overall conclusion from these results is that the speech signals are highly redundant and contain approximately 2–4 *K*bits of information only, as against 64 *K* bits of channel capacity used in a log-PCM codec.

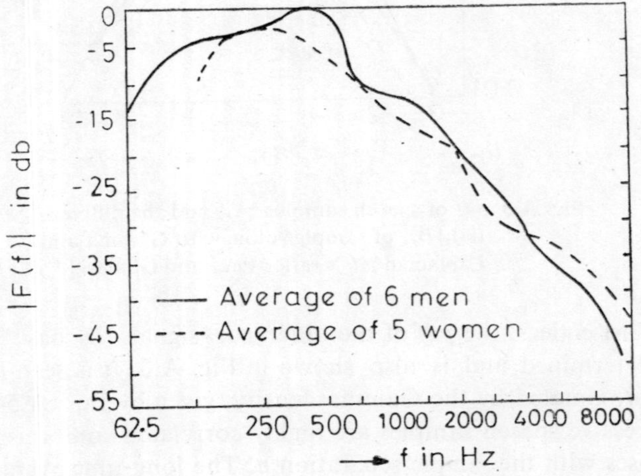

Fig. A.5 Long-time power-density spectrum for continuous speech

A.5.1 Image Signals [1, 5]

Video signals are generated by interlaced scanning of a scene through video camera, and a complete image scan, called a frame, is divided into even and odd fields. A frame contains 525/625 scan lines, the frame time is 1/30 (or (1/25) sec. and each field scan period is 1/60 (or 1/50) sec. The continuous output for many frames repeated at the frame frequency, $f_v = \omega_v/2\pi$, may be formulated as:

$$e_0(t) = \sum_{q=0}^{\infty} \left\{ \sum_{k=0}^{\infty} \frac{E_k E_q}{2} [\cos(k\omega_h t + q\omega_v t + \theta_k - \theta_q) + \cos(k\omega_h t + q\omega_v t + \theta_k + \theta_q)] \right\} \quad \text{(A.57)}$$

where, $\omega_h = 2\pi/t_h$, t_h = horizontal scan time.
The spectrum $F(f)$ of $e_0(t)$ is discrete, rather than continuous, because of the line and frame scanning processes. It has been shown that for ideal resolution, 1300×1300 pixels per frame are required and the effective spectrum spread of $e_0(t)$ is approximately 22 MHz. However, for commercial system, 512×512 pixels per frame are sufficient, and the approximate signal bandwidth is now 5 MHz. Even, this bandwidth is poorly utilized as the spectral amplitudes fall rapidly (at the rate of 12 dB/octave) after 1 MHz, as shown in Fig. A.6. This property leads to large correlation between successive pixels and the consequent redundancy reduction techniques, discussed in Chapter 3, have been immensely successful.

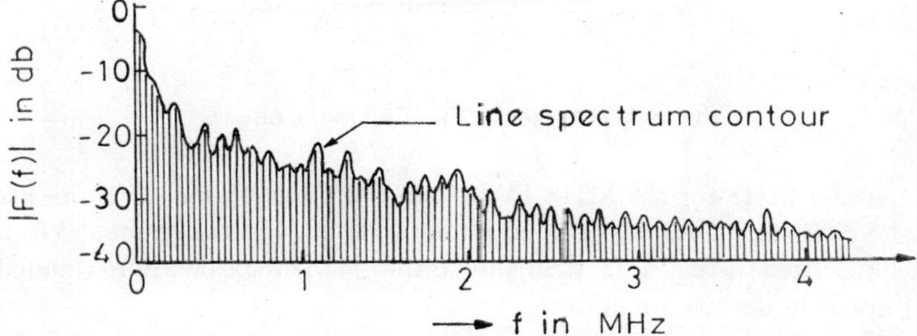

Fig. A.6 Approximate contour of the Luminance signal spectrum of a typical image

Because of the correlation between finite image samples, an image field may be modelled as a sample of a first-order Markov process. The sampled pixel values are then characterised by the convariance matrix

$$[K_{ik}] = \sigma_{px}^2 \begin{bmatrix} 1 & \rho_{px} & \rho_{px}^2 & \rho_{px}^3 & \cdots & \rho_{px}^{N-1} \\ \rho_{px} & 1 & \rho_{px} & \rho_{px}^2 & \cdots & \rho_{px}^{N-2} \\ \vdots & & & & & \\ \rho_{px}^{N-1} & \cdots & & & \cdots & 1 \end{bmatrix} \quad \text{(A.58)}$$

where the adjacent pixels along a row/column have a correlation of ρ_{px}, $0 \leqslant \rho_{px} \leqslant 1$, and a variance of σ_{px}^2. The measured values of correlation along a line, between the lines and along the diagonal of a typical image relate the values of of ρ_{px} as:

(a) $\rho_{px} = 0.953$ for pixels along a line
(b) $\rho_{px} = 0.965$ for pixels between lines
(c) $\rho_{px} \simeq 0.92$ for pixels along the diagonal.

This correlation property has been used in designing the predictors for DPCM codecs, as discussed in Chapter 3.

Frei *et al.* [6] have measured the statistics of the amplitude differences $D_r(1) = (X_r - X_{r-1})$ between consecutive samples in a busy picture. Their results for $p(D_r)$ are shown in Fig. A.7, where different sampling rates f_r of

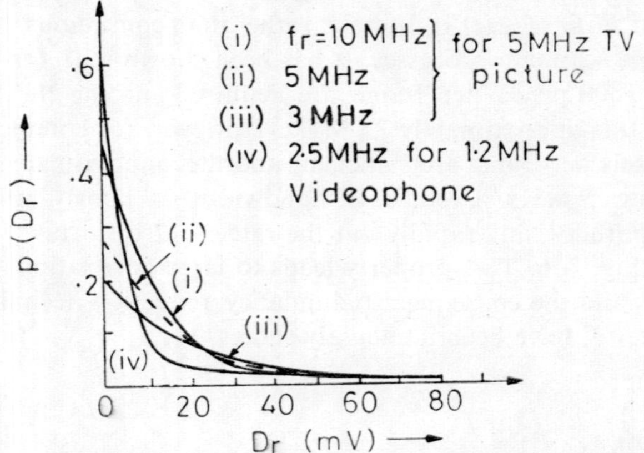

Fig. A.7 $p(Dr)$ for different sampling rates fr

10, 5, and 3 MHz for a 5 MHz TV system (625 lines/frame, 25 frames/sec.), and 2.5 MHz for a 1.2 MHz video-phone system (315 lines/frame, 25 frames/sec) have been used. It is then shown that $p(D_r)$ may be approximated by the Laplacian density function:

$$p(D_r) = \frac{1}{2V_0} \exp\left(-\frac{|D_r|}{V_0}\right) \tag{A.59}$$

where V_0 is the characteristic voltage. Further, $V_0 = 1.1$ for $f_r = 10$ MHz, 1.85 for $f_r = 5$ MHz, and 2.5 for $f_r = 3$ MHz. The above pdf and the cumulative distribution $P(D_r \leqslant V_T)$,, V_T = threshold level, indicate that, even in a DPCM codec, the generated codewords are not equiprobable, and there is a scope for further entropy coding of the DPCM words.

REFERENCES

1. Das, J., Mullick, S.K., and Chatterjee, P.K., *Principles of Digital Communications*, Chapter I, Wiley Eastern, New Delhi, 1986.
2. Davenport, W.B. Jr., and Root, W.L., *Random Signals and Noise*, McGraw-Hill, NY, 1958.
3. Rice, S.O., '*Mathematical Analysis of Random Noise*', BSTJ, vol. 23, pp. 282-332, 1944 and vol. 24, pp 46-56, 1945.
4. Paez, M.D., and Glisson, T.H., 'Minimum mean-squared-error quantization in speech', *IEEE Trans. Comm.*, vol. 20, pp 225-230, 1972.
5. Pratt, W.K., *Digital Image Processing*, John Wiley, N.Y., 1978.
6. Frei, A.H., *et al.*, 'An Adaptive dual-mode coder/decoder for Television signals', *IEEE Trans. Comm. Tech.* vol. Com-19, pp. 933-944, 1971.

APPENDIX B

Advances in Matched Filter Techniques

The basic theory of matched filters (MF) has been discussed in Sec. 2.6, and it has been shown that an MF is the most important hardware in a correlation receiver [1]. Practical MF's are generally of two types:

(a) Analogue delay lines, using all-pass networks, surface-wave devices, charged transfer devices;
(b) Digital delay lines, using discrete shift-registers, integrated-circuit semi-conductor memory using bipolar transistors or MOS-FETs.

To optimize accuracy, resolution and ambiguity of detected signals, the constraint of physical realizability and practical relalizability have to be kept in mind. As such, the usual approach in the MF design is to synthesize such networks whose impulse response is the optimized desired signal and to use the same in a complimentary way to decode the signal. Since the complimentary functions of signal-generation and signal-reception are performed by the same network, the problems of realizability are automatically satisfied. A simple example of signal generation and reception of an arbitrary waveform has been shown in Fig. 2.35, where an impulse is fed at the lefthand side and the summed weighted-output becomes a signal. If this signal is now fed at the righthand side of the delay line, then the summed weighted-output will again be an impulse giving the required cross-correlated output.

B.1 RECENT ADVANCES IN MF DEVICE TECHNOLOGY

The classical MF's for pulse-compression applications have been designed using Bridged-T LC networks in cascade. In practice, more than 150 Bridged-T sections are required for compressing a linear FM signal having a TW product of 80. As such, a considerable amount of effort is involved in the measurement and alignment procedures.

Because of this complexity and size of lumped-constant approach, Dispersive ultrasonic delay lines [2] have been developed for FM – MF's. The operation of these devices is based on the guided wave propagation properties of thin wires and strips of metals which exhibit dispersive properties in a restricted frequency region. In a typical example, for a compression ratio of

200(B = 1 MHz, T = 200 μ sec), the length of the strip delay line becomes 700 inches with a midband delay = 700 μ sec. A more recent application of ultrasonic techniques to pulse-compression systems is based on the realization of a frequency-dependent delay function with grating or array structures. However, these delay structures cannot be used as a tapped delay line for decoding discrete coded signals.

B.1.1 Acoustic Surface-wave Filters

Acoustic surface-wave (SAW) filters/delay lines [3] have been recently developed for use as MF's for both analogue and digital signals, bandpass filters and recirculating memory in UHF range. The dispersive SAW delay line utilizes surface-wave dispersive propagation in a substrate on which a layer with a low acoustic velocity is deposited. These waves travel along the surface of a solid material and can be tapped and controlled along the path of propagation. The SAW filters have a number of advantages:

(a) numerous kinds of filters can be designed with this technique
(b) they are small, planar and their manufacture is compatible with integrated-circuit technology
(c) they are expected to be reliable, reproducible and inexpensive devices.

A simple SAW filter is shown schematically in Fig. B.1, where one transducer (a comb structure) is used for launching the wave and the other for

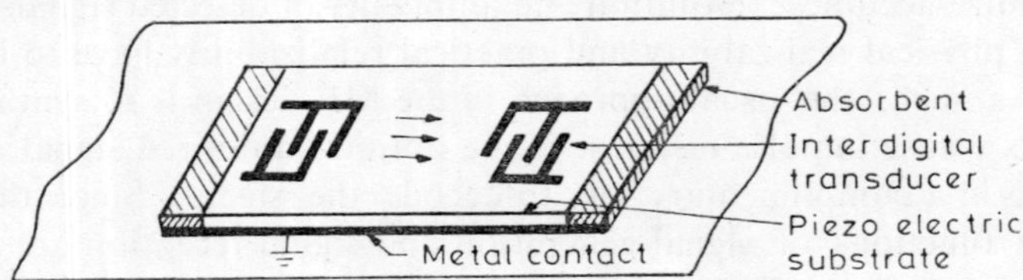

Fig. B.1 Schematic diagram of a SAW filter

receiving. The device acts as a filter, because the frequency response is determined by the structures of the input and output transducers. These transducers consist of inter-digital fingers whose spacing and overlap may be controlled to achieve the desired frequency response. The transducers are easily fabricated by standard photo-lithographic technology for frequencies in the UHF range. The overall frequency response is merely the product of the transducer responses. The mechanical design is simple and the device is very small; its dimensions being typically a few tens of acoustic wavelengths. The shaded patches at the ends represent absorbent material to prevent unwanted reflections.

For pulse compression filters using apodized comb structures constructed on $LiNbO_3$ substrate, the linearity of the group delay for the frequency range of 40 to 80 MHz has been found to be excellent. Bandpass filters and Transversal filters for the same frequency range have also been obtained with different comb structures. Large-time-bandwidth-product SAW pulse compressors [4] having BT-products in excess of 1500 have also been developed. These devices use the reflection of elastic surface waves from sputter-etched gratings on $LiNbO_3$. This reflective-array compressor is relatively free of the spurious signals which limit the performance of inter-digital transducer array. Time delays up to 3 msec. in passive systems and up to 20 msec with the addition of loss compensation by means of internal surface-wave amplifiers, have been observed in wrap-around surface-wave delay lines constructed of $Bi_{12}GeO_{20}$ and operating in a recirculating mode at 52 MHz [5].

For applications in digitally coded systems, SAW TDL's, as shown schematically in Fig. B.2, have given excellent characteristics as MF's. It has

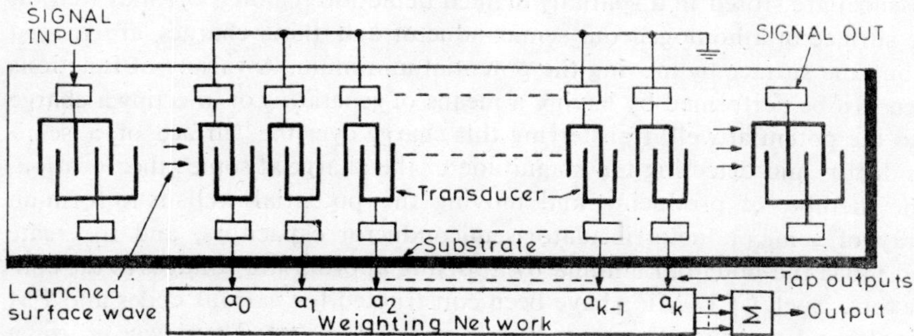

Fig. B.2 A SAW TDL

been reported that a thousand bit, at 10 *M* bit/s, SAW delay line MF on s.t.-cut quartz has been successfully implemented [6]. The centre frequency was 60 MHz and taps were spaced at 0.1 μ sec. The expected correlation peak of 30 dB above the side-lobes was also obtained.

A further report [7] has described an acoustic surface-wave memory operating at a bit rate of 220 Mb/s and a storage capacity of 1280 bit per recirculation loop. The transducers are coded using orthogonal pairs of Golay complimentary sequences to obtain pulse-in/pulse-out behaviour. The recirculation electronics use standard commercial logic for both the amplifier and the WRITE, READ, INHIBIT and RECLOCKING functions. It is expected that using internal amplification and Bismuth Germanium oxide as the substrate, it will be possible to have memories having a capacity of 100,000 bits, if both the diffraction losses and dispersion of the system can be kept low. An excellent account of SAW applications has been given by Milstein and Das [8].

B.1.2 Charge Transfer Devices [9]

Charge Transfer Devices (CTD's) which include both charge-coupled devices (CCD's) and Bucket Brigade devices (BBD's) have unique applications to many analogue signal processing systems as they are inherently analogue in operation. One of the most important signal processing functions for which

CTD's can be used is the time delay of analogue signals. When CTD's are used in this application, the signal to be delayed is first sampled at a rate greater than twice the highest frequency in the signal. Analogue samples are then clocked down the CTD shift register and appear at the output after a delay time T_d. The delayed signal is finally reconstructed by passing the samples through an appropriate filter. Alternatives to CTD's for analogue time delays are acoustic delay lines for short-time delay, or digital delay preceeded by A/d converter and followed by D/A converter. Since CTD's can achieve hundreds of milliseconds of delay, that look very attractive for a large number of signal delay functions.

The CCD's operate on the principles that minority carriers (or their absence) are stored in a spatially defined depletion region (potential well) at the surface of a homogeneous semiconductor and these charges are moved about the surface by moving the potential minimum. A variety of functions then can be performed by having a means of generating or injecting a charge into the potential well, transferring this charge over the surface of a semiconductor and detecting the magnitude of the charge at some other location. One method of producing and moving the potential wells is to form an array of semiconductor-insulator-semiconductor capacitors, and to create and move the potential minima by applying appropriate voltages to the conductors. Such CCD MF's have been constructed for various codes and FM signals, and a particular structure is shown in Fig. B.3. The device utilizes a 3-phase single-level metallisation p-channel CCD structure fabricated on $\langle 111 \rangle$ 20 Ω cm silicon. A 1000 Å gate oxide was used with a resulting threshold voltage of 3.9 V. The electrode length was about 15 μm (0.6 mil) and, with the exception of the metallisation definition, the devices were produced on standard MOS production lines. The drives ϕ_1 and ϕ_2, as seen in Fig. B.3(a) and (b), shift the charges from one electrode to the next. The non-destructive sampling, weighting and summing of the shifted charges are accomplished by splitting the ϕ_3-electrode into two sections connected to two separate $\phi_3(+)$ and $\phi_3(-)$ drive lines. The weighting coefficients h_k are controlled by the dimensions of the split electrode and the summed output is proportional to $\sum_k h_k Q_k$, Q_k = the signal charge. In Fig. (c), the correlation response of the 13-chip Barker-code device is shown.

Analogue matched filters for long digital codes and also for linear FM signals (chirp filter) have bsen experimentally made and their performances have been almost ideal. Large transversal filters have also been designed using this device. Computer simulations have shown that 100 element filters are practical with a charge transfer efficiency of 99.9 percent, while a c.t.e. of 99.99 percent would allow a filter of thousand elements to be constructed with clock rates of 20 MHz. Presently 512-length programmable CCD-TDL's are available commercially, and these have a c.t.e of 99.9999. They can be clocked at 20 MHz rate and are being used in high reliability, low-noise analog signal processing applications [10, 11].

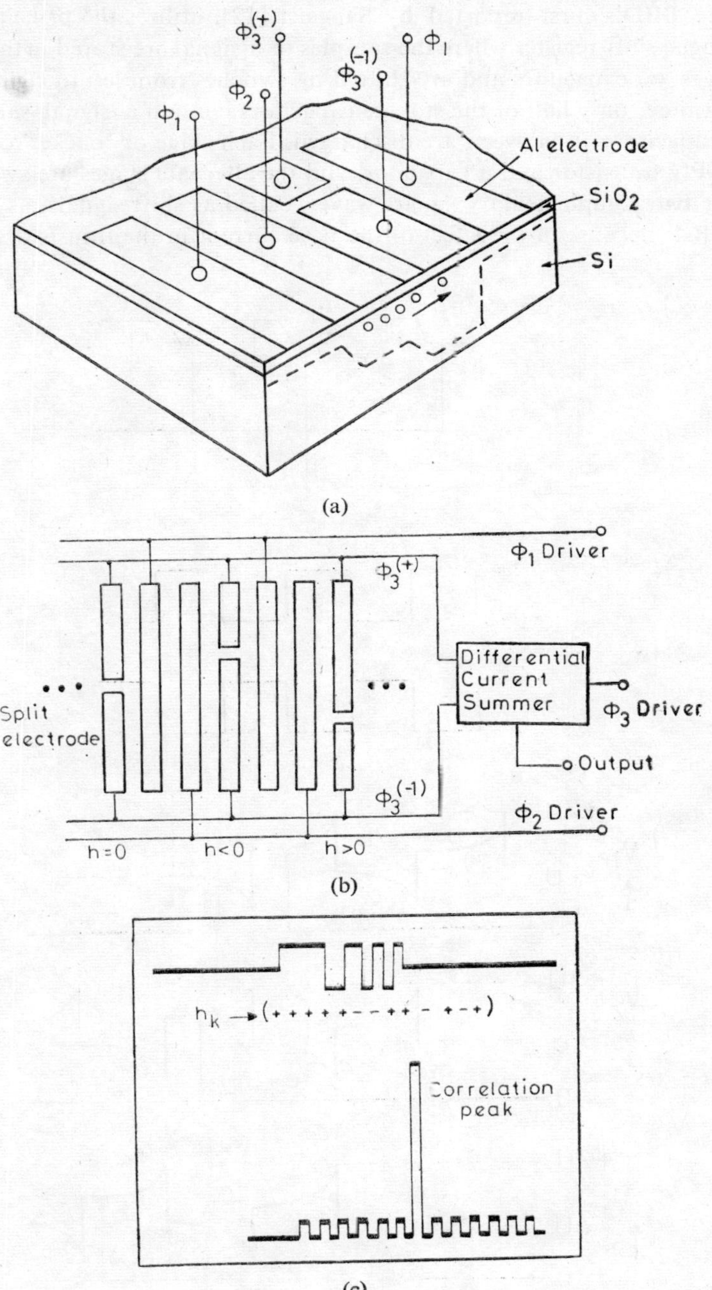

Fig. B.3 (a) A section of the CCD device; ϕ_1, ϕ_2, ϕ_3 are 3-phase drive, (b) CCD-schematic showing weighting 'h_k' and summing through ϕ_3 for a non-binary code, (c) Response of a 13-chip Barker code MF (after Buss *et al.* [9])

The BBD's, first reported by Sangster [12], utilizes the principles of an analogue shift-register where the samples of a signal are stored in the form of charges on capacitors and are shifted by switches from left to right. At any given time, only half of the storage capacitors contain a signal sample and the capacitors in between are discharged. Each stage or 'bucket' consists of an NPN transistor and a capacitor, and the alternate stages are switched by using two complementary square waves, called as shift signals, as shown in Fig. B.4. Because of the effect of the base current in the transistors, there is

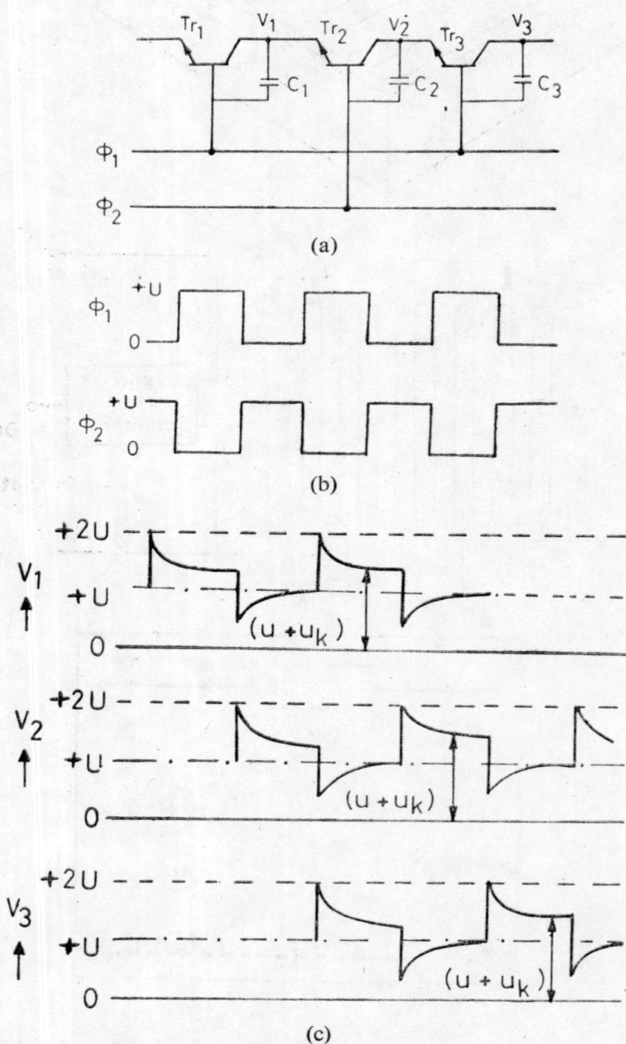

Fig. B.4 (a) Three stages of BBD circuits, (b) Shift signals ϕ_1 and ϕ_2, (c) Shows the voltages V_1, V_2, V_3 on storage capacitors during the transfer of a signal sample u_k.

certain attenuation of the signal samples as the signal travels along the line. This loss is made up by using amplifiers along the line. The BBD's using NPN transistors have been built with a length of about 300 stages in a monolithic chip. This could be used upto a sampling frequency of 30 MHz and the delay can be arranged from 5 μs to 5 msec. This delay can be continuously varied by varying the shift frequency.

The BBD's using *p*-channel MOS transistors have shown better performance because of the absence of the gate current and the simpler input-output circuits. With a large number of MOS stages, the signal source becomes distorted because of a small feedback of the drain voltage to the source in individual transistors. This feedback can be reduced by adding to each stage a second MOS transistor biased so that it always operates in saturation. These extra transistors together with the original ones may form MOS Tetrodes. The BBD's have been successfully utilized in delaying television pictures through 12 delay lines of 72 stages, with a total delay of 12×14 μs. The final picture was excellent.

In MF applications, the CTD filters can achieve a desired spectral characteristic, when the impulse response of the filter is chosen to be the Fourier Transform of the desired frequency characteristic. The CTD MF filters have been tested in spread-spectrum receivers over a wide temperature range (-60°C to $+80$°C) and it was found that their characteristics were as predicted. In communication systems utilizing signals which are too long to be processed using surface-wave filters (T_d greater than 20 μs), CTD's are the only alternative to a digital computer. When CTD's can be used, their advantages in cost, power, size and weight are very promising. The main advantages of CTD transversal filters in spectral filtering are tunability and flexibility in selecting the spectral characteristic. However, it is difficult to achieve high Q filters, because of errors in weighting coefficients and finite duration of the impulse response.

B.2 DIGITAL MATCHED FILTERS (DMF) [13]

For large length signalling waveforms, as used in M-ary orthogonal signalling and in spread spectrum techniques, the BT product of the MF required is necessarily large and such large length MF's are difficult to fabricate in production quantities. But with recent availablility of CCD's/BBD's and surface-acoustic-wave delay lines (SAWDL's), the situation has changed considerably; even then, it is more economical and reliable to use digital MF's using LSI, when the situation permits. Moreover, Turin [13] has shown that under certain jamming environments, DMF outperforms the corresponding along MF for binary-coded signals. It is, therefore, necessary to understand the operation of DMF's as well.

Under large access environment (e.g., in SSMA) and in saturated transponders (e.g., in mobile/satellite networks), the input to the receiver is necessarily limited through a bandpass limiter (BPL). Such a network with BPL, coherent detection, a lowpass limiter and a DMF (using a series of shift registers), is shown in Fig. B.5(a). Devenport [14] has shown that for low CNR, BPL gives 1 dB loss in CNR and lowpass limiter gives 2 dB loss; but the coherent detection provides 3 dB gain in SNR. Thus, the overall SNR at the input to DMF is the same as the input CNR to the receiver, and it is sufficient to study the DMF performance at the baseband only. If coherent detection is not possible, an equivalent receiver is structured with *I–Q* channels and two DMF's, whose squared outputs are added to form the desired correlated signals, as shown in Fig. B.5(b).

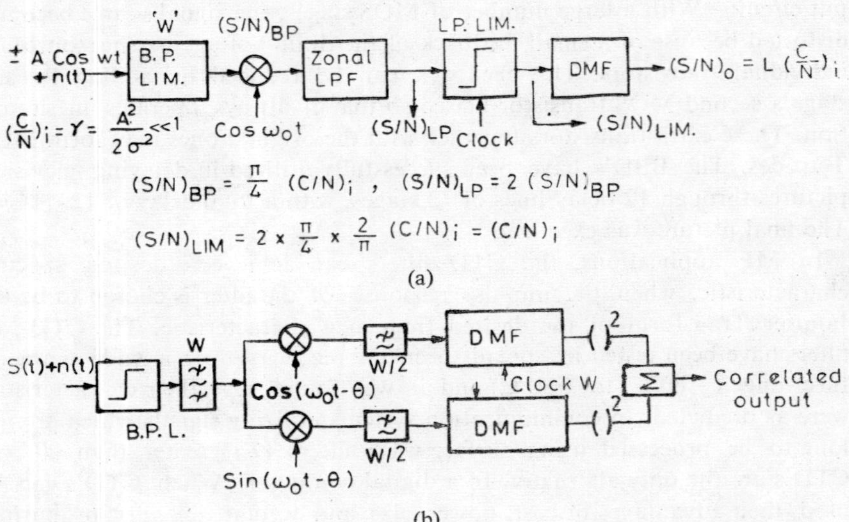

Fig. B.5 (a) Receiver with limiters and DMF, (b) Incoherent DMF receiver

For low input CNR, the error probability of the decisions (of ± polarity) at the DMF input is given by [15]:

$$p_e = 1 - \sqrt{\gamma/\pi}, \qquad \gamma = \text{CNR} = \frac{A^2}{2\sigma^2} \text{ at the receiver input.}$$

The probability of error at the DMF output is now given by the Binomial distribution and is equal to:

$$P_e \text{ at DMF output} = \sum_{k=0}^{(L-1)/2} \binom{L}{k} (1 - p_e)^k \, p_e^{(L-k)} \tag{1}$$

when L is the length of the DMF.

Cahn [16) has shown that the SNR at the DMF output is now given by

$$(S/N)_{pk} = L \frac{(1 - 2p_e)^2}{4p_e(1 - p_e)} \approx L(1 - 2p_e)^2 \tag{2}$$

where $p_e \to 0.5$ for $\gamma \ll 1$.

For an ideal correlator with the BPL, the output SNR is:

$$(S/N)_o = 2(\pi/4)\, L(C/N)_i \tag{3}$$

whereas for the ideal correlator without the BPL,

$$(S/N)_0 = 2L(C/N)_i,$$

thus showing 1 dB ($= \pi/4$) loss in the BPL.

Cahn has further shown that for worst case interference, i.e., for square wave interference with $S_I \geqslant S_i$, $(S/N)_o$ loss can be very large. But adding controlled dither to the interference, according to the scheme of Fig. B.6,

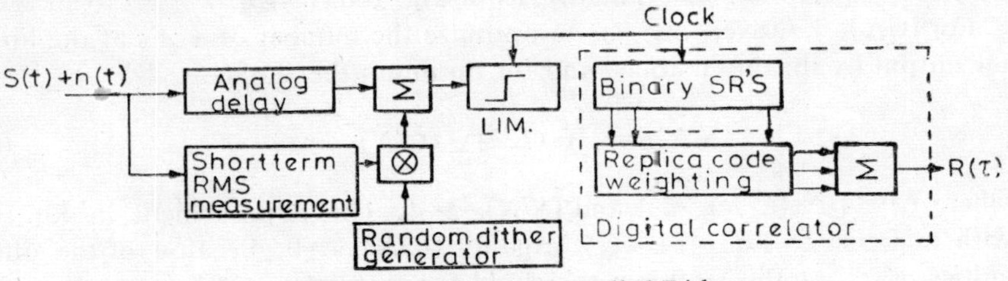

Fig. B.6 DMF with controlled Dither

the SNR loss in DMF can be reduced to 4.8 dB only. With 4 level quantization (and using 4 parallel DMF's), the loss is reduced to 2.3 dB only.

Turin (13) has made extensive studies of the DMF and expressions have been obtained for output SNR's, using binary input signals and 1-bit quantization, for the cases where the channel interference is additive and (a) white Gaussian, (b) incoherent and constant amplitude (ICAJ), and (c) coherent constant amplitude (CCAJ). Improvements in DMF performance obtained by threshold biasing and by dithering have been investigated and it has been shown that DMF's can outperform analog MF's (AMF) by proper use of the former technique. The SNR results for zero-threshold DMF's are:

(a) For WGN, $(S/N)_0 = \frac{4}{\pi} L(C/N)_i$, showing 2 dB loss in DMF

(b) For ICAJ, $(S/N)_0 = \frac{4}{\pi^2} L(C/N)_i$, showing 7 dB loss in DMF

(c) For CCAJ,

$$(S/N)_0 = \begin{cases} \frac{32}{\pi^2} \sqrt{(C/N)_i} & \text{for } A/J \ll 1 \\ \infty & \text{for } A/J \gg 1 \end{cases} \tag{4}$$

It is seen that for CCAJ, the interference captures the DMF for $A/J \ll 1$, and the signal captures the DMF when $A/J > 1$. An optimum predigitizing filter, having a notch at dc (equivalently at the carrier frequency f_0), would fully reject the interference. Such a situation exists in case of AMF's where the transfer function has a notch at f_0, and the resultant *SNR* is:

$$(S/N)_0 = 8 \frac{L^2}{Q^2} (C/N)_i \tag{5}$$

where Q = excess of the number of $+A$'s over that of $-A$'s
= 1 in case of *PN* sequences.

For balanced code with $Q = 0$, $(S/N)_0 \rightarrow \infty$. However, for ICAJ, this filter would have to be adaptive.

Biased Thresholds and Dither

Since CCAJ produces an effective DC bias at the DMF input, it is reasonable to expect that shifting the threshold bias proportionately would restore the limiter output to the $\pm$ values as given by the transmitted code. The change of the bias converts the capture of the DMF by CCAJ into capture by the signal, and $(S/N)_0 \rightarrow \infty$. However, the channel measurement is necessary to set the bias optimally, as in Fig. B.6.

For ICAJ, however, one has to optimize the number of $+A$'s at the limiter output by shifting its bias, and for optimum thersholds,

$$(S/N)_0 = \frac{32}{\pi^3}(L/2)\sqrt{(C/N)_i} \tag{6}$$

where $L \leqslant 10^3$, $(C/N)_i \ll 1$ and $(S/N)_0 \gg 1$. The improvement in Eq. (6) with reference to Eq. (4) is very large for $(C/N)_i \rightarrow 0$. Because of the difficulties of setting the optimum threshold for unknown ICAJ, it is easier to set a suboptimal threshold through the measurement of the power of J only, and the $(S/N)_0$ is now 3 dB worse as compared to Eq. (6).

Alternatively non-parametric Adaptive thresholds may be used and here, threshold of the quantizer is adjusted adaptively to always give a compromise value of 25 percent $+1$'s at the output, instead of the ideal 50 percent $+1$'s for random codes. This avoids the sensitivity problems of the optimum threshold quantizers, but at the same time degrades the ideal performance somewhat. The results now are:

$$\begin{aligned} &\text{(a) For ICAJ,} \quad (S/N)_0 = \frac{16}{\pi^3}(L)(C/N)_i \\ &\text{(b) For CCAJ,} \quad (S/N)_0 = 2(L)^2 \\ &\text{(c) For WGN,} \quad (S/N)_0 = 0.8L(C/N)_i \end{aligned} \tag{7}$$

As Cahn has shown, dither can restore the partial correlation in case of ICAJ and CCAJ interference, by shifting the thresholds randomly. Turin has also discussed this and further recommends that dither spectrum can be shaped to concentrate at the nulls of the transfer function of the DMF i.e., at the multiples of the chip rate, so that the correlator will reject the major part of the dither spectrum. That is, the dither will exert its control on the digitizer and having served its purpose, will be suppressed by the correlator. One possible way to do this is to sample a sawtooth waveform having a period $(L - 1)^{\delta/L}$, where δ is the chip time and use these samples as the dither.

Results

Simulation results of a DMF with $L = 127$ and different types of interference are shown in Figs. B.7(a), (b), (c) and (d) [13]. It is seen in Fig. B.7(a), that for ICAJ, optimum-threshold DMF performs better than AMF for low $(C/N)_i$ and in Fig. B.7(b), with CCAJ, the optimum threshold and adaptive threshold—DMF is shown to out-perform AMF by a very large margin. Effect of dither is shown in Figs. B.7(c) and (d), where Cahn's

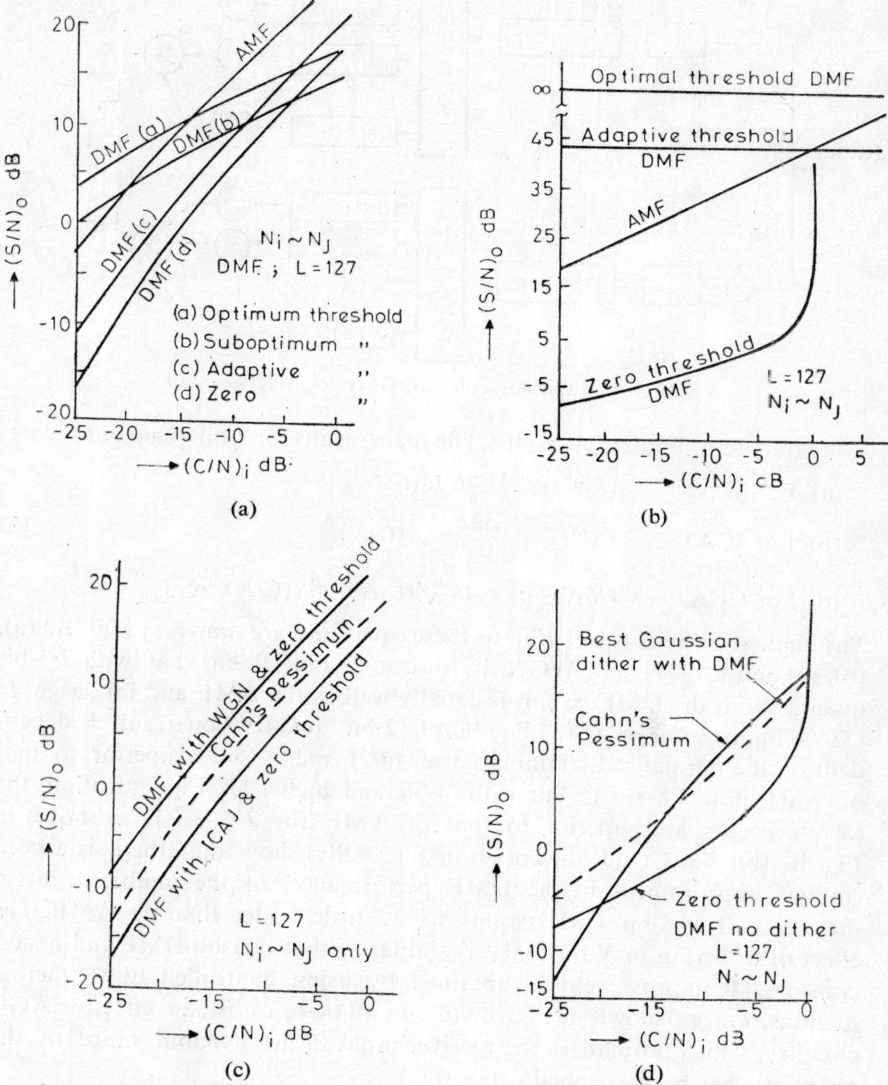

Fig. B.7 SNR curves for DMF with ICAJ and CCAJ and effect of Dither (after Turin [13]) (a) SNR with ICAJ. (b) SNR with CCAJ, (c) Effect of Dither for WGN and ICAJ, (d) Effect of Dither for CCAJ

technique of using controlled amount of dither has found to be quite efficient and the spectrally-shaped dither gives 2-3 dB improvements over Cahn's results.

Lim [17] has studied in the detail the multibit-quantized noncoherent DMF and the hardware configuration for realizing such M-bit DMF is shown in Fig. B.8. As an extention of Turin's results, SNR's for additive channel noise, using WGN. ICAJ and CCAJ, have been calculated and

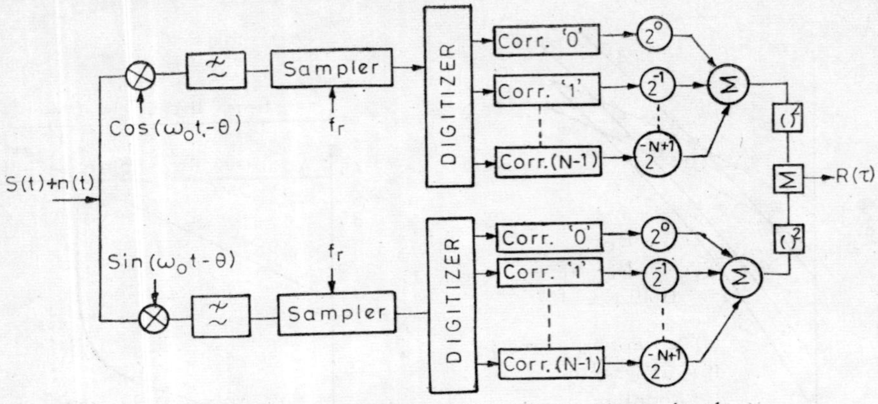

Fig. B.8 Multibit non-coherent DMF for binary signals

compared with simulation results. The main results for 2-bit quantization are:

$$\begin{aligned} &\text{(a) For WGN:} && (S/N)_0 = 1.76\, L\, (C/N)_i \\ &\text{(b) For ICAJ:} && (S/N)_0 = \frac{32L}{\pi^3} \sqrt{(C/N)_i} \\ &\text{(c) For CCAJ:} && (S/N)_0 = 8.48 \sqrt[4]{(C/N)_i} \cdots (C/N)_i \ll 1. \end{aligned} \tag{8}$$

The simulation results, relating to these equations, are shown in Figs. B.9(a), (b), (c) and (d) [17]. For WGN, it is seen in Fig. B.9(a) that with 2-3 bit quantization, the DMF performs equally well as the AMF and for large L, $(S/N)_0$ improves as in AMF. For ICAJ, 2-bit quantization result is dependent on the normalized jamming power ($J/2$) and is only superior to that of AMF for $J/2 \simeq 1$; but with 4-bit and higher level quantization, the $(S/N)_0$ results are superior to that of AMF for $J/2 \leqslant 1$, as shown in Fig. B.9(b). For CCAJ, the curves of Fig. B.9(c) show that there is a continuous improvement in the DMF performance, as the number of bits is increased. The 8-bit DMF seems to do little better than the AMF. The effect of dithering in N-bit DMF is similar to that in 1-bit DMF and hence, considerable improvement is obtained by using controlled dither before quantization, as shown in Fig. B.9(d). In all these cases, an effective AGC circuitry, which normalizes the receiver input to the dynamic range of the quantizer, has been assumed.

B.3 MULTI-SAMPLED DMF [18]

The mechanism of SNR deterioration in DMF is as follows: Large random noise, while limited in the 1-bit quantizer, creates holes and spurious pulses in the signal bit stream of ± 1's, and while taking single samples per chip at the DMF input, many of the chip values are misrepresented. The effective BW of the limiter output is also increased and the output has a correlation function [19] of the form:

$$R(\tau) = A_0^2 \exp(-1.16\, W \mid \tau \mid) \tag{9}$$

where W = bandwidth of the receiver.

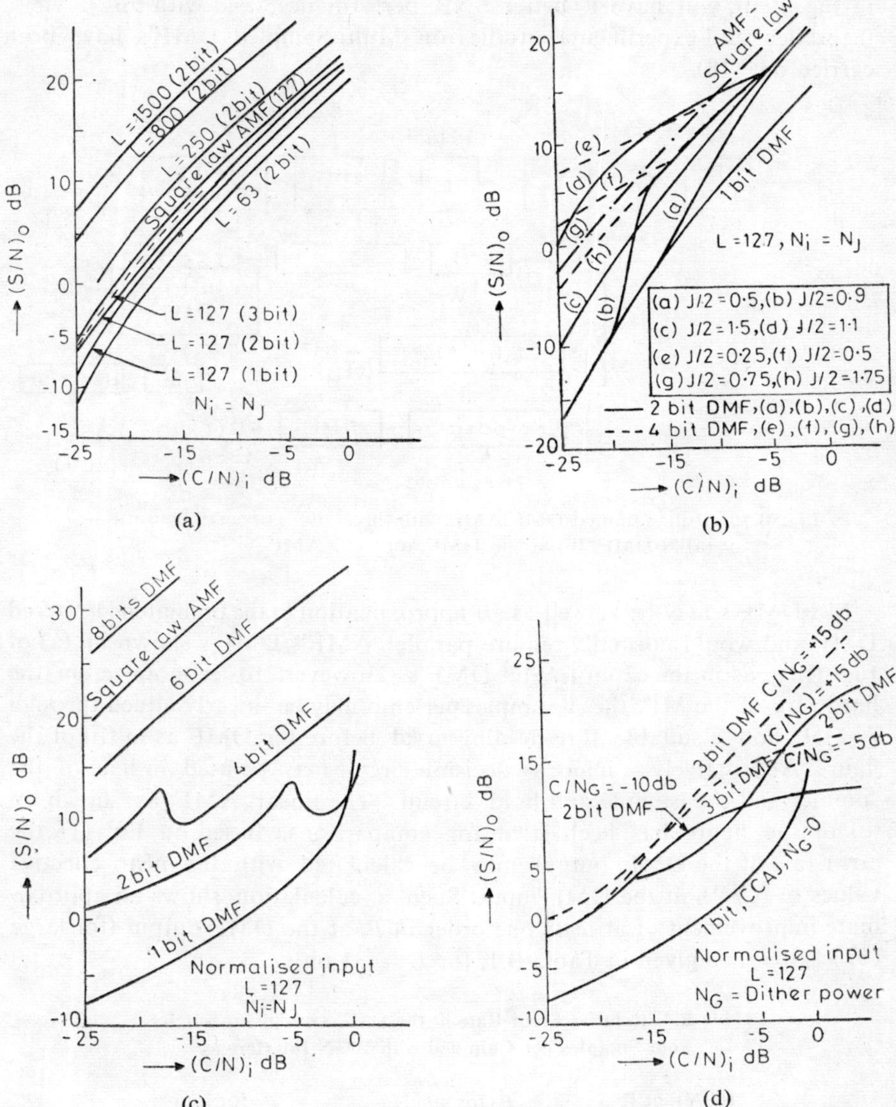

Fig. B.9 SNR curves for multibit DMF with ICAJ, CCAJ, and Dither (after Lim [17]) (a) SNR for DMF with WGN, (b) SNR with ICAJ, (c) SNR with CCAJ, (d) SNR for 2/4 bit quantization with CCAJ and Dither

Since the effective bandwidth of the limiter output is now doubled, it calls for a sampling rate $\geqslant 2W$, instead of the rate W ($\approx 1/\tau_0$, τ_0 = chip time) as is normally used in the DMF discussed before. Moreover, a pre-matched-filtering of the limited signal plus noise before the DMF would improve the $(C/N)_i$ at the DMF input by approximately 2 dB. From these arguments, it is expected that a multi-sampled DMF (MS-DMF), as shown in Fig. B.10, will have a better SNR performance, and with this in view, theoretical and experimental studies on Multi-Sampled DMF's have been carried out (18).

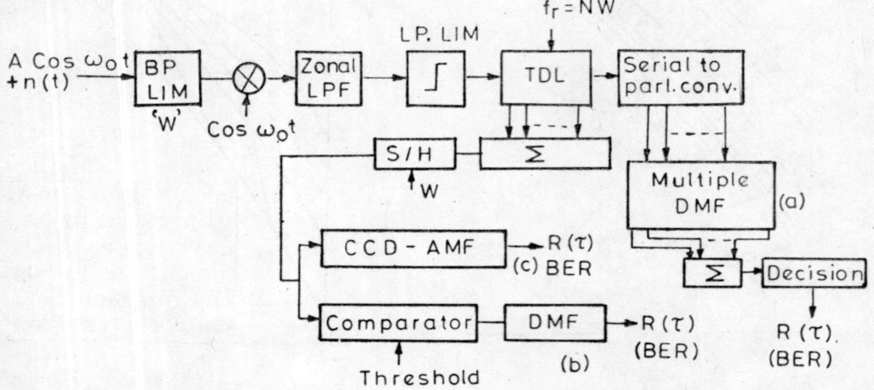

Fig. B.10 Multi-sampled DMF/AMF with three alternative configurations; (a) N-DMF, (b) Single DMF, (c) CCD-AMF

MS-DMF's may be viewed as an approximation to the prematched filtered DMF and would normally require parallel AMF's/DMF's shown in (a) of Fig. B.10, as in the case of N-bit DMF's. However, to economize on the hardware of the MF, the N-samples per chip may be linearly added through a TDL and a suitable thresholding used before the DMF as in (b) of the figure. Alternatively, a majority decision circuit may be used in lieu of the summer and the 'sample and hold' circuits. The linear AMF as shown in (c) of the figure has been used for comparisons. Based on Eq. (1), the error-rate at the DMF output may be calculated with the near effective values of $(C/N)_i$ at the DMF input. Such a calculation shows an approximate improvement of at least one order in P_e at the DMF output (for large $(S/N)_0$ only) as given in Table B.1, for $L = 31$ only.

Table B.1 Relative Error-Rate at the DMF Output for Single/ Four Samples per Chip and with WGN Interference

$(C/N)_i$ at IF input in dB	P_e for single sample/chip	P_e for four sample/chip
−3	6.56×10^{-4}	3.62×10^{-5}
−4	3.68×10^{-3}	3.76×10^{-4}
−5	9.30×10^{-3}	1.87×10^{-3}
−6	2.18×10^{-2}	6.62×10^{-3}
−8	7.72×10^{-2}	3.08×19^{-2}

Experimental verification of the performances of the MS-DMF was done with a system having a processing gain $P_G = 10^3$ and multiple samples up to $N \leqslant 8$. A concatenated 31×31 PNS code [20] mixed with WGN and ICAJ was used as the receiver signal. The receiver consisted of a saturating amplifier, TDL-summer, and AMF/DMF made of (31×31)—CCD MF's as shown in Fig. B.10. The clock rate was 1 MHz, noise BW $\simeq$ 500 kHz, mean frequency of ICAJ $\simeq$ 700 kHz. Controlled Gaussian dither was also used for experiments. Results are shown in Figs. B.11(a), (b), (c) and (d). The relative performances of AMF's and DMF's with $N = 1, 2, 4$ and 8, are shown in Fig. B.11(a) for WGN-interference. It is seen that with $N = 4$ or 8, the AMF gives $P_e \simeq 10^{-4}$ for $(C/N)_i = -20$ dB, as is theoretically expected for a $P_G = 30$ dB, and the result is equivalent to that of an ideal analog MF. DMF with single sample requires 2 dB more C_i to give the same P_e as the single-sampled AMF. Both for AMF and DMF, there is a progressive improvement with $N > 1$ and in the limit N large, DMF is still 1 dB worse than the corresponding AMF. For the ICAJ interference, the results in Fig. B.11(b), show progressive improvements with $N > 1$, and DMF is again 2 dB worse with $P_e \simeq 10^{-4}$ for $(C/N)_i = -15$ dB. But with optimum dither, error rates with ICAJ improve considerably, as shown in Figs. B.11(c) and (d), and 4-sampled AMF and DMF behave equally well; thus the results are almost ideal for $N = 4$, as $P_e = 10^{-4}$ at $(C/N)_i = -21$ dB. Here $N_i = N_j +$ dither noise power N_G, and the optimum dither is adjusted for each $(C/N)_i$. It was found that $N_J = N_G$ was optimum, and the actual value of $(C/N_J)_i = -18$ dB for $P_e = 10^{-4}$.

These results show that MS-DMF's will efficiently replace AMF's for large hardware realization, and will find wide applications where saturated transponders are encountered, a situation which is unavoidable in the cases of mobile radio communication, and SSMA systems with a central repeater.

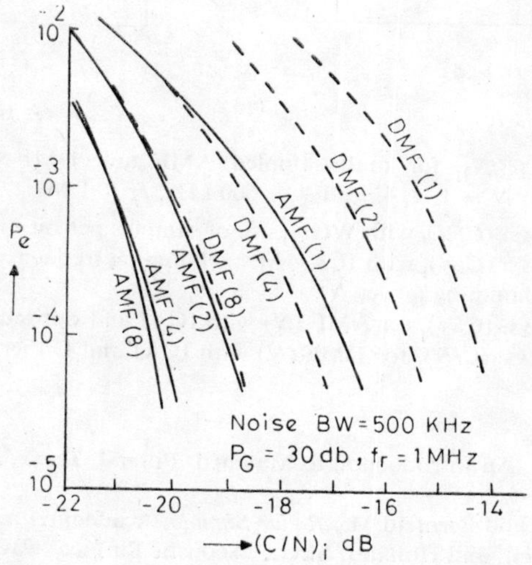

Fig. B.11 (a)

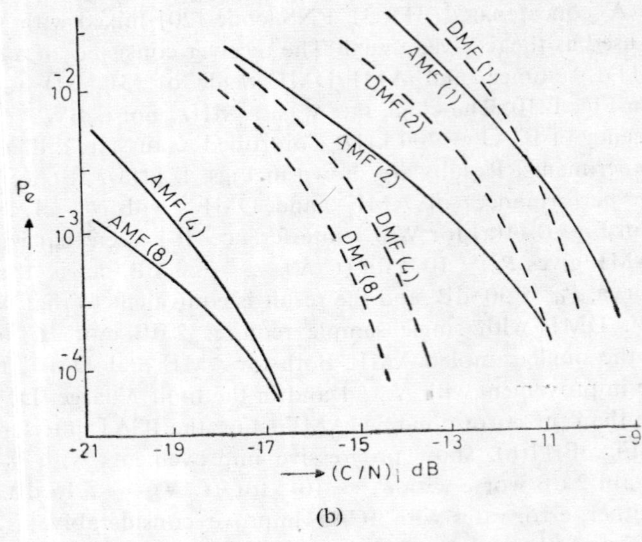

(b)

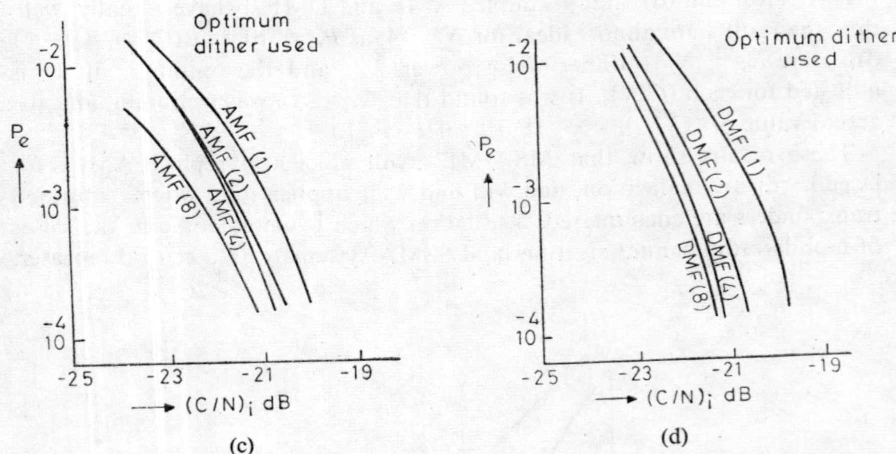

(c) (d)

Fig. B.11 P_e vs $(C/N)_i$ for multi-sampled AMF and DMF with WGN, ICAJ and Dither; $N = 1, 2, 4, 8$; BW = 500 kHz, $f_r = 1$ MHz, $P_G = 30$ dB [18]

(a) P_e vs. $(C/N)_i$ with WGN, no. of samples per bit indicated as (N),

(b) P_e vs. $(C/N)_i$ with ICAJ, Mean Jamming frequency = 700 kHz; N_i = Jamming power N_j,

(c) P_e vs. $(C/N)_i$ for AMF (N) with ICAJ and optimum Dither,

(d) P_e vs. $(C/N)_i$ for DMF (N) with ICAJ and Dither

REFERENCES

1. Turin, G.L., 'An Introduction to Matched Filters', *Trans. IRE,* vol. IT-6, p 310, June 1960.
2. Cook, C.E., and Bernfeld M., *Radar Signals*, Academic Press, NY, 1967.
3. Tancrell, R.H., and Holland, M.G., 'Acoustic Surface wave filters', *Proc., IEEE,* vol. 59, p 393, 1971.

4. Williamson, R.C., and Smith, H.J., 'Large-time-bandwidth-product Surface-wave pulse compressor employing reflective gratings', *Electronics Letters*, vol. 8, p 401, 1972.
5. Reeder, T.M., *et al.*, 'Multimillisecond time delays with wrap-around surface-acoustic-wave delay lines', *Electronics Letters*, vol. 8, p. 359, 1972.
6. Vasile, C.F., and LaRosa, R., '1000-bit surface-wave matched filter', *Electronics Letters*, vol. 8, p 479, 1972.
7. Van de Vaart, H., and Schissler, L.R., 'Acoustic surface wave recirculating memory', *Electronics Letters*, vol. 8, p 333, 1972.
8. Milstein, L.B., and Das, P.K., 'Spread-spectrum receiver using surface acoustic wave technology', *IEEE Trans.*, vol. Com-25, p 841, 1977.
9. Buss, D.D. *et. al.*, 'Transversal filtering using charge transfer devices', *IEEE Jour. of Solid State Circuits*, vol. Sc-8, April 1973.
10. Dutta Roy, S.C., *et. al.*, 'Signal Processing applications of charged coupled devices', *J. IETE*, vol. 24, p 400, 1978.
11. Gricco, D.M., 'The application of charged-coupled devices to spread-spectrum systems' *Trans. IEEE*, vol. Com-28, p. 1693, 1980.
12. Sangster, F.F.J., The 'bucket-brigade delay line', a shift register for analog signals, *Phil. Tech. Rev.*, vol. 31, p 97, 1970.
13. Turin, G.L., 'An introduction to digital matched filters', *Proc. IEEE*, vol. 64, p. 1092, July 1976.
14. Devenport, W.B. (Jr), 'Signal-to-noise ratios in bandpass limiters', *Jour. App. Physics*, vol. 24(6), p 720, 1953.
15. Aein, J.M., 'Multiple access to a hardlimiting communication satellite repeater', *IEEE Trans. on Space Electronics and Telemetry*, vol. 10, p 159, Dec. 1964.
16. Cahn, C.R., 'Performance of digital MF correlator with unknown interference', *IEEE Trans. Com.*, vol. COM-19, p 1163, 1971.
17. Lim, T.L., 'Noncoherent digital MF multibit quantization', *IEEE Trans.*, vol. COM-26(4), p 409, 1978.
18. Das, J. and Shanmugavel, S., 'Performance of multisampled digital matched filtes in noise and interference', *Proc of International Conf., on Computer, Systems and Signal Processing*, Bangalore, Dec. 9, 1984.
19. Rice, S.O., 'Mathematical analysis of random noise', *BSTJ*, vol. 23, p. 282, July 1944, and vol. 24, p 46, Jan. 1945.
20. Maskara, S.L., and Das. J., 'Concatenated sequences for spread spectrum systems', *IEEE Trans.* vol. AES-17, pp 344-349, 1981.

APPENDIX C

Codes and Sequences

In the main text, specially in Chapters 5, 8 and 9, various codes and sequences have been mentioned for the purpose of signal design and error correction. Some of the important ones are Orthogonal codes, PN sequences, Gold codes, minimum polynomials; these are discussed below. Algebra of codes is also discussed briefly.

C.1 ALGEBRA OF CODES [1, 2]

In conformity with the switching circuit algebra, the modulo-2 arithmetic is used in the discussion of codes and polynomials. Thc mod-2 addition and multiplication is defined as:

$X \backslash Y$	0	1
0	0	1
1	1	0

(a) $X \oplus Y$

$X \backslash Y$	0	1
0	0	0
1	0	1

(b) $X \odot Y$

(C.1)

The binary alphabet, 0 and 1, is called a Field of two elements (or a binary field), which is denoted by the Galois Field GF (2). In binary arithmetic, $-X = X$, and $X - Y = X + Y$. Polynomials $f(X)$ with 1/0 as the coefficients are manipulated using these relations.

Example 1

(*a*) The product of $f_1(X) = X + 1$ and $f_2(X) = X^3 + X + 1$, is given by:

$$f_1(X) \odot f_2(X) = X^4 + X^3 + X^2 + 1.$$

(b) Similarly, the division of

$$f_2(X) = X^6 + X^5 + X^3 \text{ by } f_1(X) = X^3 + X + 1,$$

gives the result:

$$f_2(X)/f_1(X) = (X^3 + X^2 + X + 1) \text{ with the remainder } 1.$$

(c) If $f(X) = (X^4 + X + 1,)$ then

$$[f(X)]^2 = X^8 + X^2 + 1 + 2X^5 + 2X^4 + 2X = f(X^2).$$

(d) Roots of polynomials may also be calculated by using the conditions that: if $f_1(1) = 0$, then $(X + 1)$ is a factor of $f_1(X)$ and the factors of the polynomials may be calculated in the normal way.
For $f_1(X) = X^4 + X^3 + X^2 + 1$, $f_1(1) = 0$, then, $(X + 1)$ is a factor of $f_1(X)$.

Galois Fields

A polynomial $f(X)$ of degree m is said to be irreducible over GF(2), if $f(X)$ does not have a factor of degree less than m, and greater than zero. The polynomials, e.g., $X^3 + X + 1$, $X^4 + X + 1$, $X^5 + X^2 + 1$, etc., are irreducible polynomials.

Fields with 2^m symbols are called Galois Fields, GF(2^m), and its arithmetic is derived as follows:

Consider a polynomial $p(X)$ of degree m (with binary coefficients) and then using a new variable α such that $p(\alpha) = 0$, a Table of powers of α is developed. If $p(X)$ is a primitive polynomial, then $\alpha^0, \alpha^1, \ldots, \alpha^{(q-2)}$, $q = 2^m$, will have distinct values and are called the set of $q(= 2^m)$ field elements. But $\alpha^{q-1} = \alpha^{2m-1} = 1$. The element α is called a primitive element of the field GF(2^m) and the polynomial $p(X)$ is called a primitive polynomial.

Example 2

GF (2^3) may be formed from GF(2) as the ground field by using the third degree irreducible polynomial $p(X) = X^3 + X + 1$. Then α is the root of $X^3 + X^2 + 1$, i.e., $(X + \alpha) = 0$, and $\alpha^3 + \alpha^2 + 1 = 0$, giving $\alpha^3 = a^2 + 1$. The seven non-zero field elements of GF (2^3) are now given in Table C.1. The representation of GF (2^4,) using $p(X) = X^4 + X + 1$, is given in Table C.2.

Table C.1 Elements of GF (2^3): Modulo Polynomial $= X^3 + X^2 + 1$

Sl. No.	Element as a power of a Primitive element	Binary Representation
0	0	000
1	$\alpha^0 = 1$	001
2	$\alpha^1 = \alpha$	010
3	$\alpha^2 = \alpha^2$	100
4	$\alpha^3 = \alpha^2 + 1$	101
5	$\alpha^4 = \alpha^2 + \alpha + 1$	111
6	$\alpha^5 = \alpha + 1$	011
7	$\alpha^6 = \alpha^2 + \alpha$	110
8	$\alpha^7 = \alpha^3 + \alpha^2 = 1 = \alpha^0$	001

Table C.2 Elements of GF (2^4): Modulo Polynomial = $X^4 + X + 1$

Sl. No.	Element as a power of a primitive element	Binary representation
0	0	0000
1	$\alpha^0 = 1$	0001
2	$\alpha^1 = \alpha$	0010
3	$\alpha^2 = \alpha^2$	0100
4	$\alpha^3 = \alpha^3$	1000
5	$\alpha^4 = \alpha + 1$	0011
6	$\alpha^5 = \alpha^2 + \alpha$	0110
7	$\alpha^6 = \alpha^3 + \alpha^2$	1100
8	$\alpha^7 = \alpha^3 + \alpha + 1$	1011
9	$\alpha^8 = \alpha^2 + 1$	0101
10	$\alpha^9 = \alpha^3 + \alpha$	1010
11	$\alpha^{10} = \alpha^2 + \alpha + 1$	0111
12	$\alpha^{11} = \alpha^3 + \alpha^2 + \alpha$	1110
13	$\alpha^{12} = \alpha^3 + \alpha^2 + \alpha + 1$	1111
14	$\alpha^{13} = \alpha^3 + \alpha^2 + 1$	1101
15	$\alpha^{14} = \alpha^3 + 1$	1001
16	$\alpha^{15} = 1 = \alpha^0$	0001

As discussed in Sec. 8.4.3, the generator polynomials for BCH codes are obtained from the minimum polynomials, $M_i(X)$, where $M_1(X)$ is the primitive polynomial $p(X)$ over GF(2^m), The successive polynomials $M_3(X)$, $M_5(X)$ etc. are the minimum polynomials of α^3, α^5, etc. for the given m and $p(X)$. Table C.3 gives the minimum polynomials for m = 3, 4, 5 and 6, and is

Table C.3 Minimum Polynomials Over GF (2^n)

m	$M^i(X)$	
3	$p(X) = M_1(X) = X^3 + X + 1$;	$M_3(X) = X^3 + X^2 + 1$.
4	$p(X) = M_1(X) = X^4 + X + 1$:	$M_3(X) = X^4 + X^3 + X^2 + X + 1$:
	$M_5(X) = X^2 + X + 1$;	$M_3(X) = X^4 + X^3 + 1$.
5	$p(X) = M_1(X) = X^5 + X^2 + 1$:	$M_3(X) = X^5 + X^4 + X^3 + X^2 + 1$;
	$M_5(X) = X^5 + X^4 + X^2 + X + 1$;	$M_7(X) = X^5 + X^3 + X^2 + X + 1$;
	$M_{11}(X) = X^5 + X^4 + X^3 + X + 1$;	$M_{15}(X) = X^5 + X^3 + 1$.
6	$p(X) = M_1(X) = X^6 + X + 1$;	$M_3(X) = X^6 + X^4 + X^2 + X + 1$;
	$M_5(X) = X^6 + X^5 + X^2 + X + 1$:	$M_7(X) = X^6 + X^3 + 1$;
	$M_9(X) = X^3 + X^2 + 1$;	$M_{11}(X) = X^6 + X^5 + X^3 + X^2 + 1$;
	$M_{13}(X) = X^6 + X^4 + X^3 + X + 1$;	$M_{15}(X) = X^6 + X^5 + X^4 + X^2 + 1$;
	$M_{21}(X) = X^2 + X + 1$;	$M_{23}(X) = X^6 + X^5 + X^4 + X + 1$.
	$M_{57}(X) = X^3 + X + 1$;	$M_{31}(X) = X^6 + X^5 + 1$.

based on the Table of irreducible polynomials given in Ref. [1]. It is to be noted that the reciprocal polynomial of an irreducible polynomial is also irreducible, and the reciprocal of the primitive polynomial is also primitive.

C.1.1 Vectors and Matrices

The polynomial $f(X)$ with binary coefficients can be represented as an ordered sequence of binary symbols, such as,

$$(1 + X + X^2 + X^5) \leftrightarrows (111001) \tag{C.2}$$

with the highest-power coefficient at the rightmost location (since in actual practice, the higher-order coefficients are transmitted first). This order sequence of binary digits is gererally called an 'n-tuple' over GF(2), and may be considered as a vector. The set of all binary n-tuples $\{V_n\}$ is called a vector space over GF(2). For n binary digits in each vector, the vector space consists of 2^n vectors. As for example, for $n = 3$,

$$\{V_n\} = \{000, 001, 010, 011. 100, 101, 110, 111\} \tag{C.3}$$

The vectors obey the following postulates:

(a) For every pair of vector α and β in $\{V_n\}$, there is a vector γ in $\{V_n\}$ such that $\alpha \oplus \beta = \gamma$
(b) For all α β, γ in $\{V_n\}$, $(\alpha \oplus \beta) \oplus \gamma = \alpha \oplus (\beta \oplus \gamma)$
(c) For all α, β in $\{V_n\}$, $\alpha \oplus \beta = \beta \oplus \alpha$
(d) There is an all-zero vector 0 in $\{V_n\}$ such that:

$$\alpha \oplus 0 = 0 \oplus \alpha = \alpha \text{ every } \alpha \text{ in } \{V_n\}$$

(e) For every α in $\{V_n\}$, there is a vector $(-\alpha)$, such that:

$$\alpha \oplus (-\alpha) = (-\alpha) \oplus \alpha = 0$$

The vector $(-\alpha)$ is called the additive inverse of α.
(f) For all scalars $\{C\}$ drawn from GF(2), and for all α in $\{V_n\}$, the scalar product $\{C\alpha\}$ is a vector in $\{V_n\}$, e.g.,

$$\{u\} = \{C_1v_1 + C_2v_2 \ldots + C_kv_k\} \in \{V_n\} \tag{C.4}$$

where C_i's are from GF(2) and are called the coefficients of v_i.

The inner product or dot product of two vectors is defined as:

If $\quad v_1 = \{x_1, x_2, x_3\}$ (triple)

$\quad v_2 = \{y_1, y_2, y_3\}$ (triple)

Then,

$$v_1 \odot v_2 = \{x_1y_1 \oplus x_2y_2 \oplus x_3y_3\} \tag{C.5}$$

where $\oplus$ and $\odot$ are mod-2 operations. If $v_1 \odot v_2 = 0$, then v_1 and v_2 are said to be orthogonal.

A set of vectors $\{v_1, v_2, \ldots v_n\}$ is said to be linearly dependent if and only if, there are scalars $\{C_1, C_2, \ldots, C_n\}$ from GF(2), not all zero, such that

$$C_1v_1 + C_2v_2 + \ldots + C_nv_n = 0 \tag{C.6}$$

If the set is not linearly dependent, then they are said to be independent.

Example 3
Consider the set of vectors v_1, v_2, v_3, v_4 for $\{V_4\}$. Then for

$$v_1 = 0011, \quad v_2 = 0101, \quad v_3 = 0100, \quad v_4 = 0010,$$

The sum $v_1 \oplus v_2 \oplus v_3 \oplus v_4 = (0000)$.
Thus the set of vectors is linearly dependent.
But the set of vectors, $v_1 = 0111$, $v_2 = 1011$, $v_3 = 1101$, $v_4 = 1110$, are seen to be linearly independent. Any vector in $\{V_4\}$ is a linear combination of these vectors.

Matrices
The code vectors, e.g., those shown in Ex. 3, may also be arranged in the form of a matrix and the usual matrix manipulation may be carried out. In general, a rectangular array of $m \times m$ elements is called a Matrix, but the matrices with binary elements only will be discussed here. Consider a matrix $[A]$ with elements a_{ij} as follows:

$$(A) = \begin{bmatrix} a_{11} & a_{12} & \cdots & a_{1n} \\ a_{21} & a_{22} & \cdots & a_{2n} \\ \cdot & & & \cdot \\ \cdot & & & \cdot \\ \cdot & & & \cdot \\ a_{m1} & a_{m2} & \cdots & a_{mn} \end{bmatrix} \tag{C.7}$$

where $a_{ij} = 0$ or 1. The matrix is over GF(2), where each row is a binary n-tuple and each column is a binary m-tuple. The following are the useful properties of the matrix:

(a) The rows and columns are known as row and column vectors.
(b) The row rank of a matrix is equal to the maximum number of independent rows of the matrix.
(c) The row rank of a matrix is invariant under elementary operations, e.g., interchange of any two rows, multiplication of any row by a non-zero field element, and addition of any scalar multiple of one row to another.
(d) Elementary row operations can be used to simplify a matrix to a standard form, say, the Echelon Canonical form or a triangular matrix.
(e) If the inner product of two vectors $v_i \odot v_j$ vanishes ($=$ zero), then v_i and v_j are said to be orthogonal.
(f) If $[A_1]$ and $[A_2]$ are two matrices, and if $[A_1 A_2^T]$ has all-zero elements, then the two matrices are said to be orthogonal.

C.2 SWITCHING CIRCUITS (1, 3)

The encoder and deeoder circuits arc usually constructed with a number of adders (mod-2), memory devices (usually shift registers) and constant multipliers. Multiplication of two polynomials $g(X)$ and $f(X)$, and division of one by the other are also mechanised using mod-2 adders and shift registers

(SR). The product of two polynomials $A(X)$ and $B(X)$ yields the polynomial $C(X)$ given by:

$$\begin{aligned} C(X) &= A(X) \odot B(X) \\ &= (a_0 + a_1X + \ldots + a_kX^k)(b_0 + b_1X + \ldots + b_nX^n) \\ &= a_0b_0 + (a_1b_0 + a_0b_1)X + (a_0b_2 + a_1b_1 + a_2b_0)X^2 + \ldots \\ &\quad + (a_{k-1}b_n + a_kb_{n-1})X^{(k+n-1)} + a_kb_nX^{k+n} \end{aligned} \tag{C.8}$$

This may be realized with the circuit of Fig. C.1(a) or C.1(b) where $A(X)$ is input and the coefficients of $B(X)$ are given as the weighting factors connected

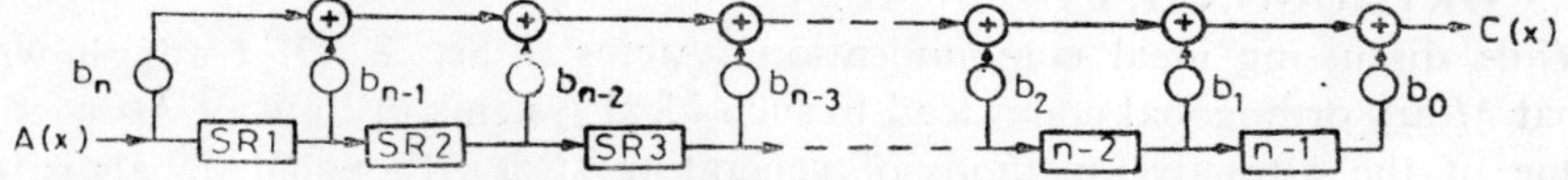

Fig. C.1(a) A circuit for multiplying $A(X)$ by $B(X)$

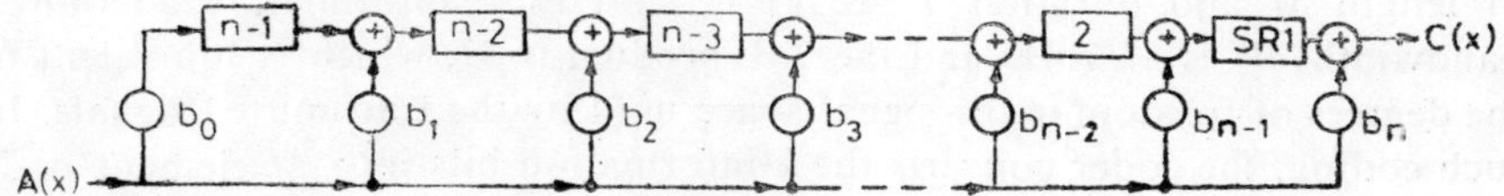

Fig. C.1(b) An alternative circuit for multiplication

to the mod-2 adders. Using the convention that the higher-order coefficients are transmitted first, the highest-order coefficient a_kb_n is obtained first at the output of the multiplier. The rest of the coefficients follow sequentially, and a total of $(k + n)$ shifts are required to complete the process. It may also be noted that the impulse response of the SR circuits is $h(X) = B(X)$, and hence the product $C(X) = A(X)*B(X)$. It is also possible to combine two multiplication operations into one by having another set of coefficients connected to the same mod-2 adders in the circuit of Fig. C.1(b).

The polynomial divider circuit is organized by using feedback connections (instead of feed-forward connections used in the multiplier) through the multiplier coefficients as shown in Fig. C.2. In the circuit for division of

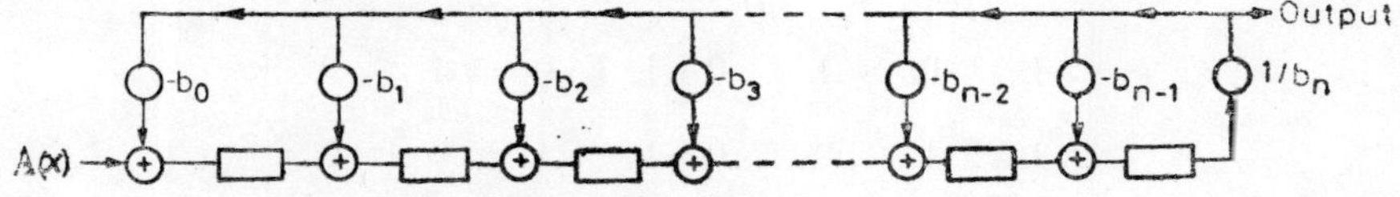

Fig. C.2 A divider circuit

$A(X)$ by $B(X)$, $k \geqslant n$, the entire quotient appears at the output after k shifts and the remainder is stored in the SR's. It is also possible to combine a divider circuit with a multiplier by combining Figs. C.1(b) and C.2 suitably.

In certain coders/decoders, a random sequence generator, also known as PN-sequence generator, is utilized. Such a generator is realized by using feedback to a series of SR's through mod-2 adders. With certain feedback conditions, the length of lhe sequence is maximum and is given by $(2^m - 1)$, where m SR's are used in cascade. Figure C.3 shows a PNS generator of length $(2^m - 1) = 255$, and its generator polynomial is

$$f(X) = 1 + X^3 + X^5 + X^6 + X^8 \tag{C.9}$$

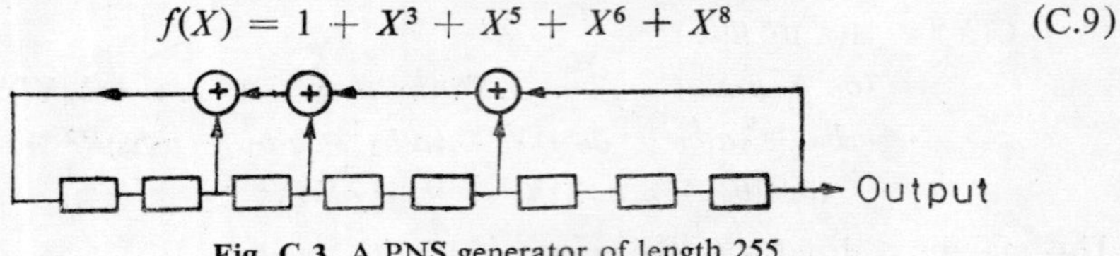

Fig. C.3 A PNS generator of length 255

C.3 ORTHOGONAL CODES (4)

While discussing ideal communication systems in Sec. 8.7.2, it was shown that M-ary orthogonal codes lead to such ideal systems in limit of $M \to \infty$. One of the attractive methods of generating such orthogonal signals is to phase-modulate a carrier by a binary orthogonal sequence. For a sequence of length M and duration T sec ($T = kT_b$, T_b = bit time), the required bandwidth W is M/T Hz and the TW product is M, which is equivalent to the degrees of freedom in the signal space used by the transmitted signals. In such coding, the coder converts the k-information bits into M-element code vectors ($M = 2^k$), where the chip time $= T/M$ sec. The binary sequences are said to be orthogonal if the cross correlation of M-vectors satisfy the conditions:

$$\rho_{ij}(m_i, m_j) = \begin{Bmatrix} M, & \text{for} & i = j \\ 0, & \text{for} & i \neq j \end{Bmatrix} \tag{C.10}$$

Such orthogonal sequences may be generated from the Hadamard matrix using the recurrence formula*.

$$[H_4] = \begin{bmatrix} H_2 & H_2 \\ H_2 & \bar{H}_2 \end{bmatrix} \quad \text{where } [H_2] = \begin{bmatrix} 1 & 1 \\ 1 & 0 \end{bmatrix}$$

and the Mth order matrix is generated from the $(M/2)$th order one. As an example, as eighth order matrix is given as

$$[H_8] \to \left[\begin{array}{cccc|cccc} 1 & 1 & 1 & 1 & 1 & 1 & 1 & 1 \\ 1 & 0 & 1 & 0 & 1 & 0 & 1 & 0 \\ 1 & 1 & 0 & 0 & 1 & 1 & 0 & 0 \\ 1 & 0 & 0 & 1 & 1 & 0 & 0 & 1 \\ \hline 1 & 1 & 1 & 1 & 0 & 0 & 0 & 0 \\ 1 & 0 & 1 & 0 & 0 & 1 & 0 & 1 \\ 1 & 1 & 0 & 0 & 0 & 0 & 1 & 1 \\ 1 & 0 & 0 & 1 & 0 & 1 & 1 & 0 \end{array}\right]$$

*The orthogonal code set may be generated also from the Walsh functions. The codes are also known as the first order Reed-Muller codes.

where $[H_8] = \begin{bmatrix} H_4 & H_4 \\ H_4 & \bar{H}_4 \end{bmatrix}$

Based on the I/O representation of the code vectors, the codes are considered orthogonal if the number of symbol positions in which they are similar equals the number of positions in which they are dissimilar. The test is satisfied by the above set of codes. By interchanging the rows and columns of the derived matrix, the orthogonal codes may be converted to cyclic one, which is easily genesated as the maximal-length linear shift-register sequence (also known as PRBS or PNS).

Biorthogonal codes are generated by taking a set of orthogonal codewords and adding to it the complements of each word. Thus, $2M$ signals are generated by M-elements and the total bandwidth requirement is halved. The cross-correlation of these codes is given by

$$\rho_{ij}(m_i, m_j = \left\{ \begin{array}{ll} 0, & i \neq j \\ \pm M, & i = j \end{array} \right\} \tag{C.11}$$

One of the best known biorthogonal codes is the first-order Reed-Muller code, an example of which for $M = 16$ is given now:

$$\begin{array}{cccccccc c cccccccc} 1 & 1 & 1 & 1 & 1 & 1 & 1 & 1 & \quad & 0 & 0 & 0 & 0 & 0 & 0 & 0 & 0 \\ 1 & 0 & 1 & 0 & 1 & 0 & 1 & 0 & & 0 & 1 & 0 & 1 & 0 & 1 & 0 & 1 \\ 1 & 1 & 0 & 0 & 1 & 1 & 0 & 0 & & 0 & 0 & 1 & 1 & 0 & 0 & 1 & 1 \\ 1 & 0 & 0 & 1 & 1 & 0 & 0 & 1 & & 0 & 1 & 1 & 0 & 0 & 1 & 1 & 0 \\ 1 & 1 & 1 & 1 & 0 & 0 & 0 & 0 & & 0 & 0 & 0 & 0 & 1 & 1 & 1 & 1 \\ 1 & 0 & 1 & 0 & 0 & 1 & 0 & 1 & & 0 & 1 & 0 & 1 & 1 & 0 & 1 & 0 \\ 1 & 1 & 0 & 0 & 0 & 0 & 1 & 1 & & 0 & 0 & 1 & 1 & 1 & 1 & 0 & 0 \\ 1 & 0 & 0 & 1 & 0 & 1 & 1 & 0 & & 0 & 1 & 1 & 0 & 1 & 0 & 0 & 1 \end{array} \tag{C.12}$$

Orthogonal set — Complementary set

A modified orthogonal code set, known as the simplex code, is generated by deleting the first row and the first column of an orthogonal set. The code set $[H_8]$ is now modified for $M = 7$ as:

$$(H_7) \rightarrow \begin{bmatrix} 0 & 1 & 0 & 1 & 0 & 1 & 0 \\ 1 & 0 & 0 & 1 & 1 & 0 & 0 \\ 0 & 0 & 1 & 1 & 0 & 0 & 1 \\ 1 & 1 & 1 & 0 & 0 & 0 & 0 \\ 0 & 1 & 0 & 0 & 1 & 0 & 1 \\ 1 & 0 & 0 & 0 & 0 & 1 & 1 \\ 0 & 0 & 1 & 0 & 1 & 1 & 0 \end{bmatrix}$$

The codes may be made cyclic by suitably interchanging some rows and columns, and their cross-correlation function is:

$$\rho_{ij}(m_i, m_j) = \begin{cases} M, & \text{for} \quad i = j \\ -1, & \text{for} \quad i \neq j \end{cases} \tag{C.13}$$

The pseudo-random binary sequences (PRBS or PNS), generated by feedback shift registers, are also used as semiorthogonal codes with low ρ_{ij}.

It is also observed from the orthogonal code sets that if a codeword is decoded incorrectly, then the number of bits in error is $M/2$, since the distance between the codewords is $M/2$. Thus, the bit error probability (BER) is related to the word (symbol) error probabillty $P_e(M)$ by the equation:

$$\text{BER} = P_b = 1/2P_e(M) \tag{C.14}$$

C.3.1 Optimum Waveforms for M-ary Signalling

Application and detection of M orthogonal codes have been discussed in Sec. 8.7.2 where it was assumed that an ideal synchronization at the receiver was achieved through auxiliary signals. It is, however, more desirable to have in-built synchronizing capability of the codes used for signalling. Thus the code set $\{C_i\}$ should have the dual properties given by

$$\rho_{ij}(C_i, C_j) = \begin{cases} N, & \text{for} \quad i = j \\ 0, & \text{for} \quad i \neq j \end{cases} \tag{C.15}$$

and

$$\rho_{ii}(C_i, C_{i+\tau}) = \begin{cases} N, & \text{for} \quad \tau = 0 \\ 0, & \text{for} \quad \tau \neq 0 \end{cases}$$

where ρ_{ii} = autocorrelation of the codes (ACF), ρ_{ij} = cross-correlation of the codes (CCF), and N = length of the codes. Such codes are not practically realizable, but by relaxing the second property as

$$\rho_{ii} = \begin{cases} N, & \tau = 0 \\ \epsilon, & \tau \neq 0 \end{cases}$$

where $\epsilon \ll N$, realizable codes may be generated.

Consider the bi-orthogonal code set for $M = 16$ given by the (16×8) matrix of Eq. (C.12). Let the code veetors be multiplied (mod-2), element by element, by a relaxed Barker code B, given by (in bipolar form):

$$\{B_{1i}\} = \{1, 1, -1, 1, -1, -1, -1, 1\}$$

whose

$$\{\rho_{ii}\} = \{8, -1, 0, -1. 0, -3, 0, 1\} \tag{C.16}$$

Then multiplying the code set $\{C_i\}$ by $\{B_{1i}\}$, the modified code set$\{C_i'\}$ is obtained as:

$$\begin{bmatrix} C'_0 \\ C'_2 \\ C'_4 \\ C'_6 \\ C'_8 \\ C'_{10} \\ C'_{12} \\ C'_{14} \end{bmatrix} = \begin{bmatrix} 1 & 1 & 0 & 1 & 0 & 0 & 0 & 1 \\ 1 & 0 & 0 & 0 & 0 & 1 & 0 & 0 \\ 1 & 1 & 1 & 0 & 0 & 0 & 1 & 0 \\ 1 & 0 & 1 & 1 & 0 & 1 & 1 & 1 \\ 1 & 1 & 0 & 1 & 1 & 1 & 1 & 0 \\ 1 & 0 & 0 & 0 & 1 & 0 & 1 & 1 \\ 1 & 1 & 1 & 0 & 1 & 1 & 0 & 1 \\ 1 & 0 & 1 & 1 & 1 & 0 & 0 & 0 \end{bmatrix} \tag{C.17}$$

Only the even numbered codes are shown and the odd numbered codes $\{C'_1, C'_3, \ldots, C'_{15}\}$ are the complements of this above set. It will be noted that all the code vectors are orthogonal to each other and they have reasonably acceptoble auto-correlation ρ_{ii} with small sideloads. As for example,

$$\{\rho_{ii}(C'_6)\} = (8, -1, 0, 3, 0, 1, 0, 1\}$$

and

$$\{\rho_{ii}(C'_{14})\} = \{8, 1, 0, -3, 0, -1, 0, -1\}$$

Thus the correlation peaks obtained at the MF output for each of the M signals transmitted may be used to synchronize the receiver clock and the sampling instants for the detection of codewords.

Similarly, for $M = 32$, bi-orthogonal codes (Read-Muller codes), one may use the synchronizing code $\{B_{2i}\}$ given by

$$\{B_{2i}\} = \{1, -1, -1, 1, -1, 1, -1, -1, 1, 1, -1, -1, -1, -1, -1, 1\}$$

where

$$\{\rho_{ii}\} = \{16, -1, 2, 1, -2, 1, 1, -1, 0, 1, 2, -1, 2, -1, -2, 1\}$$

The resultant code vectors for $M = 32$ are given in Table C.4. The higher-order codes, (say, for $M = 64, 128$ etc.) may be generated by combining higher-order Hadamard matrices with relaxed Barker codes of required length.

An alternative method of synchronization is to use a suitably modified Barker code in lieu of the all-zero code vectors, and to transmit this code repeatedly at intervals of nT sec. or whenever there is an all-zero message vector at the coder input. Continuous transmission of the synchronizing code is also used in some systems, but this requires additional signal power/ extra bandwidth for efficient performance.

The functional diagram of M-16 bi-orthogonal code generator is shown in Fig. C.4, where the binary message vector, say, 1 0 1 1, is fed to the AND gates in parallel, and the gates are controlled by the outputs of the binary dividers as shown in the figure. The gate outputs are now summed through MOD-2 adders, resulting into the waveforms shown in Fig. C.4(b). It can be easily checked that the summed output is the code $\{C_{11}\} = \{10100101\}$ and by multiplying with $\{B_{1i}\}$, the desired code $\{C'_{11}\} = \{01110100\}$ is obtained. A suitable MF, say, using a tapped delay line (TDL) detects the code in an optimum manner.

Table C.4 Bi-orthogonal Code Set and the Modified Vectors Incorporating the Synchronous Signal

Code No.	Basic Biorthogonal codes	Modified Reed-Muller codes
C_0	0 0 0 0 0 0 0 0 0 0 0 0 0 0 0 0	1 0 0 1 0 1 0 0 1 1 0 0 0 0 0 1
C_2	0 1 0 1 0 1 0 1 0 1 0 1 0 1 0 1	1 1 0 0 0 0 0 1 1 0 0 1 0 1 0 0
C_4	0 0 1 1 0 0 1 1 0 0 1 1 0 0 1 1	1 0 1 0 0 1 1 1 1 1 1 1 0 0 1 0
C_6	0 0 0 0 1 1 1 1 0 0 0 0 1 1 1 1	1 0 0 1 1 0 1 1 1 1 0 0 1 1 1 0
C_8	0 0 1 1 1 1 0 0 0 0 1 1 1 1 0 0	1 0 1 0 1 0 0 0 1 1 1 1 1 1 0 1
C_{10}	0 1 0 1 1 0 1 0 0 1 0 1 1 0 1 0	1 1 0 0 1 1 1 0 1 0 0 1 1 0 1 1
C_{12}	0 1 1 0 0 1 1 0 0 1 1 0 0 1 1 0	1 1 1 1 0 0 1 0 1 0 1 0 0 1 1 1
C_{14}	0 1 1 0 1 0 0 1 0 1 1 0 1 0 0 1	1 1 1 1 1 1 0 1 1 0 1 0 1 0 0 0
C_{16}	0 0 0 0 0 0 0 0 1 1 1 1 1 1 1 1	1 0 0 1 0 1 0 0 0 0 1 1 1 1 1 0
C_{18}	0 1 0 1 0 1 0 1 1 0 1 0 1 0 1 0	1 1 0 0 0 0 0 1 0 1 1 0 1 0 1 1
C_{20}	0 0 1 1 0 0 1 1 1 1 0 0 1 1 0 0	1 0 1 0 0 1 1 1 0 0 0 0 1 1 0 1
C_{22}	0 0 0 0 1 1 1 1 1 1 1 1 0 0 0 0	1 0 0 1 1 0 1 1 0 0 1 1 0 0 0 1
C_{24}	0 0 1 1 1 1 0 0 1 1 0 0 0 0 1 1	1 0 1 0 1 0 0 0 0 0 0 0 0 0 1 0
C_{26}	0 1 0 1 1 0 1 0 1 0 1 0 0 1 0 1	1 1 0 0 1 1 1 0 0 1 1 0 0 1 0 0
C_{28}	0 1 1 0 0 1 1 0 1 0 0 1 1 0 0 1	1 1 1 1 0 0 1 0 0 1 0 1 1 0 0 0
C_{30}	0 1 1 0 1 0 0 1 1 0 0 1 0 1 1 0	1 1 1 1 1 1 0 1 0 1 0 1 0 1 1 1

Note: The odd numbered codes are formed by complementing the preceding even numbered codes, e.g., C_{15} is the complement of C_{14}. The synchronization signal = 1001010011000001. For PSK transmission, 1/0 in the table stands for ± 1, i.e., phase reversal of the carrier.

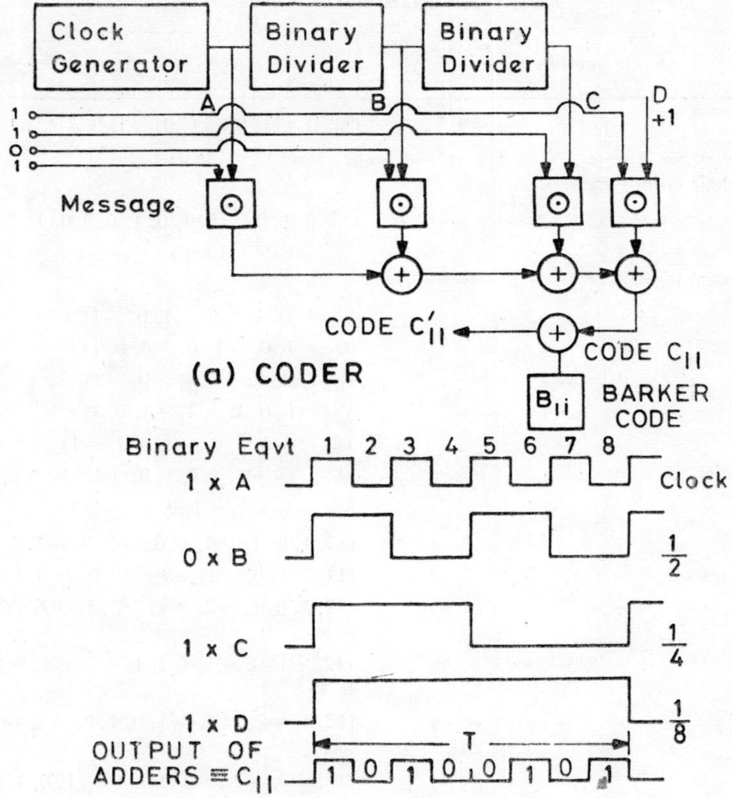

Fig. C.4 Functional diagram of M-16 code generator

C.3.2 Barker Codes [5]

Ideal Barker codes with ACF $= (0, \pm 1)$ for $\tau \neq 0$, are known for $N = 1$, 2, 3, 4, 5, 7, 11 and 13 only. But by relaxing the above condition, it is possible to generate many more quasi-ideal codes. Some of these codes for $N \geqslant 8$ with their ACF is given in Table C.5. More details are given in Ref. [5].

It is possible to generate longer codes by using dot products of smaller codes, e.g., the code $C_{11} \times C_{11}$, generated by the dot product of Barker code with $N_1 = N_2 = 11$, has an ACF peak of 121 and sidelobes 0 or -11. Thus, many suitable combinations of codes may be found for the specific application.

C.4 PN SEQUENCES AND GOLD CODES

For spread-spectrum systems, the address codes are generated by using:

(a) Long-quasi-orthogonal sequences giving code-division multi-accessing (CDMA),

Table C.5 Relaxed Barker Codes [5]

N	Code vectors	ACF $[\rho_{ii}(\tau)]$
11	(+ + + − − − + − − + −)	(11, 0, −1, 0, −1, 0, −1, 0, −1, 0, −1)
	(− + − − + − − − + + +)	(ideal)
	and their complements).	
13	(+ + + + + − − + + − + − +)	(13, 0, 1, 0, 1, 0, 1, 0, 1, 0, 1,0, 1)
	(+ − + − + + − − + + + + +)	(ideal)
	and complements.	
8	(+ + + − − − + −)	(8, 1, 0, −3, 0, −1, 0, −1)
	(+ + − + − − − +)	(8, −1, 0, −1, 0, −3, 0, 1)
	(+ + + + − − + +)	(8, 1, 0, 1, −2, −1, 0, −1)
	(+ + + − + − − +)	(8, −1, 0, 1, −2, −1, 0, 1)
	(+ + + − − + − −)	(8, 1, −2, 1, 0, −1, −2, −1)
	(+ + − − − − + −)	(8, 1, 0, −1, −2, 1, 0, −1)
	(+ − + + − − − +)	(8, −1, −2, −1, 0, 1, −2, 1)
12	(+ + + − − − − + − − + −)	(12, 1, 0, 1, −4, 1, 0, −1, 0, −1, 0, −1)
	(+ + + − − − + − + + − +)	(12, −1, 0, −1, −4, −1, 0, 1, 0, 1, 0, 1)
	(+ + + + − − − + − − + −)	(12, 1, 0, 1, −2, −1, −2, 1, −2, −1, 0, −1)
	(+ + + − − − + − − + − +)	(12, −1, 0, −1, −2, 1, −2, −1, −2, 1, 0, 1)
	(+ + + − − − + − − + − −)	(12, 1, −2, 1, 0, −1, 0, 1, 0, −1, −2, −1)
	(+ − + + − + + + − − − +)	(12, −1, −2, −1, 0, 1, 0, −1, 0, 1, −2, 1)
16	(+ − − + − + − − + + − − − − − +)	(16, −1, −2, 1, −2, 1, 2, −1, 0, 1, 2, −1, 2, −1, −2, 1)
	(+ + + − + + + − − − + − + + −)	(16, . . . , sidelobes $\leqslant$ 2)
	(+ + + + + − − + + − + − + − −)	(16, . . . , sidelobes $\leqslant$ 2)
32	(+ + + + + + − + + + − − + + + − − − + − + + − + + − + − − − +)	(32, . . . , sidelobes $\leqslant$ 3)

(b) Time-frequency (TF) matrix/frequency hopping (FH) codes giving Pulse address multiaccessing (PAMA)

(c) Hybrid techniques.

PNS Codes [6]

For CDMA, maximal length Pseudo-noise sequences (PNS) (also known as PRBS), generated by feedback shift registers (SR), are generally used, since the acyclic ACF of long PNS has small side-lobes and many quasi-orthogonal PNS may be generated if N is sufficiently large. In general, a sequence length of $N = 2^k - 1$, is generated by connecting k-stage SR in tandem and giving a feedback to the input from two or more SR-outputs through a MOD-2 adder, as shown in Fig. C.5. Maximal-length PNS (or m-sequences)

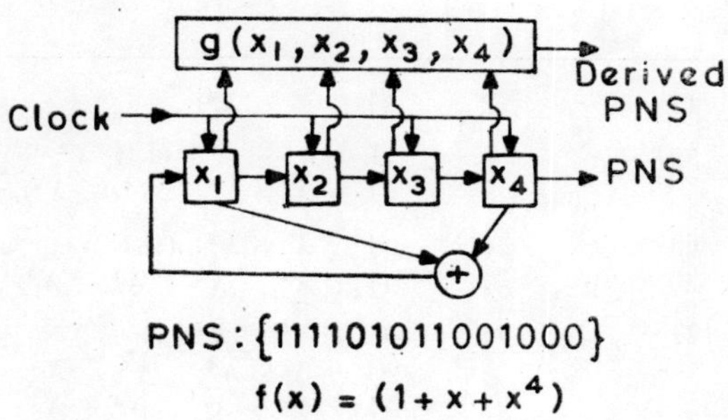

Fig. C.5 A PNS generator

are generated by certain primitive polynomials (refer to Sec. C.1), and the complements of the polynomials also give valid feedback connections. In the case of the 4-stage SR, shown in Fig. C.5, the primitive polynomial used is $f(X) = (1 + X + X^4)$ and its complement is: $f(X) = X^4(1 + X^{-1} + X^{-4}) = (X^4 + X^3 + 1)$. Thus the feedback connection may be taken either from $(X_1 \oplus X_4)$ or from $(X_3 \oplus X_4)$, and two distinct 15-length m-sequences are obtained as

$$\{0001\ 00110\ 101111\} \quad \text{or} \quad \{000\ 1111\ 010\ 11001\}$$

By combining the outputs of the SR's through a linear/non-linear switching circuit $g(X_1, X_2, X_3, X_4, \ldots)$, a derived sequence may also be obtained. Maximum number of m-sequences that can be generated by k-SR's is given by

$$M(m) = \phi(N)/k \tag{C.18}$$

where $\phi(N)$ is Euler function. A short list of m-sequences, along with their primitive polynomials is given in Table C.6, where the complements are not given and only a few of the polynomials shown. For more details, refer to [6].

The PN sequences, although generated by a known SR-configuration, exhibit random properties for $N \rightarrow$ large. These pseudorandom properties are:

(a) The balance property: In each complete period of the sequence, the number of 1's differs from the number of O's by at most one.
(b) The run property: Among the runs of consecutive 1's and 0's in one complete period of the sequence, a half of the runs of each kind are of length 1, a quarter of each kind are of length 2, an eighth are of length 3, and so on, as long as these fractions give meaningful numbers of runs.

Table C.6 *m*-sequences and their polynomials

SR-length $=k$	$N = 2^k - 1$	$M(m)$	Primitive polynomials $f(X)$
4	15	2	$(1 + X + X^4)$
5	31	6	$(1 + X^2 + X^5)$, $(1 + X^2 + X^3 + X^4 + X^5)$, $(1 + X + X^2 + X^4 + X^5)$
6	63	6	$(1 + X + X^6)$, $(1 + X + X^2 + X^5 + X^6)$ $(1 + X^2 + X^3 + X^5 + X^6)$
7	127	18	$(1 + X^3 + X^7)$, $(1 + X + X^2 + X^3 + X^7)$, $(1 + X + X^2 + X^4 + X^5 + X^6 + X^7)$, $(1 + X^2 + X^3 + X^4 + X^7)$ $(1 + X + X^2 + X^3 + X^4 + X^5 + X^7)$, $(1 + X^2 + X^4 + X^6 + X^7)$, $(1 + X + X^7)$, $(1 + X + X^3 + X^6 + X^7)$, $(1 + X^2 + X^5 + X^6 + X^7)$
8	255	16	$(1 + X^2 + X^3 + X^4 + X^8)$, $(1 + X^3 + X^5 + X^6 + X^8)$, $(1 + X + X^2 + X^5 + X^6 + X^7 + X^8)$ $(1 + X + X^3 + X^5 + X^8)$, $(1 + X^2 + X^5 + X^6 + X^8)$, $(1 + X + X^5 + X^6 + X^8)$, $(1 + X + X^2 + X^3 + X^4 + X^6 + X^8)$, $(1 + X + X^6 + X^7 + X^8)$
10	1023	60	$(1 + X^3 + X^{10})$, $(1 + X^2 + X^3 + X^8 + X^{10})$, $(1 + X^3 + X^4 + X^5 + X^6 + X^7 + X^8 + X^9 + X^{10})$, $(1 + X^3 + X^4 + X^5 + X^6 + X^7 + X^8 + X^{10})$, etc.
11	2047	176	$(1 + X^2 + X^{11})$, $(1 + X^2 + X^5 + X^8 + X^{11})$, $(1 + X^2 + X^3 + X^7 + X^{11})$, $(1 + X^2 + X^3 + X^5 + X^{11})$, $(1 + X^2 + X^3 + X^{10} + X^{11})$, $(1 + X + X^3 + X^8 + X^9 + X^{10} + X^{11})$, $(1 + X^3 + X^6 + X^8 + X^{11})$, $(1 + X^3 + X^4 + X^6 + X^8 + X^{10} + X^{11})$, etc.
15	32,767	1800	$(1 + X + X^{15})$, $(1 + X + X^5 + X^{10} + X^{15})$ $(1 + X + X^3 + X^{12} + X^{15})$, $(1 + X + X^2 + X^4 + X^5 + X^{10} + X^{15})$, $(1 + X + X^2 + X^6 + X^7 + X^{11} + X^{15})$, $(1 + X + X^2 + X^3 + X^6 + X^7 + X^{15})$, $(1 + X^2 + X^3 + X^{12} + X^{15})$, etc.
20	1,048,575	24,000	$(1 + X^3 + X^{20})$, $(1 + X^3 + X^5 + X^9 + X^{20})$, $(1 + X^2 + X^3 + X^6 + X^8 + X^{11} + X^{20})$, etc.
21	2,097,151	84,672	$(1 + X^2 + X^{21})$, $(1 + X^2 + X^7 + X^{14} + X^{21})$, $(1 + X^2 + X^5 + X^{13} + X^{21})$, etc.

(c) The correlation property: The cyclic correlation of a PNS is given by Eq. (C.13), and in normalized form

$$\rho_{ii}(\tau) = \begin{cases} 1 & , \quad \text{for} \quad \tau = 0 \\ -1/N & , \quad \text{for} \quad \tau \neq 0 \end{cases} \tag{C.19}$$

Based on this the power spectrum of a PNS is given by

$$G(\omega) = \frac{N-1}{N}\left(\frac{\sin \omega/2f_r}{\omega/2f_r}\right)^2 \sum_{\substack{-\infty \\ n\neq 0}}^{\infty} \delta\left(\omega - \frac{2\pi n f_r}{N}\right) + \frac{1}{N^2}\delta(\omega) \tag{C.20}$$

where $N = 2^k - 1$ and f_r = clock frequency of the PNS generator. The spectrum thus has a contour given by $(\sin x/x)^2$ and the equivalent noise bandwidth $B_n \simeq f_r$. Although the cyclic ACF's of m-sequences are ideal, the acyclic ACF of PNS's have considerable side-lobes and for $N \to$ large, the peak sidelobes have magnitudes $\simeq \sqrt{N}$. The CCF between pairs of m-sequences of the same length N is then quite large and is given by:

$$\rho_{ij}(C_i, C_j) > \sqrt{N-1} \tag{C.21}$$

Some values of peak $|\rho_{ij}|$ are given in Table C.7 and it is seen that all such codes cannot be used as the signalling waveforms for SS systems, unless N is very large, say, $N \geqslant 10^4$.

Table C.7 Maximum Cross-Correlation Values Between Pairs of m-sequence

k	N	Cyclic ρ_{ii}	Max. $\vert\rho_{ij}\vert$	Normalized $\rho_{ij} = \vert\rho_{ij}\vert/N$
5	31	31, −1	11	0.35
6	63	63, −1	23	0.36
7	127	127, −1	41	0.32
9	511	511, −1	113	0.22
10	1023	1023, −1	383	0.37
11	2047	2047, −1	287	0.14
12	4095	4095, −1	1407	0.34

C.4.1 Gold Codes

Gold [7] has, however, shown that certain pairs of m-sequences have CCF bounded as

$$\rho_{ij}(C_i, C_j) \leqslant 2^{(k+1)/2} + 1 \quad \text{for} \quad k \text{ odd}$$

and

$$\rho_{ij}(C_i, C_j) \leqslant 2^{(k+2)/2} + 1 \text{ for } k \text{ even and not divisible by 4.} \tag{C.22}$$

As for example, using two ($k = 6$) — PNS generators with polynomials,

$$f_1(X) = X^6 + X + 1$$

$$f_2(X) = X^6 + X^5 + X^2 + X + 1,$$

is is calculated that $\rho_{ij} \leqslant 2^4 + 1 = 17$. Such pairs of PNS are called 'preferred' sequences. From a pair of preferred sequences, say C_i and C_j, a new sequence C_l may be generated by taking MOD-2 sum of C_i with N cyclically shifted versions of C_j or vice versa, as shown in Fig. C.6. Thus N

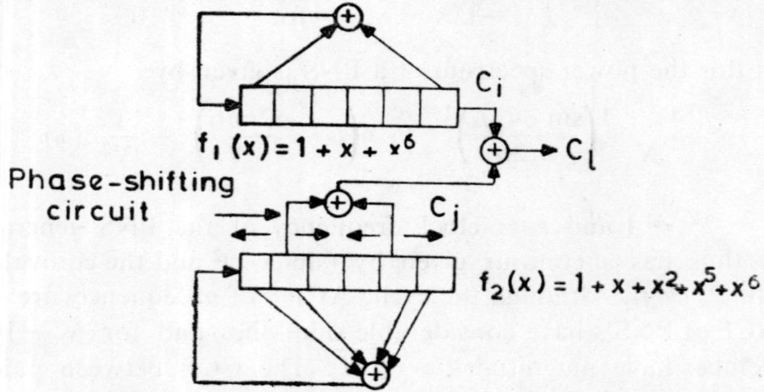

Fig. C.6 Generator for Gold code; cyclic shift of C_j are controlled by the phase-shifting circuit

new sequences are generated with the CCF bound given by Eq. (C.22) and including C_i and C_j, the total number of such sequences are $(2^k + 1)$. These sequences are known as Gold sequences (codes) whose ACF values are the same as the CCF values. Thus, the codes have poorer ACF than *m*-sequences, but better CCF values. Further, it has been shown that the Gold codes have three-level CCF's and Table C.8 gives their values and relative frequencies of occurrence. It is seen that even values of k (not divisible by 4) give better CCF values for these Gold codes.

Table C.8 CCF Values and Their Relative Frequencies for Gold Codes

k = SR-length	N	Normalized CCF	Relative frequency
k odd	$2^k - 1$	$-1/N$	0.50
		$-(2^{(k+1)/2} + 1)/N$	0.25
		$(2^{(k+1)/2} - 1)/N$	0.25
k even but not	2^{k-1}	$-1/N$	0.75
divisible by 4		$-(2^{(k+2)/2} + 1)/N$	0.125
		$(2^{(k+2)/2} - 1)/N$	0.125

The above Gold codes or *m*-sequences with $N \to$ large, are used as the addresses for the CDMA (SS) systems, and also in digital systems using *M*-ary signalling.

C.5 TF MATRIX CODES [3]

In *TF*-matrix coding shown in Fig. C.7, the bit-time T is subdivided into n slots and the available RF bandwidth B is subdivided into m slots, giving

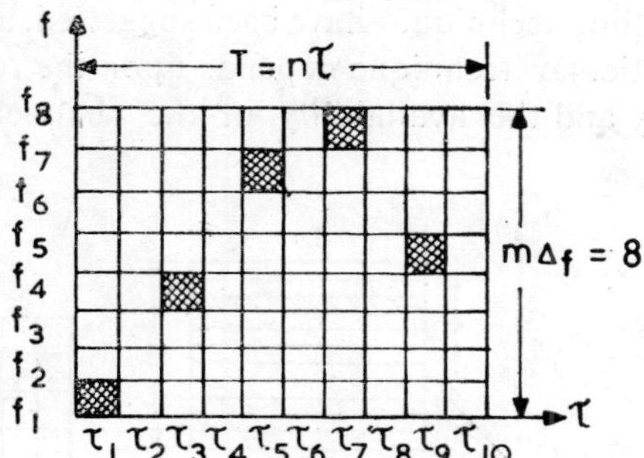

Fig. C.7 TF matrix code/FFH

rise to $(m \times n)$ slots in the *TF*-plane, where $T = n\tau$, $B = m \cdot \Delta f$, and $\Delta f \geqslant 1/\tau$. In the transmitter, the information bits control an address matrix, such that a few short pulses (carrier bursts) out of $(m \times n)$ available slots are transmitted per bit. In the receiver, the pulses are received through narrow bandpass filters and the resulting video pulses are either counted or summed through corresponding delay lines for subsequent decision on 1/0. It is seen that using only a few slots in the *TF*-plane per address, a large number of distinct addresses are possible. To make the addresses unique, assume that the first time slot of each is always occupied and in general, the same time slot and the same frequency are not used more than once in an address. Using these restrictions, the number of unique addresses M_A is given by:

$$M_A = \frac{m!(n-1)!}{(m-q)!\,(n-q)!q!} \tag{C.23}$$

where q is the number of pulses transmitted per bit of information. As for example, using $m = 4$, $n = 6$ and $q = 3$, $M_A = 80$, but using $m = 8$, $n = 10$ and $q = 5$, as in Fig. C.7, $M_A = 169{,}344$. Thus for a given matrix size, a large number of addresses may be generated, under the control of a PNS generator, whose output at any instant would select the addresses in a random manner. These codes have been also discussed in Section 5.11.2.

This signalling technique is generally known as Fast frequency hopping (FFH), since q frequencies are transmitted per signalling bit/symbol. In contrast to this, there are slow FH schemes, where the carrier frequency is randomly hopped once in every signalling interval T, under the control of a PNS code generator. The 1/0 signals are transmitted by FSK and the receiver demodulates the signal incoherently after charging the random frequencies to a constant IF by mixing the input with the output of a frequency synthesizer controlled by a synchronized PNS-code generator. When collisions with other transmitter signals occur, then one bit/symbol of information is lost; but this may be retrieved if suitable FEC codes are used in tendem.

Many methods of FH signalling have been used and it is also possible to use TH (time hopping) signalling, analogous to FH, to introduce randomness in SS-signals. Combining the different techniques of FH, TH and CDMA, many hybrid signalling techniques have been suggested, as shown in Fig. C.8. The use of a particular technique depends upon the requirements of anti-jamming capability and the availability of the channel banwidth for the link.

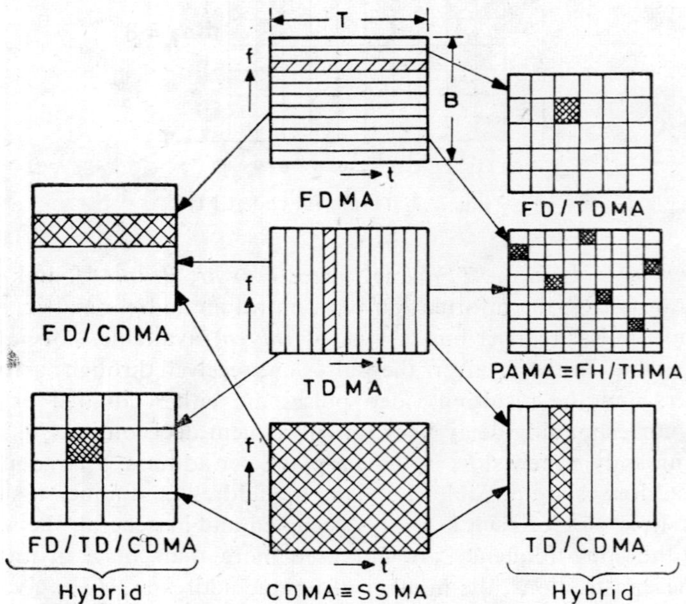

Fig. C.8 Generation of hybrid SS-codes

REFERENCES

1. Peterson, W.W. and Weldon, E.J., *Error-Correcting Codes*, MIT Press, Cambridge. Mass. 1972.
2. Clark, G.C., and Cain, J.B., *Error-Correcting coding for digital Communication*, Plenum Press, N.Y., 1981.
3. Das, J., Mullick, S.K., and Chatterjee, P.K., *Principles of Digital Communication*, Wiley Eastern, New Delhi, 1986.
4. Veterbi, A.J., *Principles of Coherent Communication*, McGraw-Hill, N.Y., 1966.
5. Moharir, P.S. and Selvarajan, A., 'Sequence with good autocorrelation for pulse compression and synchronisation' *Tech. Report* CIP/O2P/1.1, EE Deptt., IIT Kanpur, Jan. 1974.
6. Golomb, S.W., *Shift Register Sequences*, Holden-Day, San Francisco, 1967.
7. Gold, R., 'Optimal Binary Sequences for SS multiplexing', *IEEE Trans. Inf. Th.* vol. IT-13, pp 619-62, 1967.

Index